Industrial Pollution Prevention Handbook

Other McGraw-Hill Environmental Engineering Books

American Water Works Association
WATER QUALITY AND TREATMENT

Baker
BIOREMEDIATION

Chopey
ENVIRONMENTAL ENGINEERING FOR THE CHEMICAL PROCESS INDUSTRIES

Corbitt
STANDARD HANDBOOK OF ENVIRONMENTAL ENGINEERING

Freeman
HAZARDOUS WASTE MINIMIZATION

Freeman
STANDARD HANDBOOK OF HAZARDOUS WASTE TREATMENT AND DISPOSAL

Jain
ENVIRONMENTAL IMPACT ASSESSMENT

Harris and Harvey
HAZARDOUS CHEMICALS AND THE RIGHT TO KNOW: AN UPDATED GUIDE TO COMPLIANCE WITH SARA TITLE III

Levin and Gealt
BIOTREATMENT OF INDUSTRIAL AND HAZARDOUS WASTE

Kolluru
ENVIRONMENTAL STRATEGIES HANDBOOK

McKenna and Cunneo
PESTICIDE REGULATION HANDBOOK

Majumdar
REGULATORY REQUIREMENTS OF HAZARDOUS MATERIALS

Seldner
ENVIRONMENTAL DECISION MAKING FOR ENGINEERING AND BUSINESS MANAGERS

Waldo and Hines
CHEMICAL HAZARD COMMUNICATION GUIDEBOOK

Willig
ENVIRONMENTAL TQM

Industrial Pollution Prevention Handbook

Harry M. Freeman

McGraw-Hill, Inc.

New York   San Francisco   Washington, D.C.   Auckland   Bogotá
Caracas   Lisbon   London   Madrid   Mexico City   Milan
Montreal   New Delhi   San Juan   Singapore
Sydney   Tokyo   Toronto

Library of Congress Cataloging-in-Publication Data

Freeman, Harry
    Industrial pollution prevention handbook / Harry M. Freeman.
        p.   cm.
    Includes bibliographical references and index.
    ISBN 0-07-022148-0
    1. Industry—Environmental aspects. 2. Pollution. I. Title.
TD194.F74   1995
363.73'1—dc20   95-7979
                                              CIP

1 2 3 4 5 6 7 8 9 0   DOH/DOH   9 0 9 8 7 6 5 4

ISBN 0-07-022148-0

The sponsoring editor for this book was Gail F. Nalven, the editing supervisor was Kimberly A. Goff, and the production supervisor was Donald F. Schmidt. This book was set in Palatino. It was composed by McGraw-Hill's Professional Book Group composition unit.

Printed and bound by R. R. Donnelley & Sons Company.

 This book is printed on recycled, acid-free paper containing a minimum of 50% recycled de-inked fiber.

This book was edited by Harry M. Freeman in his private capacity. No official support or endorsement by the Environmental Protection Agency or any other agency of the federal government is intended or should be inferred.

Contents

Contributors

KONSTADINOS ABELIOTIS, *New Jersey Institute of Technology, Newark, New Jersey*

VITAL AELION, *New Jersey Institute of Technology, Newark, New Jersey*

FRANK ALTMAYER, *Scientific Control Laboratories, Incorporated, Chicago, Illinois*

K. P. ANANTH, *Battelle Memorial Institute, Columbus, Ohio*

PAUL T. ANASTAS, *Office of Pollution Prevention and Toxics, U.S. Environmental Protection Agency, Washington, D.C.*

CHARLES S. APPLEGATE, *Radian Corporation, Milwaukee, Wisconsin*

ELAINE APPLEY, *Concurrent Technologies, Johnstown, Pennsylvania*

JOHN ATCHESON, *U.S. Department of Energy, Washington, D.C.*

GARY E. BAKER, *Science Applications International Corporation, Cincinnati, Ohio*

DENISE P. BEHAYLO, *General Motors Corporation, AC Rochester Division, Flint, Michigan*

CAROLE BELL, *Science Applications International Corporation, Newport, Rhode Island*

RONALD L. BERGLUND, *The M.W. Kellogg Company, Houston, Texas*

JUNE BOLSTRIDGE, *GAIA Corporation, Silver Spring, Maryland*

KRISZTINA BORDACS-IRWIN, PH.D., P.E., *Newtown, Pennsylvania*

JOSEPH J. BREEN, *Office of Pollution Prevention and Toxics, U.S. Environmental Protection Agency, Washington, D.C.*

SANDRA S. BREWER, *General Motors Corporation, Warren, Michigan*

R. SCOTT BUTNER, *Shapiro and Associates, Incorporated, Seattle, Washington*

MICHAEL S. CALLAHAN, P.E., *Jacobs Engineering Group Incorporated, Pasadena, California*

LAURIE CASE, *Illinois Hazardous Waste Research and Information Center, Champaign, Illinois*

SATYA P. CHAUHAN, *Battelle Memorial Institute, Columbus, Ohio*

JAMES CRAIG, *U.S. Environmental Protection Agency, Washington, D.C.*

DALE DENNY, *Concurrent Technologies Corporation, Johnstown, Pennsylvania*

RANDY D. DOWN, P.E., *The Sear-Brown Group, Rochester, New York*

STEPHEN P. EVANOFF, P.E., D.E.E., *Lockheed Fort Worth Company, Fort Worth, Texas*

DAVID P. EVERS, *Battelle Memorial Institute, Columbus, Ohio*

HARRY M. FREEMAN, *U.S. Environmental Protection Agency, Cincinnati, Ohio*

BRIAN FREWERD, *Concurrent Technologies, Johnstown, Pennsylvania*

CARL H. FROMM, *Jacobs Engineering Group Incorporated, Pasadena, California*

ARUN R. GAVASKAR, *Battelle Memorial Institute, Columbus, Ohio*

MADELINE M. GRULICH, *Pacific Northwest Pollution Prevention Research Center, Seattle, Washington*

STEVEN M. HASSUR, *Office of Pollution Prevention and Toxics, U.S. Environmental Protection Agency, Washington, D.C.*

JOSHUA M. HELTZER, *Virginia Department of Environmental Quality Office of Pollution Prevention, Richmond, Virginia*

JACK R. HOPPER, *Lamar University, Beaumont, Texas*

JOHN HOULAHAN, *Science Applications International Corporation, Hampton, Virginia*

RICHARD L. HUBLER, *General Motors Corporation, AC Rochester Division, Flint, Michigan*

GARY E. HUNT, *North Carolina Department of Environment, Health and Natural Resources, Raleigh, North Carolina*

JOHN A. JAFFURS, *General Motors Corporation, AC Rochester Division, Flint, Michigan*

PAMELA G. JENKINS, P.E., *Science Applications International Corporation, Olympia, Washington*

MURALI KALAVAPUDI, *ENERGETICS, Incorporated, Columbia, Maryland*

GREGORY A. KEOLEIAN, PH.D., *National Pollution Prevention Center, School of Natural Resources and Environment, University of Michigan, Ann Arbor, Michigan*

LYNN KNIGHT, *Eastern Research Group, Incorporated, Lexington, Massachusetts*

DIETER S. LEIDEL, *Tanoak Incorporated, Barrie, Ontario, Canada*

JOSEPH C. LIELLO, *Radian Corporation, Milwaukee, Wisconsin*

SUBIR K. MALLICK, *New Jersey Institute of Technology, Newark, New Jersey*

JENNIFER MARRON, *American Forest & Paper Association, Washington, D.C.*

MICHAEL MELTZER, *Lawrence Livermore National Laboratory, Livermore, California*

LAURA MENDICINO, *Illinois Hazardous Waste Research and Information Center, Champaign, Illinois*

P. N. MISHRA, PH.D., *General Motors Corporation, Warren, Michigan*

VLASTA MOLAK, PH.D., *GAIA Unlimited, Incorporated, Cincinnati, Ohio*

FREDERICK L. MOORE, *Union Carbide Corporation, Danbury, Connecticut*

PEGGY MORGAN, *Hazardous Waste and Toxics Reduction Program, Washington Department of Ecology, Olympia, Washington*

FERNANDO J. MUZZIO, *Rutgers University, Piscataway, New Jersey*

THOMAS E. NATAN, JR., PH.D., *Hampshire Research Associates, Incorporated, Alexandria, Virginia*

RICHARD A. OSANTOWSKI, *Radian Corporation, Milwaukee, Wisconsin*

PRAKASH T. PALEPU, *Battelle Memorial Institute, Columbus, Ohio*

KEVIN J. PALMER, *Science Applications International Corporation, Falls Church, Virginia*

EDWARD L. PAUL, *Merck Sharp and Dohme Research Laboratories, Rahway, New Jersey*

TRACY HOEFLING PAVA, *Concurrent Technologies, Johnstown, Pennsylvania*

DEMETRI PETRIDES, *New Jersey Institute of Technology, Newark, New Jersey*

RUSSELL W. PHIFER, *Environmental Assets, Incorporated, West Chester, Pennsylvania*

ROBERT B. POJASEK, PH.D., *GEI Consultants Incorporated, Winchester, Massachusetts*

LINDA GIANNELLI PRATT, *San Diego County Department of Environmental Services, San Diego, California*

TERRY SCIARROTTA, *Southern California Edison, Los Angeles, California*

PETER SHERMAN, ESQ., *Science Applications, International Corporation, Falls Church, Virginia*

J. D. SHOEMAKER, *Lawrence Livermore National Laboratory, Livermore, California*

LESLEY J. SNOWDEN-SWAN, *Battelle Pacific Northwest Laboratories, Richland, Washington*

DAVID THOMAS, *Illinois Hazardous Waste Research and Information Center, Champaign, Illinois*

JAMES THURBER, ESQ., *Science Applications, International Corporation, Falls Church, Virginia*

PAUL S. TOBIN, *Office of Pollution Prevention and Toxics, U.S. Environmental Protection Agency, Washington, D.C.*

REBECCA TODD, *New York University, New York, New York*

O. WARREN UNDERWOOD, *General Motors Corporation, Warren, Michigan*

E. S. VENKATARAMANI, *Merck & Company, Incorporated, Rahway, New Jersey*

BRUCE VIGON, *Battelle Memorial Institute, Columbus, Ohio*

JOHN WARREN, *Research Triangle Institute, Research Triangle Park, North Carolina*

DAVID T. WIGGLESWORTH, *Pollution Prevention Office, Alaska Department of Environmental Conservation, Anchorage, Alaska*

ALLEN L. WHITE, *Tellus Institute, Boston, Massachusetts*

STEVEN L. WHITE, *Jacobs Engineering Group Incorporated, Pasadena, California*

TODD A. WILLIAMS, *General Motors Corporation, Warren, Michigan*

AZITA YAZDANI, P.E., *Pollution Prevention International, Incorporated, Brea, California*

Preface

*Pollution prevention is the use of materials, processes,
or practices that reduce or eliminate the creation of
pollutants or wastes at the source. It includes
practices that reduce the use of hazardous and
nonhazardous materials, energy, water, or other
resources as well as those that protect natural
resources through efficient use.*

USEPA

The Administrator of the U.S. Environmental Protection Agency has stated, "I
have four priorities for the Agency. First, pollution prevention...." President
Clinton in his 1993 Earth Day speech stated, "Our long-term strategy invests
more in pollution prevention...."

This book is all about the techniques, technologies, regulations, and strategies that define pollution prevention, a term that people involved with environmental improvement in the public and private sectors are using extensively
these days.

As you will see from a review of the contents list and chapters of the book,
the subject is addressed from many perspectives by prominent experts. There is
much "learning" to be had within the covers of this book through easy access to
the results of so much experience.

In many ways pollution prevention, rather than being a speciality field itself,
is actually a convergence of fields drawing upon knowledge in a wide variety
of more typical fields of expertise. This book may be seen as a bridge, since
there is a lot of bridging material to facilitate crossfield learning.

While the book is designed to be used as a reference handbook and not necessarily intended to be read from cover to cover, there is a world of interesting information within its covers. For the reader inclined to browse, treats await you.

The 78 contributing authors and I hope you find exploring the results of our work rewarding.

Harry M. Freeman

Acknowledgments

I want to acknowledge the contributors, whose willingness to take time from their busy careers and lives to share their advice, guidance, observations, and experiences have made the book possible. For those who have not yet contributed to a book such as this one, please know that these authors are motivated by something other than financial gain. We are all enriched by their commitment to furthering the pollution prevention agenda.

And on a larger scale, I would like to acknowledge the efforts of all those individuals in the private and public sectors marching in the pollution prevention parade who have made possible over the past decade the significant environmental benefits that people in countries around the world are coming to enjoy as we move toward an environmentally sustainable economy. You know who you are. Thanks.

Industrial Pollution Prevention Handbook

1

Pollution Prevention

A New Agenda

Harry M. Freeman

U.S. Environmental Protection Agency
Cincinnati, Ohio

1.1   Introduction

"An ounce of prevention is worth a pound of cure." A trite way of saying something everybody knows? Maybe, but this proverb is coming to be quoted by individuals around the world as a rallying cry for programs to enhance the quality of the world's environment. The proverb is at the heart of the pollution prevention movement.

Pollution prevention is a term used to describe production technologies and strategies that result in eliminating or reducing waste streams. The U.S. Environmental Protection Agency (EPA) defines pollution prevention as "the use of materials, processes, or practices that reduce or eliminate the creation of pollutants or wastes at the source. It includes practices that reduce the use of hazardous materials, energy, water, or other resources and practices that protect natural resources through conservation or more efficient use."[1] The idea underlying the promotion of pollution prevention is that it makes far more sense for a generator not to produce waste than to develop extensive treatment schemes to ensure that the waste poses no threat to the quality of the environment.

While *pollution prevention* is coming to be the most widely accepted term in the United States for such strategies and processes, there are other similar terms that have been used in the past and are to varying degrees still in use in the United States and elsewhere. Van Weenen compiled the exhaustive list of similar terms shown in Table 1-1.[2]

That pollution prevention and the related concept, *sustainable industrial development*, are ideas whose times have clearly come is illustrated by the selection of quotes contained in Fig. 1-1.

1

Table 1-1. Van Weenen's Waste Reduction Terms

Organization and Policy

Strategy, management, research, procedures, activities:
  "Reflection"
  Anticipate-and-prevent strategies
  Avoidance strategy
  Front-end resource management
  Waste prevention research
  Product assessment procedure
  Preventative activities
  Humane chemistry
  Source reduction
  Source control

Technology

Technology, technologies:
  New technologies
  Environmental technology
  Prevention-aimed environmental technology
  Process-integrated environmental technology
  Appropriate technology
  Clean technologies
  Cleaner technologies
  Nonwaste technology
  Low- and nonwaste technologies
  Low-waste technology
  Low-polluting technology
  Pollution control technology
  Add-on technologies
  End-of-pipe technologies
  Recycling technologies
  Waste treatment technologies
  Purification treatment
  Cleaning up technology

Waste and Pollution

Prevention, avoidance, minimization, reduction:
  Waste prevention
  Waste avoidance
  Waste minimization
  Waste reduction
  Pollution prevention
  Pollution reduction
  Recycling

"Our long term strategy invests more in pollution prevention, energy efficiency and solar energy:..."[3]

> PRESIDENT BILL CLINTON
> APRIL 1993

"I have four priorities for the Agency. First, pollution prevention.... Pollution prevention is our best hope for the future of environmental protection: not just environmental regulation, but environmental protection."[4]

> CAROL BROWNER
> ADMINISTRATOR U.S. EPA
> APRIL 1993

"We have learned the inherent limitations of treating and burying wastes. A problem solved in one part of the environment may become a new problem in another part. We must curtail pollution closer to its point of origin so that it is not transferred from place to place."[5]

> WILLIAM REILLY
> ADMINISTRATOR U.S. EPA
> DECEMBER 1990

"EPA should shift the focus of its environmental protection strategy from end-of-the-pipe controls to preventing the generation of pollution."[6]

> U.S. EPA SCIENCE ADVISORY BOARD
> SEPTEMBER 1988

"Sustainable development can be achieved only through a new and strong partnership between governments, business, citizens' groups and the broader society that comprises consumers, voters, and those who through their daily actions are the planet's true environmental decision makers."[7]

> STEPHEN SCHMIDHEINY
> CHAIRMAN, BUSINESS COUNCIL FOR SUSTAINABLE
> DEVELOPMENT

"New Jersey has already enacted restructuring measures that will be hallmarks of a new era of environmental protection, where agencies will promote pollution prevention, public access, efficiency, and a 'holistic' environmental awareness."[8]

> SCOTT A. WEINER
> COMMISSIONER, NEW JERSEY DEPARTMENT OF
> ENVIRONMENTAL PROTECTION AND ENERGY

Figure 1-1. Selected quotes about pollution prevention.

"The proactive pollution prevention approach seeks more lasting and complete solutions to environmental problems. The proactive approach to environmental management is the essence of sustainable development."[9]

ROBERT P. BRINGER AND DAVID M. BENFORADOR
3M COMPANY

"By working with governments and the environmental community in a voluntary, proactive, productive and cooperative manner, we can make sustainable development a reality."[10]

FRANK P. POPOFF
CHAIRMAN AND CEO
DOW CHEMICAL COMPANY

"P2. Onward."

HARRY FREEMAN
U.S. EPA

Figure 1-1. *Continued*

1.2 Benefits of Pollution Prevention

Answering the question "Why should you undertake pollution prevention?" in a manual to help the industrial generator make cost comparisons on the basis of costs and benefits of pollution prevention, the authors of the EPA's *Pollution Prevention Benefits Manual* state:

> Pollution prevention can help you achieve the following:
> - improve your firm's "bottom line"
> - make compliance with environmental regulations easier
> - demonstrate a proactive commitment to genuinely pursuing a pollution prevention program.[11]

Pollution prevention, nicknamed "P2," is a compelling strategy for many reasons. If no pollution is generated, there are no pollutants to be managed. Thus, future problems are avoided, such as the problems which occur when previously accepted land disposal methods are discovered to be major sources of environmental contamination. Preventing pollution before it occurs also prevents situations that might endanger not only members of the community, but workers involved in the management of pollution.[12]

One of the significant benefits of P2 is that it is often an economical approach. When wastes are reduced or eliminated, cost savings in materials result—more product is produced from the same starting materials. The close examination of manufacturing processes needed to plan a successful pollution prevention approach can produce a number of side benefits as well, such as significant improvements in energy and water conservation, and improved, or more consistent, product quality.

P2 can also lead to large savings in regulatory and compliance costs, which are lowered as less pollution is produced. Frequently the dominant cost savings come from reduced future liability for the pollution. Ever since passage of the federal Resource Conservation and Recovery Act (RCRA), with its mandate that manufacturers have "cradle to grave" responsibility for the wastes that they generate, and enactment of the joint and several liability provisions of the federal Comprehensive Environmental Response, Compensation, and Liability Act (better known as the Superfund law), waste producers have been subject to the possibility of unlimited liability for any harm caused by their wastes. This liability includes even future problems caused by wastes managed using the best current practices. Because waste site cleanups can cost hundreds of millions of dollars each, these liabilities can dwarf all other costs associated with waste generation, which makes pollution prevention options even more compelling.[12]

The environmental advantages of P2 approaches include improving effectiveness, minimizing uncertainty, avoiding cross-media transfers, and protecting resources.

1.3 Pollution Prevention in Other Countries

Pollution prevention is receiving widespread emphasis internationally within multinational organizations and within individual countries. The driving force behind this emphasis is the concept of sustainable development and the hold that this concept has over planning strategies and long-term solutions to global limits and north-south economic issues. Participants in the June 1992 United Nations Conference on Environmental Development (UNCED) in Brazil spent a great deal of time on pollution prevention.

The European Community (EC) has designed some of its rules and programs around pollution prevention. The Organization for Economic Cooperation and Development (OECD) has just completed a major assessment on pollution prevention. The United Nations Environmental Programme has a clean technologies program and the United Nations Industrial Development Office (UNIDO) held major conferences on sustainable development. Joining this group of international bodies is the North Atlantic Treaty Organization (NATO). NATO has a nonmilitary Committee on Challenges of Modern Society that has just begun a multiyear pilot study called "Pollution Prevention Strategies for Sustainable Development" in which 14 countries are involved in an information exchange program on pollution prevention policy, education, and technology.[13]

Individual countries have taken their own initiatives in developing pollution prevention programs. Canada has the Green Plan and The Netherlands has the

NEPP (National Environmental Policy Plan). Denmark and The Netherlands are extensively studying life-cycle accounting applied to a host of consumer and commercial products.

A very interesting approach to using the regulatory system is being pursued by Germany to encourage P2. The 1986 Waste Act empowers the federal government to bring its influence to bear on waste generation prior to the production and use of products.

The options, if necessary to increase recycling and to simplify waste treatment, include

- Subjecting certain products to mandatory labeling or separate handling

- Requiring the manufacturers to reclaim their products once they become waste

- Imposing bans or restrictions on marketing

Although these regulations are valid for all wastes, the two main aims are

- To reduce the pollutant content of waste and thereby enable more recycling of these pollutant-free wastes

- To reduce the amount of household waste by reducing all kinds of packaging material

In carrying out the second of the above options, the German government has passed an ordinance that required, by January 1993, that commercial dealers accept for recycling all returned packaging in or near shops.[14]

1.4 Industrial Programs

Reflecting both an interest in saving money and avoiding increasingly stringent end-of-the-pipe environmental regulations, and responding to the desire on the part of the consuming public for more environmentally friendly activities, industries around the world have adopted pollution prevention with a vengeance and have initiated many broad programs.

In the United States, the Chemical Manufacturing Association (CMA) reports a "quiet revolution" that the chemical industry is conducting within its own operation to improve its performance. As part of the CMA's Responsible Care Program, the industry has adopted a Waste and Release Reduction and Management Code that contains 10 management practices that provide a framework for reducing waste generation and releases to the environment. The CMA notes that "improved performance will take time, money, and hard work. As we move down this road, we invite others to pick up the challenge and join us."[15] If you can believe the wealth of information being produced on all fronts, the challenge is being picked up very well indeed.

1.5 Where Do We Go from Here?

Two decades of environmental regulation in the United States have shown notable results in addressing air pollution, water pollution, and solid and hazardous waste disposal. Today, to quote a 1988 EPA report,

> There is no question that the air in most of our cities is far cleaner and healthier than it was in the 1960's. Thousands of miles of rivers and streams and thousands of acres of lakes have been restored and protected for fishing and swimming. In addition, we have taken extraordinary steps to improve the management of hazardous wastes, toxic chemicals, and pesticides. Nevertheless, looking at the current roster of environmental problems, they appear to be as formidable as ever. They include global concerns such as climate change and ozone depletion; cross-media pollutants such as lead and heavy metals, solvents, and pesticides; small and dispersed source concerns such as agricultural runoff and mobile sources; shortages of waste disposal capacity and massive waste cleanup bills; and a growing public concern over the basic state of the environment in which we live. Of course, our concerns in the United States differ little from the concerns of citizens of countries throughout the world.
>
> As we have achieved success in improving our environment, it has become increasingly clear that there is only so much that can be achieved through strategies based upon developing more effective pollution control technologies. There appears to be much more to be gained from strategies based on eliminating polluted streams through improvements in the production processes themselves....Admittedly, this is a much longer term strategy, but the apparent benefits are impressive.[16]
>
> Zero discharge of wastes is not possible—but as we continue to enjoy the benefits from the efforts of creative individuals around the world who are developing new approaches for encouraging cleaner products and production technologies, we can see that a world with much less waste, and therefore a lessening of the problems arising from treating and disposing of that waste, is certainly possible.
>
> To quote a former head of the EPA's Office of Pollution Prevention, "As progress is made on the technological side of pollution prevention, there is a growing recognition of the need for prevention to become an integral part of our basic philosophy of environmental protection. Pollution prevention must become the strategy of first choice in addressing any environmental problem. Creating this new "pollution prevention ethic" requires a shift in the perspectives of those whose activities affect the environment. Without question, this is a massive undertaking, and one that will continue to challenge society in the years ahead."[17]

References

1. U.S. Environmental Protection Agency, *Environmental Protection Agency Pollution Prevention Directive,* May 13, 1990.

2. J. C. Van Weenen, *Waste Prevention: Theory and Practice,* Castricum Publishers, Delft, The Netherlands, 1990.

3. William J. Clinton, 1993 Earth Day Address, April 21, 1993.

4. Carol Browner, "Carol Browner on EPA's Priorities," *EPA Journal,* vol. 19, no. 2, April–June 1993.

5. U.S. Environmental Protection Agency, *Meeting the Environmental Challenge: EPA's Review of Progress and New Directions in Environmental Protection,* EPA, 21K-2001, December 1990.

6. U.S. Environmental Protection Agency Science Advisory Board, *Future Risk: Research Strategies for the 1990's,* SAB-EC-88-40, September 1988.

7. Stephan Schmidheiny, "Looking Forward: Our Common Enterprise," in *Environmental Strategies Handbook: A Guide to Effective Policies and Practices,* McGraw-Hill, New York, 1994.

8. S. A. Weiner, "Environmental Protection in New Jersey," in *Environmental Strategies Handbook: A Guide to Effective Policies and Practices,* McGraw-Hill, New York, 1994.

9. Robert P. Bringer and David Benforador, "Pollution Prevention and Total Quality Environmental Management," in *Environmental Strategies Handbook: A Guide to Effective Policies and Practices,* McGraw-Hill, New York, 1994.

10. Morton L. Mullins, "Industry Perspective, Environmental Health and Safety Challenges and Social Responsibilities," in *Environmental Strategies Handbook: A Guide to Effective Policies and Practices,* McGraw-Hill, New York, 1994.

11. U.S. Environmental Protection Agency, *Pollution Prevention Benefits Manual* (Draft), vol. 1, EPA Report WAM-1, October 1989. Available from Pollution Prevention Information Clearinghouse, Falls Church, Va.

12. U.S. Environmental Protection Agency, *Pollution Prevention 1991: Progress on Reducing Industrial Pollutants,* EPA 21P-3033, October 1991.

13. NATO Committee on the Challenges of Modern Society, *Pollution Prevention Strategies for Sustainable Development Newsletter,* NATO Brussels, Winter 1991.

14. Tellus Institute, *The Tellus Institute Packaging Study,* Boston, Mass., prepared for the Council of State Governments, Washington, D.C., November 1991.

15. "Improving Performance in the Chemical Industry: Ten Steps for Pollution Prevention," Chemical Manufacturers Association, Washington, D.C.

16. Harry M. Freeman and Gary E. Hunt, "Industrial Pollution in the United States," presented at Environment 90 Conference and Exhibition, Jyvasklya, Finland, May 23–25, 1990.

17. G. Kotas, "A Look Back and a Look Ahead," *Pollution Prevention News,* U.S. EPA, Office of Pollution Prevention, Washington, D.C., October 1991.

2

Overview of Waste Reduction Techniques Leading to Pollution Prevention

Gary E. Hunt
North Carolina Department of Environment,
Health, and Natural Resources
Raleigh, North Carolina

2.1 Introduction

Liquid, solid, and/or gaseous waste materials are always generated during the manufacture of any product. In addition to creating environmental hazards, these wastes represent losses of valuable materials and energy from the production process and require a significant investment in pollution control. Traditionally, pollution control relies on "end-of-the-pipe" and "out-the-back-door" management approaches that require labor hours, energy, materials, and capital expenditures. Such an approach removes pollutants from one source, such as wastewater, but places them somewhere else, such as in a landfill.

More regulations, higher disposal expenses, increased liability costs, and increased public awareness have caused industrial and governmental leaders to begin critical examinations of end-of-the-pipe control technologies. The value of reducing waste during the manufacturing process has become apparent to many industries. These companies are looking at broader environmental management objectives, rather than concentrating solely on pollution control.

Waste reduction not only is very often economically beneficial for an industry, it also improves the quality of the environment.

Waste reduction is not a new concept; it has been around as long as people have been manufacturing products. One electroplating manual from the late eighteenth century states that "nothing whatever should be allowed to go to waste in well-conducted works."[1] The manual includes many waste reduction methods that are applicable and discussed in the literature even today. These "modern" methods point up the fact that many waste reduction techniques are relatively "low tech." In fact, many industries find that simple operational changes, increased training, and improved inventory management can significantly reduce waste generation rates.

Waste reduction techniques can be applied to any manufacturing process, from one as simple as making a paper clip to one as complex as assembling an F-16 fighter aircraft. Available techniques, which range from easy operational changes to state-of-the-art recovery equipment, can be broken down into four major categories: managing inventory, modifying production processes, reducing waste volume, and recovering waste. Because the classifications are broad, there is some overlap. In actual application, waste reduction techniques generally are used in combination so that the maximum effect is achieved at the lowest cost.

One point must be stressed: technology alone will not reduce waste; it must be coupled with trained, motivated, and empowered workers. Unlike traditional end-of-pipe management, successful waste reduction strategies rely heavily on worker involvement rather than on "black box" treatment technology. Thus, the level of corporate, committee, and worker involvement will have as much impact on the success of a program as will the selection of the appropriate technologies. This critical factor must be kept in mind when you read the following sections on waste reduction methods.

A useful method of categorizing waste reduction options is shown in Table 2-1.

2.2 Managing Inventory

Proper control over raw materials, intermediate products, final products, and the wastestreams associated with production is an important waste reduction technique. In many cases, waste is just raw materials that are out of date, "off-spec," contaminated, or unnecessary; spilled clean-up residues; or final products that are damaged. The cost of disposing of these materials includes not only the actual disposal costs but also the cost of the lost raw materials or product. This waste can represent a very large economic burden on any company. For example, one furniture company spent 70 percent of the purchase price to properly dispose of two years' worth of unused coating materials.

Table 2-1. Categories of Waste Reduction Techniques

Inventory management
   Inventory control
   Material control

Production process modification
   Operation and maintenance procedures
   Material change
   Process equipment modification

Volume reduction
   Source segregation
   Concentration

Recovery
   On-site recovery
   Off-site recovery

Any effective inventory management program must include process waste. Handling waste as if it were a product will help reduce waste and increase its potential for recovery. Many of the techniques discussed in this section can be applied to waste material as well as raw materials and finished products.

2.2.1 Inventory Control Methods

Methods for controlling inventory range from simply changing the ordering procedures to implementing just-in-time (JIT) manufacturing techniques. Many companies can help reduce waste generation by tightening up and expanding their current inventory management programs. This increased control will significantly affect the three major sources of waste resulting from improper inventory control: excess, out-of-date, and no-longer-used raw materials.

1. The purchase of only the amount of raw materials needed for a production run or a set period of time is one of the keys to proper inventory control. Often, manufacturers must dispose of excess raw materials simply because they are out of date. This problem can be reduced by better applying existing inventory management procedures and by educating purchasing personnel on the problems and costs of disposing of excess materials. Also, staff environmental engineers or chemists should evaluate the set expiration dates of raw materials, especially stable compounds, to see if those dates are too short.

2. The development of review and approval procedures for all raw materials purchased is another step in establishing an inventory control program. The approval process means that all production materials are evaluated to determine if they contain hazardous constituents and, if so, the alternative nonhazardous substitutes that are available. These review procedures can be devel-

oped by one person with the necessary background in chemistry or by a committee comprised of people with a variety of backgrounds. Often, the information needed for review is on the Material Safety Data Sheets (MSDS) provided by the chemical supplier. Two companies that have established successful material review programs are IBM[2] and Hewlett-Packard.[3]

3. The ultimate in inventory control procedures is JIT manufacturing. In JIT, raw materials go from the receiving dock to the manufacturing area for immediate use, and the finished product is shipped out without any intermediate storage. The result is no inventory of either raw materials or completed products. JIT techniques are complex to implement and cannot be used by all facilities, but they can reduce waste significantly. Using JIT techniques, the 3M Company reduced waste generation by 25 percent to 65 percent in its individual plants.[4]

2.3 Modifying Production Processes

Improving the efficiency of a production process can significantly reduce waste generation at the source. In fact, some of the most cost-effective waste reduction techniques are simple and relatively inexpensive changes in production procedures. Available techniques include (1) improving current operation and maintenance procedures, (2) changing the materials used in production, and (3) modifying existing equipment or purchasing more efficient—and more cost-effective—equipment.

2.3.1 Improving Operational Procedures

Improved operational procedures are quite simply methods that make optimum use of the raw materials used in the production process. Such methods are neither new or unknown, are usually inexpensive to institute, and involve little or no capital outlay. For example, a producer of breaded foods instituted a number of operational changes such as dry cleanup, installation or modification of drip trays under processing equipment, and better systems for collection and handling of waste material. These relatively modest changes resulted in the following:

- Decreased water usage for cleanup, by about 30 percent
- Elimination of the landfilling of waste solids
- Reduced organic load of wastewater, by almost 80 percent
- Increased revenues, as the company sold 2,359,000 kg (5.2 million lb) a year of solids to recovery firms[5]

For any business, the first step in instituting improved operations geared to pollution prevention is to examine the current production process for ways to improve its efficiency. A review would include all segments of the process from the delivery of raw materials through production to final product storage.

One important area commonly overlooked or not given proper attention in many manufacturing facilities is material handling. Proper material handling ensures that raw materials reach the production area without loss from spills, leaks, or contamination. Proper procedures also ensure that materials are efficiently handled in the production process.

Another critical area is *maintenance*. One company found that one-fourth to one-half of its excess waste load resulted from poor maintenance.[6] A strict maintenance program that stresses corrective and preventive maintenance can reduce waste generation caused by equipment failure. Such a program will help spot potential sources of release and correct the problem before any material is lost. A good maintenance program is important because the benefits of the best waste reduction program can be wiped out by just one leak or equipment malfunction.

A maintenance program can include maintenance cost tracking and preventive maintenance scheduling and monitoring. To be effective, a maintenance program should be developed and followed for each operational step in the production process, with special attention given to potential problem points. A strict schedule and accurate records on all maintenance[7] activities should be maintained. Computerized maintenance scheduling and tracking programs are available from a variety of vendors. A comprehensive program should also include predictive maintenance. This approach provides the means to schedule future repairs or replacement of equipment relative to its current condition. A number of nondestructive testing technologies are available for making these evaluations.[8]

Once proper operating and maintenance procedures are established, they must be fully documented and made part of the employee training program. In fact, a comprehensive training program is a key element of any effective waste reduction program. Industry case studies show that, through training, a dairy plant,[9] a semiconductor manufacturer,[10] and a furniture plant[11] reduced waste by 14 percent, 40 percent, and 10 percent, respectively.

For a program to be effective, all levels of personnel from the line operator to the corporate executive officer should be included. The goal of any program is to make every employee aware of waste generation, its impact on the company and the environment, and ways it can be reduced. Written materials should be prepared and used in conjunction with hands-on training. The training should be an ongoing process with review updates and interaction between employees and supervisors on a regular basis.

Table 2-2 lists some good operating practices that have been found to lead to pollution prevention.

Table 2-2. Pollution Prevention through Good Operating Practices

Good operating practice	Program ingredients
Waste segregation	Prevent mixing of hazardous wastes with nonhazardous wastes.
	Store materials in compatible groups.
	Segregate different solvents.
	Isolate liquid wastes from solid wastes.
Preventive maintenance programs	Maintain equipment history cards on equipment location, characteristics, and maintenance.
	Maintain a master preventive maintenance (PM) schedule.
	Keep vendor maintenance manuals handy.
	Maintain a manual or computerized repair history file.
Training and awareness-building programs	Provide training for
	■ Operation of the equipment to minimize energy use and material waste
	■ Proper materials handling to reduce waste and spills
	■ Awareness of the importance of pollution prevention by explaining the economic and environmental ramifications of hazardous waste generation and disposal
	■ Detecting and minimizing material loss to air, land, or water
	■ Emergency procedures to minimize lost materials during accidents
Effective supervision	Closer supervision may improve production efficiency and reduce inadvertent waste generation.
	Centralize waste management. Appoint a safety/waste management officer for each department. Educate staff on the benefits of pollution prevention. Establish pollution prevention goals. Perform pollution prevention assessments.
Employee participation	"Quality circles" (free forums between employees and supervisors) can identify ways to reduce waste.
	Solicit and reward employee suggestions for waste reduction ideas.
Production planning and scheduling	Maximize batch size to reduce clean-out waste.
	Dedicate equipment to a single product.
	Alter batch sequencing to minimize cleaning frequency (light-to-dark batch sequence, for example).
Cost allocation and accounting	Charge direct and indirect costs of all air, land, and water discharges to specific processes or products.
	Allocate waste treatment and disposal costs to the operations that generate the waste.
	Allocate utility costs to specific processes or products.

SOURCE: Ref. 1.

2.3.2 Changing to Less Hazardous Materials

Hazardous materials used in either a product formulation or a production process may be replaced with a less hazardous or nonhazardous material. Reformulating a product to contain a less hazardous material will reduce the amount of hazardous waste generated both during formulation of the product and in its end use. A less hazardous material used in a production process will generally reduce the amount of hazardous waste produced and, in turn, reduce the cost of capital equipment needed to meet environmental regulatory limits.

Although product reformulation is one of the more difficult waste reduction techniques, it can be very effective. Examples of product reformulation include eliminating pigments containing heavy metals from ink, dyes, and paint formulations; replacing phenolic biocides with other less toxic compounds in metalworking fluids; and developing new paint, ink, and adhesive formulations based on water rather than on organic solvents.

Hazardous chemicals used in the production process can also be replaced with less hazardous or nonhazardous materials. Changes can range from switching to purer raw materials to replacing solvents with water-based products. The latter is a widely used waste reduction technique and is applicable in many industries. For example, a diesel engine remanufacturing facility that switched from cleaning solvents and oil-based metalworking fluids to water-based products reduced its coolant and cleaning costs by about 40 percent. Also, the company was able to eliminate one cleaning step, and it found that since machine filters lasted twice as long, material and labor costs were reduced.[12]

A word of warning: one important area that is sometimes overlooked in material changes is the impact on the total wastestream. Switching from a solvent-based to a water-based product can increase wastewater volumes and concentrations. These increases could adversely affect the current wastewater treatment system, cause effluent limits to be exceeded, and possibly increase the amount of wastewater treatment sludge produced. Thus, before any change is made, its impact on all discharges must be evaluated.

2.3.3 Modifying or Changing Equipment

Waste generation may be reduced by installing more efficient equipment or by modifying existing equipment to take advantage of better production techniques. Not only can new or updated equipment process materials more efficiently, it produces less waste. Also, by reducing the number of rejected or off-specification products, high-efficiency systems reduce the amount of material that must be reworked or discarded.

Modification of existing equipment can be a highly cost-effective method to reduce waste generation. In many cases, relatively simple and inexpensive modifications can help ensure that materials are not wasted or lost. Such modifications can be as easy as redesigning parts racks to reduce drag-out in electroplating operations, installing better seals on process equipment to eliminate leakage, or installing drip pans under equipment to collect leaking process material for reuse. One chemical company reduced the waste from a sump in a production area from 31,750 kg per year to 1360 kg per year by installing a sight glass, using better pump seals, and purchasing a broom for dry cleanup.[13]

Installing new, more efficient equipment, and in some cases, modifying current equipment, will require capital investment in equipment, facility modifications, and employee training. The extent of the investment will vary greatly with the type of equipment used. These investments, however, can have a rapid payback. For example, a power tool manufacturer replaced a spray solvent paint system with a water-based electrostatic immersion painting unit. This new equipment paid for itself in just over one year by reducing raw material costs by $600,000 a year and waste disposal costs by 97 percent and greatly increasing productivity.[14]

2.4   Reducing Waste Volume

Volume reduction methods include techniques to separate toxic, hazardous, and/or recoverable wastes from the total wastestream. These techniques are usually used to increase material recoverability; reduce waste volume, and thus disposal costs; or increase management options. The available techniques range from simple segregation of wastes at the source to complex concentration technology.

2.4.1   Segregating Waste at the Source

Segregation of wastes is, in many cases, a simple and economical technique for waste reduction. For example, by segregating wastes at the source of generation and handling the hazardous and nonhazardous wastes separately, the volume of waste, and thus the cost of managing it, are reduced. Also, the uncontaminated or undiluted wastes may be reusable in the production process or sent off site for recovery.

This technique is applicable to a wide variety of wastestreams and industries and usually involves simple changes in operational procedures. For example, in metal-finishing facilities, wastes containing different types of metals can be treated separately so that the metal values in the sludge can be recovered. Spent solvents or waste oils that are kept segregated from other solid or liquid waste may be recyclable. If wastewater containing toxic material is kept separate from uncontaminated process water, the volume of water that must be treated is reduced.

A commonly used waste segregation technique is to collect and store wash-water or solvents used to clean equipment (such as tanks, pipes, pumps, or printing presses) for reuse in the production process. This technique is used by paint, ink, and chemical formulators as well as by printers and metal fabricators. For example, a printing firm segregates and collects toluene used for press and roller cleanup operations. By segregating the used toluene by color and type of ink contaminant, the company can reuse it later for thinning the same type and color of ink. Recovery of 100 percent of the waste toluene has totally eliminated a hazardous wastestream.[15]

2.4.2 Concentrating Waste

Various techniques are available to reduce the volume of a waste through physical treatment. Such techniques usually remove a component of the waste such as water. Available concentration methods include gravity and vacuum filtration, evaporation, ultrafiltration, reverse osmosis, freeze vaporization, filter press, heat drying, and compaction. Many of these methods are actually recovery techniques and will be discussed further in the next section.

Unless the material can be recycled, to concentrate a waste so that more of it can be "fit into a drum" is not waste reduction. But in some cases, concentration of a wastestream may also increase the likelihood that the material can be reused or recycled. For example, filter presses or sludge driers can increase the concentration of metals in electroplating wastewater treatment sludges to such a level that they become valuable raw material for metal smelters. A printed circuit board manufacturer uses a filter press to dewater its sludge and reduce it to 60 percent solids. The company receives $7200 a year from the sale of the dewatered sludge for copper reclamation.[16]

2.5 Recovering Waste

The U.S. Environmental Protection Agency (EPA) does not consider waste recovery to be pollution prevention unless it is accomplished within the production process producing the wastestream. However, recovering waste is a highly cost-effective waste management alternative, and is certainly within the spirit of reducing the environmental impact of industrial operations. Waste recovery techniques can help eliminate waste disposal costs, reduce raw material costs, and possibly provide income from a salable waste.

The effective use of recovery depends on the segregation of the recoverable waste from other process wastes or extraneous material. This segregation ensures that the waste is uncontaminated and that the concentration of recoverable material is maximized. Some companies[17] have assigned the responsibility for the handling, collection, and scheduling for recovery of waste material to one individual to ensure that the maximum value of the waste can be recovered. Waste recovery can take place either on site or at an off-site facility.

2.5.1 On-Site Recovery

In most cases, the best place to recover process wastes is at the production facility. Waste can most efficiently be recovered at the point of generation, since it is there that the possibility of contamination by other waste material is lower, and most of the risk involved in handling and transporting waste materials has not yet been incurred. Wastes that are simply contaminated versions of the raw materials used in the manufacturing process are good candidates for in-plant recycling.

Some wastestreams can be recycled directly back to the original production process as raw material. This redirection is usually accomplished when the waste material is lightly contaminated or is excess raw material. Examples include cleaning waste from printers, coaters, and chemical or product formulators; electroplating drag-out solutions; process solutions from filter changes; and dust collector residue from pesticide formulators. Lightly contaminated waste can sometimes be reused in operations that do not require materials of the higher quality. For example, spent high-purity solvents generated during the production of microelectronics can be reused in less critical metal degreasing operations, or a caustic waste material can be reused to treat an acid wastestream.

Some waste may have to undergo some type of purification before it can be reused. A number of physical and chemical techniques available on the market can be used to reclaim a waste material. These techniques range from simple filtration to state-of-the-art techniques such as freeze crystallization. The method of choice will depend on the physical and chemical characteristics of the wastestream, the economics of recovery, and the operational requirements. Most on-site recovery systems will generate some type of residue (contaminants removed from the recovered material). This residue can either be processed for further recovery or properly disposed. Economic evaluations of any recovery technique must include the management of these residues.

2.5.2 Off-Site Recovery

Wastes may be recovered at an off-site facility when the equipment is not available to recover them on site, when not enough waste is generated to make an in-plant system cost effective, or when the recovered material cannot be reused in the production process. Off-site recovery usually entails the recovery of a valuable portion of the waste through chemical or physical processes. Some materials that are commonly reprocessed off site are oils, solvents, electroplating sludges and process baths, lead-acid batteries, scrap metal, food-processing waste, plastic scrap, and cardboard. The cost of off-site recycling will depend on the purity of the waste and the market for the waste or recovered material.

In some situations, a waste may be transferred to another company for use as a raw material in its manufacturing process. This exchange can be economically advantageous to both firms, as it will reduce the waste disposal costs of the generator and reduce the raw material costs of the user. For example, an x-ray film manufacturer found that it could produce a salable product from waste film stock. The company installed equipment that flakes and bales waste polyester-coated film stock, which is then sold as raw material input to another firm. The more than 9 million kg (20 million lb) of film stock exchanged each year represent $200,000 in annual savings in collection, transport, and processing costs and an annual profit of $150,000 from the sale of the materials.[18]

The upgrading of a waste into a product requires a strong commitment from the generator to find markets, both inside and outside the company, for the waste material. In some cases, the production process or the waste may have to undergo some modification to make a more salable product. Regional waste exchanges have been set up by a number of states to help companies find markets; these exchanges act as information clearinghouses for wastes available and materials wanted. The service usually offered is a listing of wastes available from generators and wanted by users in a catalogue or computer database form.

2.6 Summary

In the final analysis, waste reduction depends on looking at waste in a different way: not as something that inevitably must be treated and disposed of but as what it really is—a loss of valuable process materials, the reduction of which can have significant economic benefits. Thus, waste reduction is not an environmental issue, but a competitive one. In these days of global economies, companies must implement techniques which reduce manufacturing costs yet maintain high product standards. Manufacturing methods which also eliminate or reduce waste generation will have a further advantage. Waste reduction is also much more than just a technology; innovative business management methods are also a critical part of any waste reduction effort. Technology alone will not reduce waste—it must be coupled with significant employee involvement.

Table 2-3 is a checklist useful for identifying pollution prevention and recycling methods in a wide variety of industrial operations. Tables 2-4 and 2-5 are more specific checklists for the printing and printed circuit board industries, respectively.

Table 2-3. Checklist for All Industries

Waste origin	Waste type	Pollution prevention and recycling methods
Materials receiving	Packaging materials, off-spec materials, damaged containers, inadvertent spills, transfer hose emptying	Use just-in-time ordering system.
		Establish a centralized purchasing program.
		Select quantity and package type to minimize packing waste.
		Order reagent chemicals in exact amounts.
		Encourage chemical suppliers to become responsible partners (e.g., accept outdated supplies).
		Establish an inventory control program to trace chemical from cradle to grave.
		Rotate chemical stock.
		Develop a running inventory of unused chemicals for other departments' use.
		Inspect material before accepting a shipment.
		Review material procurement specifications.
		Validate shelf-life expiration dates.
		Test effectiveness of outdated material.
		Eliminate shelf-life requirements for stable compounds.
		Conduct frequent inventory checks.
		Use computer-assisted plant inventory system.
		Conduct periodic materials tracking.
		Properly label all containers.
		Set up staffed control points to dispense chemicals and collect wastes.
		Buy pure feeds.
		Find less critical uses for off-spec material (that would otherwise be disposed).
		Change to reusable shipping containers.
		Switch to less hazardous raw material.
		Use rinsable/recyclable drums.
Raw material and product storage	Tank bottoms; off-spec and excess materials; spill residues; leaking pumps, valves, tanks, and pipes; damaged containers; empty containers	Establish Spill Prevention, Control, and Countermeasures (SPCC) plans.
		Use properly designed tanks and vessels only for their intended purposes.
		Install overflow alarms for all tanks and vessels.
		Maintain physical integrity of all tanks and vessels.
		Set up written procedures for all loading and unloading and all transfer operations.

Table 2-3. Checklist for All Industries (*Continued*)

Waste origin	Waste type	Pollution prevention and recycling methods
Raw material and product storage	Tank bottoms; off-spec and excess materials; spill residues; leaking pumps, valves, tanks, and pipes; damaged containers; empty containers	Install secondary containment areas.
		Instruct operators to not bypass interlocks, alarms, or significantly alter setpoints without authorization.
		Isolate equipment or process lines that leak or are not in service.Use sealless pumps.
		Use bellows-seal valves.
		Document all spillage.
		Perform overall material balances and estimate the quantity and dollar value of all losses.
		Use floating-roof tanks for control of volatile organic compounds.
		Use conservation vents on fixed-roof tanks.
		Use vapor recovery systems.
		Store containers in such a way as to allow for visual inspection for corrosion and leaks.
		Stack containers in such a way as to minimize the chance of tipping, puncturing, or breaking.
		Prevent concrete "sweating" by raising the drum off storage pads.
		Maintain Material Safety Data Sheets to ensure correct handling of spills.
		Provide adequate lighting in the storage area.
		Maintain a clean, even surface in transportation areas.
		Keep aisles clear of obstruction.
		Maintain distance between incompatible chemicals.
		Maintain distance between different types of chemicals to prevent cross-contamination.
		Avoid stacking containers against process equipment.
		Follow manufacturers' suggestions on the storage and handling of all raw materials.
		Use proper insulation of electric circuitry and inspect regularly for corrosion and potential sparking.
		Use large containers for bulk storage whenever possible.
		Use containers with height-to-diameter ratio equal to 1 to minimize wetted area.
		Empty drums and containers thoroughly before cleaning or disposal.
		Reuse scrap paper for notepads; recycle paper.

Table 2-3. Checklist for All Industries (*Continued*)

Waste origin	Waste type	Pollution prevention and recycling methods
Laboratories	Reagents, off-spec chemicals, samples, empty sample and chemical containers	Use micro or semimicro analytical techniques.
		Increase use of instrumentation.
		Reduce or eliminate the use of highly toxic chemicals in laboratory experiments.
		Reuse/recycle spent solvents.
Operation and process changes	Solvents, cleaning agents, degreasing sludges, sandblasting waste, caustic, scrap metal, oils, greases from equipment cleaning	Recover metal from catalyst.
		Treat or destroy hazardous waste products as the last step in experiments.
		Keep individual hazardous wastestreams segregated, segregate hazardous waste from nonhazardous waste, segregate recyclable waste from nonrecyclable waste.
		Assure that the identity of all chemicals and wastes is clearly marked on all containers.
		Investigate mercury recovery and recycling.
		Maximize dedication of process equipment.
		Use squeegees to recover residual fluid on product prior to rinsing.
		Use closed storage and transfer systems.
		Provide sufficient drain time for liquids.
		Line equipment to reduce fluid holdup.
		Use cleaning system that avoids or minimizes solvents, and clean only when needed.
Operation and process changes	Sludge and spent acid from heat exchanger cleaning	Use countercurrent rinsing.
		Use clean-in-place systems.
		Clean equipment immediately after use.
		Reuse cleanup solvent.
		Reprocess cleanup solvent into useful products.
		Segregate wastes by solvent type.
		Standardize solvent usage.
		Reclaim solvent by distillation.
		Schedule production to lower cleaning frequency.
		Use mechanical wipers on mixing tanks.
		Use bypass control or pumped recycle to maintain turbulence during turndown.
		Use smooth heat exchange surfaces.
		Use on-stream cleaning techniques.
		Use high-pressure water cleaning to replace chemical cleaning where possible.
		Use lower-pressure steam.

Source: Ref. 1.

Table 2-4. Checklist for the Printing Industry

Waste origin	Waste type	Pollution prevention and recycling methods
Image processing	Empty containers, used film packages, outdated material	Recycle empty containers. Recycle spoiled photographic film.
Image processing	Photographic chemicals, silver	Use silver-free films, such as vesicular, diazo, or electrostatic types. Use water-developed litho plates. Extend bath life. Use squeegees to reduce carryover. Employ countercurrent washing. Recover silver and recycle chemicals.
Plate making	Damaged plates, developed film, outdated materials	Use electronic imaging, laser plate making.
Plate making	Acids, alkali, solvents, plate coatings (may contain dyes, photopolymers, binders, resins, pigment, organic acids), developers (may contain isopropanol, gum arabic, lacquers, caustics), and rinse water	Use electronic imaging, laser print making. Recover silver and recycle chemicals. Use floating lids on bleach and developer tanks. Use countercurrent washing sequence. Use squeegees to reduce carryover. Substitute iron-EDTA for ferrocyanide. Use washless processing systems. Use better operating practices. Remove heavy metals from wastewater.
Finishing	Damaged products, scrap	Reduce paper use and recycle waste paper.
Printing	Lubricating oils, waste ink, cleanup solvent (halogenated and nonhalogenated), rags	Prepare only the quantity of ink needed for a press run. Recycle waste ink and solvent. Schedule runs to reduce color change-over. Use automatic cleaning equipment. Use automatic ink leveler. Use alternative solvents. Use water-based ink. Use UV-curable ink. Install web break detectors. Use automatic web splicers. Store ink properly. Standardize ink sequence. Recycle waste ink.
Printing	Test production, bad printings, empty ink containers, used blankets	Install web break detectors. Monitor press performance. Use better operating practices. Use alternative fountain solutions. Use alternative cleaning solvents. Use automatic blanket cleaners. Improve cleaning efficiency. Collect and reuse solvent. Recycle lube oils.
Finishing	Paper waste from damaged product	Reduce paper use. Recycle waste paper.

source: Ref. 1.

23

Table 2-5. Checklist for the Printed Circuit Board Industry

Waste origin	Waste type	Pollution prevention and recycling methods

Printed circuit board manufacture	General	Product substitution
		    Surface-mount technology
		    Injection-molded substrate and additive plating

Cleaning and sur-face preparation	Solvents	Materials substitution
		    Use abrasives
		    Use nonchelated cleaners
		Increase efficiency of process
		    Extend bath life, improve rinse efficiency, countercurrent cleaning
		Recycle/reuse
		    Recycle/reuse cleaners and rinses

Pattern printing and masking	Acid fumes and organic vapors, vinyl polymers spent resist removal solution, spent acid solution, waste rinse water	Reduce hazardous nature of process
		    Aqueous processable resist
		    Screen printing versus photolithography
		    Dry photoresist removal
		Recycle/reuse
		    Recycle/reuse photoresist stripper

Electroplating and electroless plating	Plating solutions and rinse wastes	Eliminate process
		    Mechanical board production
		Materials substitution
		    Noncyanide baths
		    Noncyanide stress relievers
		Extend bath life; reduce drag-in
		    Proper rack design and maintenance, better precleaning and rinsing, use of demineralized water as makeup, proper storage methods
		Extend bath life; reduce drag-out
		    Minimize bath chemical concentration, increase bath temperature, use wetting agents, proper positioning on rack, slow withdrawal and sample drainage, computerized/automated systems, recover dragout, use airstreams or fog to rinse plating solution into the tank, collect drips with drain boards
		Extend bath life; maintain bath solution quality
		    Monitor solution activity
		    Control temperature
		    Mechanical agitation
		    Continuous filtration/carbon treatment
		    Impurity removal

Table 2-5. Checklist for the Printed Circuit Board Industry (*Continued*)

Waste origin	Waste type	Pollution prevention and recycling methods

Electroplating and electroless plating	Plating solutions and rinse wastes	Improve rinse efficiency
		    Closed-circuit rinses
		    Spray rinses
		    Fog nozzles
		    Increased agitation
		    Countercurrent rinsing
		    Proper equipment design and operation
		    Deionized water use
		Turn off rinse water when not in use
		Recovery/reuse
		    Segregate streams
		    Recover metal values

Etching	Etching solutions and rinse wastes	Eliminate process
		    Differential plating
		    Dry plasma etching
		Materials substitution
		    Nonchelated etchants
		    Nonchrome etchants
		Increased efficiency
		    Use thinner copper cladding
		    Pattern versus panel plating
		    Additive versus subtractive method
		Reuse/recycle
		    Reuse/recycle etchants

source: Ref. 1.

References

1. U.S. Environmental Protection Agency, *Facility Pollution Prevention Guide*, EPA/600/R-92/088, May 1992.
2. "Governor's Award for Excellence in Waste Management—1987," Governor's Waste Management Board, Raleigh, N.C., 1988.
3. Dadak, "Waste Minimization: The Hewlett-Packard Experience," in *Waste Minimization Manual*, Government Institutes, Inc., Washington, D.C., 1987.
4. Hunter, "Minimizing Waste by Source Segregation and Inventory Control," presented at Technical Strategies for Hazardous Waste Prevention and Control Seminar, Government Institutes, Inc., Washington, D.C., 1987.
5. Waynick, Carawan, and Tarver, "A Breaded Foods Processor Does It Too!" in *Proceedings of the Conference on Waste Reduction–Pollution Prevention: Progress and Prospects within North Carolina*, Pollution Prevention Program, North Carolina Department of Natural Resources and Community Development, Raleigh, N.C., 1988, pp. 30.1–30.9.

6. Shober, "Water Conservation and Waste Load Reduction in Food Processing Facilities," in *1988 Food Processing Waste Conference Proceedings,* Georgia Tech Research Institute, Georgia Institute of Technology, Atlanta, Ga., 1988.

7. Hunter, *op. cit.*

8. Mobley, "Turning Maintenance Dollars into Bottom-Line Profits," *CPI Equipment Reporter,* May–June 1988, p. 21.

9. Case study files from the Pollution Prevention Program, North Carolina Department of Natural Resources and Community Development, Raleigh, N.C.

10. T. Kalenowski and M. Keon, *Waste Generation and Disposition Practices and Currently Applied Waste Minimization Techniques within the Semiconductor Industry,* State of California Department of Health Services, Sacramento, Calif., 1987.

11. J. Koho et. al., *Managing and Recycling Solvents in the Furniture Industry,* North Carolina Board of Science and Technology, Raleigh, N.C., 1986.

12. Johnson, "Experiences in Getting Rid of Solvent-Based Degreasing in a Diesel Engine Remanufacturing Plant," in *Proceedings of the Conference on Waste Reduction–Pollution Prevention: Progress and Prospects within North Carolina,* Pollution Prevention Program, North Carolina Department of Natural Resources and Community Development, Raleigh, N.C., 1988, pp. 15.1–15.5.

13. Beck, "Waste Minimization—A Plant Approach to Getting Started," in *CMI Waste Minimization Workshop Proceedings,* vol. I, Chemical Manufacturers Association, Washington, D.C., 1987.

14. D. Huisingh, *Profits of Pollution Prevention: A Compendium of North Carolina Case Studies,* North Carolina Board of Science and Technology, Raleigh, N.C., 1985.

15. Ibid.

16. Ibid.

17. J. Kohl and J. Currier, *Managing Waste Oils,* Industrial Extension Service, North Carolina State University, Raleigh, N.C., 1987. See also Waynick et. al., op. cit.

18. Huisingh, op. cit.

Further Reading

Detailed discussions on developing a waste reduction program are covered in the following: H. Freeman, *Hazardous Waste Minimization,* McGraw-Hill, New York, 1990; U.S. Environmental Protection Agency, *Waste Minimization Opportunity Assessment Manual,* EPA 625/7-88/003, Hazardous Waste Engineering Research Laboratories, Cincinnati, Oh., 1988; Ontario Waste Management Corporation, *Industrial Waste Audit and Reduction Manual,* Toronto, 1987.

A list of trade associations can be found in *National Trade and Professional Associations of the United States,* Columbia Books, Washington, D.C., 1986.

3

Pollution Prevention Requirements in United States Environmental Laws

James Thurber, Esq.

Peter Sherman, Esq.

*Science Applications
International Corporation
Falls Church, Virginia*

3.1 Introduction

This section discusses federal legal authority to require pollution prevention. It presents an overview of the authority to implement pollution prevention that exists in the following major environmental statutes:

- The Pollution Prevention Act (PPA)
- The Resource Conservation and Recovery Act (RCRA)
- The Clean Air Act (CAA)
- The Clean Water Act (CWA)
- The Toxic Substances Control Act (TSCA)
- The Federal Insecticide, Fungicide, and Rodenticide Act (FIFRA)

Not all of the major environmental statutes are discussed. This is because many environmental statutes focus on the control, management, and disposal of pollutants, rather than on their prevention. Of those environmental statutes that do not directly authorize or mandate pollution prevention, many do indirectly promote pollution prevention through establishing regulatory programs that increase the cost, potential liability, and public scrutiny associated with managing hazardous materials. Two good examples include the Comprehensive Environmental Response, Compensation, and Liability Act (CERCLA) and the Emergency Planning and Community Right-to-Know Act (EPCRA).* However, the bulk of this section focuses on explicit pollution prevention authority and requirements.

This section does not focus extensively on the regulations or policies promulgated pursuant to the statutory provisions examined, except where such provisions have major significance. Nor does this section address state laws that mandate pollution prevention. Rather, state pollution prevention laws are discussed in Chap. 4.

For purposes of this section, pollution prevention is being defined in a manner consistent with the Pollution Prevention Act of 1990 (42 U.S.C. §13101 et seq.). That is, pollution prevention is being defined as source reduction, which, under the PPA, includes any practice that

> reduces the amount of any hazardous substance, pollutant or contaminant entering any waste stream or otherwise released into the environment (include fugitive emissions) prior to recycling, treatment or disposal; and reduces the hazard to public health and the environment associated with the release of such substances, pollutants or contaminants.

Source reduction specifically includes equipment and technology modifications, process or procedure modifications, reformulation or redesign of products, substitution of raw materials, and improvement in housekeeping, maintenance, training, or inventory control. It does not include practices that are not integral to the production of the product.[1]

3.2 Summary of Federal Pollution Prevention Statutory Provisions

Table 3-1 presents a summary of the pollution prevention authority provided in the environmental laws reviewed.

---

*For example, CERCLA is not a pollution prevention statute, but it promotes pollution prevention through its pervasive liability scheme. Similarly, EPCRA is a reporting and public right-to-know law—yet it is one of the primary drivers of pollution prevention due to the publicity it generates in this new era of environmentalism.

Table 3-1. Summary of Federal Pollution Prevention Provisions

Statutory provision	Pollution prevention authority/requirement
	Pollution Prevention Act
Findings and policy (42 U.S.C. §13102)	Source reduction is more desirable than waste management and pollution, yet opportunities for source reduction are often not realized.
	Mandates a national policy creating a hierarchy of preferred waste management approaches: source reduction, recycling, treatment, and disposal, all to be conducted in an environmentally safe manner.
EPA activities (42 U.S.C. §13103)	Directs the EPA to create an office (Office of Pollution Prevention at EPA Headquarters, Washington, D.C.) to implement this statute and develop a comprehensive pollution prevention strategy.
State grants for technical assistance programs (42 U.S.C. §13104)	The EPA administrator must make available to states matching grants for programs designed to promote the use of source reduction by businesses. Grants based on: availability of technical assistance to businesses; providing targeted assistance to businesses for whom lack of information is a major impediment to source reduction; and the extent to which training is made available to businesses.
	The administrator must also develop appropriate means by which to measure the effectiveness of these state grant programs.
Source Reduction Clearinghouse (42 U.S.C. §13105)	The EPA must establish the Source Reduction Clearinghouse to compile information on management, technical, and operational approaches to source reduction in a computerized database format.
Source reduction and recycling data collection (42 U.S.C. §13106)	Owners and operators of facilities subject to the annual toxic chemical release form filing requirements of SARA Sec. 313 must include with that filing a toxic chemical source reduction and recycling report for the preceding calendar year. This report must address: the amounts of chemicals released and recycled, source reduction practices aimed at the reported chemicals, methodologies used to identify source reduction opportunities, and future chemical production figures.
EPA Biennial Report (42 U.S.C. §13107)	The EPA must submit a biennial report to Congress that includes an assessment of the effectiveness of the Source Reduction Clearinghouse and the state grant program, analyzes the data generated under Sec. 6607 of the PPA to identify such topics as current trends in source reduction programs in various SIC code industries, regulatory and nonregulatory barriers to source reduction programs, opportunities to use existing federal regulatory programs to overcome existing impediments and promote source reduction in the United States, and identifies and makes recommendations with regard to innovative research and development programs targeting source reduction.

Table 3-1. Summary of Federal Pollution Prevention Provisions (*Continued*)

Statutory provision	Pollution prevention authority/requirement
	Resource Conservation and Recovery Act
Statutory objectives (42 U.S.C. §6902(a)(6))	An express objective of the RCRA is to minimize the generation of hazardous waste and the land disposal of hazardous waste by encouraging process substitution, materials recovery, properly conducted recycling and reuse, and treatment.
National policy (42 U.S.C. §6902(b))	Establishes as national policy that, to the extent feasible, the reduction or elimination of hazardous waste generation should be achieved as expeditiously as possible. Where this cannot be achieved, such wastes should be treated, stored, or disposed so as to minimize the present and future threat to human health and the environment.
Waste minimization requirements for manifests (42 U.S.C. §6922(b))	Hazardous waste generators are required to certify that they have a program in place to reduce the volume or quantity and toxicity of the materials they manage. Such programs must exist to the extent that they are economically practical.
Waste minimization requirements for permits (42 U.S.C. §6925(h))	Hazardous waste treatment, storage, and disposal facilities are required to certify that they have a program in place to reduce the volume or quantity and toxicity of the materials they manage. Such programs must exist to the extent that they are economically practical.
	Clean Air Act
Air toxics (42 U.S.C. §112)	In developing standards to control air toxics, the EPA has authority to require pollution prevention measures, including the installation of control equipment as well as process changes, the substitution of materials, changes to work practices, and operator training and certification.
Mobile source (42 U.S.C. §§241–250, 211)	Requires increasing percentage of fleet vehicles to use alternative fuels and establishes alternatively fueled vehicle pilot program in California. Requires the development of reformulated gasoline and oxygenated fuels to reduce air pollutants.
New source review (42 U.S.C. §§170–178)	Requires that new sources located in nonattainment areas use most stringent controls and provide offsets (i.e., emissions reductions from some other source that compensates for residual emissions). Such offsets may be achieved through pollution prevention.
Acid rain (42 U.S.C. §§401–406, 416)	Imposes stringent sulfur dioxide emissions limits and creates system of tradable emissions allowances. Forces power plants to either reduce their emissions to the specified rates or develop or buy emissions allowances, thereby creating market-driven incentives to reduce emissions.
Chlorofluorocarbons (42 U.S.C. §§602–605, 609, 610)	Requires the phaseout of production and sale of CFCs and several other chemicals that have been shown to contribute to the destruction of the stratospheric ozone layer. Also requires the imposition of various controls on CFC-containing products.

Table 3-1. Summary of Federal Pollution Prevention Provisions (*Continued*)

Statutory provision	Pollution prevention authority/ requirement
Clean Water Act	
Effluent limitations (33 U.S.C. §1311)	The EPA has the authority to develop technology-based, industry-specific national limits (implemented through NPDES permits) on the amounts of regulated pollutants a facility is allowed to discharge into the nation's waters. These standards may recommend, and, in some cases, mandate, in-plant controls.
National Pollutant Discharge Elimination System (NPDES) (33 U.S.C. §1342)	Subjects point-source discharges of contaminated stormwater to regulation, including pollution prevention plans.
Toxics Substances Control Act	
Manufacturing and processing notices (15 U.S.C. §2604)	Provides authority for the EPA to issue an order prohibiting or limiting the manufacture, processing, distribution in commerce, use, or disposal, of a chemical substance.
Hazardous chemical substances and mixtures (15 U.S.C. §2605)	Provides the EPA with the authority to promulgate rules to prohibit or limit production or to impose labeling or other requirements (i.e., record keeping, testing requirements, restrictions on the commercial use and disposal, notice of hazards to distributors, users, and the public) when the manufacturing, processing, distribution, use, or disposal of an existing chemical substance or mixture presents an unreasonable risk of injury to health or the environment.
Reporting and record keeping (15 U.S.C. §2607)	Requires the submittal of information about the hazards posed by chemical substances to facilitate regulatory and nonregulatory activity to mitigate known problems. Also supports voluntary submissions of chemical data.
Federal Insecticide, Fungicide, and Rodenticide Act	
Registration of pesticides (7 U.S.C. §136a)	All pesticides that are distributed or sold must be registered with the EPA, unless they are the subject of experimental use permits or an exemption. The administrator has the authority to limit the distribution, sale, or use of any pesticide that is not registered, subject to an experimental use permit, or exempted in order to prevent unreasonable adverse effects on the environment.
Administrative review and suspension (7 U.S.C. §136d)	For pesticides that are not in compliance, or if their use causes unreasonable adverse effects on the environment, the EPA may issue a notice of intent to cancel the pesticide's registration, change its classification, or hold a hearing on these issues. The EPA has authority to supercede the classification process and immediately suspend the registration if necessary to prevent an imminent hazard.
Stop sale, use, removal, and seizure (7 U.S.C. §136k)	Authorizes the EPA to cancel a pesticide if it is causing unreasonable effects on human health and the environment, or suspend its use immediately in order to prevent an imminent hazard. The administrator has the authority to seize pesticides, or to issue "stop sale, use, or removal" orders if the pesticide is in violation of any FIFRA provision, or if the registration has been cancelled.

3.3 Discussion

3.3.1 Pollution Prevention Act (PPA)

Congress responded to the growing national concern with waste generation and management practices by enacting the Pollution Prevention Act of 1990 (42 U.S.C §§13101–13109)(PPA) in October 1990. In its findings, Congress stated that source reduction opportunities often went unexploited because of a variety of factors: existing regulations and industrial resources were focused on treatment and disposal, the applicable regulations did not require or address a multimedia approach to pollution prevention, and there was a lack of essential information on source reduction technologies that industry needed to overcome institutional barriers to source reduction. Congress went on, declaring that "source reduction is fundamentally different and more desirable than waste management and pollution control."[2]

Further, Congress went on to say that it was to be the national policy that pollution should be prevented or reduced at the source whenever feasible. Pollution that cannot be prevented should be addressed through recycling programs, and if these options cannot be pursued, then the pollution should be treated and disposed of in an environmentally protective manner.

The Environmental Protection Agency has been directed to establish a source reduction program that collects and disseminates information, provides fiscal assistance to the state, and is the primary agency responsible for implementing the various provisions of the PPA. The EPA issued the Pollution Prevention Strategy (56 *Fed. Reg.* 7649, February 1991) to help clarify its pollution prevention mission and objectives to be accomplished. The strategy is designed to accomplish two main goals: (1) to provide guidance and focus for current and future efforts to incorporate pollution prevention principles and programs in existing EPA regulatory and nonregulatory programs, and (2) to set forth a program that will achieve specific pollution prevention objectives within a set, reasonable timeframe.[3]

The PPA has five major provisions that address developing and implementing a national source reduction program. Section 6604 of the PPA sets out a lengthy, comprehensive list of activities that the EPA Administrator is to develop as part of a strategy to promote source reduction. Some of the main activities include the following:

- Developing standardized methods of measuring source reduction
- Coordinating source reduction activities within the EPA as well as with other federal agencies
- Facilitating the adoption of source reduction programs by businesses using the Source Reduction Clearinghouse and the state matching grant program
- Identifying measurable source reduction goals and a strategy to successfully implement those goals

- Establishing source reduction training programs for all EPA program offices

- Identifying current barriers (regulatory, technological, policy) to achieving source reduction and making recommendations to Congress on overcoming these barriers

- Developing source reduction auditing procedures to help identify source reduction opportunities in the public and private sectors

Section 6605 of the PPA directs the EPA administrator to establish a matching grant program for states to promote the use of source reduction by business and industry. Grant requests to the EPA are evaluated for (1) the availability of specific technical assistance to businesses seeking to establish source reduction programs; (2) targeted assistance to those businesses for whom paucity of relevant information and data is a barrier to achieving some measure of source reduction program success; and (3) providing training opportunities for potential businesses in source reduction techniques.

The administrator, under Sec. 6606 of the PPA, will establish a Source Reduction Clearinghouse to compile information on management, technical, and operational approaches to source reduction in a computerized format. The Clearinghouse will serve as a center for source reduction technology transfer, develop and implement outreach and source reduction programs to encourage states to adopt source reduction practices, and collect and compile information on the operation and success of state source reduction programs operated under the matching grant program of Sec. 6605.

Section 6607 requires each owner and operator of a facility required to comply with the reporting requirements of SARA Sec. 313 (toxic chemical) to file an annual toxic chemical source reduction and recycling report with the EPA. This report must address such topics as (1) the quantity of chemical entering any wastestream; (2) the amount of chemical that is recycled and the process used; (3) any source reduction activities associated with the specific chemical(s); (4) projected amounts of the chemical(s) that will be reported for the next two calendar years; (5) a comparison of chemical production figures from the previous and current reporting years; (6) any techniques used to identify source reduction opportunities; (7) the quantity of chemicals released as a result of catastrophic events, remedial actions, or other one-time events; and (8) the quantity of chemical that is treated during the reporting year, and a comparison with similar data from the previous reporting year.

Finally, in Sec. 6608, the EPA must provide a biennial report to Congress that summarizes the data collected under the provisions of PPA Sec. 6607. The report must address the following topics:

- Data analysis on an industry-specific basis for a minimum of five Standard Industrial Codes (SICs), evaluating source reduction trends by industry, firm size, production, or other categories deemed appropriate by the EPA

- Usefulness and validity of the data in measuring trends in source reduction, and the adoption of source reduction programs by businesses

- Identification of regulatory and nonregulatory barriers to source reduction, and opportunities to use existing regulations and programs to encourage source reduction

- Identification of both industries and pollutants that require assistance in multimedia source reduction

- Identification of incentives needed to encourage research and development in source reduction technologies as well as existing opportunities to conduct such research

- Evaluation of the technical feasibility and associated costs of source reduction, and the identification of those specific industries for which there exist significant barriers to source reduction

3.3.2 Resource Conservation and Recovery Act (RCRA)

The Resource Conservation and Recovery Act (RCRA) addresses the management of solid waste, hazardous waste, and underground storage tanks that contain petroleum or hazardous substances. The RCRA establishes a comprehensive, or "cradle-to-grave," regulatory scheme applicable to hazardous wastes. However, the RCRA's hazardous waste provisions regulate wastes after they are generated; they generally do not authorize the EPA to regulate in-process materials. Hence, the RCRA does not provide extensive authority to mandate pollution prevention. Yet the RCRA does provide some authority for addressing pollution prevention. In 1984, the Hazardous and Solid Waste Amendments (HSWA) added several new provisions to the RCRA, some of which do require pollution prevention. These provisions make it clear that pollution prevention is a fundamental element of United States hazardous waste management policy.

The HSWA established the prevention of the generation of hazardous waste as the national policy of the United States. Under RCRA, this policy states that "...wherever feasible, the generation of hazardous waste is to be reduced or eliminated as expeditiously as possible...."[4] This policy has been adopted and expanded upon in the Pollution Prevention Act (1990).[5] In addition, the HSWA states that one statutory objective under the RCRA is to minimize "the generation of hazardous waste and the land disposal of hazardous waste by encouraging process substitution, materials recovery, properly conducted recycling and reuse, and treatment."[6]

The HSWA also mandated that hazardous waste generators and treatment, storage, and disposal facilities have waste minimization programs in place. Under RCRA Sec. 6923(b) and Sec. 6925(h), hazardous waste generators and facilities that treat, store, or dispose of hazardous waste generated on site are required to certify that they have a program in place to reduce the volume or quantity and toxicity of the materials they manage. Such programs must exist to the extent that they are economically practical. Generators must include such certifications on every hazardous waste manifest. Treatment, storage, and disposal facilities must have a requirement for such a program as a condition of their RCRA permit. The EPA has issued interim final guidance to assist hazardous waste generators and owners and operators of hazardous waste treatment, storage, and disposal facilities in complying with these requirements.[7]

3.3.3 Clean Air Act (CAA)

The Clean Air Act (CAA), originally passed in 1967 and most recently amended in 1990, seeks to protect our nation's air quality through imposing emission standards on stationary and mobile sources of air pollution. Compliance with the requirements imposed under the CAA has generally relied upon the use of end-of-pipe controls (i.e., air pollution control devices or APCDs). However, several provisions under the act do require or provide authority for pollution prevention. These include requirements addressing air toxics, mobile sources of air pollution, the review of new sources of air pollution, acid rain emission allowances, and ozone protection. All of these provisions have been significantly amended under the 1990 CAA amendments. Each is discussed following.

The CAA air toxics program provides controls over the large numbers of toxic substances that are not covered by national ambient air quality standards. For a variety of reasons, prior to the 1990 CAA amendments, EPA had only established emission standards for seven toxic substances.* Concerned about this lack of regulatory activity regarding toxics, Congress significantly strengthened and revised Sec. 112 in 1990. Under the amended Sec. 112, the EPA is required to regulate 189 substances presumed to merit regulation as air toxics. Regulation of these substances will consist of two phases: application of control technologies and subsequent addressing of residual risk. The first phase incorporates pollution prevention.

Initially, the EPA will have to establish maximum achievable control technology (MACT) standards governing selected categories of industrial facilities that emit the target pollutants. MACT standards must achieve "the maximum degree of reduction in emissions of hazardous air pollutants" determined by the EPA to be achievable (including prohibition where achievable) through application of specified processes, systems, or techniques.† These processes,

---

*Arsenic, asbestos, benzene, beryllium, mercury, radionuclides, and vinyl chloride. (See 40 CFR Part 61.)

†Such standards must take into consideration costs, energy requirements, and non-air-quality health and environmental benefits (See §112(d)(2).)

systems, or techniques may rely upon a wide range of control measures, including the installation of control equipment as well as process changes, the substitution of materials, changes to work practices, and operator training and certification.[8] While such requirements clearly provide an opportunity to promote pollution prevention, it is not clear how these provisions will be implemented. The first MACT standards were to be published by November of 1992; however, this schedule has been adjusted under a consent decree. Standards for 25 percent of the listed categories must be published by November of 1994.

The second CAA program that mandates pollution prevention is the mobile source program. This program focuses on reducing pollution from cars and other vehicles, as well as on the development of clean fuels (methanol, ethanol, mixtures of the two, reformulated gasoline, natural gas, liquified petroleum gas, and electricity). In addition to making the emissions standards for cars and other vehicles more stringent, the CAA amendments impose two clean-fueled-vehicles requirements, one for vehicle fleets and a pilot program applicable to vehicles in California. Both these and the clean-fuel provisions are discussed as follows.

Under the fleet requirements, cars and trucks that operate in fleets (generally 10 or more vehicles owned by one person and that are capable of being centrally fueled) in serious, severe, and extreme ozone nonattainment areas must comply with tailpipe emission standards that are more stringent than those applicable to nonfleet vehicles. These standards effectively require the use of alternative (i.e., clean) fuels. The percentage of each fleet that must meet the more stringent standards increases over time, thus achieving further reductions.[9]

The California clean-fueled-vehicle pilot program requires a portion of the vehicles sold in California to meet standards that are substantially more stringent than nationally applicable emission standards. The number of clean-fueled vehicles increases from 150,000 in 1996 to 300,000 in 1999 and continues to increase each year thereafter.[10]

The clean-fuels requirements of the CAA create programs mandating the use of reformulated gasoline and oxygenated fuels in highly polluted metropolitan areas. The gasoline program requires that in the nine cities with the highest ozone levels, reformulated gasoline is to be phased in beginning in 1995. The reformulated gasoline must meet specified limits for oxygen content, aromatic hydrocarbons, and benzene, as well as meeting restrictions on the formation of volatile organic compounds (VOCs) and hazardous air pollutants. The oxygenated-fuels program basically requires that in carbon monoxide nonattainment areas the oxygen content of fuels be at least 2.7 percent, starting in 1992.[11]

The third CAA program area that promotes pollution prevention is the new source review requirements. These requirements seek to minimize additional air pollution in attainment areas and prevent additional pollution in nonattainment areas. The latter requirement is achieved through imposition of the most stringent emission controls on new or modified sources and requirements that such sources provide offsets for any emissions that do result from the new or modified source. These offsets constitute emissions reductions that go beyond what would normally be required, and may be achieved in a variety of ways, including through the use of standard pollution prevention practices.[12]

The fourth program area that promotes pollution prevention is the acid rain provisions. These provisions focus on reducing the emissions of sulfur dioxide from power plants, since such plants account for approximately 80 percent of sulfur dioxide emissions to the atmosphere. The acid rain requirements are designed to be implemented in two phases: imposing reduced sulfur dioxide emissions rates in 1995 (2.5 lbs/mmBtu in Phase I) and further decreasing allowed emissions in the year 2000 (1.2 lbs/mmBtu in Phase II). The act also establishes, for the first time in a major federal environmental statute, a scheme of tradable emissions allowances. Effectively, these requirements force power plants to either reduce their emissions to the specified rates or to develop or buy emissions allowances. If a plant reduces its emissions below the specified level, it will generate allowances, which can be banked, used in Phase II, used at another plant, or sold to other facilities or industries. Each of the regulated plants must hold one allowance for every ton of sulfur dioxide (SO_2) emitted each year, beginning January 1, 1995. This approach diverges from the EPA's traditional command and control regulation by allowing power plants to generate whatever level of emissions they determine to be appropriate, provided they possess the allowances necessary. This approach also creates a market incentive to examine all methods of reducing air emissions, including pollution prevention.[13]

Finally, the ozone protection requirements of the CAA also rely upon pollution prevention to reduce the depletion of stratospheric ozone. The amended CAA requires the phaseout of production and sale of CFCs and several other chemicals that have been shown to contribute to the destruction of the stratospheric ozone layer. In addition, the provisions also require the imposition of various controls on CFC-containing products (motor vehicle air-conditioning units, motor vehicle refrigerant, nonessential CFC-containing products).[14]

3.3.4 Clean Water Act (CWA)

The Clean Water Act (CWA) seeks to restore and maintain the chemical, physical, and biological integrity of the nation's waters. The act contains five main components that work to accomplish this goal: (1) technology-based, industry-specific minimum national effluent (i.e., water discharge) standards; (2) water quality standards; (3) a permit program for discharges to United States water bodies; (4) specific provisions applicable to certain toxic and other pollutant discharges (i.e., oil, hazardous chemicals); and (5) a revolving Publicly Owned Treatment Works (POTW) construction loan program.

The primary focus of these provisions is to ensure that toxic levels of pollutants are not discharged to the nation's waters. This is achieved through restricting the types and amounts of pollutants that are discharged. Such restrictions are imposed through the use of enforceable effluent standards specified in National Pollution Discharge Elimination System (NPDES) permits. By its nature, the NPDES permit program primarily relies upon treatment as the principal means of achieving compliance with discharge restrictions. In this sense, the CWA does not focus primarily on pollution prevention. However, two CWA programs include pollution prevention components. These programs include effluent standards development and stormwater regulation (part of the NPDES's permitting of non-point-source pollution).

The most significant program components that encourage pollution prevention are the effluent discharge standards. These standards, which are developed for major industries and subcategories within these industries, force regulated industries to either reduce the amount of waterborne pollution they generate or pay the cost of treatment. To facilitate waste reduction, the agency publishes in-plant controls in each effluent standard development document. In-plant controls include recommended changes to process engineering, process management, equipment, and manufacturing or processing systems. For example, food processing often uses large quantities of water, and several in-plant controls focus on using alternative technologies that reduce the amount of water needed and/or the degree to which that water is contaminated. In many instances, such controls are similar to the pollution prevention techniques recommended today. Yet, in some cases, such control may be out of date. It is important to note that, in most cases, in-plant controls are not enforceable. Rather, they are suggested controls that will help the industry comply with the effluent standards. Following the enactment of the Pollution Prevention Act, there is a renewed emphasis on fostering source reduction opportunities through the development of effluent guidelines.

The second CWA program that emphasizes source reduction is the stormwater program. This program seeks to prevent contaminated stormwater runoff from polluting United States waters. Under the stormwater regulations, targeted facilities are required to develop stormwater pollution prevention plans. Implementation of these plans prevents stormwater runoff from becoming contaminated, thereby preventing pollution and avoiding the need to treat contaminated runoff.

3.3.5 Toxic Substances Control Act (TSCA)

The Toxic Substances Control Act (TSCA) of 1976 provides the EPA with the authority to test chemicals for their potential health and environmental effects prior to their introduction into commerce and to regulate these substances where they pose an unreasonable threat to health or environment. The act provides the authority to regulate the manufacturing, processing, distribution in commerce, use, and disposal of chemical substances and mixtures. The term *chemical substance* is broadly defined under the TSCA; however, several exclusions are provided, including pesticides, tobacco and tobacco products, nuclear materials and by-products, food, food additives, drugs, cosmetics, and devices.* TSCA Secs. 5, 6, and 8 provide the basic authority to prevent unreasonable harm to human health or the environment.

TSCA Sec. 5 addresses chemical manufacturing and processing notices. Section 5(a) of the TSCA prohibits any person from manufacturing or importing a new† chemical substance [i.e., one that does not appear on the chemical substance list (or "Inventory") established under Sec. 8(b)] without notifying the EPA at least 90 days prior to commencing the activity (i.e., manufacturing or importing). Notification is accomplished through the submittal of a Premanufacture Notice (PMN). Submittal of a PMN enables the EPA to screen new chemicals before commercial production or importation begins. The review period provides the EPA with an opportunity to review and evaluate information pertaining to the substance to determine whether manufacture, processing, distribution in commerce, use, or disposal should be limited or prohibited because of insufficient data with which to determine health and environmental effects of the substance or because the substance will present an unreasonable risk of injury to health or the environment. After determining the adequacy of the data submitted and/or the degree of risk posed by the chemical, the EPA may take several actions to regulate the chemical substance, including issuing an order or promulgating a rule prohibiting or limiting the manufacture, processing, distribution in commerce, use, or disposal of such a substance.

If the administrator determines that the information available is insufficient to permit a reasonable evaluation of the health and environmental effects of a chemical substance *and* that the manufacture, processing, distribution in commerce, use, and disposal of such a substance may present an unreasonable risk to health or the environment (*or* the substance will be produced in substantial amounts that may either enter the environment or cause substantial or significant human exposure to the substance), the administrator may issue a proposed order (i.e., Sec. 5(e) order) prohibiting or limiting the manufacture, processing, distribution in commerce, use, or disposal of such a substance. Such an order becomes effective upon the expiration of the PMN notification period.

If it is determined that there is a reasonable basis to conclude that the manufacture, processing, distribution, use, or disposal of a chemical substance presents or will present an unreasonable risk of injury to health or environment, the administrator may either issue a rule under Sec. 6 limiting the amount of such substance that may be manufactured, processed, or distributed (or take other actions under Sec. 6) or issue an order under Sec. 5(f). Under a Sec. 5(f) order, the administrator may prohibit the manufacture, processing, or distribution of a chemical substance or may seek an injunction to prohibit the manufac-

---

*For specific definitions of each exemption from the term *chemical substance* under TSCA, see §3 of TSCA ("Definitions").

†Significant new uses of existing chemicals may also be subject to similar regulation. See TSCA §5(a)(2).

ture, processing, or distribution of the chemical. A proposed rule issued under Sec. 5(f) (implemented through Sec. 6) will take effect immediately upon publication in the *Federal Register*. A Sec. 5(f) order becomes effective on the expiration of the notification period.

Section 6 of the TSCA addresses the regulation of hazardous chemical substances and mixtures. When the manufacturing, processing, distribution, use, or disposal of an existing chemical substance or mixture presents an unreasonable risk of injury to health or the environment, the administrator has broad authority under TSCA Sec. 6 to promulgate rules to prohibit or limit production or to impose labeling or other requirements. Such other requirements include imposing record keeping and testing requirements, imposing restrictions on the commercial use and disposal of the chemical, and requiring that notice of potential hazards be provided to distributors, users, and the public. Any such action undertaken by the administrator must be the least burdensome of the alternatives specified under the act.

Section 6(e) prohibits the manufacture, processing, and distribution in commerce of polychlorinated biphenyls (PCBs) except where totally enclosed. The administrator may authorize by rule the manufacturing, processing, distribution, or use of PCBs in a manner other than totally enclosed if it can be demonstrated that such activities will not present an unreasonable risk of injury to health or the environment. In addition, persons can petition the administrator for an exemption to the ban and the administrator may grant such petitions where it is found that an unreasonable risk of injury to health or the environment would not result and efforts have been made to develop safer substitutes. Such exemptions are limited to one year from the date they are granted.*

TSCA Sec. 8 consists of five subsections that establish reporting and record-keeping requirements. Chemical manufacturers, importers, processors, and, in certain cases, distributors, may be required to submit reports and/or maintain records. These requirements provide the EPA with considerable leverage for obtaining detailed data about chemical substances, which supports actions to restrict or prohibit chemical use.

Section 8(e) of the TSCA requires that "...any person who manufactures, processes, or distributes in commerce a chemical substance or mixture and who obtains information which reasonably supports the conclusion that such substance or mixture presents a substantial risk of injury to human health or the environment shall immediately inform the Administrator of such information unless such person has actual knowledge that the Administrator has been adequately informed of such information." The EPA's Office of Pollution Prevention and Toxics evaluates this information to determine whether specific

---

*The EPA has promulgated regulations prohibiting the production and use of PCBs. These regulations also address PCB marking, storage, disposal, spill clean-up, record keeping, and reporting. (See 40 CFR Part 761.) In addition, the EPA imposes restrictions on the manufacturing, processing, and distribution of fully halogenated chlorofluoroalkanes and on the manufacturing, importation, or processing of asbestos. The asbestos regulations also address the use of asbestos in schools and abatement. (See 40 CFR Parts 762, 763.)

chemical substances pose an unreasonable risk to human health or the environment. Response actions may include labeling and MSDS changes, processing changes, notifying workers and customers, or discontinued chemical processing or use. In addition, voluntary industry submissions are encouraged under Sec. 8(e). Since 1980, over 5000 such submissions have been received by the EPA and in over 500 cases, the businesses have acted to reduce releases or to slow or stop production of hazardous chemicals.*

3.3.6 Federal Insecticide, Fungicide, and Rodenticide Act (FIFRA)

The Federal Insecticide, Fungicide,and Rodenticide Act (FIFRA), and the FIFRA amendments of 1975, 1978, 1980, and 1988 establish the federal authority to regulate the distribution, sale, and use of pesticides in the United States. FIFRA's major provisions include product registration, pesticide use, and removal of pesticides from the market.

Before a pesticide can be lawfully sold or distributed in the United States, the product must first be registered by the EPA.† Registration allows a pesticide product to be sold and distributed for specified uses in accordance with specified use instructions, precautions, and other conditions. A pesticide product may be registered or remain registered only if it performs its intended functions without causing unreasonable adverse effects on the environment.[15] Additionally, any establishment that produces pesticide products also must be registered with the EPA.

Pesticide use is controlled mainly through labeling requirements and restricting the use and application of certain types or classes of pesticides. Registered pesticides are classified into *general* and *restricted* categories. General use pesticides are available to any member of the general public, but restricted use pesticides can only be sold and used by a certified pesticide applicator. Every pesticide product must bear a label or labeling stating what precautions are required and the directions for proper and safe use. Application and use must be consistent with the directions on the label.

Whenever the EPA receives information indicating that a registered pesticide product may cause unreasonable adverse effects on human health or the environment, the EPA has the authority to begin a special review process. During the course of this process, the EPA may specify particular modifications in the terms and conditions of registration, such as the deletion of particular uses or labeling revisions, as an alternative to cancellation of the registration; a de facto cancellation of the pesticide. This review process may result in a cancellation

---

*The EPA imposes testing and reporting requirements on certain dibenzo-para-dioxins and dibenzofurans under 40 CFR Part 766.

†As of 1991, there were approximately 1200 registered active ingredients in over 20,000 different pesticide products. (EPA, *Pollution Prevention 1991: Progress on Reducing Industrial Pollutants,* Office of Pollution Prevention, Washington, D.C., EPA 21P-3033, October 1991, p. 141.)

order being issued by the EPA to establish risk reduction measures to avoid outright cancellation. If, during the review process, the EPA determines that continued availability and use constitutes an imminent hazard to human health or the environment, the EPA may immediately suspend a pesticide's registration, eliminating its production, sale, and use. The EPA also has the authority to require the recall of any pesticide product that has been suspended and cancelled if the EPA administrator finds the recall is necessary to protect human health and the environment.

This ability of the EPA to impose restrictions on the use of a pesticide, to suspend or cancel its registration, or to recall it from the marketplace provides a powerful incentive to industry to reduce or eliminate activities leading to pollution of our environment. Restricting or preventing the use of a particular chemical in a pesticide because of adverse environmental risks is likely to lead to industry's use of less harmful ingredients since the alternative is to not be able to sell the product at all and risk financial difficulties. Since 1985, over 30 chemicals have been cancelled or restricted under FIFRA for posing unreasonable risks to human health or the surrounding environment.*

3.4 Executive Orders Promoting Federal Pollution Prevention

In addition to federal statutory law, there are several executive orders that also require or promote pollution prevention. Generally, these executive orders are binding on the federal government and affiliated entities. Table 3-2 summarizes the content of these orders. A more detailed discussion follows.

3.4.1 Executive Order 12902— Energy Efficiency and Water Conservation at Federal Facilities (March 8, 1994)

Executive Order 12902 outlines several different energy conservation programs to be designed and implemented at most federal facilities.† The Deparment of Energy is the lead agency for this effort, implementing it through the Federal Energy Management Program (FEMP). These programs supersede some simi-

---

*7 U.S.C. §136(a). These chemicals include aldrin, carbon tetrachloride, dinoseb, cadmium, chlordane, and captan.

†The Energy and Policy Conservation Act (Act) (42 U.S.C. 6201 et seq.) exempts certain types of facilities and associated operational activities (industrial or energy intensive) from the energy and water conservation requirements of the Act. However, this EO does require each agency to develop a plan to reduce energy and water waste at all other non-exempt operations occurring at these exempt facilities. The EO requires the agencies to revise their designation of "exempt facilities" so that only individual buildings in which industrial or energy intensive operations are occurring remain designated as exempt under the Act.

Table 3-2. Summary of Executive Orders Addressing Pollution Prevention

Executive order	Summary of major provisions
EO 12902 (3-9-94) *Energy Efficiency and Water Conservation at Federal Facilities*	Require federal agencies to develop and implement programs to reduce energy consumption and increase energy efficiency at their facilities and buildings by using prioritization studies, facility audits, and technologies for energy efficiency, water conservation, and renewable energy
	Develop a program to signficantly increase the use of solar power and other renewable energy sources
	Develop and implement programs to use cleaner, less-polluting fuels and energy sources instead of petroleum-based products, and reduce the use of petroleum where such alternatives are not practical or cost efficient
EO 12856 (8-3-93) *Federal Compliance with right-to-know laws and pollution prevention requirements*	Ensure that all federal facilities operate in a manner that reduces the amount of toxic chemicals entering any waste-stream through source reduction and recycling activities, ensuring compliance with the provisions of the Pollution Prevention Act
	Comply with EPCRA statutory provisions, particularly toxic chemical and hazardous substance reporting requirements, to ensure public knowledge and awareness
	Revise existing procurement process to reflect source reduction principles, and use and test innovative pollution prevention technologies on site to encourage market development and wider use of more effective pollution prevention methodologies
EO 12845 (4-21-93) *Requiring agencies to purchase energy-efficient computer equipment*	Encourage energy efficiency and reduced emissions by power plants such as coal-fired utilities by purchasing and using power-conserving computer equipment that meet the criteria of the EPA's "Energy Star" computer purchasing program
	Educate federal users about the energy-saving and environmental benefits of using less energy and generating fewer pollutants, thereby encouraging wider use and future benefits
EO 12844 (4-21-93) *Federal use of alternative-fueled vehicles*	Mandates use of alternatively fueled vehicles by federal agencies to substantially reduce toxic and hazardous air pollutants
	Directs all federal agencies to increase their purchases of alternatively fueled vehicles by 50 percent over levels specified by Energy Policy Act for fiscal years 1993 through 1995
EO 12843 (4-21-93) *Procurement requirements and policies for federal agencies for ozone-depleting substances*	Directs all federal agencies to revise procurement regulations and policies to conform with the provisions of Title VI of the Clean Air Act Amendments addressing stratospheric ozone layer protection
	Agencies must maximize their use of alternatives to ozone-depleting substances, evaluate present and future needs for ozone-depleting substances, and develop recycling initiatives to reduce and prevent further ozone-layer degradation
	Agencies must modify procurement specifications and practices to substitute non-ozone-depleting substances for ozone-depleting substances currently used

Table 3-2. Summary of Executive Orders Addressing Pollution Prevention (*Continued*)

Executive order	Summary of major provisions
EO 12780 (10-31-91) *Federal agency recycling and the Council on Federal Recycling and Procurement Policy*	Directs federal agencies to promote cost-effective waste reduction and recycling activities, providing a positive forum for the development and study of policy options and procurement practices that enhance environmentally sound and protective waste reduction and recycling practices
	Requires all federal agencies to develop an affirmative procurement program designed to purchase products with recycled content
	Creates the Council on Federal Recycling and Procurement Policy whose broad mission is to encourage federal agencies to purchase products that reduce waste generation, assist in the development of waste reduction and recycling programs, and collect and disseminate federal agencies' information concerning waste reduction methodologies, recycling program costs and savings, and current market prices of recycled content products as well as those that reduce wastes.
EO 12759 (4-17-91) *Federal energy management*	Encourages wise energy management practices by all federal agencies in a variety of ways including using alternative, less-polluting fuel, minimizing use of petroleum products, and encouraging employee outreach programs.

lar programs contained in EO 12759 (4-17-91) since Sec. 701 of EO 12902 specifically revokes all but three programs authorized by EO 12759.*

This executive order requires federal agencies to develop and implement a program to reduce energy consumption at their facilities 30 percent by 2005, using 1985 data as a baseline for calculations. Agencies must also increase energy efficiency at all industrial facilities 20 percent by 2005, using 1990 data as a baseline. The EO also requires agencies to conduct a prioritization survey of facilities that will be used to establish objectives for the comprehensive facility audits also mandated under this executive order. Within 180 days of completing the comprehensive facility audit, agencies must begin implementing the audit's recommendations for the installation of energy efficient, water conserving, and renewable energy technologies at individual facilities.

DOE is required to develop a program to significantly increase the use of solar and other renewable energy resources, and will take the lead in working with other federal agencies to implement the program. All agencies are required to develop and implement programs to reduce the use of petroleum at buildings and facilities by switching to a less polluting and nonpetroleum-based energy source, such as natural gas, solar, or other renewable energy sources. If the use of such alternative fuels is not practical or cost effective, the agency shall

*Please see Sec. 3.4.7 referencing the specific pollution prevention programs that are still viable under EO 12759.

improve the efficiency with which petroleum is used. Finally, each agency involved in the construction of a new facility must design and construct it to use energy efficiently, conserve water, and employ renewable energy technologies.

3.4.2 Executive Order 12856—
Federal Compliance
with Right-to-Know Laws
and Pollution Prevention Requirements
(August 3, 1993)

Executive Order 12856 (EO) is the most comprehensive of the executive orders, and sets out to ensure federal facility compliance with the chemical-reporting requirements of the Emergency Planning and Community Right-to-Know Act of 1986 (42 U.S.C. §§11001–11050)(EPCRA) and the pollution prevention require- ments of the Pollution Prevention Act of 1990 (42 U.S.C. §§13101–13109)(PPA). The overall objectives of EO 12856 are (1) to ensure that all federal facilities con- duct their facility management and acquisition practices in such a manner as to reduce the amount of toxic chemicals entering any wastestream through source reduction and recycling activities; (2) to require federal agencies to report and make available to the public information on any toxic chemicals entering any wastestream from their facility, and to improve local emergency planning, response, and accident notification; and (3) to encourage markets for clean tech- nologies and safe alternatives to toxic chemicals and hazardous substances through revisions to standards and practices of the federal procurement process and the testing of innovative pollution prevention technologies at these facili- ties.[16] This order applies to all federal agencies that either own or operate a "facility" as defined in Sec. 329(4) in EPCRA, and supplements, but does not replace any other existing obligations to which these federal facilities are already subject pursuant to EPCRA and PPA provisions.

More specifically, EO 12856 requires the head of each federal agency to develop a written pollution prevention strategy to achieve these three objec- tives. Each agency is encouraged to involve the public in forming the strategy and monitoring the relative success of its implementation. The strategy should include a pollution prevention statement delineating specific development, implementation, and assessment responsibilities for the strategy, and a strong commitment to achieve pollution prevention using source reduction as the pri- mary means.[17]

Executive Order 12856 also directs each federal agency to establish voluntary goals to reduce the agency's total releases of toxic chemicals to the environ- ment, and to reduce the off-site transfer of such toxic chemicals for treatment and/or disposal, by 50 percent by December 31, 1999. The agency is specifically requested to use source reduction principles and activities to achieve these reductions whenever practicable.[18] Alternatively, the federal agency may choose to implement a plan that achieves a 50 percent reduction in its toxic pol- lutants. Each federal agency must develop a written pollution prevention plan by December 31, 1995, and conduct follow-up assessments as required to ensure the development of such plans and the pollution prevention programs.[19]

Each federal agency must establish a plan to eliminate or reduce the purchasing of products containing toxic or extremely hazardous substances. This effort to use the procurement process to achieve pollution prevention must also include an effort to reduce each agency's own use, manufacturing, or processing of extremely hazardous substances and toxic chemicals. The Department of Defense (DOD) and the General Services Administration (GSA) will be reviewing their specifications and standards over the next two years to identify additional opportunities to reduce the use of these two types of substances, and any required changes to the Federal Acquisition Regulation (FAR) will be made within two years of EO 12856 publication. Federal agencies are strongly encouraged to identify and use innovative pollution prevention technologies at their facilities to aid the development of a strong market for these technologies.

Federal agencies are specifically directed to comply with the provisions in Sec. 313 of EPCRA (reporting requirements), Sec. 6607 (source reduction and recycling data collection) of the PPA, and all implementing regulations and guidance issued by the Environmental Protection Agency. Each facility will comply with these provisions without regard to SIC delineations that apply to the facility, and each report will address all releases, transfers, and wastes at the facility site.

Finally, each federal agency is directed to comply fully with all of the chemical and hazardous substance reporting, record keeping, and additional requirements as set forth in EPCRA, Secs. 301–312, and all implementing regulations. Federal facilities must provide all data required under Sec. 303 (comprehensive emergency response plans) to the local emergency planning committee in the area within one year of EO 12856 publication.

3.4.3 Executive Order 12845—
Requiring Agencies to Purchase
Energy-Efficient Computer Equipment
(April 21, 1993)

Executive Order 12845 aims to encourage energy-efficient facility management, a reduction in the generation of toxic pollutants by energy-producing power plants such as coal-fired utilities, and the procurement and market development of pollution prevention technologies. The purchasing of computer equipment was chosen because, according to the executive office issuing EO 12845, the United States government is the largest purchaser of computer equipment in the world. This is an extremely powerful position from which to encourage pollution prevention principles through responsible procurement practices.

Under the authority of the Energy Policy and Conservation Act (42 U.S.C. 6361), Sec. 152 of the Energy Policy Act (PL-102-486), and Sec. 205 of the Federal Property and Administrative Services Act (40 U.S.C. 486), federal agencies must currently ensure that all computer equipment purchased meets "EPA

Energy Star" requirements for energy efficiency. Case-by-case exemptions are allowed, with the following taken into account: commercial availability, cost differential, agency missions, and agency performance requirements. Agencies are directed to educate federal computer users concerning the economic and environmental benefits, including pollution prevention, of using this energy-efficient, low-power feature as outlined in the EPA Energy Star program.[20]

3.4.4 Executive Order 12844—
Federal Use
of Alternatively Fueled Vehicles
(April 21, 1993)

Executive Order 12844, under the authority of the Energy Policy and Conservation Act (42 U.S.C. §6201 et seq.), the Motor Vehicle Information and Cost Savings Act (15 U.S.C. §1301 et seq.), and the Energy Policy Act of 1992 (PL 102-486), mandates the use of alternatively fueled vehicles by federal agencies in order to substantially reduce toxic and hazardous air pollutants and reduce pollution associated with energy development activities. Each federal agency is directed, given fiscal constraints, to purchase 50 percent more alternatively fueled vehicles during the period of 1993 through 1995 than currently specified in the Energy Policy Act of 1992. The federal fleet acquisition program implementation will be overseen by the interagency Federal Fleet Conversion Task Force convened by the Secretary of Energy.[21]

3.4.5 Executive Order 12843—
Procurement Requirements and Policies
for Federal Agencies
for Ozone-Depleting Substances
(April 21, 1993)

This executive order recognizes the importance of addressing the current depletion of the protective ozone layer caused by the worldwide use of various ozone-depleting substances. The Montreal Protocols call for a phaseout of the production and consumption of these substances, and the United States, as a signatory, is using EO 12843 as one tool in achieving this laudable goal.

Agencies are directed to accomplish several important objectives. Procurement regulations and policies must be revised to conform with the requirements of Title VI of the Clean Air Act Amendments addressing stratospheric ozone protection. Agencies must maximize their use of alternatives to ozone-depleting substances, evaluating current and future uses of ozone-depleting substances and identifying opportunities for recycling. Procurement specifications and practices must be modified, whenever economically practicable, to substitute non-ozone-depleting substances for those ozone-depleting substances currently purchased and used. Agencies were directed to submit a report summarizing efforts to implement the specific provisions of this executive order to the Office of Management and Budget by October 23, 1993.[22]

3.4.6 Executive Order 12780—
Federal Agency Recycling
and the Council on Federal Recycling
and Procurement Policy
(October 31, 1991)

This executive order, under the authority of the Resource Conservation and Recovery Act (RCRA)(42 U.S.C. §6901 et seq.), directs federal agencies to promote cost-effective waste reduction and recycling activities, providing a positive forum for the development and study of policy options and procurement practices designed to promote environmentally sound waste reduction and recycling of precious national resources. Agencies also are encouraged to integrate these waste reduction and recycling programs into all waste management programs.

In implementing EO 12780, federal agencies must develop a program to promote cost-effective waste reduction and recycling programs that contain waste-reducing elements and recycling of specific commodities such as glass, paper, and used oil. Within six months of the order's effective date, each federal agency must develop and adopt an affirmative procurement program to increase the purchase of materials and products with recycled content. The Council on Federal Recycling and Procurement Policy was created, consisting of representatives from several federal agencies and departments. This council has several broad objectives. First, it is to develop incentives for federal agencies to purchase products that reduce waste and/or contain recycled content. Second, it is to assist federal agencies in developing guidelines for cost-effective waste reduction and recycling programs. Third, the council is directed to review all agency specifications and standards and, where necessary, recommend changes that will enhance the procurement of recycled-content products. Fourth, it is to collect and disseminate to the federal agencies information concerning waste reduction methodologies and economic benefits associated with waste reduction and recycling. Finally, the council is charged with the responsibility of developing guidelines for agency waste reduction and recycling programs and assisting federal agencies in developing long-range goals for federal waste reduction and recycling programs.[23]

3.4.7 Executive Order 12759—
Federal Energy Management
(April 17, 1991)

The prime objective of this executive order, under the authority of the Energy Policy and Conservation Act (42 U.S.C. §6201 et seq.), is to encourage wise energy management practices by federal agencies through a wide variety of energy conservation practices, use of alternative, less-polluting fuels. However, a significant number of EO's 12759 provisions were later revoked by Executive Order 12902, *Energy Efficiency and Water Conservation at Federal Facilities* (March 8, 1994). The specific pollution prevention programs that remain in effect are:

- Minimize the use of petroleum as a primary energy source, either by substituting a cleaner alternative fuel such as natural gas, or, as an alternative, by developing dual fuel usage capabilities where practicable.

- Require each agency to implement outreach programs such as ride-sharing and employee awareness programs in order to reduce petroleum fuel usage.

- For agencies operating 300 or more commercial vehicles, develop a plan that will reduce motor vehicle gasoline and diesel consumption by a minimum of 10 percent by 1995 in comparison to 1991 fuel usage data.

References

1. 42 U.S.C. §13102.
2. 42 U.S.C. §13101(4).
3. EPA, *Pollution Prevention 1991: Progress on Reducing Industrial Pollutants*, Office of Pollution Prevention, EPA 21P-3003, October 1991, p. 130.
4. 42 U.S.C. §6902(b).
5. 42 U.S.C. §13101(b).
6. 42 U.S.C. §6902(a)(6).
7. 58 *Fed. Reg.* 31114; May 28, 1993.
8. See 42 U.S.C. §7412(d)(2).
9. See 42 U.S.C. §§7581–7586.
10. See 42 U.S.C. §7589.
11. See 42 U.S.C. §7545.
12. See 42 U.S.C. §§7501–7507.
13. See 42 U.S.C. §§7651a–7651e, 7651o.
14. See 42 U.S.C. §§7671a–7671d, 7671h, 7671i.
15. 7 U.S.C. §136(a).
16. 59 *Fed.Reg.* pp. 41981–41982, August 6, 1993.
17. Ibid., p. 41983.
18. Ibid.
19. Ibid., p. 41984.
20. 58 *Fed.Reg.* 21887, April 21, 1993.
21. 58 *Fed.Reg.* 21885, April 21, 1993.
22. 58 *Fed.Reg.* 21881, April 23, 1993.
23. 56 *Fed.Reg.* 56289, October 31, 1991.

4

Expanding the Pollution Prevention Framework

Roles for State and Local Programs

David T. Wigglesworth

Pollution Prevention Office
Alaska Department of Environmental
Conservation
Anchorage, Alaska

Linda Giannelli Pratt

San Diego County Department of
Environmental Services
San Diego, California

4.1 Introduction

Most successful state and local pollution prevention programs share several important characteristics:

- Development of a program vision and plan
- Existence of program champions
- Top-management support
- Strong technical assistance services
- Established trust with industry

- Development of partnerships with the private sector
- Incentives for industry to consider pollution prevention
- Incorporation of pollution prevention into the base operating budgets
- Development of new approaches to measuring success

This chapter explores these issues by taking a look at the variety of roles that state and local programs play in moving environmental protection programs toward pollution prevention. Understanding the nature and the purpose of these roles is critical to developing methods to evaluate program effectiveness. This chapter also suggests that the desire by program evaluators to measure pollution prevention program effectiveness solely by actual reductions in pollution, while important, may

1. Underestimate the value of many state and local program services
2. Ignore the range of important and necessary roles for state and local programs

The authors want to emphasize that this chapter presents only a brief analysis of these issues in hopes of spurring additional discussion on the topic.

4.2   Local Government Perspective

Leveraged success through coordination at various levels of government is essential for effective business assistance programs. This can be best accomplished when there is clarification as to the responsibilities and capabilities of each government agency.

One of the roles of federal and state government is to provide direction through policies and legislation that mandate or otherwise foster the development of pollution prevention programs, and provide resources for implementing those programs. Complementing that effort is the ability of local agencies to act as the link between federal and state programs and the intended "customers." Local governments understand the interconnectedness of the various sectors of their community. In other words, they are keenly aware of the regional economy, pressing environmental issues, sensitive political matters, and the established infrastructure. It is this knowledge that enables better communication with businesses and the general public. Another benefit to garnering the support of local agencies is that they are accepted as a part of the community, whereas federal and state governments may be viewed as outsiders.

Federal and state agencies advocate and support pollution prevention programs primarily by emphasizing technology and information *development*. At this time, there needs to be a shift to technology and information *transfer*. In many cases, outstanding assistance programs and technologies have been developed, but information about the availability of those programs has not been adequately disseminated. "Reality checks" are critical to ensure that policies, programs, and technologies are meeting the needs of the "customer."

Representatives from local agencies can provide that function more effectively than other levels of government because they are easily accessible to businesses and can evaluate the success of the programs.

Government-sponsored pollution prevention programs typically include a variety of educational outreach efforts such as workshops, on-site consultations, resource clearinghouses, and written material. It is essential that there be established a long-term commitment between the agency and the intended customers, whether they are businesses, elected officials, or the general public. The goal is to bring about changes in the knowledge, behavior, commitment, and vision of the community.

Using the terms *marketing* and *government* in the same sentence may challenge conventional wisdom. However, the decade of the nineties can be characterized as "business as unusual" for government agencies. While there are some differences between marketing in the public sector and marketing in the private sector, the fundamental objectives are the same:

Understanding and satisfying the needs of your customers

Creating an awareness (positive image and value) of your service

Generating (or leveraging) revenue

It seems obvious that a pollution prevention program coordinator must identify the significant industries within a community and have at least a general understanding of the effect of those businesses on the environment. It is also critical to look for other agencies, industry associations, economic development organizations, business support groups, and environmental advocacy coalitions that can help "market" the concept of pollution prevention. First of all, cooperation with such groups provides an opportunity to leverage resources. Pollution prevention information can be included in established newsletters and brochures from many sources rather than being viewed as a "stand-alone" program. Secondly, the message may be perceived as more credible when stated in a cohesive way by many sources. Collectively, educational outreach activities are magnified, and the sphere of influence is expanded.

Many local agencies have in recent years begun to implement pollution prevention programs. These programs have become a component of city and county health departments, publicly owned treatment works, fire departments, and other agencies. For the most part, representatives are actively involved with a variety of business assistance programs that offer technical and management support information.

4.2.1 The TEAM Project

The Technical and Educational Assistance Model (TEAM) Project provided an opportunity to test strategies for multiagency pollution prevention integration at the local government level. The environmental health departments of San Bernardino, San Diego, and Ventura counties were utilized as the lead agencies for this project. Funding was received from the U.S. Environmental Protection

Agency's (EPA's) Office of Pollution Prevention, and the grant was administered by the California State Department of Toxic Substances Control. The specific findings are described in the report and reflect the unique political, economic, and institutional makeup of each county. However, common themes emerged, and encompass the following issues:

- Pollution prevention is the bridge between environmental quality and economic competitiveness and can be successfully advocated as a *tool* for encouraging business retention within a community.

- "Nontraditional partnerships" which unite environmental regulatory agencies with other sectors of the community, such as economic development organizations, public utilities, industry and trade associations, environmental advocacy coalitions, and universities, can magnify the effectiveness of educational outreach efforts.

- Compliance issues must be adequately addressed before businesses are receptive to information about pollution prevention.

- The recognized value of coordinating the pollution prevention educational outreach activities of several environmental regulatory agencies is that it provides more opportunities to identify and address other needs, such as permit streamlining.

- Resource limitations have resulted in tremendous competition for existing staff and funding, and may compel management to require quantifiable justification of program benefits.

- Regulatory agency staff are struggling with their dual roles of regulator and educator, and this conflict may limit the scope of technical and administrative information provided to businesses.

As a means to enhance the quality of their programs, many local agencies like those that participated in the TEAM project have become part of regional committees. For example, California has three regional pollution prevention committees, which are located in southern California, the San Francisco Bay area, and the Sacramento Valley. These regional committees provide a mechanism for exchange of information about ongoing pollution prevention activities, and facilitate coordination of these efforts when possible. They also offer a forum for ongoing education of members by utilizing speakers who are experts on specific topics. Statewide pollution prevention roundtables have been formed throughout the nation which include representatives from federal, state, and local agencies. Typically, the roundtables' two-day meetings focus attention on selected areas of common concern, staff training is initiated, and there is a review of pollution prevention clearinghouse activities.

Pollution prevention is a tool that can be used to build the foundation for environmental quality, economic competitiveness, and business retention. Ultimately, this will lead to a sustainable quality of life that is valued by the community. Local government representatives can fulfill an important role in this endeavor.

4.3 State Government Perspective

While spending on pollution control and cleanup will remain paramount, investment in pollution prevention is becoming more widespread in both the private and public sectors. This is being driven by a number of factors, including

- Increase in the number of state laws which require facilities to develop pollution prevention plans
- Implementation of voluntary pollution prevention partnerships
- The EPA's increasing interest in supporting state programs by ensuring flexibility for pollution prevention within operating grant guidance
- Technical assistance from state and local pollution prevention programs
- Industry-initiated activities

Investment in pollution prevention is approaching a critical juncture. New resources for pollution prevention programs often result from disinvestment in other activities. Increasing demands are being placed on pollution prevention initiatives to demonstrate results. Those evaluating state and local programs often focus solely on whether or not program activities result in actual pollution reductions. While this is obviously a very important desired outcome, this type of evaluation may

1. Underestimate the value of state and local program services that have indirect linkages to quantitative reductions in pollution
2. Ignore other important and necessary roles for state and local programs
3. Require additional field-testing and development of specific methodology to be truly reliable

4.3.1 State Program Roles

A central role of a state program is to provide direct technical assistance to industry to foster pollution prevention activities. Often this role is nonregulatory; however, this is changing as more states implement facility-planning laws with regulatory requirements.

While state programs are service providers, they also have many other roles that need to be factored into any analysis of their effectiveness. These roles include that of (1) leader, (2) educator, (3) facilitator, (4) supporter, (5) innovator, and (6) partner.

These other roles demonstrate that advancement of prevention strategies requires social, economic, and behavioral change, in addition to providing technical information on waste reduction techniques and/or requiring facility plans. Some of these roles cannot be linked directly to actual reductions in waste but are central to overall efforts to sustain statewide and/or industry-specific prevention activities. State program roles are very state-specific.

Effectiveness studies need to consider the needs of the state pollution prevention framework. This calls for a new appreciation of qualitative evaluation methods and results when determining whether investment in pollution prevention is warranted.

Role as Leader.   State programs (and local programs) have taken a leadership role in the development of a pollution prevention framework in the United States. In 1981, there were fewer than five state programs operating in the United States. Now, virtually every state in the country has a pollution prevention program.[1]

The National Roundtable for State Pollution Prevention Programs was established in 1985, and has been instrumental in developing state programs, securing EPA and congressional support of pollution prevention funds such as the Source Reduction and Recycling Technical Assistance grants, and the Pollution Prevention Incentives for States grant program. The Roundtable (in addition to regional and state roundtables) continues to have an evolving leadership role in the development of national and state pollution prevention policy.

Individual states such as Colorado, Vermont, and Alaska are taking leadership in advancing new efforts to secure base program funding support through the EPA's operating grant flexibility initiatives.

Increasing emphasis on pollution prevention today can be directly linked to this historic leadership role of state programs. Each state program is doing its part to foster pollution prevention, some by their presence alone. This role cannot be easily connected to actual reductions in waste but is essential to sustaining program resources, developing new policy, and increasing understanding of pollution prevention options.

Role as Educator.   Pollution prevention program staff are educators. This role is part of all other roles described in this chapter. State programs educate businesses and communities about pollution prevention technologies and techniques through workshops, newsletters, technical assistance, and other program services. These efforts support changes in behavior by suggesting an alternative prevention-based framework for environmental problem solving.

Role as Facilitator.   The cross-media, and often nonregulatory, nature of a state program enables it to play an important role as facilitator within and outside a state agency. A key role of pollution prevention staff is to establish a dialogue and develop mutual trust between themselves and media program staff, and/or a facility operator and CEO. Pollution prevention programs often facilitate collaboration among diverse groups to revisit current pollution control approaches. These facilitated discussions often result in a variety of outcomes, including

A new level of understanding between regulators and industry about each other's needs

A breakdown in compartmentalized environmental management and a new understanding of cross-media shifts in pollution and multimedia strategies

Increased dialogue between individuals rather than between media programs, or between a regulatory agency and an industry

Brainstorming discussions that result in unanticipated, beneficial outcomes which emphasize pollution prevention. For example, the Pacific Northwest Pollution Prevention Research Center and the Alaska Department of Environmental Conservation sponsored a meeting with the fish-processing industry to explore pollution prevention research needs. At a break in the meeting, the state program was able to inform an industry representative of the availability of pollution prevention matching grants. The industry then applied for and received matching funds to evaluate recycling solvent used in routine testing for fish meal. The project resulted in a 50 percent reduction in solvent consumption (in a particular process) and cost savings for the industry. The industry is now evaluating other pollution prevention options in other parts of their operation.[2]

This example brings to light two important issues. First, it is very difficult to measure reductions in waste when a state program assumes the role of facilitator. However, states do know that providing a nonthreatening forum, or opportunity, for discussion on pollution prevention is critical to (1) fostering the behavioral change that emphasis on pollution prevention requires and (2) developing a relationship with industry. Second, these forums provide an opportunity for *initiating* projects resulting in quantifiable reductions in waste. In the case cited, the state program supported industry efforts to reduce consumption of a particular solvent. However, it is the industry that deserves the credit for reducing the waste. Two important question to consider about this case study are:

1. If the meeting had not been held and the state program had not been able to provide information about this grant program, would the successful project ever have been initiated?

2. If the state grant program had not been available, would the project ever have been initiated?

These questions probably do not have any one right answer. The important point is that the facilitator role of a state program can have a variety of potential positive short- and long-term outcomes that need to be legitimized in program evaluations.

Role as Supporter.   As previously mentioned, state programs also play the role of supporter of pollution prevention activity. Support activities include

- Providing telephone technical assistance
- Conducting on-site assessments
- Sponsoring training programs

In this role, state programs are more directly linked with audiences that are typically those interested in making changes to reduce waste. This provides an easier opportunity to assess quantitative reductions in pollution. For example, some state on-site assessment programs and matching grant programs include follow-up evaluation that links the state activity to measurable reductions in waste. Technical assistance hotlines can also include follow-up to determine if the assistance was useful and what if any waste reduction activity occurred as a result of the telephone assistance.

It is important to note that quantifiable results may not be readily forthcoming even when states assume this role. For example, extremely valuable information may be provided, but the company may not act on the advice until a later time because it needs to budget for the changes. It is difficult then to link state program involvement to measurable pollution reductions, unless the program consistently tracks company efforts. In addition, the technical assistance provided may be only one of many factors influencing a company to invest in pollution prevention. This underscores the difficulty of making inferences about the relationships between reductions in waste and state program activities. However, many states can demonstrate that program support services are valued by their "customers" (businesses). This demonstrates the importance of these services and the inherent value of state pollution prevention programs.

Role as Innovator.   The role of innovator is one of the more exciting ones a state program can play. The current pollution prevention framework allows ample opportunity for state programs to test pollution prevention approaches.

Many states are exploring options for incorporating pollution prevention concepts into environmental agency functions (e.g., permits, inspections, enforcement actions). Others are identifying linkages with economic development agencies. For example, the state of Louisiana piloted the Environmental Scorecard program, which strives to reduce waste by linking environmental performance with a facility's tax exemptions.[3] The state of New Jersey is attempting to use pollution prevention as a tool for racheting down end-of-pipe permit limits.[4] The state of Alaska has fostered pollution prevention activities through media grant "flexibility" negotiated during the state/EPA agreement (SEA) process.[5] The states of Oregon, Washington, and Massachusetts are piloting innovative approaches to facility pollution prevention planning.

Innovation is a necessary condition for change. Many of these innovation efforts can be and are being evaluated for measurable reductions in waste. However, these efforts play equally important roles as catalysts in the process of incorporating pollution prevention concepts into social, political, and economic systems.

Role as Partner.   A state program role as partner in pollution prevention is becoming increasingly important. Limited staff, and grant requirements encouraging collaboration, are driving state programs to assume this role. In addition, many states understand the importance of linking up with trade associations, chambers of commerce, environmental groups, and education organizations in order for pollution prevention to take hold as an environmental protection strategy of first choice.

The partnership roles state programs play can vary widely. It is often associated with whether the partnership is merely an idea or an ongoing effort. For example, the state of Alaska Department of Environmental Conservation works extensively with the Anchorage Chamber of Commerce and the Alaska Center for the Environment to develop and implement the Green Star program—a partnership advancing pollution prevention in the small business community. The program received the EPA Administrator's Award for Excellence in Pollution Prevention in 1992. More recently, the program was highlighted in proposed legislation reauthorizing the Clean Water Act. In this bill, the EPA is directed to consult with the Anchorage Green Star program as it develops voluntary pollution prevention programs.

The state pollution prevention program has played (and continues to play) a number of roles in this innovative program, including

- Initiating a dialogue with the Chamber and the Center to implement this voluntary program
- Identifying fund sources and helping to develop proposals for funding
- Legitimizing the program (The partnership has gained legitimacy in the community due to the fact that business, government, and the environment are all active players.)
- Providing technical input on program materials and voluntary standards
- Evaluating business efforts to achieve the Green Star standards
- Building support for the program both within and outside state government
- Assisting with efforts to transfer the program to other communities in the state

The Green Star program has resulted in measurable pollution reductions. This analysis is based on business-specific information provided as a part of the Green Star achievement forms used in the program. However, a close look at the role of "partner" and the role of a state program within a partnership effort underscores the importance of justifying investment in this program for many reasons—not just waste reduced.

The Green Star program has heightened community awareness about pollution prevention and increased the ability of the state, business, and others to collaborate on other prevention projects such as developing an Alaska Materials Exchange and developing a Green Star for schools program.

The program has also provided a means for chambers of commerce to assume a leadership role in pollution prevention. The Kauai, Hawaii, Chamber of Commerce recently began efforts to implement the program islandwide. Other chambers of commerce in Alaska are initiating efforts to replicate the voluntary program in their communities.

These outcomes emphasize the multiplier effect that state program involvement in partnerships can have. Other partners in the Green Star program have equally important roles. It is safe to say that without the involvement of any one of the Green Star partners, the program would not have achieved its current level of success, and may not have been implemented.

4.4 Conclusions

Central themes emerging from this glance at the role of state and local pollution prevention programs include:

1. Pollution prevention programs do not have a single role to play. Program evaluation must consider this fact and take into account that state and local program activities have many desired outcomes in addition to reductions in pollution. These may include customer satisfaction, increasing participation in partnerships, and leveraging resources to support prevention programs.

2. Program roles are extremely time-intensive and require experienced staff with considerable interpersonal skills. Traditional program evaluation, based on "beans" delivered, does not give credit for skillful facilitation and partnership activities conducted by a state program. These functions need to be legitimized in program evaluations.

3. Investment in pollution prevention must be widespread, both within and outside government. Investments must strive toward incorporating pollution prevention into traditional activities. Investments must also support core pollution prevention programs to implement and sustain pollution prevention emphasis.

4. Considerable time and attention needs to be given to linking program evaluation to specific needs of state programs and to various state program roles. These may vary from state to state. This is the only true way to capture program effectiveness.

5. State and local programs should be proactive in developing methods to measure their effectiveness.

As the state and local pollution prevention agenda continues to unfold, it is important that federal, state, and local program evaluators spend time understanding the roles of a pollution prevention program. State and local programs are extremely interested in developing successful methods to measure progress and are open to collaboration. A closer consideration of evaluation methods used to measure preventive health programs may provide insight into appropriate methods to assess the effectiveness of pollution prevention programs at the state and local level.

References

1. Roger N. Schecter, "Profile of State Waste Reduction Programs," in *Hazardous Waste Minimization*, Harry M. Freeman (ed.), McGraw-Hill, New York, 1990, p. 225.

2. Joe Frazier, "Final Report: Use of Re-Distilled Toluene for Fat Analysis of Fish Meal," prepared for the Alaska Department of Environmental Conservation, July 12, 1993, unpublished.

3. Brenda S. Davis and Barbara M. Greer, *Pollution Prevention—State Strategies for Industrial Change*, Princeton University Press, Princeton, N.J., 1993, p. 80.

4. Ibid, p. 95.

5. U.S. Environmental Protection Agency, Office of the Administrator, *Pollution Prevention Media Grant Guidance*, 1993, p. 15.

5

Voluntary Pollution Prevention Programs

John Atcheson

U.S. Department of Energy
Washington, D.C.

5.1 Genesis of Voluntary Programs

The early advocates of pollution prevention subscribed to a philosophy that was rooted in traditional market theory—a view that still dominates many pollution prevention programs today. Voluntary programs grew out of this tradition.

Joel Hirshorn, in *Serious Reduction of Hazardous Waste,*[1] articulated this perspective as follows: there were a great number of cost-effective opportunities to reduce the generation of waste; these opportunities were not being realized due to market failures and impediments. The primary market failures were seen as a disparity in the availability of information; the primary impediments were seen as regulatory barriers to the introduction of new and innovative technologies.* These two phenomena essentially combined to form a negative synergy—rules based on "best available technology" created a culture that focused research, capital, and corporate policy on end-of-pipe solutions, rather than solutions embedded in the products, processes, procedures, or practices of manufacturing—pollution prevention.

The solution, according to Hirshorn and Oldenfeldt: remove the impediments, provide the information, and the market would unleash a tremendous

---

*Amory Lovins of the Rocky Mountain Institute had been making similar arguments about energy-efficient technologies for over a decade, and by the mid-eighties national efficiency improvements were proving him right.

reservoir of technological and process-based potential to reduce waste in a cost-effective manner. Finding the right mix of policies and programs to aid this became the Holy Grail of public policy makers in the states and the federal government. Voluntary programs became one of the primary strategies used to unleash and stimulate the market potential.

5.2 Voluntary Programs of the EPA

The U.S. Environmental Protection Agency (EPA) established its Pollution Prevention Office in the fall of 1988, and the new office began investigating voluntary programs and awards—which had been used effectively in private industry and a few states—as one tool to foster innovation, stimulate information exchange, and, ultimately, change the culture surrounding environmental compliance.

5.2.1 The Dye Project— EPA's First Voluntary Program

The agency's first voluntary pollution prevention program was a cooperative effort with the Ecological and Toxicological Association of Dyestuffs (ETAD), a trade association representing dye manufacturers. In many ways it remains the most comprehensive voluntary program to date. By design, it focused on the entire industry, and on all aspects of facility operations. The goal of this cooperative effort is "to develop and implement a comprehensive, industrywide pollution prevention program that arrives at quantifiable and documented results."

Participating ETAD member companies agreed to identify pollution prevention opportunities in all media and implement comprehensive prevention programs. In ETAD's words, pollution prevention is "more than just a `technological fix'; pollution prevention provides an integrated framework for addressing all aspects of a company's activities."

Incentives. At the time the project was being considered, the dye-manufacturing industry was facing a potential increase in regulatory controls. Wastes from azo-dye manufacturing were being considered for listing as a hazardous waste under the Resource Conservation and Recovery Act (RCRA), the Clean Water Act was coming up for reauthorization, and some states were considering new and more stringent laws which had the potential to affect the industry. At the same time, the EPA was devoting considerable research to the risks posed by the wastes from dye manufacturing, and was compiling information on the pollution prevention opportunities in batch-processing industries. It was against this backdrop that the two parties sat down to negotiate a joint, voluntary program.

Results.   At this writing, the survey data from this project are still being compiled and processed, so quantifiable results are not available. Preliminary indications, however, show that the companies participating in the project have substantially reduced emissions to *all* media, and they have outperformed reductions achieved by companies reported in the general toxics release inventories. However, some other measures of success can be assessed. One measure is the degree of participation: all major dye manufacturers in the United States participated. The industry has developed a comprehensive survey that looks at all aspects of waste generation on a multimedia basis and it has catalogued an extensive set of successful case studies of pollution prevention opportunities. From an industry perspective, this has facilitated benchmarking, and raised the performance level of each of the participants, enabling the industry at large to most expeditiously benefit from the increased efficiencies inherent in pollution prevention. The industry has also developed a manual that provides guidance specific to the dye manufacturers on starting up corporate pollution prevention programs. The manual addresses opportunities in administrative, fiscal, marketing, research, and supplier and customer networks, as well as process- and manufacturing-based strategies.

Issues.   Working through a trade association offered distinct advantages. The EPA was able to work with an entire industry, participants were able to share information, and there was ownership and acceptance of the approaches developed, since they were developed collectively. Moreover, in contrast to many voluntary programs, it appears that there has been *real and substantive change in corporate culture, and that the commitment to prevention and the results will persist and expand long after the project ends.*

The approach also has some disadvantages. The consensus and industry ownership approach—while valuable for long-term change—has made the process slow and often incremental.

5.2.2   The 33/50 Program

The 33/50 Program was based on a startlingly simple concept: challenge industry to reduce emission of 17 of the most ubiquitous and toxic chemicals by 33 percent in 1992 and 50 percent in 1995. In exchange, participants would receive assistance, be given recognition, and be eligible for awards. Reductions would be measured against 1988 baseline data as reported in the Toxics Release Inventory (TRI).*

*The Toxics Release Inventory is a publicly available database that contains information on the release of 312 chemicals from the manufacturing sector. It was established under the Emergency Planning and Community Right to Know Act of 1986 (EPCRA). Under EPCRA, manufacturers who meet certain thresholds must report releases and transfers of the covered chemicals. The Pollution Prevention Act required these manufacturers to also report on source reduction and recycling activity.

Incentives.   The program relies on the TRI for more than just baseline data. Both corporate and public reaction to the first round of TRI data was dramatic. The first survey (based on 1987 data) showed that although releases were subject to regulatory and permit requirements, the cumulative total was staggering. Individual companies and the general public reacted strongly. The public exposure of emission data by facilities and companies proved a powerful motivator to reduce releases.

Results.   The program has been dramatically successful in meeting its objectives. In 1991 the TRI reporting indicated that releases and transfers of the 17 chemicals and classes of chemicals covered by the program declined by 34 percent—surpassing the 1992 goal of 33 percent a full year ahead of schedule. This translates into a 500 million pound reduction in releases of the 17 chemicals. In the assessment of the role of the 33/50 Program one measure of its success is how much better participating companies performed than nonparticipating companies and facilities.

Participating companies achieved a 40 percent reduction in releases and transfers between 1988 and 1991 for the 17 chemicals; nonparticipating companies reported only a 25 percent decline in those chemicals over the same period. The difference between participants and nonparticipants was even more marked when looking at on-site releases: participants reported reductions of 36 percent, nonparticipants only 20 percent. Numbers like these speak for themselves. Although public exposure through the TRI and competitive pressures have stimulated reductions in nearly all facilities, the 33/50 Program has clearly extended reductions beyond what we could expect from these forces alone.

There are other, less tangible, results from the program. Within some participating corporations, it has begun to spawn a culture of prevention that extends beyond the specific goal of cutting releases of the 17 chemicals. In addition, the program has stimulated activity by states, trade associations, and other "leveraging" organizations.

Issues.   Several issues have been raised as the program developed and was implemented. The 33/50 Program is in the process of conducting a thorough survey and analyses to learn how effective it was and how the EPA might improve future voluntary efforts. Three of the primary criticisms follow. First, critics suggested that, given corporate planning cycles, participants had already put plans in motion, and the program was often not stimulating new activity, merely rewarding ongoing efforts. Second, the ambitious time frames in the goals made substantive changes internal to manufacturing processes difficult, thereby pushing participants to achieve results by the most expeditious means possible—which often meant off-the-shelf treatment technologies rather than pollution prevention. Third, focusing on a specific set of chemicals allowed companies to meet the objectives of the program without fundamentally changing corporate policies in a way that would encourage a more enduring commitment to pollution prevention, and possibly without substantive environmental benefit. For example, companies reported meeting program objectives through "chemical swapping" without knowing whether substitutes were safer, and without getting at corporate policies which could have effected a more broadly based pollution prevention ethic.

The issues raised are based on anecdotal information; the validity of these concerns and the extent to which they affected the program cannot be measured until the EPA's evaluation and survey is completed. In the meantime, the numbers are compelling. It may be that the concept can be improved on, but the results are incontrovertible, and impressive.

5.2.3 Green Lights, Energy Star, and Golden Carrots

Nearly two decades ago, Amory Lovins, founder of the Rocky Mountain Institute, began a debate on energy policy that is a true David and Goliath story. Lovins' premise was simple: it was cheaper to provide additional electrical capacity through improved efficiency than by building new capacity. In a series of publications and articles, Lovins threw down a guantlet.* Utilities and power interests responded with their own articles, editorials, and lobbying efforts. One by one they capitulated. Today, it would be hard to find a utility without an aggressive demand-side management (DSM) program aimed at generating capacity through reduced demand, not additional capacity—along with the associated pollution it would create.

Meanwhile, the EPA and the state environmental agencies were finding that the environmental ground was shifting. During the 1980s we began to learn that our environmental programs were not addressing the most serious environmental risks, and that they were not particularly efficient at alleviating those risks they did address. In particular, there was a set of emerging problems associated more with the ubiquitous use of materials, energy, and land than with releases from relatively large point sources. Lovins' prescriptions offered a strategy for addressing these endemic problems, and as the eighties closed, the EPA's Office of Air and Radiation developed a series of innovative voluntary programs aimed at reducing energy use through efficiency that built on Lovins' work. These programs were aimed at, in Lovins' words, "abating global warming at a profit."

Incentives. The Green Lights program, and the Energy Star and Golden Carrot programs that followed it, combine information, recognition, and technical assistance that enable companies to recognize cost-effective opportunities, "sell" the capital investment within the corporation, and locate the vendors and sources of finance needed to realize these opportunities. The Energy Star and

---

*See, for example, A. B. Lovins, "Energy Strategy: The Road Not Taken?" *Foreign Affairs*, **55**(1):65–96, October 1976.

Golden Carrot programs also provided incentives to manufacturers to design, produce, and market highly efficient products, as well as stimulating demand on the part of consumers of these products.

The underlying incentive is the integration of economic and environmental goals which occurs when environmental objectives are realized through more efficient use of materials and energy. In the Green Lights program, companies are asked to sign an agreement to upgrade 90 percent of their square footage, to the extent the upgrade is profitable and does not compromise the quality of lighting. Participants must do so within five years of signing the agreement.

Results. We will look primarily at the Green Lights program, since the Energy Star and Golden Carrot programs are relatively new. Green Lights, after two years, had over 1000 corporations and other participating organizations, including more than 100 of the Fortune 500. The program covers more than 3 percent of U.S. commercial floor space. Internal rates of return (IRRs) for participating companies have often far exceeded the "prime plus 6 percent" figure that represents the investment floor used by the program. In many cases, IRR is approaching and even exceeding 30 percent above prime rates.

Translating the economic potential to environmental results is relatively easy. For each kilowatt-hour not used, participants prevent formation of 1.5 lb of carbon dioxide, 5.8 g of sulfur dioxide, and 2.5 g of nitrogen oxides, as well as mitigating the habitat loss, acid mine drainage, methane releases, and air emissions of toxics (particularly heavy metals) associated with the generation and transmission of power. With lighting consuming an estimated 20 to 25 percent of electricity nationally, and commercial and industrial space accounting for some 80 to 90 percent of lighting demand, the "environmental rate of return" is impressive.

Perhaps more impressive are the indirect benefits the program has created. First, the economic focus of Green Lights—expressed in the cold hard facts of corporate financial language—has done a great deal to change the way corporations look at costs and benefits of environmental strategies.

Second, it has profoundly affected public policy. As this book goes to press, the nation is implementing an aggressive Climate Action Plan aimed at addressing the specter of global climate change. Voluntary partnerships built on the Green Lights model form the basis of most of the programs. The strategies being implemented will abate global warming *and* enhance the competitiveness of U.S. industries. No other voluntary program has been so broadly replicated nor applied, and none has so dramatically pointed to the potential synergy between environmental goals and economic objectives when pollution prevention is the strategy adopted.

Issues. One of the criticisms leveled against the program is that it has overstated the economic gains and environmental returns for participants. While many of the assumptions are based on best-case scenarios, the experience of Green Lights partners is generally confirming the predictions.

5.3   Conclusion

Voluntary programs have been effective, particularly when economic self-interest supports the voluntary efforts. The most effective programs combine information, recognition, and technical assistance. These examples come from the federal government, but the approach has been used within corporations as well as at the state and local level. One of the key public policy questions in designing these programs centers on how far pure volunteerism can take you, and how the programs can be integrated with more traditional command and control regulations to foster pollution prevention, when they do not take you far enough.

Reference

1. Joel Hirshorn and Kirsten Oldenburg, *Serious Reduction of Hazardous Waste*, Office of Technology Assessment, Washington, D.C., 1987.

6

State Facility Planning Requirements

Peggy Morgan

Hazardous Waste and Toxics Reduction Program
Washington Department of Ecology
Olympia, Washington

Beginning in 1989 with passage of legislation in Oregon and Massachusetts, a new direction in state-initiated environmental policy was established. Although the details of the two laws vary considerably, they have a common theme: the requirement that certain hazardous waste generators and toxic substance users prepare plans outlining how they intended to reduce their use of toxic substances and their generation of hazardous wastes. Since 1989 over one-third of the states have enacted legislation with a similar planning requirement.

Passage of these laws was due to a variety of reasons, including pressure from citizen groups to reduce chemical use in the workplace and community, public awareness of the types and amounts of toxic emissions into the environment, and a strong move within state government to work in partnership with industry to try new approaches to solve problems.

The plans, which are often referred to as facility plans, generally contain four main components:

1. A comprehensive review of all industrial processes that use, generate, or release toxic or hazardous materials

2. The identification of pollution prevention opportunities in all processes in which toxic or hazardous materials are handled

3. A ranking for each of these opportunities and a schedule for their implementation

4. The implementation of these options, including some measure of their success*

Generally speaking, the rationale behind this requirement is that, although pollution prevention activities can often provide a competitive advantage, a company may not pursue such activities unless they conduct a thorough analysis of their processes and develop a plan detailing how they intend to institute pollution prevention activities.

6.1 Planning Requirements

Although the specific planning requirements vary from state to state, the following are the elements most often required by facility planning laws:

- A policy statement of management support for pollution prevention, the facility plan, and its implementation

- A statement of reduction goals, the reasoning behind them, and a schedule for meeting those goals

- A description of efforts initiated in the past that qualify as pollution prevention and an assessment of those efforts' successes and failures

- A detailed, numeric description of current processes in which toxic chemicals are used and hazardous wastes generated (usually produced by teams reviewing and assessing those processes)

- Identification of pollution prevention options in specified areas, including (at a minimum) changes in a product or its formulation, substitution of raw materials in existing processes and products, equipment modification or modernization, and changes in operating and maintenance procedures

- Detailed financial and technical analyses of identified and practical options in light of current operating conditions

- Detailed criteria or rationale for choosing or discarding identified options for implementation

- A detailed schedule for implementing selected options and a procedure for measuring and monitoring progress in achieving reductions

- A description of opportunities for employee involvement and training

- Certification by responsible corporate officers or facility managers.†

---

*SOURCE: *Pollution Prevention Review* vol. 2, no. 4, Executive Enterprises Publications Co., Inc., New York, 1992, pp. 479–490.

†SOURCE: Ibid.

Table 6-1 lists some of the key components from the legislation of the states that have mandatory facility planning requirements. As can be seen from the information in the table, many similarities as well as many differences exist in the states' laws. This flexibility is seen as a positive aspect by many states as it allows them to address the specific needs and political climates of their states.

6.1.1 Who Needs to Complete a Plan?

A few states require only hazardous waste generators to complete plans and several others require only TRI reporters (those facilities required to report their toxic releases under the Emergency Planning and Community Right-to-Know Act) to complete plans. However, most of the states require both types of facilities to prepare plans.

6.1.2 What Needs to Be Planned for?

As shown in Table 6-1, the focus of the plan varies considerably from state to state. California, Georgia, New York, and Tennessee focus on hazardous waste reduction, whereas Massachusetts and New Jersey emphasize hazardous substance use reduction, and Minnesota concentrates on the reduction of TRI releases. Several states require planning facilities to evaluate reduction in all of these categories. By broadening the scope beyond hazardous waste reduction, the focus is expanded to more of a multimedia approach, with less potential for simply shifting risks from one type of media to another.

6.1.3 Reduction Goals

Another component in which the state legislation varies considerably is in the area of reduction goals. For the most part, the reduction goals that are legislatively established are done so on a statewide basis rather than as actual targets for specific facilities. The belief is that there are too many variables between facilities, such as past reduction accomplishments and changing production levels, to uniformly apply specific reduction goals. Several states require facilities to set their own goals, but these are considered targets and not mandatory levels to achieve.

6.1.4 Progress Reports

All of the states, except for Iowa, require planning facilities to submit periodic progress reports. Most states require that the reports be submitted annually except for Georgia and Massachusetts, which require them every two years, and California, which requires them every four years. Pennsylvania requires

Table 6-1. Key Components of State Facility Planning Laws

State	Planning facilities	Plan focus	Plan due dates	Reduction goals
Arizona	—TRI reporters —certain LQGs	—HW —toxic substances used —releases to air and water	12/92	none

California	—certain LQGs	—HW	9/91 (larger generators)
			9/93 (smaller generators)
				statewide:
				  5% of all HW per year from 1993–2000
				facilities:
				  set their own goals

Georgia	—LQGs —out-of-state LQGs using Georgia TSDs	—HW	3/92 LQGs	facilities: set their own goals
Iowa*	—TRI reporters —LQGs	—HW —toxic substances used —TRI releases	No mandatory schedule	statewide: 25% reduction of HW by 7/94

Maine	—LQGs
	—SQGs
	—toxics users (excludes some LQGs, POTWs)
		—toxic substances used
		—HW
			1/93	statewide:
				  (toxics use reduction)
				  10% by 7/93
				  20% by 7/95
				  30% by 7/97
				facilities:
				  set their own goals

Massachusetts	—large-quantity toxics users
	—small-quantity toxics users
		—toxic substances used	7/94	statewide:
				  50% waste reduction
				facilities:
				  set their own goals

Minnesota	—TRI reporters plus some additional SIC codes	—TRI releases	7/92 additional SIC codes: 7/95	none
Mississippi	—LQGs —SQGs —TRI reporters	—HW —toxic substances used —releases to air and water	1/92	statewide: 25% waste reduction by 1/96

New Jersey	—TRI reporters plus priority facilities in selected SIC codes	—toxic substances used	1/94	statewide:
				  50% reduction in releases by 1996
				facilities:
				  set their own goals

*Under Iowa law, facilities are encouraged, not required, to complete plans.

Plans or summaries submitted?	Public review	Plan updates	Progress reports	Penalties
yes	available	no	annual	various
no	available	every four years	every four years	up to $1000/day
full plans	available	every two years	every two years	none
yes	available	no	no	none
neither, unless a facility does not meet its goals then state can request that a summary be submitted	available, unless a facility requested confidentiality	every two years	annually	$2,000–10,000 per day
summaries	plans are public, except for trade secrets	every two years	every two years	normal administrative processes
no	progress reports only	every two years	annually	not set
certified report submitted annually	certified reports are available	annually	annually	none
yes	summaries, updates, and annual progress reports except for trade secrets	every five years, unless processes change	annually	yes, not determined yet

Table 6-1. Key Components of State Facility Planning Laws (*Continued*)

State	Planning facilities	Plan focus	Plan due dates	Reduction goals
New York	—HW generators of >25 tons/year —on-site TSD permit holders and TSD permit applicants	—HW	phased schedule: 7/91–7/96	statewide: 50% hazardous waste reduction by 1996
North Carolina	—HW generators —air and water permit holders	—HW —releases to air and water	not yet determined	statewide: none facilities: not yet determined
Oregon	—TRI reporters —LQGs —SQGs	—toxic substances used —HW	—TRI reporters/ LQGs: 9/91 —SQGs: 9/92	none
Pennsylvania	—LQGs —large residual waste generators	—HW —residual wastes	—residual wastes: 7/93 —hazardous wastes: 1/94	none
Tennessee	—LQGs —SQGs	—HW	LQGs: 1/92 SQGs: 1/94	statewide: 25% by 6/95
Texas	—LQGs —TRI reporters	—HW —TRI releases	7/93 for certain facilities; 1/97 for remainder	statewide: 50% reduction of TRI releases and/or HW by 2000
Vermont	—LQGs —SQGs —TRI reporters	—HW —air and water releases —toxic substances used	—LQGs: 1/92 —SQGs: 7/93 —TRI: 7/95	facilities: for wastestreams that exceed 10% of total waste
Washington	—LQGs —TRI reporters	—HW —toxic substances used	phased schedule 9/92–9/94	statewide: 50% HW reduction by 1995 facilities: numeric goals encouraged

Plans or summaries submitted?	Public review	Plan updates	Progress reports	Penalties
full plans are submitted	nonconfidential portions are available	every two years	annually	generator is not allowed to sign manifest; therefore, cannot transport waste off site
not yet determined	not yet determined	not yet determined	not yet determined	not yet determined
no	a few are available	annually	annually	public hearing
only with permit application or request to process or dispose of waste	yes	every five years or if process changes	when plan is updated	can deny permits
no, can be reviewed by inspectors	no	annually	annually	fines up to $10,000/day
executive summaries	executive summaries only	required as program is modified by the facility	annually	fines up to $25,000 per day
executive summaries	executive summaries only	updated on a three-year cycle	annually	deficient plans can be reviewed by public if not adequately modified
executive summaries	executive summaries only	every five years or if significant changes are made to the facility's processes	annually	fines: the greater of $1000 or three times the amount of the previous year's fee (up to $30,000)

them when the plans are updated, which occurs every five years, or when processes change significantly. In the states that require facility-specific goals, information from these reports is used to measure progress against the goals. For other states, the reports are used in more general ways to evaluate success of the states' planning and program efforts.

6.2 Compliance Issues and Penalties

With the exception of Iowa (which simply encourages facilities to write plans) and North Carolina (which is not currently implementing its facility planning law), all of the states listed in Table 6-1 require that the designated facilities comply with their planning requirements. In addition, almost all of those states review the plans for adequacy. The laws in several states (Arizona, New York, Texas, Pennsylvania, and California) also contain provisions for mandatory implementation of the plans.

Penalties for noncompliance with both the planning and implementation requirements include fines (up to $25,000 a day in Texas), denial of permits, public hearings, and in New York the prohibition for the facilities to sign hazardous waste manifests, which in effect deny them the ability to send hazardous waste off site for treatment, storage, or disposal.

6.3 Measurement and Normalization Issues

Measurement in general and normalization in particular are issues that have received considerable attention from a number of states as well as from the EPA. In regard to reporting progress, the states handle the normalization issue in a variety of ways. Some require that waste generation or toxic substance usage be normalized through the use of a specific unit (e.g., the amount of hazardous waste generated per employee, per "widget" produced, or per production hour). Others allow the facility to decide on the most appropriate normalization measure. Yet again, others are more interested in aggregate reduction and therefore do not require normalization at all, but instead ask for total volumes of toxic substance used, wastes generated, or releases emitted.

6.4 Protection of Confidential Business Information

States allow for the protection of confidential business information in a number of ways. Some states, although requiring that entire plans or summaries of the plans be submitted to the state, allow a facility to designate certain information as confidential and therefore not available for public review. Other states allow the plans to remain at the facility, only accessible to state personnel. Massachusetts takes this one step further and allows businesses to shield confidential business information from state agency personnel as well.

6.5 Technical Assistance Activities and Funding Mechanisms

All of the states provide technical assistance to facilities required to prepare plans. These technical assistance activities include on-site assessments, workshops, classes, telephone assistance, written guidance materials, videotapes, teleconferences, clearinghouses, resource centers, and newsletters. Maine provides facilitators for a company's pollution prevention team. Washington has found that a five-session "step-by-step" class was extremely successful. Facility planners were encouraged to complete portions of their plans over the course of the class. This allowed the facilities to receive individualized assistance and allowed the state to review and comment on portions of the plans before the actual due date for the plans, resulting in better plans in the long run.

State technical assistance programs are funded through a variety of mechanisms, including hazardous waste generator fees, disposal fees, hazardous substance usage fees, solid waste tipping fees, state general funds, and EPA grants.

6.6 Grants

Several states also provide grants to businesses to assist them in a variety of pollution prevention activities. Examples include:

Pennsylvania, which provides funds for the purchase or lease, as well as the installation, of equipment that will result in waste reduction or recycling

Vermont, which administers a "Pollution Prevention Incentives Grants" program to encourage small businesses to invest in pollution prevention projects

6.7 Pollution Prevention Barriers and Reasons for Success

Although exact circumstances differ from facility to facility, a number of barriers to achieving pollution prevention have been identified by states that have worked with planning facilities. Some of the most common barriers identified include lack of technical information, lack of funds, and lack of management support. Fear of future regulations, shortage of skilled staff, and reluctance to change are other barriers that have also been observed.

Overwhelmingly, the most common reason cited for success in pollution prevention is corporate and upper management commitment and support. Teamwork and employee participation are other important elements of successful pollution prevention programs.

6.8   State-Specific Components

As mentioned previously, each state's planning requirements are somewhat unique, distinguished by the types of facilities or materials covered, or the penalties applied for failure to comply, etc. The following are examples of some of the distinctive components from a selection of states.

6.8.1   Arizona

Facilities that have an approved plan are assessed only one-half the regular hazardous waste fee of $10 for each ton of waste generated and shipped off site. In addition, one of the enforcement possibilities for failure to complete a plan is for the state to write the plan and then bill the facility for their time.

6.8.2   Iowa

Facilities are encouraged, not required, to prepare a plan.

6.8.3   Minnesota

Enforcement staff frequently require a permittee or enforcement client to prepare a facility plan as part of the conditions of an enforcement settlement or permit.

6.8.4   New Jersey

New Jersey's law contains very comprehensive reporting and goal-setting requirements. Companies are required to publicly report numeric goals for the reduction in the use of each hazardous substance and nonproduct output (waste prior to treatment or recycling) for each of their production processes.

6.8.5   New York

Facilities that fail to prepare or implement a plan are not allowed to sign their hazardous waste manifests, which in effect denies the facility the ability to send hazardous waste off-site for treatment, storage, or disposal.

6.8.6 Oregon

Industry has played a strong role in policing itself to ensure that high compliance rates are achieved.

6.8.7 Pennsylvania

Planning facilities are required to include residual wastes (industrial, nonhazardous wastes) in their plans. Implementation is mandatory for the hazardous waste portions of the plan, voluntary for the residual waste portions. Permits can be denied to facilities that fail to meet the planning and implementation requirements.

6.8.8 Texas

Plan implementation is mandatory.

6.9 Planning Follow-up

California and Washington have both surveyed their planners to help determine the effectiveness of their planning processes.

Table 6-2 lists some of the questions and responses from California's survey.

Table 6-2. Selected Questions and Responses from California's SB 14 Survey

Of the approximately 2000 questionnaires that were mailed out, 580 were returned. Of those that responded, 225 were generators.

1. Was completing a plan a worthwhile experience?

   YES   74%   NO   25%

2. Will cost savings due to identified waste minimization opportunities counterbalance the cost of producing a plan?

   YES   41%   NO   59%

3. Indicate the hazardous waste minimization as a percentage of total generation at your site that was identified by completing the plan.

% reduction	Number of generators
0	24
<10	63
10–25	57
25–50	17
50–75	7
>75	9

Washington's survey included questions about the planning process and the assistance provided by the state.

One of questions on the survey was "What was the single most useful part of preparing your plan?" Examples of responses to this question include: "Having a goal to aim at from the start," "Formalization of costs/benefits," "Employee awareness through organizing a team to prepare a plan," and "The opportunity to fully evaluate all opportunities." Facilities were also asked what was the most difficult part of preparing a plan. The following aspects were commonly identified: understanding what was wanted, collecting data, researching opportunities, the time needed, and financial analysis of opportunities.

Washington has also conducted a preliminary analysis of 85 percent of their first-wave plans (those that were due on September 1, 1992). The analysis revealed that the plans identified over 4800 pollution prevention opportunities. Of these, 1860 were selected for implementation and 707 will be analyzed further. For plans which listed numeric reduction goals, compared to 1991 baseline data:

97 facilities plan to reduce hazardous substance use by 48 million pounds per year by 1997

150 facilities plan to reduce hazardous waste generation by 75 million pounds per year by 1997

30 facilities plan to reduce toxic emissions by 11.9 million pounds per year by 1997

6.10 Summary

Due to the limited time that the facility planning requirements have been in place, it is too soon to tell how effective these requirements may be in reducing the generation of hazardous waste, the use of hazardous substances, or the release of hazardous emissions. However, hope is high, as these requirements are based on the following sound principles:

- Cooperation between business and government is essential.
- Pollution prevention is a win-win activity.
- A structured planning process helps businesses help themselves.
- Voluntary measures help businesses set goals.

Further Reading

Davis, Brenda S., and Barbara M. Greer, *Pollution Prevention: State Strategies for Industrial Change,* Princeton University, Princeton, N.J., 1993.

Edmunds, Cynthia, "States Legislate Hazardous Waste Reductions," *Hazmat World*, December 1990, pp. 10–11.

Foecke, Terry, "A New Mandate for Pollution Prevention," *Pollution Prevention Review*, vol. 1, no.1, Winter 1990–1991, pp. 91–97.

Foecke, Terry, and Robert Style, "An Update on State Facility Planning Legislation and Programs," *Pollution Prevention Review*, vol. 2, no. 4, Autumn 1992, pp. 479–490.

Innes, Al, "Scope and Approach of State Pollution Prevention Legislation: An Overview," *Pollution Prevention Review*, vol. 1, no.4, Autumn 1991, pp. 369–387.

National Advisory Council for Environmental Policy and Technology, "Building State and Local Pollution Prevention Programs," U.S. EPA, EPA-130-R-93-001, December 1992.

Shah, Arvind, "California's Pollution Prevention Strategy," *Pollution Prevention Review*, vol. 1, no.1, Winter 1990–91, pp. 55–65.

Willis, Dennis G., "Pollution Prevention Plans—A Practical Approach," *Pollution Prevention Review*, vol. 1, no. 4, Autumn 1991, pp. 347–355.

Wise, Marian, and Hillel Gray, "Toxics Use Reduction: New Jersey's Approach to Pollution Prevention," *Pollution Prevention Review*, vol. 2, no. 1, Winter 1991–92, pp. 31–50.

7

Pollution Prevention Incentives, Barriers, Regulations, and State Programs

Murali Kalavapudi
ENERGETICS, Incorporated
Columbia, Maryland

7.1   Introduction

Environmental policy over the past twenty years has evolved through many statutes and regulations that specify limits to the release of pollutants into the various environmental media. So far the regulatory emphasis has been on controlling the volume of pollutants at discharge (end-of-pipe), through the stack, or disposed of or buried in the ground. Although this approach has assisted in protecting our environment, it has also imposed greater costs on industrial pollution control and its management, therefore increasing costs to consumers and taxpayers who pay for pollution control either directly or indirectly.

Earlier regulations and environmental guidelines emphasized handling and treating liquid and solid waste discharges rather than recycling these wastes. During this period, only a few industries introduced the concept of waste reduction, treatment, and reuse into their manufacturing operations solely due to cost benefits. Typical examples of this approach were found in the electroplating and photographic industries, where valuable inorganic materials were extracted from process wastewaters. As a result of this waste recycling and extraction operation, the process simultaneously reduced discharge concentrations. Recently, pollution prevention awareness has been widespread and there is more emphasis on reducing waste generation than on end-of-pipe treatment.

Waste reduction is defined as an appropriate means of improving environmental quality beyond that achieved under the traditional "command and control" strategies characteristic of many past and current environmental regulations. Implementing waste reduction helps to conserve raw materials as waste materials are recycled into other products, and treatment and disposal activities are thus significantly reduced. Today, waste reduction is considered a viable economic tool for preventing pollution. It has been proven that waste reduction results in lower product cost to the consumer while simultaneously minimizing environmental impact.

Both industry and government have noticed the cost benefits of pollution prevention and are supporting waste reduction as the preferred environmental management technique. Federal agencies have also established pollution prevention programs in most of their operations, while the United States Environmental Protection Agency (EPA) has integrated pollution prevention laws and programs into its regulatory framework. The agency has also realized that pollution prevention laws alone have directed the large and successful corporations in establishing pollution prevention programs, primarily as a cost-benefit tool for waste treatment.

The economic benefits of waste reduction measures vary among specific industrial applications. However, the advantages of using these measures over conventional pollution control and regulatory compliance have been widely demonstrated. Hundreds of firms have implemented waste reduction measures and have reported significant cost savings. Although pollution prevention incentives have emphasized cost benefits and waste reduction, several barriers exist and need to be addressed.

This section discusses potential incentives and barriers that prevail in the regulatory and management framework. The status of current regulations and management principles that may either promote or inhibit pollution prevention programs within industry or the government are also discussed in this section.

7.2 Pollution Prevention Incentives

Incentives for pollution prevention can be an advantage to management as well as to industry in general. Case studies[1-4] have shown that implementation of pollution prevention within the industry is successful when employees are involved and receive incentives for their efforts in pollution prevention programs.

Incentives are needed to implement pollution prevention. The particular incentive mechanisms selected can crucially influence how these issues are addressed. Escalating regulatory and economic consequences have forced companies to consider establishing pollution prevention programs to eliminate more stringent future regulations and the limited treatment options. Incentives may be classified into the following categories:

- Economic benefits
- Enhanced public image and relations
- Regulatory compliance
- Reduction in liability

Following is a description of each category.

7.2.1 Economic Benefits

Waste reduction minimizes the costs associated with treating or handling wastes. The costs of transportation, disposal, and/or treatment will be lowered as the volume of waste generated decreases. Current trends indicate a steady increase in waste disposal costs in the future. In addition, waste reduction will result in income that can be derived through the sale, reuse, or recycling of certain waste materials. The not-so-obvious gains of a waste reduction program include reductions in safety and health costs of employees and other beneficiaries, insurance costs, raw material purchase costs, reporting costs, manifesting costs, and permit costs.

The EPA and state and local agencies can influence companies to investigate pollution prevention techniques through offers of reductions in fines and penalties. A pollution prevention program can result in total regulatory compliance while the permit fees are structured in a way that promotes pollution prevention activities. Offers of reductions in fines and penalties are provided through the Supplemental Environmental Projects (SEP) program established by the EPA and discussed under Sec. 7.2.5.

7.2.2 Enhanced Public Image and Relations

A growing awareness of the importance of protecting the environment at different levels of society has resulted in increased emphasis on environmental issues. Political campaigns have made the environment a top-priority issue on their agendas as a result of special concerns of the public. In addition, the public accessability to the EPA's Toxic Release Inventory under SARA Title III has revealed factual information concerning industrial actions that have resulted in decreases in some companies' overall sales as well as diminished public image.

7.2.3 Regulatory Compliance

Implementation of a pollution prevention program resulted in the successful reduction of other regulatory compliance issues involving industry. In cases where an industry is subject to meeting other compliance requirements, the EPA has considered reducing penalties and regulatory pressures on it.

7.2.4   Reduction in Liability

Both short- and long-term liabilities can be reduced with pollution prevention programs. Short-term liabilities, such as releases to the environment, can be significantly reduced through reductions in overall waste generation and other process modifications. Worker exposure liabilities are also eliminated through pollution prevention programs. Long-term liabilities of disposal problems associated with wastes will also be eliminated. In addition, if an industry is considered a potentially responsible party under CERCLA actions, liabilities may be significantly reduced as confidence between the industry and the EPA is enhanced.

7.2.5   Supplemental Environmental Projects (SEP)

A new program established by the EPA enables companies confronted with fines for violating environmental regulations to consider supplemental environmental projects (SEPs) in order to minimize their fines. The SEPs have the potential to improve the relationship between business and government regulators, advance pollution prevention efforts, and provide cost advantages, while the companies can still be held liable for their actions. SEPs have also included environmental activities that a company uses to negotiate with the EPA in exchange for a reduction in fines. A settlement that includes SEP implementation will still require the company to pay for the economic benefit that it gained through noncompliance. The settlement, generally in the form of a fine reduced below the normal maximum potential fine, is an incentive for a company to agree to a SEP. Costs for implementing the SEP may often be more expensive than the reduction in the potential fine, and they may not always be practically implementable. However, a SEP may be implemented when the maximum fine is large and in situations wherein the SEP results in considerable savings.[2]

SEPs also provide a variety of cost benefits, although on certain occasions the cost of implementation may be high. SEPs are economically advantageous for several reasons.[2]

Savings Through Process Changes.   Cost savings noticed in the SEP are a result of decreasing waste through production process changes. After identifying possible pollution prevention opportunities, companies can negotiate with the EPA to include in the settlement those SEPs that will result in greater cost savings.

Tax Incentives. In establishing a SEP, the industry benefits because capital invested is tax deductible as a business expense. Since fines are not tax deductible, negotiating a reduced fine in combination with a costly SEP could be more cost-effective than a high fine. If the SEP involves the purchase of equipment, capital costs can be depreciated, resulting in additional tax savings.

Decreased Transaction Costs. In establishing a SEP, the industry can reduce its legal fees. The SEP is negotiated through EPA regional offices and is treated as an administrative settlement, so it does not involve court action or approval by the Department of Justice. Thus, a settlement can be resolved more quickly than through traditional litigation.

The EPA and Congress have explored a broad range of incentive-based approaches. The Final Emissions Trade Policy set up by the EPA indicates federal rules authorizing emitters to create, store, and substitute inexpensive emissions reductions for costly ones. This has saved American industry nearly a billion dollars over the cost of uniform stack-by-stack controls, with uniform or better environmental results. This policy has also facilitated flexible compliance and environmental integrity through stringent safeguards against "nonsurplus" reductions, adverse ambient effects, or incidental toxins increase. The policy also provides needed safety valves allowing for adjustments of national rules to site-specific conditions. It also offers states methods to protect sources which bank or trade against loss of credits, if further reductions to meet health standards are required.[2]

A SEP may be initiated through discussions with the regional EPA offices either through the legal council divisions of enforcement activities or through communications with the compliance officers for a specific medium of concern, e.g., air, water, groundwater, soil, or solid waste. More information on SEP policy can be obtained from the EPA's Pollution Prevention Information Clearinghouse (PPIC).

7.3 Pollution Prevention Barriers

Although several incentives for pollution prevention are available, sometimes implementation can be inhibited by barriers. Barriers may be of three types: regulatory, economic, or operational and administrative.[5] Each of these barriers is discussed as follows.

7.3.1 Regulatory Barriers

Regulatory barriers are restraints causing delays and depletion of financial resources in meeting current regulatory compliance requirements imposed by federal, state, and local agencies. For example, an industry may be limited in resources for exploring pollution prevention opportunities and R&D efforts due to penalties and other investments involved in meeting regulatory compliance requirements. These may be related to noncompliance of applicable regulations, penalties imposed on the industry for cleanup, and delays in environmental remedial actions. These issues require reiterating priorities and cumbersome negotiations with the regulators on one hand, and creating financial resources to enable employees to participate in training and awareness programs on the other. Barriers may also involve loss of time and delays on meeting existing regulatory commitments, due to the absence of key personnel, such as environmental managers and facility or site supervisors.

Such regulatory barriers may be overcome by increased communications and constant negotiations between the industry and the agency. This will help resolve industry-specific issues on meeting current stringent regulatory standards and penalties imposed. Such negotiations may emphasize the need for decreased penalties in other areas of regulatory compliance, financial support to R&D efforts, and informational support and guidance via technology clearinghouses that are currently available through various federal, state, and local agencies. Most states and EPA regional offices support an industry in establishing and accomplishing its pollution prevention goals.

7.3.2 Economic Barriers

Economic barriers involve restraints caused by lack of additional funding for achieving future R&D goals and commitments. Economic barriers may be due to the combined stress of dealing with the remaining management system, budget shortfalls, trade deficits, and shortfalls in projected sales due to international competition. Economic barriers may be diminished by establishing a constant communication system with the federal, state, and local agencies. Interacting with the regulator to assist in meeting current compliance requirements and seeking financial incentives are further possibilities.

7.3.3 Operational and Administrative Barriers

Operational and administrative barriers are caused by the enormous capital investments required for process modifications and related in-plant process changes. Changes to the existing sizing of units, design modifications, and the purchase of new and/or innovative equipment demand large capital commitments. Associated time delays during process modification mean further delays in implementing the overall pollution prevention program. In addition, consulting costs may increase due to a lack of appropriate technical resources in-house. An industry caught in such a situation may result in further disappointment on the part of the management. Typical management concerns and administrative barriers to implementing pollution prevention programs may be caused by the following:

- Management impassivity
- Deficiency in financial commitment
- Deviations from production concerns
- Expanded research, development, and design considerations
- Failure to monitor pollution prevention program success
- Middle management priorities and decisions
- Lack of agreement within the corporation
- Disorientation on regulatory compliance
- Perplexity on economic benefits
- Bureaucratic resistance to change
- Deficient or inadequate multimedia management structure

Following is a description of each of these barriers and suggestions for possible solutions.

Management Impassivity.   This may be caused by an indifferent attitude to pollution prevention programs. Since administrators in many companies lack technical orientation, many upper-level managers have difficulty realizing the enormous benefits of the pollution prevention program. These attitudes may be changed by educating upper management, through training courses and workshops, about the financial gains of a pollution prevention program.

Deficiency in Financial Commitment.   This is due to the nonavailability of financial and technical resources within the industry. Generally, this is a problem in small or large industries that are experiencing a financial downturn. Such deficiencies may be eliminated by a search for external financial resources provided by government agencies and/or negotiations which can lead to the release of financial pressures on those trying to conform to current regulatory requirements.

Deviations from Production Concerns.   These are generally caused by supervisors and middle management staff spending additional time and effort educating themselves about pollution prevention programs and their incentives, which in turn, may result in the absence of a manager when needed on the job. A company can eliminate such a concern by providing incentives and encouragement for such staff to attend workshops and training courses during noncritical hours of the day or in the evenings.

Expanded Research, Development, and Design Considerations.   These considerations are generally discouraged by management due to supervisory and management concerns about diverting the efforts of their staff from meeting production goals and assigned deadlines. A very common misconception on a supervisor's part is that any pollution prevention activity leads to delays on projects and costs more money. A company may eliminate such a concern by providing incentives and encouragement to the work force to pay particular attention to pollution prevention programs through workshops and after-work meetings and interactions. Workers may be rewarded and recognized for their increased efforts and interests in pollution prevention programs.

Inability to Monitor Success. Periodically, this is a serious constraint on a pollution prevention program that is implemented. Lack of a monitoring structure that periodically displays the results of employee efforts on a pollution prevention program leads to disaffection and lack of interest in the team's continued efforts. Personnel responsible for monitoring and communicating the success of a pollution prevention program should be held accountable for a failure to display or communicate the results to the team.

Middle Management Priorities and Decisions. These are sometimes a hurdle to pollution prevention programs. Middle managers have a serious influence on upper management's efforts and corporate policies. Frequently, middle management's priorities emphasize overall profit reporting through sales and production efforts. Because this is the primary objective of middle management, many middle managers are not totally receptive to pollution prevention programs or ideas. To eliminate this obstacle, middle managers should be assigned the responsibility of periodically reporting pollution prevention ideas to upper management. This should be done in a controlled manner and should include the provision of incentives for meeting pollution prevention goals.

Lack of Agreement within the Corporation. This is caused by a poor understanding of the advantages of pollution prevention. The corporation may eliminate such a problem by assessing proper pollution prevention areas within the process operations and by setting short-term goals and prioritizing them. A proper program for pollution prevention requires a consensus of key corporate personnel, the efforts and ideas of upper and middle managers, and the creativity and interest of all support staff. A lack of agreement within a corporation may be counteracted by requirements for establishment of short- and long-term commitments towards meeting pollution prevention goals.

Disorientation on Regulatory Compliance. This may result from multiple violations involving various areas of the industry. The difficulty of tracking regulatory updates and revisions may lead to lack of compliance, which results in penalties and serious liabilities. In smaller corporations, lack of technical resources on environmental issues leads to penalties and fines. Permit review processes are doomed to fail if corporations and responsible individuals are not kept updated and informed about particular regulations.

Perplexity on Economic Benefits.   Until recent times, cost tracking has not been emphasized by an industry—particularly costs associated with environmental management and compliance issues. Corporations have not fully realized that environmental compliance must be a significant and integral part of their infrastructure and that it requires more attention to stay in compliance with regulations for generating, treating, and disposing of wastes. Due to the changes in regulations, treatment and disposal costs have increased dramatically, and liability concerns will continue to increase in the future. Today, overall corporate acumen about this issue is possible through creative management and competitiveness.

One of the biggest challenges in implementing a pollution prevention program is to develop an effective cost tracking and accounting system that accurately reflects all the costs and benefits of prevention as opposed to treatment. Currently, only a few simple projects are seen as cost-effective when analyzed by traditional accounting methods that assign all environmental costs to overhead costs.

Total cost accounting (TCA) systems may make companies nervous, because they require projections of future liability. Meanwhile, activity-based accounting systems, which break down accounts by product or business unit, based on function, take too long and cost too much to implement. Many companies with proactive pollution prevention programs already incorporate some limited total cost accounting for environmental projects by including raw material savings and avoiding treatment and disposal costs. The most progressive of them are combining elements of both total cost-based and activity-based accounting in their systems, assigning costs back to their original product lines, and factoring in more of the hidden costs of pollution prevention.

Frequently neglected in the early analysis of pollution prevention programs are costs for handling and moving materials, record keeping, occupancy and overhead, all supplies, quality inspections, scheduling, labeling and manifesting, and utilities. If all these factors have been weighed and the project still has a negative *net present value* (NPV), cost tracking through a checklist can be used to evaluate the more tangible of the "soft" liability issues.

Bureaucratic Resistance to Change.   This involves serious resistance on the part of large corporations to implementing pollution prevention through process modification and change. Frequently, corporations do not realize that the more resistant their attitude, the greater are the risks of failure in global markets, which may result in corporate downfalls or downsizing.

Deficient or Inadequate Multimedia Management Structure.   This problem is commonly found in the industry today. A multimedia approach may be applied to a product, system, or process, and requires analysis of process streams, utilities, and other inputs, as well as an understanding of multimedia waste discharges for the overall facility.[5] Such an analysis is mandatory for every unit in the facility, with environmental compliance, energy conservation, and future expansions of the facility taken into consideration.

In addition to the foregoing administrative barriers, pollution prevention may be almost impossible to achieve in situations where technological changes are mandated to outdated equipment that is still in operation. In current times, industries have found it difficult to operate while seeking process changes, due to fierce international competition and the difficulties of survival in the global market. As a result, priorities are more focused on survival in the industry rather than on pursuing pollution prevention programs related to process modifications. However, it should be noted that an industry should not rule out ideas related to pollution prevention—rather, it should consider recycle and reuse options that may provide considerable reductions in manufacturing costs. Also, an industry should consider how manufacturing ancillary products from waste by-products can generate additional financial resources for expansion and marketability.

7.3.4   Federal, State, and Local Barriers

Presently, federal and state organizations are considering removing bureaucratic obstacles to encourage pollution prevention and to provide an economic boost to the industries. Following are examples of such barriers prevalent within the regulatory system:

- Incomplete guidance and information about additional feasible alternatives for controlling the multitude of diverse, changing products and processes that contribute to pollution.

- Rising control costs, based on changes to environmental regulations and emission standards that have exceeded industrial investments by over half a trillion dollars.

- Delayed government response to new knowledge and changing industrial conditions because of centralized rules tailored to individual processes or chemicals. These rules require volumes of feasibility data, take years to complete reviewing, and often halt existing or past control technologies in place rather than stimulating new ones.

- Diminishing resources for complex environmental problems due to economic constraints and sluggish market conditions, with expanding statutory conditions.

- Lack of motivation for regulated dischargers to provide more than the minimum required, since no benefits are offered. Also, dischargers are subject to further scrutiny and regulation.

Today's complex environmental compliance problems are dominated by issues relating to underground storage tanks, non-point-source runoff, smog, or groundwater contamination. The effective control of pollution is associated with local or site-specific factors, land use or related changes, and global problems such as stratospheric ozone depletion, acid rain, or pollution of international waterways. These problems dominate the environmental protection system and are not amenable to centralized regulation or enforcement. Whether incentives are financial, informational, or technical, a shift from regulation toward incentives is possible when individual private interests improve in meeting environmental goals established by the industry. A number of private pollution prevention "self-help" networks exist on the state and local level, through state-imposed pollution prevention reporting requirements.

Furthermore, technology transfer programs and applied R&D in pollution prevention are provided by industrial associations. The EPA plans to improve its pollution prevention information clearinghouse and to effectively coordinate cross-media programs to promote new and innovative pollution prevention technologies. This is possible with the currently available communications technologies and networking systems.

7.4 Regulatory Framework Encouraging Pollution Prevention

There are several regulations employing language that encourages pollution prevention. These include the Resource Conservation Recovery Act (RCRA), the Comprehensive Environmental Response Compensation Liability Act (CERCLA), the Clean Air Act (CAA), the Clean Water Act (CWA) and other point- and non-point-source discharge regulations, the Pollution Prevention Act, the Hazardous Materials Transportation Act, the Omnibus Trade and Competitiveness Act, and other laws affecting waste reduction. These regulations are discussed in Chap. 3.

Some further legislative initiatives that rigorously encourage pollution prevention are the Emergency Planning and Community Right-to-Know Act (EPCRA) and other laws influencing waste reduction, such as the Hazardous Materials Transportation Act (HMTA), the Occupational Safety and Health Act (OSHA), the Omnibus Trade and Competitiveness Act, the Tax Reform Act, the Revenue Reconciliation Act, and the Revenue Act.

7.4.1 Other Federal Laws Influencing Pollution Prevention

In addition to the major environmental legislation, a number of laws that either directly or indirectly influence waste-reduction activities exist. Typically, the authority for promulgating regulations and enforcing these laws has been the responsibility of various federal departments other than the EPA. These include the Department of Transportation, Department of Labor, Department of Commerce, and Department of Treasury.

7.5   State Programs Promoting Pollution Prevention

While reduction and recycling of municipal and solid wastes has been addressed by many states, laws addressing reduction and/or recycling of hazardous wastes are relatively recent. Several states have passed pollution prevention into their legislation. With only a few exceptions, most states are focusing on methods that encourage and aid development of pollution prevention strategies instead of mandating waste reduction standards, thus leading the way as promoted by the EPA in several guidance documents and policy statements.

State programs are increasing pollution prevention activities as well. States are primarily concerned about land disposal and, as a result, are promoting pollution prevention. In fact, some states have found it practical to take a non-regulatory approach to promote pollution prevention. Due to limited financial resources and limited labor resources, few attempts have been or are likely to be made to measure the effectiveness of state programs. Although states have led the federal government in actively promoting pollution prevention, it is clear that a parallel federal effort is needed to raise pollution prevention to a stature comparable to that of pollution control.[4]

State programs often deal primarily with waste management, not reduction of waste at its source. States have not explored possibilities in dealing with non-RCRA wastes and multimedia issues in their programs, as their regulations emphasize waste problems of small-quantity generators and the accompanying disposal issues.

Pollution prevention has been viewed as less important and urgent than siting and as representing a diversion of resources. This was due to the uncertainty that pollution prevention introduces, which clouds the market's need for new waste management facilities. But pollution prevention can be viewed as a means of alleviating the need for site waste-management facilities and of assuring the public that only truly necessary facilities will be sited. There are indications that some siting programs are now taking a positive view of pollution prevention rather than seeing it as a threat. Overall, the pressure associated with siting difficulties has probably played a positive, though indirect, role in stimulating interest in pollution prevention by industry and the public.

So far, a few states view pollution prevention as contributing to safe and publicly acceptable industrial development. Many more may do so as the economic and industrial benefits of pollution prevention are better understood. Highlights from selected state programs are discussed in Chap. 6. A variety of information sources under state programs on pollution prevention is available. Listed here are some information sources that are available through most state environmental agencies.[5]

- Information clearinghouses
- Waste exchanges

- Technical assistance programs, including waste audits
- Grants and tax incentives
- Demonstration projects
- Merit awards
- Education for industry and schools
- Training programs for employees
- Research and development opportunities

References

1. Resource Conservation and Recovery Act, as amended, Sec. 3005(h), 1984.
2. "Growing Exchange of Information Spurs Pollution Prevention Efforts," *Chemical and Engineering News*, 26 July 1993, pp. 8–25.
3. U.S. Department of Energy, Office of Environmental Analysis, *The Impact of State Hazardous Waste Minimization Programs on Operational Practices in the Petroleum Refining Industry*, December 1989.
4. U.S. Department of Energy, *Federal Legislative and Regulatory Incentives and Disincentives for Industrial Waste Reduction* (Final Report), October 1991.
5. L. Theodore and Y. C. McGuinn, *Pollution Prevention*, Van Nostrand Reinhold, New York, 1992.

Further Reading

Environmental Defense Fund, *Approaches to Source Reduction: Practical Guidance from Existing Policies and Programs*, Berkeley, Calif., 1986.

The Hazardous and Solid Waste Amendments of 1984, Public Law No. 98-616, 42 U.S.C.A. 6902 et. seq., 1984.

Resource Conservation and Recovery Act, Public Law No. 94-580, Stat. 2795, 42 U.S.C.A. § 6901 et. seq., 1976. Also, *RCRA*, as amended, §§ 3002(b), 3002(a)(6)(C–D), 8002(r), 1003(a)(6), 1984.

U.S. Environmental Protection Agency, *Report to Congress: Minimization of Hazardous Waste*, EPA/530-SW-86-033, October 1986.

Water Environment Federation, *Industrial Wastewater, NPDES Stormwater Program Provides Leverage for Pollution Prevention*, Alexandria, Va., August/September 1993, p. 39.

8

Developing and Maintaining a Pollution Prevention Program

Laurie Case

Laura Mendicino

David Thomas
*Illinois Hazardous Waste Research and
Information Center
Champaign, Illinois*

Industrial waste generation in the United States averages billions of pounds daily. Discharged to our air, water, or land, this waste represents a significant loss of raw materials and a potential threat to human health and the environment. To be responsible guardians of environmental quality, waste generators must review their production processes and business operations as well as consider the economic and the environmental benefits of implementing a pollution prevention program.

Due to increasing environmental concerns associated with industrial waste, companies must now incorporate waste management and prevention strategies into their industrial processes. Adopting a pollution prevention program as a way of doing business can provide a number of significant benefits to a company. Figure 8-1 provides a brief overview of what pollution prevention means,

---

What

Pollution prevention—any in-plant practice that reduces or eliminates the amount and/or toxicity of pollutants which would have entered any wastestream or would otherwise have been released into the environment prior to management techniques such as recycling, treatment, or disposal. Pollution prevention includes the design of products and processes that will lead to less waste being produced by the manufacturer or the end user.

Who

Any business that

- Generates waste
- Uses hazardous materials
- Emits or discharges waste into the air, water, or land
- Wants to save money through reducing waste-handling costs, raw material costs, and production costs
- Wants to operate in an environmentally sound and responsible manner

Why

Businesses that implement a pollution prevention program

- Avoid rising costs of waste disposal
- Save money in other areas such as purchasing of raw materials
- Increase their industrial efficiency
- Maintain or increase competitiveness
- Decrease long-term liability
- Follow state and federal policy guidelines
- May reduce present and future regulatory burdens
- Improve environmental and workplace conditions
- Ensure community safety
- Maintain good corporate image

Figure 8-1. An introduction to pollution prevention.

who should be considering developing a pollution prevention program, and why it is important.

By decreasing the amount of waste generated or released, a company can reduce waste disposal costs, improve worker safety, and reduce long-term liability. In addition, pollution prevention methods should lead to increased efficiency of the production line and decreased costs associated with the purchase of raw materials, inventory control, and so forth. Any resulting changes in efficiency or expenditures may help the company to maintain or improve its competitiveness in the marketplace.

A pollution prevention program develops from a thorough understanding of the facility from both a production and human resources standpoint. This knowledge, combined with commitment to implement pollution prevention techniques or technologies, can help a company meet, exceed, or even eliminate its regulatory responsibilities.

This chapter focuses on the establishment of a pollution prevention program, and methods to sustain such a program once in place. Other steps needed to implement a pollution prevention program are discussed only briefly, as many are covered elsewhere in this handbook or in the publications listed at the end of this chapter. There is no simple formula that every company can follow to establish a program; however, there are some basic steps every company should examine and undertake in developing its unique approach to pollution prevention.

8.1 Benefits of and Obstacles to a Pollution Prevention Program

Pollution prevention can be viewed as business planning with environmental benefits. The most common benefits of and incentives for establishing a pollution prevention program are presented in Table 8-1. Some of the obstacles that may hinder implementation or program development are presented in Table 8-2.

Table 8-1. Incentives for Pollution Prevention

Reduced operating cost	Money will be saved in the long term.
	Many projects have good return on investment and short payback period.
	Savings in disposal and raw materials costs reduce overall operating costs.
Improved worker safety	Reducing toxics improves working environment and decreases personal protective requirements.
Reduced compliance costs	Pollution prevention projects can reduce regulatory exposure.
	Some projects may eliminate need for permits, manifesting, monitoring, and reporting, saving time and money.
Increased productivity	Pollution prevention can result in more efficient use of raw materials due to improved processes and operations.
Increased environmental protection	Pollution prevention reduces waste generation, assuring improved long-term environmental protection and reduced future liability costs.
Continuous improvement	Pollution prevention can be an integral part of a company's TQM or continuous improvement program.

Table 8-2. Obstacles to Pollution Prevention

Capital requirements	Projects may require capital investments.
Specifications	Production materials may be toxic and/or hazardous and could be replaced with more environmentally sound alternatives, but contracts may specify that the toxic compounds be used.
Regulatory issues	New or modified permits may be necessary.
Product quality issues	Pollution prevention projects may change product quality.
Customer acceptance	Anything that affects quality or perceptions of quality may affect customer acceptance.
Immediate production concerns	Companies are often hesitant to admit that the "old way" may not have been the best way.
Available time and technical expertise	Time or sufficient expertise may be lacking.
Inertia	The "If it ain't broke, don't fix it" attitude may prevail.

8.2   Definition of Pollution Prevention Program, Plan, and Project

There is considerable confusion about the meanings of the terms *pollution prevention program, plan,* and *project.* Many companies have compiled a list of projects and called the list a plan—such a list is not a plan. Creating a *pollution prevention program* involves developing and implementing a continuous strategy to reduce all waste generated by a facility, in addition to procedures for prioritizing and systematically reducing these wastes. A *pollution prevention plan* is a written guide to how the program will be implemented and is used to chart the progress of the program. It reiterates management support, lists reasons for the program, identifies the pollution prevention team, describes how waste will be characterized, provides a strategy and schedule for pollution prevention assessments, institutes a cost allocation system, indicates how technology transfer will take place, addresses training needs, and discusses how the program and projects will be evaluated and implemented. The plan must be updated periodically to reflect the continuous nature of a pollution prevention program. *Projects* are the specific activities undertaken to reduce or eliminate waste.

8.3   Overview of Steps to Implementing a Pollution Prevention Program

Many companies initiate only portions of a pollution prevention program, undertaking some specific projects, but then finding it difficult to maintain momentum. Often this is because there is not full commitment to a program. Unless pollution prevention is integrated into the company's business practices, early successes with pollution prevention projects may not be enough to sustain the program. By developing a program based on the following steps, a company can build and maintain a vigorous pollution prevention program. The concepts presented are applicable to the reduction of all waste regardless of medium, quantity, or toxicity. Suggestions are targeted at middle-management personnel. To apply these suggestions, each industry or facility will need to interpret them so they are made specific for their situation.

A manual developed by the Illinois Hazardous Waste Research and Information Center (HWRIC), *Pollution Prevention: A Guide to Program Implementation* (Illinois Hazardous Waste Research and Information Center, 1992), details each of the eight steps (see Fig. 8-2). The steps to establish and maintain a pollution prevention program are:

1. Obtain support from top management.

2. Get the program started by beginning to incorporate changes throughout the company, developing a written pollution prevention plan, and training employees in pollution prevention.

3. Review and describe in detail the manufacturing processes within the facility to determine the raw materials used and the sources of waste generation and to define a baseline inventory to be used to set goals and evaluate progress.

4. Identify potential pollution prevention opportunities for the facility.

5. Determine cost of current waste generation and establish a system of proportional waste management charges for those departments that generate waste.

6. Select the best pollution prevention options for the company and implement these choices.

7. Evaluate the pollution prevention program on a companywide basis, as well as evaluating specific pollution prevention projects.

8. Maintain and sustain the pollution prevention program for continued growth and benefits to the company. Reevaluate the program as economic situations change and/or process equipment requires upgrading.

This chapter will specifically examine steps 1, 2, and 8. Other chapters should be examined for in-depth information on the other topic areas.

8.4 Top-Management Support

Management commitment to the philosophy and the long-term process of implementing and maintaining a pollution prevention program provides the

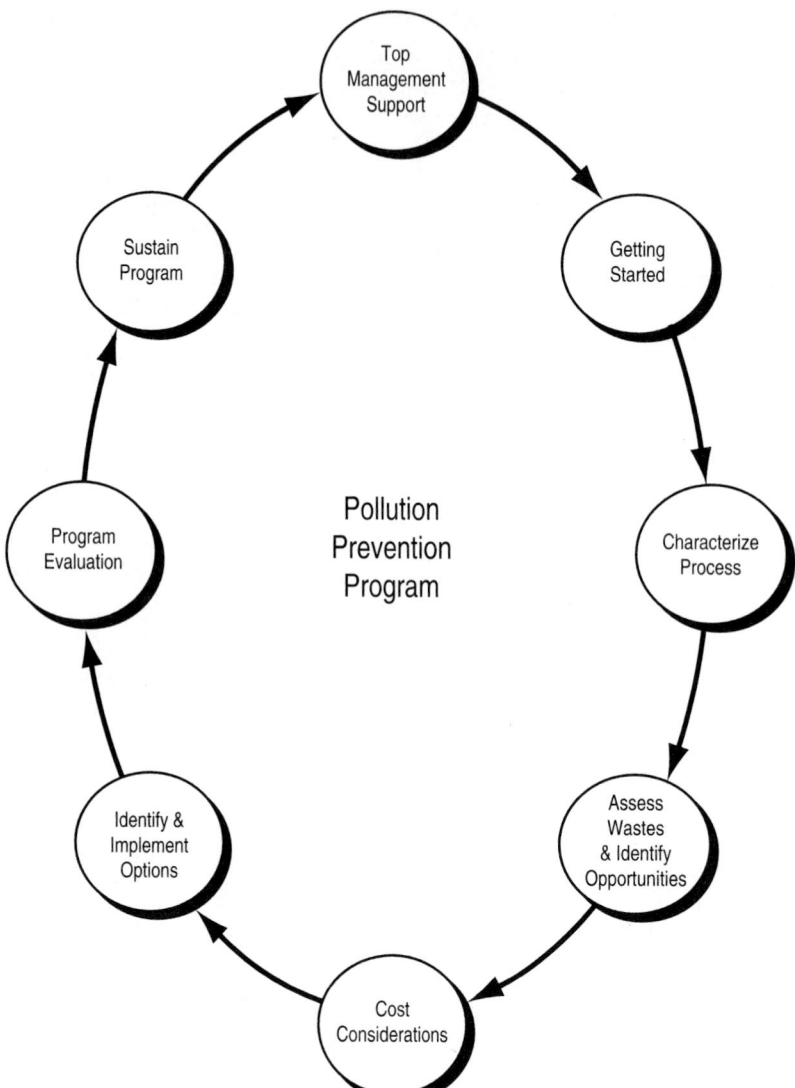

Figure 8-2. The continuous pollution prevention loop.

internal framework upon which a solid and enduring program can be built. The initial groundwork includes developing awareness and understanding of pollution prevention among top management. Providing a concise oral or written presentation about pollution prevention and how such a program can benefit the company can serve as an introductory overview. Figure 8-3 provides a checklist of topic suggestions.

Cost savings through reduced raw material usage, and reduced waste-handling, transportation, and storage costs

Increased productivity

Improved product quality

Regulatory compliance

Worker health and safety

Reduction of potential long-term liability

Examples of what other similar companies have achieved

Improved public image for the corporation

Emphasis of EPA policy on pollution prevention as a priority for the 1990s

Current state policies regarding pollution prevention planning

Figure 8-3. Suggested pollution prevention topics to discuss with management.

One way to follow up such initial presentations is to highlight pollution prevention case studies from companies using similar industrial processes. Seeing other successes can demonstrate the practicality of pollution prevention and provide tangible bottom line results. Case study examples are available from state pollution prevention agencies and the U.S. Environmental Protection Agency (EPA). Trade associations may also have case studies available specific to their industries.

Seeking assistance from state environmental agencies, both regulatory and nonregulatory, that offer pollution prevention services can be considered a next step. Staff from these agencies, consultants, and industry pollution prevention experts can be invited to talk to management about developing a pollution prevention program.

Once management approves a pollution prevention program, the foundation is complete and the first floor is ready to be built. Management's role is far from over, however. A brief written policy statement in support of a pollution prevention program needs to be drafted. Endorsements of the policy by all management levels are needed, and then the policy should be distributed to all employees. This statement can be a useful tool used for increasing employee awareness of pollution prevention.

Depending on the company's organization and structure, developing a corporate policy statement can be a lengthy process. Rather than allow this procedure to delay proceeding with the program, an interim policy or area-specific policy can be developed. This gets the program started; the corporate policy can follow later. Figure 8-4 shows an example of a management policy statement.

We, [company name], are committed to excellence and leadership in protecting the environment. In keeping with this policy, our objective is to reduce waste generation and emissions. We strive to minimize adverse impact on the air, water, and land through excellence in pollution prevention. By successfully preventing pollution at its source, we can achieve cost savings, increase operational efficiencies, improve the quality of our products and services, and maintain a safe and healthy workplace for our employees.

[Company name]'s environmental guidelines include the following:

- Environmental protection is everyone's responsibility. It is valued and displays commitment to [company name].

- Preventing pollution by reducing and eliminating the generation of waste and emissions at the source is a prime consideration in research, process design, and plant operations. [Company name] is committed to identifying and implementing pollution prevention opportunities through encouragement and involvement of all employees.

- Technologies and techniques which substitute nonhazardous materials and utilize other source reduction approaches will be given top priority in addressing all environmental issues.

- [Company name] seeks to demonstrate its corporate citizenship by adhering to all environmental regulations. We promote cooperation and coordination between industry, government, and the public toward the shared goal of preventing pollution at its source.

Figure 8-4. A management policy statement. (*From Minnesota Office of Waste Management, 1991.*)

Additional support needed from management includes assigning responsibility for developing the pollution prevention program, developing a means for evaluating and tracking progress, allocating adequate time and budget, and recognizing achievements. Continuity of the pollution prevention program is important. From the initial stages, the program should be set up in such a way that one step can flow naturally into the next in a continuous cycle.

When a pollution prevention program flows from one step to the next, it comes to be viewed by all personnel at the facility simply as their way of doing business. One way to introduce pollution prevention into the company psyche is through incorporating it within a total quality management (TQM) program. It also builds nicely on a health or environmental safety program because it can benefit the company in a number of ways as shown in Fig. 8-5. Whether pollution prevention becomes integrated into another program or stands alone, it is a group effort with all employees working together toward meeting company pollution prevention goals.

Pollution prevention reduces the amount and/or toxicity of chemicals in the workplace.

Pollution prevention reduces short- and long-term exposure of employees, visitors, and contractors.

Pollution prevention reduces or eliminates monitoring requirements.

Pollution prevention reduces reporting requirements.

Pollution prevention reduces toxics, thereby reducing HVAC requirements needed.

Pollution prevention reduces or eliminates the need for personal protective equipment.

Figure 8-5. Why pollution prevention can be incorporated into a company's health and safety program.

8.5 Getting Started

Once a policy statement has been written, approved, and distributed, the process of incorporating a pollution prevention program into daily company activities begins. The steps involved include designating a pollution prevention coordinator, developing a pollution prevention team, increasing employee awareness and involvement, establishing a recognition program, training employees, goal setting, and developing a written pollution prevention plan.

8.5.1 Designate a Pollution Prevention Coordinator

While a pollution prevention program needs top-down support and commitment, it also needs bottom-up input and implementation. This means that teamwork and participation from all levels within the company are essential. A key element for success is to find a good advocate and coordinator for the pollution prevention program.

The pollution prevention coordinator will be responsible for establishing the pollution prevention team(s), conducting meetings, and making sure the company is working toward its pollution prevention goals. More than likely, the coordinator will come from a middle-management position. He or she needs to be well organized, an advocate for the program, enthusiastic, and a motivator of people. If the coordinator has top-management support and the confidence of supervisors and others on the team, he or she will likely develop a very successful program. The coordinator will act as the key liaison to top management. This helps to ensure that the best pollution prevention ideas in terms of need, feasibility, and benefit to the company are delivered to top management for consideration. Also, the coordinator will need to obtain interdepartmental cooperation and resources on a continuing basis.

8.5.2 Develop a Pollution Prevention Team

A pollution prevention team needs to be organized prior to beginning the characterization and assessment processes. The team should not be assigned from any one department. Some suggested key personnel to include are representatives (both supervisors and line workers) from maintenance, production, environmental, health and safety, purchasing, shipping and receiving, legal, and engineering departments. Plant and executive managers should also be included. Not every company will have all these designations; the point is to include those individuals knowledgeable about the processes generating wastes and involve them from the beginning.

The initial pollution prevention team meeting should be an informal session to discuss the concept of pollution prevention, how the company can be expected to benefit, and where and how to begin. General information about the company's processes and operational procedures should be reviewed. The team will be responsible for developing the formal pollution prevention plan as discussed later.

In addition to those individuals assigned duties on the pollution prevention team, others may wish to help. Do not turn away volunteers—everyone should be encouraged to participate in the pollution prevention program. All volunteers should be commended in some way (in articles in the in-house newsletter, for example) for their ideas and interest in helping the company, their coworkers, and the environment. Employee suggestions should continually be encouraged—supervisors need to listen carefully because innovative ideas can come from any employee. Pollution prevention must continue for the life of the facility; establishing a sound, cooperative program from the start will be beneficial in future years.

8.5.3 Develop a Written Pollution Prevention Plan

After the pollution prevention team has been organized, development of a written plan should be its first official task. (Some states now require facility pollution prevention plans.) The written plan should include all the ideas developed by the team, such as the statement of support from management; the pollution prevention team's structure, organizational guidelines, and statement of purpose; the methods for fostering participation by all employees; the company's general goals; the structure of an incentive and/or award program; the procedures for conducting process assessments; selection of projects; procedures, criteria, and schedule for implementing pollution prevention projects; and provisions for employee training. Again it should be emphasized that the plan must contain clear guidelines for employee participation on the team, as well as criteria for selecting and implementing projects. There should also be provision for replacing team members as necessary.

Maintaining a pollution prevention program requires that the plan be an accepted part of company procedure. This plan should be presented and agreed to by management so that they understand how the pollution prevention team will proceed and what resources and support will be required. Copies of the plan can be made available for all employees to read if they so desire. The plan should be modified on an annual basis as pollution prevention experience is gained and goals are reached. A company should strive to continually improve the entire program.

8.5.4 Set Goals

There are different types of goals a company should set when beginning its pollution prevention program. Some goals will be waste-specific, such as replacing organic cleaning solvents with water-based alternatives or replacing a wet paint line with powder coating. Other goals will be activity oriented such as establishing a preventive maintenance program, incorporating pollution prevention into performance evaluations of all management staff, installing a revised accounting system that charges the cost back to the production line generating the waste, training all employees in pollution prevention, or holding monthly team meetings.

The team should discuss what types of goals are appropriate for the company. For example, a company may want to set an ultimate goal of "zero percent waste generation" to acknowledge the fact that pollution prevention is a continuous quality management program. This is very similar to company goals like "zero product defects" or "zero lost workdays." Another goal may be to replace some or all toxic substances used with less toxic substances and thus reduce risk to employees, the public, and the environment. Specific goals will also be based on the type of production process undergoing change, for example, replacing hexavalent with trivalent chromium in a plating line.

In addition to specific goals, more general goals should be set. These could include improving worker health and safety in the facility or improving the company image and attractiveness to investors.

It is a good idea to set a number of measurable goals to track progress over a given period. Numerical goals for waste reduction may be established once the wastes are characterized. Goals should be continually updated as they are achieved. Do not remain static. Build on the successes achieved.

8.5.5 Increase Employee Awareness and Involvement

One method of increasing pollution prevention knowledge is through a corporate and/or facility awareness program. Supervisors should discuss the status of the pollution prevention program at weekly meetings. They should encourage employees to bring pollution prevention ideas to them so they can be forwarded to the facility pollution prevention team meetings. Some companies may already have "quality circles" in place to improve product quality and production efficiency. The team should work with these groups to develop ideas for pollution prevention initiatives. Figure 8-6 provides suggestions for the pollution prevention team to use in developing its awareness program.

8.5.6 Train Employees

As pollution prevention strategies are identified, the training requirements must be considered by the pollution prevention team prior to implementation. Pollution prevention training programs should be tailored for management, line, and maintenance staff and should be incorporated into company procedures. This training can be incorporated into other training programs, such as process operation training or health, safety, and environmental training. Consolidated training for different groups can also stimulate discussion between employees who would not interact otherwise. Additional personnel training may be needed if materials-handling or accounting changes are made.

Provide a definition and explanation of the primary components of pollution prevention: source reduction and in-process recycling.

State company policies and guidelines clearly.

Mention any state government requirements for pollution prevention planning.

Identify company goals to reduce waste generation and to improve operations.

Stress that pollution prevention is essential and benefits the company.

Encourage employee participation as extremely important to improve facility and environmental conditions.

Distribute information on pollution prevention and the company's commitment to all employees.

Make management and pollution prevention team members available to respond to employee suggestions and new ideas.

Present facts on safety improvement that occurs when a pollution prevention program is implemented.

Stress the relationship between the cost of generating waste and company competitiveness.

Equate savings from pollution prevention with the company's fiscal health (i.e., increasing job security to encourage employee involvement).

Figure 8-6. Building a pollution prevention awareness program.

The facility or company may want to include a pollution prevention orientation program for all new employees, regardless of their job function. Employees will need thorough training on any new technologies or techniques added to unit processes. Depending upon the size of the facility, this may require training on more than one shift.

8.5.7 Reward Pollution Prevention Successes

To generate interest and participation in pollution prevention, establish an employee award or recognition program. Competition at larger plants may motivate participation. Shifts, departments, or even individuals can be encouraged to compete against their own past year's performance. Recognition in the form of an awards ceremony, a bonus, company logo merchandise, a special parking place, or added vacation time provides a tangible reward to individuals and departments who have reached their pollution prevention goals. Further recognition may be promoted in a regular column in the company newsletter. When a company newsletter is not available, a one-page fact sheet on pollution prevention could be started that acknowledges employee participation and accomplishments.

8.6 Out in the Plant— A Learning Process

This chapter focuses on the first two steps, top-management support and getting started, and the last step, sustaining a program, from the eight-step program used by HWRIC. A brief overview of the remaining five steps is provided below. In-depth procedures on approaching each aspect of a pollution prevention program are found elsewhere in this handbook.

8.6.1 Step 3: Characterize Process

To effectively implement a pollution prevention program, it is necessary to develop a complete understanding of the various unit processes and points in these processes where waste is being generated. The team should characterize all processes in the facility and keep these characterizations updated as process changes are made.

Two general approaches to characterize processes and waste generation are used. One technique involves gathering information on total multimedia (air, land, and water) waste releases at the end of each process, and then backtracking to determine waste sources. The other technique utilizes tracking of materials from the point where they enter the plant until they exit as wastes or products. Both techniques provide a baseline for understanding where and why wastes are generated, and a basis to measure waste reduced after implementation of pollution prevention projects. The steps involved in these characterizations include gathering background information, defining a production unit, general process characterization, understanding unit processes, and completing a material balance.

Developing process flow diagrams or updating existing diagrams can be useful for determining where, why, and how much waste is generated. The information assembled in the process characterization and flow diagrams will be used to help identify pollution prevention opportunities.

8.6.2   Step 4: Assessing Waste and Identifying Pollution Prevention Opportunities

Before conducting an assessment to identify what pollution prevention opportunities are present, wastestreams and unit processes should be prioritized to determine which should be examined first. When establishing priorities for pollution prevention, all input and output streams should be ranked—beginning with those which require immediate attention, followed by those which are less urgent. The team should maintain this list of priorities, so that once a project is completed they can move on down the list. The team should also update this list, as priorities may change with new regulations, changes in business direction, and other factors. It is also important, however, for the company to identify easy-to-do projects that can reduce waste and save money. Implementing these projects first can yield early successes in the program. Many of these types of changes can be started at the same time as more involved projects.

Once wastestreams are identified and prioritized, the assessment of specific pollution prevention opportunities can begin. This procedure involves looking at the specific processes contributing to the candidate streams. The team needs to observe the process in operation—ideally during all shifts, and if possible during a shutdown/cleanout/start-up period—to identify materials used and wastes generated by these procedures. Figure 8-7 lists what the assessment teams should be looking for.

Generate Reduction Options.   Once the process assessments are complete, the data should be reviewed for thoroughness by all pollution prevention team members. This review will aid the next activity: brainstorming for ideas to reduce waste and/or eliminate toxic and hazardous materials. Team members should also solicit ideas from personnel at all levels throughout the facility. Many times these personnel already have ideas for reducing waste but have never had the opportunity to express them. All options should be written down and given serious consideration.

> Observe operation procedures of line workers.
>
> Determine quantities and concentrations of materials (especially wastes).
>
> Observe collection (including exact sources) and handling of waste (note if wastes are mixed).
>
> Examine any recordkeeping; obtain copies of records if this has not been done already.
>
> Create flow diagram and follow through actual process.
>
> Look for any leaking lines or poorly operating equipment.
>
> Look for any spill residues.
>
> Examine any damaged containers.
>
> Determine physical and chemical characteristics of the waste or release.

Figure 8-7. Process assessment observations.

8.6.3 Step 5: Cost Considerations

Before pollution prevention projects are evaluated for economic feasibility, the full cost of waste generation must be determined. This full cost is necessary to develop the economics of pollution prevention techniques and technologies, including calculating the cost savings and payback periods.

Determine Full Cost of Waste. The full cost of waste generation includes more than treatment or disposal costs; it includes all the costs incurred in producing and handling waste. All of the expenditures associated with the wastestream, both direct and indirect, should be identified. These include, but are not limited to, the following: purchasing, storage and inventory, and in-process use of materials; air and water emissions, solid waste collection, waste storage, and on-site treatment or recycling; waste disposal; waste transportation; lost raw materials; labor costs. A pie chart showing the typical cost for waste generation is shown in Fig. 8-8. Often, wasted raw material costs are three-fourths of the full cost of generating waste. Waste disposal costs are typically less than half the total costs (Selman and Czarnecki, 1988). Many pollution prevention options will not appear to be justified if only half, or less, of the likely savings are considered.

Develop Economics. Once the full costs of the wastestreams are determined, an economic analysis of each pollution prevention project can be conducted. This analysis will provide management information on the costs and benefits associated with the techniques and/or technologies so that management can decide whether it is economically feasible to proceed with implemen-

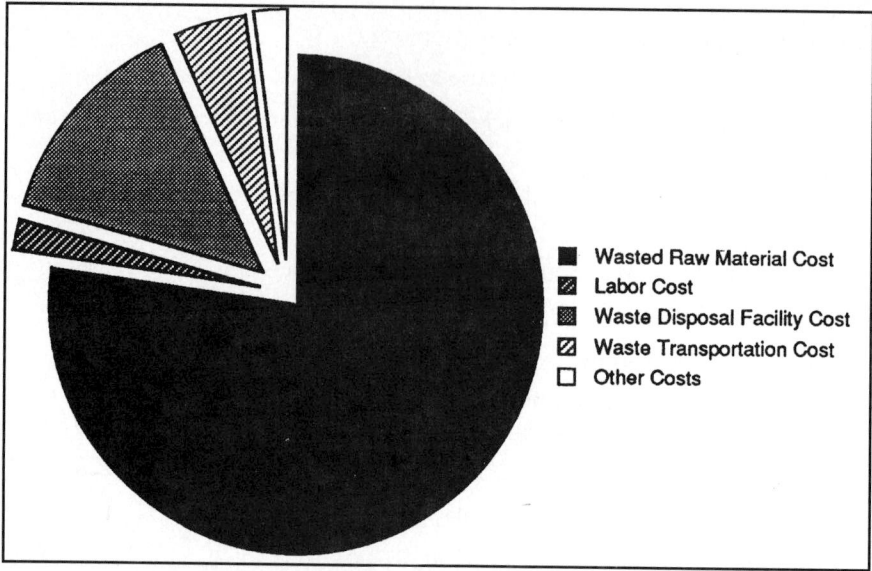

Figure 8-8. Typical cost distribution for waste generation.

tation. Certain benefits, such as reduced long-term liability, reduced worker exposure to toxic chemicals, and improved community relations, will be difficult to quantify.

Establish Cost Allocation System.   A cost allocation system is an important element of a pollution prevention program. In this system, each department or process is charged for the total waste management costs of the wastes it generates. This cost allocation system should lower the total overhead cost, as most companies charge waste disposal costs to overhead (i.e., the environmental department), and may provide a more accurate idea of the total cost of waste generation at the facility. It will also provide incentives for employees associated with the departments or processes that are charged for the waste handling to reduce their waste generation and subsequently their costs.

8.6.4   Step 6: Identifying and Implementing Pollution Prevention Options

Once suggestions for pollution prevention options are gathered and the costs associated with these options calculated, they should be reviewed by the pollution prevention team and the least beneficial options eliminated. These options may, however, be reviewed in the future, as what is less beneficial now may work better later. Each remaining option should then be examined in more detail to determine its overall benefits. Technical and economic feasibility of each option, based on the company's requirements for these criteria, should be studied. Those options found to be consistent with the company's goals can then be scheduled for implementation. There may even be cases in which certain benefits of a project override low economic return.

Some projects may be worth doing for reasons other than pollution prevention. For example, preventive maintenance programs can reduce the risk of leaks, spills, or possible explosions. Increasing process efficiency may improve economic competitiveness. These types of projects may also result in waste reductions. Regardless of the criteria for selection, any pollution prevention benefits associated with a project should be identified and documented.

Technical Evaluation. There are many factors which should be considered when one determines if a project is technically feasible. Personnel that will be directly affected by implementing the project should be consulted and included in the decision-making process. They typically have knowledge of process details that may inhibit project success and are essential in proper implementation. For projects that involve a new technology and/or technique, a bench-scale or pilot test may be required to assess technical feasibility. At this point, if it is determined that an option is not feasible by these criteria, the option should be deferred for consideration at a later time when the circumstances may be different.

Economic Evaluation. Once a pollution prevention project has been found to be technically feasible, the economics of the project should be examined.

Any project that yields a cost savings, i.e., savings in annual waste-handling or annual operating costs, has potential for profitability. If there are no initial costs involved, then a project can be considered economically feasible if there is a cost savings. Options such as better operating practices may be the most practical to implement first since they do not require an initial capital investment.

Implementing Projects. Once the pollution prevention team selects the projects to be implemented, management approval must be obtained. If management support was obtained as described previously, the approval process should not be difficult providing the project benefits, profitability, and feasibility are acceptable. The pollution prevention coordinator should present the details of the project along with the budget and project justification—particularly economics—to management. Once approval is received and implementation begins, it is important for team members to stay involved in the project. They should closely monitor all pollution prevention activities to make sure any problems are addressed immediately.

8.6.5 Step 7: Program Evaluation

The pollution prevention program should be continuously evaluated and updated to improve overall effectiveness. This periodic review by the pollution prevention team should be conducted for all stages of the program, from management support and team selection to project implementation.

The progress of the pollution prevention program can be determined by looking at the individual activities and projects. One way of measuring progress is quantitative. For example, look at actual waste reduction, both in terms of actual change in quantity and change in hazard level. The actual change in quantity is the difference between the waste per production unit reported in the current year and the waste per production unit reported in the previous year. The change in hazard level is based on toxicity, reactivity, ignitability, and corrosivity of the waste and industrial hygiene/employee exposure–type measurements. This comparative measurement is most useful when one evaluates an alternative material substitution, such as switching from an organic solvent to a water-based solvent. (Switching to water may even eliminate an OSHA sampling requirement.) These measures of waste reduction may not be appropriate for all facilities and wastes. Other quantitative measurements are adjusted quantity change and throughput ratio. Additional guidelines and detailed descriptions on measuring pollution prevention progress can be found in other chapters of this handbook.

When evaluating the elements of the program, it is important to identify those strategies and techniques which have been very successful, marginally successful, or have failed. If possible, the reasons why these projects were or were not successful should be determined. This information will be beneficial for modifying the program and redefining goals.

8.7 Sustain the Pollution Prevention Program

A pollution prevention program, once initiated, should have no "last step." Rather, the challenge lies in continuing and increasing the forward momentum cycle after cycle. Specific projects may be completed, but new projects should always be under development. The pollution prevention program cycle involves maintaining a strong commitment to the program from all operating levels. Some specific ideas for sustaining the program while carrying employee participation forward include summarizing the economic benefits of the program, bringing new personnel into the pollution prevention team, providing additional training, and publicizing success stories.

8.7.1 Reemphasize Economic Benefits

To sustain momentum in the pollution prevention program, the economic benefits realized through the activities undertaken should be summarized and presented not only to management personnel but to all employees. This information will also be important to include in the annual report to stockholders.

Emphasize cost savings resulting from pollution prevention projects over what would have been spent prior to implementation. Continue to track cost savings on implemented projects, as this may be the primary driving force toward sustaining the program.

8.7.2 Rotate Pollution Prevention Team

To maintain the flow of fresh ideas and introduce new perspectives, the pollution prevention team members should be rotated. With an ongoing pollution prevention program, there may be new employees who join the company over the years that want to participate. It is especially important that if the primary advocate and/or team leader for the pollution prevention program steps down, management find a suitable replacement who will give the program high priority and active guidance. A new team leader should step in with high energy, enthusiasm, and creativity. As members step down, they should be encouraged to serve as "consultants" to their replacements. There may also be team members who wish to remain; this should be encouraged, as they have gained valuable field experience. Team composition should always include employees from all levels and departments. The importance of having a written pollution prevention plan in place can be better understood once there is turnover of team members. This document provides the blueprint for operating the pollution prevention program and provides necessary continuity when there is a shift in team members.

8.7.3 Provide Refresher Training

Pollution prevention awareness and training should be conducted on a periodic basis so that all new or reassigned employees understand the company's commitment to the program and their individual responsibilities. As discussed earlier, pollution prevention training can be incorporated into already existing training programs (health and safety, environmental, processes, etc.) if that works best for the company.

8.7.4 Publicize Success Stories

Publicity is one of the most effective means to sustain the pollution prevention program. Internal publicity raises employee awareness of activities at the facility and encourages further participation. The results of the various projects should be relayed through bulletin board postings, newsletters, interoffice memos, and so forth. The names of the pollution prevention team members, as well as those employees offering suggestions, should be included in these publications. Hold periodic special events such as a free lunch for all employees contributing pollution prevention ideas during a specified month. If individual successes are recognized, other employees may wish to join in to receive the same recognition. Award ceremonies for employees or teams that have met their pollution prevention goals will also help publicize successes. Cost savings, waste reductions, and product quality improvements due to pollution prevention activities and projects should be highlighted.

The pollution prevention program can be a key public relations tool. Any reduction in waste is a benefit to employees, the community, and the environment, and should be publicized. News releases should be prepared for local and state news media documenting the project and the benefits to the company and the surrounding community. Reporters could also be invited to the facility for a demonstration of a new technology. Community forums such as city council meetings or local chamber of commerce meetings are excellent avenues to promote the company's environmental efforts.

Further public recognition can be facilitated through state, county, and local award programs. Pollution prevention award programs can be found in many states. These awards are presented to industrial facilities, trade organizations, vendors, community groups, and educational institutions that demonstrate significant achievements in pollution prevention. Check with the agency in your state responsible for overseeing pollution prevention activities on whether or not an award program exists.

Trade association meetings and publications are another good avenue for promoting a company's pollution prevention program. Case studies can be submitted which demonstrate the company's progressive stance in environmental protection while describing the use of innovative technologies and techniques to reduce waste. These case studies should emphasize the benefits accruing to the company—not only waste reduction but also cost savings, quality improvements, safety improvements, regulatory compliance, and better community relations.

8.8 Summary

The development of a pollution prevention program can involve a significant change in a company's traditional business practices. Whereas regulatory compliance has usually been the responsibility of an environmental department or coordinator, pollution prevention must involve employees from all areas of the plant and levels of management, including hourly workers. For companies developing TQM or continuous quality improvement programs, pollution prevention can become a part of these programs. A pollution prevention program can also become the primary strategy a company uses to achieve and maintain regulatory compliance. In fact, it may allow companies to get out of the regulatory system altogether.

As discussed, a pollution prevention program requires a firm commitment from management to survive. A written plan which affirms this commitment, provides guidance for conducting the program, lists criteria for evaluating and selecting projects, and gives procedures for documenting successes and failures is essential. A key component in every successful program is the pollution prevention advocate. The company must choose a program leader who understands and believes in a pollution prevention approach and who has the ability to work with people at all levels of the organization to implement the program.

A fully implemented pollution prevention program will affect a company's relationship with its suppliers, customers, and neighbors. A facility may even consider its neighbors to be its customers and thus inform them of the facility's activities and solicit their input. Community relations can be an important part of a company's pollution prevention program and can help sustain the program.

A company can also apply pressure to its suppliers and vendors to have them adopt pollution prevention programs and can encourage them to develop pollution prevention solutions for products and processes that may create significant amounts of hazardous and/or toxic wastes. The company can also look into providing "green" products and marketing them as such. For example, one company abandoned its line of organic solvents and now manufactures solvents with an aqueous base that do not generate hazardous waste. These cleaner solvents are now in great demand by many companies around the country. Companies with "green" products may find their pollution prevention approach to doing business will be sustained as it becomes an essential part of their business practice. This is the ultimate key to maintaining a successful pollution prevention program.

Bibliography

Illinois Hazardous Waste Research and Information Center, *Pollution Prevention: A Guide to Program Implementation*, TR-009, Champaign, Ill., 1992, 44 pp.

Minnesota Office of Waste Management, *Minnesota Guide to Pollution Prevention Planning*, Minneapolis, Minn., 1991, 125 pp.

Belle F. Selman and Charles A. Czarnecki, *Industrial Waste Prevention: Guide to Developing an Effective Waste Minimization Program*, Waste Advantage, Inc., Southfield, Mich., 1988.

Further Reading

U.S. Environmental Protection Agency, *Facility Pollution Prevention Guide*, EPA/600/R-92/088, 1992, 143 pp.

U.S. Environmental Protection Agency, *Pollution Prevention Information Exchange System (PIES) User Guide Version 2.1*, EPA/600/R-92-213, November 1992, 91 pp.

U.S. Environmental Protection Agency, *Pollution Prevention Resources and Training Opportunities in 1993*, EPA/560/8-92-002, 1993.

U.S. Environmental Protection Agency, *Total Cost Assessment: Accelerating Industrial Pollution Prevention through Innovative Project Financial Analysis*, 1992, 168 pp.

Washington State Department of Ecology, *Pollution Prevention Planning, Guidance Manual for Chapter 173-307 WAC*, Publ. 91-2, Olympia, Wash., 1992.

9

Pollution Prevention and Total Quality Management

Gary E. Baker

Science Applications International
Corporation
Cincinnati, Ohio

9.1 The Paradigms: Their Basis and Substance

Don't run to the dictionary—a paradigm is not as frightening or nebulous as the term appears. It is simply a pattern, example, or model, according to Webster. And both total quality management and pollution prevention are exactly that—a pattern, example, or model for their respective concepts to be followed and put into effect in United States industrial production. They each represent a framework for making things better, doing things smarter, and for the examination and evaluation of current practices. Both pollution prevention and total quality management welcome innovative thought and the mandate to question the validity of practices conducted simply because "that's how we've always done it." They strive to disrupt the inherent human resistance to change and the stubborn management mindset that considers product count to be the only measurement that matters.

This chapter investigates the striking similarities of purpose and content of pollution prevention and total quality management techniques. It examines the reasons and directions that United States industry is following to make both paradigms integral parts of conducting business.

9.1.1   The Background and Essence of TQM

> The basic cause of sickness in American industry and resulting unemployment is failure of top management to manage.... This book is an attempt to do something about productivity; to improve productivity, not just to measure it.... Everyone doing his best is not the answer. It is necessary that people know what to do. Drastic changes are required. The responsibility for change rests on management. The first step is to learn how to change.[1]

These words are printed in the preface to Dr. Deming's book on total quality management. They have served to awaken many top managers, and to help bring a new ethic to American business: the two basic foundations of any successful business are service and customer satisfaction.

Total quality management concepts were first presented to the public around 1982. They were reportedly put into practice in Japan by their author, W. Edwards Deming, decades before that time; Deming was awarded a medal from the Emperor of Japan in 1960. Significant controversy and debate have ensued since the publication of *Quality, Productivity, and Competitive Position,* but few argue about the promise and predictability of implementing Deming's ideas. In this book, and subsequent ones, Deming argued for acceptance of change, benchmarking the competition, and always looking for better and more productive ways of doing things. He summarized his management treatise with 14 points:

1. Create constancy of purpose toward improvement of product and service, with a plan to become competitive and to stay in business. Decide to whom top management is responsible.

2. Adopt the new philosophy. We are in a new economic age. We can no longer live with commonly accepted levels of delays, mistakes, defective materials and defective workmanship.

3. Cease dependence on mass inspection. Require, instead, statistical evidence that quality is built in, to eliminate need for inspection on a mass basis. Purchasing managers have a new job, and must learn it.

4. End the practice of awarding business on the basis of price tag. Instead, depend on meaningful measures of quality, along with price. Eliminate suppliers that can not qualify with statistical evidence of quality.

5. Find problems. It is management's job to work continually on the system (design, incoming materials, composition of material, maintenance, improvement of machine, training, supervision, retraining).

6. Institute modern methods of training on the job.

7. Institute modern methods of supervision of production workers. The responsibility of foremen must be changed from sheer numbers to quality. Improvement of quality will automatically improve productivity. Management must prepare to take immediate action on reports from foremen concerning barriers such as inherited defects, machines not maintained, poor tools, fuzzy operational definitions.

8. Drive out fear, so that everyone may work effectively for the company.

9. Break down barriers between departments. People in research, design, sales, and production must work as a team, to foresee problems of production that may be encountered with various materials and specifications.

10. Eliminate numerical goals, posters, and slogans for the work force, asking for new levels of productivity without providing methods.

11. Eliminate work standards that prescribe numerical quotas.

12. Remove barriers that stand between the hourly worker and his right to pride of workmanship.

13. Institute a vigorous program of education and retraining.

14. Create a structure in top management that will push every day on the above 13 points.[2]

Deming's points define a new approach in American industry, one that happened to coincide with the recognition that United States industry was, for the first time, beginning to lose competitive ground to the Japanese and even to the Europeans. Deming's words were taken quite seriously, and became the foundation of total quality management.

To manage the whole, it is not only inherent, but essential, that there be a complete examination of why and how we do things. Every person in the organization is important. Every person must feel that he or she is responsible for the quality of the product. Employees' ideas for improvements and changes to the process must be welcomed by management and given a fair and serious evaluation. Staff must play a part in the evaluation of change and use their collective creativity and ingenuity to overcome barriers and hurdles.

Obstacles to change must similarly be identified, categorized, and evaluated. Typically, obstacles fall into two major categories: technical or organizational. Technical obstacles require adaptation to physical limitations such as availability of space, chemical or mechanical limitations imposed by the processes themselves, or analysis of the specific qualities of the product either desired by the customer or necessary to the next step in the production array. Obstacles to technical change are repeatedly frustrated in today's industry because the specific characteristics or qualities of a process are not fundamentally known; they are done that way because that's how it has always been done. Evaluation of the characteristics of a surface coating imparted to a part by electroplating of chrome, for example, would typically include such factors as hardness, lubricity, wear, fatigue, microstructure, and adhesion. Many other coatings, applied by alternative technologies or processes that may enhance productivity, decrease worker exposure to toxic chemicals, and actually decrease rework frequency, may be deficient in some of the characteristics of chrome. What is important to remember is that the evaluation of technical factors cannot stop at metallurgical qualities; the specific service environment of the part in the component should be what determines the essential characteristics of the coating.

The type of change required puts old-style managers in a mode of survival for their jobs. It questions standard operating procedures and rules and regulations set in place by these types of controlling managers, and it opens worker vitality to the point of mutiny. In a total quality management framework, there is an organizational—better yet, an *institutional*—team working toward common goals.

Total quality management means total process consciousness and the flexibility in mindset to pursue and make radical change. A man named Chester Barnard, according to Peters and Waterman,[3] may have said it best in his book called *The Functions of the Executive,* in which he describes the role and responsibility of the business executive in managing the whole:

> The common sense of the whole is not obvious, and in fact is often not effectively present. Control is dominated by a particular aspect—the economic, the political, the religious, the scientific, the technological—with the result that [top performance] is not secured and failure ensues or is perpetually threatened. No doubt the development of a crisis due to unbalanced treatment of all the factors is the occasion for corrective action on the part of executives who possess the art of sensing the whole. A formal and orderly conception of the whole is rarely present, perhaps even rarely possible, except to a few men of executive genius, or a few executive organizations, the personnel of which are comprehensively sensitive and well integrated.[4]

Since the system, represented by all of these interplaying components, is so complex, it is logical that the advent of a team approach to problem solving (dubbed the skunkworks, or brainstorming session) is essential to bring the pieces together to make the whole. Teams of individuals knowledgeable of the impacts of a decision on purchasing, production, the environment, quality, and productivity are integral to making the correct decisions. No individual should presume to know enough about all of the aspects of a major process change or management direction to make the decision alone. The best-run companies in Peters and Waterman's book effectively use skunkworks, constantly strive to improve product quality and service, and evaluate what their customers want from their products or services. They use a myriad of tools to measure the effectiveness of their actions, and another host of tools to effect the proper management mindset, or culture, in every employee.

9.1.2   Soft and Hard TQM

Since the advent of TQM, there has emerged the notion of two distinct, but integral, aspects of the method. *Soft TQM* refers to establishment of the corporate and employee culture in a company to be innovative, to constantly strive for customer satisfaction and product quality improvement, and to make a difference at the individual level. These concepts require promotion of the entrepreneurial spirit and employee awareness and appreciation of other parts of the system whole. Not every individual or company is capable of these kinds of cultural adjustments. Indeed, some companies have no specific culture that is identifiable as anything other than total autocratic control.

Promoting soft TQM cultural changes can be done by inviting employee input, by carefully selecting skunkwork teams to consider as many aspects of a decision as possible, and by designing cross-feed of the factors into others. This latter method may involve practices such as those employed by the Japanese. A good friend of mine spent a designated part of a lengthy orientation at a local Honda plant on the production floor assembling vehicles long before commencing his job as director of safety and environment. This kind of hands-on perspective cannot be learned or appreciated in any other way. Again, Peters and Waterman have aptly summarized the four basic human needs in an organization: "(1) people need meaning; (2) people need a modicum of control; (3) people need positive reinforcement; and (4) people need to have their actions and behaviors shape attitudes and beliefs rather than vice versa."[5]

People must have the ability to try new things or methods, and to fail without retribution. Failure must be an established criterion for the pursuit and accomplishment of innovation. These kinds of thoughts cannot exist in a company disciplined by the profit motive. In a recent article by Drs. Michael and Timothy Mescon,[6] an analogy is drawn between today's successful corporations and children's toys called "Transformers"™ ("exceptional creatures that metamorphose to better adapt to their changing environment").[7] They describe the advent of the corporate "chameleon," who "adapt[s] to the needs of the marketplace while staunchly adhering to a vision, maintaining a strategic direction, and keeping everyone always informed." They describe the tactics of Motorola, where line employees participate in production and product design, cost management, and sales forecasts. It is called a "total-employment environment" and reportedly maximizes employee contributions and has made ongoing and dramatic improvements to the critical bottom line. They have documented ten characteristics of successful corporations:

1. Long-term thinking and planning are definitely in.
2. Stars are completely customer-driven. Simply stated: no customers, no profit. Research indicates that a two percent improvement in customer retention equals a ten percent cut in costs.
3. The stars build organizations where all stakeholders prosper.
4. Inspiration and leadership must emanate from the top.
5. Companies must be fast, focused, and ferocious.
6. Nanosecond response times are no longer silly science fiction.
7. Bundle up when necessary. Bundling up into task-oriented projects gives management cost relief from creating unnecessary infrastructures and the capability to deliver products and services that might not be plausible without these alliances.
8. There is no substitute for training.
9. The dinosaur is still dead. In a global market in which we all compete, bigness is goodness only if it allows for the flexibility to do what must be done when it must be done.
10. Change is the norm.[8]

The Mescons caution that each company must build a corporate culture that is best suited for itself. This is best done by "developing a system and culture from top to bottom, inside out, that recognizes, reinforces, and rewards meaningful change..."[9]

Hard TQM refers to the measurement methods employed to effect and ensure quality in a product. They are different types of statistical tools that enable the company to focus on what is wrong much faster than without them. Although they are used for specific problem identification, some common types are the Pareto chart, histograms, scatter diagrams, stratification of data, check sheet (tabulation of results or decision matrix), Shewart control chart (analysis of variation and control limits), and the Ishikawa diagram (used to examine cause and effect relationships). For example, a Pareto chart is a bar graph used to prioritize data. Figure 9-1 shows a Pareto chart for the cost of disposal of wastes generated at a facility in a given year. The data in this chart helps identify the magnitude of costs for each type of waste generated, and therefore allows for prioritization of effort.

A decision matrix or check sheet allows for instant evaluation of implementing change relative to several important variables of importance. In Table 9-1, a decision matrix allows the user to examine cost versus waste generation along with perceived technical difficulty, community concerns, worker exposure, and the toxicity of the waste. In these categories, a qualitative ranking of the factor has been assigned. In this case, although Process 2 is more costly than Process 3 or 4, the ease of implementation, the ability to reduce the impact on natural resources, and the potential that this process will be regulated in the future make Process 2 the one to pursue.

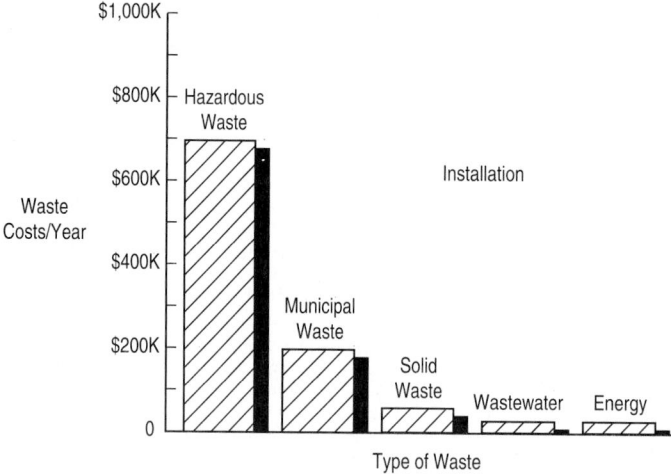

Figure 9-1. Pareto chart for waste data at a hypothetical installation.

Table 9-1. Decision Matrix (SAIC)

Criteria	Process 1	Process 2	Process 3	Process 4
Haz. waste disposal costs/yr	$325,000	$256,500	$235,500	$183,000
Releases to all media/yr (lbs)	2000	4600	3000	1500
Perceived ease of changing process (1–5)*	2	5	1	4
Community concerns with waste type (1–5)	1	5	2	1
Natural resource impact (1–5)	2	4	1	1
Potential for future regulations (1–5)	2	4	2	3

*Ranking 1 to 5, with 5 the highest.

9.1.3 The Pollution Prevention Ethic

So, what is this pollution prevention paradigm, anyhow? Increasingly stringent regulations and statutes in the environmental arena have finally caused people to integrate the concept of reducing or eliminating the generation of waste at its source. Another more comprehensive definition of pollution prevention is

> Any action that reduces the impact that an operation or activity may have on the environment, including impacts to the air, surface waters, ground waters and soils, through the reduction or elimination of wastes, and more efficient use of raw material or energy.[10]

In this definition, pollution prevention applies to solid (municipal, nonhazardous) waste, as well as to hazardous waste, air emissions, and wastewater discharges, and it involves the efficient use of all materials, including energy and water. Its basis is the Pollution Prevention Act of 1990, and the ensuing National Strategy, which presented a hierarchy of management for waste. This hierarchy is presented in Fig. 9-2.

Although simple to understand, implementing changes that reduce the very generation of waste requires a lot of work and evaluation. The source reduction level is the most important step in the hierarchy, and should not be bypassed. It is this source reduction aspect that has caused parallel management models of TQM and pollution prevention to actually merge. But more on this later.

As a contractor to the EPA, Science Applications International Corporation (SAIC) has conducted scores of process assessments at numerous facilities across the country. The assessment process allowed the teams to question even the most fundamental elements of why things were being done in the manner in which they were being done. At one facility, an equally capable nondestructive inspection method was not being used because the electrical power at a new facility had yet to be installed. Instead of forcing the installation of power and relocation of the unit, the staff was content to use a fluorescent-based system that generated significant wastes from the developer, dye, and rinsing steps. The specific frequency of testing that they employed was based on a written procedure that no one questioned. There were new dry developer system modifications available

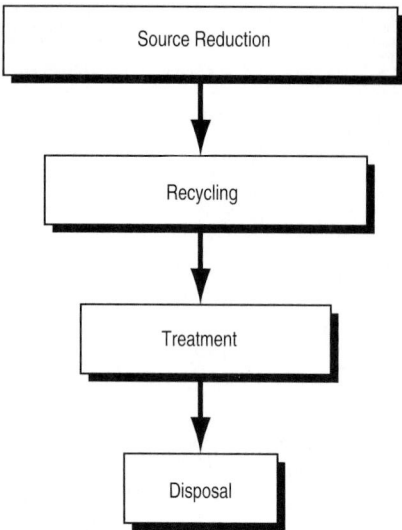

Figure 9-2. The waste management hierarchy.

from the original vendor that could have eliminated the liquid wastestream. The specifications of the dye and developer clearly required that the developer be handled as hazardous waste; they were currently disposing of the dye only as a hazardous waste, although they really didn't know why.

Pollution prevention has both a soft and hard component. The soft side of pollution prevention requires fostering a fundamental caring attitude, not only for the environment, but for the quality of the product being made. It requires top management commitment and buy-in, but also across-the-board acceptance and employee participation. It means that, finally, environmental impacts can be fully considered as another factor in the process of making decisions. It must not continue to be an overhead expense that is tolerated simply because there are regulatory inspections, fines, or worse in the event the rules are not obeyed. Beyond this, pollution prevention offers the ultimate compliance strategy to American industry.

A recent article by Olin Jennings called "A Practical Approach to Improving Quality"[11] looks at the issues of using TQM in environmental services companies. Although significant, the problems of this adaptation are overcome by using a nine-step approach:

1. Top management commitment, training, and planning

2. Branch-level operations analysis or review

3. Customer interviews (including lost customers and prospects)

4. Branch-level brainstorming, problem solving, and training

5. Personal improvement objectives for all employees

6. Improvement program action plans

7. Documentation and communication of the program

8. Implementation

9. Ongoing monitoring and improvement

Does this resemble the steps of the pollution prevention facility plan? Jennings describes these steps as bringing TQM to this type of services company.

The process used to conduct pollution prevention projects is startlingly similar to those requirements of TQM; rethinking why things are done the way they are is fundamental to fostering accountability in employees, forcing management to respond positively and fairly to change, and ultimately effecting better product quality and customer satisfaction. The evaluation team must decide what aspects of environmental protection are most important to them. The basis for conducting pollution prevention efforts can be direct, such as need to achieve and maintain compliance with a specific provision, or intangible, such as reducing the impact to the community landfill, and thereby becoming more of a good neighbor.

9.2 Convergence of the Paradigms

Although it arrives nearly a decade later to American industry than TQM, pollution prevention is clearly based on the principles of TQM. Like TQM, pollution prevention means total process consciousness and the flexibility to pursue and make radical change. It is founded on a new American management ethic that involves employees and management together in the decision-making process.

Let's examine Deming's 14 points for TQM, this time relative to pollution prevention.

1. *Create constancy of purpose toward improvement of product and service. ...Decide to whom top management is responsible.* The ethical change of this point is most evident in source reduction efforts since the product (or service) must be rethought and reevaluated as to why and how things are done.

2. *Adopt the new philosophy.* Companies that are successful using TQM methods will aggressively adopt those of pollution prevention as integral to their success.

3. *Cease dependence on mass inspection.* Both TQM and pollution prevention use the same statistical methods to build quality into the option or proposal being considered.

4. *End the practice of awarding business on the basis of the price tag.* The cost of proper environmental compliance today is such a significant component that it can no longer be overlooked. Those companies that refuse to incorporate com-

pliance (and therefore pollution prevention) into their culture will not be able to compete for long.

5. *Find problems.*   This point is fundamental to deciding what environmental elements are most important to a facility, and in focusing on those that need the first and most attention to their solution.

6. *Institute modern methods of training on the job.*   Training is essential for all employees to achieve the right mindset for pollution prevention, to know how to properly conduct assessments and involve other team members, and to apply the implemented options to the process and make it the new way of doing business.

7. *Institute modern methods of supervision of production workers.*   Supervisors and other managers must respect and appreciate the input of those employees on the production line, in the laboratory, or responsible for purchasing and supplying materials.

8. *Drive out fear, so that everyone may work effectively for the company.*   This point is strongly related to the previous one, and essential for pollution prevention to work. Concepts, ideas, and even experimental results from anyone connected with a given process are critical to option identification, and for proper evaluation of the success of the options.

9. *Break down barriers between departments.*   This point speaks of improved communication throughout an organization, and of the ability to propose ideas with retribution.

10. *Eliminate numerical goals, posters, and slogans for the work force, asking for new levels of productivity without providing methods.*   These kinds of motivations are condescending and corny. Involve the worker in detailing ideas he has had over the course of years with the same process or product, and you will generate the best pollution prevention options.

11. *Eliminate work standards that prescribe numerical goals.*   Pollution prevention methods invite everyone to think and not be tied to producing a set number of widgets per shift.

12. *Remove barriers that stand between the hourly worker and his or her right to pride of workmanship.*   Whether it is an invitation to a skunkwork for a product issue or for a pollution prevention issue, breaking down the artificial walls between managers and workers will yield the best and the desired results.

13. *Institute a vigorous program of education and retraining.*   Both TQM and pollution prevention require the right culture. Education and retraining of managers and workers collectively will instill these values into their daily mindset more rapidly than by management edict.

14. *Create a structure in top management that will push every day on the preceding 13 points.*   Like pollution prevention, TQM requires constant, dynamic change.

Criteria for Improved Profitability.   It is truly beginning to happen. The environmental responsibilities of American industry are achieving equal, or occasionally even higher, status than responsibility to the shareholder. As TQM helps modern managers separate from the profit motive toward management of the whole, the environmental ethic has emerged as an integral part of the whole. Environmental responsibility cannot be slighted, overlooked, or outright ignored. The commitment to staffing environmental compliance groups will similarly not be ignored.

Change in Culture: Rewarding Flexibility and Innovation in the United States Workplace. A recent project in a research facility illustrates the change of culture mandated by both TQM and pollution prevention. This project was initiated by one person, who, in the definition of Peters and Waterman, is both a pioneer and an "egomaniac with a mission"—a true champion.

John, as he will be called here, believed that there were many pollution prevention opportunities for his facility, but he did not know which process or organization to start with. He did realize, however, that he needed a plan. Not another rigid plan typically used to level the short leg of the conference table, but one that would be used by those challenged to find solutions to problems that were identified, and one that his management would accept and support. The plan that was developed did just that. It was based on the personal interviews of staff who were engaged in conducting processes and activities daily. It compiled the basic data on raw materials, energy, water, and waste generation. It assigned accountability by identifying the parts of the organization responsible. It actively incorporated the top management criteria for environmental action, and assigned numerical values to these criteria. Finally, the plan identified 58 pollution prevention options, and prioritized 15 for the first year's implementation plan.

John presented the plan to his management. He was rewarded by their praise for the effort, additional staff to support the program and its implementation, and he secured the assistance of professionals to get the first-year plan implemented.

Similarly, the President's Council on Environmental Quality conducted a recent initiative that is documented in a report called *Total Quality Management—A Framework for Pollution Prevention.*[12] This initiative reports the knowledge and experience gained by 12 facilities in using TQM principles and tools to reduce or prevent pollution. While the best the council would do was to describe TQM and pollution prevention as complementary concepts, they went on to say

> The heart of TQM is the systematic analysis of processes or services by empowered, cross-functional, multidisciplinary teams. The same is true regarding pollution prevention. Emission or waste reduction opportunities are most successful when groups of employees with diverse skills and experiences are fully empowered to identify sources of pollution and to make innovative, cost-effective recommendations for addressing identified sources. TQM tools are useful at every step in this process.[13]

Like TQM, the council recommends a flexible approach to both the cultural and the statistical tools as applied to any individual industry or company. Pollution prevention, like TQM, can be achieved without large capital investments: "Success often depends as much on the creativity and energy of the employees involved as on the amount of capital invested."[14] In summary, the council stated, "The projects demonstrated that TQM principles are effective in achieving pollution prevention. Increased communication and further integration with facilities not involved in the QEM [quality environmental management] process were considered necessary to ensure that the QEM ethic becomes the predominant philosophy and business approach for companies."[15]

Cam Metcalf, a lecturer at the University of Tennessee,[16] has compiled a list of barriers to pollution prevention, that, this author submits, is identical to the barriers of companies to TQM. They are:

Institutional

Lack of top management support

Lack of clear communication of priorities or support

Organizational structures separating environmental decisions from production decisions

Habit and inertia inhibiting change

Lack of involvement of affected workers

Reward system not focusing on pollution prevention

Firms lacking the technical ability to apply preventive methods and technologies

Frequent changes to output, product design, and other factors making implementation more difficult

Lack of information about sources of waste releases, alternative strategies, and resources

Preventive applications not currently available

Lack of consumer environmental awareness

Economic

Inaccurate market signals

Incomplete cost/benefit analysis

Inappropriately short time horizons

Fear of market share loss and/or consumer pressure

Inappropriate product or process specifications

Fear of production interruption

Limited access to necessary resources

Worker fear of job loss

Regulatory

End-of-pipe focus

Media-specific focus

Regulatory program evaluation criteria

Regulatory inflexibility

Regulatory uncertainty

Pollution fees

Data gathering and management

For the institutional barriers to be those of TQM, one need only insert "quality" for "pollution prevention." From an environmental perspective, quality and pollution prevention through source reduction are synonymous. The economic barriers plaguing pollution prevention are identical to those to be overcome by TQM culture and statistical tools. Regulatory barriers are a specific subset of variables that apply to environmental analysis.

9.3 The Future Factory

The cultural change required for TQM and pollution prevention requires new blood, with creative and probing personalities. It is no longer acceptable to rest on management laurels gained decades ago, any more than it is appropriate to say you know pollution prevention because twenty years ago you implemented wastewater or air pollution control systems. Both TQM and pollution prevention require the pioneering assertiveness of a champion. The concepts are even easier to facilitate if the champion is in a position of authority or management.

The future factory will fully integrate the concept of pollution prevention into its TQM program. The Department of Energy has published presentations such as "Environmentally Conscious Manufacturing," by Barry Granoff.[17] Such presentations espouse that the solution to environmental problems is treating waste as a "quality defect," and they reinforce the idea that the total systems integration concepts of TQM are the key to getting the right people, information, and communication together. This has proven effective.

The future factory will spend considerable time benchmarking the competition. This concept refers to a thorough evaluation of products or services that are made by a competitor and are considered by the consumer to be better. A company with the right approach to TQM will be flexible enough and adventurous enough to make changes of its own to improve its own product or service. The DOE is currently applying such benchmarking to the private sector, in an attempt to improve its own products and services to preclude privatization.

The future factory will consider the environmental implications of a new product, service, or workload. It will design pollution prevention into its research experiments. It will insist on pollution prevention in the acquisition of new components or systems from others so that waste generation in the life cycle of its product will be minimized. The future factory will constantly use TQM tools to find problems and seek solutions. It will inculcate the TQM culture into every person in the company.

References

1. W. Edward Deming, *Quality, Productivity, and Competitive Position,* Massachusetts Institute of Technology—Center for Advanced Engineering Study, Cambridge, Mass., 1982.

2. Ibid.

3. Thomas Peters and Robert Waterman, Jr., *In Search of Excellence: Lessons from America's Best Run Companies,* Warner Books, New York, N.Y., 1982.

4. Chester Barnard, *The Functions of the Executive,* Harvard University Press, Cambridge, Mass., 1938.

5. Peters and Waterman, *op. cit.*

6. Michael H. Mescon and Timothy S. Mescon, "Rising Stars," *Sky Magazine,* June 1993, pp. 20–24.

7. *Ibid.*

8. *Ibid.*

9. *Ibid.*

10. Scientific Applications International Corporation, *Army Pollution Prevention Plan Manual: A Guide for Army Installations,* Army Environmental Policy Institute, Champaign, Ill., May 1993.

11. Olin R. Jennings, "A Practical Approach to Improving Quality," *Environmental Protection.*

12. President's Council for Environmental Quality, *Total Quality Management—A Framework for Pollution Prevention,* Quality Environmental Management Subcommittee, Washington, D.C., January 1993.

13. *Ibid.,* p. 39.

14. *Ibid.,* p. 39.

15. *Ibid.,* p. 41.

16. Cam Metcalf, "Parallel Paradigms," lecture material, University of Tennessee, transparency hard copies.

17. Barry Granoff, "Environmentally Conscious Manufacturing," presented at the *DOE/DP Pollution Prevention and Integrated Technologies Workshop,* Albuquerque, N.M., March 1993.

10

Examples of Successful Pollution Prevention Programs

Thomas E. Natan, Jr., Ph.D.

Hampshire Research Associates, Inc.
Alexandria, Virginia

10.1 Introduction

Motivation for implementing a pollution prevention program can come from many sources: federal, state, or local mandates; community involvement; cost analyses; or companies' own desires for superior environmental management. However, the decision to apply pollution prevention techniques is often part of a larger process of reducing environmental releases and transfers of toxic chemicals. When corporations are given the opportunity to choose pollution prevention as a part of total environmental management, the result can be an extremely successful program. The U.S. Environmental Protection Agency's (EPA's) 33/50 Program, begun in 1991, provides an example of a diverse population of companies in various industries creating pollution prevention opportunities in response to an invitation to reduce their environmental releases and transfers of 17 priority chemicals 33 percent by 1992 and 50 percent by 1995, using 1988 Toxic Release Inventory (TRI) data as a baseline. Figure 10-1 lists the chemicals targeted by the 33/50 Program. Companies were asked to consider accomplishing their reduction goals primarily through source reduction before considering options such as recycling and treatment.

As of August 1993, 1172 parent companies had agreed to participate. An examination of the 1991 TRI data for 33/50 Program chemicals shows a 34 percent reduction from 1988 levels, exceeding the program's interim goal of a 33

Benzene

Cadmium and cadmium compounds

Carbon tetrachloride

Chloroform (trichloromethane)

Chromium and chromium compounds

Cyanide and cyanide compounds

Dichloromethane (methylene chloride)

Lead and lead compounds

Mercury and mercury compounds

Methyl ethyl ketone

Methyl isobutyl ketone

Nickel and nickel compounds

Tetrachloroethylene (perchloroethylene)

Toluene

1,1,1-Trichloroethane (methyl chloroform)

Trichloroethylene

Xylene

Figure 10-1. 33/50 Program chemicals.

percent reduction by 1992. Participating companies showed a total decrease of 40 percent in targeted chemicals from 1988, while the decrease for nonparticipating companies was 25 percent. These same participating companies accounted for 62 percent of the total releases and transfers of program chemicals in 1988, but only 57 percent in 1991. Response from companies making commitments to the 33/50 Program was almost uniformly positive, with the most frequent comment being an appreciation of the EPA's willingness to allow companies to choose their own methods of setting and achieving reduction goals. Of those companies providing numerical goals (not necessarily the program targets) and offering insight as to methods for achieving reductions, 60 percent have pledged to accomplish their goals by source reduction alone, while an additional 19 percent have pledged to use a combination of source reduction, recycling, and a change of chemicals (changes where no nontoxic or aqueous-based substitute is mentioned), indicating that pollution prevention is clearly the way these companies prefer to accomplish reductions in releases and transfers. Since the program began in 1991, it is interesting to compare the 1990 and 1991 TRI data to see the effect of the program on targeted chemicals in its first year of existence. An examination of all facilities reporting the use of

33/50 Program chemicals to TRI (not just those participating in the 33/50 Program) reveals that total releases and transfers of program chemicals declined 21.2 percent from 1990 to 1991—an amount greater than the 16.3 percent reduction from 1988 to 1990, suggesting that voluntary reduction accomplished primarily through pollution prevention has had a significant effect on releases and transfers of targeted chemicals, even though only 15 percent of eligible companies have thus far agreed to join the 33/50 Program.

Participating companies were also asked to provide progress updates and descriptions of techniques used to accomplish reductions. Responses range from simple restatements of the companies' TRI data to detailed plans for each targeted chemical. This chapter profiles three companies participating in the 33/50 Program that have provided details of interesting and successful pollution prevention programs resulting in varying degrees of reduction of environmental releases and transfers. The companies, Grumman Corporation of Bethpage, New York; Panel Processing of Alpena, Michigan; and Avondale Industries of New Orleans, Louisiana, were chosen for the range of industries represented: manufacture of transportation equipment, coating and lamination of wood products, and shipbuilding. Examples of other interesting pollution prevention plans will also be discussed.

10.2   Overview of Reduction Techniques

It is difficult to generalize a specific set of techniques for companies engaged in pollution prevention activities, or even generalize for certain industrial classifications. Nonetheless, certain basic patterns emerge, as described by companies participating in the 33/50 Program, one for the solvents and another for the metals.

Targeted solvents are used primarily in painting, coating, or degreasing operations when they are not integral formulation components. Substitution of aqueous-based materials is the primary technique used by companies to reduce targeted solvent use. Under a materials use/waste management hierarchy, the strategy is first to seek an acceptable substitute—that is, one which will accomplish the task without sacrificing product quality, price, or customer acceptance—by working with materials suppliers and conducting tests over the range of use of the material in different products and processes. When a suitable substitute cannot be found, attempts are made to use the material more efficiently in processing. Tests to determine minimum use levels are conducted, as well as surveys of equipment and handling procedures to prevent unnecessary losses. Recovery systems are considered only after these other measures have been exhausted.

33/50 Program metals are used in a variety of applications, ranging from use as key formulation components in metal products to use as coating materials, additives in polymers, or processing aids. Again, substitution is examined but is usually not an option when the metal is used as a formulation component. Reductions in releases and transfers of these metals are accomplished primarily through improved material-handling techniques to increase efficiency in their use, followed by recycling and treatment.

10.3   Grumman Corporation

Grumman Corporation operates nine facilities which reported releases and transfers of 33/50 Program chemicals to TRI for the years 1988–1991. Grumman's facilities report to TRI under Standard Industrial Classifications (SIC) codes 3711, 3713, and 3721 as manufacturers of various forms of transportation equipment, including aerospace and ground transport. The company pledged a 64 percent reduction in total releases and transfers of 33/50 Program chemicals by 1995, based on a 1988 TRI baseline. This baseline figure of 2,460,467 lb includes releases and transfers of chromium and chromium compounds, dichloromethane, methyl ethyl ketone, methyl isobutyl ketone, tetrachloroethylene, toluene, 1,1,1-trichloroethane, trichloroethylene, and mixed xylenes.

In pledging to participate in the 33/50 Program, Grumman described its ongoing activities in joint programs with other companies to research and develop source reduction techniques relevant to the transportation manufacturing industry. With regard to specific pollution prevention activities at Grumman, the company referred to several specific projects:

- Development of a new materials use reporting system designed to identify and accurately track source reduction opportunities.

- Replacement of certain chromium-plating operations with a sulfuric acid anodization process or aluminum deoxidation.

- Substitution of a blasting technique using fine, reusable plastic particles for dichloromethane paint-stripping operations, and use of alkaline and aqueous cleaners to replace trichloroethylene vapor degreasing.

- Implementation of solvent recovery for tetrachloroethylene in milling operations, reclaiming 90 percent of total use.

The company provides a breakdown of its reduction activities which reveals that specified source reduction techniques (closed-loop solvent recovery, reformulation, nontoxic substitution, and process changes) account for 33 percent of its total reduction, with chemical substitution (not necessarily considered source reduction, since the replacement chemicals are not specified) comprising 25 percent, while the remaining reductions come from unspecified source reduction, chemical substitution, and unspecified solvent recovery (possibly recycling) techniques. The company's TRI releases and transfers of 33/50 Program chemicals decreased 25 percent between 1988 and 1990 and 42 percent between 1988 and 1991, well ahead of a linearized reduction schedule to achieve a 64 percent reduction by 1995. Grumman Aerospace, which accounts for over 70 percent of Grumman Corporation's total 33/50 Program releases and transfers, decreased its own releases and transfers 24 percent between 1988 and 1990, and 37 percent between 1988 and 1991. Grumman committed to the 33/50 Program in 1991, and the decrease in releases and transfers between 1990 and 1991 is greater than the average of the previous two years' reductions for both Grumman Aerospace and Grumman Corporation (17 percent versus 12 percent), making it probable that the implementation of the 33/50 Program reduction techniques detailed above has had an early effect. Off-site transfers of chromium and chromium compounds decreased 64 percent between 1988 and 1991, in keeping with Grumman's efforts to reduce chromium use.

Grumman Corporation's data, provided in the company's 1991 TRI Form R submissions, projected a 49 percent decrease in 33/50 Program releases and transfers by 1993. When combined with reductions already achieved, this amount brings Grumman's total projected 33/50 reduction to 66 percent by 1995. Tables 10-1 and 10-2 compare the data by chemical for Grumman's

Table 10-1. Grumman Aerospace Corporation, Bethpage, New York, Facility, TRI Data, 1991 (in Pounds)

Chemical	Recycling	Energy recovery	Treatment	Releases and disposal	Total wastes
Methyl chloroform	5,851	0	11,658	133,102	150,611
Chromium compounds	0	0	15,500	541	16,041
Dichloromethane	0	0	2,079	29,603	31,628
Methyl ethyl ketone	0	32,248	100	57,563	89,911
Perchloroethylene	17,031	0	1,570	347,558	366,159
Toluene	0	11,598	928	15,635	28,161
Trichloroethylene	44,540	0	486	188,674	233,700
Total	67,422	43,846	32,321	772,676	916,265
Percent of total wastes	7.4	4.8	3.5	84.3	

Table 10-2. Grumman Aerospace Corporation, Bethpage, New York, Facility, 1991 TRI Data Projection for 1993 (in Pounds)

Chemical	Recycling	Energy recovery	Treatment	Releases and disposal	Total wastes
Methyl chloroform	4,100	0	8,200	93,200	105,500
Chromium compounds	0	0	11,000	380	11,380
Dichloromethane	0	0	1,155	16,466	17,601
Methyl ethyl ketone	0	21,499	67	38,375	59,941
Perchloroethylene	346,000	0	1,000	18,000	365,000
Toluene	0	10,431	835	14,072	25,345
Trichloroethylene	14,847	0	450	62,891	78,188
Total	364,947	31,937	22,707	243,364	622,955
Percent of total wastes	55.0	4.8	3.4	36.8	

Bethpage, New York, facility for 1988 and 1991, projecting a large increase in on- and off-site recycling of tetrachloroethylene, consistent with the company's projection in committing to the 33/50 Program. Wastes containing chromium compounds are expected to continue decreasing as other materials are used to reduce corrosion in metal parts. The strongest indicator of Grumman's commitment to pollution prevention arises from a comparison of the percentage of total waste attributable to each category of data. Recycling, which consists primarily of the closed-loop solvent recovery process for perchloroethylene, is expected to increase from 7.4 to 55.0 as the percentage of total waste between 1991 and 1993, while total wastes will decline by 32 percent over the same period. Treatment and energy recovery are expected to maintain their relative use on a per-pound basis.

10.4   Panel Processing, Incorporated

Panel Processing, Incorporated, produces finished, wood-fiber-based panels for the store fixture and display industries in six plants. Two facilities reported releases and transfers of 33/50 Program chemicals to TRI for the years 1988–1991, reporting under SIC code 2499 for wood products. The company has committed to a reduction of 90.2 percent in releases and transfers of targeted chemicals based on 1988 TRI data. The 1988 baseline of 440,499 lb consists of methyl ethyl ketone, methyl isobutyl ketone, toluene, and xylene.

The 33/50 Program solvents used by Panel Processing are ingredients in epoxy-coating and laminating processes. The company reports that laminating operations, previously done on older equipment using solvent-based glues, were moved to a new facility in February 1992, enabling the company to investigate new aqueous-based glues and modified processes. The move continues a trend toward high-solids, water-based emulsions which Panel Processing has been pursuing since the early 1980s. Reformulation is a multistage process which requires consumer acceptance of new materials as well as competitive pricing for the resulting products. Conversion to aqueous materials indicates that the cost of solvent recovery and high-temperature curing of "cheaper" organics is higher than the seemingly expensive up-front price of the aqueous stocks, although factors such as worker safety play a part in such decisions. Panel Processing reports an 85.4 percent decrease in total releases of 33/50 Program chemicals from 1988 to 1992, despite a 25 percent increase in sales volume. Remaining solvents are released as stack emissions or transferred in a liquid state for energy recovery. Specific pollution prevention measures include

- Reformulation of laminating processes to eliminate dichloromethane and trichloroethane.

- Development of high-gloss coating finishes using catalyzed aqueous emulsions with low-temperature curing. Production of one particular product was suspended for nearly four years because readily available substitutes created unacceptable product.

Figure 10-2 illustrates changes in releases and transfers of 33/50 Program chemicals for Panel Processing from 1988 to 1991 (see also Table 10-3). Total releases and transfers have been reduced from 440,499 to 47,559 lb, a reduction of 89 percent and nearly equaling the 1995 goal of 90.2 percent. Releases and transfers of methyl isobutyl ketone and xylene have been eliminated, and off-site transfers of methyl ethyl ketone and toluene are no longer taking place.

Panel Processing's TRI data as reported in 1991 project that total releases and disposal of 33/50 Program chemicals will drop from 47,559 lb in 1991 to 34,881 lb in 1993. This drop translates to a 92 percent decrease from 1988, slightly

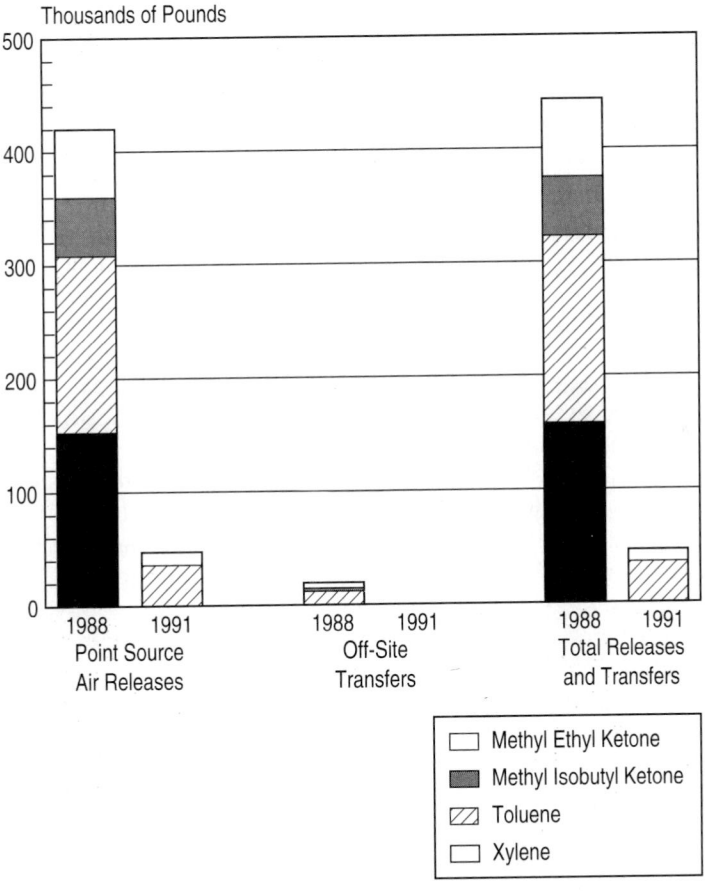

Figure 10-2. Panel Processing, Incorporated, releases and transfers of 33/50 Program chemicals by medium, 1988 and 1991 (in pounds).

Table 10-3. Panel Processing, Incorporated, Releases and Transfers of 33/50 Program Chemicals by Medium, 1988 and 1991 (in Pounds)

Chemical	Point-source air releases		Off-site transfers		Total releases and transfers	
	1988	1991	1988	1991	1988	1991
Methyl ethyl ketone	61,366	10,751	4,021	0	65,387	10,751
Methyl isobutyl ketone	51,020	0	1,522	0	52,542	0
Toluene	155,705	36,808	12,903	0	168,608	36,808
Xylene	153,962	0	0	0	153,692	0
Total	422,053	47,559	18,446	0	449,229	47,559

greater than the 1995 reduction goal. Since most of Panel Processing's source reduction has been accomplished by materials substitution, the slight decrease projected for 1991–1993 indicates that the company's pollution prevention measures are for the most part in place, and, whether through lack of suitable aqueous replacements or customer demand, little additional reduction is anticipated.

10.5 Avondale Industries, Incorporated

Avondale Industries, Incorporated, is a shipbuilding company with three facilities which reported 33/50 Program chemicals to TRI for 1988 and 1991 under SIC code 3731 for shipbuilding and repairing. Since 1991, one facility has been sold; however, the data presented here represent Avondale Industries as the parent company. The company has committed to reducing the total volume of releases and transfers of all 33/50 Program chemicals 33 percent per ton of steel processed by 1992, and 50 percent per ton of steel processed by 1995, using 1988 TRI data as a baseline. This particular commitment was chosen because the amount of chemicals used in any given year is a function of the number and size of ships built in that year. Weight of steel used rather than number of ships built is considered a better indication of total reduction. The company has also provided chemical-specific commitments for methyl isobutyl ketone, toluene, 1,1,1-trichloroethane, and xylene independent of production, representing a 54 percent decrease from a 1988 baseline of 1,558,614 lb.

Since the largest use of 33/50 Program chemicals occurs in the coating process, Avondale has concentrated on specifying coating materials with low amounts of volatile organic compounds (VOCs) and modifying the coating process itself. Some coatings are applied before assembly and others during or after assembly, depending on location and strategic importance. These factors also dictate the composition of the coating materials. For those materials for which an acceptable substitute could not be found, solvent recovery and recycling were instituted to avoid unnecessary releases. Specific projects which have resulted in reductions in VOC use include:

- Consultation with paint suppliers to develop a low-VOC, low-zinc preconstruction primer. The primer had to meet rigorous standards for weldability, quick drying, ease of application, and durability. While the new primer still uses VOCs, they are present in a much smaller volume ratio than previously. In addition, the new primer has reduced the amount of solvents needed to clean the painting equipment.

- Revision of painting processes. Previous painting sequences included using one primer before construction, assembling the ship hull pieces, and removing the primer because it was incompatible with the final coating system. This primer was necessary to protect the hull pieces before and after assembly. To complete the process, a compatible primer would then be applied to the completed hull before the final coating. The new painting sequence uses compatible primers before assembly, saving primer, cleaning compounds, and thinners.

- Revision of the painting process to include reusable bulk paint totes, avoiding losses in transfer of paint, as was common with smaller containersn.

- Development of a new flushing protocol. Avondale changed its procedure for flushing paint lines, pumps, and guns, replacing the old "clean solvent" procedure with reused solvent for initial flushing. Clean solvent is used only for the final pass, reducing the amount of clean solvent used by at least 50 percent.

- Consultation with the Navy, Avondale's major client, to change coating specifications to allow for reductions in VOC use.

Table 10-4 lists Avondale's total releases and transfers of 33/50 Program chemicals for 1988 and 1991. Since the company has not provided any data to index their reductions to 1988 production levels, it is not possible to relate the overall reduction of 83 percent to the company's goal. The data submitted with

Table 10-4. Avondale Industries, Incorporated, 33/50 Program Total Releases and Transfers, 1988 and 1991 (in Pounds)

Chemical	Total releases and transfers, 1988	Total releases and transfers, 1991
Chromium	0	45
Methyl isobutyl ketone	401,681	11,917
Nickel	0	5
Toluene	588,000	17,381
1,1,1-trichloroethane	16,983	31,960
Xylenes	551,950	199,373
Total	1,558,614	260,681

the company's 1991 Form Rs indicate that production declined 12 percent between 1990 and 1991, while total wastes fell from 383,953 to 354,592 lb, a decrease of 8 percent, and total releases and transfers decreased by 6 percent. Avondale has made an interim switch to use of 1,1,1-trichloroethane in coatings while it seeks other non-VOC alternatives. The data projections for 1993 indicate that solvent recovery systems will not be in place by that date. The 1992 TRI data with projections for 1994 will give a better indication of when solvent recovery will be instituted.

10.6   Other Pollution Prevention Achievements

The three companies described above represent individual areas of manufacturing outside the chemical and petroleum industries. Companies in these two industries also have developed pollution prevention plans; in general, however, their plans focus on longer-term goals and take future products into account. For example, Mobil Corporation, headquartered in Fairfax, Virginia, operates thirteen 33/50 Program facilities. These facilities fall under the supervision of Mobil Oil Corporation, Mobil Chemical Corporation, and the Netches River North Regional Treatment Plant. The company has pledged to reduce total releases and transfers of 33/50 Program chemicals 50 percent by 1995 based on a 1988 baseline of 3,802,781 lb. Mobil plans to integrate its 33/50 Program goal into its reformulation of gasolines. As this is a long-lead-time activity requiring new construction, the company anticipates that its greatest reductions in releases and transfers of targeted chemicals will come closer to 1995. The company's TRI data for 1988–1991 show a 20 percent decrease in total releases and transfers, which is on track for a linearized reduction schedule. The 1991 TRI data projects for 1993 indicate that while total releases and disposal corporatewide will decrease 9 percent, Mobil Chemical's decrease will be 43 percent, the treatment plant's decrease will be 94 percent, and Mobil Oil's decrease will be 1 percent, consistent with the company's long-term reduction plans.

Another industry of interest is pulp and paper production, which produces large amounts of chloroform in the bleaching of wood pulp. Two approaches are used for pollution prevention. The first is more efficient use of chlorine through usual methods of vigilance in processing and handling. The second approach involves development of chlorine-free paper products, and has been utilized by Louisiana Pacific's bleached kraft pulp mill in Samoa, California. The plant reduced the release of chloroform by 59 percent between 1988 and 1991 and provided news of its achievement on chlorine-free paper.

One final example of pollution prevention comes from The Bass Plating Company of Bloomfield, Connecticut. Wastes from the electroplating industry have long been considered intractable, since few alternatives to the traditional cyanide plating processes have produced acceptable results in terms of coating thickness and ductility. Bass Plating has succeeded in using noncyanide processes where appropriate in zinc-plating lines and in installing a low-temperature distillation process for the rinse effluent from cadmium-plating lines. Reusing the rinse water is predicted to reduce the amount of cadmium and cyanide wastes generated by 50 percent while a similar installation for the zinc-plating lines will reduce the cyanide waste by another 35 percent by 1995.

10.7 Conclusions

The company profiles in the previous sections are not necessarily definitive examples of successful pollution prevention plans. The profiles do reveal how individual circumstances such as supplier input and customer feedback can result in dramatic reductions in environmental releases and transfers. The TRI data relating to the 33/50 Program for 1991 are encouraging since they reveal that pollution prevention activities can have a dramatic effect on total releases and transfers from the outset. Cost data has not been provided, but a study of a Halstead Industries facility in Wynne, Arkansas, gives some insight into cost savings resulting from simple materials-handling changes in the use of trichloroethylene as a degreaser of copper tubing. The facility was able to amass annual savings of $30,000 from a one-time investment of $15,000 in simple equipment changes and enhancement of operator training. As companies exhaust their immediate options of improved materials handling and substitution, additional examples of successful pollution prevention programs will emerge from companies participating in the 33/50 Program.

11

Agile Manufacturing

Dale Denny

Concurrent Technologies Corporation
Johnstown, Pennsylvania

11.1 Introduction

The following description of "agile manufacturing" is taken almost completely from material prepared by Steven Goldman of the Agile Manufacturing Enterprise Forum (AMEF).[1] Key components to *agility* in this context are the ability to make information flow freely, the ability to form teams quickly, and the ability to move products to the market rapidly. Such an environment will require innovative technologies for communications between companies and new product and process design methods. Institutional procedures for government execution contracts, antitrust determinations, and issuance of environmental operating permits may also have to change dramatically.

Participants in the AMEF who developed the Agile Manufacturing Enterprise Vision foresee representation of the U.S. Environmental Protection Agency (EPA), the Occupational Safety and Health Administration (OSHA), and the general public on product design teams. At the conclusion of a design, the team members would be empowered to issue operating permits and to represent the project to the public as one which was in the overall best interest of the community. The concepts of design for the environment, pollution prevention, and life-cycle environmental responsibility as espoused in this handbook are clearly supportive of the transition to an agile manufacturing environment.

11.2 Manufacturing Background

In *Turing's Man*,[2] J. David Bolter developed the notion of "defining technologies"—the clock, the steam engine, today the computer—technologies that perfuse their cultures, shaping institutions, ideas, values, sensibilities, consciousness itself. The late nineteenth-century fusion of mass production manufacturing and the modern industrial corporation was just such a defining technology. By synthesizing existing production, transportation, and communication technologies into a centrally administered, hierarchically structured, vertically integrated managerial organization, this fusion created a new system of manufacturing. The influence of this system on life in the twentieth century has been, and continues to be, profound.

Explicitly, mass production manufacturing defined a system for the creation, production, and distribution of goods and services. Implicitly, it defined patterns for the organization of society to create, distribute, and consume the kinds of goods and services distinctive of this system of production. Today, a new system for the production of goods and services is emerging, one that coordinates new production technologies with a new organizational structure for the industrial corporation. This agile manufacturing system threatens to displace mass production corporations from the century-long dominance they have enjoyed. By analogy with mass production, agile manufacturing, too, will define patterns for the organization of society to create, distribute, and consume the goods and services distinctive of its mode of operation. In the process, it will evoke from society new institutions, new values, and new sensibilities, reflecting the spreading influence of this new system of production.

11.3 Agile Manufacturing History

The essence of this new production system was first described in a report, *21st Century Manufacturing Enterprise Strategy: An Industry Led View*,[3] published by Lehigh University's Iacocca Institute late in the fall of 1991. In a little over a year, more than 15,000 copies of this report were sold, and an industry-supported organization (the AMEF) was created to develop and disseminate the ideas contained in the report. Agile manufacturing has become a new "buzzword" for Fortune 500 company senior management. It has won powerful congressional supporters, found a place in Clinton administration industrial policy circles, and has begun to attract the attention of leading management consulting firms.

The initial conceptualization of agile manufacturing was the result of a 1991 study of requirements for global U.S. commercial manufacturing competitiveness that was jointly funded by 13 corporations and the U.S. Navy Manufacturing Technology Program. In the process of identifying the technological, the human, and the social resources necessary to accomplish this goal, the study participants realized that incremental improvement of the current organization of manufacturing would not be enough. The kinds of computer-based information and production technologies that were becoming available to industry opened up the possibility of an altogether new *system* of manufacturing.

11.4 How Does an Agile Company Operate?

An agile company functions in the way that a first-rate, state-of-the-art hospital emergency room does. A hospital emergency room consists of a pool of resources—human, technological, and institutional—that can be configured very quickly to meet the specific requirements of people with largely unanticipated medical problems who appear in the emergency room in no particular order.

On demand, emergency room personnel must form themselves into teams in response to the needs of individual patients. From experience, medical administrators can accurately predict the overwhelming majority of emergency room requirements. The medical resources immediately at hand are, however, only a small fraction of the total resources available to the emergency room staff. When called for, emergency room personnel can quickly link remote resources to a newly formed patient response team. This process begins with ambulance or helicopter paramedics who establish voice and data telephone links with personnel and equipment located at the hospital in order to begin responding to the patient on site or en route, if that is necessary.

In addition, routine access to remote medical resources includes

- Accessing specialized knowledge possessed by physicians associated with the hospital or, increasingly routinely, by real and "virtual" experts (medical databases) located almost anywhere in the world
- Accessing specialized expertise or equipment located elsewhere in the hospital, or in other hospitals
- Using extensive hospital facilities not dedicated to the emergency room, but available to it, from the pathology lab to beds

The emergency room patient is the analogue of the customer of an agile producer of goods and services. The primary objective of management in such an enterprise must be to create an organization capable of quickly configuring subsets of its resources in response to the particular requirements of individual customers. To do so, the organizational "walls"—and the accompanying lines of control from above—that traditionally circumscribe functional resources in a company must be routinely reconfigurable.

11.5 What Capabilities Exist in an Agile Manufacturing Company?

The answer to this question is: the capabilities required to support the ability to successfully compete in an environment of continuous and unanticipated change, and to respond quickly to rapidly changing markets driven by customer-based valuing of products and services.

Agility entails the capability to achieve

Rapid, cost-effective development of products and production facilities

Continuous improvement of product development and production processes in ways that are easy to understand and easy to maintain

Rapid, efficient changes in production volume (up/down)

Low variability of volume-to-unit cost ratio with volume changes

Full electronic access for product data, change orders, and performance and status reporting

Full, location-independent interactive communications among all participants in the product and/or service design, development, production, marketing, and service process

Scalable, modular production facilities that can be organized into dynamic confederations of manufacturing modules with different capabilities, depending on their configuration

Distributed control decisions are made at relevant functional knowledge points, not centralized in dedicated managers

An agile company employs processes for designing, manufacturing, marketing, and supporting its products and services that are integrated into a customer-centered whole committed to pursuing sharply lower product realization times and costs.

Concurrently, production in an agile company is pulled by sales of products that can be customer-configured at a negligible cost premium, and that can serve as platforms for a continually variable mix of new value-adding hardware options, information, and services.

11.6 What Are the Characteristics of an Agile Company?

An agile company has a customer-centered organizational structure keyed to rapid time to market for new, high-quality, easily customized products. It routinely engages in intensive collaboration with suppliers and customers in designing, manufacturing, and servicing new products. It practices a total product life-cycle design philosophy. It is committed to such collateral customer values as optimal environmental, health, safety, and community impact of products, services, and company activities.

The products of agile companies are designed to evolve and to be long-lived in the face of constant technological and market change. As the needs of users change, as improvements are introduced, users can readily reconfigure or upgrade what they have bought instead of replacing it. Concurrently, the production systems and business processes of an agile company must be reprogrammable, reconfigurable, and continuously changeable. They must be capable of being integrated into a new, information-intensive manufacturing system that makes the lot size of an order irrelevant. The cost of production is the same for 10,000 units of one model as for one unit each of 10,000 different configurations of all of the models of a single product.

Because of the longevity of its evolutionary product lines, an agile company develops strategic relationships with its consumer, as well as with its commercial customers. In place of the "sale and limited warranty" relationship, it offers customers a continuously variable mix of products, services, and value-adding information. To facilitate this relationship, it communicates with its products while they are in their users' possession. Information is exchanged, software upgraded, diagnostic servicing performed, and individual product histories maintained.

Agile businesses are strategically focused, seeing opportunities for growth and profit in constant change because their production technologies and organization are highly flexible and able to exploit change. To accomplish this in an agile company, authority is diffused, not concentrated in a chain of command. Where technology, under the mass production system, is perceived as the key to solving manufacturing and marketing problems, under agile manufacturing, people creatively utilizing technology, are the problem solvers.

The technologies required to do this are foreseeable: flexible, programmable machine tools grouped in reconfigurable, modular, and scalable manufacturing cells; "intelligent" manufacturing-process controllers; closed-loop monitoring of manufacturing processes employing sensors, samplers, and analyzers coupled to intelligent diagnostic software; the computer power, and the manufacturing process knowledge base, to design complex products digitally, to simulate their properties and behaviors reliably, and to model the processes of their manufacture accurately.

The rapid creation, development, and manufacture of new products requires making these technologies available to a work force capable of using them and organized in a way that empowers workers to improve existing products and create new ones on their own initiative. An agile company must therefore be one in which the work force shares a common vision with management, a common commitment to shared values for joint success, and shared responsibility for creating that success. In return, the work force must be nurtured by management as the central asset of the organization, and its continuous training counted as an investment, not a cost.

Agile companies maintain an open information environment, one in which information flows seamlessly among manufacturing, engineering, marketing, purchasing, finance, inventory, sales, and research personnel, and is routinely shared with collaborating companies as well. The development of new products and of the processes for manufacturing and marketing them take place concurrently, not sequentially, as is the norm today. Design is no longer the province of engineering, not even of engineering and manufacturing jointly. Instead, representatives of every stage in a product's life cycle, from materials employed in its manufacturing to its ultimate disposal—including representatives from supplier, collaborator, and customer companies—participate in setting product design specifications.

The effectiveness of agile organizations is dependent on the creation of an atmosphere of mutual trust, within and between companies, the public, government, academia, and the investment community, based on the need to make cooperation a first-choice approach to problem solving. Trust and shared responsibility together bear on the capability for sharing information and for localized decision making that is a major determinant of agility.

The ultimate expression of trust, given the proprietary attitudes toward information that prevail today, is the routine formation of "virtual" companies. If speed to market with new products offers major competitive advantage, the quickest route to the introduction of a new product often lies in selecting the human and physical resources and capabilities needed to create it from different companies that already possess them. These resources, if they are compatible with one another, can then be synthesized into a single, electronic business entity. Such an entity accomplishes functional integration via bytes, not bricks. This is what is meant by a virtual company.

The advantages of the virtual company are much lower sunk costs, much lower risk, and much shorter development time for new products and services. Personnel and facilities dedicated to specific products are minimized. For as long as a market opportunity can be exploited profitably, the virtual company continues in existence. When the opportunity passes, the company dissolves and the personnel and facilities pursue other projects.

11.7   What's New in Agility?

No one of the features of agility, taken by itself, is an innovation unique to it. Flattening the managerial hierarchy, cross-functional work teams, work force empowerment, the customer-focused organization, and intensive collaboration with other companies have all been proposed in other contexts. So have "smarter" production technologies, capable of uncoupling volume and unit cost, and so have concurrent engineering, enterprise integration, and the total product life-cycle design philosophy.

Each of these innovations was being pursued prior to the publication of the report that launched the concept of agility. Companies serving rapidly changing global markets have for years been attempting to push beyond the increasingly manifest limitations of mass production manufacturing, following lines of development opened up by programs such as just-in-time inventory management, total quality control and total quality management, and lean production.

The situation of agile manufacturing today is strikingly reminiscent of the situation of the industrial corporation 100 years ago. It is all too easy to forget that the modern industrial corporation has not always been with us and did not just happen. It had to be invented, as a new way of doing business, one that promised dramatic competitive advantage from a new, systematic coordination of production, marketing, and distribution.

Back in the 1880s and 1890s, when Gustave Swift was creating the Swift Meat Packing Company and J. P. Morgan and John D. Rockefeller were creating U.S. Steel and Standard Oil of New Jersey, neither the factory system nor mass production nor the steam-power-based Industrial Revolution were new, or any longer newsworthy. The factory system was by then 200 years old. Mass production, in textiles for example, was 100 years old, as were improved steam engines, which had for a number of years been used in ships and railroad locomotives, as well as factory machinery. Even the telegraph was "old hat."

What Swift, Morgan, and Rockefeller saw was the possibility of integrating existing production, transportation, and information technologies into a new organizational structure: a centrally administered, vertically integrated, hierarchically managed corporation. They envisioned the unprecedented competitive power that such an organization unleashed. And they were right. Companies that adopted this structure, or adapted to it, survived. Few of the companies that had to compete against U.S. Steel, Standard Oil, General Electric, Du Pont, General Foods, Ford, Goodyear, among many other still familiar names, and that refused to adapt, survived for long, even if there was nothing absolutely new in what these corporations were doing. The news about agility is that it promises to be the twenty-first-century analogue of the industrial corporation's version of the mass production manufacturing system. That is, agility, too, promises to create new business strategies.

11.8   Why Agility?

Agility is becoming a condition of survival in a growing number of commercial and consumer markets. An intensely competitive environment for advanced industrial goods and services has already begun to take shape. Markets are fragmenting and changing rapidly as a steady flow of new products appears in a wide range of continually changing models. Stimulated by the variety of models available, commercial customers and consumers alike are becoming more demanding. At the same time, more and more products are being designed to compete in highly diversified global markets. Their design, manufacture, distribution, and servicing are being managed by transnational corporations using globally distributed assets and resources.

If competitive pressures are "pulling" a transformation of manufacturing into existence, it is also the case that social pressures are driving manufacturing in the same direction. Environmental concerns are the most obvious source of pressure, ranging from anxiety over the global impact of manufacturing to local pollution issues. Energy concerns, natural resource depletion, manufacturing workplace safety, and the social impact of manufacturing are drivers second only to environmental ones.

The posture of industrial management toward these social issues has, until recently, been reactive, treating them as negative externalities to be coped with as they arose. The implementation of a new manufacturing system, more explicitly coupled to a broad range of social institutions and values than mass production manufacturing was, offers an opportunity to adopt a new, proactive posture. In an agile manufacturing environment, it is natural for management to assimilate into the managerial decision-making process the total impact of manufacturing. The concept of manufacturing expands from a narrow focus on production of consumable and/or disposable goods to the comprehensive process of creating, developing, selling, and maintaining products over their entire life cycles, which for many products will be highly extended by reconfiguration and upgrade capabilities.

With this expansion comes a complex set of interdependencies among manufacturers, suppliers, customers, and public institutions that make it advantageous for manufacturers to internalize social drivers, rather than have them constrain manufacturing from the outside—for example, by political responses to special interest activism. The routine formation of virtual companies, as well as of smaller-scale electronic alliances, diminishes the attractiveness of single-enterprise vertical alliances. The complex of small suppliers, machine shops, and specialty manufacturers that is an internationally acknowledged strength of American industry can in this way be knit into a new national resource. Accomplishing this will require equal access to comprehensive-coverage, broad-band, high-speed computer communication channels, and the creation of a national industrial database, with uniform data exchange standards for sharing digital engineering drawings and three-dimensional parts descriptions. Not yet identified, but analogous, processes will take place in the bulk materials industries, such as the chemical industry.

The task of effecting the transition to a new era in American manufacturing is a social task, one that must be led by industry if it is to be accomplished at all, but one in which society is a stakeholder along with industry. The rise of mass production manufacturing profoundly affected a wide range of institutions in all the societies that adopted that mode of industrialization. The rise of agile manufacturing will exert an analogous social influence. Some of the changes that will take place can be foreseen, at least in part, because they will be deliberate. Some of the changes, perhaps the most profound, cannot be foreseen. But the competitive challenge that American industry confronts today implicates the future prosperity, quality of life, and security of American society.

References

1. Steven L. Goldman, *Agile Manufacturing—A New Production Paradigm for Society*, Agile Manufacturing Enterprise Forum, Iacocca Institute, Lehigh University, Bethlehem, Pa., 1993.

2. J. David Bolter, *Turing's Man: Western Culture in the Computer Age*, University of North Carolina Press, Chapel Hill, N.C., March 1984, pp. xiii and 264.

3. Roger Nagle and Rick Dove, *21st Century Manufacturing Enterprise Strategy*, vols. 1 and 2, Iacocca Institute, Lehigh University, Bethlehem, Pa., 1991.

12

Facility Pollution Prevention Planning

David P. Evers

Battelle Memorial Institute
Columbus, Ohio

12.1 Developing a Pollution Prevention Program

Pollution prevention planning is a comprehensive and continual evaluation of business operations. The resulting pollution prevention program affects many functional areas within a company, such as the production line, accounting practices, and management. It has much in common with the planning already conducted for other aspects of business operations, but it looks at the facility as an integrated whole instead of as a series of disjoint parts or operations.

Figure 12-1 is a flowchart illustrating the major elements in the pollution prevention program. This section describes the elements of pollution prevention program planning and design. These elements include building support for pollution prevention throughout the company, organizing the program, setting goals and objectives, performing a preliminary assessment of pollution prevention opportunities, and identifying potential problems and their solutions.

12.1.1 Establish the Pollution Prevention Program

Executive Level Decision. In some companies, the initiative for setting up a pollution prevention program will be taken at the executive level. In others, lower-level managers or employees will be the catalysts. In either case, it may

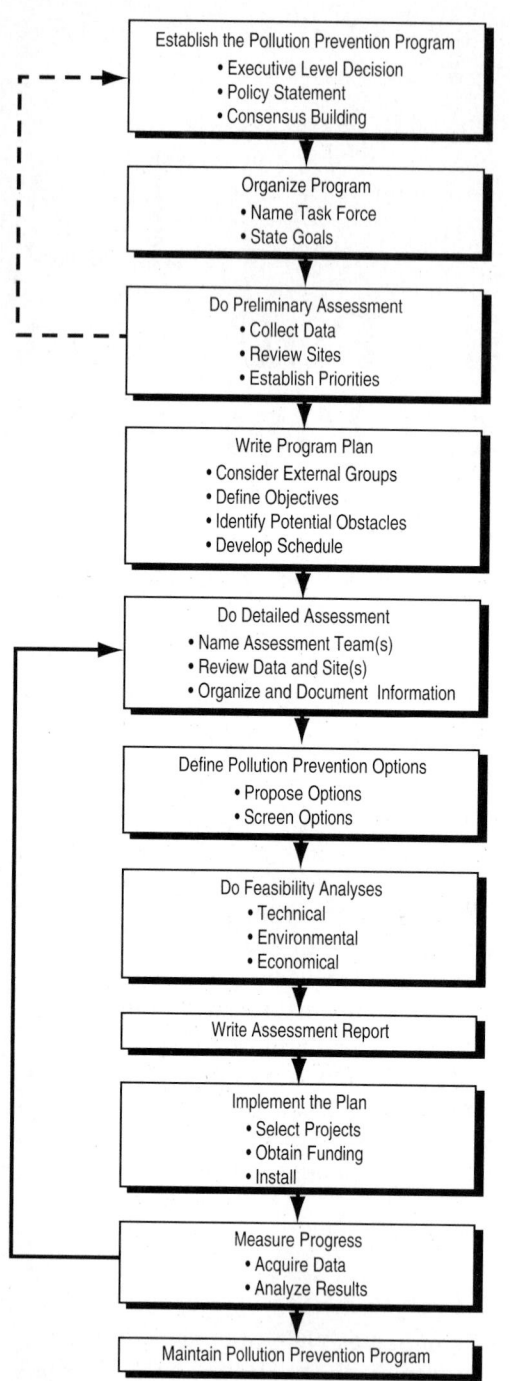

Figure 12-1. Major elements of a pollution prevention program.

be necessary to gather preliminary information to demonstrate that pollution prevention opportunities exist and should be explored. This preliminary information will be used by company executives as they weigh the potential value of pollution prevention and decide whether to commit the resources necessary to develop and implement the program.

One way to gather this information is to perform a preliminary assessment of only one or two areas of the facility and, perhaps, even identify several low-cost, quick-payoff pollution prevention techniques that can be implemented readily. A preassessment is part of the formal program design effort and is, therefore, described later in this section.

Once the decision to establish a pollution prevention program has been made it should be conveyed to all employees through a formal policy statement. This establishes a framework for communicating the formal commitment throughout the organization.

Policy Statement.   As with other company policy statements, the pollution prevention policy statement should state:

- Why a program is being established
- What is to be accomplished in qualitative terms
- Who is responsible for pollution prevention

Consensus Building.   After the pollution prevention policy statement has been developed, consider how it should be presented to employees so that they will see it as an ongoing, companywide commitment. While the priorities and tone of the pollution prevention program will be set by the management team, the attitude of production-level employees will have a significant effect on its success. Since it is through their daily activities that waste is generated, their support of the program is essential.

Publication or distribution of the policy depends on the size and the culture of your company. Suggestions include a general meeting or several small group meetings, distribution via a company newsletter, a special insert with a paycheck, or other types of publicity that have been found effective.

Employee incentives for continued participation and to keep interest from waning are necessary. Bonuses or other awards might be offered to employees who suggest ways to prevent pollution. Announcing awards in newsletters or on bulletin boards provides additional incentive to employees and further publicizes the program. Finally, pollution prevention should be included in job objectives and performance evaluations for all employees.

The company's commitment to pollution prevention cannot be overemphasized and neither can the need for employee participation. This will help to establish a positive atmosphere, will reassure employees concerned with resulting changes, and will also elicit worthwhile pollution prevention suggestions.

Employees' participation should be encouraged at all stages of the pollution prevention program including:

- Defining company goals and objectives

- Reviewing processes and operations for use and generation of pollutants or emissions

- Recommending ways to eliminate or reduce waste production at the source

- Designing or modifying forms and records to monitor materials used and waste produced

- Finding ways to involve suppliers and customers

- Acknowledging and rewarding employee contributions to the pollution prevention effort

12.1.2 Organize the Pollution Prevention Program

The program will be directed by the pollution prevention task force. Its first task will be to delineate program goals.

Name the Pollution Prevention Task Force.   The people who will direct the pollution prevention program should be selected carefully. They will have overall responsibility for developing the plan and directing its implementation. Their capabilities and their attitudes toward the effort will be major determinants of how successful it is.

As with other areas of your operation, successful program execution will require integration and continuity through planning, implementation, modification, and maintenance. Therefore, all individuals named to this task force should have substantial technical, business, and communication skills as well as thorough knowledge of the company. The responsibility and authority of each individual should be established during this organizational stage.

Once the task force has been established, it will be a valuable resource within the company both now and in the future. When plans are being made to expand the facility or to design or redesign products, the task force can review the plans to determine whether pollution prevention opportunities have been evaluated thoroughly.

Program Leader.   The role of the leader is to facilitate the flow of information among all levels in the company. The program leader's responsibilities include keeping the program on track, and ensuring that pollution prevention becomes an integral part of the overall corporate plan. The program leader's attributes should include a high degree of authority within the organization and the personal qualities necessary to elicit broad-based support from the company's employees.

Champion.   The task of a "champion" is to overcome possible resistance to proposed changes in operations. Champions will be the task force members who serve as liaisons and should be visible within the production areas and respected and trusted at all levels.

Team Members.   Other task force members might be selected for their specific technical or business expertise. Potential task force members include environmental and plant process engineers, production supervisors, health and safety personnel, experienced line workers, and purchasing and quality-assurance staff.

In some cases, it may be necessary or desirable to retain outside consultants to work with the in-house task force when suitable skills are not available within the company.

State Goals.   The task force members, with input from concerned employees and management, will need to establish goals that state the long-term direction for the pollution prevention program. Goals, which may have been stated in general terms in the policy statement, now need to be stated more specifically. This will help to focus effort and build consensus.

A starting point for establishing pollution prevention goals is the zero-discharge perspective. This ideal situation would involve 100-percent utilization of resources, eliminating disposal costs and regulatory compliance needs. However, like zero-defect production goals, zero-discharge goals are not achievable given the current technology, but serve to encourage an attitude of continually striving for improvement.

Goals can be *qualitative,* such as to "achieve a significant reduction of toxic substance emissions to the environment," or *quantitative,* which are more difficult to develop but are worth the extra effort. Quantitative goals spell out your pollution prevention commitment and give all participants and observers a yardstick for measuring progress.

Finally, goals should be flexible and adaptable. Conditions change in actual practice and as information is acquired. As your pollution prevention program becomes more focused, the task force and program more mature, and the pollution-specific aspects of the operation become better known, the goals can be refined. Periodic review of the goals and the achievements by the task force, management, and the employees will keep your program active and visible within the company.

12.1.3   Do the Preliminary Assessment

Even though some aspects of the preliminary assessment may have been completed as input to the executive decision to develop a pollution prevention program, a deeper examination is needed. Data collection as part of the preassessment serves two purposes: (1) to help the task force review the data that are already available, and (2) to begin defining ways to process that data. These data, along with site visits, will enable the task force to establish priorities and procedures for detailed assessments. Chapter 13 describes the detailed assessment phase and the more in-depth data collection and analyses that will be done at that stage.

Collect Data.   The extent and complexity of the system used for collecting pollution prevention data should be consistent with the goals of the pollution prevention program and the needs of the company. The objective of the program is to prevent pollution, not to collect data. Use the simplest system that is consistent with the goals and needs. Depending on the nature and size of the firm, much of the data needed for a pollution prevention program may already have been collected as a normal part of plant operations or as part of a response to existing regulatory requirements.

An all-media approach, which considers all air, water, and solid waste emissions and releases, will be the most effective. For each wastestream, the sources are identified, the methods for pollution abatement are evaluated, and the true costs of abatement, treatment, and waste disposal are quantified. A number of information sources to consider are outlined in Fig. 12-2.

Regulatory Reports. National Pollutant Discharge Elimination System (NPDES) and SARA Title III reports document the volume, composition, and degree of toxicity of wastewater discharged. The toxic substance release inventories required by SARA Title III, Sec. 313, may provide information on emissions into all environmental media.

Engineering and Operating Data.   Shipping manifests will provide quantities of hazardous waste shipped during a given period, but may not provide information on the chemical analysis, specific source, and the time period during which the waste was generated. The plant design documents and equipment operating manuals and procedures may yield specific data for streams inside of the plant.

Plant Business Records.   Records available from inventory control, purchasing, records management, accounting, marketing, and training can provide data needed for the preassessment and may themselves present opportunities for pollution prevention. For example, improved inventory control and judicious purchasing can significantly reduce the volume of raw materials that must be disposed of because they become outdated. In reviewing existing data, you may find that current accounting practices are not appropriate for placing the burden of pollution and pollution control at the point of generation. These findings should be taken into account when costs of pollution control measures are analyzed. (See Chap. 16.)

Visit Sites.   Part of the information the task force will need in order to prioritize the processes, operations, and wastes that will be studied during the detailed assessment phase should be garnered by a site visit by the task force. During this phase, the most important waste problems should be targeted, with lower-priority problems dealt with as resources of time, staff, and money per-

Data Sources for Facility Information

Regulatory information:

- Waste shipment manifests
- Emission inventories
- Biennial hazardous waste reports
- Waste, wastewater, and air emissions analyses, including intermediate streams
- Environmental audit reports
- Permits and/or permit applications
- Form R for SARA Title III, Sec. 313

Process information:

- Process flow diagrams
- Design and actual material and heat balances

Raw material/production information:

- Product composition and batch sheets
- Material application diagrams
- Material safety data sheets
- Product and raw material inventory records
- Operator data logs
- Operating procedures
- Production schedules

Accounting information:

- Waste handling, treatment, and disposal costs
- Water and sewer costs, including surcharges
- Costs for nonhazardous waste disposal, such as trash and scrap metal
- Product, energy, and raw material costs

Figure 12-2. Information sources for preliminary assessments.

Typical Considerations for Prioritizing Wastestreams for Further Study

- Compliance with current and anticipated regulations
- Costs of waste management (pollution control, treatment, and disposal)
- Potential environmental and safety liability
- Quantity of waste
- Properties of the waste (including toxicity, flammability, corrosivity, and reactivity)
- Other safety hazards to employees
- Potential for pollution prevention
- Potential for removing bottlenecks in production or waste treatment
- Potential recovery of valuable by-products
- Available budget for the pollution prevention assessment program and projects
- Minimization of wastewater discharges
- Reduction of energy use

Figure 12-3. Considerations for prioritizing wastestreams.

mit. The preassessment site visits will provide the information needed to accomplish this prioritization and to designate the detailed assessment teams, which will be selected on the basis of expertise in particular areas.

Establish Priorities.   Assigning priorities (see Fig. 12-3) to processes, operations, and wastestreams will focus the remainder of the pollution prevention plan development effort. The priorities set in this stage will guide the selection of areas for detailed assessment.

12.1.4 Prepare the Program Plan

With the information collected during the preassessment, the task force can develop a detailed program plan. This plan will address the extent to which external organizations will be involved, define pollution prevention program objectives, identify potential obstacles and solutions, and define the data collection and analysis procedures that will be used. A summary of the points that should be addressed in a program plan appears in Fig. 12-4.

Contact External Groups.   At this point, the task force should consider soliciting input from outside the company. Inclusion of the surrounding com-

---

Elements of the Formal Written Pollution Prevention Plan

- Corporate policy statement of support for pollution prevention.

- Description of your pollution prevention planning task force makeup, authority, and responsibility.

- Description of how all of the groups (production, laboratory, maintenance, shipping, marketing, engineering, and others) will work together to reduce waste production and energy consumption.

- Plan for publicizing and gaining companywide support for the pollution prevention program.

- Plan for communicating the successes and failures of pollution prevention programs within your company.

- Description of the processes that produce, use, or release pollutants, including clear definition of the amounts and types of substances, materials, and products under consideration.

- List of treatment, disposal, and recycling facilities and transporters currently used.

- Preliminary review of the cost of pollution control and waste disposal.

- Description of current and past pollution.

Figure 12-4. Elements for a pollution prevention plan.

munity in the pollution prevention planning process can create a new forum for communication. Valuable technical information can also be exchanged with technical, trade, or professional organizations.

Legislative and executive officials can provide their perspectives on environmental protection issues and information on their planning processes. In return, they can gain information that will help them make decisions on future public issues related to the environment.

Community involvement is a good way to build credibility and focus pollution prevention efforts on the discharge paths that most concern your neighbors. However, it may be wise to wait until the program is established before seeking to involve the community. Having a few pollution prevention projects underway will demonstrate your good faith. Positive community involvement can be encouraged through holding open meetings, granting interviews to the media, advertising, circulating direct-mail surveys, and conducting opinion polls.

Other businesses can be sources of information on technical issues and suppliers, either because they are in the same geographical area or because they have similar technical areas of interest. Local business groups are a good way of locating resources in the immediate area, while trade and professional associations can provide contacts in other parts of the country or the world. Of course, the companies with the most similar interests may be competitors, but it should be possible to interact without risking disclosure of business-sensitive information.

Define Objectives.   During the preliminary assessment phase, the task force will have identified opportunities for pollution prevention and will have worked with the executive group to establish priorities. These will be the starting point for defining short- and long-range objectives.

Objectives are the specific tasks that will be necessary to achieve goals. For example, in order to reach a goal of reducing waste, the objectives might be defined as reducing solvent, paper, and packaging wastes by specific amounts over a stated period of time.

Objectives can be defined at the facility or the department level, depending on the size and diversity of your company. A small company could decide to develop a single set of objectives to cover all of its operations. A larger company with many facilities or products might develop an overall corporate plan describing goals and objectives, supplemented by facility- or product-specific goals. In any case, the management at each location must understand and support its objectives if the pollution prevention program is to be successful.

Objectives should be stated in quantitative terms and should have target dates. These two attributes make objectives effective tools for directing effort and measuring progress.

Identify Potential Obstacles.   As the task force begins to develop and implement a pollution prevention program, a number of factors that will complicate the process will likely be encountered. These need to be recognized, and the means for overcoming them needs to be defined. Apparent obstacles will be less likely to impede the process if everyone understands that there is a mechanism for addressing them in a later stage.

The mix of factors and the relative degree of difficulty each presents will vary from company to company. Those that are likely to be encountered by most businesses are discussed here. They fall into four broad categories: economic, technical, regulatory, and institutional.

Economic Obstacles.   The task force should recognize that some complex economic factors may need to be addressed later. Broadly defining procedures now for dealing with them will help prevent economic concerns from stifling the creative process of defining options.

Cost-benefit analysis procedures should be defined. Many proposed pollution prevention options will have start-up costs. For example, additional or replacement equipment may need to be purchased, staff training may be required, or alternative raw materials may cost more. Some of these additional costs can be justified readily because they clearly will be cost-effective and will have short payback times. However, many will not be so clear-cut and will need more sophisticated analysis as described in Chap. 15.

Limited financial resources for capital improvements may also be a problem, even for options that will ultimately be profitable. The task force should investigate the availability of and conditions for funding assistance or low-interest loans from state or local agencies. Chapter 53 provides some additional information on whom to contact.

Technical Obstacles.   Information will be needed on alternative procedures that should be considered, how to integrate them in the production process, and what side effects are possible.

Information resources could be a problem. As a small or medium-sized business, you may not have ready access to a central source of information on pollution prevention techniques. There are several ways to deal with this problem. Contact appropriate agencies listed in Chap. 53 for assistance. Encourage employees to watch for information in the technical journals and newsletters they read and to pass it on to the task force. Those who belong to professional societies may get ideas from other members. Metropolitan or university library reference departments can provide assistance in locating sources of published information as well as names of people who might be able to provide information in specific areas. If the scope of the technical problem and resources permits, it may be appropriate to retain a consultant.

Limited flexibility in the manufacturing process may pose another technical barrier. A proposed pollution prevention option may involve modifying the work flow or the product, or installing new equipment; implementation could require a production shutdown, with loss of production time. The new operation might not work as expected or might create a bottleneck that could slow production. In addition, the production facility might not have space for pollution prevention equipment. These technical barriers can be overcome by having design and production personnel take part in the planning process and by using tested technology or setting up pilot operations.

Product quality or customer acceptance concerns might cause resistance to change. For example, in some printing and publishing operations it is possible to minimize waste by substituting a water-based ink for a solvent-based ink. But for some products, quality suffers when water-based ink is used. Avoid potential product quality degradation by contacting customers and verifying customer needs, testing the new process or product, and increasing quality control during manufacture.

Regulatory Obstacles.   Regulations may be a barrier to some pollution prevention options. For example, changing to another feed material may require changing the existing permits. In addition, it may be necessary to learn what regulations might apply to proposed alternative input materials.

Working with the appropriate regulatory bodies early in the planning process will help overcome this barrier. The EPA and the state environmental agencies have developed a number of documents to facilitate pollution prevention efforts by industry; many are listed in Chap. 53.

Local health department and city and county waste disposal and treatment offices may also provide assistance. Industry task forces and consultants might also be contacted.

Institutional Obstacles.   As with any other new program, general resistance to change and friction among elements within the organization may arise. These can result from many factors, such as lack of awareness of corporate goals and objectives, individual or organizational resistance to change, lack of commitment, poor internal communication, requirements of existing labor contracts, or an inflexible organizational structure.

Analyze these barriers from different perspectives in order to understand the concerns. Management is concerned with production costs, efficiency, productivity, return on investment, and present and future liability. Workers are concerned about job security, pay, and workplace health and safety. The extent to which these issues are addressed in the pollution prevention program will affect the success of the program.

Institutional barriers can be overcome with education and outreach programs. As was pointed out earlier, it is vital to gain the support of staff at all levels very early in the pollution prevention effort.

Develop Schedule.   The final aspect of planning the pollution prevention program is to list the milestones within each of the stages from detailed assessment through implementation and then assign realistic target dates. The execution of these stages should follow this schedule closely. Significant deviations may cause the program to falter because certain steps are not completed. Adherence to the schedule will also help control the start-up or implementation costs of the program.

12.2 Developing and Implementing Pollution Prevention Projects

This section outlines how to execute the pollution prevention program plan that resulted from the activities already outlined.

As with the other stages, the degree of formality should be tailored to the size of the company and the diversity of its product lines. Thus, a small company may need to do only one detailed assessment and prepare one implementation plan, while a larger, more diverse company might require several in order to address all production processes. If multiple plans are developed, it will be necessary to examine how they fit together, resolving any conflicts and prioritizing them to fit available resources.

12.2.1 Detailed Assessment Phase

As part of the program design, a preliminary assessment of the facility identified areas of opportunity for pollution prevention. Now, detailed assessments will focus on specific areas targeted by the preliminary assessment.

Assessment teams will be assigned to each operational area of the facility to gather data for later analysis. As was the case during the preliminary assessment, existing written materials and site evaluations will be used. However, teams will delve much more deeply into each production process, interviewing workers and compiling necessary data that may not have been collected before.

During this process, the team may identify some options that can be implemented quickly and with little cost or risk. It is likely, however, that many options will be more complex and will require in-depth analysis later.

Detailed Assessment Team(s). The detailed assessment phase should be started by a member of the task force, which was identified during program design. Unless the company is small enough that the task force and the detailed assessment team are the same, additional staff (three to six individuals) will need to be named to form one or more detailed assessment teams which will focus on specific pollution prevention opportunities.

Ideally, one member of the task force will be included on each team to facilitate communication. The additional team members should be people with direct responsibility for and knowledge of the wastestreams and/or areas of the facility under consideration. To the extent practical, engineers, supervisors, and production workers as well as finance and accounting, purchasing, and administrative staff should be considered when the team members are selected, much the same as when the task force was chosen. A multidisciplinary team is likely to be more successful in achieving a comprehensive assessment and providing the best input possible to the data analysis and option definition stages. Specialists can be consulted as needed.

Aside from field of expertise, consider a candidate's ability to work on a team, apparent interest in and commitment to the program, and capacity for looking at situations from new perspectives and for thinking creatively. Some example teams are given in Table 12-1. Note that for each team, the team leader is someone who has day-to-day operations responsibility and experience.

Table 12-1. Examples of Detailed Assessment Teams

Metal Finishing Department in a Large Defense Contractor
Metal finishing department manager
Process engineer responsible for metal finishing processes
Facilities engineer responsible for metal finishing department*
Wastewater treatment department supervisor
Staff environmental engineer

Small Pesticide Formulator
Production supervisor*
Environmental engineer
Maintenance engineer

Cyanide Plating Operation
Environmental engineer*
Electroplating facility engineering supervisor
Plant chemist

*Recommended detailed assessment team leader.

Review Data and Sites.   Numerous data sources probably exist for a given site. Many of these may have been identified during the preliminary assessment. The detailed assessment team for that site will search for additional sources of data that will be useful in studying the targeted processes, operations, or wastestreams.

However, most of their effort will be directed toward performing a thorough site review and interviewing workers. This will help them understand the data already collected and identify factors that are not well documented and for which data will need to be collected. Site review guidelines are outlined in Fig. 12-5.

A careful site review will provide the assessment team with a systems perspective of the process, operation, or wastestream in question and of how it fits into the overall facility operation. This perspective is a prerequisite for thorough assessment of options in later phases of the pollution prevention plan development cycle. If consultants are on the assessment team, the site review enables them to become familiar enough with the facility to utilize their expertise effectively.

The site review should not be performed perfunctorily, even though the assessment team members who are employed at the facility will all be familiar to some extent with the work site being reviewed. Those who are not involved in the day-to-day operation in that area will see factors that otherwise would be overlooked. Furthermore, personnel assigned to that specific site will often see it in a new

---

Site Review Plan

- Prepare an agenda in advance that covers all points that still require clarification. Provide staff contacts in the area being assessed with the agenda several days before the inspection.

- Schedule the inspection to coincide with the particular operation that is of interest (e.g., makeup chemical addition, bath sampling, bath dumping, start-up, or shutdown).

- Monitor the operation at different times during all shifts, and if needed, during all three shifts, especially when waste generation is highly dependent on human involvement (e.g., in painting or parts cleaning operations).

- Interview the operators, shift supervisors, and work leaders in the assessed area. Discuss the waste generation aspects of the operation. Note their familiarity with the impacts their operation may have on other operations.

- Photograph or videotape the area of interest, if warranted. Pictures are valuable in the absence of plant layout drawings.

Figure 12-5. Site review guidelines.

Typical Questions to Ask During Site Reviews

- What is the composition of the wastestreams and emissions generated in the company? What is their quantity?

- From which production processes or treatments do these wastestreams and emissions originate?

- Which waste materials and emissions fall under environmental regulations?

- What raw materials and input materials in the company or production process generate these wastestreams and emissions?

- How much of a specific raw or input material is found in each wastestream?

- What quantity of materials is lost in the form of volatile emissions?

- How efficient is the production process and the various steps of that process?

- Does mixing materials produce any unnecessary waste materials or emissions which could otherwise be reused with other waste materials?

- Which good housekeeping practices are already in place to limit waste generation?

Figure 12-6. Questions to ask during site reviews.

light when performing a pollution prevention assessment. Some of the information that can be gathered through site reviews is summarized in Fig. 12-6.

Site visits should be well planned to ensure that maximum benefit is obtained without excessive expenditures of time. While multiple visits to check or supplement data will usually be required, good planning can minimize such repetitions. Several suggestions for preparing for site visits are given here.

Review existing documentation, such as operators' manuals and purchasing and shipping records. This will enable the team to focus on the topics to be investigated. Decide on data collection formats to ensure that the data collection will be rigorous and compatible with the compilation and analysis stage described later. In particular, it is worthwhile to predetermine the boundaries and bases for calculating the energy and material balances that will be worked out during that stage. Doing a preliminary balance during the data collection phase can help identify data gaps and determine sampling requirements. Photographs and videos are an excellent means of capturing extensive detail quickly and accurately.

Prepare an agenda and make sure that all team members and supervisors at the site receive it in advance.

Schedule site visits by contacting the staff in the area to be visited. Ask when they will be performing the operations you are particularly interested in assessing.

Observe operations as they are actually performed by different shifts and under various circumstances. Process units may be operated differently from the methods described in their operating manuals, or the equipment may have been modified without being so documented in the flow diagrams or equipment lists.

Interview workers and supervisors to determine how aware they are of what wastes are generated by their operation. They may have suggestions for reducing these wastes.

Follow the process from beginning to end, from the point where input materials enter the work site to the point where products and wastes exit. This will help identify all suspected sources of waste. Waste sources to inspect include the production process; piping; maintenance operations; and storage areas for raw materials, finished product, and work-in-process. Examine housekeeping practices and the waste treatment area, as well. Make follow-up visits as missing or unclear data are identified during the analysis stage.

Organize and Document Process Information.   Analyzing process information involves preparing material and energy balances as a means of analyzing pollution sources and opportunities for eliminating them. Such a balance is an organized system of accounting for the flow, generation, consumption, and accumulation of mass and energy in a process. In its simplest form, a material balance is drawn up according to the mass conservation principle:

$$\text{Mass in} = \text{Mass out} - \text{Generation} + \text{Consumption} + \text{Accumulation}$$

If no chemical or nuclear reactions occur and the process progresses in a steady state, the material balance for any specific compound or constituent is as follows:

$$\text{Mass out} = \text{Mass in}$$

The first step in preparing a balance is to draw a process diagram, which is a visual means of organizing the data on the energy and material flows and on the composition of the streams entering and leaving the system. Such a diagram shows the system boundaries, all streams entering and leaving the process, and points at which wastes are generated.

Boundaries should be selected according to the factors that are important for measuring the type and quantity of pollution prevented, the quality of the product, and the economics of the process. The amount of material input should equal the amount exiting, corrected for accumulation and creation or destruction.

A material balance should be calculated for each component entering and leaving the process. When chemical reactions take place in a system, there is an advantage to performing the material balance on the elements involved.

The limitations of material and energy balances should be understood. They are useful for organizing and extending pollution prevention data and should be used whenever possible. However, the user should recognize that most balance diagrams will be incomplete, approximate, or both.

- Most processes have numerous process streams, many of which affect various environmental media.
- The exact composition of many streams is unknown and cannot be easily analyzed.
- Phase changes occur within the process, requiring multimedia analysis and correlation.
- Plant operations or product mix change frequently, so the material and energy flows cannot be accurately characterized by a single balance diagram.
- Many sites lack sufficient historical data to characterize all streams.

These are examples of the complexities that will recur in the analysis of real-world processes.

Despite the limitations, material balances are essential to organize data, identify gaps, and permit estimation of missing information. They can help calculate concentrations of waste constituents where quantitative composition data are limited. They are particularly useful if there are points in the production process where it is difficult or uneconomical to collect or analyze samples. Data collection problems, such as an inaccurate reading or an unmeasured release, can be revealed when "mass in" fails to equal "mass out." Such an imbalance can also indicate that fugitive emissions are occurring. For example, solvent evaporation from a parts cleaning tank can be estimated as the difference between solvent put into the tank and solvent removed by disposal, recycling, or dragout.

12.2.2 Define Pollution Prevention Options

Once the sources and nature of wastes generated have been described, the assessment team enters the creative phase. In a two-step procedure, they will propose and then screen pollution prevention options. Their objective is to generate a comprehensive set of options, ranked as to priority, that merits detailed feasibility assessment.

Propose Options. As with other planning efforts, the best results will be achieved in an environment that encourages creativity and independent thinking by each assessment team member. Brainstorming sessions are useful for encouraging creative thought because they provide a nonjudgmental, synergistic atmosphere in which ideas can be shared. Then, these ideas can be developed by means of group decision-making techniques. This approach will enable the assessment team to identify options that the individual members might not have come up with on their own.

Structuring option definition sessions according to the EPA pollution prevention hierarchy will encourage the team to look first at true source reduction options, such as improved operating procedures and changes in technology, materials, and products. Then, options that involve reuse, or closed-loop recycling, would be examined. The team would next consider off-line and off-site recycling and, finally, alternative treatment and disposal methods.

Screen Options.   Many proposed options may result from the previous step. Since detailed technical, economic, and environmental feasibility analysis can be costly, the proposed options should be screened by the assessment team. Some options will be found to have no cost or risk attached; these can be implemented immediately. Others will be found to have marginal value or to be impractical; these will be dropped from further consideration. The remaining options will generally be found to require feasibility assessment.

This screening does not require detailed and costly study. Screening procedures can range from an informal review, with a decision made by either the program manager or a vote of the team members, to the use of quantitative decision-making tools. Figure 12-7 shows questions to be considered in option screening.

The informal review is a procedure by which the assessment team selects the options that appear best after discussing and examining each option. As is the case when the team is proposing options, its approach to screening should employ group decision-making techniques whenever possible.

In more complicated situations, the team may need to use a formal decision analysis technique designed for use in complex decision-making situations.

Questions for Option Screening

- Which options will best achieve the goal of waste reduction?
- What are the main benefits to be gained by implementing this option (financial, compliance, liability, workplace safety, etc.)?
- Does the necessary technology exist to develop the option?
- How much does it cost? Does it appear to be cost-effective, meriting in-depth economic feasibility assessment?
- Can the option be implemented within an reasonable amount of time without disruptions in production?
- Does the option have a good track record? If not, is there convincing evidence that the option will work as required?
- What other areas will be affected?

Figure 12-7. Considerations for option screening.

12.2.3 Feasibility Analyses

The final product of the option definition phase is a prioritized list of pollution prevention options. These options now should be examined to determine which are technically, environmentally, and economically feasible and to prioritize them for implementation.

Depending on the resources currently available, it may be necessary to postpone feasibility assessments for some options. However, all options should be evaluated eventually.

Technical Evaluation. The assessment team will perform a technical evaluation to determine whether a proposed pollution prevention option is likely to work in a specific application. Technical evaluation for a given option may be relatively quick or it may require extensive investigation. The list in Fig. 12-8 suggests some criteria that could be used in a technical evaluation. Some of these are more detailed versions of questions asked during the option screening phase.

All groups in the facility that will be affected directly if the option is adopted should contribute to the technical evaluation. This might include people from production, maintenance, QC/QA, and purchasing. In some cases, customers may need to be consulted and their requirements verified. Prior consultation and review with these groups will ensure the viability and acceptance of an

---

Typical Technical Evaluation Criteria

- Will it reduce waste?

- Is the system safe for our workers?

- Will our product quality be improved or maintained?

- Do we have space available in our facility?

- Are the new equipment, materials, or procedures compatible with our production operating procedures, work flow, and production rates?

- Will we need to hire additional labor to implement the option?

- Will we need to train or hire personnel with special expertise to operate or maintain the new system?

- Do we have the utilities needed to run the equipment? Or must they be installed at increased capital cost?

- How long will production be stopped during system installation?

- Will the vendor provide acceptable service?

- Will the system create other environmental problems?

Figure 12-8. Typical technical evaluation criteria.

option. If the option calls for a change in production methods or input materials, carefully assess the likely effects on the quality of the final product. If, after the technical evaluation, the option appears impractical or can be expected to lower product quality, drop it.

For options that do not involve a significant capital expenditure, the team can use a "fast-track" approach. For example, procedural or housekeeping changes can often be implemented quickly, after the appropriate review, approvals, and training have been accomplished. Material substitutions also can be accomplished relatively quickly if there are no major production rate, product quality, or equipment changes involved.

Equipment-related options or process changes are more expensive and may affect production rate or product quality. Therefore, such options require more study. The assessment team will want to determine whether the option will perform in the field under conditions similar to the planned application. In some cases, they can arrange, through equipment vendors and industry contacts, visits to existing installations. Experienced operators' comments are especially important and should be compared with vendors' claims. A bench-scale or pilot-scale demonstration may be needed. It may also be possible to obtain scale-up data using a rental test unit for bench-scale or pilot-scale experiments. Some vendors will install equipment on a trial basis, with acceptance and payment after a prescribed time, if the user is satisfied.

Environmental Evaluation.   In this step, the detailed assessment team will weigh the advantages and disadvantages of each option with regard to the environment. Often the environmental advantage is obvious—the toxicity of a wastestream will be reduced without generating a new wastestream. Most housekeeping and direct efficiency improvements have this advantage. With such options, the environmental situation in the company improves without new environmental problems arising.

Unfortunately, the environmental evaluation is not always so clearcut. Some options require a thorough environmental evaluation, especially if they involve product or process changes or the substitution of raw materials.

For example, the engine rebuilding industry is dropping solvent and alkaline cleaners to remove grease and dirt from engines prior to disassembly. Instead, they are using high-temperature baking followed by shot blasting. This shift eliminates waste cleaner but presents a risk of atmospheric release because small quantities of components from the grease can vaporize.

To make a sound evaluation, the team should gather information on the environmental aspects of the relevant product, raw material, or constituent part of the process. This information would consider the environmental effects not only of the production phase and product life cycle but also of extracting and transporting the alternative raw materials and of treating any unavoidable waste.

Energy consumption should also be considered. To make a sound choice, the evaluation should consider the entire life cycle of both the product and the production process.

Economic Evaluation. Estimating the costs and benefits of some proposed pollution prevention projects is straightforward, while others prove to be complex. Despite the ease with which the cost calculations may be done for some options, it is advisable to document all that are adopted and to estimate the economic effects of each. This will help ensure that these real accomplishments of your pollution prevention program will not be overlooked when you measure the program's progress, as discussed in Chap. 16.

There are a number of factors that make pollution prevention costs and benefits difficult to calculate for many proposed projects. The total costs of continuing to pollute are not discernible in most corporate accounting systems. Furthermore, many of these costs are probabilistic—although the risks are real, it is difficult to predict the cost and even the occurrence date from past experience. The long-term need to avoid the spiraling costs of waste treatment, storage, and disposal, as well as future regulatory and liability entanglements, are likely to be major elements of your pollution prevention project economic evaluation.

Chapter 16 discusses an approach and gives an overview of the types of cost and benefit factors that should be examined when proposed pollution prevention projects are studied. It suggests some approaches to calculating indirect and probabilistic costs so that their full impact can be included in economic feasibility assessments. It also discusses ways to track the economic effects of pollution prevention projects after they are implemented.

12.2.4 Assessment Report

The task force should write a report that summarizes the results of the detailed pollution prevention assessments. The minimum contents are summarized in Fig. 12-9. The report will provide a schedule for implementing prevention projects and will be the basis for evaluating and maintaining the pollution prevention program. It may also be needed to secure internal funding for projects that

---

Discussion Points for Each Proposed Project Report

- Its pollution prevention potential

- The maturity of the technology and a discussion of successful applications

- The overall project economics

- The required resources and how they will be obtained

- The estimated time for installation and start-up

- Possible performance measures to allow the project to be evaluated after it is implemented

---

Figure 12-9. Contents of an assessment report.

require capital investment, if the members of the pollution prevention assessment task force do not have the authority to commit funds.

It may be tempting to omit this step if the company has an owner-manager and only a few employees. A summary assessment report may not be needed to resolve pollution prevention project conflicts among different areas, and funding approvals probably are not a formal procedure requiring cost justifications. However, an assessment report will help focus subsequent pollution prevention efforts and will be useful as a record of what aspects of the business have been examined for pollution prevention opportunities.

Input of the Assessment Teams. In a company that has several assessment teams, the task force will need to evaluate the results and resolve any conflicts that might exist among the teams as to approach and resources required for the projects proposed.

As input to this integration effort, each assessment team should prepare a summary report, presenting the results of its investigations and listing the options it screened. Each report should describe in some detail the options that the team has determined are feasible and propose a schedule for implementing them. The options recommended for immediate implementation should then be described in detail as proposed projects.

These proposals should evaluate each project under different scenarios. For example, the profitability of each could be estimated under both optimistic and pessimistic assumptions. Where appropriate, sensitivity analyses indicating the effect of key variables on profitability should be included. Each should outline a plan for adjusting and fine-tuning the initial projects as knowledge and experience increase. The proposals should include a schedule for addressing those areas and wastestreams with lower priorities than the ones selected for the initial effort.

Preparing and Reviewing the Assessment Report. The task force will use the assessment teams' reports and project proposals to prepare the summary assessment report and implementation plan. The report should include a qualitative evaluation of the indirect and intangible costs and benefits of a pollution prevention plan to the company, its employees, and the community. It will provide the basis for obtaining funding of pollution prevention projects. Pollution prevention projects should not be sold on their technical merits alone; a clear description of both tangible and intangible benefits can help a proposed project obtain funding.

Before the report is issued in final form, managers and other experienced people in the production units that will be affected by the proposed projects should be asked to review the report. Their review will help to ensure that the projects proposed are well defined and feasible from their perspectives. While they probably were involved in the site reviews and other early efforts of the task force, they may spot inaccuracies or misunderstandings on the part of the assessment teams that were not apparent before.

In addition to ensuring the quality of the assessment report and implementation plan, this review will help ensure the support of the people who will be responsible for the success of the project.

12.2.5 Implement the Pollution Prevention Plan

Select Projects for Implementation. Final decisions on which projects will be implemented and what the schedule will be are made at this point. If the task force or company executives question aspects of some projects, the assessment teams or pollution prevention program champions may be asked to produce additional data. The task force should be flexible enough to develop alternatives or modifications. The members should also be willing to do background and support work, and they should anticipate potential problems in implementing the options. Above all, they should keep in mind that an idea will not sell if the marketers are not convinced.

Obtain Funding. The task force will seek to secure funding for those projects that will require expenditures. There will probably be other projects, such as expanding production capacity or moving into new product lines, that will compete with the pollution prevention program for funding. If the task force is part of the overall budget decision-making procedure, it can make an informed decision that a given pollution prevention project should be implemented right away or that it can wait until the next capital budgeting period. The task force will need to ensure that the project is reconsidered at that time.

Some companies will have difficulty raising funds internally for capital investment. If this applies, look to outside financing. Private sector financing includes bank loans and other conventional sources of financing. Financial institutions are becoming more cognizant of the sound business aspects of pollution prevention.

Government financing is available in some cases. It may be worthwhile to contact your state's department of commerce or the U.S. Small Business Administration for information regarding loans for pollution control. Some states can provide financial assistance.

Install the Selected Projects. Many pollution prevention projects will require changes in operating procedures, purchasing methods, or materials inventory control. Company policies and procedures documents and employee training will also be affected by the changes.

For projects that involve equipment modification or new equipment, the installation of a pollution prevention project is essentially the same as any other capital improvement project. The phases of the project include planning, design, procurement, construction, and operator training. As with other equipment acquisitions, it is important to get warranties from vendors prior to installation of the equipment.

Training and incentive programs may be needed to get employees used to the new pollution prevention procedures and equipment.

Measure Progress Against the Goals.   By reviewing the program's successes and failures, managers at all levels can assess the degree to which pollution prevention goals at the facility and production unit levels are being met and what the economic results have been. The comparison identifies pollution prevention techniques that work well and those that do not. This information will help guide future pollution prevention assessment and implementation cycles.

Quantitative evaluation also enables comparison with similar units in the company and with data from other companies. This knowledge is needed to plan enhancements to the current pollution prevention program, to select technologies for transfer from other operations, and to help identify new pollution prevention options.

Review and Adjust.   The pollution prevention process does not end with implementation. After the pollution prevention plan is implemented, track its effectiveness versus the claims made—technical, economic, etc. Options that do not meet your original performance expectations may require rework or modifications. Above all, reuse the knowledge gained by continuing to evaluate and fine-tune pollution prevention projects. Chapter 17 provides details on measuring progress after implementation and evaluating it against goals. Chapter 8 deals with ways to maintain and enhance a program after it is implemented.

13

Descriptive Approach for Pollution Prevention Assessments

Robert B. Pojasek, Ph.D.

GEI Consultants Inc.
Winchester, Massachusetts

13.1 Introduction

The practice of industrial pollution prevention has been entrusted to the environmental specialists. This came about largely as a result of congressional delegation of the Pollution Prevention Act of 1990 to the U.S. Environmental Protection Agency (EPA). Congress has also included waste minimization, resource conservation, and source reduction in previous legislation, all of which was assigned to the EPA.

This view of pollution prevention as an environmental issue has created a fundamental problem for those who practice pollution prevention. Incorporation of pollution prevention in the federal regulations, as well as special state initiatives requiring pollution prevention plans, create a "command and control" focus. If something is not done, enforcement can be expected. Regulatory personnel use checklists, questionnaires, and worksheets to ensure compliance with the regulations fashioned from the legislation. In other words, their approach is quite prescriptive.

Since the rapid proliferation of environmental regulations began in the early 1970s, environmental specialists have become increasingly involved in regulatory compliance audits. Protocols for these activities are necessarily prescriptive. With the large number of applicable regulations for any industrial operation, one would be foolhardy not to use this approach. When pollution prevention became fashionable, these practitioners were convinced that this prescriptive approach should be the method of choice.

Recently the EPA, along with a large number of industrial concerns in the regulated community, endorsed the concept of quality improvement. A number of useful tools became available to those involved in these programs. These tools include process mapping, brainstorming, storyboarding, dendrograms, cause and effect diagrams, functionality analysis, force-field analysis, and Pareto diagrams. Environmental specialists have been slow to adopt these tools. However, with the growing interest of production personnel in nonregulatory programs like pollution prevention, quality improvement tools are beginning to gain acceptance in environmental programs. Some firms now claim to practice "total quality environmental management."

Many companies now view pollution prevention as an exercise in making a process or operation more efficient. They try to control materials use and the loss of materials by making changes or improvements to the process. This form of loss control is accomplished through the many iterations which are the hallmark of the continuous improvement element of quality improvement programs.

This chapter will explore some of the fundamental differences between a prescriptive and a descriptive approach to pollution prevention assessments. There is a reluctance on the part of some environmental specialists to change. However, pollution prevention is all about change—getting people to change the way they are doing things in order to be more efficient. The reader will see how the quality improvement tools mentioned above are readily incorporated into the *descriptive approach* to pollution prevention discussed below.

13.2   Prescriptive Approach

The EPA encouraged the *prescriptive approach* to pollution prevention assessments with the publication of the document entitled *Waste Minimization Opportunity Assessment Manual*[1] in July 1988. This manual presented a collection of 32 pages of checklists, questionnaires, and worksheets. A more recent EPA manual entitled *Facility Pollution Prevention Guide*[2] and a series of industry-specific pollution prevention manuals have perpetuated the use of these prescriptive tools. Let's examine some of the shortcomings of this approach.

Prescriptive forms are meant for individuals to fill out. Let's examine the "Process Information" worksheet (see Fig. 13-1) to illustrate this point. Picture if you would, three chemical engineers who are asked to view a particular process flow diagram (PFD) and determine whether it is "current" (Y/N as indicated on the form). The first engineer notes that the PFD is two years old. Having seen many PFDs older than this, this person enters a "Y" in the worksheet. A second engineer notices all the "minor" changes that have been made

Firm _____	**Waste Minimization Assessment**	Prepared By _____
Site _____		Checked By _____
Date _____	Proj. No. _____	Sheet _1_ of _1_ Page __ of __

| WORKSHEET **6** | **PROCESS INFORMATION** | ☸ **EPA** |

Process Unit/Operation: _____

Operation Type: ☐ Continuous ☐ Discrete
☐ Batch or Semi-Batch ☐ Other _____

| Document | Status | | | | | |
	Complete? (Y/N)	Current? (Y/N)	Last Revision	Used in this Report (Y/N)	Document Number	Location
Process Flow Diagram						
Material/Energy Balance						
Design						
Operating						
Flow/Amount Measurements						
Stream						
Analyses/Assays						
Stream						
Process Description						
Operating Manuals						
Equipment List						
Equipment Specifications						
Piping & Instrument Diagrams						
Plot and Elevation Plan(s)						
Work Flow Diagrams						
Hazardous Waste Manifests						
Emission Inventories						
Annual/Biennial Reports						
Environmental Audit Reports						
Permit/Permit Applications						
Batch Sheet(s)						
Materials Application Diagrams						
Product Composition Sheets						
Material Safety Data Sheets						
Inventory Records						
Operator Logs						
Production Schedules						

Figure 13-1. Prescriptive worksheet from EPA publication.

in the piping and the relocation of several tanks and enters an "N" on the worksheet. Because of the lack of suitable detail on the PFD, the third engineer refuses to vote (i.e., this is not really a qualified PFD). How then do we fill out the worksheet when we have perhaps one yes, one no, and an abstention?

Prescriptive forms request more information than might be necessary to determine suitable opportunities for pollution prevention. Note all the information requested on this form. Will all this material be gathered, read, and understood by the auditor? Probably not. Probably, the requestor will simply use the information to check off the appropriate boxes in the form for completeness' sake. Meanwhile, gathering all this information for this purpose is very disruptive and costly to the facility.

Use of prescriptive forms entails a series consistency issue. If the form is given to the three engineers described above for use in reviewing the same facility at different times, it is possible that there will be serious differences between their forms. Even a single person is not likely to be consistent given the differences between a Monday morning and a Friday afternoon just before a three-day weekend.

The spirit of exploration is discouraged because of the highly structured format of the prescriptive forms. If a question is not on the form, it is much less likely to be asked. And this approach often does not allow penetration to the source of the problem because it focuses on an overall process and not on the individual functional steps within the operation. For example, inputs are not divided substantially into individual steps but rather are allocated to "streams" or production units consisting of many steps.

Overall, the prescriptive approach places an emphasis on readily identified wastes instead of identifying all losses from the process. The need for industry-specific booklets as the source of suitable options stifles innovative solutions to mitigation of these process losses.

There are some appealing aspects of the prescriptive approach, especially to those who use it. There is a feeling that a lot of training will not be necessary to use this approach and that it will in fact ensure completeness. The industry-specific forms are quite useful to orient and prepare audit teams for a particular assignment. Finally, it creates a uniform format that allows all reports to be written in a consistent form.

13.3   Descriptive Approach

Noting the shortcomings of the prescriptive approach, the author developed the descriptive approach. It was first described in a book chapter[3] and subsequently in a number of magazine and journal articles.[4,5] This approach replaces the checklists, questionnaires, and worksheets with PFDs.

One of the most important tools utilized in quality improvement programs is *process mapping*. An example of this is the exercise to determine the number of steps it takes to get an invoice issued. A quality improvement team carefully investigates the procedure by interviewing those involved and observing what is actually happening. Using a consensus-building approach, the team can prepare a map of the process. This process mapping can be utilized to describe any operation or process. Please note that it does *not* utilize checklists, questionnaires, and worksheets. Engineers refer to these process maps as *process flow diagrams* (PFDs), as described above. Our hypothetical engineers would have been asked to prepare a process map together and not to fill in a form individually.

Functionality is a key concept in the use of the descriptive approach. For any process or operation there is a functional sequence of events or actions. This is the sequence that is captured in the process mapping. Understanding this sequence is based on the premise that one action initiates others, which in turn initiate still others, until the process has completed its overall function with some kind of product (i.e., throughput) or result.

Consider the following statement: "Important functional is sequence very." Of course, there is more immediate understanding of what this message is trying to convey if it is stated as follows: "Functional sequence is very important." It is not good enough to know that a certain process makes a given product. It is not even good enough to know how this is done procedurally. One must understand the functionality of each step and the functionality of every material used in those steps.

Using the descriptive approach in the assessment is like becoming an explorer. Author Roger von Oech[6] provides a number of tips on becoming a good explorer. First is to be curious and ask lots of questions. This means not being bound to a script of questions as provided by a questionnaire. Second, one should look to other industries and contexts to understand what one is observing in a given location. Like the use of analogies, this will help one understand the item better. Third, one must generate lots of ideas. It is well understood that the only way to have a good idea is to have lots of ideas. One of them is more likely to be a good idea. Fourth, don't overlook the obvious. Often the means to understanding or a great idea for preventing pollution is right in front of you. Fifth, look for ideas in "avoided" places. There are many things assessors do not wish to consider because of their background or inclinations. The understanding may lie in that area. Sixth, consider the "big picture." All little things take place in a bigger sequence. What happens in any given area may have impacts not only on what happens after but also what happens before. Seventh, pay attention to a variety of information. If you are an engineer, you should not disregard sociological information and vice versa. Finally, never explore alone. The descriptive approach to pollution prevention is a team activity.

The product of the descriptive approach is a PFD like the one shown in Fig. 13-2. This PFD provides a template which tracks materials used in each step of the operation. It also tracks materials loss by unit. Numbering the boxes allows the corresponding text to describe the functionality of the materials used in each step as well as the functionality of the overall process. Verification of the PFD is the major focus of a pollution prevention assessment. Please note that the PFD is used in place of the prescriptive tools described above.

Every loss from the process or operation becomes an opportunity not to have that loss. As one can see, the descriptive approach identifies many more opportunities than the prescriptive approach—so many more in fact, that the pollution prevention team will have to rank-order these opportunities in order to effectively deal with them in a feasibility study. Usually a set of criteria are selected in order to place the opportunities in categories of high, medium, and low interest and importance. These criteria will incorporate elements of effectiveness, implementability, and cost/benefit. It is important not to have all the

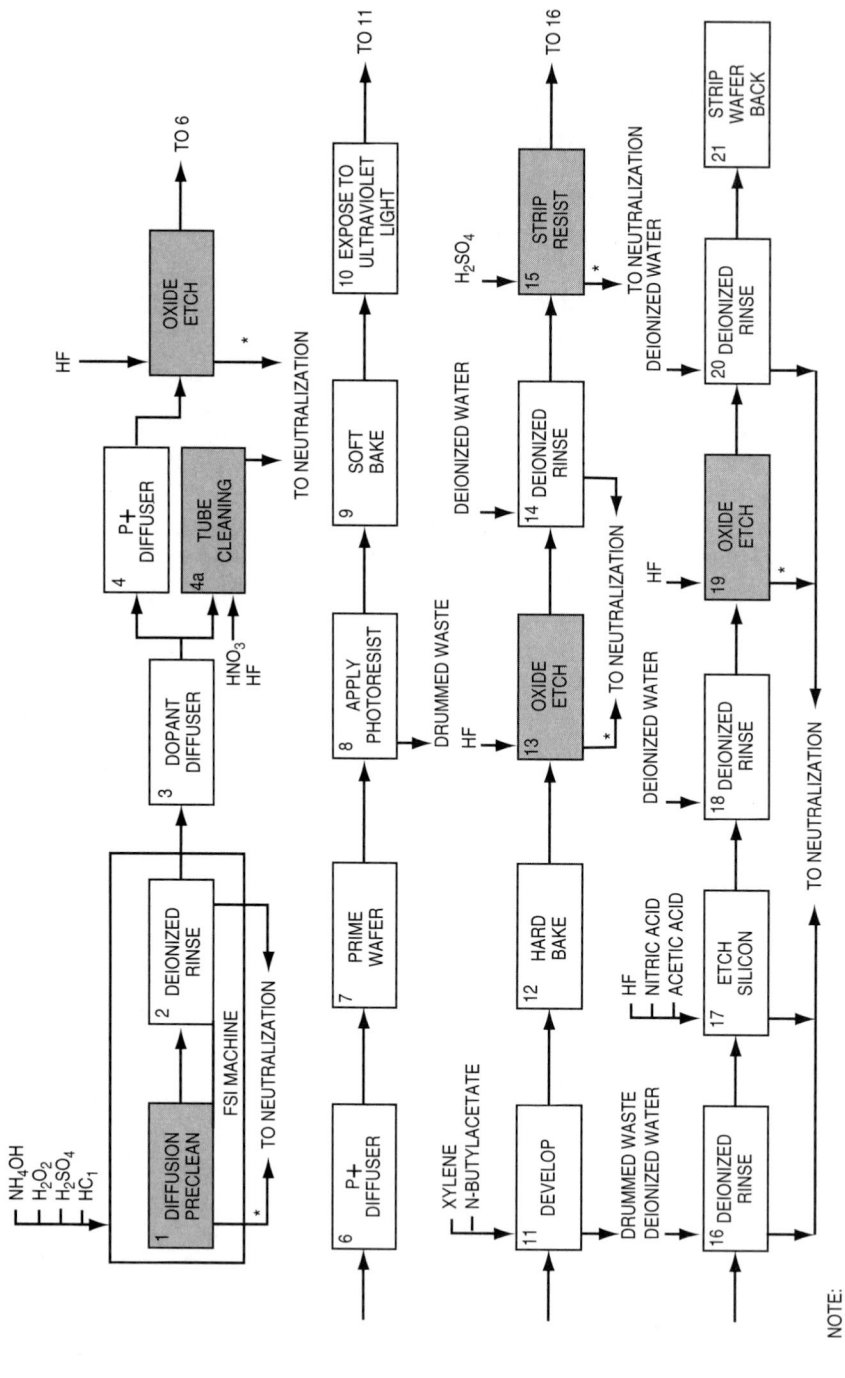

Figure 13-2. Process flow diagram for semiconductor device fabrication.

most difficult opportunities in the high-priority category. Building on small successes[7] is much more effective at getting management and employee "buy-ins" on the pollution prevention program.

Now the pollution prevention team must derive *all* the possible alternatives for each primary opportunity. By paying attention to von Oech's "artist," the team will derive many ideas to accomplish the reduction. Quality improvement tools are particularly helpful. These ideas will include the following pollution prevention program categories: changes, in operating practices, materials substitution, process and/or product changes, and recycling and/or reuse. Every attempt is made to avoid premature adoption of the "right answers." This is when someone on the team lobbies hard for a single way to accomplish pollution prevention. The group needs lots of ideas to be brought forth. Ideas must not be dismissed with a "killer phrase"[8] like "We've tried that before but it didn't work" or "That's too expensive."

Next there is a feasibility study to determine which idea will be utilized. Note that the prescriptive approach focuses on "right answers" and this diminishes the need for a feasibility study.[9] The descriptive approach emphasizes the feasibility study, which von Oech describes in terms of being a good "judge." This is a good analogy since the team that conducts a good feasibility study will show the same impartiality and willingness to consider all evidence as a prudent jurist does.

Finally, the pollution prevention team seeks means of implementing the most feasible alternative. This is the tricky phase because there are barriers to change built into any operation. Roger von Oech likens pressing for implementation to being a "warrior." It may be necessary to fight for your programs.

13.4 Conclusions

The descriptive approach to pollution prevention uses quality improvement tools to solve the problems associated with inefficient operations, i.e., the loss of materials to the workplace and the environment. No matter what operation or process is studied, the approach is valid. There is no need to customize checklists, questionnaires, and worksheets because they are not used.

Individuals do not use the descriptive approach. Just as quality improvement teams are empowered to solve problems in a manufacturing or service operation, pollution prevention teams need to have the same outlook. However, it is important that these teams receive proper training in the use of group dynamics practices and that they be encouraged to use creative problem-solving techniques.[10]

In contrast, the prescriptive approach to pollution prevention used regulatory compliance auditing tools. These tools are customized for various industries. This practice does not allow for cross-pollination between industries (e.g., two industries with a similar waste but generated in a different manner). Also it does not account for cultural differences between companies in the same industry or multiple facilities in the same company.

EPA contractors continue to utilize the prescriptive approach. However, this does not make their information worthless to those firms that now use the descriptive approach. By utilizing this information within the framework described in this chapter, the pollution prevention team will get a head start on its deliberations.

The author has been training students at Tufts University[11] since 1988 to utilize this descriptive approach. Contrary to the prescriptive adherents, this approach is not more difficult to use. It replaces the rigid logic of checklists, questionnaires, and worksheets with a commonsense approach which really works on student projects conducted during the semester.

An interesting exercise was conducted by the President's Commission on Environmental Quality[12] to demonstrate that quality improvement and pollution prevention are one and the same. Once companies and those who seek to provide them with technical assistance finally recognize this fact, the practice of pollution prevention can advance more significantly.

References

1. U.S. Environmental Protection Agency, *Waste Minimization Opportunity Assessment Manual*, EPA/625/7-88-003, Cincinnati, Oh., July 1988.

2. U.S. Environmental Protection Agency, *Facility Pollution Prevention Guide*, EPA/600/R-92/088, Washington, D.C., May 1992.

3. R. B. Pojasek, "Waste Reduction Audits," in Eric B. Rothenberg and Dean Jeffery Telego (eds.), *Environmental Risk Management—A Desk Reference*, RTM Communications, Arlington, Va., 1991.

4. R. B. Pojasek and L. J. Cali, "Contrasting Approaches to Pollution Prevention Auditing," *Pollution Prevention Review*, vol. 1, no. 3, Summer 1991, pp. 225–235.

5. R. B. Pojasek, "For Pollution Prevention: Be Descriptive Not Prescriptive," *Chemical Engineering*, September 1991, pp. 136–139.

6. R. von Oech, *A Kick in the Seat of the Pants*, Harper & Row, New York, 1986.

7. R. B. Pojasek, "Getting Your Pollution Prevention Program Off to a Quick Start," *Pollution Prevention Review*, vol. 3, no. 3, Summer 1993, pp. 351–357.

8. C. Thompson, *What a Great Idea*, Harper Perennial, New York, 1992.

9. R. B. Pojasek, "Reviving the Feasibility Study for Use in Pollution Prevention Programs," *Pollution Prevention Review*, vol. 2, no. 4, Autumn 1992, pp. 107–111.

10. von Oech, *op cit.*

11. R. B. Pojasek, *Prevention Pollution: Theory & Tools for Implementing the Practice*, textbook in preparation.

12. President's Commission on Environmental Quality, *Total Quality Management: A Framework for Pollution Prevention*, PCEQ, Washington, D.C., 1993.

14

Application of Risk Analysis to Set Pollution Prevention Priorities

Vlasta Molak, Ph.D.

GAIA Unlimited, Inc.
Cincinnati, Ohio

14.1 Overview

14.1.1 Introduction

Risk analysis is a collection of methods that evaluates the probability of an adverse effect of an agent, industrial process, technology, or natural process. An *adverse effect* is usually understood as an adverse effect to human health (death or disease), but it may also be defined as economic loss (economic risk analysis) or an effect on ecological systems. Risk analysis of a toxic substance often does not derive a probability of a disease or death (except for cancer), but establishes if a given exposure to a specific chemical substance is above or below a presumably safe level. Risk analysis is an inexact science that cannot be used in accurate predictions of actual or absolute risks. However, it can be used to help determine an actual or potential threat. *Risk assessment* is a term frequently used interchangeably with *risk analysis*. However, in this chapter *risk assessment* denotes an actual application of and results of performing risk analysis in a particular situation.

Industrial pollution can be defined as the presence of toxic substances in air, water, or soil, often resulting from inefficiencies in production processes. The presence of these substances can present a health risk to humans or ecological systems. These risks can be estimated and compared using risk analysis methods. Therefore, risk analysis can serve to establish a priority ranking of pollution problems based on the magnitude of risk that they pose either to human health or ecological systems.

Pollution can also be regarded as resources distributed in wrong places. Therefore, pollution prevention at the source can be regarded as saving on resources. Since economic risk analysis can indicate economic losses resulting from pollution, it can be used to encourage pollution prevention at the source as a means of improving the "bottom line." *Risk communication* explains the data derived from toxicological risk analysis and economic risk analysis to decision makers in the most compelling manner in order to encourage pollution prevention at the source. This chapter will demonstrate some of the applications of risk analysis in dealing with industrial pollution.

14.1.2   Modern Risks Associated with Pollution

The following are examples of the changes in a modern industrial society that must be factored into risk analysis and risk management associated with industrial pollution:

- A shift in the nature of risks from infectious diseases to degenerative diseases
- Increased new risks such as from nuclear plant accidents, radioactive waste, pesticides and other chemicals released, oil spills, chemical plant accidents, ozone depletion, acid rain generation, and global warming
- Increased ability of scientists and equipment to measure contamination
- Increased number of formal risk analysis procedures capable of predicting a priori risks
- Increased role of governments in assessing and managing risks
- Increased participation of special interest groups in societal risk management (industry, workers, environmentalists, scientific organizations), which increases necessity for public information
- Increasing level of citizen concern and demand for protection

Risk analysis can help manage technology in a more rational way to promote sustainability of desirable conditions of societies and eliminate those conditions that are detrimental to the well-being of humans and ecosystems. Figure 14-1 gives a schematic presentation of technological activities that can result in environmental pollution, which in turn results in risks to human and ecological health. Since many of these human and ecological health risks are associated with industrial pollution (air, soil, and water contamination), risk analysis can play a role in industrial pollution prevention by establishing the magnitude of

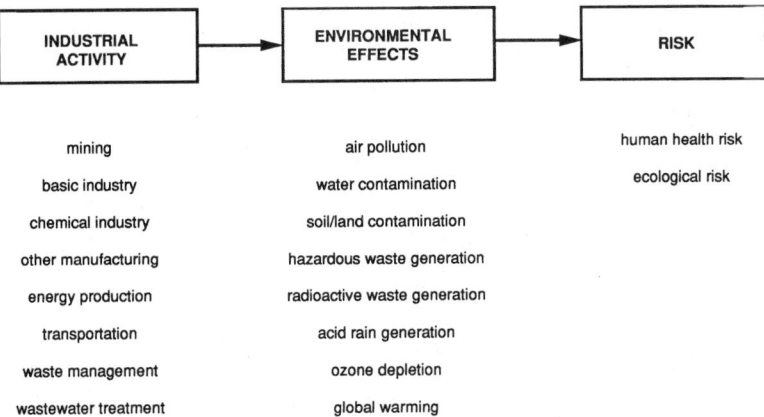

Figure 14-1. Technological activities that can result in environmental pollution.

risk associated with each pollution case and indicating which pollution prevention option will result in highest reduction of risk (Shorthouse, 1990–1991).

14.2   Types of Risk Analysis

Formal risk analysis can be characterized as shown in Fig. 14-2. For noncarcinogenic chemicals, it is assumed that an adverse effect occurs only if an exposure to the chemical exceeds a threshold. Risk analysis is used both for establishing criteria and standards for chemicals in the environmental media and for evaluating risk in particular cases of exposures to toxic chemicals (such as to contaminated water, soil, or air in the vicinity of a pollution source or when evaluating Superfund sites).

Risk analysis for toxic substances is probabilistic only in the case of carcinogens. The probability of developing cancer or a cancer potency slope as a result of exposure to a particular chemical is derived by modeling from animal data. Depending on the model applied, a variety of results may be obtained (Johannsen, 1990; Casarett and Doull, 1986).

Probabilities of developing cancer or other diseases can also be obtained by epidemiological research correlating exposures to toxic substances with development of cancer or other type of disease. Epidemiological risk analysis deals with the establishment of correlations or causal relationships between exposure to a chemical and disease (Stayner, 1992). Most frequently, retrospective cohort mortality studies of occupational groups are used for assessing cancer risk. Standard morbidity or mortality ratios can be regarded as denoting increase in probability of a health risk with exposure. However, because of the large uncertainty in estimating exposure, the results of the epidemiological studies are

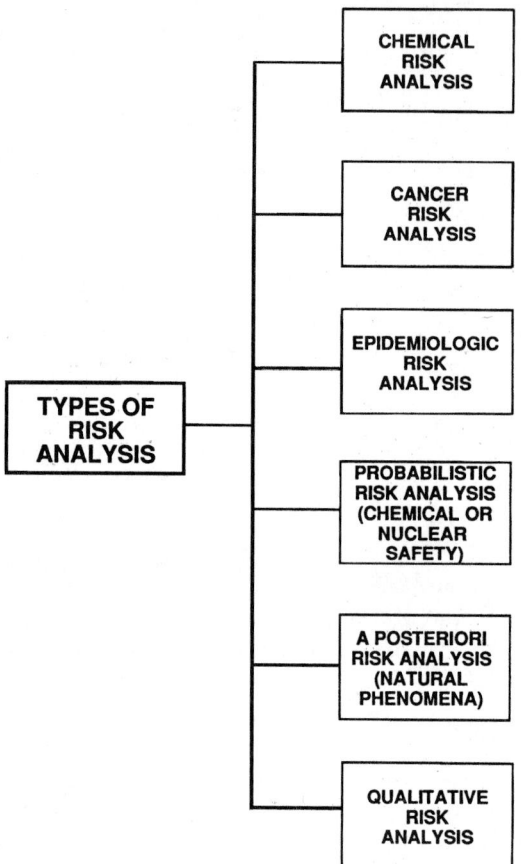

Figure 14-2. Risk analysis characterization.

combined with studies in animals in order to confirm the causal relationship between exposures to an agent (carcinogen) and cancer.

Probabilistic risk analysis is applied to industrial process safety and nuclear plant safety (fault-tree and failure-tree analysis). The probability of an adverse outcome (failure of a component or a system) of a series of interconnected events is obtained through evaluation of probabilities of failures of individual components. These probabilities are obtained either based on historical data or on assumptions of failure. Once a probability of failure of a chemical process is established, one can apply chemical risk analysis to establish severity of consequences of a release of a particular toxic substance.

Based on historical data, one can establish probabilities of adverse effects from natural phenomena (e.g., lightning, floods), or types of human activities (e.g., transportation accident rates). This type of risk analysis is used extensively by the insurance industry to establish insurance rates. Economic risk analysis can be also regarded as belonging to this category, because adverse economic effects are obtained from known prices of wasted chemicals and other costs associated with pollution (cost of cleanup of hazardous waste sites, legal costs, medical costs to society, etc.).

Some recent phenomena are not yet quantifiable. For example, risks from acid rain generation are not yet easily amenable to numerical analysis, nor are the risks from global warming (Gore, 1993). Therefore, one can only establish qualitative risks until more data are obtained to perform quantitative risk analysis.

14.3 Chemical Risk Analysis and Pollution Prevention

Most environmental problems that concern the public have to do with some kind of chemical exposure (by inhaling air, by ingestion of water or food, or dermal exposure) to pollutants originating from the chemical or other industries, power plants, road vehicles, or agricultural chemicals. Chemical risk analysis does not generally determine the probability of an adverse effect. Rather, it establishes levels of chemicals in our food, water, or air that could be tolerated by most people without experiencing adverse health effects. These levels (either concentrations of chemicals in environmental media or total intake of a chemical by one or all routes of exposure) are then established as "criteria," which serve for establishing standards. In a particular pollution situation, one can measure or estimate exposures to a contaminant and compare these to previously established criteria. The likelihood of harm increases if the exposure levels exceed the derived "safe" levels.

Chemical risk analysis is generally divided into four steps (NAS, 1983):

1. *Hazard identification.*   Identification of potentially toxic chemicals.

2. *Determination of dose-response relationships.*   Determination of toxicological properties' dependence on amounts ingested, inhaled, or otherwise entering the human organism. Dose-response relationships are usually determined from animal studies.

3. *Exposure assessment.*   Determination of the fate of the chemical in the environment and its consumption by humans. Ideally, by performing environmental fate and transport studies, and by evaluating food intakes, inhalation, and possible dermal contacts, one can assess total quantities of toxic chemicals which may cause adverse health effects in an exposed individual or population.

4. *Risk characterization.*   Risk characterization consists of evaluating and combining data from steps 2 and 3. For establishing criteria and standards, assumptions are made about "average exposures" and the criteria are set at the concentration at which it is believed that no harm would occur. For example, the reference dose (RfD) and "health advisory" levels (for 1-day,

10-day, and subchronic exposures) are derived for many chemicals with use of safety (uncertainty) factors to protect most individuals. If an actual exposure to an environmental pollutant (or pollutants) exceeds limits set by the criteria, efforts should be made to decrease the concentrations of pollutant. The magnitude of risk can be estimated by comparing the particular exposure to derived criteria or reference doses.

14.3.1 Toxicological Bases of Toxic Substances Risk Analysis

Over 110,000 chemicals are used in commerce in the United States. The Registry of Toxic Effects of Chemical Substance (RTECS) database, maintained by the National Institute for Occupational Safety and Health (NIOSH), contains updated information on toxicity of those chemicals (RTECS, 1993). Since the number of chemicals appearing in the environment is large, and the toxicological effects are very complex and are different depending on the chemical and conditions of exposure, it is sometimes difficult to determine "how toxic is toxic." Risk analysis helps determine which chemicals are dangerous and under what circumstances. It can also help establish relative risks from various chemicals (ranking risks). If, for example, in a particular industrial setting, the derived health risk from a pollutant A is higher than from a pollutant B, action should first be taken to decrease the pollution by A. In order to be able to use information on such a large number of substances, toxicologists have developed a classification of chemicals by their acute, subchronic, and chronic toxicity (Casarett and Doull, 1986). This classification is described below.

Acute Toxicity.   Acute toxicity is the most obvious and easiest to measure and is generally defined by the LD_{50} ("lethal dose 50 percent"). This is the dose, expressed in milligrams per kilogram of body weight, which causes death within 24 hours in 50 percent of exposed individuals after a single treatment, either orally or dermally. LD_{50} is usually derived from animal studies (mice and rats). The measure of acute toxicity for gases is LC_{50} (lethal concentration of chemical in air that causes death in 50 percent of animals if inhaled for a specified period of time, usually 4 hours). Based on that definition, chemicals are characterized as practically nontoxic, moderately toxic, very toxic, extremely toxic, and supertoxic (see Table 14-1).

In the sixteenth century the Swiss physician and alchemist Paracelsus stated that "the dose makes the poison"; chemicals could be very useful at small doses and poisonous at high doses. For example, selenium, oxygen, and iron are nontoxic or even essential at certain doses but can be lethal at higher doses. Generally, we are concerned with chemicals which are very, extremely, or supertoxic. Unless the chemical is a carcinogen or has some other chronic health or environmental effects (such as PCBs or heavy metals), there is little concern about those chemicals in the moderately toxic and practically nontoxic groups.

Table 14-1. Toxicity Characteristics for Chemicals

Toxicity rating	Probable lethal oral dose for humans*	Example Chemical	LD_{50} (animals)
1 = practically nontoxic	> 15 g/kg		
2 = slightly toxic	5–15 g/kg	Ethanol	10 g/kg
3 = moderately toxic	0.5–5 g/kg	Sodium chloride	4 g/kg
4 = very toxic	50–500 mg/kg	Phenobarbital	150 mg/kg
5 = extremely toxic	5–50 mg/kg	Picrotoxin	5 mg/kg
6 = supertoxic	< 5 mg/kg	Dioxin	0.001 mg/kg

*Doses given in grams or milligrams per kilogram of body weight.

Subchronic and Chronic Toxicity.   In some instances, chemical substances can heave very low acute toxicity but can cause cancer (e.g., PCBs), birth defects (thalidomide), or ecological effects (DDT) (Casarett and Doull, 1986). Long exposures to even relatively low concentrations of chemicals can cause specific organ damage or cancer. Therefore, chemicals are also evaluated for their subchronic and chronic systemic toxicity, carcinogenicity potential, or reproductive and developmental toxicity. Data are usually obtained from animal studies and sometimes from epidemiological studies.

Cancer Risk Assessment Models and Cancer Potency.   Various cancer models can serve to determine a cancer potency slope for a particular chemical (Johannsen, 1990; Casarett and Doull, 1986). While for health effects other than cancer a threshold dose is assumed, for cancer it is assumed that any exposure may potentially cause cancer. However, the probability of getting cancer at low exposure concentrations may be so low as to be of no practical concern. The U.S. Environmental Protection Agency (EPA) defines negligible risk for cancer as the one smaller than 1 in 1 million (U.S. EPA, 1980), and for OSHA risk of less than 1 in 1000 is "acceptable" (OSHA, 1989). The EPA has used a multistage linear model to establish potency slopes for approximately 140 cancer-causing chemicals, which can serve to establish the risks of pollutants in the air, water, and food (U.S. EPA, 1988a). Since most of these potency slopes are derived from animal data, there is an uncertainty associated with their numerical values. An additional uncertainty is posed by extrapolation from high to low doses, because animal studies are, for practical reasons, performed at relatively high doses.

14.3.2   Dose-Response Relationships

For each chemical there are dose-response relationships for different types of toxicological effects (see Fig. 14-3). With increasing dose the percentage of individuals with some type of health effect increases. For noncarcinogens, a threshold dose is assumed which defines a "no observable adverse effect" level. It is

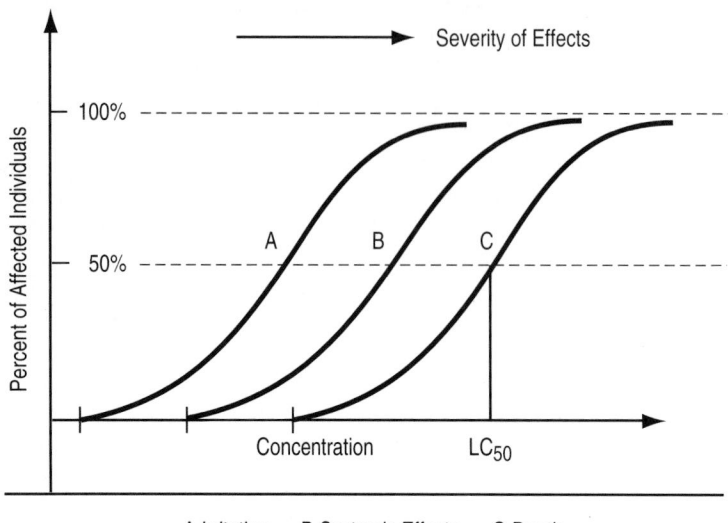

Figure 14-3. Dose-response relationships for different types of toxicological effects.

assumed that exposure to a chemical that results in a dose smaller than the threshold dose is handled by the organism and no adverse health effects result. For carcinogens, however, it is assumed that no threshold exists and that even a small number of molecules of carcinogen could potentially cause alterations in DNA resulting in cancer (Upton, 1988). A similar curve applies for dose-effect relationship, in which the severity of effect in an individual increases with dose (Casarett and Doull, 1986; OSHA, 1989).

14.3.3   Exposure Assessments

Exposures are determined by determining the concentration of the chemical in a particular environment and then establishing average amounts of a chemical consumed by an exposed person or population by ingestion of food and water, inhalation, or dermal contact.

In deriving criteria for particular pollutants, one assumes an average consumption of food and water and derives criteria to result in a dose which would have no adverse effects. Based on exposure assessment in a particular situation, one can derive total dose to an individual and compare it with existing criteria. Therefore, for chemicals with existing criteria, one only has to perform exposure assessments to establish possible adverse effects of pollution.

Without exposure to a particular pollutant, there is no risk. Therefore, to use risk analysis in pollution prevention, the most important task is to establish or estimate true potential exposures and then estimate risk for the "maximally exposed individual" or for an average exposure, for example, based on food surveys. The EPA's guidelines for exposure assessment (U.S. EPA, 1986*b*) are useful for deriving real-life exposures. If a company has reliable monitoring data on their pollutants, it should be relatively simple to estimate exposures to potentially exposed individuals. To perform a proper exposure assessment, one needs either to measure the environmental concentrations and/or be able to model the chemical's fate and transport in the environment (bioaccumulation, degradation in the environment, chemical transformation, etc.). For each particular chemical or situation, different sets of parameters may apply. For better exposure assessment, it is also useful to know environmental pharmacokinetics. Substances that easily degrade and do not bioaccumulate are probably of less consequence than persistent compounds, such as DDT, dioxins, and heavy metals.

14.4 Examples of Chemical Risk Analysis

Most of the chemical risk analysis methodologies in use in the United States were developed by the EPA. The National Institute for Occupational Safety and Health (NIOSH), the Occupational Health and Safety Administration (OSHA), and the Food and Drug Administration (FDA) have subsequently also started to use risk analysis for their evaluation of toxic substances (DHHS Committee to Coordinate Environmental and Related Programs, 1985). The EPA has developed methods for dealing with toxic substances that contaminate the environment in general, and NIOSH, OSHA, and the FDA deal with occupational contaminants and food contaminants, respectively.

14.4.1 EPA Risk Analysis

The EPA has a long tradition of dealing with environmental pollutants and has developed criteria and standards for drinking water, ambient water, air total intake reference dose (RfD), reportable quantities (RQs), and levels of concern (LOCs) for many environmental pollutants from various lists of toxic chemicals. These lists, sometimes overlapping, contain over 1200 chemicals and/or chemical categories. Various lists appear in the text of the Resource Conservation and Recovery Act (RCRA), the Comprehensive Environmental Response, Compensation, and Liability Act (CERCLA), and Title III (302 and 313) of the Superfund Amendments and Reauthorization Act (SARA) (U.S. EPA 1992*a*). Based on risk analysis for those pollutants, several types of criteria and standards for various media were derived using EPA-developed guidelines for carcinogen risk assessment, mutagenicity risk assessment, health risk assessment of chemical mixtures, suspect developmental toxicants, estimation of exposures, and systemic toxicants (U.S. EPA, 1986*a*).

Criteria and Standard Deviation.   Initially, the EPA's risk analysis for chemicals was developed in order to derive criteria and standards for chemicals that were polluting waters in the United States (U.S. EPA, 1980). Gradually, risk analysis methods were expanded to all environmental media (U.S. EPA, 1986a). Most of the criteria value are derived from extrapolation from animal studies using assumptions about inhalation, water consumption, food consumption, and weight of average human. The details for criteria derivations and corresponding assumptions are available from the EPA (U.S. EPA, 1986a,b,c).

Ambient water quality criteria (AWQC) were derived in 1980 for priority pollutants (U.S. EPA, 1980). In derivation of these criteria, toxicity in fish and other aquatic organisms, as well as bioaccumulation, was considered.

Health advisories for drinking water (HA) indicate "safe" concentrations of particular chemicals in drinking water for 1-day, 10-day, and subchronic consumption. Usually, these are derived from short-term drinking water studies in rats and mice and application of a proper uncertainty factor (U.S. EPA, 1988b).

Reference dose (RfD), previously known as *acceptable daily intake* (ADI), is defined as total daily dose of the chemical (in milligrams per kilogram of body weight) that would be unlikely to cause adverse health effects even after a lifetime exposure (Barnes and Dourson, 1988). RfDs are established from all available toxicological data for several hundred chemicals, particularly those associated with Toxic Release Inventories (TRIs). The RfDs and risk assessment methodologies used for their derivation are available on-line from the Integrated Risk Information System (IRIS, 1993).

Level of concern (LOC) is defined as concentration of a toxic chemical in air that the general public could endure for up to an hour without suffering from an irreversible health effect (U.S. EPA, FEMA, and DOT, 1987). LOCs are derived from IDLH (immediately dangerous to health and safety) values by dividing them by a factor of 10, or from LD_{50}s by dividing them by 100.

Reportable quantities (RQs) are derived for possible spill reporting. The RQ for a particular substance is 1, 100, 500, 1000, or 5000 lb, depending on its acute toxicity, carcinogenicity, fate and transport in the environment, and reactivity (U.S. EPA, 1987).

Cancer potency slopes (q* slopes) are derived from animal studies using linear multistage analysis (U.S. EPA, 1986c). Cancer potency slope is an indication of magnitude of cancer threat; however, there is a great uncertainty in the accuracy of this number, because of various assumptions made in its derivation (U.S. EPA, 1986c, 1988a).

Reference concentrations (RfCs) for chronic inhalation from air were developed for some chemicals on IRIS (U.S. EPA, 1989). Although for many chemicals air criteria are established based on risk analysis, only six air standards exist (CO, SO_2, O_3, NOX, lead, and particulates) (Casarett and Doull, 1986).

Standards for a chemical in air, water, or soil are derived considering criteria and other factors, such as cost, policy issues, and public perception. Generally, cost/benefit analysis is performed and alternative risks are considered. For example, although chlorination may cause cancer in a small number of individuals, chlorination removes a certain risk of infectious diseases. An outbreak of cholera in Peru led to death from cholera of more than 300 people because officials decided that they did not want to expose the population to chlorine (Anderson, 1991). However, in order to prevent a hypothetical risk of death of 1 in 1 million the officials have introduced far greater risk of cholera, that resulted in an actual death rate of 1 in 1000.

14.4.2 Risk Analysis by Other Institutions

For regulating chemicals in the workplace, OSHA uses *permissible exposure limits* (PELs), that are generally derived from *threshold limit values* (TLVs) developed by the Association of Governmental Industrial Hygienists. Although in 1989 (OSHA, 1989), OSHA had established PELs for over 600 substances, they were thrown out of court and only old, less protective values are now in effect. NIOSH has similarly developed *recommended exposure limits* (RELs) for the same substances (NIOSH, 1990). There was no formal risk assessment initially applied in derivation of either TLVs (and PELs) or RELs, and the numbers were derived based on expert committees (qualitative and semiquantitative risk analysis). Frequently, such TLVs were a compromise between technology and human health protection, not necessarily always protecting human health. The last several years have seen development of epidemiological risk assessment at NIOSH and cancer risk assessment at OSHA, similar to that at the EPA (Stayner, 1992; OSHA, 1989).

14.5 Application of Risk Analysis to Industrial Pollution Problems

In an ideal world, with unlimited resources, we could technically eliminate all industrial pollution and develop closed-loop production systems with no waste generated. However, since our resources are limited, we should concentrate on the pollution problems causing most human health problems and environmental damage. Application of risk analysis to industrial pollution provides an opportunity to derive neutral sets of data and compare the magnitudes of problems in terms of human and environmental health and costs of alternatives. Until recently most decisions regarding pollution were made based on public perception. For example a heavy smoker might tolerate a risk of 1 in 2 of emphysema, cancer, or heart disease but would not tolerate a risk of 1 in 10,000 imposed on him by a factory releasing pollutants. Generally, the public accepts much greater voluntary risks than imposed (involuntary) risks (Slovic, 1987). In the last few years, the EPA and other federal agencies have emphasized risk analysis as a basic tool to determine priority of environmental problems, rather than being pushed into action by the perception of risks (U.S. EPA, 1990).

Figure 14-4 depicts how a particular industrial plant can apply risk analysis at various points of production. Risks due to transportation of raw materials, their

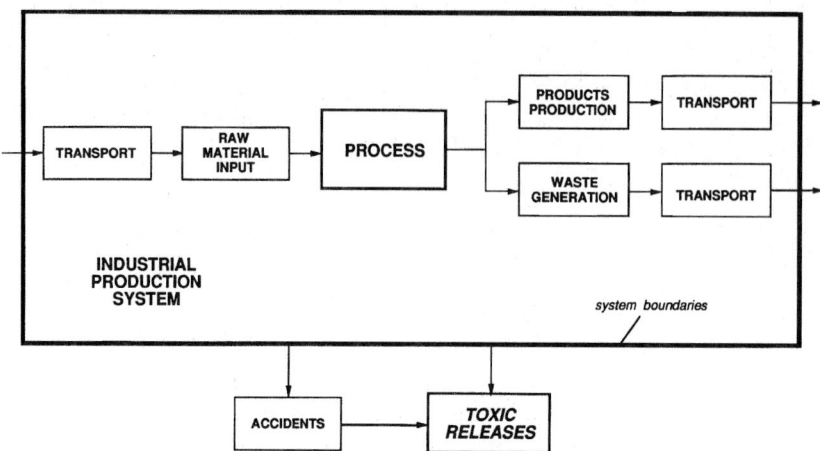

Figure 14-4. Applying risk analysis to industrial production.

input into process, the production process itself, toxic releases and products leaving the production process, and the transport of process outputs can be evaluated using historical data and measures could be taken to decrease risks of accidents and resulting pollution. Within the plant, one can evaluate process safety (and probability of spill and/or accident that would result in pollution) by applying probabilistic risk analysis. Based on this analysis, additional levels of safety may be introduced and/or a change of equipment and process implemented. Based on probable severity of release resulting in air, water, and/or soil/land contamination, one can perform risk assessment of effects of pollutants on human and ecological health. Similarly, "routine" releases can be evaluated using chemical risk analysis or carcinogen risk analysis.

Each industry filing a TRI report has an opportunity to evaluate its practices using risk analysis. From measured concentration of pollutants in air, water, and/or soil one can perform comparative risk assessments for each particular chemical or process, and rank them according to the magnitude of risk (either probability of cancer, or magnitude by which releases exceed RfDs, air criteria, or water criteria). Although the company may not violate any permits, it is conceivable that the potential exposures to toxic chemicals released from the plant exceed safe levels. Each company can perform its own risk analysis. Numerous computer programs exist to simulate and evaluate health risks from chemicals (RISKWARE, 1992). A good starting point for performing chemical risk analysis may be TRI reports, since they represent an inventory of pollution, even if incomplete.

Chemical risk analysis and economic risk analysis can be applied to evaluate various waste disposal practices. Based on such analysis one can set priorities in pollution prevention. A description of risk assessments that could be performed by industries is given next.

14.5.1 Chemical Industry

Since the chemical industry deals with the largest number of toxic chemicals and processes, the application of risk analysis is most appropriate.

1. One could study exposure of workers (from the existing data or from modeling routine releases) to raw materials, intermediates, releases, and final products (occupational safety standards). Based on these studies one could devise reduction of releases of pollutants into the environment, with priority given to those posing greatest risk. Probabilistic risk analysis of the process safety (fault-tree analysis) could be used to pinpoint weak points in the system and thus decrease unintentional catastrophic releases. An attempt should be made to minimize loss of chemicals (releases) during the process and establish safety measures to minimize risk of other types of injuries (e.g., from explosions or falls).

2. Hazardous wastes and/or toxic releases to air, water, and/or soil are other targets for risk analysis because of potential exposure of surrounding populations and ecosystems.

3. Products themselves may be highly toxic and thus may cause problems to the users that may outweigh benefits. However, it may be difficult to persuade a producer to stop producing a chemical that may be harmful and not really necessary (although the market has gotten used to it for various reasons) but that brings financial rewards to the producer.

14.5.2 Other Manufacturing

One can perform risk assessments for any industry involving chemical pollution in the same manner as is done for the chemical industry. Additional analysis may be performed on packaging material from a waste reduction point of view, or using economic risk analysis.

14.5.3 Hazardous Materials Handling and Pollution Control

Another argument for industrial pollution prevention could be made by performing risk analysis of various options for disposal of pollutants, since such disposal has in many cases led to environmental problems of great magnitude (soil contamination at Love Canal, dioxin-contaminated soil in Missouri). Risk analyses of numerous hazardous waste sites indicate the magnitude of the problem, which was caused by the choice of inappropriate industrial waste disposal options. Superfund risk assessments are part of a standardized procedure used in evaluating sites (U.S. EPA, 1992). Potential health and ecological risks associated with hazardous waste handling and disposal methods are

1. Exposure of workers and the public to toxic chemicals (groundwater and surface water contamination).
   a. *Superfund sites.* Numerous studies have been performed both by the EPA and by industry to determine necessity and priority of cleanup of the site (U.S. EPA, 1992*a*). The EPA study is termed a *remedial investigation/feasibility study* (RI/FS).
   b. *Other sites.* Industry was often storing the waste on the premises or partially degrading the toxic waste by using lagoons. Such treatment of hazardous waste often results in site contamination and groundwater and surface water contamination.
2. Potential for generation of other more potent toxic chemicals by municipal solid waste incinerators. One of the by-products of incineration performed at temperatures under 900°C is dioxine, a potent human toxicant and carcinogen (Mukerjee and Cleverly, 1987). Also, incineration of other chlorinated organics results in generation of chlorine, which in itself is highly toxic.
3. Contamination of wastewater systems and waterways. Frequently the largest number of toxic chemicals from a TRI list are released into sewers, eventually reaching publicly owned wastewater treatment works (Molak, 1989). If such releases are done over a short period of time, they may cause die-off of microorganisms and subsequent pollution of waterways.

When pollution is evaluated using a systems approach, one may find that previously used options for treatment of hazardous waste are more risky (from a human health point of view, and often from an economic point of view) than pollution prevention.

14.5.4 Energy Production

Energy production is unfortunately often connected with industrial pollution. An increase of efficiency in energy consumption by industry can be regarded as pollution prevention. Different types of energy production, however, introduce different risks that could be evaluated and quantified.

Although in most cases manufacturing plants do not have an option as to which energy source to use, in some instances they could opt for oil or gas versus coal based either on chemical risk assessment (lowest pollution option) or economic risk analysis or both.

14.5.5 Transportation

Transportation of raw materials, finished products, and, very often, waste by industry also contributes to industrial pollution in the form of

- *Air pollution:* 60 to 70 percent of total air pollutants in any large city are generated by transportation (major pollutants are CO, NOX, organics causing ozone development).

- *Routine and catastrophic spills of toxic chemicals during transport from production to consumers:*   Most frequent risks are posed by overturned trucks carrying gasoline or other oil products and other toxic chemicals.

- *Routine and catastrophic oil spills* (*water and land*):   The *Exxon Valdez* spill alerted the public to the ecological consequences of oil transportation. However, the results of routine spills are even more hideous because total oil released by routine spills is twice that of the catastrophic spills.

Because of the pollution potential of transportation, design of more efficient flow of raw materials, products, and waste that will minimize amounts of materials carried and types of carrier used is a useful form of pollution prevention. For example, per ton of material used, the least polluting and least expensive alternative to trucks in many cases may be trains and/or boats.

14.6 Application of Risk Analysis to a Particular Pollution Source

For each industrial pollution source, one can perform an environmental exposure assessment for the pollutants coming from that source. By comparing the particular exposure with the previously established criteria for that chemical (or criteria derived using the risk assessment guidelines), one can establish possible human health and ecological risks from that industrial source. However, even in cases where the pollutant concentration in air, water, and/or soil does not exceed criteria, one could make a case for pollution prevention based on economic risk analysis. The scheme presented in Fig. 14-4 can be applied to a chemical plant that produces organic chemicals. For example, if a plant X was found to use five processes that may result in catastrophic releases, a fault tree can be constructed for each process and for the plant overall (Shorthouse, 1990–1991). Probabilistic risk analysis can evaluate the likelihood of an accident. Also it can indicate amounts of toxic chemicals potentially released from such an accident. Chemical risk analysis can then evaluate possible adverse effects of such a spill on humans and ecological systems.

The human and ecological risks from routine releases can be evaluated separately based on known data on exposures to and properties of the chemical. One can study both human health risk and economic risk. In many cases it would be difficult to do complete health risk analysis (air modeling, exposure modeling, etc.) for TRI chemicals since TRI data only include total yearly releases. However, in cases where patterns of release are known, one can perform a valid chemical risk analysis and set priorities for pollution prevention.

Economic risk analysis based on TRI data can be used to decrease pollution around industrial facilities, because the TRI enables us to see inefficiencies in the process (i.e., forces the company to evaluate its materials flow). Economic risk analysis has been used with data from TRI to demonstrate economic losses in Hamilton County, Ohio, due to pollution (Molak, 1989, 1991).

Based on a premise that pollution is resources distributed in wrong places, economic losses of wasted resources in Hamilton County were calculated from total TRIs to amount to $25 million (Molak, 1989). Since TRIs report only 5 to 10 percent of released toxic chemicals, the estimated real value of loss is $250 to $500 million, or $250 to $500 per person (Molak, 1991). A rough estimate for the United States based on TRI data is $60 to $120 billion. These numbers indicate that the cost analysis argument for waste reduction and pollution prevention may be compelling (Molak, 1990).

The major objection to that argument is that it would take much more money to recapture pollutants from the smokestacks, wastewaters, or soil than would be saved. This argument is valid only in the old paradigm of pollution control, where pollution is controlled at the end of the tailpipe. It is too late to capture pollutants at the end of the pipe, since at that point the entropy has already increased and large amounts of energy are needed to concentrate chemicals. However, prevention of losses of chemicals in the process itself or recycling of materials would bring about savings.

Often it is difficult to prove that exposure to a given concentration of a particular pollutant causes harmful health effects, but since it cannot do any good, and companies are losing money on raw materials, it is prudent to prevent the pollution at the source.

14.7 Use of Chemical Informatics in Pollution Prevention

In a complex industrial production plant, one way to understand the system under observation is to define its boundaries as well as input and output material flows (Olbina, 1991). A flowchart diagram helps establish points of potential human and ecological health risks due to release of toxic chemicals. A variety of risk assessment software is available which incorporates the risk analysis method described or similar methods (RISKWARE, 1992). Based on the flowchart diagrams and the risk assessments performed, it is possible to find those processes where changes can be made to decrease risk and subsequently pollution. Pollution prevention strategies could be based on priorities obtained by risk analysis. All material and energy inputs and outputs could be accounted for and efficiency in production maximized. Reductions in material losses through releases to air, water, and/or soil could then be regarded as improvement in efficiency or process optimization.

14.8 Conclusion

Application of risk analysis to industrial pollution has demonstrated that

1. Risk analysis can help establish priorities for pollution prevention by first dealing with chemicals and processes that result in the highest human and ecological health risks.

2. Pollution can be regarded as resources distributed in wrong places. The old paradigm of pollution control is economically less desirable than pollution prevention, as can be demonstrated by the application of economic risk analysis.

3. Both an economic case, and often a public health case, for introduction of waste minimization and pollution prevention can be made by using risk analysis.

Bibliography

Anderson, C., "Cholera Epidemic Traced to Risk Miscalculation. Peru Outbreak of Cholera as a Consequence of Faulty Risk Miscalculation," *Nature*, **354**(6351):255, Nov. 28, 1991.

Barnes, D. G., and M. Dourson, "Reference Dose (RfD): Description and Use in Health Risk Assessments," *Regulatory Toxicology and Pharmacology*, **8**:471–486, 1988.

Casarett and Doull, *Casarett and Doull's Toxicology*, New York, Macmillan, 1986.

Covello, V. T., and J. Mumpower, "Risk Analysis and Risk Management: A Historical Perspective," *Risk Analysis*, **5**(2):103–120, 1985.

DHHS Committee to Coordinate Environmental and Related Programs, *Risk Assessment and Risk Management of Toxic Substances. A Report to the Secretary*, Department of Health and Human Services, April 1985.

Felton, W. T., et al., *Man, Medicine, and Work. Historic Events in Occupational Medicine*, Public Health Publ. no. 1044, 1964.

Freeman, H., et al., "Industrial Pollution Prevention: A Critical Review," *Journal of the Air Waste Management Association*, **42**(5):618–656, 1992.

Gore, A., *Earth in the Balance: Ecology and Human Spirit*, 1993.

IRIS, on-line Integrated Risk Information System, 1993. User support telephone: (513) 569-7254.

Johannsen, F. R., "Risk Assessment of Carcinogenic and Non-Carcinogenic Chemicals," *Critical Reviews in Toxicology*, **20**(5):341–366, 1990.

Molak, V., "Waste Minimization and Community-Right-to-Know Law," in *Proceedings of 1st International Conference on Waste Minimization and Clean Technology*, Geneva, May 29–June 1, 1989, ISWA, pp. 404–410.

Molak, V., "Pollution as a Terminal Disease," *Risk Analysis*, **10**(4):605–607, 1990.

Molak, V., "Over $100 Billion/Year Wasted by Industry into the Air, Down the Drain, and into the Countryside of the United States," *Journal of Clean Technology and Environmental Sciences*, **1**(2):155–157, 1991.

Mukerjee, D., and Cleverly, D. H., "Strategies for Assessing Risk from Exposures to Polychlorinated Dibenzo-p-dioxins and Dibenzofuranes Emitted from Municipal Incinerators," *Waste Management and Research*, **5**:269, 1987, p. 28.

National Academy of Sciences (NAS), *Risk Assessment in the Federal Government: Managing the Process*, National Academy Press, Washington, D.C., 1983.

National Institute for Occupational Safety and Health (NIOSH), *NIOSH Pocket Guide to Chemical Hazards*, U.S. Department of Health and Human Services, 1990.

Occupational Safety and Health Administration (OSHA), "Air Contaminants; Final Rule" (Codified at 29 *CFR* 1910), *Federal Register* **54**:2332–2983, Jan. 19, 1989.

Olbina, R., "Computer Supported Modelling of Hazardous Waste Management," Ph.D. dissertation, University of Ljubljana, Faculty of Science and Technology, Ljubljana, Slovenia, 1991, p. 271.

RISKWARE SOFTSTRACTS, presented at Society for Risk Analysis 1992 Annual Meeting. Telephone: (703) 790-1745.

RTECS, Registry of Toxic Effects of Chemical Substances—on-line Toxicology Information Program, National Library of Medicine.

Shorthouse, B. O., "Using Risk Analysis to Set Priorities for Pollution Prevention," *Pollution Prevention Review*, **1**(1):41–53, 1990–1991.

Slovic, P., "Perception of Risk," *Science*, **236**:280–285, 1987.

Stayner, L., "Methodology Issues in Using Epidemiologic Studies for Quantitative Risk Assessment," in *Proceedings of the Conference on Chemical Risk Assessment in the DoD: Science, Policy and Practice*, H. J. Clewell (ed.), ACGIH, Cincinnati, Oh., pp. 43–51, 1992.

Upton, A. C., "Are There Thresholds for Carcinogenesis? The Thorny Problem of Low Level Exposure," *Annals of the New York Academy of Science*, **534**:863–884, 1988.

U.S. Environmental Protection Agency (U.S. EPA), "Water Quality Criteria Documents Availability, Appendix C. Guidelines and Methodology Used in Derivation of the Health Effect Assessment Chapter of the Consent Decree Water Criteria Document," *Federal Register*, **45**(231):79347–79379, 1980.

———, *The Risk Assessment Guidelines of 1986*, EPA/600/8-87/045, August 1987, 1986*a*.

———, "Guidelines for Estimating Exposures," *Federal Register*, **51**:34042, 1986*b*.

———, "Guidelines for Carcinogen Risk Assessment," *Federal Register*, **51**:33992, 1986*c*.

———, Environmental Criteria and Assessment Office, *Health and Environmental Effects Profile for Hexachlorocyclohexanes*, NTIS PB89126585XSP, 1987.

———, Office of Health and Environmental Assessment, *Evaluation of Potential Carcinogenicity of Acrylonitrile*, NTIS PB93181631XSP, 1988*a*.

———, Office for Cooperative Management, *Development of Maximum Contaminant Levels under the Safe Drinking Water Act*, NTIS PB89225619XSP, 1988*b*.

———, *Interim Methods for Development of Inhalation Reference Doses*, EPA/600-8-88/066F, Research Triangle Park, N.C., August 1989.

———, *Reducing Risk: Siting Priorities and Strategies for Environmental Protection*, SAB-EC-90-02, 1990.

———, *List of Lists. Consolidated List of Chemicals Subject to Reporting under the Emergency Planning and Community Right-to-Know Act*, NTIS PB92500792XSP, 1992*a*.

———, Office of Emergency and Remedial Response, *Guide for Conducting Treatability Studies under CERCLA*, NTIS PB93126787XSP, 1992*b*.

———, Federal Emergency Management Agency (FEMA), and U.S. Department of Transportation (DOT), *Technical Guidance for Hazard Analysis*, U.S. Government Printing Office, Washington, D.C., 1987.

15

Profitability Analysis of Pollution Prevention Investments Using Total Cost Assessment

The Bottom Line

Allen L. White
Tellus Institute
Boston, Massachusetts

15.1 Introduction

This chapter focuses on approaches to estimating the profitability of pollution prevention investments. Whether large or small, multiple or single product, complex or simple, all manufacturing firms undertake some form of analysis to evaluate the best use of limited capital resources. For some firms, the process is a formal, annual capital budgeting cycle involving multiple layers of management review and fairly rigid financial performance standards for project approval. For other firms, the process is more ad hoc and episodic, with few, if any, formal approval standards.

Whatever the degree of formality and complexity, sound capital budgeting requires some level of measurement, data compilation, analysis, interpretation, and communication of financial information. How these activities are conducted is critical to the ability of pollution prevention projects to compete on equal footing with alternative projects vying for limited capital resources. These activities may occur through inadequate physical characterization of current processes (that is, materials flows), incomplete cost inventory or allocation procedures, and/or choices of time horizon and profitability indicators which fail to capture all the relevant benefits of a prevention project. If any one or a combination of these conditions prevail, firms will tend to direct investment funds away from prevention toward end-of-pipe projects, or projects for which environmental improvement plays little or no role.

Is there reason to believe that pollution prevention projects do, in fact, confront such obstacles, thereby leading to systematic underinvestment in prevention technologies? A first approximation suggests the answer is yes, in part owing to the historical context of environmental regulation and in part to the nature of prevention investments themselves.[1] To date, the dominant form of industrial environmental projects have been of a compliance, or "must-do," type driven by national or state regulation. This is to be expected, given the historical emphasis on pollution control using prescribed technology standards aimed at controlling the release of pollutants *after* their generation. Such is the case with air pollution control equipment such as scrubbers, filters, and precipitators, as well as wastewater pretreatment and treatment systems. These examples represent one end of a technology continuum. In the middle of this continuum are recycle and reuse technologies, with on-site and in-process recycle and reuse moving in the direction of pollution prevention. At the other end of the continuum are approaches at the core of prevention strategies: process changes, product redesign, and materials substitution.

In practice, of course, boundaries along this technology continuum are blurry. Changes in current practices within a single process line, for example, may involve a mix of process changes and on-site recycling, combined with upgraded wastewater pretreatment prior to discharge to a publicly owned treatment facility. Nonetheless, as firms move further upstream in the production process, and along the technology continuum toward core forms of prevention, a number of repercussions for project financial analysis begin to emerge.[2]

First, upstream changes linked to prevention techniques create, by definition, complex and disparate repercussions midstream and downstream. Changing material inputs from more- to less-toxic substances, for example, necessitates careful consideration of how such changes will affect equipment performance and product quality. In the same vein, product redesign likely requires equipment modifications within a process line. Accompanying these technology adjustments will be a set of capital and operating costs (and savings) which may be far-reaching, indirect, and not immediately evident. Thus, in evaluations of the profitability of such investments, compiling and analyzing operational and capital costs becomes that much more complicated and prone to error and/or omission of potentially significant items. Accurate costing quickly moves beyond the capabilities and data sources of a single staff person such as the environmental engineer, the materials manager, the production engineer, and the financial officer. Contributions from multiple, rather than single, departments will be necessary to assemble such data.

Second, prevention technologies are associated with certain contingent benefits which are difficult to predict and quantify. Avoided liability is the prime example of this situation. The firm examining a potential project aimed at eliminating a hazardous wastestream or emission may be motivated by a desire to eliminate certain risks of litigation linked to personal or property damages. However, such risks are probabilistic in nature; that is, they may materialize if and when an accident occurs or a claimant sues the firm for such damages. This could occur next week, next month, next year, or never. Thus, incorporating the future monetary benefits of such an investment into a project profitability analysis is problematic. They depend on whether, when, and how much liability cost is avoided.

Third, profits from prevention projects may materialize well beyond the two- to five-year time frame commonly applied in investment analysis. Though the savings of some practices—such as improved inventory control—may accrue in the short term, major and more costly changes to processes, materials, and products are likely to extend well beyond the two- to five-year time period. Analytical methods which fail to capture future streams of savings contain inherent biases against clean technology investments. In the competition for limited capital resources, such investments are likely to be rendered noncompetitive with more traditional pollution control projects as well as with those which are nonenvironmental in nature.

A priori, improved cost accounting cannot ensure the improved profitability and competitiveness of pollution projects in the capital budgeting process. However, the preceding suggests the strong likelihood that conventional cost accounting practices are prone to certain biases and may well systematically diminish the competitiveness of such investments as they go head-to-head with other investment options under consideration by the firm. Such conditions are often exacerbated by a historical view of environmental projects in general.

In setting priorities for capital allocation, moderate- and large-size enterprises typically place projects into one of three categories*:

1. Market expansion projects
2. Profit-adding projects (includes cost-reduction projects)
3. Profit-sustaining projects (includes compliance, infrastructural, and maintenance projects)

Typically, market expansion projects are the first priority because they contribute most directly to growth of the firm. The stronger the incentives to create new markets and/or expand existing ones, the more dominant market expansion projects will be in the firm's capital allocation process. In general, this category of projects represents relatively large investments and, in the case of new products, higher levels of risk.

Profit-adding (or cost-reducing) projects are typically projects which enhance the competitiveness of the firm through efficiency gains and cost reduction.

---

*The name and number of categories vary widely across firms; however, differentiating investments either formally or informally according to their strategic value is shared by virtually all businesses.

Equipment purchases which reduce labor costs, upgrade product quality, reduce line losses, or eliminate costly wastestreams fall into this category. Such investments are likely to enhance the market position of the firm for a product it already sells. Though favored by plant managers, profit-adding investments are less favored by growth-oriented managers at higher levels since they serve to enhance competitiveness in existing markets rather than creating new markets for the firm.

Profit-sustaining projects are those targeted at continuing production at existing levels through maintenance, compliance, and/or replacement projects. Managers typically see this category as must-do, or nondiscretionary, projects. It is in this category that environmental investments have traditionally been placed, spurred by the need to comply with a performance or technology standard and perceived as a diversion from new market development or profit-adding alternatives.

The treatment of profit-sustaining projects as net losers is mirrored, and reinforced by, the project justification methods applied to such investments. Under the presumption that returns on the investment are negative, standard profitability analysis is often set aside in favor of simple, short-term analysis of capital and possibly operating cost comparisons across technological options. For example, such a comparison may be developed to determine the least-cost option for complying with a new air toxics standard or a new pretreatment standard imposed by a publicly owned treatment facility to which a plant discharges effluents. Some of these options may be end-of-pipe; others may be prevention oriented. It certainly is the case that prevention-oriented approaches historically have received inadequate attention as an approach to compliance-driven projects.

In recent years, conventional wisdom and, to a much lesser extent, project analysis methods, have shown signs of adjusting to the new environmental reality. This new reality is reflected in a slow shift away from the presumption that every environmental investment is a net loser. The reasons for this shift are several, ranging from the continued price escalation at off-site waste management facilities to the growing consumer demand for clean processes and "green products." In particular, the idea that environmentally friendly processes and products in and of themselves may enhance the market position of the firm has begun to move some managers beyond the narrow confines of the must-do, minimalist, and compliance-driven approach to pollution management, toward a broader vision of translating environmentalism into concrete economic payoffs.

15.2 Conventional versus Total Cost Assessment (TCA)

Progress toward more prevention-friendly project financial analysis methods will require several adjustments to current practices[3]: identification and calculation of a wider range of project costs and savings beyond those conventionally computed, rigorous allocation of costs and savings to processes and products instead of overhead accounts, a longer time horizon for calculation of financial indicators, and utilization of indicators which capture the impact of such horizons. Taken as a package, these changes are referred to as *total cost assessment,* or TCA. TCA is a broader concept than full-cost accounting, a term used by the accounting profession to describe accounting systems to assign all costs to a process, product, or product line, most often for purposes of pricing and cost control. TCA, in contrast, is an investment concept which incorporates full-cost accounting within one of its four elements, cost allocation. Total cost *assessment* is preferable to total cost *accounting* in recognition of the substantial qualitative and discretionary aspects of investment decision making.

A move in the direction of TCA first requires organizational awareness that permits prevention projects to enter the firm's capital budgeting process. Once in the process, the fate of such projects may then hinge to a large degree on the method by which their profitability is calculated. While more rigorous quantitative analysis is no substitute for management's discretion in allocating capital resources, such analysis is a key ingredient to enhancing the competitiveness of prevention projects as they move through the capital budgeting process.

The TCA approach comprises four elements: expanded cost inventories, extended time horizon, use of multiple financial indicators, and rigorous allocation of project costs to processes, products, or product lines.

15.2.1 Cost Inventory

While conventional cost analysis practices generally include only the capital costs directly associated with the investment plus obvious operational costs and savings such as waste disposal and labor, TCA considers a broader range, including certain probabilistic (or contingency) costs and savings. These may be grouped into four categories, given as follows with examples. Table 15-1 contains a more comprehensive list.

1. Direct Costs
   - capital expenditures
     buildings
     equipment
     utility connections
     equipment installation
     project engineering
   - operation and maintenance expenses/revenues
     labor
     waste disposal

Table 15-1. List of Potential Capital Costs

Purchased equipment
  Equipment
  Delivery
  Sales tax
  Price for initial spare parts
  Process equipment
  Monitoring equipment
  Preparedness/protective equipment
  Safety equipment
  Storage & materials handling equipment
  Laboratory/analytical equipment
  Insurance

Materials
  Piping
  Electrical
  Instruments
  Structural
  Insulation
  Building construction materials
  Painting materials
  Ducting materials

Utility connections and new utility systems
  Electricity
  Steam
  Sewerage
  Water
  Refrigeration
  Fuel
  Plant air
  Inert gas
  General plumbing
  Cooling water
  Process water
  Gas connection
  Oil connection

Site preparation
  Demolition and clearing
  Old equipment/rubbish disposal
  Walkways, roads, and fencing
  Grading, landscaping

Construction/installation
  In-house
  Contractor
  Vendor
  Labor
  Supervision
  Taxes
  Insurance
  Equipment rental

Engineering/contractor
  In-house planning
  In-house engineering
  Procurement
  Contractor/consultant
  Design
  Drafting
  Accounting
  Supervision

Start-up/Training
  In-house
  Vendor/contractor
  Trials/manufacturing variances
  Training

Contingency

Permitting
  Fees
  In-house
  Contractor/consultant
  Labor
  Supervision
  Environmental impact studies

Initial charge for catalysts and chemicals

Working capital (raw materials, materials and supplies, product inventory)

Salvage value

Direct materials
  Raw materials
  Catalysts and solvents
  Wasted raw materials costs
  Transport
  Storage

Waste management (materials & labor)
  Predisposal treatment
  On-site handling
  Storage treatment
  Hauling
  Insurance
  Disposal

Utilities
  Electricity
  Steam
  Water
  Sewerage
  Water (cooling or process)
  Refrigeration
  Fuel (gas or oil)
  Plant air and inert gas

Table 15-1. List of Potential Capital Costs (*Continued*)

Direct labor
  Operating labor
  Operating supervision
  Manufacturing clerical labor
  Inspection (QA/QC)
  Worker productivity changes
	Regulatory compliance (materials & labor)
	  Right-to-know training
	  Record keeping
	  Inspection
	  Closure/postclosure care
	  Generator fees/taxes

Other
  Miscellaneous indirect labor
  Maintenance (materials & labor)
  Medical surveillance
	Insurance
	
	Revenues
	  Sale of product
	  Marketable by-products

Regulatory compliance (materials & labor)
  Manifesting
  Reporting
  Monitoring
  Testing
  Labeling
  Training
  Permitting
  Repermitting
	Change in manufacturing throughput
	  Change in sales from:
	    Market share/consumer acceptance
	    Corporate image
	
	Future liability
	  Fines/penalties
	  Personal injury
	  Real property damage
	  Natural resource damage

     utilities: energy, water, sewerage

     value of recovered material

2. Indirect or Hidden Costs

    ■ compliance costs

      permitting

      reporting

      monitoring

      manifesting

    ■ insurance

    ■ on-site waste management

    ■ operation of on-site pollution control equipment

    ■ revenues from sale of pollution credits, e.g., SO_2, NOX*

Costs are indirect or hidden in that they frequently are allocated to overhead rather than to their source (production process or product), or are omitted entirely from the project financial analysis.

*Under provisions of the Clean Air Act Amendments of 1990, emitters of SO_2 may purchase or sell pollution permits, thereby creating a marketplace which, in theory, should lead to efficiencies in meeting national targets. Emitters which can reduce pollution less expensively will do and sell their surplus permits to those facing higher costs to reduce. Markets for NOX and other pollutants also may be created in the next decade as the federal government and the states attempt to establish market mechanisms for meeting air quality standards. The implications for prevention investments thus become clear: projects which reduce tradable pollutants ought to be credited with a revenue stream generated by the creation of salable permits.

3. Liability Costs
   - penalties and fines
   - personal injury and property damage
   - natural resource damages

Liability costs originate from two principal sources: penalties and fines for non-compliance; and legal claims, awards, and settlements for remedial action, personal injury, and property damage due to routine or accidental hazardous releases. Pollution prevention, by definition, reduces or eliminates potential liability costs by reducing or eliminating the source of the risk from the production process. From a financial analysis perspective, liability conceptually is no different from any other operational cost except that it is nonrecurrent; that is, it occurs at a discrete point(s) in time over the life of an investment. For example, an accident or remediation settlement may occur eight years into an investment time horizon *if* a prevention project is not undertaken. The fact that this same prevention project eliminates (or at least greatly reduces) the possibility of a liability cost translates into a savings of $x in Year 8 of a financial analysis for the subject project.

Liability costs are by nature difficult to estimate and equally difficult to locate at a point in the life cycle of a project. By including estimates of future liability directly into a financial evaluation, the analyst introduces considerable uncertainty, which top management may be unaccustomed, or unwilling, to accept as part of a project justification.

Firms currently use several alternative approaches to considering liability costs in project analysis. For example, in the narrative accompanying a profitability calculation, a firm may include a calculated estimate of liability reduction, cite a penalty or settlement which may be avoided (based on a claim against a similar company using a similar process), or qualitatively indicate, without attaching dollar value, the reduced liability risk associated with the pollution prevention project. Alternatively, some firms have chosen to loosen the financial performance requirements (e.g., by raising the required payback period from three to four years, or lowering the required internal rate of return from 15 to 10 percent) of the project to account for liability reductions.[4]

For publicly traded companies in the United States, liability estimation is controversial because the Securities and Exchange Commission requires firms to report liabilities to stockholders and accrue assets to cover these future costs. Also, a liability estimate may be damaging to a firm if it is made public in a legal proceeding. For all these reasons, if firms consider liability costs in project analysis, they normally exercise substantial caution in assigning a quantitative estimate of liability to a specific investment.

4. Less Tangible Benefits
   - increased revenue from enhanced product quality
   - increased revenue from enhanced company and product image

- reduced health maintenance costs from improved employee health
- increased productivity from improved employee relations

A pollution prevention project may deliver substantial benefits from an improved product and company image or from improved employee health. These benefits, like liability, are difficult to predict and estimate. A TCA analyst may find a qualitative analysis more appropriate and salable to management.

15.2.2 Cost Allocation

A firm's cost accounting system traditionally serves as a tool for tracking and allocating costs to a product or process line, principally for operational budgeting, cost control, and pricing. In a TCA context, careful allocation serves to illuminate exactly which processes and products are responsible for generating pollutants and, as a result, create the aforementioned wide spectrum of direct and indirect costs whose reduction is the goal of pollution prevention.

For purposes of investment analysis, the ideal cost accounting system has two primary features.[5] First, the system should allocate all costs to the processes which are responsible for their creation. This is a perennial challenge to financial officers and cost accountants who decide on the placement of costs into either overhead or product or process accounts. Waste disposal costs, for example, are often placed in overhead accounts, while a process or product allocation would assign such costs based on some activity or component of the manufacturing process.

Second, it is not enough to simply allocate costs to appropriate processes. Costs should be allocated in a manner that is reflective of the way in which costs are actually incurred. For example, waste disposal costs in some companies are allocated across operating centers—administrative, research and development, and manufacturing—on the basis of floor space rather than on the quantity and type of waste generated by each. This impedes a rigorous estimation of the financial benefits of reduced waste generation. Thus, effective cost allocation directs management attention to the sources of waste generation and to the potential savings of waste-avoidance measures.

15.2.3 Time Horizon

In addition to a more comprehensive cost inventory and more precise allocation of costs, a third feature of TCA is its longer time horizon. The reason this is critical is tied to the nature of prevention technologies themselves—by nature, they take longer to begin showing profitability than conventional end-of-pipe management strategies. Conventional project cost analysis typically considers the investment in a three- to five-year time period. Retaining this horizon runs the risk of losing sight of the costs and benefits which TCA is designed to capture. For this reason, an extended time horizon of five to ten years or more is integral to the effective application of TCA.

15.2.4 Financial Indicators

To consistently provide corporate decision makers with accurate and comparable project financial assessments, indicators of financial performance must meet two essential criteria: (1) they must consider all cash flows (positive and negative) over the life of the project; and (2) they must consider the time value of money, i.e., they must appropriately discount future cash flows. The net present value (NPV), internal rate of return (IRR), and profitability indicator (PI) methods meet both these criteria.

Where projects compete for limited resources, the NPV method is preferred because there are certain conditions under which the IRR or PI methods fail to identify the most advantageous project. The payback method, commonly used by small companies, does not meet either of these criteria. The following describes each of these indicators in more detail.*

Net Present Value (NPV).   Under the NPV method, the present value of each cash flow, both inflows and outflows, is calculated and discounted at the project's cost of capital. The sum of the discounted cash flows is the project's NPV. A positive NPV means a project is worth pursuing; a negative NPV indicates it should be rejected. If the availability of capital is constrained (as it usually is) or several projects are competing with one another, other things being equal, the project or combination of projects with the highest positive NPV should be chosen. The NPV method, particularly as applied to long-term projects with significant cash flows in later years, is very sensitive to the level of the discount rate. Thus, for a project with most of its cash flows in the early years, its NPV will not be lowered much by increasing the discount rate. On the other hand, the NPV of a project whose cash flows come later will be substantially lowered, rendering the project a much less attractive investment opportunity.

Internal Rate of Return (IRR).   The IRR method calculates the discount rate which equates the present value of a project's expected cash inflows to the present value of the project's expected costs. Thus, the basic formula to calculate the IRR is the same as that for the NPV; for the IRR, the NPV is set to zero and the discount rate is calculated; for the NPV, the discount rate is known and the NPV is calculated. A project is worth pursuing when the calculated IRR is greater than the cost of capital to finance the project. Where several projects are

---

*These are conventionally calculated in most project financial spreadsheets. Two sources which describe and provide guidance for calculation of financial indicators specifically for pollution prevention projects are: *P2/FINANCE User's Manual*, Tellus Institute, Boston, 1993; and Northeast Waste Management Officials Association (NEWMOA), *Costing and Financial Analysis of Pollution Prevention Projects: A Training Packet*, Boston, 1992.

vying for limited resources, all else being equal, the project with the highest IRR should be pursued.

Profitability Index (PI).   The profitability index is also known as the *benefit/cost ratio*. The PI is simply the present value of benefits (cash inflows) divided by the present value of costs (cash outflows), and shows the relative profitability of a project, or present-value benefits per dollar of costs. Projects with profitability indices greater than 1.0 should be pursued, and the higher the PI, the more attractive the project.

Payback.   Payback is the simplest of the techniques for evaluating capital project investments. It provides a quick, "back-of-the-envelope" appraisal of the financial prospects of a project. While the payback calculation may suffice for a preliminary assessment, it should not be relied upon as the sole method for project evaluation. The payback period is the expected number of years required to recover the original project investment. The payback period can be calculated before or after taxes, and serves as a type of "break-even" calculation. If cash flows come in at the expected rate until the payback year, then the project will break even from a dollar standpoint. However, the regular payback does not account for the cost of capital, so that the cost of the debt and equity used in the investment is not reflected in the cash flows or the calculation. Another major drawback of the payback method is that it does not take account of cash flows beyond the payback year. The payback period does, however, provide an estimate of how long funds will be tied up in a project and is therefore often used as an indicator of project liquidity.

15.3   Three Examples

How do the various elements of TCA affect the bottom-line profitability for proposed pollution prevention investments? Three projects, drawn from actual projects recently considered by three firms, illustrate how TCA provides a clearer picture of the financial viability of such investments. These projects are: (1) in a paper coating mill, conversion from a solvent/heavy metal coating technology to one which is aqueous/heavy metal free; (2) in a metal fabrication firm, installation of a paint-water separator to reduce materials purchases and hazardous waste generation; and (3) in a diversified chemical company, a by-product recovery project to recover and reuse waste materials. In each case, the conventional (company) analysis is compared to the TCA for the same project, thereby demonstrating how the two differ.

15.3.1   Cost Inclusion

While cost categories considered in a financial analysis will tend to differ according to the nature and scale of the project, Table 15-2 presents an overview

Table 15-2. Cost Inventory: Company versus TCA Analysis

X = cost(s) included
P = cost(s) partially included

	Project 1[1] Company	TCA	Project 2[2] Company	TCA	Project 3[3] Company	TCA
Capital costs						
Purchased equipment	X	X	X	X	X	X
Materials (e.g., piping, elec.)					X	X
Utility systems		X				
Site preparation						
Installation				X		
Engineering/contractor	X	X			X	X
Start-up/training	X	X		X		
Contingency					X	X
Permitting						
Initial chemicals						
Working capital			X	X		
Salvage value						
Other:						
Project audit					X	X
Operating costs						
Direct costs:[4]						
Raw materials/supplies	P	X	X	X	X	X
Waste disposal	P	X	P	X	X	X
Labor	X	X	X	X	X	X
Revenues—general					X	X
Revenues—by-products						
Other						
Transportation					X	X
Indirect costs:[5]						
Waste management						
Hauling		X		X	X	X
Storage		X				
Handling		X				
Waste-end fees/taxes		X				
Hauling insurance				X		
Utilities						
Energy		X	X	X	X	X
Water		X		X		
Sewerage (POTW)		X		X		
Pollution control/solvent recovery		X				
Regulatory compliance		X		X		
Insurance						
Future liability		X				X

1. Solvent/heavy metal to aqueous/heavy metal-free coating conversion at paper coating company.
2. Paint/water separator at metal fabrication company.
3. By-product recovery project at diversified chemical company.
4. We use the term *direct costs* to mean costs which are typically allocated to a product or process line (i.e., not charged to an overhead account) and are typically included in project financial analyses.
5. We use the term *indirect costs* to mean costs which are typically charged to an overhead account and typically not included in project financial analyses.

of the costs included in three sample projects for which detailed comparisons of company versus TCA analyses are made. The TCA column represents a complete picture of the internal costs and revenues affected by the project. By comparing the company analysis column with the TCA column, one can see a picture of the firm's project costing approach emerging.

Direct and Indirect Costs.   In the case of Project 1, the paper coating firm omitted all nondisposal waste management costs, utilities (energy, water, and sewerage), solvent recovery, and regulatory compliance costs from its analyses of the aqueous conversion project. The firm also omitted several costs associated with the storage needs and shorter shelf life of aqueous coatings—namely, a steam heating system for the coating storage shed, lost raw material value, and cost to dispose of spoiled coatings.

The engineer at the metal fabrication company did not include installation and training in her capital estimate for the paint/water separator, and omitted waste hauling, insurance, water, and sewerage costs from her annual operating cost estimate. While she considered the avoided cost of disposing of the paint/water and oil/water waste, she did not include the cost to dispose of the sludge generated by the separator.

The financial analyses developed by the firm for Project 3 were extremely detailed and comprehensive, leaving little room for improvement for a TCA analysis. In other words, the firm demonstrates project analysis methods which, from a cost inventory standpoint, conform well with a TCA approach. Since the wastestream targeted for recovery is not a regulated (RCRA) hazardous waste, it is not subject to manifesting, monitoring, reporting, and other regulatory requirements. Therefore, regulatory and nondisposal waste management costs are not relevant to either the company or the TCA analysis.

Future Liability Costs.   Liability may occur in two general forms: liability from personal injury or property damage (e.g., from a leaking landfill), and penalties and fines for violation of environmental regulations. None of the companies included an estimate of avoided future liability costs in their own financial analyses. In the case of Project 1, the paper coating firm alluded to this benefit in a qualitative way in its appropriations request, as follows: "...major reductions in levels of fugitive emissions, and amounts of solid hazardous waste going to landfill, are very positive from a regulatory and community standpoint."

The environmental engineer at the metal fabrication company did not include an estimate or mention of future liability in her analysis of Project 2. By complying with state and federal hazardous waste regulations and sending all waste to an incinerator, she feels that the company is safe from future financial liabilities from fines, personal injury, and property damage. In a previous project, she analyzed the direct economic benefits of a new paint procurement policy in light of a state regulation which contained a set of fines for improper shipping of hazardous materials. Since the company was not meeting these requirements, she clearly stated that they could be fined up to $25,000 after two offenses. If incurred, the fine would far outweigh the long-term savings in operating costs for the new system. The decision to suspend the new procurement practice was based in large measure on the threat of fines.

The diversified chemical company financial analysts are cautious in valuing indirect and intangible benefits of a project, choosing instead to calculate only—and thoroughly—conventional, direct costs while leaving TCA-type benefits, such as avoided liability, for qualitative consideration in a project appropriations request. Although the company has developed a procedure for estimating costs associated with Superfund liability, it did not include either an estimate or mention of avoided future liability in any financial analyses for Project 3. As Table 15-2 shows, this has been added to the TCA analysis in a fashion viewed as realistic by the firm's managers.

While no firm included estimates of avoided future liability in their quantitative project analyses, liability avoidance does remain a major concern of United States firms (and a growing concern of European firms as well). Many of the companies interviewed are looking to pollution prevention to minimize the likelihood of liability associated with hazardous waste generation. However, the availability of incineration as a disposal option clearly gives many of these firms a level of comfort which may act as a disincentive to pollution prevention.

Less Tangible Benefits.   Less tangible benefits from pollution prevention investments, such as increased revenue from enhanced product quality, improved company or product image, increased worker productivity, and reduced worker health maintenance costs are among the most difficult to predict and quantify. None of the three company analyses or TCA analyses contain estimates of less tangible benefits. In the case of Project 1, the coated paper product is sold domestically, on the basis of cost, visual appearance, and performance durability, to book publishers and other intermediate product manufacturers. Although the company expects some quality improvements using aqueous coating, it does not anticipate an increase in market value. Therefore, it expects no increase in domestic sales as a result of the conversion to the aqueous/heavy metal-free coating. The company hopes to improve its competitive advantage in the European market if the European Economic Community implements lead-free packaging standards (which would apply to books) as expected. However, it would not speculate on the potential revenue effects associated with increased European market share.

Neither the metal fabrication company nor the diversified chemical company expects to increase market share or product value as a result of Projects 2 and 3, respectively. Both of these companies, as well as the paper coating firm, are manufacturers of intermediate, rather than consumer, products and cannot directly market their products on the basis of environmental performance in the same way that a consumer products manufacturer (such as Procter & Gamble) can.

A reduction in solvent use at the paper coating firm will certainly reduce worker exposure to fugitive solvent emission and the elimination of nitrocellulose from the coating mixture will reduce flammability and explosivity hazards. While reduced solvent exposure may result in a lower incidence of worker illness over the long term, and the elimination of nitrocellulose may result in fewer worker injuries, we had neither the information nor the resources to estimate the potential impact of these benefits on either the company's health care costs or long-term worker productivity. In this case, this issue was dealt with qualitatively in a section of an appropriations request called "Safety/Health Impact of Converting from Solvent to Aqueous Coating." The section listed specific project benefits that will improve safety and industrial hygiene, such as:

"Reduce risk of fire in chemical storage, mixing and coating areas."

"Minimize employee physical activity and fire risk when loosening and removing nitrocellulose from drums."

"Minimize employee exposure to organic vapors, reducing health risks and need for IH monitoring and record keeping."

"Minimize odor complaints in Mill and the administration building when retained solvents are released during converting or solvents are used to clean converting equipment."

Although many company representatives believe that project benefits are more persuasive if they are monetized and included in the project financial analysis, when costs are difficult or impossible to monetize, a qualitative approach may be more credible to management.

Discovery of Previously Omitted Nonenvironmental Costs. In developing the TCA analyses for the three projects analyzed in-depth, we added to the company analyses any capital or operating costs or savings which could be attributed to the project and reasonably estimated. While our focus was on environmental costs typically omitted from project analyses, the process of developing a more comprehensive list of costs unearthed, in the case of Projects 1 and 2, other, "nonenvironmental" costs which were not originally included by the company. For example, in Project 1, all previous analyses of the aqueous/heavy metal-free conversion ("aqueous coating conversion") had omitted the costs of heating system installation, the energy needed to prevent the aqueous coating from freezing, and the additional energy needed to dry aqueous versus solvent-based coating. While the latter cost was acknowledged by several production engineers and managers, it had never been estimated nor included in previous analyses. In the case of Project 2, equipment installation and operator training costs (while admittedly small) were not included in the company's initial analysis. These items, which tended to increase the cost of the prevention project, were included in the TCA analysis. While probably not surprising to most project analysts, the TCA analyses of Projects 1 and 2 led to a general finding that nonenvironmental direct and indirect costs may also be left out of project analyses.

15.3.2 Profitability Comparisons

A comprehensive TCA should include direct costs, indirect costs, liability costs, and less tangible benefits. It requires evaluation of project costs and savings over a longer time horizon and the use of profitability measures which reflect the long-term profitability of the project. The following considers how each of these elements affects the financial analyses of the three projects in a conventional versus TCA context.

Effect of Cost Inclusion on Financial Indicators.   In all three project analyses, the inclusion of waste management, regulatory compliance, future liability, and other previously unquantified costs in the TCA analyses results in a net improvement in project cash flows and financial indicators as compared to the company analyses. The degree of improvement varies widely, depending on the complexity and cost of the project, as well as the degree to which TCA practices are already operational. Table 15-3 summarizes these results.

The magnitude of the effect, as illustrated by the percent change in IRR (years 1 to 15) ranged from a low of −5 percent for Project 1 to 3 percent for Project 2. The 15-year NPV difference is more dramatic, ranging from a low of −94 percent for Project 1 to a high of 33 percent for Project 2. Only in the case of Project 2 is payback significantly reduced, from 4.3 to 3.8 years, moving the project closer to the range of hurdle rates normally sought by firms who use this financial indicator. Overall, the degree of change is determined principally by the magnitude of the aggregated annual operating costs and benefits added to the TCA, and to a lesser degree by the addition of capital costs which were not accounted for in the company analysis. With respect to future liability, the impact of this cost on a financial indicator depends on the size of the estimated liability, the projected year in which it will occur, and the discount rate used in the calculation of NPV and IRR. Each of these is discussed as follows.

Future Liability Costs.   A savings of $35,000 in avoided liability costs appears in Project 1. In computing this figure, we used a component of a method first developed by General Electric (GE) for their internal project evaluation process.[6] One key variable in this method—adjustment of the liability cost estimate according to the waste management technology used—causes all waste destined for incineration to be heavily advantaged from a cost reduction standpoint under the assumption that incinerated waste is unlikely to result in downstream remedial action costs. Since the paper coating firm incinerates its waste, the modest savings which appears in this project analysis flows from this assumption as well as from the diluting effect of discounting future savings (liability exposure is assumed to occur in year 13). The net effect is to yield only a modest 1 percent difference in the 10- versus 15-year IRR in the TCA analysis.

Table 15-3. Summary of Financial Data for Projects 1, 2, and 3

	Project 1			Project 2			Project 3		
	Company analysis	Total cost assessment	Difference	Company analysis	Total cost assessment	Difference	Company analysis	Total cost assessment	Difference
Total capital costs	$893,449	$923,449	3%	$19,659	$19,733	0%	$4,961,251	$4,961,251	0%
FINANCIAL INDICATORS									
Annual savings (BIT)†	$118,112	$79,127	-33%	$4,583	$5,234	14%	$2,293,600	$2,293,600*	0%
Net present value—years 1–10	($314,719)	($480,512)	-53%	$3,860	$6,227	61%	$6,651,492	$8,250,053	24%
Net present value—years 1–15	($203,643)	($395,625)	-94%	$9,332	$12,436	33%	$10,035,274	$11,633,835	16%
IRR—years 1–10‡	6%	0%	-6%	17%	20%	3%	38%	39%	1%
IRR—years 1–15‡	11%	6%	-5%	20%	23%	3%	40%	41%	1%
Simple payback (years)	7.6	11.7	54%	4.3	3.8	-12%	2.2	2.2	0%

*Does not include estimated future liability cost.
†Before interest and taxes.
‡Differences are expressed in percentages.

Only liability distinguishes the company and TCA analyses for Project 3.[7] For this project, the firm's own liability estimation procedure was used. Though the liability estimate was not insubstantial—$4,615,065—the result shows the 15-year IRR increases by a mere 1 percent, and the 15-year NPV by 15 percent, again a demonstration of how discounting over a long time horizon dilutes the effect of even substantial benefits from liability avoidance.

Both the methodology developed by GE and that used by the diversified chemical company base their estimation of avoided liability on the quantity of waste reduced by the pollution prevention project. While this approach provides a rational basis for linking magnitude of financial liability to waste quantity, it does not necessarily reflect the way in which liability costs for waste disposal sites are assigned among potentially responsible parties (PRPs) under the joint and several liability provision of CERCLA. Under this provision, a PRP may be assessed a share of the cleanup cost that is disproportional to the amount or degree of hazard of the waste sent to the site. This may occur if a particular PRP has sufficient financial resources and others do not.

The GE liability estimation procedures assume that incineration operates as a powerful safety net, minimizing future damages against the firm. This is based on the underlying assumption that because incineration facilities do not directly bury waste (they do produce ash which must be buried), potential liability for generators is virtually eliminated "up the stack." While there are few cases of generator liability stemming from second-party incineration, there have been cases where generators were held responsible for damage resulting from improper storage of wastes at commercial incineration facilities. Most incineration sites include waste storage systems. In addition, advances in risk assessment, and the evolution of regulations and legal precedents may eventually lead to long-term personal injury or property damage claims on behalf of incinerator neighbors against the generators that supplied the waste.

Not considered in our analyses are the questions and potential cost implications of future availability of incineration capacity and continued allowance of interstate shipping of hazardous waste. In view of recent waste market conditions, ongoing site investigations, and court rulings which have upheld taxes and surcharges on waste imports, capacity shortfall and interstate transportation barriers are plausible potential constraints facing any waste generator during the next decade. Given this situation, inclusion of some measure of uncertainty into the liability avoidance calculation to reflect these conditions could alter the outcome and may lead to a more substantial savings from any proposed pollution prevention investment in which off-site waste disposal is a current cost.

Less Tangible Benefits.   Less tangible benefits were not included in any of the TCA project analyses. We earlier defined these benefits as gains in market share owing to enhanced corporate or product image, increased worker productivity owing to a safer workplace, and similar long-term payoffs reasonably ascribed to pollution prevention investments. These may stand on their own or be tied to other more concrete benefits such as liability avoidance. In this case, "staying out of the newspaper," to quote one company, may be one measurable gain from such an investment.

Impact of Previously Omitted Nonenvironmental Costs. In the previous discussion of the qualitative effects of TCA analysis, we mentioned that the process of expanding the list of environmental costs may actually turn up previously omitted nonenvironmental costs. The effect of such costs on the project's financial performance depends on whether the item represents a cost or a savings for the project. In the case of Project 1, nonenvironmental costs tended to reduce the profitability of the project by adding to capital and operating costs. For example, capital requirements (a direct cost) for the Project 1 TCA were 3 percent higher than for the company analysis. Including steam costs in the TCA (more steam will be needed to dry aqueous-coated paper under the pollution prevention scenario) contributes to reducing the IRR (years 1 to 10) by 94 percent.

While this finding is probably no surprise to those who prepare project analyses, it is important to point out that *the TCA process may actually reveal additional costs as well as savings for the project*. If the financial benefits identified in a TCA analysis by including regulatory compliance or waste management activities are marginal, they may be negated by the addition of one or two previously omitted nonenvironmental costs.

Are All Conventional Costs Created Equal? By reducing waste, fewer drums need to be handled, labeled, and stored, and fewer manifests need to be prepared and managed. While these cost savings can be estimated, they are only realized "on the books" of the company if, in the case of labor, the payroll is reduced or if a worker is given productive activities as opposed to nonproductive waste management labor. In the case of waste storage, estimated cost reductions are realized if the company can either put that storage space into productive use, reduce space heating and lighting costs, or reduce rental costs on rented storage space. In contrast to waste disposal costs, which are often visibly reduced when waste generation is reduced, estimated labor, waste storage, and other similar cost reductions may or may not be realized on the firm's ledgers. As the director of the division technical center at the diversified chemical company noted, business managers tend to be skeptical of the real value of these "paper" savings until they can be convincingly documented.

This issue has clear implications for TCA. Where a majority of savings identified by a TCA analysis may be labor-related, the results of a TCA analysis are likely to be less persuasive than for a project with a majority of savings in disposal, raw material, and energy cost reductions. For example, the environmental engineer at the metal fabrication company, a firm of approximately 250 people, saw no opportunity for real cost reductions through reduced person-hours spent labeling and inspecting fewer drums for incorporation into Project 2. As a salaried employee that regularly works 60 hours a week, she knows that the only marginal savings from these effects will be in the amount of overtime she will need to work. If hourly labor were responsible for such tasks, however, reduced labor requirements for on-site waste management may be a significant benefit to a pollution prevention investment.

Financial Indices and Time Horizon. To examine the effect of the choice of financial indicators and time horizon, we create two functional categories of indices: discounted cash flow methods which consider a stream of future cash flows for the investment (e.g., NPV and IRR), and those that do not (e.g., simple payback period and return on investment).

By extending the time horizon of the investment analysis to 10 or more years, the effect of long-term costs or savings can be analyzed. Discounted cash flow methods are well-suited to long-term investment analysis since they consider the time value of money, and can easily include tax, depreciation, and inflation effects. On the other hand, payback period and ROI typically consider only the average annual or net annual cash flow and capital costs for the investment, thereby overlooking costs or savings which may be realized in out-years.

While discounted cash flow measures and reasonably long time horizons for investment analysis are decidedly advantageous to evaluating pollution prevention investments, their actual effect depends on whether the analysis includes costs or savings in years after the initial investment. If long-term costs are included in the analysis, an extended time horizon and use of discounted cash flow indicators are essential. For example, the payback periods calculated for the Project 3 company analysis and for the TCA analysis are identical—2.2 years—because the payback calculation ignores the avoided future liability cost entered in year 10 of the financial analysis.

Even if a 10- or 15-year time horizon is chosen, and discounted cash flow indicators are used, the outcome of the financial analysis will be greatly influenced by the magnitude of future costs (relative to first-year costs), the year in which future costs accrue, and the discount rate used in the analysis. If the magnitude of future costs is small relative to first-year costs, then, despite the use of a long time horizon and NPV or IRR, the effect of these long-term costs on financial indices will be small. Thus, the inclusion of an avoided liability cost of $35,000 in the project 1 TCA had virtually no effect on 10- versus 15-year IRR. Even if the future cost is high, by discounting the cost to present-year dollars, the impact of this future cost is diluted. In the case of Project 3, though the estimate for future liability cost is significant ($4.6 million), because it is heavily discounted relative to first-year costs, the effect on IRR is negligible—only a 3 percent gain over the company analysis.

By performing sensitivity analyses on the TCA for Project 3, we can test the effect of financial indicators on the timing of future costs and choice of discount rate. If liability costs were predicted to materialize in year 5 rather than year 10, the IRR (years 1 to 15) for the TCA would be 44 percent rather than 40 percent. And if a discount rate of 10 percent rather than 12 percent is used, the NPV would be $13.6 rather than $11.6 million.

Once a firm accepts in principle the idea of including liability in its financial analysis, prudence would suggest the execution of this type of sensitivity analysis in order to establish reasonable ranges of savings from a pollution prevention investment. This task is made more necessary by the many uncertainties in waste management conditions, and by the assumptions built into the various liability estimation methodologies. Of course, this is one reason why firms have been reluctant to either quantify liability or include such savings items in their financial analyses. A firm willing to quantify, but stopping short of incorporation of a figure into a project financial analysis, will still have taken a step in the direction of prudence.

15.3.3   Cost Allocation

When costs for waste management and regulatory compliance are properly allocated to the production processes or product lines which are responsible for their creation, managers have a clearer picture of the full costs of running a process or producing a product and project analysts have a better source of cost data for project financial analysis. Reallocation of these costs from overhead accounts to cost or profit centers, however, comes at a price. The price is both the one-time administrative expense of reorganizing the firm's cost accounting system, and the recurrent expense of charging these costs to several or many different accounts as opposed to one central overhead account.

Even when firms do allocate waste management costs to facility operations (versus overhead accounts), their procedures may not reflect accurately how such costs are incurred.[8] The paper coating firm, for example, allocates waste solvent recovery costs to product lines according to quantity of coating used, not according to quantity of waste solvent produced. So, within the existing internal costing system, the product line is not credited with reductions in solvent recovery costs which are realized through reductions in waste solvent generation. A pharmaceutical company (not included in these project evaluations) allocates waste disposal costs to administrative, R&D, and manufacturing facilities at one site according to square footage rather than according to the types and quantities of wastes produced in each operating area. Therefore, the economic signals and incentives that could be provided by an accurate allocation process are not recognized.

One large firm attempts to allocate all costs—including waste management—to product lines according to labor. Environmental staff have initiated efforts to include waste disposal costs into raw material acquisition pricing such that procurement decisions are made considering both purchase and disposal costs.

15.4   Conclusions

The basic appeal of TCA stems from a simple fact: it offers the opportunity to translate subjective or discretionary information on pollution prevention benefits into the language that management understands and will respond to.

At the most general and intuitive level is the desire to protect the firm from future regulatory actions, the scope of which can never be fully predicted in the rapidly evolving and often unpredictable world of environmental regulation. For managers with even a medium-term outlook, pollution prevention makes sense as a vehicle for "getting out from under" the uncertainties and costs of new and more stringent regulations. For these managers, TCA is a tool for fortifying the case for prevention investments as they move through the firm's capital budgeting process.

Closely coupled with this perspective is consistent mention of liability and liability avoidance as a powerful force in reducing or eliminating waste generation. Virtually all firms recognize the potential costs which waste generation and disposal represent. For large firms involved in Superfund cases, the knowledge is first-hand and the costs already may have amounted to millions of dollars. For others yet to be identified as responsible parties in site cleanup, the mere prospect is daunting and, in at least one case, has led the environmental manager to ship all waste to incineration facilities as a protective measure. The continued tension between the realities of liability and the omission of quantitative estimates of liability avoidance benefits in project financial analysis is a strong selling point for a TCA approach.

For our three sample projects, comparing the financial indicators from the company versus the TCA analyses shows that projected project profitability is improved by using the TCA approach. Using a mix of a more expansive cost inventory, extended time horizons, and financial indicators more sensitive to a longer time frame, the TCA analysis reveals that TCA produces advantages in the sample pollution prevention projects.

Are these advantages enough to move managers to adopt TCA methods? What role do organizational issues play in this process? This will depend on many factors including the size of the firm, the organizational structure of the firm, and on a number of regulatory barriers and incentives from federal and state governments. Our discussion of existing capital budgeting practices in case study firms indicates that formalized processes used in capital decision making at larger firms are favorable for the adoption of TCA approaches. Where several or many projects are competing for capital, where projects are evaluated over relatively long time horizons, and sensitive financial indicators are regularly employed, the conversion to a TCA approach could be relatively straightforward. On the other hand, small and medium-sized firms which tend to rely on informal capital budgeting processes and short-term analyses of investment options may find only one or two aspects of TCA useful (e.g., expanding the cost inventory).

From an organizational perspective, firms with strong top-level mandates to eliminate pollution, effective champions at the staff level and, for intermediate product manufacturers, close linkages to other companies with strong environmental records are likely to be most receptive to TCA concepts. Furthermore, the trend toward increasing levels of external pressure from both predictable and unpredictable regulatory and courtroom developments, and opportunities to exploit emerging "green" markets may exercise influence over a firm's view of the benefits of TCA.

References

1. A. White, *Accelerating Corporate Investment in Clean Technologies through Enhanced Managerial Accounting Systems,* Report prepared by Tellus Institute for the Organization for Economic Co-Operation and Development (OECD), Programme on Technology and Environment, January 1993.

2. A. White, M. Becker, and J. Goldstein, *Alternative Approaches to the Financial Evaluation of Industrial Pollution Prevention Investments,* Report prepared by Tellus Institute for the New Jersey Department of Environmental Protection and Energy, November 1991, Revised Executive Summary, June 1993; A. White, M. Becker, and J. Goldstein, *Total Cost Assessment: Accelerating Industrial Pollution Prevention through Innovative Project Financial Analysis, With Application to the Pulp and Paper Industry,* Report prepared by Tellus Institute for U.S. EPA, Office of Pollution Prevention, December 1991, Revised Executive Summary, June 1993.

3. A. White, M. Becker, and D. Savage, "Environmentally Smart Accounting: Using Total Cost Assessment to Advance Pollution Prevention," *Pollution Prevention Review,* Summer 1993, pp. 247–259. U.S. Environmental Protection Agency Risk Reduction Engineering Laboratory, *A Primer for Financial Analysis of Pollution Prevention Projects,* EPA/600/R093/059, April 1993.

4. U.S. Environmental Protection Agency, Hazardous Waste Engineering Research laboratory, *Waste Minimization Opportunity Assessment Manual,* EPA/627/7-88/003, July 1988.

5. R. Cooper and R. Kaplan, "Profit Priorities from Activity-Based Costing," *Harvard Business Review,* May–June 1991, pp. 130–135.

6. General Electric Company, Corporate Environmental Programs, *Financial Analysis of Waste Management Alternatives,* Stamford, Conn., 1987.

7. A. White, M. Becker, and J. Goldstein, *Alternative Approaches to the Financial Evaluation of Industrial Pollution Prevention Investments, op. cit.*

8. R. Cooper and R. Kaplan, *op. cit.*

16

Accounting for the Environment

Environmental Cost Profiling

Rebecca Todd
New York University
New York, New York

16.1 Introduction

From the point of view of business, environmental wastes represent lost resources, forgone profits, and increased risk to employees, suppliers, investors, and creditors. However, the range of environment-related costs greatly exceeds those traditionally associated with end-of-pipeline waste and effluents. For example, in many firms, most if not all employees spend at least some of their time with environment-related matters including environmental monitoring, regulatory compliance and reporting, environment-driven legal and legislative affairs, engineering and research and development, marketing and public relations, and auditing of suppliers' environmental activities. Companies, in general, regard environmental vigilance as a proper and desirable part of every employee's job profile. However, in recent years, managers are finding that such efforts are consuming increasingly large portions of employees' time, supplanting their primary profit-generating activities. To the extent that this is the case, the opportunity cost to the firm may be large.

Management information systems, of which accounting systems are a part, provide information useful to managers for decision support, monitoring of operations, and control and motivation of employees. Environmental concerns cut across all of these functions. However, traditional accounting systems are not generally designed to isolate environmental costs and information in a routine and systematic manner. It is frequently observed that if managers don't "see" the costs, they will not be able to control them. This is particularly crucial for environmental costs because the ultimate objective of managers of firms is to reduce or eliminate environmental losses and attendant costs and risks.

This chapter reviews the purposes of managerial accounting information, inadequacies of current accounting systems to provide useful managerial information, and proposals for environmental information enhancements to current systems to better reflect the environmental cost burden to the firm. The benefits for many firms are likely to be improved profitability, reduced embedded firm risk, and a stronger market competitive position.

16.2   Accounting Information Support for Managerial Decision Making

This section reviews the various roles of accounting information within the firm and the importance of environmental cost information in this context.

Business activities require the use of several kinds of scarce resources. Typically, production requires the acquisition of *additional resources* in the form of materials, labor, or additional equipment. The cost of these additional resources is the additional cash which must be expended. In addition, production will require the use of *existing long-term resources* or plant capacity. Although the firm already possesses these assets, a cost is associated with their use, commonly termed an *opportunity cost*, the value of other production forgone. Finally, firms frequently employ the use of *rights resources*, intangible assets which may be very valuable to a firm. For example, patents for a manufacturing firm and copyrights for a publishing firm may make it possible to conduct business at all. Similarly, permits to emit environmental effluents may make it possible for the firm to remain in business, providing breathing room while the firm develops new processes or transitions to new markets. Consequently, such permits are likely to have large economic values. It is important to observe that managers are constrained by the resources available and that adequate information support for the various managerial functions becomes crucial to the success of the firm.

16.2.1   Managerial Decision Support

Managers must make a wide variety of important decisions in the routine course of business. In many firms, nearly all such decisions have environmental implications. For example, the product mix a firm chooses to manufacture and sell has a profound effect on the profitability and risk of the firm. Moreover, such choices are of utmost importance in defining the company's environmental profile. Individual product choices determine both the set of manufacturing processes which are available and the exposure to costs associated with hazardous waste, regulatory action, and potential environmental liability.

Other closely related decisions include product pricing, whether to make or buy an input, capital investment choices, research and development proposals, waste treatment alternatives, and liability management strategies, among many others. Clearly, environmental costs are pervasive in many, if not all, of these decisions. A failure to accurately consider the environmental cost implications may lead to suboptimal decisions.

16.2.2 Monitoring of Operations

A major managerial function is the continual monitoring, or "tracking," of information variables which have been previously identified as important for the success of the firm. Such variables may include, for example, production volumes, production costs, sales dollars and unit volumes, product quality, and waste disposal costs. The objective of monitoring is to obtain sufficiently timely information to allow for changes and adjustments to operations, either to alter the course of action or to initiate new activities. Again, unless environmental costs are identified in a manner which will permit monitoring, action cannot be implemented to control the costs.

A second and particularly difficult aspect of monitoring is the detection of "unknown" events of material importance to the firm. By their very nature, these nonroutine changes have not been previously identified and incorporated into the information-gathering and -reporting system. Thus, an expert knowledge of managers is required to identify potential sources of change or threats to the firm. Examples of nonroutine changes are technological breakthroughs involving major shifts in resources consumed and consequent costs of operations (and the obsolescence of existing processes or products) and contingencies arising from a major new court ruling. The environmental area, where both technical knowledge and regulatory requirements are subject to rapid change, is an especially important one for monitoring.

16.2.3 Control and Motivation

In order for a firm to achieve corporate targets, use scarce resources in an optimal fashion, and otherwise become and remain a strong market competitor, managers develop systems for the control of operations and the motivation of employees. Monitoring is important for providing useful information to managers. However, the supplying of information alone is not sufficient to achieve the firm's objectives. Additional systems and mechanisms are established to move the organization in the direction which managers have previously decided upon. As part of this process, a system of rewards and incentives is usually instituted to encourage management's and other employees' compliance with the company's goals. Typical systems include periodic performance evaluations, promotion systems, and bonus and stock option awards.

Mechanisms for controlling environmental performance improvement and motivating employees to seek pollution reduction are particularly important.

The focus of most organizations is on profitability, new product development, and market expansion. However, environmental performance cuts across all such areas and usually involves costs rather than profit potential. Thus, managers may be less inclined to devote large amounts of time to pollution prevention *unless the system of rewards and incentives is focused at least partially on these efforts.*

A particular problem in the environmental area is that of *joint costs.* These are costs which are shared by a number of corporate operating units. Typical examples include wastewater treatment, incineration for hazardous wastes, and contracts for outside waste disposal. Where such contracts are entered into by the corporation as a whole, problems immediately arise as to whether the firm should attempt an equitable distribution of the costs in order to encourage waste minimization, or perhaps retain the costs in the joint facility and provide other mechanisms for control of waste generation and motivation of employees.

16.3 Environmental Cost Profiling

16.3.1 Modernizing Traditional Historical Cost Systems

Most managerial accounting and information systems are subsets of the firm's financial accounting systems. The reason is that publicly traded companies are required by the Securities and Exchange Commission to produce financial reports for investors and creditors, including financial statements prepared according to *generally accepted accounting principles* (GAAP). Even privately held companies prepare GAAP statements for the use of creditors and others who are familiar with the GAAP basis. In addition, the Internal Revenue Service (IRS) requires for-profit entities to prepare financial statements for tax purposes using a different set of IRS accounting principles. However, the IRS system is similar in many respects to the GAAP system.

Faced with this heavy burden of required external financial reporting, many firms attempt to "satisfice" by using directly or modifying portions of the GAAP or IRS accounting systems for internal managerial needs, including decision support, monitoring, controlling, and motivation. However, the objectives of the two, external reporting to investors and creditors and providing information for internal decision needs, are likely to be quite different. For example, investors and creditors are provided with highly aggregated "full cost" historical cost of goods sold numbers to permit them to evaluate operational profitability. Managers, on the other hand, may need to know which costs are arising from different activities in the firm, including environmental activities.

Thus, for managerial decision, monitoring, and control and motivation purposes, *an examination of the firm's environmental cost-generating activities should be undertaken.* A reasonably simple first approach would be to evaluate the environmental implications of each line item in the historical cost schedules currently prepared for the firm. Once the "material" (i.e., major or important) sources of environmental costs have been identified, managers in the responsible areas and the accountants who prepare the schedules can determine how best to modify the current system to accumulate and report the costs. This may require nothing more than the addition of another line item or two in the existing schedule, as would be done for any process modification. However, for major sources of environmental costs, management may consider the development of new costing schedules to report specifically on the environmental costs. The latter choice may involve essentially the addition of a new reporting dimension to the existing accounting system. Such efforts are, as always in business, subject to cost/benefit evaluation. This is why the emphasis is placed on material sources of environmental costs.

For many firms, end-of-pipeline waste is a relatively small cost. Since the late 1980s, however, many firms have begun to find that regulatory compliance monitoring and reporting requirements have grown to the point that many employees spend significant amounts of their time on such activities. Thus, monitoring of such activities may indicate areas where development of, for example, joint or automated information capture and reporting procedures may result in substantial out-of-pocket savings. At the very least, the company may wish to be aware of potential opportunity costs the firm is incurring (as a result of reduced employee time available for efficiency improvements and for exploring opportunities for profit enhancement through market expansion or new product development), and of other potential profit losses.

Another area which has grown rapidly for many firms is in the area of capital expenditures for property, plant, and equipment in order to reduce environmental effluents or otherwise comply with environmental regulations. Such costs are likely to be significant for many firms. Moreover, monitoring of such costs across divisions of a firm may indicate much redundancy of engineering and research and development activity. Thus, cost savings may be achieved, and profits enhanced, by joint efforts in these areas. Similar savings may be realized in basic research and development, legal and public relations costs, and other activities not usually associated directly with manufacturing.

16.3.2 Profiling Escapable and Nonescapable Cost Components

A second possible environmental enhancement to traditional accounting systems is separation of those costs which are "escapable" in the short term—for example, variable out-of-pocket costs—from those which are either fixed in the short run (long-term noncancellable contractual arrangements) or allocations of past historical costs (depreciation on property, plant, and equipment). In general, the full-cost financial numbers reported to investors and creditors of the firm do not distinguish between escapable and nonescapable costs. Given the financial-reporting orientation of many managerial accounting systems, such information is unlikely to be retained for managers' use in decision making, monitoring, and so forth.

However, the crux of many managerial decisions rests on the potential effects in the near or medium term of alternative business choices. Many firms, especially in highly competitive industries, are finding that profound changes are occurring at an increasing rate. And this condition is especially severe in the environmental area, where multiple factors are at work, including rapidly changing and proliferating regulations, changing technologies, and evolving legal precedents. *Firms may well consider analyzing and decomposing the environmental costs identified above into variable (escapable), and fixed and historical (nonescapable), components for those items which are material. It is also important to periodically monitor other environmental costs which have the potential to change rapidly.*

16.4   Developing Time Series for Environmental Costs

Routine financial reporting requires relatively short time horizons: three years for income statement numbers and two years for balance sheets. Thus, many financial reporting systems retain only short histories of numbers in readily accessible formats. But decision making, as well as control and motivation, frequently requires, or can benefit from, relatively long lead times in order to allow for cost-effective responses and changes to operations.

In the environmental area, where regulatory constraints may be binding upon firms, *it is imperative that sufficient time series of essential environmental cost numbers be retained to permit managers to track the costs and respond in a timely fashion.* Five years may be sufficient; however, this is likely to vary from firm to firm.

16.5   Summary

The suggested proposals, expanding the traditional historical-focused accounting system to include (1) an environmental cost dimension, (2) the ability to distinguish escapable from nonescapable costs, and (3) sufficiently long time series of important environmental cost numbers to permit timely changes to operations are both fundamental and pervasively important to many firms with large environmental exposure. It is clear, however, that such changes can prove important far beyond the prevention of pollution. Adoption of the techniques for other important aspects of businesses may well improve long-term profitability and competitiveness.

17

Measuring Pollution Prevention Progress

John Warren
Research Triangle Institute
Research Triangle Park, North Carolina

James Craig
U.S. Environmental Protection Agency
Washington, D.C.

17.1   Introduction/Overview

Corporations and organizations of all types are increasingly concerned about their environmental performance. Corporate management is pressured to prove the success of their environmental programs. The federal government, under the Pollution Prevention Act of 1990, is required to establish standard methods for measuring source reduction. The public is demanding more information on what organizations are doing to reduce the quantities and toxicities of harmful releases in their communities, and environmental groups in particular have an interest in tracking environmental performance.

The need for a tool to assess progress in pollution prevention is clearly needed. However, with such a wide array of organizations and stakeholders involved, no single measure of pollution prevention will suit everybody's needs. State programs may have different goals and objectives from industry or regulatory pollution prevention programs. Likewise, large corporations may have different technological capabilities and resources available for measuring pollution prevention progress than do small companies. Therefore, different evaluation approaches are necessary to evaluate their respective effectiveness.

This chapter discusses the importance of measuring pollution prevention and describes various measurement techniques available. The intent is not to identify the best or worst of available techniques, but to provide an overview which may serve to broaden and promote an improved understanding of ways to measure progress.

17.2   What Is Pollution Prevention Progress?

Pollution prevention is defined by the Environmental Protection Agency as

> the use of materials, processes, or practices that reduce or eliminate the creation of pollutants or wastes at the source. [It] includes practices that reduce the use of hazardous materials, energy, water or other resources, and practices that protect natural resources through conservation or more efficient use. (*U.S. EPA, 1991*)

Broadly defined, pollution prevention progress includes any action which furthers the goal of pollution prevention. Such actions range from implementing a pollution prevention program, to raising awareness of government officials, to realizing a quantifiable reduction in pollution released to the environment.

17.3   Why Measure Progress?

Pollution prevention is not a new field and the benefits of pollution prevention programs are well known and widely accepted. The reasons for measuring progress in pollution prevention, however, may be less convincing, particularly given the complex and often resource-intensive nature of the process. In the following paragraphs, we highlight several of the key reasons why tracking progress contributes to pollution prevention success.

Measuring progress allows decision makers in industry and government to evaluate program successes and failures. For industry, knowing what components of a pollution prevention program are most successful allows management to save time and resources by directing their efforts to their most efficient and effective use. Measuring pollution prevention can also help a facility identify opportunities for additional pollution prevention activities and helps management determine the cost savings attributable to their pollution prevention programs. For regulators, it may be important to evaluate pollution prevention progress in order to assess the need for regulatory adjustments or additional regulation to better meet reduction goals.

Measuring progress and being able to report on the environmental benefits of pollution prevention brings credibility to pollution prevention programs—both public and private. Making such information available to the public may also encourage other organizations and the public at large to participate in pollution prevention activities.

17.4   What to Measure?

As a relatively new field, measuring pollution prevention has not developed to the point where there are accepted techniques, indicators, or even definitions for making consistent and comparable assessments. Many of the data sources currently used in measuring pollution prevention were not designed for this purpose and it is unclear whether they adequately do the job. Also, the data available for measuring progress differs greatly from organization to organization. Thus, individual organizations need to determine what measurements will best suit their needs in assessing whether goals and objectives have been met. We will discuss some of the criteria and data options for determining a pollution prevention metric.

17.4.1   Criteria for Determining What to Measure

Goals, Reasons for Measuring.   It is critical to clearly articulate and understand program goals before a measurement technique can be selected. Goals should be set with measuring progress in mind. While it may not be desirable to set only goals that can be measured, it may be appropriate to alter goals or objectives somewhat to reflect what is possible to measure. Another approach is to stay with a goal that cannot be fully measured but to recognize that an alternative set of indicators will be used to determine if the goal has been reached.

Measurement Options.   Determining what to measure will depend on program objectives as well as state and federal requirements and should be carefully tailored to the goals of the program or project. Measurements may be needed to assess the effectiveness of a project or program or to confirm program direction or identify broad trends. Some may require detailed measurements using extensive waste sampling and highly technical equipment, while others may only need to estimate an order of magnitude. Measurement systems should be simple whenever possible.

Data Availability.   Tracking progress in pollution prevention on a year-to-year basis requires data that can be compiled into a complete and consistent database. Before an organization selects a measurement technique, it must consider its information requirements. To the extent possible, available or easily obtainable information should be used. If the necessary data is not available, organizations must consider the time and resources required to collect that information. Certainly, the allocation of resources for measuring progress should not inhibit pollution prevention programs themselves.

Data Accuracy.   Another data consideration is accuracy—how well measurement results provide a true indication of progress and how well the data help determine whether a goal has been achieved. It is important not to expect too much from the data. That is, avoid asking overly complex questions. For example, it may not be possible to show the overall progress at a plant, but it should be possible to select and assess a few objectives of the program such as CFC use reduction at a facility.

17.4.2 Data Options

Options for evaluating pollution prevention effectiveness include quantitative and descriptive measures. Quantitative measures are the traditional means of measuring the amount of pollution released into the environment. Descriptive measures provide an indication of the quality of pollution prevention programs and whether progress has been achieved, but do not quantify the degree of progress in terms of environmental impact. While quantitative and descriptive measurements should be related, to date there are no comprehensive measurement tools to link the two types of information.

Measurement selection between the two types of data should be based on the program goals being measured. Large industrial corporations already subject to release and transfer reporting requirements may find quantitative measures most effective in measuring progress. State or regional governmental programs, however, which may not have complete access to release information for their region, or for whom the volume of data may be too extensive to calculate quantitative measures, may find descriptive measures most appropriate for tracking pollution prevention progress.

Quantitative Measures.   The following are the types of quantitative data which can be used to measure the environmental impact of pollution prevention programs.

- Monitoring data
- Reporting data such as TRI and Hazardous Waste Biennial Reports
- Concentration-based measures
- Quantity of waste generated
- Quantity of waste recycled
- Indicators of level of hazard (toxicity, acidity, reactivity, ignitability)
- Waste composition

Descriptive Measures.   The following are types of descriptive data that can be measured from year to year to provide an indication of the level of pollution prevention program activity and program success.

- Number of pollution prevention projects
- Number of pollution prevention projects showing quantity reductions

- Number of pollution prevention inspectors
- Number of pollution prevention inspections completed
- Number of enforcement actions
- Level of pollution prevention training offered
- Number of staff or staff hours allocated to pollution prevention
- Level of funds budgeted for pollution prevention
- Number of requests for pollution prevention information
- Number of attendees at pollution prevention seminars or workshops

Descriptive measures may also include anecdotal information resulting from surveys and outreach services. These data provide an indication of the quality of pollution prevention program administration.

17.5 How to Measure

Measuring progress with descriptive measures is relatively uncomplicated. Once the project goals and metrics have been determined, a base year is selected, and data is gathered and analyzed on a regular basis. Measuring progress with quantitative measures, however, can be more complex.

17.5.1 Quantitative Measures of Pollution Prevention Progress

How do you measure waste that is not created? The desired measure for assessing the quantity of waste reduced is the quantity of waste that was not generated because pollution prevention programs were in place. This quantity can be difficult to measure and can be misunderstood or misrepresented. Therefore, comparison of changes in the amount of waste generated from one year to the next has been the most common method used to measure pollution prevention progress.

Many methods can be used to calculate the quantity of waste reduced from one year to the next. Three of these—the actual change method, the production-normalized method, and the throughput ratio—are discussed here along with the degree-of-hazard method, which assesses progress in terms of the change in the level of risk to human health and the environment from one year to the next.

Actual Change. The actual quantity change in waste generated is the most straightforward means of assessing waste reduction. It measures the difference between the waste generated in one year and that generated in the previous year:

$$\text{Actual change} = \text{waste, year } (n) - \text{waste, year } (n-1)$$

The advantage of this approach is its simplicity. The quantity change is relatively easy to calculate and track, and quantities can be combined to make meaningful aggregate totals. The disadvantage of this approach is that it does not account for change that was not due to factors such as changes in production levels, introduction of a new product, changing market conditions for a product, changes in product quality, or other changes in operating conditions. Actual quantity change will overstate progress if a facility's business activity has declined and understate progress if activity has increased.

Production-Normalized Change.   The production-normalized metric is a measure which accounts for changes in the level of production, service, or other business activity for the processes that generate waste. This method uses an index (known as the production index) to adjust the actual quantity change for changes in business activity. It is calculated as follows:

$$\text{Production index} = \text{production, year } n \div \text{production, year } (n-1)$$

$$\text{Production-normalized change} = \text{waste, year } (n) - (\text{waste, year } (n-1)$$
$$\times \text{production index})$$

Options for the measure of production used in the production index may include:

- Units of product
- Units of input
- Product revenue
- Product profit
- Number of employees
- Employee hours

No single production measure is appropriate for all industries. The selection made should be the measure which most highly correlates with changes in waste generation.

The advantage of this approach is that it removes the impact of production fluctuations from the waste reduction figure. The disadvantage of this approach is that it assumes that waste generation and production levels are linearly related and that no factors other than production level impact the quantity of waste generated. This assumption does not always hold true. For example, a one-time event such as an accidental spill will affect the waste reduction figure and may outweigh the quantity decrease due to pollution prevention activities.

Throughput Ratio.   The throughput ratio also adjusts for production fluctuations by calculating a mass balance for each chemical in a wastestream.

Throughput is essentially the amount of chemical processed through the manufacturing line. The throughput ratio is a ratio of the quantity of chemical generated as waste to the quantity of throughput. This calculation is illustrated as

Throughput = quantity of chemical consumed in production
+ quantity of chemical shipped off-site
+ quantity of chemical released to all media

Throughput ratio = quantity of chemical released to all media
÷ throughput quantity

The throughput ratio is essentially a measure of chemical use efficiency. A lower throughput ratio indicates that a smaller portion of the chemical used (throughput) is going to waste and a higher portion of the chemical used is being used in production.

The advantage of the throughput-normalized measure is that it can be useful for evaluating waste management practices, tracking toxic chemicals used on site, and identifying waste reduction opportunities. The drawback of this measure is the difficulty in gathering the necessary data. A facility using this measure should calculate the throughput ratio for each production unit—a time-consuming and costly procedure. In addition, a reduction in the throughput ratio does not necessarily mean that pollution prevention has occurred.

Degree of Hazard.   Assessing the change in the degree of hazard posed by wastes is an important consideration in evaluating pollution prevention progress. Ideally, a degree of hazard measure would be incorporated with each of the quantity change measurements previously discussed. However, to date, no method exists to easily and at low cost assess changes in all factors that affect the level of hazard of a waste. The types of factors which should be included in such a calculation include:

- Toxicity, ignitability, corrosivity, and reactivity of the waste
- Location of release
- Environmental medium of release
- Extent of human or environmental exposure to the waste
- Rate at which waste decomposes or dissipates

Although classifying wastes into strict degree-of-hazard categories is extremely difficult (scientifically and politically), a number of ranking systems, with broad risk categories, have been developed. An example of a ranking system used by Polaroid is discussed in the following section.

A summary of the quantitative pollution prevention measurements described in this section is provided in Table 17-1.

Table 17-1. Advantages and Disadvantages of Pollution Prevention Measurement Techniques

Measurement technique	Major advantages	Major disadvantages
Actual change	Easy to compute; aggregated totals are meaningful	Does not account for changes due to production fluctuations
Production-normalized change	Adjusts for production fluctuations	Selection of production measure is difficult; assumes linear relation between waste quantity and production level; aggregation of normalized data is less meaningful
Throughput ratio	Measures efficiency of chemical use; adjusts for production level	Extensive data requirements; comparisons across facilities are not always meaningful
Degree of hazard	Offers a way to account for changes in risk to human health and environment	Extensive data requirements; scientifically and politically difficult to classify chemicals

17.5.2 Industry Examples of Measuring Pollution Prevention Progress

Polaroid Corporation.   In 1988, Polaroid initiated its Toxic Use and Waste Reduction Program (TUWR), an example of a degree-of-hazard ranking system for evaluating waste reduction. TUWR is a voluntary corporatewide reduction program which tracks progress at all levels of the corporation. Under the program, all chemicals used by the company were placed in five categories, depending on their environmental and health risks. Chemicals having the highest risk such as carcinogens and other high-priority toxics were placed in Category 1 and chemicals with the lowest risk such as nontoxic solid wastes were placed in Category 5. The program's five-year reduction goals called for a 10-percent reduction per year in *usage* of Category 1 and 2 chemicals and a 10-percent per year reduction of *waste by-products* of Category 3, 4, and 5 materials. This approach ensures that progress is made across all categories, and not primarily the high-volume, low-toxicity materials where reductions are more easily achieved. At the end of the first four years of the program, Polaroid reported a corporatewide reduction of 27 percent (U.S. EPA, ORD, 1993).

Niagara Mohawk Corporation (NMPC).   In 1991, NMPC announced a new corporate environmental policy with five major objectives: (1) to meet or surpass the requirements of all environmental laws and regulations; (2) to promote the wise use of energy; (3) to assume a responsible stewardship for natural resources; (4) to reduce pollution and conserve raw materials; and (5) to make environmental considerations a part of corporate planning and decision making. To measure progress in achieving these objectives, NMPC developed an Environmental Performance Index (EPI). The EPI is a weighting and rating index which evaluates progress in three categories: waste/emission reductions, compliance, and enhancements. The score for the waste reductions category is based on annual percentage quantity change figures for emission levels and solid/hazardous waste generation and disposal. The compliance score measures how well NMPC is meeting state and federal environmental regulations. The enhancement score is based on NMPC's dollar investment in environmental enhancements (e.g., new equipment, research, and development).

The environmental index is used to establish annual performance objectives and assess program progress. Since the index measures widely disparate parameters, the actual score on the index for a given year has little significance, or it can be used to compare performance with other corporations. The real value of the index is to make internal performance comparisons over time. Also, as a broad-based index, employees from all divisions have an opportunity to contribute to its value and may be inspired to improve performance to meet their environmental goals (GEMI, 1992).

3M.   3M has been quantitatively measuring environmental performance since 1975. As part of its Pollution Prevention Pays (3P) program, the organization measures the quantity of pollution prevented and the monetary savings realized as a result of its prevention programs. In 1990 as an extension of the 3P program, 3M instituted a new system for measuring waste reduction. The criteria for this system were that it be simple yet accurate and reproducible; be indexed to production; measure pollution prevention according to 3M's definition of source reduction and environmentally sound reuse and recycling; and measure waste levels before treatment, control, or disposal (U.S. EPA, ORD, 1993).

The resulting system classifies the outputs from a production facility into one of three categories:

Product:	the intended output from the manufacturing facility
By-product:	the residual from operations that is later used productively through some form of recycling or reuse
Waste:	the material that is subject to waste treatment, pollution control, or is directly released into the environment

Added together, these categories equal the total output from the manufacturing facility. The facility then calculates the waste ratio, defined as:

$$\frac{\text{Waste}}{\text{Waste} + \text{by-product} + \text{product}} = \frac{\text{waste}}{\text{total output}}$$

3M calculates its waste quantities using a materials accounting approach. The weight of inputs to the production process is compared to the weight of resulting products. Any loss in weight during the production process is considered a waste.

The waste ratio metric serves as a measure of manufacturing efficiency. It is reported on a divisional basis and reflects the efforts of the entire division—not just manufacturing. Research and development scientists need to develop new products and processes that generate less waste. Engineers need to design equipment that is more efficient and that can make use of recycled or reused materials. And marketing needs to know which products generate the least amount of waste and provide information and create consumer demand for them.

The advantage of the waste ratio is its simplicity. It can be applied consistently throughout the company, across widely varied operations. Also, by focusing on total waste, this approach emphasizes efficiency in the use of resources. The disadvantage of the waste-per-unit-volume approach is that it encourages operating units to focus on those areas where they can obtain the greatest reductions in waste volume (i.e., high-volume, low-toxicity wastes). There may also be reasons other than pollution prevention or recycling for reductions in the ratio.

In 1990 when 3M's metric and measurement system was instituted, management established a waste reduction goal of 35 percent by 1995. Each division was asked to accomplish a 7 percent reduction in the rate of waste generation for each of the five years. Both 1991 and 1992 results were ahead of schedule.

17.6 Measurement Issues

There are many measurement issues or questions to ask when progress is assessed. While many of these are of primary concern in measuring national progress, they can also be important at the plant level.

17.6.1 Lack of Consistent Definitions

There is considerable variation among industry, state, and federal organizations in the types of waste measurements and definitions used. Even pollution prevention is subject to different interpretations. Many organizations define pollution prevention in terms of the Pollution Prevention Act hierarchy or RCRA waste management hierarchy, though there is much debate over whether to include recycling in the definition. Some define pollution prevention more narrowly as source reduction or toxics use reduction. And, others view pollution prevention more conceptually as any process that involves continuous improvement and movement up the environmental management hierarchy. These discrepancies in definitions, and consequently in resulting measurements, make it difficult to compare progress between companies or industrial groups. However, this does not have to pose a problem for a plant manager evaluating a specific objective, if the measurement methods used were designed to assess whether that objective had been achieved. Thus, if reducing CFCs at the source or recycling waste solvents is the objective, the focus of measurement should be on that objective.

17.6.2 Problems Defining
an Accurate Baseline

One consideration in measuring pollution prevention progress is developing an accurate baseline. Without an accurate baseline, any comparison made may not result in a reasonable conclusion. That is, based on information collected, a project may appear unsuccessful when, in fact, it is successful, or vice versa. Among the problems that complicate defining the baseline are data quality, production index, and aggregation.

17.6.3 Data Quality

If data are not reliable and accurate, the answer obtained could be left to random chance, rather than reflecting the results of the program.

17.6.4 Adjusting for Production

Calculating a production index can be a complex task. For each wastestream, facilities must calculate an index to accurately reflect changes in the level of activity that generated that waste. For some waste-generating activities, defining an appropriate measure of the activity level is difficult. Such activities include laboratory research, service activities, and cases in which many different products or processes contribute to the generation of a single wastestream (multiproduct manufacturing). Also, comparing activity levels over time is difficult if the product has changed due to improved quality or materials substitution.

Another concern with the production index is that normalizing waste measurements using the index is not always appropriate. Many other changes in operating conditions, which may not be reflected in the production index, can affect the quantity of waste generated. For example, rainfall and collection surface area can affect water pollution levels. Also, raw material quality, throughput rates, and worker productivity can affect waste generation.

17.6.5 Aggregating Increases
and Decreases

To aggregate the change in quantity of waste generated for all facilities in a region, both increases and decreases in waste generation are added together. This gives the *net change* for all facilities—those that reduced the quantity of waste generated and those that did not. If many facilities had increases in waste generation over the time period, the region as a whole might not show pollution prevention progress, even though some individual facilities experienced significant progress. To obtain an aggregate measure of pollution prevention progress, it is helpful to look separately at the aggregate quantity increases and decreases. Assessing both of these indicators reveals the true volume of activity and provides a more complete description of progress.

17.6.6 Problems Associated with Comparing Current Data to the Baseline

Among the problems associated with comparing current data to the baseline are delayed effects of pollution prevention programs, the impact of real verus paper changes, and the tendency to attribute any change to pollution prevention.

17.6.7 Delayed Effects of Pollution Prevention Programs

Pollution prevention programs require change, both technological and behavioral, which take time to implement. The effects of pollution prevention programs may not be evident for several years or such programs may have a cumulative effect that requires time to affect total releases. This makes measuring the impact of pollution prevention a difficult process, particularly if measurements are desired for a single calendar year. Tracking the results of pollution prevention over several consecutive years would provide a more accurate assessment of progress.

17.6.8 Real versus Paper Changes

Many factors contribute to changes in reported release information. These can be categorized as "real" changes and "paper" changes. Changes in production levels or product lines as well as materials substitution, procedure modifications, improved management, and other source reduction activities are considered real changes in chemical releases. Changes in measurement or estimation methods and changes in reporting requirements, or a firm's understanding of reporting requirements, are referred to as paper changes because they affect reported releases without physically reducing the quantity of chemicals released. There is currently no means of differentiating these many factors affecting releases.

17.6.9 Tendency to Attribute Any Change to Pollution Prevention

There can be many reasons for changes in release quantities. Just because there is a reduction in waste does not mean that a pollution prevention effort has been successful. This is an important factor to consider when one is setting objectives and determining a measurement system to evaluate progress.

17.7 What Have We Learned From Assessing Pollution Prevention Progress under TRI?

In 1991, Research Triangle Institute (RTI), under a cooperative agreement with the EPA, conducted a study to assess the comparative impact of real versus paper changes on TRI submissions. In particular, the study focused on the extent to which three factors—measurement/estimation changes, production changes, and source reduction activities—affected changes in TRI data between 1989 and 1990.

Data for the study were collected through telephone interviews with over 1200 facilities that were selected using statistical random sampling techniques. Each facility was asked to explain the extent to which the three study factors contributed to reported changes in their TRI data. Participants' responses were weighted and estimates were made for the number of Form Rs and facilities associated with each reason for change as well as the quantity change in pounds attributed to each reason.

The national level estimates of Form Rs and facilities associated with each reason for change are summarized in Table 17-2.

The impact of production fluctuations was the most common reason for change—69 percent of all facilities claimed an impact. Nearly 40 percent of all facilities realized a change in TRI releases and transfers due to source reduction activities. The actual percentage of facilities that have implemented source reduction programs, however, may be even higher because the results of prevention programs may not be evident for several years after implementation. A change in measurement or estimation techniques, a form of "paper" change, was the least common reason for change.

The national level quantity estimates are summarized in Fig. 17-1. The pie represents the 866-million-pound decrease in TRI submissions between 1989 and 1990 and the pieces show how this quantity change is apportioned among measurement changes, production changes, source reduction, and other factors. Other factors include changes due to reasons other than the three change variables of this study (e.g., changes due to recycling, data errors, or technical guidance).

Nearly 45 percent of the total change in TRI submissions was due to source reduction activities. An estimated 5 percent was due to production change, less than 3 percent was due to measurement change, and the remaining 48 percent was due to other factors. Thus, at least 50 percent of the total change (production change and source reduction combined) falls under the category of real change. Only 3 percent can be labeled with certainty as paper change. The remaining 48 percent decrease combines both real and paper changes.

It is important to note that the values shown in Fig. 17-1 represent net changes; both increases and decreases in TRI releases and transfers have been added together. Thus, the actual volume of activity for any given reason may be masked by large quantities of increases and decreases that have canceled each other out.

Table 17-2. Number and Percentage of Facilities and Form Rs Indicating Each Reason for Change

| | Reasons | | |
	Measurement change	Production change	Source reduction
Number of facilities	4,630	13,124	7,570
Percentage of facilities	24.4%	69.3%	39.9%
Number of Form Rs*	12,545	38,525	15,767
Percentage of Form Rs*	14.6%	44.9%	18.4%

*These estimates may be low. Facilities that submitted more than 20 Form Rs were asked to explain the reasons for change for only those chemicals making up the top 80 percent of their total change. Consequently, 4 percent of the Form R population was excluded from the interview process.

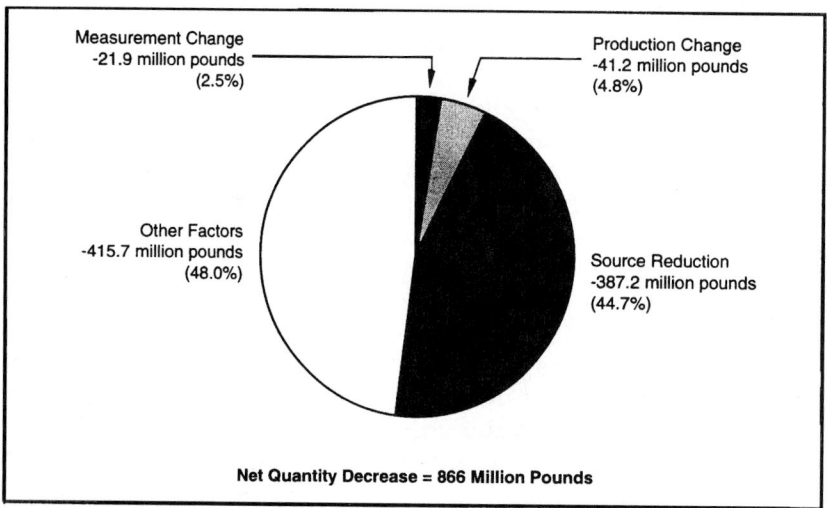

Figure 17-1. Net quantity change in TRI submissions, by reason.

Figure 17-2 illustrates how the net change in releases and transfers for each of the three change variables is derived from quantity increases and decreases. Although the net change for all three variables is a decrease, both increases and decreases were attributed to each variable. The greatest absolute quantity change was attributed to production change, supporting this variable as the most frequently cited reason for change. Yet the aggregated impact of production change is comparatively small.

The results of this study have provided insight into the reasons for changes in TRI submissions between 1989 and 1990. At least 50 percent of the total change reported represents an *actual* change in the physical quantity of chemi-

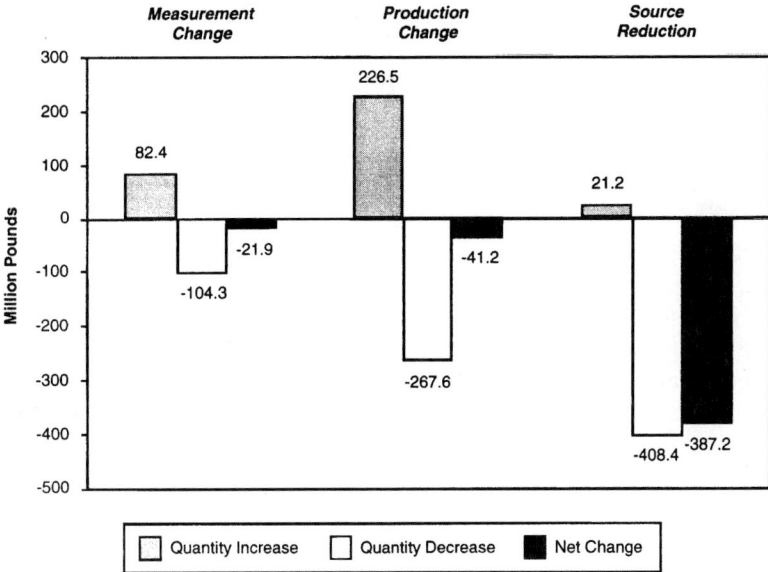

Figure 17-2. Aggregating increases and decreases in TRI submissions, by reason.

cals released (i.e., production change and source reduction). Reporting or paper changes and other factors not covered by the review also account for changes.

This study also serves as a first step in measuring source reduction progress. Results can be used as a baseline for comparison with Pollution Prevention Act data collected as part of the 1991 Form R reporting. With these new data, the EPA will be better able to track and evaluate source reduction progress from year to year. This type of longitudinal study is critical for an accurate assessment of reduction progress because the effects of many prevention activities may not be evident for several years after implementation.

17.8 Overview: Selecting a Method for Measuring Pollution Prevention Progress

Measuring progress in pollution prevention allows management to identify program successes and failures and assess the extent to which prevention goals are being met. However, measuring progress can be quite complex and requires careful consideration of industry and site-specific conditions. As stressed throughout this chapter, it is important for companies to select a measurement technique or combination of techniques that will best fit their needs and abilities. Complicated, resource-intensive techniques are not worthwhile if they inhibit a facility's ability to successfully implement and maintain their prevention programs. No single measure is practical for all industry. In the following, we highlight suggested steps for selecting and implementing an appropriate technique.

17.8.1 Conduct Waste Assessment

If a pollution prevention program is not already in place, the first step is to conduct a facility waste assessment. This involves performing site audits, making an inventory of current waste levels, identifying prevention opportunities, and prioritizing reduction options. Even if a pollution prevention program is already in place, a waste assessment may be useful to assess how well project goals are being met.

17.8.2 Determine Pollution Prevention Goals

Pollution prevention goals should focus on the wastestream priorities identified in the facility assessment. Goals should be challenging, yet reasonable and practical. Goals should also be set with measurement in mind. It is important to incorporate techniques for measuring program effectiveness at the program design stage so that an appropriate baseline can be determined and yearly comparisons made.

17.8.3 Determine Measurement Techniques and Data Requirements

Once the pollution prevention goals have been set, a facility must determine what measurement techniques exist to measure these goals and whether they are technically and financially feasible to execute. Data for assessing progress range from simple descriptive program indicators to highly sophisticated and resource-intensive quantitative measures. Facilities which select quantitative measures should also consider how to normalize the data to correct for any factors other than pollution prevention which may have influenced quantity or toxicity changes. If a facility finds that the techniques necessary to assess their program goals are too complex or resource-consuming, they may want to rethink their goals and make modifications for more feasible program evaluation.

17.8.4 Develop Baseline

Whether using quantitative or descriptive measurements, an organization must develop a baseline of data for purposes of comparing progress from year to year. Baseline figures are often drawn from data of the year prior to program implementation. An average of performance data over a period of years prior to program implementation can also be used to develop a baseline.

17.8.5 Collect Data

Data necessary for program assessment should be collected at regular intervals and compiled into a consistent and accurate database. Depending on the size and characteristics of a company, required data may already be available from reporting information.

17.8.6 Analyze Data and Assess Progress

The most common method of assessing pollution prevention progress is to make an internal comparison of annual results to baseline data. Some data such as state and federal reporting data may be appropriate for external comparisons, allowing companies to examine their progress relative to others in their industry. Progress should be reviewed frequently enough to enable division managers to make adjustments as needed to stay in line with program goals. Program successes should be recognized and program shortcomings should be reviewed and reworked to be more effective.

Bibliography

Baker, Rachel Dickstein, Richard W. Dunford, and John L. Warren, *Alternatives for Measuring Hazardous Waste Reduction*, Hazardous Waste Research and Information Center, HWRIC #89-067, 1991.

Bush, Barbara L., "Measuring Pollution Prevention Progress: How Do We Get There from Here?" *Pollution Prevention in Practice*, Executive Enterprises Publications Co., Inc., New York, 1993, pp. 175–187.

Global Environmental Management Initiative, *Corporate Quality/Environmental Management II: Measurement & Communication Conference*, Proceedings, Washington, D.C., 1992.

Harriman, Elizabeth D., Jay Markarian, Jay S. Naparstek, and James W. Stolecki, *Measuring Progress in Toxics Use Reduction*, Department of Environmental Protection, Commonwealth of Massachusetts, 1991.

Pojasek, Robert B., and Lawrence J. Cali, "Measuring Pollution Prevention Progress," *Pollution Prevention Review*, **1** (2):119–130, 1991.

Riley, Gwen J., John L. Warren, and Rachel D. Baker, *Assessment of Changes in Reported TRI Releases and Transfers Between 1989 and 1990*, U.S. EPA, Office of Pollution Prevention and Toxics, EPA Contract #CR818760-01-0, 1993.

U.S. Environmental Protection Agency, Office of Pollution Prevention, *Pollution Prevention 1991: Progress on Reducing Industrial Pollutants*, EPA 21P-3003, 1991.

U.S. Environmental Protection Agency, Office of Pollution Prevention and Toxics, *Toxics Release Inventory*, EPA 745-R-93-003, 1993.

U.S. Environmental Protection Agency, Office of Research and Development, *Facility Pollution Prevention Guide*, EPA 600-R-92-088, 1992.

U.S. Environmental Protection Agency, Office of Research and Development, *Measuring Pollution Prevention*, Conference Proceedings, 1993.

Wells, Richard P., Mark N. Hochman, Stephen D. Hochman, and Patricia A. O'Connell, "Measuring Environmental Success," *Measuring Environmental Performance*, Executive Enterprises Publications Co., Inc., New York, 1993, pp. 1–13.

18

Pollution Prevention through Life-Cycle Design

Gregory A. Keoleian, Ph.D.

National Pollution Prevention Center
School of Natural
Resources and Environment
University of Michigan
Ann Arbor, Michigan

18.1 Introduction

Product design offers tremendous opportunities for achieving pollution prevention. Through integration of environmental requirements into the earliest stages of product development, adverse environmental impacts can be reduced or eliminated in the manufacture, use, and end-of-life management of a product. Pollution prevention by design is the antithesis of "end-of-pipe" treatment or remedial action. Accordingly, it can provide significant benefits including enhanced resource efficiency, reduced liabilities, and enhanced competitiveness. Many organizational and operational changes, however, must take place both internal and external to a product manufacturer to effectively guide environmental improvement through design.

The design of a product system can be represented logically as a series of decisions and choices made individually and collectively by design participants. These choices range from the selection of materials and manufacturing processes to choices relating to shape, form, and function of the product. A design team represents a wide range of functional responsibilities including industrial design, process engineering, product development management, accounting, purchasing, marketing, human and ecosystem health, safety, and regulatory compliance. Each decision or choice made by these team members during development and implementation will shape the overall environmental profile of the product system.

Existing knowledge and experience guide individual and group design decisions. Both new information and new approaches to synthesizing and evaluating this information are essential to achieve pollution prevention through design. Recognizing that no single design method has universal appeal, this chapter offers guidelines rather than prescriptions. These guidelines are based on the life-cycle design framework developed by the author for the Pollution Prevention Branch of the U.S. Environmental Protection Agency (EPA).[1] Individual designers and design teams that recognize the benefits of pollution prevention are invited to adapt the ideas and guidelines for their own specific applications.

18.2 Definition of the Product System

18.2.1 Life-Cycle Stages

The product life cycle provides a logical system for addressing pollution prevention because the full range of environmental consequences associated with the product can be considered. By focusing on this system, designers can prevent the shifting of impacts between media (air, water, land) and between stages of the life cycle. In addition, this framework encompasses the many stakeholders (suppliers, manufacturers, consumers/users, resource recovery and waste managers) whose involvement is critical to successful design improvement. The life-cycle system is complex due to its dynamic nature and its geographical scope. Stages of the life cycle are changing continuously and changes often occur independently. Life-cycle stages are also widely distributed on a geographical basis, and environmental consequences occur on global, regional, and local levels.

Figure 18-1 is a general flow diagram of the product life cycle. As this figure shows, a product life cycle is circular. On an elementary level resources are consumed and residuals will eventually accumulate in the earth and biosphere. The product life cycle can be organized into the following stages:

1. Raw material acquisition
2. Bulk material processing
3. Engineered and specialty materials production
4. Manufacturing and assembly
5. Use and service
6. Retirement
7. Disposal

Raw material acquisition includes mining nonrenewable material and harvesting biomass. These *bulk materials* are *processed* into base materials by separation and purification steps. Examples include flour milling and converting bauxite

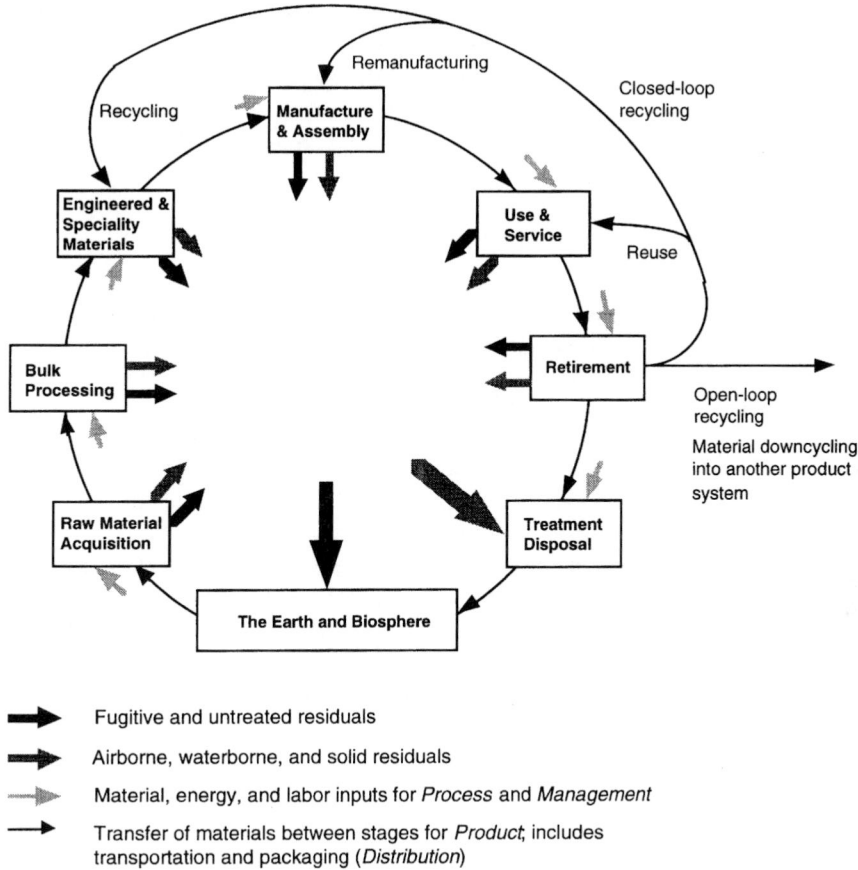

Figure 18-1. The product life-cycle system. (*Courtesy of U.S. EPA, Life Cycle Design Guidance Manual: Environmental Requirements and the Product System EPA 600/R-92/226.*)

to aluminum. Some base materials are combined through physical and chemical means into *engineered and specialty materials.* Examples include polymerization of ethylene into polyethylene pellets and the production of high-strength steel. Base and engineered materials are then *manufactured* through various fabrication steps, and parts are *assembled* into the final product.

Products sold to customers are *consumed* or *used* for one or more functions. Throughout their use, products and processing equipment may be *serviced* to repair defects or maintain performance. Users eventually decide to *retire* a product. After retirement, a product can be reused or remanufactured. Material and energy can also be recovered through recycling, composting, incineration, or pyrolysis. Materials can be recycled into the same product many times (closed loop) or used to form other products before eventual discard (open loop).

Some residuals generated in all stages are released directly into the environment. Emissions from automobiles, wastewater discharges from some processes, and oil spills are examples of direct releases. Residuals may also undergo physical, chemical, or biological *treatment*. Treatment processes are usually designed to reduce volume and toxicity of waste. The remaining residuals, including those resulting from treatment, are then typically *disposed* in landfills. The ultimate form of residuals depends on how they degrade after release.

18.2.2 Product System Components

The *product system* is defined by the material, energy, and information flows and conversions associated with the life cycle of a product. In addition to life-cycle stages, this system can be organized into four basic components: product, process, distribution, and management. As much as possible, life-cycle design seeks to integrate these components.

Product. The *product component* consists of all materials constituting the final product and includes all forms of those materials in each stage of the life cycle. For example, the product component for a wooden baseball bat consists of the tree, stumpage, and unused branches from raw material acquisition; lumber and waste wood from milling; the bat, wood chips, and sawdust from manufacturing; and the broken bat discarded in a municipal solid waste landfill. If this waste is incinerated, gases, water vapor, and ash are produced.

The product component of a complex product such as an automobile consists of a wide range of materials and parts. These may be a mix of primary (virgin) and secondary (recycled) materials. The materials invested in new or used replacement parts are also included in the product component.

The remaining three components of the product system, process, distribution, and management, each share the following subcomponents:

Facility or plant

Unit operations or process steps

Equipment and tools

Labor

Direct and indirect material inputs

Energy

Process. Processing transforms materials and energy into a variety of intermediate and final products. The *process component* includes direct and indirect materials used to make a product. Catalysts and solvents are examples of direct process materials. They are not significantly incorporated into the final product. Plant and equipment are examples of indirect material inputs for processing. Resources consumed during research, development, testing, and product use are included in the process component.

Specific process-oriented pollution prevention design strategies are addressed in Chaps. 21–26.

Distribution. *Distribution* consists of packaging systems and transportation networks used to contain, protect, and transport products and process materials. Both packaging and transportation result in significant environmental impacts. Packaging accounted for 31.6 percent of municipal solid waste generated in the United States in 1988.[2] Transportation networks include modes and routes. Trains, trucks, ships, airplanes, and pipelines are some major modes of transport. Material transfer devices such as pumps and valves, carts and wagons, and material-handling equipment (forklifts, crib towers, etc.) are part of the distribution component.

Storage facilities such as vessels and warehouses are necessary for distribution. The selling of a product is also considered part of distribution. This includes both wholesale and retail activities.

Management. The *management component* includes the entire information network that supports decision making throughout the life cycle. Within a corporation, management responsibilities include administrative services, financial management, personnel, purchasing, marketing, customer services, legal services, and training and education programs. Each of these has a strong influence on product development. In addition, significant pollution is generated and substantial resources are consumed in support of the management function.

18.3 Goals of Life-Cycle Design

The fundamental goal of life-cycle design is to promote sustainable development at the global, regional, and local level. In simple terms, sustainable development seeks to meet current needs without compromising the ability of future generations to satisfy their needs. Essential elements of sustainable development include pollution prevention, resource conservation, environmental equity, human health, and maintenance of ecosystem structure and function. Stated succinctly, life-cycle design seeks to minimize environmental impacts and utilize resources efficiently in meeting basic societal needs.

A major challenge in sustainable development is achieving environmental equity, both intergenerational and intersocietal. Enormous inequities in the distribution of resources continue to exist between developed and less-developed countries. Inequities also occur within national boundaries. Pollution and other impacts from production are also unevenly distributed.[3] Studies show that low-income communities in the United States are often exposed to higher health risks from industrial activities than are higher-income communities.[4] Inconsistent regulations in the United States also have led to different definitions of acceptable risk levels for workers and consumers.[5]

Life-cycle design goals are articulated through a corporation's environmental management system, to be discussed in the next section. This system then provides the structure for the product development team to specify environmental requirements which shape the design.

18.4   Development Activities

Figure 18-2 demonstrates the complexity in integrating environmental issues into design. The goal of sustainable development is located at the top to indicate its fundamental importance. This goal should be embraced by the entire development team. Various forces shape the creation, synthesis, and evaluation of a design by a product manufacturer, including both internal and external factors.

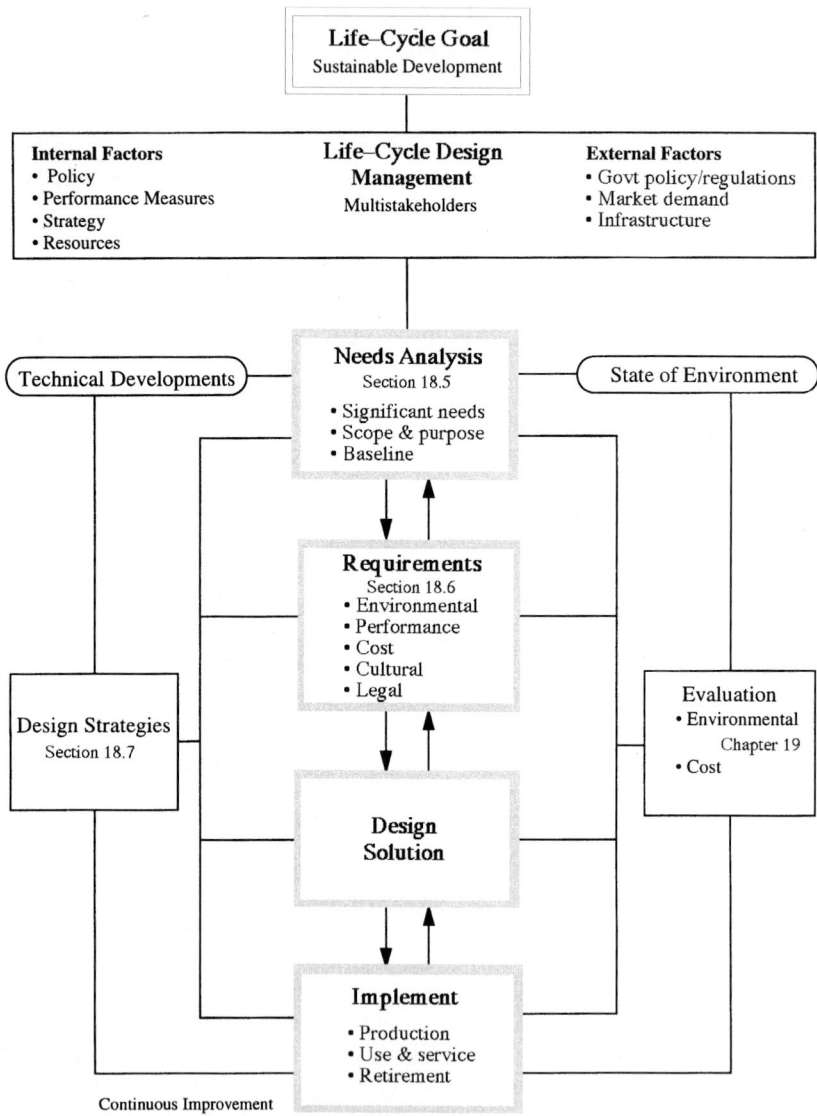

Figure 18-2. The product development process.

External factors include government regulations and policy, market demand, infrastructure, state of the economy, state of the environment, scientific understanding of environmental risks, and public perception of these risks. Many of these issues are addressed elsewhere in this handbook and are also discussed in Ref. 6. Within a company, both organizational and operational changes must take place to effectively implement life-cycle design.

Of the internal factors, management exerts a major influence on all phases of development. Both concurrent design and total quality management (TQM) provide models for life-cycle design. In addition, appropriate corporate policy, goals, and performance measures, as well as adequate resources, are needed to support design projects.

Research and technology development uncovers new approaches for reducing environmental impacts, while the state of the environment provides a context for design. Recognition and prioritization of global, regional, and local environmental problems by the scientific community and the general public should be used to guide improvement. Accordingly, current and future environmental needs are translated into appropriate designs.

A typical design project begins with a needs analysis, then proceeds through formulating requirements, conceptual design, preliminary design, detailed design, and implementation. During the needs analysis, the purpose and scope of the project are defined, and customer needs are clearly identified.

Needs are then expanded into a full set of design criteria that includes environmental requirements, which are discussed in Secs. 18.5 and 18.6. Design alternatives are proposed to meet these requirements. Strategies for satisfying environmental requirements are presented in Sec. 18.7.

The development team continuously evaluates alternatives throughout the design process. Environmental analysis tools include *life-cycle assessment* (LCA), which is outlined in Chap. 19. Several barriers and limitations must be overcome for LCA to be applied to design on a widespread basis.[7] Successful designs must ultimately balance environmental, performance, cost, cultural, and legal requirements.

18.4.1 Design Management

Environmental Management System. Successful life-cycle design projects depend on commitment from all employees and all levels of management. The result is a corporation's environmental management system, which supports environmental improvement through design. Key components of this system include an environmental policy and goals, performance measures, and a strategic plan. This system must also provide access to accurate information about environmental impacts. A well-managed environmental information system is critical to guiding the design process in the direction of environmental improvement. Ideally the environmental management system is well integrated within the corporate structure and not treated as a separate function.

Environmental Policy and Goals.   Company policies that support pollution prevention, resource conservation, and other life-cycle principles foster life-cycle design. Although a step in the right direction, vague environmental policies may not be much help. To benefit design projects, a firm's environmental policies must be specific and clearly stated. Management should offer objectives and guidelines that are detailed enough to provide a practical framework for the actions of designers and others in the company. Examples of environmental goals include phasing out the use of specific chemicals under a specific time line, reducing Toxic Release Inventory (TRI) chemicals by set targets, enhancing the energy efficiency of the product in use, and reducing packaging waste from suppliers to a specific level.

Environmental Performance Measures.   The progress of design projects should be clearly assessed with appropriate measures to help members of the design team pursue environmental goals. Consistent measures of impact reduction in all phases of design provide valuable information for design analysis and decision making. It is important to establish measures that cover efficiency of resource use (materials and energy utilization), and waste generation (multimedia), as well as measures to assess human health and ecosystem sustainability. Life-cycle assessment provides a framework for establishing corporate performance measures that address these issues.

Companies may measure progress toward stated goals in several ways. In each case, life-cycle design is likely to be more successful when environmental aspects are part of a firm's incentive and reward system. Even though life-cycle design can cut costs, increase performance, and lead to greater profitability, it may still be necessary to include discrete measures of environmental responsibility when assessing an employee's performance. If companies claim to follow sound environmental policies, but never reward and promote people for reducing impacts, managers and workers will naturally focus on other areas of the business.

Environmental Strategy.   Strategic planning is essential to manage the complex and dynamic life-cycle system. This activity can seem overwhelming given the different time cycles affecting product system components. Time scales of different events that can influence design include:

Business cycle (recovery, inflation, recession)

Product life cycle (R&D, production, termination, service)

Useful life of the product

Facility life

Equipment life

Process

Cultural trends (fashion obsolescence)

Regulatory change

Technology cycles

Environmental impacts

Shorter-term and longer-term environmental goals should be defined based on these cycles. Although challenging, understanding and coordinating time scales can be a key element in improved design. For life-cycle design to be effective corporations must make long-term investments which will also promote sustainability of the corporation. Such actions include

- Identifying and planning reduction of a company's environmental impacts
- Discontinuing or phasing out product lines with unacceptable impacts
- Investing in research and development of low-impact technology
- Investing in improved facilities and/or equipment
- Recommending regulatory policies that assist life-cycle design
- Educating and training employees in life-cycle design

Effective planning requires correctly assessing company strengths, capabilities, and resources. Many companies are under pressure to shorten development times. This is due in part to competition to continuously bring new products to market. Strategic planning must balance these factors with the need to meet and even exceed life-cycle goals.

18.4.2 Concurrent Design

Traditionally, product and process design have been treated as two separate functions. This can be characterized by a linear design sequence: product design followed by process design. In the last two decades, much progress has been made using process-oriented pollution prevention and waste minimization approaches. Product-oriented approaches are also now gaining recognition. Life-cycle design seeks to integrate product and process design functions to more effectively reduce environmental impacts associated with the entire product system.

Life-cycle design is a logical extension of *concurrent manufacturing,* a procedure based on simultaneous design of product features and manufacturing processes. In contrast to projects that isolate design groups from each other, concurrent design brings participants together in a single team.[8] By having all actors in the life cycle participate in a project from the outset, problems that often develop between different disciplines can be reduced. Product quality can be improved through such cooperation. Efficient teamwork can also reduce development time and lower costs. Table 18-1 shows how various members of the design team can participate.

18.5 Needs Analysis

A development project should first clearly identify customers and their needs. Design can then focus on meeting those needs. Ideas that lead to design pro-

Table 18-1. Role of Participants in Life-Cycle Design

Life-cycle participant	Duties and responsibilities
Accounting	Assign environmental costs to products accurately; calculate hidden, liability, and less tangible costs.
Advertising	Inform customers about environmental attributes of product.
Community	Understand potential impacts and benefits; define and approve acceptable plans and operations.
Distribution and packaging	Design distribution systems that limit packaging and transportation while ensuring protection and containment.
Environmental, health, and safety staff	Ensure occupational, consumer, and community health and safety; provide environmental information for other participants.
Government regulators and standards organizations	Develop policy, regulations, and standards that support life-cycle design goals.
Industrial designers	Create a design concept that meets environmental criteria while also satisfying all other important functions.
Legal	Interpret statutes and promote pollution prevention to minimize cost of regulation and possible future liability.
Management	Establish corporate environmental policy and translate into operational programs; establish measures for success; develop corporate environmental strategy.
Marketing and sales	Give designers feedback on existing products and demand for alternatives; promote design of low-impact products.
Process engineers	Design processes to limit resource inputs and pollutant outputs.
Procurement and purchasing	Select suppliers with demonstrated low-impact operations; assist suppliers in reducing impacts of their operations to ensure steady supply at lower costs.
Production workers	Maintain process efficiency; ensure product quality; minimize occupational health and safety risks.
Purchasers and/or customers	Provide information about needs and environmental preferences; offer feedback on design alternatives.
Research and development staff	Perform basic and applied research on impact reduction technology or product innovations.
Service	Help design product system to facilitate maintenance and repair.
Suppliers	Provide manufacturers with an environmental profile of their goods.
Waste management professionals	Offer information about the fate of industrial waste and retired consumer products and propose options for improved practices.

jects come from many sources, including customer focus groups, and research and development. Environmental assessment of existing products may uncover opportunities for design improvement. One such approach, life-cycle improvement analysis, is discussed in Chap. 19. One improvement strategy involves targeting major environmental impacts for reduction or elimination.

Life-cycle development projects properly focus on filling significant customer and societal needs in a sustainable manner. Avoiding confusion between trivial desires and basic needs is a major challenge of life-cycle design. Unless life-cycle principles such as sustainable development shape the needs analysis, projects may not create low-impact products. By including the environmental requirements in the set of customer requirements that must be satisfied, designers will be motivated to focus on environmental improvement.

Product development managers should first recognize that environmental impacts can be substantially reduced by ending production of high-impact product lines for which lower-impact alternatives are available.

18.5.1 Define Scope of Design Project

In choosing an appropriate system boundary, the development team should initially consider the full life cycle from raw material acquisition to the ultimate fate of residuals. More restricted system boundaries may be justified by the development team. Beginning with the most comprehensive system, design and analysis can focus on the full life cycle, partial life cycle, or individual stages or activities. Choice of the full-life-cycle system will provide the greatest opportunities for impact reduction.

In some cases, the development team may confine analysis to a partial life cycle consisting of several stages, or even a single stage. Stages can be omitted if they are static or not affected by a new design. As long as designers working on a more limited scale are aware of potential upstream and downstream impacts, environmental goals can still be reached. Even so, a more restricted scope will reduce possibilities for design improvement.

After a project has been well defined and is deemed worth pursuing, a project time line and budget should be proposed. Life-cycle design requires funds for environmental analysis of designs. Managers should recognize that budget increases for proper environmental analysis can pay dividends in avoided costs and added benefits that outweigh the initial investment.

18.5.2 Establish Baseline Life-Cycle Data

Comparative analysis and benchmarking are used to establish a basis for environmental improvement. "Benchmarking" is used to compare cost and performance of best-in-class competitors; in life-cycle design environmental performance is also compared.

18.6   Requirements

Formulating requirements may well be the most critical phase of design. Requirements define the expected outcome and are crucial for translating needs and environmental goals into effective design solutions. Design usually proceeds more efficiently when the solution is clearly bounded by well-considered requirements. In later phases of design, alternatives are evaluated on how well they meet requirements.

This discussion focuses on environmental requirements. Incorporating environmental requirements into the earliest stage of design can reduce the need for later corrective action. This proactive approach enhances the likelihood of developing a lower-impact product. Pollution control, liability, and remedial action costs can be greatly reduced by developing environmental requirements at the outset of a project.

Life-cycle design seeks to integrate environmental requirements with traditional performance, cost, cultural, and legal requirements. All requirements must be properly balanced in a successful product. A low-impact product that fails in the marketplace benefits no one.

Regardless of the project's nature, the expected design outcome should not be overly restricted or too broad. Requirements defined too narrowly eliminate attractive designs from the "solution space." On the other hand, vague requirements lead to misunderstandings between potential customers and designers while making the search process inefficient.[9]

When too little time is devoted to developing excellent requirements, a design project can proceed along a mistaken path. Such false starts delay the discovery of critical elements. Mistaken assumptions may also shape design until it is too late or too expensive to develop the proper product.[9,10] Surprises are unavoidable in any development project, but they are far more common and likely to be disastrous when requirements are compiled too hastily.

Activities through the requirements phase typically account for 10 to 15 percent of total product development costs.[11] Yet decisions made at this point can determine 50 to 70 percent of costs for the entire project.[11,12]

18.6.1   Requirements Matrix

Different methods are available to assist the design team in establishing requirements, including requirements matrices and design checklists. This chapter describes a matrix approach. Matrices allow product development teams to study the interactions between life-cycle requirements.

Figure 18-3 shows a multilayer matrix for developing requirements. The matrix for each type of requirement contains columns that represent life-cycle stages. Rows of each matrix are formed by the product system components described in Sec. 18.2: product, process, distribution, and management. Each row is subdivided into inputs and outputs. Elements can then be described and tracked in as much detail as necessary.

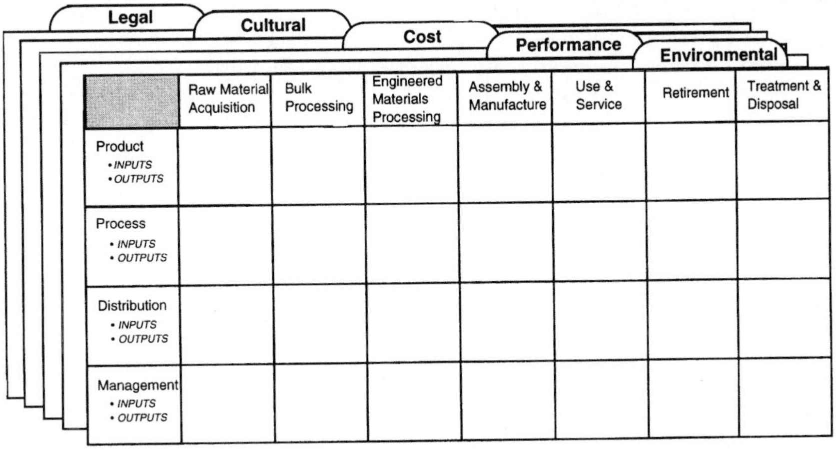
Legal
Cultural
Cost
Performance
Environmental
Raw Material Acquisition
Bulk Processing
Engineered Materials Processing
Assembly & Manufacture
Use & Service
Retirement
Treatment & Disposal
Product
• INPUTS
• OUTPUTS
Process
• INPUTS
• OUTPUTS
Distribution
• INPUTS
• OUTPUTS
Management
• INPUTS
• OUTPUTS

Figure 18-3. Conceptual requirements matrices. (*Courtesy of U.S. EPA,* Life Cycle Design Guidance Manual: Environmental Requirements and the Product System, *EPA 600/R-92/226.*)

The requirements matrices shown in Fig. 18-3 are strictly conceptual. Practical matrices can be formed for each class of requirements by further subdividing the rows and columns of the conceptual matrix. For example, the manufacturing stage could be subdivided into suppliers and the original equipment manufacturer. The distribution component of this stage might also include receiving, shipping, and wholesale activities. Retail sale of the final product might best fit into the distribution component of the use phase.

There are no absolute rules for organizing matrices. Development teams should choose a format that is appropriate for their project.

Table 18-2 is a further illustration of how categories in the matrix can be subdivided. This example shows how each row in the environmental matrix can be expanded to provide more detail for developing requirements.

18.6.2 Types of Requirements

Environmental. Environmental requirements should be developed to minimize

- Use of natural resources (particularly nonrenewables)
- Energy consumption
- Waste generation
- Health and safety risks
- Ecological degradation

Through translation of these goals into clear functions, environmental requirements help identify and constrain environmental impacts and health risks.

Table 18-2. Example of Subdivided Rows for
Environmental Requirements Matrix

Product

Inputs
  Materials
  Energy (embodied)

Outputs
  Products, coproducts, and residuals

Process

Inputs
  Materials
    Direct: process materials
    Indirect: first level (equipment and facilities)
            second level (capital and resources to produce first level)
  Energy: process energy (direct and indirect)
  People (labor)

Outputs
  Materials (residuals)
  Energy (generated)

Distribution

Inputs
  Materials
    Packaging
    Transportation
      Direct (e.g., oil and brake fluid)
      Indirect (e.g., vehicles and garages)

  Energy
    Packaging (embodied)
    Transportation (Btu/ton · mile)
  People (labor)

Outputs
  Materials (residuals)

Management

Inputs
  Materials, office supplies, equipment and facilities
  Energy
  People
  Information

Outputs
  Information
  Residuals

SOURCE: U.S. EPA, *Life Cycle Design Guidance Manual: Environmental Requirements and Product System*, EPA 600/R-92/226.

Table 18-3 lists issues that can help development teams define environmental requirements. This chapter cannot provide detailed guidance on environmental requirements for each business or industry. Although the lists in Table 18-3 are not complete, they introduce many important topics. Depending on the project, teams may express these requirements quantitatively or qualitatively. For example, it might be useful to state a requirement that limits solid waste generation for the entire product life cycle to a specific weight.

In addition to criteria discovered in the needs analysis or benchmarking, government policies can also be used to set requirements. For example, the

Table 18-3. Issues to Consider When Developing Environmental Requirements

Materials		

Amount
  Material intensiveness
 	
Type
  Direct
    Product related
    Process related
  Indirect
    Fixed capital (build-
    ing and equipment)
 	
Source
  Renewable
    Forestry
    Fishery
    Agriculture
  Nonrenewable
    Metals
    Nonmetals
	Character
	  Virgin
	  Recovered (recycled)
	  Reusable/recyclable
	  Useful life
	  Resource base factors
	    Location
	    ▪ Locally available
	    ▪ Regionally available
	    Scarcity
	    ▪ Threatened species
	    ▪ Reserve base
	    Quality
	    ▪ Composition
	    ▪ Concentration
	    Management/restora-
	    tion practices
	    ▪ Sustainability
		Impacts associated with
		extraction, processing, and use
		  Residuals
		  Energy
		  Ecological factors
		  Health and safety

Energy		

Amount
  Energy efficiency
 	
Type
  Purchased
  Process by-product
  Embodied in materials
	Source
	  Renewable
	    Wind
	    Solar
	    Hydro
	    Geothermal
	    Biomass
	  Nonrenewable
	    Fossil fuel
	    Nuclear
		Character
		  Resource base factors
		    Location
		    Scarcity
		    Quality
		    Management/restoration
		    practices
		
		Impacts associated with
		extraction, processing, and use
		  Materials
		  Residuals
		  Ecological factors
		  Health and safety
		  Net energy

Table 18-3. Issues to Consider When Developing Environmental Requirements (*Continued*)

Residuals		
Type	Characterization	Environmental fate

Solid waste
  Solid
  Semisolid
  Liquid
 	
Air emissions
  Gas
  Aerosol
  Particulate
 	
Waterborne
  Dissolved
  Suspended solid
  Emulsified
  Chemical
  Biological
	Nonhazardous
	  Constituents
	  Amount
	
	Hazardous
	  Constituents
	  Toxicity
	  Concentration
	  Amount
	
	Radioactive
	  Potency/half life
	  Amount
	  Concentration
		Containment
		Degradability (physical, biological, chemical)
		Bioaccumulation
		Mobility/transport mechanisms
		  Atmospheric
		  Surface water
		  Subsurface/groundwater
		  Biological
		Treatment/disposal
		  Impacts
		   • Residuals
		   • Energy
		   • Materials
		   • Health and safety effects

Ecological Factors		
Ecological stressors	Type of ecosystems impacts	Scale
Physical (disruption of habitat) Biological Chemical	Diversity Sustainability Rarity Sensitive species	Local Regional Global

Human Health and Safety		
Population at risk	Toxicological characterization	Nuisance effects

Workers
Users
Community
	Morbidity
	Mortality
	Exposure
	  Routes
	   • Inhalation
	   • Skin contact
	   • Ingestion
	  Duration
	  Frequency
		Odors
		Noise
		
		Accidents
		  Type

Integrated Solid Waste Management Plan developed by the EPA in 1989 targets municipal solid waste disposal for a 25 percent reduction by 1995.[13] Other initiatives, such as the EPA's 33/50 Program, are aimed at reducing toxics. It may benefit companies to develop requirements that match the goals of these programs.

It can also be wise to set environmental requirements that exceed government statutes. Designs based on such proactive requirements offer many benefits. Major modifications dictated by regulation can be costly and time consuming. In addition, such changes may not be consistent with a firm's own development cycles, creating even more problems that could have been avoided.

Performance. Performance requirements define functions of the product system. Functional requirements range from size tolerances of parts to time and motion specifications for equipment. Typical performance requirements for an automobile include fuel economy, maximum driving range, acceleration and braking capabilities, handling characteristics, passenger and storage capacity, and ability to protect passengers in a collision. Environmental requirements are closely linked to and often constrained by performance requirements.

Performance is limited by technical factors. Practical performance limits are usually defined by "best available technology." Absolute limits that products may strive to achieve are determined by thermodynamics or the laws of nature. Noting the technical limits on product system performance provides designers with a frame of reference for comparison.

Other limits on performance also need to be understood. In many cases, process design is constrained by existing facilities and equipment. This affects many aspects of process performance. It can also limit product performance by restricting possible materials and features. When this occurs, the success of a major design project may depend on upgrading or investing in new technology.

Designers should also be aware that customer behavior and social trends affect product performance. Innovative technology might increase performance and reduce impacts, but possible gains can be erased by increased consumption. For example, automobile manufacturers doubled average fleet fuel economy over the last twenty years. However, gasoline consumption in the United States remains nearly the same because more vehicles are being driven more miles.

Although better performance may not always result in environmental gain, poor performance usually produces more impacts. Inadequate products are retired quickly in favor of more capable ones. Development programs that fail to produce products with superior performance therefore can contribute to excess waste generation and resource use.

Cost. Meeting all performance and environmental requirements does not ensure project success. Regardless of how environmentally responsible a product may be, many customers will choose another if it cannot be offered at a competitive price. In some cases, a premium can be charged for significantly superior environmental or functional performance, but such premiums are usually limited.

Modified accounting systems that fully reflect environmental costs and benefits are important to life-cycle design. With more complete accounting, many low-impact designs may show financial advantages. Chapter 15 discusses methods of financial analysis that can help companies make better decisions in developing requirements.

Cost requirements should help designers add value to the product system. These requirements can be most useful when they include a time frame (such as total user costs from purchase until final retirement) and clearly state life-cycle boundaries. Parties who will accrue these costs, such as suppliers, manufacturers, and customers, should also be identified.

Cost requirements need to reflect market possibilities. Value can be conveyed to customers through estimates of a product's total cost over its expected useful life. Total customer costs include purchase price, consumables, service, and retirement costs. In this way, quality products are not always judged on least first cost, which addresses only the initial purchase price or financing charges.

Cultural. Cultural requirements define the shape, form, color, texture, and image that a product projects. Low-impact designs must satisfy cultural requirements to be successful. Material selection, product finish, color, and size are guided by consumer preferences. These choices have direct environmental consequences.

However, because customers usually do not know about the environmental consequences of their preferences, creating pleasing, environmentally superior products is a major design challenge. Successful cultural requirements enable the design itself to promote an awareness of how it reduces impacts.

Cultural requirements may overlap with those in other categories. Convenience is usually considered part of performance, but it is strongly influenced by culture. In some cultures, convenience is elevated above many other functions. Cultural factors may thus determine whether demand for perceived convenience and environmental requirements conflict.

Legal. Local, state, and federal environmental, health, and safety regulations are mandatory requirements. Violation of these requirements leads to fines, revoked permits, criminal prosecution, and other penalties. Both companies and individuals within a firm can be held responsible for violating statutes. In 1991, people convicted of violating environmental regulations served prison terms totaling 550 months.[14] Firms may also be liable for punitive damages.

Environmental professionals, health and safety staff, legal advisors, and government regulators can identify legal issues for life-cycle design. Principal local, state, federal, and international regulations that apply to the product system provide a framework for legal requirements. Laws and regulations relating to pollution prevention are discussed in Chaps. 4 and 6.

Federal regulations are administered and enforced by agencies such as the EPA, the Food and Drug Administration (FDA), and the Consumer Product Safety Commission (CPSC). In addition to such federal authorities, many other political jurisdictions enforce regulations. For example, some cities have imposed bans on certain materials and products. Regulations also vary dramatically among countries. The take-back legislation in Germany is beginning to draw more attention to end-of-life issues in product design.

Whenever possible, legal requirements should take into account pending and proposed regulations that are likely to be enacted. Such forward thinking can prevent costly problems during manufacture or use while providing a competitive advantage.

18.6.3 Example of Partial Matrix

The following example illustrates how part of a requirements matrix might be filled in. Requirements in this hypothetical example are proposed for the next generation of a consumer refrigerator. Only requirements for the use stage of the life cycle are shown in Tables 18-4 through 18-8.

This is just a sample of possible requirements. In this example, requirements are stated generally, without specific numerical constraints. An actual project would likely set more requirements in greater detail.

The requirements outlined here demonstrate some of the conflicts and trade-offs that arise in design. For example, increasing insulation in the walls and door reduces energy use, but it can also increase material use and waste at the time of disposal while reducing usable space. If cultural requirements dictate that refrigerators must fit in existing kitchens and maintain a certain usable space, energy-saving actions that increase wall thickness might be precluded. Also, CFCs are usually more efficient than alternatives that do not deplete ozone. Replacing CFCs might increase energy use.

18.6.4 Ranking and Weighing

Organizing. Ranking and weighting distinguishes between critical and merely desirable requirements. After requirements are assigned a weighted value, they should be ranked and separated into several groups. An example of a useful classification scheme follows:

1. *Must* requirements are conditions that designs have to meet. No design is acceptable unless it satisfies all must requirements.

2. *Want* requirements are desirable traits that are not mandatory. Want requirements help designers seek the best solution, not just the first alternative that satisfies mandatory conditions. These criteria play a critical role in customer acceptance and perceptions of quality.

3. *Ancillary functions* are low-ranked in terms of relative importance. They are relegated to a wish list. Designers should be aware that such desires exist. But ancillary functions should only be expressed in design when they do not compromise more critical functions. Customers or clients should not expect designs to reflect many ancillary requirements.

Once must requirements are set, want and ancillary requirements can be assigned priority. There are no simple rules for weighting requirements. Assigning priority to requirements is always a difficult task, because different classes of requirements are stated and measured in different units. Judgments based on the values of the design team must be used to arrive at priorities.

The process of making trade-offs between types of requirements is familiar to every designer. Asking "How important is this function to the design?" or

Table 18-4. Some Use and Service Requirements for Refrigerators

Environmental Matrix

Product
Material type—Based on a materials inventory of components/parts (refrigerator/freezer compartments, refrigeration system, compressor, condenser, evaporator, fans, electric components).
Eliminate high-impact materials: substitute for CFC-12 with lower ozone-depleting-potential and global-warming-potential alternatives.
Material amount
Reduce material intensiveness: specify pounds of material.
Residuals—Specified in Retirement stage.

Process
Energy
Reduce energy use: specify energy consumption for compressor, fans, antisweat heaters (average yearly energy use).
People
Noise: specify frequency and maximum loudness.
Residuals
Reduce waste: specify systems for recovering refrigerant during service; specify level of refrigerant loss during normal use and service; requirements for reuse, remanufacture, recycle of components are stated in Retirement stage.

Distribution
Material type
Reduce impacts associated with packaging materials: specify low-impact materials.
Material amount
Reduce material intensiveness of packaging: specify pounds of material.
Energy
Conserve transportation energy: specify constraints on energy associated with delivery.
Residuals
Reduce packaging waste: specify reusable, recyclable packaging.
Reduce product waste: specify maximum amount of damaged products during distribution.

Management
Information
Provide consumers with information on energy use: meet DOE labeling requirements for energy efficiency.

"What is this function worth (to society, customers, suppliers, others)?" is a necessary exercise in every successful development project.

Resolving Conflicts. Development teams can expect conflicts between requirements, as was demonstrated in the refrigerator design example. If conflicts cannot be resolved between must requirements, there is no solution space

Table 18-5. Some Use and Service Requirements for Refrigerators
Performance Matrix

Product
Material
Dimensions: H × W × D; capacity in cubic feet; shelf area; usable storage space.
Features: ice making; meat keeping; crisper humidity.

Process
Material
Identify best available technology for refrigeration system components as a practical limit to performance.
Specify useful life of product and components
Specify reliability.
Specify durability.
Energy
Identify thermodynamic limits to performance (e.g., maximum efficiency determined by temperatures inside and outside the refrigerator).
Specify temperature control: balance, uniformity, compensation.

Distribution
Material
Specify product demand.
Specify installation time and equipment requirements.
Specify packaging requirements for protection and containment.
Energy
Specify location of retail outlets relative to market.

Management
Information
Specify minimum information requirements for owner's manual.
Specify warranty period.

for design. When a solution space exists but is so restricted that little choice is possible, must requirements may have been defined too narrowly. The absence of conflicts usually indicates that requirements are defined too loosely. This produces cavernous solution spaces in which virtually any alternative seems desirable. Under such conditions, there is no practical method of choosing the best design.

In all of these cases, design teams need to redefine or assign new priorities to requirements. If careful study still reveals no solution space or a very restricted one, the project should be abandoned. It is also risky to proceed with overly broad requirements. Only projects with practical, well-considered requirements should be pursued. Successful requirements usually result from resolving conflicts and developing new priorities that more accurately reflect customer needs.

Table 18-6. Some Use and Service Requirements for Refrigerators

Cost Matrix

Product

Material
  Retail price.
  Cost for replacement parts.

Process

Material and labor
  Service costs (cost for service and parts).

Energy
  Electricity ($/kWh × kWh/yr).

Distribution

Material, energy, and labor
  Delivery and installation cost.

Residuals
  Packaging disposal cost.

Management

Information
  Manufacturer's guarantee.
  Payback period to user for purchasing more expensive energy-efficient unit.

Table 18-7. Some Use and Service Requirements for Refrigerators

Cultural Matrix

Product

Material
  Color preferences.
  Size (dependent on frequency of shopping and on convenience).
  Finishes and materials (affects cleaning, appearance).

Process

Material
  Manual vs. automatic defrost.
  Compartmentalization—ability to organize food.

Residuals
  Food spoilage—ability to control temperature.

Management

Information
  Instructions clearly written.

Table 18-8. Some Use and Service Requirements for Refrigerators

Legal Matrix

Product

Material
  Consumer Product Safety Commission.
  Montreal Protocol for discontinuing the use of CFCs.
  TSCA (Refrigerants meet regulations for use).

Process

Energy
  National Appliance Energy Conservation Act—January 1, 1993 [maximum energy consumption rate = $E = 16.0$ AV + 355 kWh/yr (AV = adjusted volume of top-mounted refrigerator)].

Distribution

Residuals
  Packaging: German take-back legislation; community recycling ordinance.

Management

Information
  FTC guidelines on environmental claims.
  DOE labeling requirements for energy efficiency.

18.7 Design Strategies

This section will focus on design strategies relating to product and distribution components of the product system. Process- and management-oriented strategies for achieving pollution prevention are addressed elsewhere in this handbook.

Appropriate strategies satisfy the entire set of design requirements, thus promoting integration of environmental requirements into design. For example, essential product performance must be preserved when design teams choose a strategy for reducing environmental impacts. If performance is degraded, the benefits of environmentally responsible design may be illusory.

General strategies that may be followed to fulfill environmental requirements are presented in Table 18-9. Most of these strategies reach across product system boundaries. Product life extension strategies can also be applied to equipment used in processing, distribution, and management. Similarly, process design strategies are not limited to manufacturing operations. They are also useful when product use depends on processes. For example, the drive train of an automobile functions like a miniature industrial plant with a reactor, storage tanks, electric power generator, and process control equipment. Process strategies can thus lower environmental impacts caused by automobile use.

The following sections present impact and risk reduction strategies. It is unlikely that a single strategy will be best for meeting all environmental require-

Table 18-9. Design Strategies

General strategy	Specific strategy
Product life extension	Appropriately durable Adaptable Reliable Serviceable Remanufacturable Reusable
Material life extension	Recycling
Material selection	Reformulation Substitution
Reduced material intensiveness	
Process improvement (see Chaps. 21–26)	Process substitution Process control Improved process layout Inventory control and material handling Facility planning (Chap. 12)
Efficient distribution	Transportation Packaging
Improved management practices	Office management (Chap. 30) Total quality management (Chap. 9) Accounting (Chap. 16)
Improved information provision	Product labeling (Chap. 20)

ments. One strategy is even less likely to satisfy the full set of requirements. For that reason, most development projects should adopt a range of strategies.

18.7.1 Product System Life Extension

Extending the life of a product can directly reduce environmental impacts. In many cases, longer-lived products save resources and generate less waste, because fewer units are needed to satisfy the same needs. Before pursuing this strategy, designers should understand the concept of useful life.

Useful life measures how long a system will operate safely and meet performance standards when maintained properly and not subject to stresses beyond stated limits.[15] Measures of useful life vary with function. Some common measures and examples are listed below:

Measures for useful life	*Product examples*
Number of uses or duty cycles	Clothes washers, switches
Length of operation (i.e., operating hours, months, years, or miles)	Automobiles, light bulbs
Shelf life	Food, unstable chemicals

Retirement is the defining event of useful life. Reasons why products are no longer in use include

- Technical obsolescence
- Fashion obsolescence
- Degraded performance or structural fatigue caused by normal wear over repeated uses
- Environmental or chemical degradation
- Damage caused by accident or inappropriate use

A product may be retired for fashion or technical reasons, even though it continues to perform its design functions well. Clothing and furniture are often retired prematurely when fashions change. Technical obsolescence is common for electronic devices.

Users may also be forced to retire a product for functional reasons. Normal wear can degrade performance until the product no longer serves a useful purpose. Repeated use can also cause structural deformation and fatigue that finally result in loss of function.

Some products are exposed to a wide variety of environmental conditions that cause corrosion or other types of degradation. Such biological or chemical stresses can reduce performance below a critical level. This type of deterioration may also cause products to be retired for aesthetic reasons, even though they continue to perform adequately.

Accidents or incorrect use also cause premature retirement. Poor design or failure to consider unlikely operating conditions may lead to accidents. Some of these events can be avoided through better operating instructions or warnings.

Understanding why products are retired helps designers extend product system life. To achieve a long service life, designs must successfully address issues beyond simple wear and tear. A discussion of specific strategies for product life extension follows.

Appropriately Durable.   *Durable* items can withstand wear, stress, and environmental degradation over a long useful life.

A durable product continues to satisfy customer needs over an extended life. Some design actions may make a product more durable without the use of additional resources. However, enhanced durability may depend on increased resource use. When this happens, design alternatives should be compared on a normalized basis (total impacts/useful life).

Development teams should enhance durability only when appropriate. Designs that allow a product or component to last well beyond its expected useful life can be wasteful.

Products based on rapidly changing technology may not always be proper candidates for enhanced durability. If a simple product will soon be obsolete, making it more durable could be pointless. In complicated products subject to rapid change, adaptability is usually a better strategy. For example, modular construction allows easy upgrading of fast-changing components without replacing the entire product. In such cases, useful life is expected to be short for certain components, so they should also not be designed for extreme durability.

Durable designs must also meet other project requirements. When least first cost is emphasized, durable products may encounter market resistance. Even so, durability is often associated with high-quality products. For example, garden tools with reinforced construction can withstand higher stresses than lower-quality alternatives and thus generally last longer. Although these tools are initially more expensive, they may be cheaper in the long run because they do not need to be replaced as frequently.

Enhanced durability can be part of a broader strategy focused on marketing and sales. For some durable products, leasing may be more successful than sale to customers. Leasing can be viewed as selling services while maintaining control over the means of delivering those services. Durability is an integral part of all profitable leasing. Original equipment manufacturers who lease their products usually have the most to gain from durable designs.

Adaptable.   *Adaptable* designs either allow continual updating or they perform several different functions. *Modular components* allow single-function products to evolve and improve as needed.

As previously mentioned, adaptability can extend the useful life of products that quickly become obsolete. Products with several parts are the best candidates for adaptable design. To reduce overall environmental impacts, a sufficient portion of the existing product must usually remain after obsolete parts are replaced.

Adaptable designs rely on interchangeable components. Interchangeability controls dimensions and tolerances of manufactured parts so that components can be replaced with minimal adjustments or on-site modifications.[15] Thus, fittings, connectors, or information formats on upgrades are consistent with the original product. For example, an adaptable strategy for a new razor blade design would ensure that blades mount on old handles so the handles don't become part of the wastestream.

Adaptable design may be particularly beneficial for processes and facilities. This strategy allows rapid response to changing conditions through continual upgrades. Such adaptable manufacturing may make it much easier to offer low-impact products that meet customer demands. A well-designed system helps save suitable plant and equipment for continued use.

Reliable.   *Reliability* is often expressed as a probability. It measures the ability of a system to accomplish its design mission in the intended environment for a certain period of time.

Environmental impacts are influenced by reliability. Unreliable products or processes, even if they are durable, may be retired prematurely. Customers will not tolerate untrustworthy performance, inconvenience, and expense for long. Unreliable designs can also present safety and health hazards.

The number of components, the individual reliability of components, and configuration are important aspects of reliability. Parts reduction and simplified design can increase both reliability and manufacturability. Simpler designs may also be easier to service. All these factors can reduce resource use and waste. Aside from environmental benefits, producers and customers can save money with reliable products.

Reliability cannot always be achieved by reducing the number of parts or making designs simple. In some cases, redundant systems must be added to provide needed backup. When a reliable product system requires parallel systems or fail-safe components, costs may rise significantly.

Reliability should be designed into products rather than achieved through later inspection. Screening out potentially unreliable products after they are made is wasteful because such products must either be repaired or discarded. In both cases, environmental impacts and costs increase.

Serviceable.   A *serviceable* system can be adjusted for optimum performance under controlled conditions. This capacity is retained over a specified life.

Many complex products designed to have a long useful life require service and support. When designing serviceable products, the team should first determine who will provide the service. Any combination of original equipment manufacturers, dealers, private business, or customers may service a product. Types of tools and the level of expertise needed to perform tasks strongly influence who is capable of providing service. In any case, simple procedures are an advantage.

Design teams should also recognize that equipment and an inventory of parts are a necessary investment for any service network. Service activities may be broken into two major categories: maintainability and repairability.

Maintainable.   The relative difficulty or time required to maintain a certain level of system performance determines whether that system can be practically *maintained*.

Maintenance includes periodic, preventative, and minor corrective actions. Proper maintenance helps to conserve resources and prevent pollution. For example, tuning an automobile engine improves fuel economy while reducing toxic tailpipe emissions. On the other hand, delaying or ignoring maintenance can damage a product and shorten its useful life.

Designers wishing to create product systems that are easy to maintain should address the following topics:

Downtime, tool availability, personnel skills

Complexity of required procedures

Potential for error

Accessibility to parts, components, or system to be maintained

Frequency of design-dictated maintenance

This is not an exhaustive list, but it identifies some key factors affecting maintenance. Most of these criteria are interrelated. If maintenance is complex, specialized personnel are required, downtime is likely to be long, and the potential for error increases. Specialty tools also make maintenance less convenient.

Similarly, if parts or components are not readily accessible, complexity and costs can increase. Spatial arrangement is the key to easy access. Critical parts and assemblies within a piece of equipment should be placed so they can be reached and the necessary procedures performed. Simpler designs are usually easier to maintain.

Maintenance schedules should balance a variety of requirements. For an automobile, changing motor oil every 500 miles would obviously be wasteful, but changing oil every 50,000 miles would damage the engine. Customers usually believe that the less often maintenance is required the better, so designs that preserve peak performance with minimal maintenance are likely to be more popular. In addition, low-maintenance designs are more likely to stay in service longer than less robust designs. Products dependent on continual readjustments for an acceptable level of performance are generally considered low-quality. Such products can be wasteful, and they are not likely to gain much market share.

Repairable.   *Repairability* is determined by the feasibility of replacing dysfunctional parts and returning a system to operating condition.

A two-step process is usually followed when a product needs repair. First, a diagnosis identifies the defect. Then, several questions critical to resource management should be asked:

Should the product be repaired or retired?

Are other components near the end of their useful life and likely to fail soon?

Should the defective component be replaced with a new, remanufactured, or used part?

Answers to these questions should take into account life-cycle consequences.

Factors relating to downtime, complexity, and accessibility are as important in repair as they are in maintenance. Easily repaired products also rely on interchangeable and standard parts. *Interchangeability* usually applies to parts produced by one manufacturer. *Standardization* refers to compatible parts made by different manufacturers. Standardization makes commonly used parts and assemblies conform to accepted design standards.[15]

Use of standard parts designed to codes established by numerous manufacturers greatly aids repair. Designs that feature unique dimensions for common parts can confound normal repair efforts. Specialty parts usually require expanded inventories and extra training for repair people. In the burgeoning global marketplace, following proper standards enables practical repair.

Cost also determines repairability. If normal repair is too expensive, practical repairability does not exist. Labor, which is directly related to complexity and accessibility, is a key factor in repair costs. When labor is costly, only items of relatively high value will be repaired. However, a substantial purchase price is not enough to promote repairability. Designs that impede repair may be retired prematurely regardless of initial investment. As with maintenance, infrequent need, ease of intervention, and a high probability of success lower operating costs, increase customer satisfaction, and translate directly into perceptions of higher quality.

Repairable designs need proper after-sale support. Firms should offer information about troubleshooting, procedures for repair, tools required, and the expected useful life of components and parts.

Remanufacturable. *Remanufacturing* is an industrial process that restores worn products to like-new condition. In a factory, a retired product is first completely disassembled. Its usable parts are then cleaned, refurbished, and put into inventory. Finally, a new product is assembled from both old and new parts, creating a unit equal in performance and expected life to the original or a currently available alternative. In contrast, a *repaired* or *rebuilt* product usually retains its identity, and only those parts that have failed or are badly worn are replaced.[16]

Industrial equipment or other expensive products not subject to rapid change are the best candidates for remanufacture. Typical remanufactured products include jet engines, buses, railcars, manufacturing equipment, and office furniture. Viable remanufacturing systems rely on the following factors:[17]

A sufficient population of old units ("cores")

An available trade-in network

Low collection costs

Storage and inventory infrastructure

Design teams must first determine if enough old units will exist to support remanufacturing. Planning for proper marketing and collection after retirement helps ensure a sufficient population of cores. To remain competitive with new products, the cost of cores must be low. Costs for collecting cores include transport and a trade-in to induce customer return.

Systems for collecting and storing the needed number of cores at competitive prices support remanufacturing. But no remanufacturing program can succeed without design features and strategies such as

Ease of disassembly

Sufficient wear tolerances on critical parts

Avoiding irreparable damage to parts during use

Interchangeability of parts and components in a product line

Designs must be easy to take apart if they are to be remanufactured. Adhesives, welding, and some fasteners can make this impossible. Critical parts must also be designed to survive normal wear. Extra material should be present on used parts to allow refinishing. Care in selecting materials and arranging parts also helps avoid excessive damage during use. Design continuity increases the number of interchangeable parts between different models in the same product line. Common parts make it easier to remanufacture products.

Reusable.   *Reuse* is the additional use of an item after it is retired from a clearly defined duty. Reformulation is not reuse. However, repair, cleaning, or refurbishing to maintain integrity may be done in transition from one use to the next.

The environmental impacts of reusable products are often contrasted with those of single-use alternatives. Examples include diapers, cameras, razors, and clothing. Which designs are environmentally superior is controversial in some cases. In others, reuse offers a clear advantage.

Reusable products are returned to the same or less demanding service without major alterations. They may undergo some minor processing, such as cleaning, between services. For example, dishware or glass bottles can be washed before reuse.

The environmental profile of a reusable product does not always depend on the number of expected uses. If the major impacts occur in manufacturing and earlier stages, increasing the number of uses will reduce total environmental impacts. However, when most impacts are caused by cleaning or other steps between uses, increasing the number of duty cycles may have little effect on overall impacts.

Convenience is often cited as a major advantage of single-use products. However, customers usually fail to consider the costs and time of purchasing, storing, and disposing single-use products. Single-use products often cost more per use than reusable products.

Several environmental comparisons between reusable and single-use products have been made. These are mostly confined to life-cycle inventories, which are discussed in the next chapter.

18.7.2 Material Life Extension

Recycling.   *Recycling* is the reformation or reprocessing of a recovered material. The EPA defines *recycling* as "the series of activities, including collection, separation, and processing, by which products or other materials are recovered from or otherwise diverted from the solid waste stream for use in the form of raw materials in the manufacture of new products other than fuel."[18]

Many designers, policymakers, and consumers believe recycling is the best solution to a wide range of environmental problems. Recycling does divert discarded material from landfills, but it also causes other impacts. Before designers focus on making products easier to recycle, they should understand several recycling basics. A discussion of types of recovered material, pathways, and infrastructure will provide a framework for understanding recycling.

Types of Recycled Material.   Material available for recycling can be grouped into the following three classes: home scrap, preconsumer, and postconsumer.

Home scrap consists of materials and by-products generated and commonly recycled within an original manufacturing process.[18] Many materials and products contain home scrap that should not be advertised as recycled content. For example, mill broke (wet pulp and fibers) is easily added to later batches of product at paper mills. This material has historically been used as a pulp substitute in paper making rather than discarded, so it is misleading to consider it recycled content.

Preconsumer material consists of overruns, rejects, or scrap generated during any stage of production outside the original manufacturing process.[18] It is generally clean, well-identified, and suitable for high-quality recovery. Preconsumer material is now recycled in many areas.

Postconsumer material has served its intended use and been discarded before recovery. Unfortunately, in many cases postconsumer material is a relatively low-quality source of input for future products.

Recycling Pathways.   Development teams choosing recycling as an attractive way to meet requirements should be aware of the two major types of pathways recycled material can follow: closed-loop pathways and open-loop pathways.

In *closed-loop systems*, recovered materials and products are suitable substitutes for virgin material. They are thus used to produce the same part or product again. Some waste is generated during each reprocessing, but in theory a closed-loop model can operate for an extended period of time without virgin material. Of course, energy, and in some cases process materials, are required for each recycling.

Solvents and other industrial process ingredients are the most common materials recycled in a closed loop. Postconsumer material is much more difficult to recycle in a closed loop, because it is often degraded or contaminated. Designs that anticipate closed-loop recycling of such waste may thus overstate the likely benefits.

Open-loop recycling occurs when recovered material is recycled one or more times before disposal. Most postconsumer material is recycled in an open loop. The slight variation or unknown composition of such material usually causes it to be downgraded to less demanding uses.

Some materials also enter a *cascade open-loop model* in which they are degraded several times before final discard. For example, used white ledger paper may be recycled into additional ledger or computer paper. If this product is then dyed or not de-inked, it will be recycled as a mixed grade after use. In this form, it could be used for paperboard or packing, such as trays in produce boxes. At present, the fiber in these products is not valuable enough to recover. Ledger paper also enters an open-loop system when it is recycled into facial tissue or other products that are disposed after use.

Infrastructure.   Types of recycled materials, and the major routes they follow, provide an introduction to recycling. Infrastructure is the key to understanding how recycling actually occurs. Suitable programs must be in place or planned to ensure the success of any recycling system. Key considerations include

Recycling programs and participation rates

Collection and reprocessing capacity

Quality of recovered material

Economics and markets

Economic and market factors ultimately determine whether a material will be recycled. Markets for some secondary materials may be easily saturated. Recycling programs and high rates of participation address only collection; unless recovered material is actually used, no recycling has occurred.

In addition, if a material is not one of the few now targeted for public collection, recovery could be difficult. It may not be possible to create a private collection and reprocessing system that competes with virgin materials. However, if demand for recovered material increases in the future, this will greatly aid collection efforts.

Design Considerations.   Recycling can be a very effective resource management tool. Under ideal circumstances, most materials would be recovered many times until they became too degraded for further use. Even so, design for recyclability is not the ultimate strategy for meeting all environmental requirements. For example, studies show that refillable glass bottles have a much lower life-cycle energy usage than single-use recycled glass to deliver the same amount of beverage.[19]

When suitable infrastructure appears to be in place, or the development team is capable of planning it, recycling is enhanced by

Ease of disassembly

Material identification

Simplification and parts consolidation

Material selection and compatibility

Products may have to be taken apart after retirement to allow recovery of materials for recycling. However, easy disassembly may conflict with other project needs. For example, snap-fit latches and other joinings that speed assembly can severely impede disassembly. In some products, easy disassembly may also lead to theft of valuable components.

Material identification markings greatly aid manual separation and the use of optical scanners. Standard markings are most effective when they are well-placed and easy to read. Symbols have been designed by the Society of the Plastics Industry (SPI) for commodity plastics. The Society of Automotive Engineers (SAE) has developed markings for engineered plastics. Of course, marked material must still be valuable and easy to recover or it will not be recycled. In addition, labeling may not be useful in systems that rely on mechanical or chemical separation, although it can be a vital part of collection systems that target certain materials or rely on source separation.

Simplification and parts consolidation can also make products easier to recycle. This is an attractive strategy for many other reasons. As previously mentioned, simple designs also ease assembly and may lead to more robust, higher-quality products.

In many design projects, material selection has not been coordinated with environmental strategies. As a result, many designs contain a bewildering number of materials chosen for combined cost and performance attributes. There may be little chance of recovering material from such complex products unless they contain large components made of a single, practically recyclable material.

Even without separation, some mixtures of incompatible or specialty materials can be "downcycled." At present, several means are available to form incompatible materials into composites. However, the resulting products, such as plastic lumber, may have limited appeal.

Designers can aid recycling by reducing the number of incompatible materials in a product. For example, a component containing parts composed of different materials could be designed with parts made from the same material. This strategy also applies within material types. Formulations of the same material might have such different properties that they are incompatible during recycling. Designers will usually have to make trade-offs when selecting only compatible materials for a product. Making single-material or compatible components may be possible in some cases but not in others.

18.7.3   Material Selection

Material selection, which is fundamental to design, offers many opportunities for reducing environmental impacts throughout a product life cycle. In life-cycle design, material selection begins with identification of the nature and source of raw materials. Then environmental impacts caused by material acquisition, processing, use, and end-of-life product management are evaluated. Finally, proposed materials are compared to determine best choices.

When modest improvements of existing products or the next generation of a line are designed, material choice may be constrained. Designers may also be restricted to certain materials by the need to use existing plant and equipment. This type of process limitation can even affect new product design. Substantial investment may then be needed before a new material can be used. On the other hand, material substitutions may fit current operations and actually reduce costs. In either case, material choice must meet all project requirements.

Reformulation is also an option when materials are selected. Most materials or products may be reformulated to reduce impacts, even when material choice is constrained.

Substitution.   Substitution is a strategy available for improvement of existing designs. The challenge with substitution is to reduce life-cycle environmental impacts without compromising performance, cost, or other requirements. These material substitutions can address a wide range of issues, such as replacing rare tropical woods in furniture with native species.

Material substitutions can be made for product as well as process materials, such as solvents and catalysts. For example, water-based solvents or coatings can sometimes be substituted for high-VOC alternatives during processing. On the other hand, materials that don't require coating, such as some metals and polymers, can be substituted in the product itself.

Reformulation.   Reformulation is a less drastic alternative than substitution. It is an appropriate strategy when a high degree of continuity must be maintained with the original product. Consumables and other products that must fit existing standards may limit design choices. Rather than entirely replace one material with another, designers can alter percentages to achieve the desired result. Some materials can also be added or deleted if characteristics of the original product are still preserved. Gasoline is one product that has undergone many reformulations to reduce fugitive emissions as well as emissions from combustion. In this case, reformulation is further complicated because it can reduce fuel economy or engine performance.

18.7.4   Reduced Material Intensiveness

Resource conservation can reduce waste and directly lower environmental impacts. A product that is less material intensive may also be lighter, thus saving energy in distribution or use. Designing to conserve resources is not always simple. Reduced material use may affect other requirements in complex ways.

In some cases, using less material affects no other requirements and thus clearly lowers impacts. When the reduction is very simple, benefits can be determined without a rigorous life-cycle assessment. However, careful study may be needed to ensure that significant impacts have not been created elsewhere in the life cycle. In addition, impacts might have been reduced further by use of another material rather than less of the current choice.

18.7.5   Efficient Distribution

Both transportation and packaging are required to transfer goods between locations. A life-cycle design project benefits from distribution systems that are as efficient as possible.

Transportation.   Life-cycle impacts caused by transportation can be reduced by several means. Approaches that can be used by designers include:

Choose an energy-efficient mode

Reduce air pollutant emissions from transportation

Maximize vehicle capacity where appropriate

Backhaul materials

Ensure proper containment of hazardous materials

Choose routes carefully to reduce potential exposure from spills and explosions

Trade-offs between various modes of transportation will be necessary. Transportation efficiencies are shown in Table 18-10. Time and cost considerations, as well as convenience and access, play a major role in the choice of the best transportation. When selecting a transportation system, designers should also consider infrastructure requirements and their potential impacts.

Packaging.   Packaging must contain and protect goods during transport and handling to prevent damage. Regardless of how well designed an item might be, damage during distribution and handling may cause it to be discarded before use. To avoid such waste, products and packaging should be designed to complement each other.

The concurrent practices of life-cycle design are particularly effective in reducing impacts from packaging. As a first step, products should be designed to withstand both shock and vibration. When cushioned packaging is required, members of the development team need to collaborate to ensure that cushioning does not amplify vibrations and thus damage critical parts.[20] Cooperation between design specialties can greatly reduce such product damage.

The following strategies may be used to design packaging within the life-cycle framework. Most of these strategies also result in significant cost savings.

Packaging reduction

- Elimination: distribute appropriate products unpackaged
- Reusable packaging
- Product modifications
- Material reduction

Material substitution

- Recycled materials
- Degradable materials

Packaging Reduction.   Shipping items without packaging is the simplest approach to impact reduction. In the past, many consumer products, such as screwdrivers, fasteners, and other items, were offered unpackaged. They can still be hung on hooks or placed in bins that provide proper containment while allowing customer access. This method of merchandising avoids use of unnecessary plastic wrapping, paperboard, and composite materials. Wholesale

Table 18-10. 1990 Transportation Fuel Requirements

	Fuel consumed per 1000 ton-miles	Energy consumed* (Btu/ton-mile)
Combination truck (tractor trailer)		
Diesel	11.8 gal	1945
Gasoline	11.8 gal	1782
Single-unit truck		
Diesel	19.1 gal	3136
Gasoline	20.8 gal	3132
Rail		
Diesel	3.1 gal	514
Barge†		
Diesel	2.0 gal	330
Residual	0.6 gal	96
Total		426
Ocean freighter†		
Diesel	0.1 gal	16
Residual	1.0 gal	173
Total		190
Pipeline—natural gas		
Natural gas	2300 ft³	2657
Pipeline—petroleum products		
Electricity	22 kWh	236
Pipeline—coal slurry		
Electricity	235 kWh	2517

*Includes precombustion energy for fuel acquisition.
†An average ratio of diesel and residual fuels is used to represent barge and ocean freighter transportation energy.

SOURCE: Franklin Associates, Ltd.

packaging can also be eliminated. For example, furniture manufacturers commonly ship furniture uncartoned. Uncartoned furniture is protected with blankets that are returned after delivery to the distribution center.

Reusable packaging systems are also an attractive design option. Wholesale items that require packaging are commonly shipped in reusable containers. Tanks of all sizes, wire baskets, wooden shooks, and plastic boxes are frequently used for this purpose.

Necessary design elements for most reusable packaging systems include

Collection or return infrastructure

Procedures for inspecting items for defects or contamination

Repair, cleaning, and refurbishing capabilities

Storage and handling systems

Unless such measures are in place or planned, packaging may be discarded rather than reused. Manufacturers and distributors cannot reuse packaging unless infrastructure is in place to collect, return, inspect, and restore packaging for another service. Producers can reduce these infrastructure needs by offering their product in bulk. Some system will still be required for reusable wholesale packaging, but it should be much less complex than that needed to handle consumer packaging. When products are sold in bulk, customers control all phases of reuse for their own packaging.

Even so, waste generation and other environmental impacts are only reduced when customers reuse their container several times. Customers who use new packaging for each bulk purchase generally consume more packaging than customers who buy prepackaged products. This is particularly true of items distributed in single-use bulk packaging.[21]

Product modification is another approach to packaging reduction. Sturdy products may require less packaging and may also prove more robust in service. Depending on the delivery system, some products may safely be shipped without packaging of any kind. Even when products require primary and secondary packaging to ensure their integrity during delivery, product modifications may decrease packaging needs. Designers can further reduce the amount of packaging used by avoiding unusual product features or shapes that are difficult to protect.

Reformulation is another type of product modification that may be possible for certain items. Products that contain ingredients in diluted form may be distributed as concentrates. In some cases, customers can simply use concentrates in reduced quantities. A larger, reusable container may also be sold in conjunction with concentrates. This allows customers to dilute the product as appropriate. Examples of product concentrates include frozen juice concentrates and concentrated versions of liquid and powdered detergent.

Material reduction may also be pursued in packaging design. Many packaging designers have already managed to reduce material use while maintaining performance. Reduced thickness of corrugated containers (board grade reduction) provides one example. In addition, aluminum, glass, plastic, and steel containers have continually been redesigned to require less material for delivering the same volume of product.

Material Substitution. As discussed, material substitution can reduce impacts in other areas of design. One common example of this strategy in packaging is the substitution of more benign printing inks and pigments for those containing toxic heavy metals or solvents. The less harmful inks are usually just as effective for labels and graphic designs. When some properties depend on toxic constituents, designers can develop new images that are compatible with sounder pigments, inks, and solvents.

Whenever possible, designers can create packaging with a high recycled content. Many public and private recycling programs currently focus on collecting packaging. As a direct consequence, firms are being encouraged to increase the recycled content of their packaging.

However, using recycled material in packaging design cannot be thought of as a complete strategy in itself. Opportunities for material reduction and packaging reduction or elimination should still be investigated. Recycling and recycled materials were discussed in more detail earlier in this chapter.

Degradable materials are capable of being broken down by biological or chemical processes or exposure to sunlight. At first glance, package designs based on degradable material appear to be an attractive solution to the mounting problem of waste disposal. But the lack of sunlight, oxygen, and water in modern landfills severely inhibits degradation. Degradable materials thus provide only limited benefits in packaging that will be properly disposed. This may change if composting of municipal waste becomes more widespread.

In any event, degradability is a desirable trait for litter deposited in aesthetically pleasing natural areas. In particular, polymers or other materials that are normally resistant to decay are less of a nuisance if they can be formulated to quickly break down. Degradable materials may also benefit some aquatic species that encounter litter. Various mammals, birds, and fish can die from entrapment in such items as six-pack rings and plastic sacks. Even so, it may be difficult to determine whether degradable packaging is an asset or just encourages irresponsible behavior.

Previously resistant materials that are now designed to decay may also cause unanticipated problems. Degradable polymers can impede recycling efforts by acting as a contaminant in recovered materials. Questions have also been raised about the environmental impacts of degraded polymers. Degradation can liberate dyes, fillers, and other potentially toxic constituents from a material that was previously inert.

18.8 Summary of Life-Cycle Design Principles

Life-cycle design principles for achieving pollution prevention and guiding the environmental improvement of the product system are summarized here.

1. Addressing environmental issues in the earliest stages of design is one of the most efficient approaches to achieving pollution prevention. Other related benefits include enhancing resource efficiency, reducing liabilities, and achieving competitiveness.

2. The ultimate goal of life cycle design is to achieve sustainable development. Sustainable development seeks to satisfy basic societal needs of today without compromising future generations' ability to meet their needs. Maintenance of ecosystem structure and function (the planet's life support system) is critical to achieving this goal.

3. The product life cycle is a useful framework for evaluating and reducing adverse environmental impacts associated with the manufacture, use, and

end-of-life management of a product. Designers can prevent the shifting of adverse impacts between media and life-cycle stages.

4. Both internal and external factors strongly influence design. Internally, the environmental management system, which includes goals and performance measures, provides the organizational structure within a company to implement pollution prevention by design. Access to accurate information about environmental impacts is also critical for achieving environmental improvement. External factors that shape design include government regulations, market forces, infrastructure, and state of the environment, as well as scientific understanding and public perception of risks.

5. The concurrent design of product system components (product, process, distribution, and information/management) is an important principle in life-cycle design management. Interdisciplinary participation is key to defining requirements that reflect the needs of multiple stakeholders: suppliers, manufacturers, consumers, resource recovery and waste managers, the public, regulators.

6. Specification of requirements is one of the most critical design functions. Requirements guide designers in translating needs and environmental objectives into successful designs. Environmental requirements should focus on minimizing natural resource consumption, energy consumption, waste generation, and human health risks, as well as promoting the sustainability of ecosystems.

7. Life-cycle design seeks to optimize environmental objectives while also optimizing cost, performance, cultural, and legal requirements. The challenge is to apply value-added design strategies that resolve conflicting requirements.

Two industry demonstration projects of the life-cycle design framework are being conducted by the EPA's Pollution Prevention Branch and National Pollution Prevention Center based at the University of Michigan. Results from demonstration projects with AT&T and Allied Signal are currently being documented and will be published by the EPA. The author has also recently completed a critical review of life cycle design.[22]

References

1. U.S. Environmental Protection Agency, Office of Research and Development, Risk Reduction Engineering Laboratory, *Life Cycle Design Guidance Manual: Environmental Requirements and the Product System,* prepared by G. A. Keoleian and D. Menerey, National Pollution Prevention Center, University of Michigan, EPA/600/R-92/226, U.S. EPA, Cincinnati, Oh., 1993.

2. U.S. Environmental Protection Agency, Office of Solid Waste, *Characterization of Solid Waste in the United States: 1990 Update,* EPA 530-SW-90-042A, U.S. EPA, Washington, D.C., 1990.

3. U.S. Environmental Protection Agency, *Environmental Equity: Reducing Risk for All Communities*, vol. 1: *Workgroup Report to Administrator*, EPA 230-R-92-008, U.S. EPA, Washington, D.C., 1992.

4. U.S. Environmental Protection Agency, *Environmental Equity: Reducing Risk for All Communities*, vol. 2: *Supporting Document*, EPA 230-R-92-008A, U.S. EPA, Washington, D.C., 1992.

5. Joseph V. Rodricks and Michael R. Taylor, "Comparison of Risk Management in U.S. Regulatory Agencies," *Journal of Hazardous Materials*, **21**:239–253, 1989.

6. U.S. Congress, Office of Technology Assessment, *Green Products by Design: Choices for a Cleaner Environment*, OTA-E-541, 1992.

7. G. A. Keoleian, "The Application of Life Cycle Assessment to Design," *Journal of Cleaner Production*, in press.

8. D. E. Whitney, "Manufacturing by Design," *Harvard Business Review*, July–August 1988, pp. 83–91.

9. Donald G. Gause and Gerald M. Weinberg, *Requirements: Quality before Design*, Dorset House, New York, 1989.

10. Mark Oakely, *Managing Product Design*, Wiley, New York, 1984.

11. Bill Hollins, *Successful Product Design: What to Do and When*, Butterworth, Boston, 1989.

12. Walter J. Fabrycky, "Designing for the Life Cycle," *Mechanical Engineering*, **109**(1):72–74, 1987.

13. Municipal Solid Waste Task Force, *The Solid Waste Dilemma: An Agenda for Action*, U.S. EPA Office of Solid Waste, Washington, D.C., 1989.

14. Frank Edward Allen, "Few Big Firms Get Jail Time for Polluting," *The Wall Street Journal*, December 9, 1991, p. B-1.

15. Marvin A. Moss, *Designing for Minimal Maintenance Expense*, Marcel Dekker, New York, 1985.

16. Robert T. Lund, "Remanufacturing," *Technology Review*, **87**:18–23, 28–29, 1984.

17. H. C. Haynsworth and R. Tim Lyons, "Remanufacturing by Design, the Missing Link," *Productivity and Inventory Management*, **28**:24–29, 1987.

18. U.S. Environmental Protection Agency, "Guidance for the Use of the Terms `Recycled' and `Recyclable' and the Recycling Emblem in Environmental Marketing Claims," *Federal Register* **56**(191):49992–50000, 1991.

19. V. R. Sellers and J. D. Sellers, *Comparative Energy and Environmental Impacts for Soft Drink Delivery Systems*, Franklin Associates, Prairie Village, Kan., 1989.

20. Frank C. Bresk, "Using a Transport Laboratory to Design Intelligent Packaging for Distribution," World Packaging Conference, Sevilla, España, January 27, 1992, Lansmont Corporation, Monterey, Calif., 1992.

21. Gregory Keoleian and Dan Menerey, "Packaging and Process Improvements: Three Source Reduction Case Studies," *Journal of Environmental Systems*, **21**(1):21–37, 1992.

22. Gregory Keoleian and Dan Menerey "Sustainable Development by Design: Review of Life Cycle Design and Related Approaches," *Journal of Air and Waste Management Association*, **44**(5):645–668, 1994.

19

Life-Cycle Assessment

Bruce Vigon
Battelle Memorial Institute
Columbus, Ohio

Life-cycle assessment (LCA) has evolved over the past 20 years to a point where designers and engineers can effectively use the approach to identify the consequences of their choices of materials, manufacturing processes, and product usage. Life-cycle assessment is both a concept and methodology for auditing and evaluating the environmental performance of a product, process, or activity throughout its entire existence from raw materials acquisition to ultimate disposition through recycling, incineration, landfilling, or composting (Fig. 19-1).[1] As currently defined, the LCA approach consists of three components: inventory analysis (LCI), impact assessment, and improvements assessment.[2] These components may be used independently or combined, not necessarily sequentially. Goal definition and scoping have been incorporated into the method to allow clear definition of the purpose and system boundaries for the analysis, as illustrated in Fig. 19-2.[3]

LCI is based on the principles of systems analysis. A system is defined as a collection of operations that perform some precisely defined function.[4] The function of a system need not be the production of a final product. A perfectly reasonable function definition is to clean a part such that it can be used as a component of an electronic assembly. A system may also have a service as its output. In some cases where LCA is applied to pollution prevention (P2), the comparison may be between a service and a product which produce the equivalent function. In general it may be said that the interest in application of LCA to pollution prevention is to allow selection of the operations associated with a system which produces its output in the most efficient manner considering the entire life cycle. Consideration of the consequences just at the point of application of the pollution prevention activity does not provide the entire or even correct answer to the question of which alternatives or practices are best.

Life-Cycle Stages

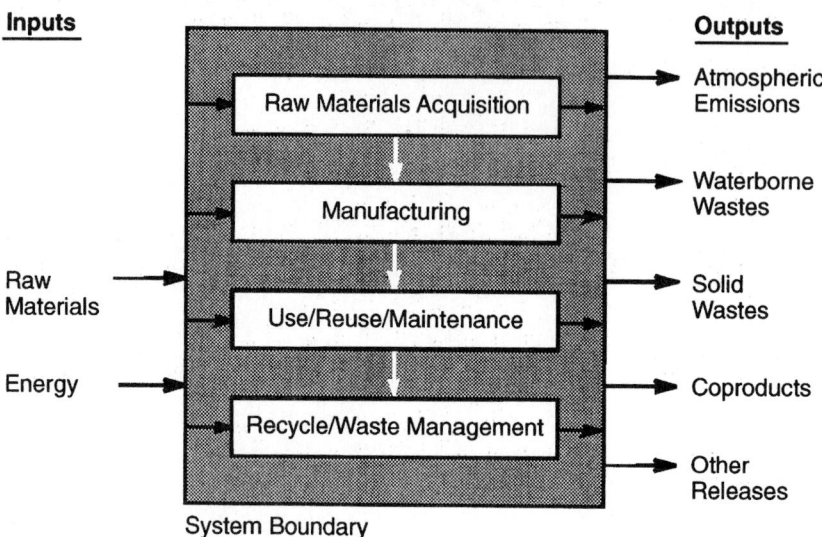

Figure 19-1. Boundary diagram illustrating typical LCA scope.[6]

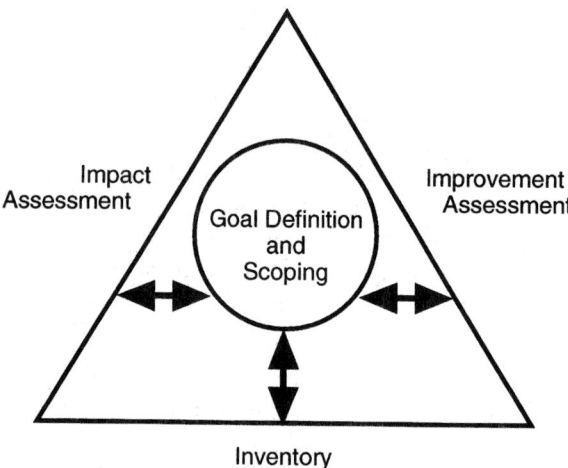

Figure 19-2. Components of a life-cycle assessment.[3]

An improvements assessment is a systematic procedure for identifying, evaluating, and selecting among these alternative opportunities for improving the energy, resource consumption, and environmental release profile of a product or process.[5] In using the life-cycle approach to undertake pollution prevention, the engineer compares the resource consumption, energy usage, and environmental burdens for the existing system to those that would characterize alternative future systems. The relationship between LCA and P2 is the degree to which P2 alternatives represent environmentally superior alternatives when considered from a broader perspective.

19.1 Concepts and Methods

The LCA concept suggests that decision making should be based on consideration of the cradle-to-grave characteristics of a product, process, or activity. This holistic process is a step beyond what most companies have used in the past and has led to the development of a new set of analytic methods and associated tools. Some of these methods are relatively well developed while others are still at a largely conceptual stage.

19.1.1 Holistic Thinking and Pollution Prevention Strategies

Limited information is presently available to manufacturers regarding product and process changes and their environmental consequences. This is especially true when the life-cycle impacts of each option are considered. Even if several alternatives, when considered from a local facility or single life-cycle stage perspective, appear to be equal in performance, cost, energy consumption, and waste generation, they may not be equal if the entire life cycle of each is examined. For example, printing operations considering volatile organic compound (VOC) controls and recycling of ink from press maintenance versus switching to low-VOC soy-based inks may be equally appealing at a local facility level. Considered on a life-cycle basis, the refining operations and disposal concerns associated with the petroleum-based product would be compared with the impacts of soybean production with regard to pesticides, fertilizers, and energy. Use of the life-cycle concept facilitates more reasonable comparisons or decisions on the relative environmental benefits of P2 projects.

19.1.2 Life-Cycle Inventory

A stepwise process has been developed for performing life-cycle inventories.[6] The inventory component is the best defined portion of an LCA from a methodology standpoint. The procedure begins with a clear definition of the purpose for conducting the inventory analysis and identification of the boundaries which define the life-cycle system. In fact, this goal definition and scoping step is important enough that it has become accepted by many as a fourth component of LCA. When the boundaries have been determined, a system flow diagram can be developed to pictorially describe the system, as illustrated in Fig. 19-3.

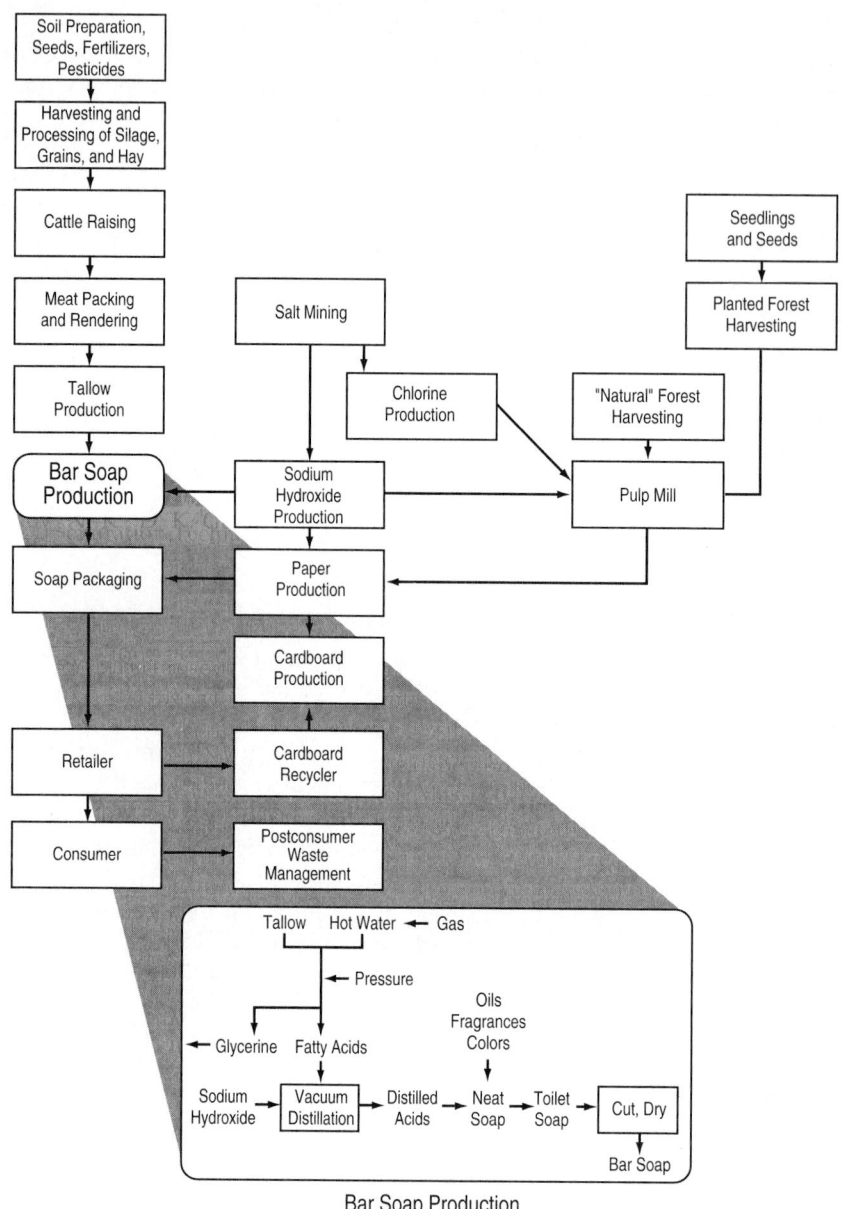

Figure 19-3. Sample product system flow diagram.[6]

LIFE-CYCLE INVENTORY CHECKLIST PART I—SCOPE AND PROCEDURES
INVENTORY OF: _____

Purpose of Inventory: (check all that apply)

Private Sector Use

Internal Evaluation and Decision Making
- ☐ Comparison of Materials, Products, or Activities
- ☐ Resource Use and Release Comparison with Other Manufacturer's Data
- ☐ Personnel Training for Product and Process Design
- ☐ Baseline Information for Full LCA

External Evaluation and Decision Making
- ☐ Provide Information on Resource Use and Releases
- ☐ Substantiate Statements of Reductions in Resource Use and Releases

Public Sector Use

Evaluation and Policy-making
- ☐ Support Information for Policy and Regulatory Evaluation
- ☐ Information Gap Identification
- ☐ Help Evaluate Statements of Reductions in Resource Use and Releases

Public Education
- ☐ Develop Support Materials for Public Education
- ☐ Assist in Curriculum Design

Systems Analyzed
List the product/process systems analyzed in this inventory: _____
_____

Key Assumptions: (list and describe)
_____
_____
_____

Define the Boundaries
For each system analyzed, define the boundaries by life-cycle stage, geographic scope, primary processes, and ancillary inputs included in the system boundaries.

Postconsumer Solid Waste Management Options: Mark and describe the options analyzed for each system.
- ☐ Landfill _____
- ☐ Combustion _____
- ☐ Composting _____
- ☐ Open-loop Recycling _____
- ☐ Closed-loop Recycling _____
- ☐ Other _____

Basis for Comparison
- ☐ This is not a comparative study.
- ☐ This is a comparative study.

State basis for comparison between systems: *(Example: 1000 units, 1,000 uses)* _____

If products or processes are not normally used on a one-to-one basis, state how equivalent function was established.
_____

Computational Model Construction
- ☐ System calculations are made using computer spreadsheets that relate each system component to the total system.
- ☐ System calculations are made using another technique. Describe: _____

Describe how inputs to and outputs from postconsumer solid waste management are handled. _____
_____
_____

Quality Assurance: (state specific activities and initials of reviewer)
Review performed on:
- ☐ Data Gathering Techniques _____
- ☐ Coproduct Allocation _____
- ☐ Input Data _____
- ☐ Model Calculations and Formulas _____
- ☐ Results and Reporting _____

Peer Review: (state specific activities and initials of reviewer)
Review performed on:
- ☐ Scope and Boundary _____
- ☐ Data Gathering Techniques _____
- ☐ Coproduct Allocation _____
- ☐ Input Data _____
- ☐ Model Calculations and Formulas _____
- ☐ Results and Reporting _____

Results Presentation
- ☐ Methodology is fully described.
- ☐ Individual pollutants are reported.
- ☐ Emissions are reported as aggregrated totals only.
  Explain why: _____
- ☐ Report is sufficiently detailed for its defined purpose.
- ☐ Report may need more detail for additional use beyond defined purpose.
- ☐ Sensitivity analyses are included in the report. List: _____
- ☐ Sensitivity analyses have been performed but are not included in the report. List: _____

Figure 19-4. Prototype checklist for a life-cycle inventory.[6]

Once the purpose and boundaries have been defined, a checklist is a useful tool to cover most decision-making areas in the performance of an inventory analysis (see Fig. 19-4). It includes the geographic scope, types of data used, how the data were gathered and developed, how the data were modeled, and how the results are presented. A tool such as this can help clarify the issues, boundaries, and conditions to be dealt with in a particular study.

Strong consideration should be given to the formation of a peer review panel. Overall, a peer review process, whether conducted internally by company staff

LIFE-CYCLE INVENTORY CHECKLIST PART II—MODULE WORKSHEET

Inventory of: _____ Preparer: _____

Life-Cycle Stage Description: _____

Date: _____ Quality Assurance Approval: _____

MODULE DESCRIPTION: _____

	Data Value[a]	Type[b]	Data[c] Age/Scope	Quality Measures[d]
		MODULE INPUTS		
Materials				
Process				
Other[e]				
Energy				
Process				
Precombustion				
Water Usage				
Process				
Fuel-related				
		MODULE OUTPUTS		
Product				
Coproducts[f]				
Air Emissions				
Process				
Fuel-related				
Water Effluents				
Process				
Fuel-related				
Solid Waste				
Process				
Fuel-related				
Capital Repl.				
Transportation				
Personnel				

(a) Include units.

(b) Indicate whether data are actual measurements, engineering estimates, or theoretical or published values and whether the numbers are from a specific manufacturer or facility, or whether they represent industry-average values. List a specific source if pertinent, e.g., "obtained from Atlanta facility wastewater permit monitoring data."

(c) Indicate whether emissions are all available, regulated only, or selected. Designate data as to geographic specificity, e.g., North America, and indicate the period covered, e.g., average of monthly for 1991.

(d) List measures of data quality available for the data item, e.g., accuracy, precision, representativeness, consistency-checked, other, or none.

(e) Include nontraditional inputs, e.g., land use, when appropriate and necessary.

(f) If coproduct allocation method was applied, indicate basis in quality measures column, e.g., weight.

Figure 19-4. (*Continued*).

not directly involved in the project, or externally, addresses the following four areas:

- Scope definition and system boundary setting
- Data compilation and quality of data
- Validity of results
- Communication of results

Peer review involvement in a study ideally occurs at several points: (1) reviewing the purpose, system boundaries, and data collection approach; (2) reviewing the compiled data and the associated quality measures; and (3) reviewing the draft inventory report, including the intended communication strategy for external studies. The peer review process is flexible to accommodate variations in the application or scope of life-cycle studies.

LCAs are data-intensive and, therefore, data quality can affect the outcome of an analysis. All data should be critically reviewed for source and content before being included.[7] Data quality goals are the required performance specifications for information in a life-cycle inventory. Establishment of these specifications is determined by the defined purpose of the life-cycle inventory. Data quality indicators are qualitative or quantitative characteristics of data. These include acceptability, bias, representativeness, and other attributes that measure data goodness and applicability (see Table 19-1).

Possible sources of data are electronic nonbibliographic databases (government and industrial), electronic bibliographic databases, electronic database clearinghouses, relevant documents (government and open literature reports and books), facility-specific industrial data (publicly accessible and nonpublicly accessible), laboratory test data, and study-specific data. The purpose, scope, and boundary of the inventory help determine the level or type of information that is required. Typically, most publicly available life-cycle documents present industry averages, while many internal private company studies use plant-specific data obtained directly from facility records.

Data on individual activities or processes are combined into subsystem modules which express the input/output characteristics for some standardized output of product, for example, 1000 kg. One major tool in life-cycle inventory analysis is the template, which is a guide that identifies the information that must be obtained for every step involved in an inventory analysis. Templates, or material and energy balance diagrams, are tools used to support data gathering and development for life-cycle inventory analyses. In an evaluation of the effectiveness of P2 activities, the template can assist in the identification of input and output information—for example, on energy or certain emissions—that would ordinarily not be collected. Energy, resources, and emissions for processes producing more than one product are allocated based on the relative weights of product output or another justifiable method. The next stage is model construction. It consists of incorporating the data and material flows from the modules into a computational framework typically using a computer spreadsheet. The systems accounting data that result from the computations of the model give the total results for the energy and resource use and environmental releases from the overall system. The analysis also allows the local effects of the pollution prevention activity to be compared with the total life cycle.

The written presentation of results should thoroughly describe the methodology employed in the analysis and explicitly define the system analyzed and the boundaries that were set. All assumptions made in performing the inventory should be clearly explained. If the inventory was conducted for purposes of product or process comparison, the basis for comparison among systems

Table 19-1. Data Quality Indicators (DQI) Relevant to LCA

DQI	Definition
Quantitative	
Accuracy	Measurement of the agreement of data with a known standard.
Bias	Measurement of systematic error that causes the values of a data set to be consistently higher or lower than the corresponding true parameter values.
Completeness	Percentage of data made available for the analysis compared with the amount of data needed.
Precision	Measurement of the spread or variability in data values or measurements compared to the mean.
Qualitative	
Acceptability	Degree to which the data source has been peer-reviewed or found adequate by colleagues in the field.
Accessibility	Ease of obtaining the data source or values.
Comparability	Degree to which different methods, data sets, or decisions agree or can be represented as similar or equivalent.
Data aggregation	Degree to which the data are nonaggregated so that trends in the data can be assessed.
Data collection method	Information that describes the method of data collection.
Limitations enumerated	Degree to which data source discusses limitations associated with the data.
Model documentation	Degree to which data source provides documentation of the associated model.
Model limitations	Degree to which data source discusses limitations associated with the model.
QA/QC	Data subject to quality assurance and/or quality control procedures.
Referenced	Data values reference the actual data source.
Representativeness	Describes how accurately the data represent what the analyst is trying to describe or depict.
Reproducible	The extent to which protocols followed in generating a data set enable another data set to be created having similar properties.
Statistical measures	Data source provides, or enables, the calculation of the mean, standard deviation, and skewness.
Verification and validation	Data source has been checked for errors and/or evaluated against an accepted method or standard to determine the accuracy of the results.

should be given and the background details should be available for review. Inventory results are presented in either tabular or graphical form. Tabular data presentation is often more comprehensive; however, graphical presentation can augment and clarify insights that result from performing an inventory. Individual atmospheric and water pollutants should be reported separately. Sometimes it is useful to report total energy results while also breaking out the contributions to the total from the process energy and energy of material resource. Energy of material resource, also known as latent or inherent energy, refers to the energy contained in a material that otherwise could have been utilized as energy. Solid wastes can be separated into postconsumer and industrial solid waste. Solid waste can be further designated, if appropriate, as nonhazardous and hazardous. A typical graphical presentation of LCI data is shown in Fig. 19-5.

19.1.3 Life-Cycle Impact Assessment

Life-cycle impact assessment is the quantitative and/or qualitative examination of potential environmental and human health effects associated with the use of resources (energy and materials) and environmental releases.[3] Unlike the inventory analysis, where much of the details of the methodology have been

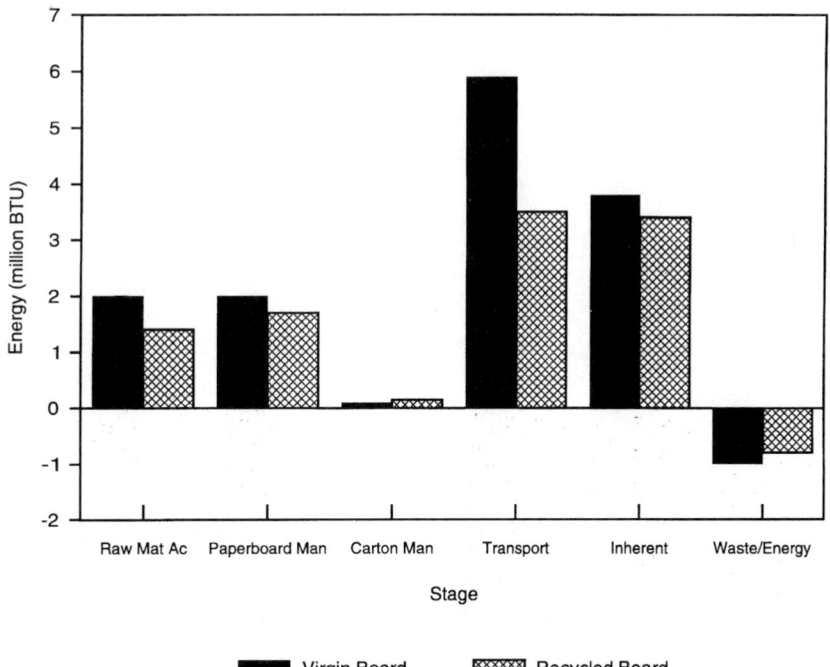

Figure 19-5. Example energy inventory data (10,000-carton basis). (*Battelle, unpublished data.*)

developed and validated over a period of time, impact assessment has emerged as a formal component of LCA only in the past several years. The conceptual framework for impact assessment consists of three phases: classification, characterization, and valuation (see Fig. 19-6).

Classification is the process of assigning inventory items to impact categories. For example, an industrial operation may emit a volatile organic com-

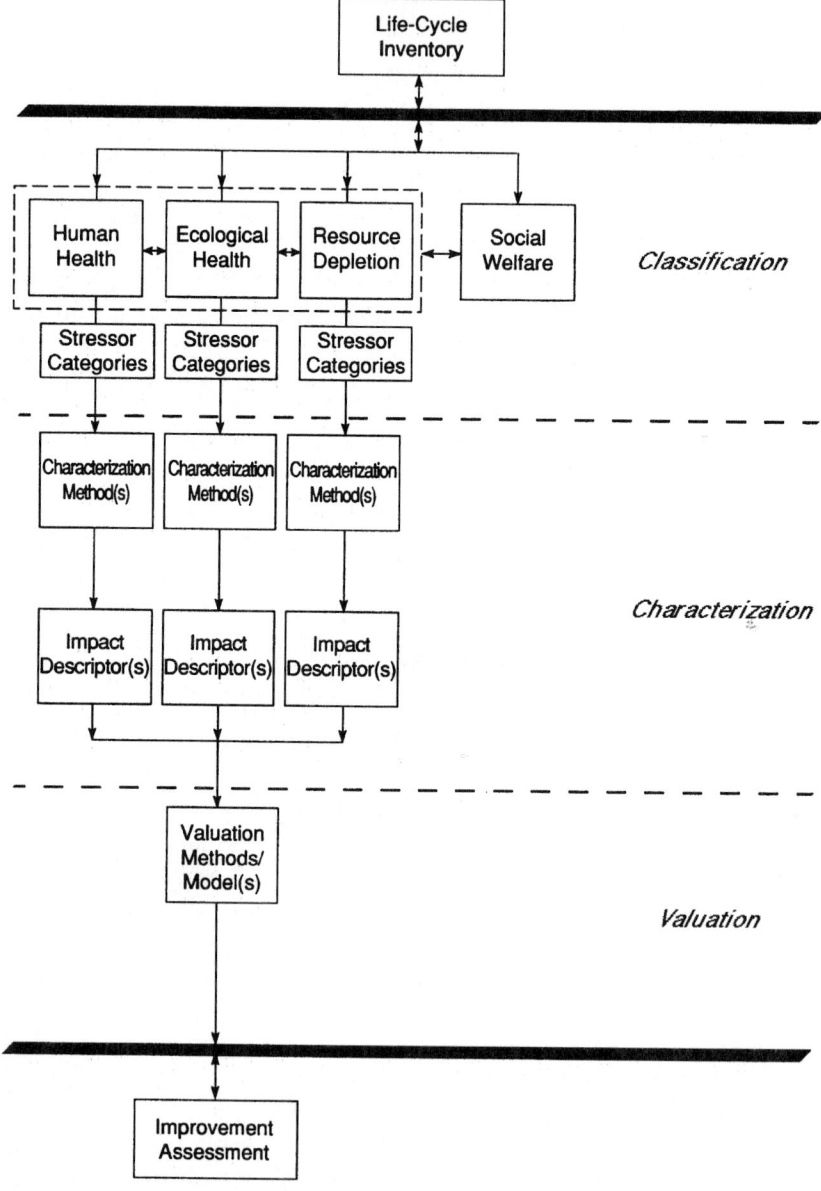

Figure 19-6. Conceptual framework for life-cycle impact assessment.[3]

pound, such as trichloroethane (TCA). This emission would be listed in the inventory as a mass loading of TCA per unit of production, say per 1000 parts cleaned. To conduct an impact assessment, the parameter TCA would be assigned to one or more categories of potential environmental stressors. Stressors are conditions that may lead to impacts. Life-cycle impact assessment is concerned with three broad types of impacts: resource depletion, ecological health, and human health. In addition, secondary impacts in any of the three primary areas related to social welfare consequences of the product or process under investigation are included. Stressor categories have been identified in each of the impact areas. Within ecological health, stressor categories include climate change gases, ozone depletion gases, toxicants, acidification precursors, photochemical oxidants, nutrients, habitat alteration, siltation, waterborne litter, soil compaction, ionizing radiation, and noise. Comparable stressors have been identified for human health impacts.

Thus, within the ecological health category, TCA would be assigned to several stressor categories, including climate change, photochemical oxidants, and toxicants. Several other inventory releases, such as CO_2, also could be assigned to a category, in this case, climate change. For some practitioners, classification also involves the aggregation of these outputs into common units based on the application of normalization or equivalency factors as discussed here.

Characterization models are applied in tiered fashion in the second phase of impact assessment to convert certain inventory values into impact descriptors. These particularly apply to impact indicators associated with ecological and human toxicants. Five levels of analysis have been identified. Beginning with a Level 1 analysis, which uses the inventory data directly as a measure of impact, parameters such as energy, which can be summed, may be added together. Level 2 aggregates loadings by equivalency factors. Emission standards have been used historically as one approach to Level 2 impact assessment. Thus, if the standard for TCA is 10 mg/m^3, the impact measure for a release of 10 mg/functional unit would be 1 m^3. Similar critical volumes can be computed for other loadings where standards exist. In principle, these equivalency transformations need not involve only a single-valued conversion factor. The transformation could be made with a linear or nonlinear function, with recognition of the existence of toxicity thresholds or the increasing scarcity value of toxic assimilation capacity as the loadings increase. One difficulty with this approach is that established regulatory levels often reflect political and economic practicalities as much as scientific considerations.

Equivalency factors also have been developed for stressors/impacts other than toxicity. In this example, TCA and CO_2 would be combined in the climate change stressor category through the application of global warming potential (GWP) equivalence factors. Since CO_2 by definition has a GWP of 1, the mass emission of TCA per functional unit would be multiplied by its 20-year GWP of 350. This results in one possible consistent measure of impact potential (CO_2 equivalents) for two different environmental loadings contributing to the same stressor.[8] In some cases, the aggregated GWP values could be taken further by relating them to specific ecological impacts of interest. For example, estimated regional temperature increases could be used to predict higher mortality for certain cold water fish species.

Level 3 impact assessment aggregates the loading parameters by inherent properties such as toxicity, persistence, and bioaccumulation potential. In many ways this approach is simply an alternative set of conversion factors to those applicable at Level 2. Levels 4 and 5 go beyond the basic set of considerations by beginning to address exposure and potential effects of exposure. In Level 4, considerations of transport and fate are treated generically while Level 5 impacts are determined based on site-specific information.

Following characterization of all relevant impacts and aggregation to the extent feasible, there may still be a need to cross-compare impacts occurring in different categories, e.g., climate change, ozone depletion, and toxicity. This stage of impact assessment is known as valuation. Valuation methods range from opinion recording methods, such as focus groups and the Delphi approach, to more formalized and structured methods, such as multiattribute utility theory and the analytic hierarchy process. Valuation based on economic factors has also been used. Economic valuation may be based on the estimated cost to control a particular impact to a predetermined level or on the estimated value of the potential damages caused by the impact. In any of these approaches, the objective is to derive a set of weighting factors to combine dissimilar kinds of impacts into a single value, recognizing that the valuation process is largely subjective.

All life-cycle impact assessment approaches are at a very early stage of development. There are significant gaps in elucidation of both the scientific principles underlying the impact characterization and valuation process and major information deficiencies that will have to be addressed before a consensus on methodology emerges. Nevertheless, the concept that inventory loadings result in potentially quantifiable impacts is important for the application of LCA to pollution prevention.

19.1.4 Making Pollution Prevention Decisions Using LCA

The LCA concept supports decision making on the selection of pollution prevention factors at several levels. In its simplest form the decision maker is faced with a set of alternative pollution prevention activities that could be undertaken. Selection criteria traditionally have been based on effectiveness of source reduction potential at the point of application. In the parts cleaning example using TCA, the reductions in VOC emissions and hazardous waste generation associated with solvent disposal would typically form the environmental basis for selecting among alternatives. Technology or best practice alternatives could include covers, vapor recycling, and switching to an alternative cleaner. Application of the LCA concept does two things: it expands the range of local parameters considered in the decision and it broadens the decision framework from the local shop level to the global level. The first modification would suggest that some nonconventional factors such as energy be added to the basic slate of local criteria. The second suggests that upstream and downstream features of the pollution prevention practice be compared to the local profile to establish the extent to which the alternative truly represents an environmental improvement.

The specific methods by which this expansion and broadening occurs are still being developed and, in any case, will need to be tailored to the scope and objectives of a particular decision. The implementation of the concept could range from a simple set of questions to be asked about the alternatives to a detailed comparison of local and global consequences. In the former, a company could establish an ad hoc working group that would attempt to identify any potentially adverse consequences of each alternative that might occur outside the boundaries of the company's facilities. In the TCA example, certain substitute cleaners might be very energy-intensive to manufacture or might involve large amounts of toxic substances during their manufacture. A checklist flagging such potential concerns would place each alternative in a more global context. Alternatives that are favorable from a local perspective as well as from a global one could be distinguished from those in which there is some tradeoff involved.

An example of a more detailed life-cycle pollution prevention assessment approach, termed the *impact analysis matrix* (IAM) is illustrated in Figs. 19-7 and 19-8, wherein TCA vapor degreasing is compared to caustic aqueous cleaning.[9] The IAM is developed in stepwise fashion by convening an expert panel. The expectation is that this group can reach a consensus on the ratings, given available background technical information.

Step 1:   For these comparisons, the IAM consisted of five columns of inventory input (resource utilization) and output (emission) parameters, and seven rows of environmental impact categories. The impact categories, selected using expert judgment, included: (1) the major global impact areas specific to the use of halogenated solvents, e.g., global warming and ozone depletion potential; (2) other important global considerations, e.g., nonrenewable resource utilization; and (3) important regional and local impact areas, e.g., air and water quality, land disposal, and transportation effects.

In application of this approach in other settings, these categories could be varied to match the particular characteristics of each comparison and could also be expanded to include other impact areas (e.g., occupational exposures). It will be important to provide sufficient rationale for the selection and exclusion of various categories given the absence of rigorous methodology or consensus to guide the selection process.

Step 2:   The next step was to determine whether some cells represent either double-counting or meaningless comparisons. In the present case, for example, it was judged that aqueous wastes have no significant impact on global warming, so this cell in the matrix was eliminated. In this way, 17 of the 35 cells were deemed meaningful.

Step 3:   Next, unweighted scores were assigned to each viable cell in the IAM. These scores were assigned relative to a particular option chosen as a

Risk Areas \\ Impacting Parameters	Material Inputs	Energy Inputs	Atmospheric Emissions	Aqueous Wastes	Solid Wastes	TOTAL
Global Warming		+1	-1			0
Ozone Depleting Potential			-1			-1
Non-Renewable Resource Utilization	-1	+1				0
Air Quality		+1	-1	+1	-1	0
Water Quality		+1		+1		+2
Land Disposal		+1		+1	-1	+1
Transportation Effects	-1	+1			-1	-1
TOTAL	-2	+6	-3	+3	-3	+1

Figure 19-7. Life-cycle impact analysis matrix method (local perspective).[9]

base; a +1 means a discernibly larger environmental impact than the base option, a −1 means a discernibly lesser impact, and a 0 means little or no perceived difference in impact. Scores were assigned based on a combination of inventory data and expert knowledge of associated impacts.

Step 4 (optional):   Next, the initial unweighted scores could have been given weightings to determine whether results will change significantly: a relatively strong environmental impact is assigned a ++ and a relatively large reduction in impact is assigned a −−.

It should be noted that the comparisons used to arrive at the weightings need not be restricted to a single impact category (i.e., row), depending on the views of the expert judges. However, the basis for assigning weights and the scope of comparison (i.e., within an impact category or across categories) should be explicitly described.

Step 5:   The pluses, minuses, and zeros were then summed to derive an overall score for each row and column; if appropriate, the summation could have been done for the entire matrix. Unweighted scores covered a potential range from +18 to −18, weighted scores from +36 to −36.

As one example of the type of information provided by the IAM approach, one can compare and contrast the scores under the energy inputs column evaluated at the user (shop) versus global level. From the perspective of the user,

Risk Areas \ Impacting Parameters	Material Inputs	Energy Inputs	Atmospheric Emissions	Aqueous Wastes	Solid Wastes	TOTAL
Global Warming		0	-1			-1
Ozone Depleting Potential			-1			-1
Non-Renewable Resource Utilization	-1	0				-1
Air Quality		0	-1	+1	-1	-1
Water Quality		0		+1		+1
Land Disposal		0		+1	-1	0
Transportation Effects	-1	0			-1	-2
TOTAL	-2	0	-3	+3	-3	-5

Notes:
a. Base alternative = TCA vapor degreasing. Proposed alternative = aqueous cleaning.
b. Shaded squares signify no basis for impact on risk area.
c. Rating on -1 represents decreased impact, 0 represents the same impact, and +1 represents an increased impact.

Figure 19-8. Life-cycle impact analysis matrix method (global perspective.)[9]

impacts arising from energy input requirements are a dominating category, and are much higher for the aqueous-based relative to the TCA-based system because of the high pumping and heating requirements of the former; in contrast, viewed from a global perspective, impacts arising from energy requirements were found to be essentially the same for the two systems.

19.2 Applications of LCA for Pollution Prevention

The LCA concept can be applied to pollution prevention in a range of applications. These may be broadly classified as either internal to the organization or external. The nature of pollution prevention activity is such that the majority of industrial applications will be internal.

19.2.1 Corporate Strategic Planning

Company policy statements often contain statements regarding the corporate commitment to producing products in an environmentally sound manner. Companies in the manufacturing sector also are committing time and resources to product stewardship programs. LCA has a potentially important strategic role to play in providing information about the optimal mix of company responses to act on policy initiatives. As shown in Fig. 19-9, the time line for resolution of environmental issues can be divided into four stages: identification, analysis, response development, and solution implementation. LCA can be used effectively in the latter three stages.

Further, once solution alternatives are identified, a company must make a decision as to their position on implementation. Implementation strategies range from the minimum legal compliance through more ambitious activities which go beyond regulatory requirements but are still economically favorable, to those where social investments are being made. Strategic use of LCA in conjunction with more traditional cost accounting can help frame the desired balance between the "bottom line" and socially oriented corporate environmental stewardship.

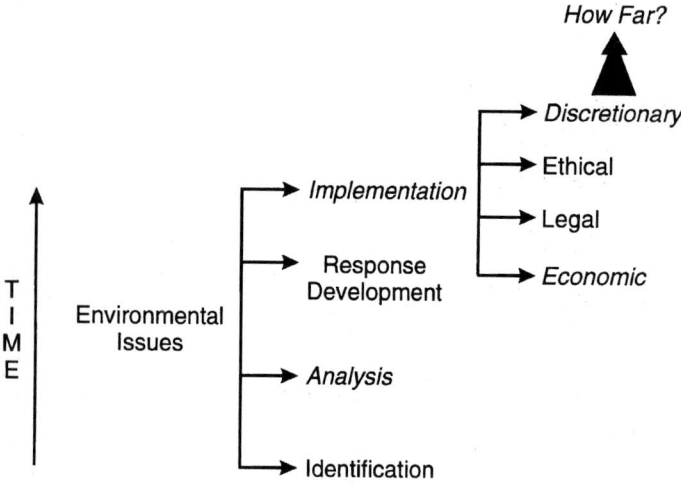

Potential Strategic Roles of Life-Cycle Assessment

Figure 19-9. Corporate strategic applications potential.

19.2.2   Product Development

Developing products in today's competitive setting requires careful weighing of product attributes, cost of manufacturing, waste generation and management, and a host of other considerations. Use of an LCA approach to design has been discussed in the literature.[10,11] This application normally would include formal pollution prevention considerations as part of the design process; i.e., selecting materials and manufacturing processes for new or redesigned products that produce less waste.

19.2.3   Process Selection and/or Modifications

LCA can be applied to select processes that have the greatest potential for waste reduction or prevention. This can be done as part of an overall product LCA, where the manufacturing stage has been or would be identified as a significant contributor to the overall system environmental loading. Alternatively, LCA can be used as a tool in a more narrowly scoped pollution prevention facility audit where the analysis is restricted to certain operations in which pollution prevention options will be identified. If the usable output of the system is the same for all process alternatives, then it may be possible to curtail the extent of the downstream system boundary and consider only the fate of environmental emissions from the process itself. In addition, any upstream effects associated with process inputs should be included. As companies build an internal database on various operations, the time, effort, and cost of evaluating alternatives will decrease.

19.2.4   Market Claims and Advertising

One of the possible applications of LCA is to support claims to the external community of pollution prevention gains made by a company. Use of LCA in this manner should be treated with extreme caution. Information on methodology, assumptions, and results should be communicated completely and transparently. Under no circumstances should the findings be selectively reported and careful understanding of the audience's communication needs should dictate the format in which the information is presented.

19.3   Facilitating Applications

While the life-cycle concept can be employed in pollution prevention in a descriptive fashion, quantitative application of LCA methodology, particularly for new users, can be information- and computationally intensive. Fortunately, software support tools and enhanced databases are becoming available, although the first few commercial offerings are not entirely satisfactory.

19.3.1   Analysis Support Software

Life-cycle inventories consist of developing and linking modules for the various unit operations involved in the overall system based on a defined system boundary and a functional analysis unit. Commercial inventory analysis software for personal computers contains a collection of standard modules for a variety of industrial operations, transportation modes, and fuel and power acquisition activities. Some LCA software programs are designed to be used exclusively for packaging applications, while other programs are more broadly capable of handling both packages and products. The number of operations modules in a given software package varies widely and little or no attempt is usually made to provide an indication to the user of the quality of the data in each module. The level of aggregation of the operations varies from module to module, depending on the information source, and not all modules will be mass- or energy-balanced, even though this is a desirable feature. For these reasons, several software vendors recommend that users consider the results as no more than a screening approach upon which a more detailed assessment can be built.

Some form of a user interface is provided to enable the user to specify the system to be inventoried as well as to enhance the embedded data by creating new operations modules or combining it in new ways. Most commercial LCA software is based on a spreadsheet concept, using either Lotus 1-2-3™ or Excel™. The user interacts with the spreadsheet either by putting data directly into cells in some of the less sophisticated implementations or indirectly through input screens and prompts for user-supplied information. A few LCA programs are organized around database software such as dBase IV™.

Output tables for some of the software packages can be presented to meet the needs of a variety of user audiences. Graphical output is generally available through the spreadsheet itself, although some programs allow file transfers of either data or output to other programs. Life-cycle impact assessment capability is available from several programs, typically consisting of simple classification and Level 2 equivalence analysis.

19.3.2   Data and Databases

The data-intensive nature of LCA makes it desirable that information be readily available to users to perform inventories and impact assessments. Most LCA studies do not involve substantial primary data collection, despite the existence of significant gaps in information and inconsistencies in the way that various secondary data sets are assembled. Secondary data sources are essential to the conduct of LCAs. Most software packages for performing LCA have data incorporated into the program. However, other sources of information are almost always necessary. Several compilations of information for preparing LCAs have been published and these may be consulted for identification of additional data sources.[12,13]

19.4 Summary

Life-cycle assessment represents a potentially useful and powerful tool for identifying and maximizing the environmental benefits of pollution prevention. LCA can be used in a variety of ways ranging from a conceptual thought process to allow companies to think more broadly about their process or materials choices to a formal analytical process for supporting decisions. As software and databases expand, the routine use of LCA will become easier, but even at the present state of development, LCA can be effective in more fully informing the decision maker of potential environmental consequences.

References

1. J. A. Fava, R. Denison, B. Jones, M. A. Curran, B. W. Vigon, S. Selke, and J. Barnum (eds.), *A Technical Framework for Life-Cycle Assessments*, Pensacola: Society of Environmental Toxicology and Chemistry, August 18–23, 1990, Smuggler's Notch, Vt., 1991, pp. 1–2.

2. Environmental Protection Encouragement Agency (EPEA), *Umweltauswirkungen von Stickoxidminderungs*, Technologien bei der Mullverbrennungsanlage Bielefeld, Hereford, Hamburg, 1990.

3. J. A. Fava, F. Consoli, R. Denison, K. Dickson, T. Mohin, and B. Vigon (eds.), *A Conceptual Framework for Life-Cycle Impact Assessment*, Pensacola: Society of Environmental Toxicology and Chemistry, February 1–7, 1992, Sandestin, Fla., 1993, pp. 4–8, 11, 19.

4. I. Boustead, and G. F. Hancock, "Theoretical Industrial Systems," *Handbook of Industrial Energy Analysis*, Ellis Horwood Chichester and John Wiley, 1979, p. 38.

5. Society of Environmental Toxicology and Chemistry, *Guidelines for Life-Cycle Assessment: A "Code of Practice,"* Pensacola: Society of Environmental Toxicology and Chemistry, Sesimbra, Portugal, March 31–April 1993, p. 25.

6. U.S. Environmental Protection Agency, 1993a. *Life-Cycle Assessment: Inventory Guidelines and Principles*, EPA/600/R-92/245, NTIS PB93-139681, ISBN: 1-55670-015-9, Prepared by Battelle and Franklin Associates Ltd. for the Risk Reduction Engineering Laboratory, Office of Research and Development, 108 pages.

7. U.S. Environmental Protection Agency, 1993b. *Life-Cycle Assessment: Guidelines for Assessing Data Quality*, Draft Report, Prepared by Research Triangle Institute for the Office of Solid Waste, September 1993.

8. D. J. Wuebbles, and J. Edmonds, *Primer on Greenhouse Gases*, Lewis Publishers, Chelsea, Mich., 1991. p. 99.

9. Source Reduction Research Partnership, *Source Reduction of Halogenated Solvents,* Final Summary Report, Appendix A, Jacobs Engineering Group, Inc., Pasadena, Calif., 1992.

10. U.S. Environmental Protection Agency, 1993c. *Life-Cycle Design Guidance Manual: Environmental Requirements and the Product System,* EPA/600/R-92/226, Prepared by the National Pollution Prevention Center, University of Michigan, for the Risk Reduction Engineering Laboratory, Office of Research and Development, 181 pages.

11. U.S. Congress, Office of Technology Assessment, *Green Products by Design: Choices for a Cleaner Environment,* OTA-E-541, U.S. Government Printing Office, Washington, D.C., 1992, 117 pages.

12. U.S. Environmental Protection Agency, 1993d. *Life-cycle Assessment: Public Data Sources for the LCA Practitioner,* Draft Final Report, Prepared by Battelle for the Office of Solid Waste, September 1993, 166 pages plus appendices.

13. SustainAbility Ltd., *The LCA Sourcebook: A European Business Guide to Life-Cycle Assessment,* Prepared for the Society for the Promotion of LCA Development and Business in the Environment, London, 1993.

Further Reading

Bailey, P. E., "Life-Cycle Costing and Pollution Prevention," *Pollution Prevention Review* **1**(1):27–39, 1991.

Guinee, J. B., H. A. Udo de Haes, and G. Huppes, "Quantitative Life-Cycle Assessment of Products 1: Goal Definition and Inventory," *J. Cleaner Production* **1**(1):3–14, 1993.

Johansson, A., *Clean Technology,* Lewis Publishers, Boca Raton, Fla., 1992, 196 pages.

U.S. Environmental Protection Agency, *Proc. International Symp. on Pollution Prevention in the Manufacture of Pulp and Paper: Opportunities and Barriers,* EPA/744/R-93-002, 1993, 339 pages.

Van Weenen, J. C., *Waste Prevention: Theory and Practice,* CIP-Gegevens Koninklijke Bibliotheek, Den Haag, The Netherlands, 1990, 418 pages.

Vigon, B. W., and M. A. Curran, "Life-Cycle Improvements Analysis: Procedure Development and Demonstration," *Proc. 1993 International Symposium on Electronics and the Environment,* Arlington, Virginia, 93CH3209-4, 1993, pp. 151–156.

20

Product Labeling

Jennifer Marron

American Forest & Paper Association
Washington, D.C.

20.1 Introduction to Environmental Labeling

When making purchasing decisions, consumers are influenced primarily by three factors: price, performance, and, in some cases, brand name. For many consumers, a fourth factor has recently come into play: environmental impact. Numerous consumer surveys show that today's shoppers are willing to pay more for products which they perceive as being good for, or less harmful to, the environment. In a rush to capitalize on this consumer willingness to spend, manufacturers have started making a variety of environmental claims about their products. These claims range from general statements, such as "environmentally friendly," to more specific declarations, like "CFC-free." Some claims, such as "recyclable," are made about the product packaging rather than the product itself. Examples of typical environmental labeling claims (also referred to as "green labeling" claims) appear in Table 20-1.

Until recently, the use of green labeling claims has been largely unregulated. This lack of oversight has resulted in consumers being confused and, in some cases, misled by manufacturers' claims about the positive environmental attributes of their products. To bring some consistency to environmental labeling in the United States, the federal government, state governments, and private organizations have begun developing standards and definitions for certain green labeling claims. Actions have been taken by the Federal Trade Commission (FTC), the coalition of State Attorneys General, and the National Advertising Division (NAD) of the Better Business Bureau to ensure that green labeling claims are truthful and useful to consumers. Many states have passed regulations governing the use of specific terms such as "recyclable" and "biodegrad-

Table 20-1. Typical Environmental
Labeling Claims

Biodegradable
CFC-free
Compostable
Degradable
Dioxin-free
Energy efficient
Environmentally friendly
Natural
Non-chlorine-bleached
Nontoxic
Organic
Oxygen-bleached
Ozone-friendly
Photodegradable
Recyclable
Recycled
Reusable
Won't harm the ozone layer

---

able." Both the FTC and the coalition of State Attorneys General have published guidance documents on the use of environmental labeling claims on product packaging and in advertising. NAD has responded to challenges brought by one manufacturer against another. NAD's evaluation of green labeling claims on a product-by-product basis has usually resulted in manufacturers agreeing to modify, or stop making, potentially misleading claims about the environmental benefits of their products.

Two private organizations in the United States, Green Seal and Scientific Certification Systems, or SCS (formerly Green Cross), are using life-cycle analysis to evaluate the environmental impact of various categories of consumer products and to award environmental seals of approval to those products judged to be environmentally preferable. The focus of this chapter is on the work being done by these two organizations to develop environmental labeling as a reliable means of guiding consumers in their purchasing decisions. Green Seal and SCS have taken a comprehensive approach to assessing the environmental impacts of consumer products and to promoting those products that are the least damaging to the environment.

In developing their programs, both Green Seal and SCS have looked to other countries for models. This chapter also provides descriptions of the environmental labeling programs in place in Germany, Canada, and Japan. All three programs are well established and have achieved some notable successes. Each provides a different example from which the green labeling efforts in the United States can draw.

Before beginning a discussion of the comprehensive environmental labeling programs being developed by Green Seal and SCS, it is useful to briefly examine the Society for the Plastics Industry's (SPI's) plastic resin coding system. The SPI coding system is considerably more narrow in scope than the private labeling programs, as it is for use exclusively on plastic bottles and containers. The confusion surrounding the implementation of the SPI coding system illustrates well how consumers can be simultaneously informed and misled by environmental labels.

20.2 Society of the Plastics Industry Resin Coding System

The primary purpose of the SPI resin code is to facilitate recycling by distinguishing plastic bottles and containers from one another according to their resin type. The SPI code combines the familiar triangular chasing arrows symbol, used to denote recycling, with a number and an abbreviation for the resin type (see Table 20-2). The number appears in the center of the triangle and the abbreviation appears below it. SPI asserts that it developed the resin coding system for use only on bottles and solely for the benefit of processors of recovered plastics. However, the resin code now appears on containers and film plastics, as well as bottles, and it is used by consumers and recycling officials, in addition to processors, as a convenient way to distinguish among resin types. In fact, 39 states now require that a manufacturer's plastic coding system be consistent with the SPI resin code.

The SPI code alleviates the need for consumers to distinguish between different types of plastics, such as polyethylene terephthalate (PET) and high-density polyethylene (HDPE). This is especially helpful for recycling programs. Rather than being told how to differentiate among plastic packages, participants in a recycling program can simply be instructed that "ones" and "twos" are being collected. Although this seems straightforward, problems have arisen due to the use of the chasing arrows to bracket the numerical resin code. When consumers see the familiar recycling symbol, they often assume that the container is recyclable, regardless of its resin type. This confusion has resulted in the con-

Table 20-2. SPI Numerical Resin Codes

Code	Resin type
1	Polyethylene terephthalate (PET)
2	High-density polyethylene (HDPE)
3	Polyvinyl chloride (PVC)
4	Low-density polyethylene (LDPE)
5	Polypropylene (PP)
6	Polystyrene (PS)
7	All other resins

tamination of plastics collected for recycling. It has also led some consumers to purchase products that they believe are better for the environment because their packaging is recyclable, only to find out that the packaging is not recyclable in their area. For example, polystyrene containers bear the chasing arrows with a number 6 in the center—yet few, if any, recycling programs accept polystyrene.

To correct these misperceptions, representatives from the National Recycling Coalition (NRC) and SPI in February 1993 formed a committee to determine whether the chasing arrows should be eliminated from the SPI code. The committee is also investigating what types of educational and legislative efforts might help to ensure the proper use of the existing, or modified, code in the future. In addition, the American Society for Testing and Materials (ASTM), the International Organization for Standardization (ISO), and the Society of Automotive Engineers (SAE), are working independently on new plastic resin codes. All three organizations have designed an alternative to the chasing-arrows symbol for bracketing the numerical resin code.

The SPI code is potentially confusing to consumers because it conveys information without any qualification. The chasing arrows bracketing the numerical resin code imply recyclability; yet, while all plastic bottles and containers are recyclable in theory, in practice, a bottle or container's recyclability varies from community to community. In addition, the symbol does not account for variations within a resin type. For example, the type of HDPE resin used to make blow-molded bottles is different from that used to make injection-molded containers. Nonetheless, HDPE bottles and containers receive the same SPI resin code. This lack of distinction often results in the contamination of HDPE packaging collected for recycling. The problems encountered in using the SPI resin code demonstrate the difficulty inherent in designing a label that conveys the right amount of information.

As can be seen in the discussion that follows, both Green Seal and SCS have responded to the need for consumers to be better informed. Neither organization permits its seal of approval to be used alone. Instead, the seals must be accompanied by a qualifying statement appropriate to the product type, explaining the basis for certification. In an attempt to provide consumers with even more information on which to base their purchasing decisions, SCS is working on an "environmental report card" that visually depicts the environmental burdens associated with the production, use, and disposal of a given product. It is then up to the consumer to compare report cards for similar products and to decide which product is environmentally preferable.

20.3   Green Seal

Green Seal is a private, nonprofit organization dedicated to environmental standard setting, product certification, and public education. The organization's stated mission is to reduce the environmental impacts associated with the manufacture, use, and disposal of consumer products to the greatest extent possible, given certain technological and economic constraints. Green Seal seeks to meet this goal by harnessing the purchasing power of consumers.

Green Seal awards its seal of approval to products which it judges to be environmentally preferable compared with other items in the same product category. To facilitate this evaluation, a different environmental standard is developed for each product category. All Green Seal environmental standards are issued in proposed form, so that comments may be received from industry, trade associations, government agencies, environmental and other public interest groups, consumers, and other interested parties. The standards are finalized following careful consideration of the comments.

Green Seal certification is available to all products for which Green Seal environmental standards have been set. Manufacturers voluntarily submit their products for evaluation. Those that comply with Green Seal's requirements may be authorized to use the Green Seal certification mark on their products and in advertising. This mark, or seal of approval, consists of a blue globe with a green checkmark crossing it and the words "Green Seal" above (see Fig. 20-1). In addition, a qualifying statement must appear next to the logo. The content of the statement is determined by Green Seal and varies from product to product. A manufacturer may use a different qualifying statement as long as it has received written permission from Green Seal to do so. Manufacturers granted the right to use the seal of approval are subject to an ongoing program of testing, inspection, and enforcement. Underwriters Laboratories conducts all of Green Seal's product testing.

To date, Green Seal has published eight environmental standards. Ten other standards currently are in various stages of development (see Table 20-3). The environmental standards for paper towels and napkins (GS-9) and for paints (GS-11) are discussed in greater detail in the following sections.

Figure 20-1. Logo for Green Seal.

Table 20-3. Green Seal Environmental Standards

Standards published	Standards under development
Tissue paper (bathroom and facial) (GS-1)	Household cleaners (GS-8)
Re-refined engine oil (GS-3)	Windows (GS-13)
Compact fluorescent lamps (GS-5)	Window films (GS-14)
Water-efficient fixtures (GS-6)	Water heaters
Printing and writing papers (GS-7)	Room air conditioners
Paper towels and napkins (GS-9)	Newsprint
Paints (GS-11)	Batteries
Major household appliances (refrigerator-freezers, freezers, clothes washers, clothes dryers, dishwashers, and cooktops and ovens) (GS-20 through GS-25)	Laundry detergents
	Luminaires
	Coated paper

20.3.1   The Green Seal Environmental Standard for Paper Towels and Napkins

The Green Seal "Environmental Standard for Paper Towels and Napkins" was issued on January 27, 1993. The focus of this standard, which presents the environmental requirements for paper towels and napkins, is on the amount of recycled paper used in making these products. The standard carefully defines the terms *recovered material* and *postconsumer material* and lays out the method by which the percentages of recovered content and postconsumer content are to be calculated. To receive a seal of approval, a given paper towel or napkin product must contain 100 percent recovered material and at least 40 percent postconsumer material by weight.

In addition to containing the specified amount of recovered and postconsumer material, paper towels and napkins must meet several other requirements to be approved by Green Seal. For example, certain chemicals may not be used (e.g., chlorine) to de-ink the recovered fiber. If the fiber is bleached using chlorine or any of its derivatives, "the adsorbable organic halogen content of the effluent from the production location may not exceed 1.0 kilogram per air-dried metric ton of pulp." The product also must perform in accordance with reasonable industry practice. That is, it should have no wrinkles or tears, no odors, no foreign materials, no ragged edges, and it must dispense properly from the container or roll. The environmental standard also regulates the use of inks, pigments, dyes, and fragrances in paper towels and napkins and the products' packaging; and it requires that the core of a roll of paper towels be manufactured from 100 percent recovered material.

Once a paper towel or napkin product is judged to be in compliance with all of these requirements, it has the right to use the Green Seal certification mark. The mark must appear on the product's packaging and can be embossed or printed on the product itself. The Green Seal certification mark cannot be used in conjunction with any modifying terms, phrases, or graphic images that could mislead consumers as to the extent or nature of the certification. However, if the product is certified to have been made without bleach, to have been bleached without chlorine, or to have been bleached using oxygen, it may bear the additional endorsement "Unbleached," "No Chlorine Bleach," or "Oxygen Bleached," respectively. In addition, whenever the Green Seal certification mark appears on a package of paper towels or napkins, the package must contain a description of the basis for the certification. The description must be in a location, style, and typeface that are easily readable by the consumer. Unless otherwise approved by Green Seal, the description must read (where XX is the certified level of postconsumer material):

> This recycled product contains 100% recovered material including XX% post-consumer materials by fiber weight, and meets Green Seal's environmental standards for bleaching, de-inking and packaging. It contains no added fragrances.

In addition to the fiber weight description, manufacturers may also add to the package of a certified product the description "XX% recovered material including YY% postconsumer material by total weight."

20.3.2 The Green Seal Environmental Standard for Paints

The Green Seal "Environmental Standard for Paints" was issued on May 20, 1993. The focus of this standard, which presents the environmental requirements for paints, is on the concentration of volatile organic compounds (VOCs), as they are defined by the U.S. Environmental Protection Agency (EPA) in 40 *CFR* Section 51.100(s), (s) (1). For the purpose of this standard, paint is defined as a

> liquid, liquefiable, or mastic composition that is converted to a solid protective, decorative, or functional adherent film after application as a thin layer. These coatings are intended for on-site application to interior or exterior surfaces of residential, commercial, institutional or industrial buildings.

This standard does not cover stains, clear finishes, or paints sold in aerosol cans. To receive a seal of approval, the VOC content of a given paint product (calculated excluding water and tinting color added at the point of sale) may not exceed certain limits established using EPA Reference Test Method 24. For interior coatings, the limit is 150 grams of VOC per liter of product minus water for nonflat paints and 50 grams for flat paints. For exterior coatings, the limit is 200 grams of VOC per liter of product minus water for nonflat paints and 100 grams for flat paints. Paints that are formulated without VOCs are designated Class A and may contain a special designation to that effect on their label.

In addition to not exceeding the specified VOC concentration, qualifying paints must meet several other requirements. Products intended for use as interior opaque topcoat must pass industry tests for scrubbability, hiding power (opacity), and washability (stain removal). Paints intended for use as exterior opaque topcoat must pass the industry test for hiding power (opacity). In addition, the products may not contain more than 1 percent by weight of the sum total of aromatic compounds (which the standard defines as hydrocarbon compounds containing one or more six-carbon benzene rings in the molecular structure). A manufacturer also must demonstrate that its paint cans are made without lead and that its paints do not contain any of the following chemical compounds:

- Halomethanes: methylene chloride
- Chlorinated ethanes: 1,1,1-trichloroethane
- Aromatic solvents: benzene, toluene (methylbenzene), ethylbenzene
- Chlorinated ethylenes: vinyl chloride
- Polynuclear aromatics: naphthalene
- Chlorobenzenes: 1,2-dichlorobenzene
- Phthalate esters: di (2-ethylhexyl) phthalate, butyl benzyl phthalate, di-*n*-butyl phthalate, di-*n*-octyl phthalate, diethyl phthalate, dimethyl phthalate
- Miscellaneous semivolatile organics: isophorone
- Metals and their compounds: antimony, cadmium, hexavalent chromium, lead, mercury
- Preservatives (antifouling agents): formaldehyde
- Ketones: methyl ethyl ketone, methyl isobutyl ketone
- Miscellaneous VOCs: acrolein, acrylonitrile

Once a paint product is judged to be in compliance with all of these requirements, its manufacturer has the right to use the Green Seal certification mark on it. The mark must appear on the product's packaging and be accompanied by a description of the basis for certification. The description must be in a location, style, and typeface that are easily readable by the consumer. Unless otherwise approved by Green Seal, the description must read:

> This product meets Green Seal environmental standards for volatile organic compounds (VOCs) and other ingredients.

The packaging must also include a brief statement discouraging the disposal of paints down drains and encouraging consultation with local authorities regarding disposal requirements and recycling opportunities.

As can be seen from the preceding descriptions, the process of establishing environmental standards such as those promulgated by Green Seal is lengthy and complex. The criteria used to judge products such as paper towels and paints must be clearly defined. Although Green Seal's environmental standards are based primarily on one characteristic, such as recovered material content or the use of VOCs, many other aspects of a product are considered when decisions are made as to whether or not to award a seal of approval. It is this effort to take into account the many facets of a product's impact on the environment that makes Green Seal's approach a form of life-cycle analysis.

20.4   Scientific Certification Systems

The stated mission of SCS is "to spur the private and public sectors toward more environmentally sustainable policy planning, product design, management, and production through programs based on sound scientific principles." The organization was founded in 1984, when it launched its NutriClean program, a third-party certification system for testing for pesticide residues in fresh produce. In 1990, SCS initiated its two-part environmental labeling program: (1) the Environmental Claims Certification Program, which verifies specific environmental claims, and (2) the "environmental report card," which provides an environmental profile of products and packaging based on a life-cycle analysis.

The Environmental Claims Certification Program verifies the accuracy of manufacturers' environmental claims about their products. This verification effort has focused on the use of phrases such as "recycled content" or "biodegradability" to promote certain products. SCS also works with U.S. retailers to monitor the environmental claims made by their vendors. It is the goal of SCS to educate manufacturers about the proper use of green labeling terms. To date, claims for more than 1000 stock-keeping units (SKUs) have been certified. Manufacturers whose claims have been certified are permitted to display an authorized certification emblem accompanied by a precise description of the claim that has been verified. The SCS logo is comprised of a green cross superimposed on the right side of a blue globe, with the words "Scientific Certification Systems" in blue appearing below the cross (see Fig. 20-2). The emblem and accompanying description may appear on the product itself, the product packaging, product shelf signs, or in special educational materials prepared by manufacturers or retailers.

The second part of SCS's green labeling program centers around the environmental report card, which graphically depicts the environmental burdens associated with a product's production, use, and disposal. These burdens include the depletion of natural resources, energy consumption, the release of pollutants to the air and water, and the generation of solid waste. These burdens are the key elements in the life-cycle analysis that SCS performs for each product. By showing consumers the results of this analysis, the report card label tells people that all products have negative impacts on the environment but that it is possible for them to choose one product over another to minimize these impacts. SCS hopes that its labeling system will encourage consumers to take a more active role in bringing about environmental change by helping them to make better-informed purchasing decisions.

Figure 20-2. Logo for Scientific Certification Systems.

Manufacturers who would like to have their environmental claims certified and/or would like to use the report card label pay the cost associated with auditing their products. All services are provided by SCS on a direct fee-for-service basis, with no licensing fees for use of its certification mark. Payment of fees to undergo evaluation does not guarantee that certification will be granted. The certification process begins with the signing of a contract between a manufacturer and SCS, in which the manufacturer commits to providing detailed information related to the environmental claim and SCS provides confidentiality assurance. SCS then reviews the information, conducts a site visit, and performs several tests to confirm the accuracy of the claim. Independent databases are consulted to compare the results of SCS's review with industry norms. Following this evaluation process, certification is either granted or denied. The certification records for certified products are updated quarterly by SCS to ensure that the products are continuing to meet the required standards.

SCS is currently conducting 15 life-cycle analyses, which it expects to result in the issuance of 12 environmental report cards. To date, fewer than twenty of the thousands of products that have received SCS certification have been certified to use the report card.

20.5 Labeling Programs around the World

Many countries have taken steps toward instituting environmental labeling programs, including Germany, Canada, Japan, the United States, Austria, France, and Portugal, as well as Norway, Sweden, and Finland (the Nordic Council Programme). No two programs are exactly alike, yet they all have some key elements in common. All the programs follow the same basic steps for product category and criteria selection and refuse to consider any inherently dangerous products. Each program tries to include in its assessments the environmental impact of a product's production, distribution, use, and disposal. These programs are voluntary and all charge a fee for use of their label. Countries trying to establish green labeling programs have found the effort to be more difficult and expensive than they anticipated; consequently, most are government financed. The three countries with the most well established environmental labeling programs are Germany, Canada, and Japan, whose programs are discussed in greater detail below.

20.5.1 Germany

Germany's "Blue Angel" Program is generally regarded as the model for all other environmental labeling programs. The program awards its Blue Angel label to consumer products that are clearly more beneficial to the environment than others in the same product category, as long as a product's primary function and safety are not significantly impaired. The German program's seal of approval is comprised of the Blue Angel symbol of the United Nations Environmental Programme (UNEP) and, above the symbol, the word *umweltzeichen*, which means "environmental label." The Blue Angel Program is administered by three agencies: the Federal Environment Agency (FEA), the Environmental Label Jury (ELJ), and the Institute for Quality Assurance and Labeling (RAL). The ELJ, comprised of a broad spectrum of interested parties, has the ultimate say in the selection of product categories.

The German program awarded its first environmental label in 1978. By 1991, the program had issued over 3600 labels in 64 product categories. The Blue Angel Program has been credited with increasing Germany's consumption of recycled paper and low-VOC paints, lacquers, and varnishes. In 1981, the program required that recycled paper have a waste paper content of 51 percent; today, recycled paper must be made from 100 percent recovered paper to be awarded the Blue Angel.

Germany's environmental labeling program has been criticized for having awarded over half of its labels in only four (low-pollutant varnishes and coatings, low-emission gas burners, pH-neutral stripping agents for wastewater treatment, and recycled paper) out of sixty-four product categories. Examples of the product categories appear in Table 20-4. It has been difficult for the German program to get manufacturers to apply in all product categories. Manufacturers sometimes decline to apply, even if they meet the criteria, for fear that receipt of an environmental seal of approval for one product will draw attention to the fact that their other products do not carry the seal. The program also has been criticized for achieving only partial success with the awarding of the Blue Angel to low-VOC paints, varnishes, and lacquers. While purchases of these products have increased among do-it-yourselfers, they have not increased among professionals, who believe the low-VOC formulations are less effective.

The German government stresses that environmental labeling is a "soft" market instrument and that it functions best as a complement to regulations,

Table 20-4. Examples of Product Categories and Products Approved by Germany's Blue Angel Program

Product category	Products
Recyclables	Retreaded tires, returnable bottles, recycled paper, reusable crates for food products, building materials made from waste paper, building materials made from recycled glass, building materials made from recycled gypsum, recycled cardboard
Low in hazardous substances	Low-waste hair sprays, deodorants, and shaving creams, asbestos-free brake linings, zinc-air batteries, insecticide-free products for indoor pest control, low-formaldehyde wooden products (for indoor use), unbleached paper filters
Low emissions	Motor vehicles with exhaust emission control, low-emission and waste-reducing copiers
Low noise	Low-noise lawnmowers, durable low-noise automobile mufflers, low-noise compost choppers
Water efficient	Wastewater treatment for car washes, electronically operated shower facilities, water-saving flushing valves
Energy efficient	Gas-fueled combination boilers and circulating-water boilers, solar energy products and mechanical watches, highly heat-insulating multilayer window glass
Biodegradable	Soil nutrients and conditioners made from compost

taxes, and subsidies. The German government believes that its labeling program is an important tool for protecting the environment, but that even without green labeling, manufacturers would be lessening the environmental impacts associated with the production, use, and disposal of their products.

20.5.2   Canada

Canada started its Environmental Choice Program in 1988, and products bearing the program's seal of approval first appeared in stores in March 1990. The program's logo is a maple leaf (representing Canada's environment) comprised of three doves, which symbolize the three major partners joined to protect the environment: government, industry, and commerce. Accompanying the logo is a brief statement explaining why the product has received Canada's environmental seal of approval. The Environmental Choice Program is administered by the Canadian government. A 16-member board, comprised of representatives from environmental, manufacturing, health care, retailing, and consumer groups, acts as a final advisor to the Environment Minister on the selection and certification of products. The Canadian Standards Association (CSA), an independent testing and standards-setting organization, verifies products against guidelines and licenses companies on behalf of the Environmental Choice Program.

In selecting products for certification, the staff of Canada's Environmental Choice Program acts according to four principles: (1) The program should address significant long-term environmental issues rather than short-term issues that will be addressed by regulations; (2) a product's full life cycle should be considered before criteria are established, even if the criteria are based on just a few aspects of the product; (3) the program should be directed at educating consumers about environmental trade-offs and the fact that even a labeled product may not be absolutely safe for the environment; the program should also make clear that the label certifies a product, not the company; (4) the label should promote industry leadership by identifying environmentally superior goods.

At present, Canada's Environmental Choice Program has set its criteria so that 10 to 20 percent of the products in a given category can qualify for the environmental seal of approval. During the first two years of operation, the program has focused on recycled and low-pollution products. To date, 58 products in 18 product categories have been certified. Fourteen certified products are listed in Table 20-5. The board is in the process of considering the more than 400 product categories that have been recommended for certification. Canada had hoped that its program, which is currently administered by the government, would be totally self-sufficient in two to five years. While it has been unable to meet this goal, Canada still plans for its program to be self-financing once it generates more revenue from product licensing fees. Canada's program is the only government-run program that aspires to such self-sufficiency; neither Germany nor Japan intend to make their programs self-financing.

20.5.3 Japan

After two years of study, Japan initiated its EcoMark Program (also called the Project for the Promotion of Ecologically Safe Merchandise) in February 1989. By January 1991, the EcoMark Program had issued over 850 labels in 31 different product categories. The EcoMark logo is comprised of two arms embracing the

Table 20-5. Examples of Product Categories and Products Approved by Canada's Environmental Choice Program

Product categories	Products
Recyclables	Fine paper from recycled paper, craft forms from recycled paper, newsprint from recycled paper, insulation from recycled wood-based cellulose fiber, products made from recycled plastic, re-refined motor oil, reusable cloth diapers, composting systems for residential waste, reusable shopping bags
Low-pollution products	Solvent-based paint, water-based paint, zinc-air batteries, ethanol-blended gasoline
Miscellaneous	Heat-recovery ventilators

world, which symbolizes the protection of the earth with our own hands. The two arms form the letter "e," which stands for "environment," "earth," and "ecology."

Japan's green labeling program seeks to encourage environmental conservation by certifying products and services that are judged to be relatively beneficial to the environment. In awarding the EcoMark, the Japanese program looks for products that meet four criteria: (1) minimal environmental impact from use; (2) significant potential for improvement of the environment by using the product; (3) minimal environmental impact from disposal after use; (4) other significant contributions to improve the environment. Certified products must also demonstrate that (1) appropriate environmental pollution control measures are employed during production, (2) treatment for disposal of product is relatively easy, (3) energy or resources can be conserved by using the product, (4) the product complies with laws, standards, and regulations pertaining to quality and safety, and (5) price is not extraordinarily higher than comparable products.

The daily operations of Japan's EcoMark Program are handled by the secretariat of the Japan Environment Association. The job of awarding environmental labels is shared by two committees: the nine-member EcoMark Promotion Committee comprised of representatives from consumer and industry groups, the Environmental Agency, the National Institute for Environmental Studies, and local governments; and the five-person Committee for Approval, comprised of representatives from consumer protection organizations, environmental science experts, and technical experts from the Environment Agency and the National Institute for Environmental Studies. The EcoMark Promotion Committee advises on day-to-day operations, approves the guidelines for the operation of the program, and selects appropriate product categories and criteria. The Committee for Technical Approval decides whether or not products qualify for the label.

Compared with other countries' environmental labeling programs, Japan's EcoMark Program is less expensive to administer, in part because it provides for more rapid approval of products and product categories. Japan's program is less complex than Germany's and provides for less public participation than Canada's. To date, most of the products approved by the EcoMark program fall into three categories: products designed to reduce kitchen waste, products that foster recycling, and products that conserve energy. Examples of some of these products appear in Table 20-6.

Japan has a second environmental labeling program called the Greenmark Program, which has been in operation since 1981. The Greenmark program is sponsored by the Ministry of International Trade and Industry (MITI) on behalf of the Japan Paper Recycling Promotion Center (JPRPC). The Greenmark is used only to identify paper products with recycled content. Certain school and community groups are eligible to collect the Greenmarks: The groups purchase labeled paper products and send the Greenmarks to the JPRPC. In return, the JPRPC sends these groups coupons that can be used to buy seedlings at All Japan Tree Growers Association stores.

Table 20-6. Examples of Product Categories and Products Approved by Japan's EcoMark Program

Product categories	Products
Kitchen waste	Strainers for use in domestic kitchen sinks, blotting paper for used cooking oil, home composters
Recyclables	Recycled paper, soap made from used cooking oil, cloth diapers, recycled rubber tires
Energy efficient	Solar heating systems, thermal insulation for buildings, products that use solar batteries
Miscellaneous	Cellulose sponges, cans with stay-on tabs, water-saving faucets and valves, blast furnace cement

20.6   Conclusion

The U.S. EPA estimates that Americans discarded 4.3 pounds per person per day of municipal solid waste in 1990 and that this rate will likely increase to 4.5 pounds per person per day by the year 2000. At this rate, the American public will be generating 222 million tons of solid waste per year. In addition to filling up scarce landfill space, the generation and management of this large quantity of municipal solid waste can result in the pollution of our soil, water, and air. One way to significantly reduce the amount and toxicity of our discarded waste is to purchase only those products whose production, use, and disposal do the least possible harm to the environment. To do so requires that manufacturers take into account the environmental impacts of their products and inform consumers about these impacts.

Green labeling is one of the best mechanisms for providing consumers with the information they need to select environmentally preferable products from among essentially equivalent items. Purchasing decisions made by environmentally concerned consumers can help to push manufacturers toward decreasing the environmental impacts associated with the production, use, and disposal of their products, by rewarding manufacturers that do so with consumer dollars. For the power of consumer spending to be effectively harnessed, consumers must have complete confidence in the accuracy of the environmental labels that guide their purchasing decisions. The work being conducted by Green Seal and SCS, and by countries such as Germany, Canada, and Japan, to establish a reliable system of environmental labeling is essential to building this confidence.

Further Reading

John Elkington, Julia Hailes, and Joel Makower, *The Green Consumer*, Viking Penguin, New York, 1990.

Federal Trade Commission, *Guides for the Use of Environmental Marketing Claims: The Application of Section 5 of the Federal Trade Commission Act to Environmental Advertising and Marketing Practices*, FTC, Washington, D.C., July 28, 1992.

Organization for Economic Cooperation and Development, *Environmental Labelling in OECD Countries*, OECD, Paris, 1991.

Thompson Publishing Group, *Environmental Packaging: U.S. Guide to Green Labeling, Packaging, and Recycling*, Washington, D.C., 1993.

U.S. Congress, Office of Technology Assessment, *Green Products by Design: Choices for a Cleaner Environment*, OTA-E-541, U.S. Government Printing Office, Washington, D.C., October 1992.

U.S. Environmental Protection Agency "Green Advertising Claims," brochure doc. no. 530-F-92-024, October 1992. (Call 1-800-424-9346 to obtain a copy.)

Organizations to Contact for Additional Information

Coalition of Northeastern Governors (CONEG)
400 North Capitol Street, NW, Suite 382
Washington, DC 20001
(202) 624-8450

Green Seal
1250 23rd Street, NW, suite 275
Washington, DC 20037-1101
(202) 331-7337

Federal Trade Commission
Consumer Protection Bureau
Division of Advertising Practices
6th Street and Pennsylvania Avenue, NW
Washington, DC 20580
(202) 326-3090

National Recycling Coalition/
Recycling Advisory Committee
1101 30th Street, NW, Suite 305
Washington, DC 20007
(202) 625-6406

Northeast Recycling Council (NERC)
139 Main Street, Suite 401
Brattleboro, VT 05301
(802) 254-3636

Scientific Certification Systems
1611 Telegraph Avenue, Suite 1111
Oakland, CA 94612-2113
(510) 832-1415

U.S. Office of Consumer Affairs
1620 L Street, NW, Suite 700
Washington, DC 20036
(202) 634-4140

21

Pollution Prevention in Process Development and Design

R. Scott Butner

Shapiro and Associates, Inc.
Seattle, Washington

21.1 Introduction

Process development describes the refinement of a process concept from early conceptual stages (articulation of process objectives, selection of process steps, determination of constraints) through preliminary engineering (development of preliminary economic analysis, piping and instrumentation diagrams, and process flow diagrams). Process development is also understood to include experimental studies at the laboratory, bench-scale, and pilot-plant level. These studies are focused on obtaining key process information required to guide the remainder of the design effort and often take place concurrently with the conceptual design. Process development efforts can be oriented around new products, improvement of existing products, or increased utilization of available capital equipment assets (i.e., retrofit design).

The primary reason for incorporating pollution prevention into process development and design is one of cost effectiveness. Decisions made early in the development process often determine later development activities, including laboratory and pilot-plant studies, equipment and materials selection, and project economic analysis. By addressing environmental issues early in the development cycle, one can anticipate unforeseen technical, regulatory, and economic consequences of design choices. The net result is a reduction in the technical and economic risk associated with environmental issues. Additional benefits may also be obtained when concurrent consideration of environmental issues with other engineering factors leads to quicker time-to-market, process innovation, improved quality of products, or increased efficiency.

21.1.1 Linkages Between Product and Process Design

It is important to understand the strong linkage that exists, particularly in early stages of development, between product and process. Because product design issues were dealt with at length in Chaps. 18–20, they will not be given much attention here. However, it bears repeating that the process development team should be represented in the product development and specification efforts. Conversely, having a representative of the product team represented during process development efforts may make it easier to challenge assumptions regarding product requirements when pollution prevention options appear to be in conflict.

Product specifications developed by the product development team play an important role in establishing process design options. It is important to identify any product specifications which are likely to result in significant environmental issues, since changing the product specifications at this stage of development will be easier than during later phases, when significant time and resources have been devoted to process development. Some of the questions which should be raised early in the process development cycle include:

Were all opportunities for reducing the toxicity of the product itself explored? In a well-designed process plant, the product generally represents the largest single release to the environment.

What is the basis for product purity specifications, if any? Many times, purity specifications are used as a proxy for some other performance-based specification (such as color, specific gravity, or melting point). If purity specifications can be relaxed in favor of meeting performance specifications, additional latitude may be gained in designing separation and recycle systems.

What related products are anticipated? Can off-specification material be used to formulate one of these related products? Can process equipment and production runs be coordinated to minimize cleaning-related wastes?

How will the product be packaged? Will containers be returned to the plant for recycle? Packaging operations can be the source of waste from broken or leaky containers, transfer wastes, etc. Ideally, design of the packaging and design of the process will be coordinated.

21.1.2 Defining Design Objectives and Constraints

A critical first step in implementing pollution prevention in the process development cycle is to clearly and explicitly identify the environmental-related process development objectives and constraints. This should be done as part of the initial

Table 21-1. Examples of Environmental Design Constraints and Objectives

Constraints	Objectives
Compliance with all applicable environmental regulations	Minimal use of toxics in-process
	Minimize life-cycle impact within acceptable financial parameters
Compliance with existing permit requirements for discharge, emissions	Implement all pollution prevention options meeting investment hurdles
Process loadings not to exceed existing treatment capacity	Maximize use of recycled raw materials
Zero discharge of regulated wastes	

scope of work for the process development effort in the same way that other process requirements (product specifications, technology constraints, process throughput, equipment constraints, economic hurdle rates, etc.) are defined.

While the establishment of environmental constraints and objectives seems like an obvious first step, it is often overlooked. The consequence of this oversight is that incorporating pollution prevention into the existing process development framework becomes difficult. Lacking clear objectives and constraints, the process development team must make assumptions which may not hold up as the project moves into the engineering phase.

It is a fact of life that design objectives may conflict with one another. For example, cooling of process streams prior to storage is a useful strategy to reduce air emissions from storage tanks*; however, cooling of these streams is likely to increase overall cooling requirements for the plant. Optimizing the trade-offs between conflicting objectives is probably most readily accomplished by fully costing the environmental objectives into economic terms. When data is not available to determine the full costs of waste and emissions, or when the environmental objectives include so-called externalized costs which are difficult to assign economic value to (e.g., emission of nonregulated greenhouse gases), then other techniques must be used.[2]

21.2   The Process Development Cycle

The process of moving a product concept from initial concept to a working process is often referred to as the process development cycle. Despite the implication that process development activities are cyclic in nature, most process development occurs in a project setting. In other words, for any given project (a debottlenecking project, for example), work proceeds in more or less distinct

---

*With the cooling of the process stream prior to storage, the vapor pressure of the stored material is reduced. This can reduce the amount of leakage from tank vents, valve packings, etc. Nelson[1] provides several strategies for implementing this concept.

phases, complete with milestones, decision points, and interim products (e.g., reports, drawings, specifications, or other documents). Thus, much of the discussion in this chapter will refer to the development activities in project terms.

A typical process development project can include several more or less distinct phases, including bench-scale testing to validate process chemistry, conceptual design to determine economic feasibility, pilot-scale testing to determine engineering issues for process scale-up, and preliminary engineering, which typically results in specifications for detailed (preconstruction and construction-phase) engineering.

The process development project represents a gradual accumulation of information about the process, during which the process constraints become increasingly well defined. During early stages of development, much of the information available is incomplete or uncertain, making quantitative evaluation of pollution prevention options difficult. During these preliminary stages, qualitative tools including risk-assessment, root-cause analysis, and heuristic (e.g., "checklist") approaches to design are generally most appropriate. As more data is collected, quantitative tools including a wide variety of optimization techniques, life-cycle impact assessment, and total cost accounting, can be effectively applied. Many of these techniques are already employed in traditional process development, and it is generally more effective to integrate environmental considerations into existing tools and workflows than to completely restructure the development process.

21.2.1 Bench-Scale Development

As opposed to product development research, bench-scale development activities are generally focused on obtaining information needed to design the manufacturing process. Especially on projects which involve new or untried chemistry, bench-scale development efforts play an important role in better defining potential environmental impacts of the process. During bench-scale development, process chemistry is typically the major focus of investigation. Reaction stoichiometry, yields, and rates are determined for the system or systems of interest. Calorimetric studies may be undertaken to determine the amount of heat generated during reaction. Determination of catalyst activity, selectivity, and lifetime is generally carried out at this stage of development. In some cases, thermodynamic equilibrium data for key separation steps must be collected as well.

Much of the data which is needed to support pollution prevention decisions later in the development project should be collected during the bench-scale testing. Careful consideration should be given to the use of statistical experimental design and quality function deployment (QFD) techniques in both the scoping and conduct of bench-scale studies.[3] Simplified life-cycle inventory techniques, root-cause analysis, "what-if," or other screening techniques for identifying potential sources of pollution can also be useful in scoping bench-scale studies. In the absence of more specific scoping requirements, Table 21-2 describes some typical types of data which should be considered during experimental design and data analysis.

Table 21-2. Pollution Prevention Data from Bench-Scale Testing

Unit operation	Data required
General	Determine corrosion rates of candidate construction materials
	Screen for catalytic effects of candidate materials, corrosion products, feed impurities
Reactors	Determine reaction stoichiometry
	Determine equilibrium yield
	Measure catalyst selectivity, activity, and lifetime
	Identify and characterize reaction by-products
	Determine kinetics of major side-reactions
	Determine effects of recycle
Distillation and other separations processes	Obtain vapor pressure data for products and intermediates
	Obtain vapor-liquid equilibrium (VLE) data for potential entrainers, diluents, and trace compounds
	Determine loading capacity, regenerative properties of absorbents

21.2.2 Conceptual Design

Process conceptual design often occurs at several stages during the development of a new process, usually concurrent with the bench-scale testing. During the process conceptual design, fundamental decisions are made regarding the desired chemistry or processing operations to be used, the sequencing of unit operations, the relationship of the process with other operations, and whether batch or continuous processing will be employed. Often, these decisions must be made preliminary to the collection of any engineering data regarding actual process yields, generation of reaction by-products, or the efficacy of any needed separation steps. However, it is still possible even at this early stage of development to begin including some pollution prevention strategies, as shown in Fig. 21-1.

From a pollution prevention perspective, probably the most critical need at this early stage of process development is identification of key potential pollution sources which arise from the preliminary process flow diagram. These can be used to focus the environmental constraints and refine the data collection efforts which will take place as the project proceeds.

21.2.3 Pilot-Scale development

By the time a process reaches pilot-scale development, a major commitment to the process chemistry has typically been made. The emphasis thus shifts to the study of equipment design issues. The objective of pilot-scale development is generally to accomplish the following:

- Avoid adsorptive separations where adsorbent beds cannot be readily regenerated.

- Provide separate reactors for recycle streams, to permit optimization of conversions.

- Consider low-temperature distillation columns when dealing with thermally labile process streams.

- Consider high-efficiency packing rather than conventional tray-type columns (this reduces pressure drop and decreases reboiler temperatures).

- Consider continuous processing when batch cleaning wastes are likely to be significant (e.g., with highly viscous, water-insoluble, or adherent materials).

- Consider scraped-wall exchangers and evaporators with viscous materials to avoid thermal degradation of product.

Figure 21-1. Pollution prevention strategies for conceptual design.

- Demonstrate the technical and economic viability of a process at a scale which is commercially meaningful

- Obtain a better determination of key process parameters and quantify relationships between these parameters and process performance and scale-up

- Confirm results of laboratory studies and study process variables that are not readily scaled (for example, heat and mass transfer effects and flow distribution in process vessels.)

By the time a process has reached pilot-scale testing, some equipment design decisions will have been made. Depending on the specific objectives of the pilot studies, the type of reactor model (fluidized bed, packed bed, CSTR, etc.) will often be selected. Certainly, by the end of pilot-scale testing, data should have been gathered on heat exchange coefficients, fouling rates, and required power inputs for agitation and mixing.

Some issues should be thoroughly explored during pilot-stage testing, to help direct pollution prevention efforts during preliminary engineering. These include:

- *Effect of reactor mixing and feed distribution on by-product formation.*   Pilot testing often provides the first opportunity to conduct the reaction stages of the process in equipment which approximates the reactors to be used commercially. Since by-products can be formed due to residence time distributions caused by poor mixing or feed distribution, hot spots on catalysts, impurities not present in lab-grade chemicals, or catalytic effects of construction materials, be sure to inspect and analyze products for unwanted by-products.

- *Fouling rates in heat exchange equipment.*   Cleaning wastes resulting from routine exchanger maintenance and thermal decomposition products formed by increased wall temperatures forced by reductions in heat exchange coefficients are both common causes of process waste. Use pilot-scale studies to fully investigate these phenomena.

- *Corrosion studies.*   Although these should have been initiated during bench-scale testing, it is important to monitor corrosion rates during pilot-scale testing as well.

- *Sedimentation rates and product stability.*   Pilot-scale studies are a good opportunity to look for sedimentation or sludge formation in process vessels, and to determine product stability during storage.

21.2.4 Preliminary Process Engineering

A key objective of the preliminary engineering phase of development is to provide a preliminary cost estimate (typically ±30 percent) and economic analysis of the process. By this time, the process chemistry has been firmly established, and data is available to support assumptions underlying the capital and operating costs estimates. The emphasis of the design has shifted almost completely from the qualitative aspects of the process to the quantitative. Major process vessels are sized, and initial valve counts are often completed. By the end of the preliminary engineering stage, a preliminary piping and instrumentation diagram (P&ID) will typically be complete, and broad considerations of facility site design will have been concluded. Opportunities for major process changes are few, but there are still many opportunities for incorporating pollution prevention into the process. In fact, as the emphasis shifts from conceptual to operational aspects of the proposed project, it may prove easier to implement suggestions for pollution prevention because the uncertainties surrounding the suggestions can be more readily evaluated.

Table 21-3 lists some of the strategies for preventing pollution during the preliminary engineering phase of the project.

21.3 Key Tools and Techniques for Incorporating Pollution Prevention

The preceding material illustrates the opportunities for incorporating pollution prevention at each stage of the process development project. While many of the concepts are relatively simple or intuitive, incorporating them into the existing development process is often challenging. Simply including pollution prevention considerations into a postdesign review is unlikely to be effective. A more pragmatic strategy is to incorporate these considerations into existing process development tools and techniques. The following section describes this approach.

Table 21-3. Strategies for Pollution Prevention during Preliminary Engineering

Activity	Strategy
Vessel design	Ensure easy access to storage tanks, reactors, etc., to simplify cleaning of vessels
	Design tank and vessel drains to ensure complete draining of vessels
Piping design	Recover wastestreams separately
	Minimize length of piping runs (reduces material inventory)
	Minimize valve and flanges (reduce valve counts, use welded fittings)
	Route drains, vents, and relief lines to recovery or treatment
	Specify bellow-seal or zero-emission valves
Instrumentation design	Select in-line process analyzers to reduce sampling wastes
	Use closed-loop (purge style) sampling ports
	Install preventative maintenance monitoring equipment (e.g., vibration monitors, torque sensors) to optimize maintenance schedules
	Instrument heat exchangers to permit real-time monitoring of fouling, leakage
	Consider advanced control strategies such as model-based control
Materials selection	Consider costs of waste disposal when selecting maximum allowable corrosion rates
	Consider foul-resistant materials (e.g., Teflon) on heat exchange surfaces and vessels requiring frequent cleaning
	Consider glass- or polymer-lined vessels where frequent cleaning is required
Cost estimation	Incorporate "hidden" waste costs (handling, regulatory tracking, and reporting, treatment, disposal, liability), in cost equations
	In absence of detailed waste costs, consider "penalty functions" for releases based on environmental objectives

A number of tools and techniques exist for incorporating pollution prevention into the development process. Many of these are simply adaptations of existing invention, decision, and optimization techniques already in common use by process development teams. Some, such as the various components of life-cycle analysis (inventory, impact assessment, etc.) have been developed largely outside the context of process design. In addition, a number of design heuristics (rules of thumb) have been developed by various authors[4,5] and can provide a great deal of useful guidance to members of the development team.

Design heuristics have the advantage of being easily communicated and incorporated into existing design practices. A common way of incorporating these heuristics into the design process is through the use of design checklists, as described in the following section.

21.3.1 General Principles

Just as general principles such as designing for disassembly, recyclability, and life extension can be applied to product design,* similar principles can be effectively applied to process design. The use of applicable general principles reduces the amount of new information that the development team needs to assimilate, and can often lead to new insights regarding process alternatives. These principles are particularly useful in early stages of process development, when the initial scoping of the process development effort is being established.

A number of general principles have been reported in the literature, and are summarized in Table 21-4.

21.3.2 Checklist Approaches

Checklists are commonly used in process design as a means of structuring the analysis of process options, conveying commonly used expertise or knowledge, and promulgating "informal" corporate design standards. They are useful in pollution prevention because they force the evaluation of pollution prevention options early and continuously throughout the development project.

Checklists should be developed specifically to fit within the design procedures used by a company. Checklists are commonly used in screening (e.g., to evaluate the relative risk of potential waste or process streams), in equipment specification and design (to ensure that all options have been considered), and during process economics (to ensure that indirect costs related to the environment have been included). Figure 21-2 illustrates a typical equipment design checklist, used prior to final specification of a distillation column.

21.3.3 Application of Life-Cycle Inventories During Process Development

Life-cycle analysis is a methodology for inventorying, assessing, and ultimately valuing the environmental impacts across the entire life cycle of a product, including resource extraction, manufacture, shipping and storage, use, and ultimate disposition. It has been widely discussed in the literature[6] and is covered extensively elsewhere in this volume. This discussion focuses on specific application of LCA techniques to process development.

---

*A very well articulated set of product design principles has been developed by the American Electronics Association. Up-to-date white papers on this program can be obtained by contacting the AEA's offices in Washington, D.C.

Table 21-4. Process Design Heuristics for Pollution Prevention

Principle	Rationale	Examples
Seek to minimize the number of process steps	Reducing number of process steps also reduces the potential for leaks, spills, contamination	Minimize interim storage where possible Seek simpler process chemistry
Minimize potential for leaks	Non-point-source and fugitive emissions can represent major sources of environmental impact from a chemical process plant	Specify sealless pumps for process fluids Use low- or zero-emission valves and valve packings Precool volatile fluids prior to storage Use welded fittings on piping Reduce the number of flanges, valves, and sample points Reduce relief venting by increasing margin between operating and design pressure
Maximize process selectivity at each unit operation	Eliminating production of by-products reduces load on downstream separations, waste treatment	Avoid thermal decomposition by reducing temperatures at heat exchange surfaces Choose catalyst on basis of selectivity and durability rather than activity Optimize reactor design, feed distribution, and temperature profiles for maximum selectivity Screen materials for unwanted catalytic properties
Minimize process utility requirements	Process utilities such as steam, cooling water, and electricity often represent significant sources of waste	Recycle process cooling water where possible Use heat integration to reduce overall heating load Use variable speed drives on pumps and compressors
Segregate process streams where possible	Mixing of wastestreams can hinder reuse, recycling, and treatment efforts	Avoid use of common process sewers or drains Avoid contamination of storm water by use of covered process equipment, drip pans, segregated sewers
Design for operability	Many wastes are result of process upsets, off-spec product, or maintenance and cleaning	Design heat exchangers with smaller operating margins to reduce fouling Design piping, pumps, process vessels for capture of drainage Design for easier cleaning of vessels (e.g., material and finish specifications, geometry, access ports) Provide interim storage for reclamation and recycling of cleaning fluids

- Is there likelihood of thermal decomposition in the materials being separated?

- If yes, have you considered:

    Reducing pressure drop in the column?
    Using high-efficiency packings rather than trays?
    Increasing the vapor line size?
    Using a falling-film reboiler?
    Reducing column pressure?
    Increasing insulation on column?

- Is the column adequately drained to permit cleaning and maintenance?

- Were alternative separation (e.g., membranes, crystallization, adsorption) techniques evaluated?

- If the column is provided with relief valves, are they vented to recovery or treatment devices?

- Can the operating margin between column working pressure and design capacity be increased in a cost-effective manner? (This can help prevent relief events.)

- Can feed streams be preheated with waste heat to reduce utility loads and prevent thermal decomposition?

Figure 21-2. Example equipment design checklist (distillation column).

In a strict sense, life-cycle analysis tends, by definition, to include a much broader scope of impacts than is typically considered in process development, which emphasizes the manufacture of the product, rather than its use. In this light, it is perhaps a more appropriate technique for product development efforts than for process development. However, the general concept of life-cycle analysis—the systematic evaluation of environmental impacts—can be applied successfully to process development activities.

Scoping of a life-cycle assessment is perhaps the most important step in applying the technique, since the establishment of process boundaries and the environmental issues to be considered determines both the usefulness of the analysis, as well as the amount and type of data required to support the effort. Since data collected throughout the process development effort is likely to support the LCA, the assessment should be seen as an iterative process. During early phases of the process development project, it may be enough to simply identify the LCA system boundaries so that the full range of environmental impacts can be flagged for later study. In later stages of the project, as more data on process performance is available, quantitative assessment of impact may be attempted.

As a screening tool, LCA can be used to protect the development team from "surprises" later in the project. These surprises often arise from the realization that environmental concerns related to nonroutine or indirect wastes were overlooked in preliminary project planning. Typical issues identified in LCA scoping include:

Wastes related to production and extraction (i.e., mining) of raw materials and intermediate products

Emissions resulting from production of energy for the process (including off-site generation)

Wastes resulting from packaging, storage, and transportation of raw materials and products

Wastes from decommissioning of process facilities

Cleaning, maintenance, start-up, and shutdown wastes

Non-point-source emissions, including contamination of storm water, trash, and soils in processing areas

Secondary wastes generated during waste treatment operations (ash, sludges, biosolids, spent adsorbents, etc.)

Direct release of product to the environment during use

Of course, the range of environmental "vectors" to consider is also significant. Although a full life-cycle analysis may consider such "nonwaste" issues as resource depletion, habitat destruction, social disruptions, and the like, the emphasis in this handbook is on pollution prevention. Appropriate screening vectors might include:

- Persistent or bioaccumulative toxins (heavy metals, dioxins, etc.)
- Acute toxins or other materials requiring special handling
- Specifically regulated materials (e.g., Toxics Release Inventory chemicals)
- Greenhouse gases
- Ozone-depleting chemicals
- Materials specifically identified in existing or anticipated permits

Ideally, life-cycle analysis should be used throughout the product development cycle and become a basis for making product deployment decisions. Realistically, much of the data required to conduct these analyses is not available during early stages of the product and process development. A practical strategy is to begin the analysis during the early stages and to both periodically update the available data as well as review the study boundaries and vectors of concern.

21.3.4 Use of Design Specifications

Design specifications are the primary means for communicating design constraints and performance standards and a key means of capturing collected experience and expertise of the design team. Because project design specifications are often derived from "generic" specifications which establish the general practices and standards of the company, incorporation of pollution prevention considerations in these specifications can have a widespread impact, helping pollution prevention become part of standard practice in the development effort.

References

1. K. E. Nelson, "Process Modifications that Reduce Waste," Paper presented at the *American Institute of Chemical Engineers (AIChE) 1991 Summer National Meeting*, Pittsburgh, Pa.

2. Pacific Northwest Laboratory, *Project Opportunity and Benefit Model Methodology Guide*, Pacific Northwest Laboratory, Richland, Wash., 1993.

3. C. M. Overby, "QFD and Taguchi for the Entire Lifecycle," *Transactions of the 1991 ASQC Quality Congress*, American Society for Quality Control, Minneapolis, Minn., 1991.

4. Chemical Manufacturer's Association, *Designing Pollution Prevention into the Process*, Chemical Manufacturer's Association, Washington, D.C., 1993.

5. R. L. Berglund, and G. E. Snyder, "Minimize Waste During Design," *Hydrocarbon Processing*, April 1990, pp. 39–42.

6. Society of Environmental Toxicologists and Chemists (SETAC), *A Technical Framework for Lifecycle Assessment*, SETAC Foundation for Environmental Education, Washington, D.C., 1991.

Further Reading

Allenby, B. R., and A. Fullerton, "Design for Environment—A New Strategy for Environmental Management," *Pollution Prevention Review*, Winter 1991–1992, pp. 51–61.

Early, W. F., and M. A. Edison, "Design for Zero Releases," *Hydrocarbon Processing*, August 1990, pp. 47–49.

Fromm, C. H., "Pollution Prevention in Process Design," *Pollution Prevention Review*, Autumn 1922, pp. 389–401.

Jacobs, R. A., "Design Your Process for Waste Minimization," *Chemical Engineering Progress*, June 1991, pp. 55–59.

K. E. Nelson, "Use These Ideas to Cut Waste," *Hydrocarbon Processing*.

Pojasek, R. B., and L. J. Cali, "Measuring Pollution Prevention Progress," *Pollution Prevention Review*, Spring 1991, pp. 119–130.

Pojasek, R. B., "Looking Beyond the Manufacturing Process for Opportunities to Prevent Pollution," *Pollution Prevention Review*, Autumn 1993, pp. 469–473.

U.S. Environmental Protection Agency, "Life Cycle Design Guidance Manual," EPA600/R-92/226, U.S. EPA-ORD/RREL, Cincinnati, Ohio, January 1993.

22

Pollution Prevention Through Reactor Design

Jack R. Hopper

Lamar University
Beaumont, Texas

22.1 An Overview of Reaction Engineering and Pollution Prevention

Generation of waste in the chemical processing industries has its beginning in the heart of the process—the reaction system. Pollution prevention will have the greatest impact in minimizing the generation of waste through the design and operation of chemical reactors by reducing generation at the source—source reduction.

Pollution prevention by modification of reaction parameters is defined as changing the selectivity of the reaction so that undesirable reactions which produce waste products are minimized while at the same time producing the desirable products.

Process reactor design is an extensive subject which requires mastery of all of the fundamentals of chemical engineering: material balances, energy balances, heat transfer, mass transfer, and momentum transfer. Moreover, few rules are generally applicable to the process design of chemical reactors because of the diversity of the behavior of chemical reactions. However, the well-established general rules for mechanical design of pressure vessels, heat exchangers, agitators, and many other pieces of equipment may be applied to mechanical design of reactors.

22.2   Basic Terminology

A *homogeneous reaction* is one that involves only one phase. A *heterogeneous reaction* involves more than one phase, and reaction usually occurs at or very near the interface between the phases. An *irreversible reaction* is one that proceeds in only one direction and continues in that direction until the reactants are exhausted. A *reversible reaction,* on the other hand, can proceed in either direction, depending on the concentrations of reactants and products relative to the corresponding equilibrium concentrations. An irreversible reaction behaves as if no equilibrium condition exists.

Batch reactors are designed on the basis of actual time of reaction (in addition to time to heat up and cool down). Continuous reactors are designed on the basis of time in which the reactants stay in the reactor (residence time). True residence time is difficult to determine and an apparent residence time is used as the independent variable.

Take as the basis species A, which is one of the reactants that is disappearing in the general reaction:

$$aA + bB \rightarrow cC + dD \tag{22.1}$$

The rate of disappearance of A, $-r_A$, depends on temperature and composition. It can be written as the product of a reaction rate constant k and a function of the concentrations of the various species involved in the reaction, i.e.,

$$-r_A = [k(T)][fn(C_A, C_B, \ldots)] \tag{22.2}$$

The quantity k is referred to as the specific reaction rate (constant). It is strongly dependent on temperature. The temperature dependence of the specific reaction rate k is typically correlated by the Arrhenius equation:

$$k(T) = Ae^{-E/RT} \tag{22.3}$$

Multiple reactions can be classified into three general reaction types: series, parallel, and combination. There are two kinds of products that can be obtained from multiple reactions: desired products and by-products or undesired products. The desired product is the product that is produced on purpose and the by-product is the product that is formed as a second component in a single reaction or is a product formed in a competing or side reaction which may become a wastestream. Modification of reaction parameters to change the selectivity of the reaction so that undesirable reactions are minimized is pollution prevention technology. This selectivity may be affected by all of the reaction conditions. These reaction conditions include reactor temperature, reactor pressure, catalyst, reactor mixing conditions, feed temperature, feed pressure, ratio of feed concentrations, feed mixing conditions, reaction type, and reactor type and design options.

Multiphase reactions or heterogeneous systems are classified by: (1) two-phase systems (gas-liquid, liquid-liquid, gas-solid, and liquid-solid) and (2) three-phase systems (gas-liquid-solid). Moreover, the two-phase and the three-phase systems may be either catalytic or noncatalytic.

In heterogeneous reaction systems, the overall rate expression becomes complicated because of interaction between physical and chemical processes. The complication occurs because the reactants have to be transported to a common phase where the reactions take place. Heterogeneous systems include a series of physical resistances combined with the chemical kinetic resistance.

22.3 Classification of Reactors

Reactors may be classified by mode of operation (batch, semibatch, or continuous) and by the kinds of phases that are being contacted (homogeneous or heterogeneous). A matrix showing one concept for classification of reactors on this basis is shown in Fig. 22-1. Design considerations are given for homogeneous, heterogeneous fixed-bed-catalytic, and heterogeneous gas-liquid-solid catalytic reactors.

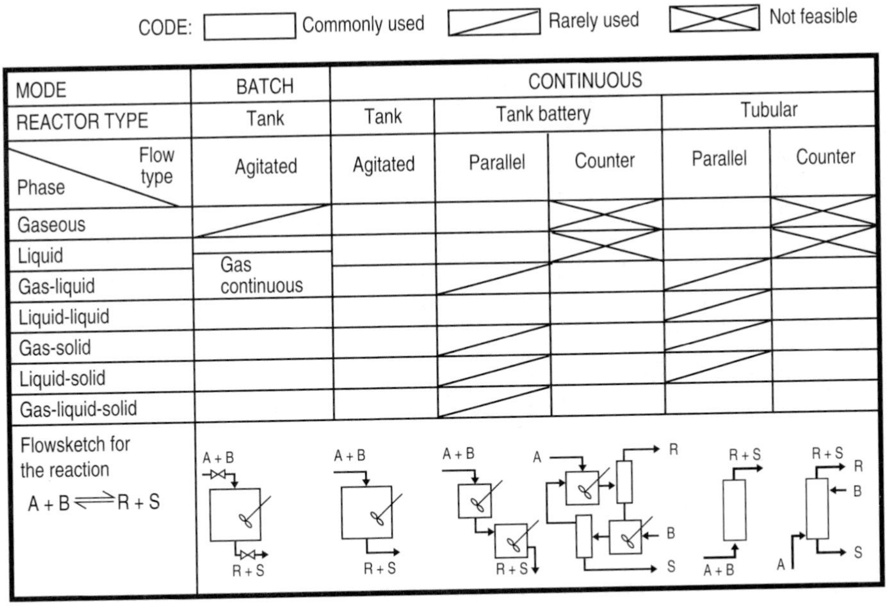

Figure 22-1. Classification of reactors according to the mode of operation and the kinds of phases involved. (*Reprinted with permission from S. M. Walas, Chemical Process Equipment, Butterworths Publishers, Boston, 1988.*)

22.3.1   Homogeneous Reactors

Homogeneous (all gas or liquid) reactors are generally classified as stirred tank (batch or continuous) or tubular. A description of these reactors follows:

Homogeneous Tank Reactors

Batch Reactors.   Stirred tanks are the most common type of batch reactor. Typical dimensions are shown in Fig. 22-2. This type of reactor is filled at the beginning of the operation and no material is fed into or removed from the reactor during the reaction period (Fig. 22-3*a*). Stirring is used to initially mix the ingredients, to maintain complete mixing during reaction, and to increase heat transfer. Batch processing is used when an extended period of reaction time is required or when the volume of product is small and/or many different products are produced.

Continuous Stirred Tank.   Stirred tanks are also commonly used in continuous processing in the chemical industry as shown in Fig. 22-3. Reactants are fed to the reactor and products withdrawn in a continuous flow through the reac-

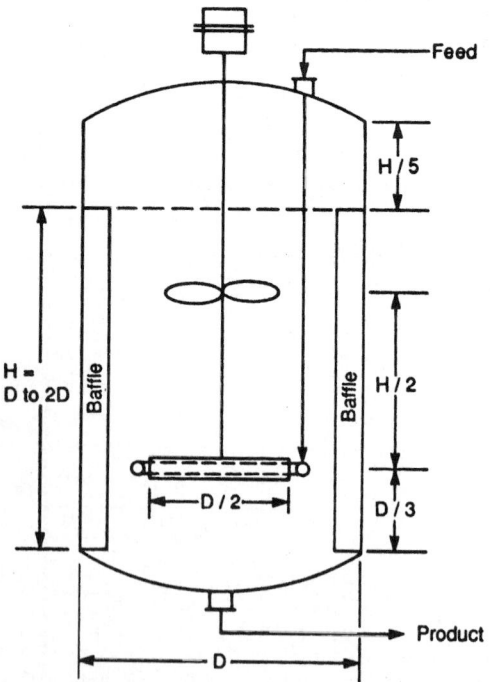

Figure 22-2. Typical proportions of a stirred tank reactor with radial and axial impellers, four baffles, and a sparger feed inlet. (*Reprinted with permission from S. M. Walas,* Chemical Process Equipment, *Butterworths Publishers, Boston, 1988.*)

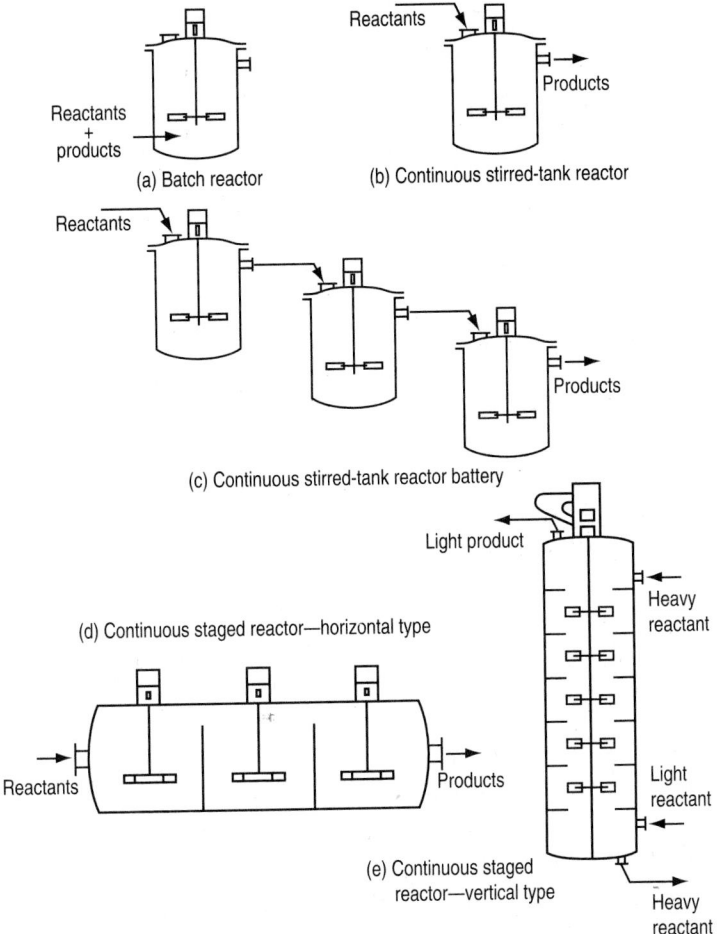

Figure 22-3. Stirred tank reactors. [*Reprinted with permission from R. H. Perry et al. (eds.), Chemical Engineers Handbook, 6th ed., McGraw-Hill, 1984.*]

tor. The processing may be in a single tank (Fig. 22-3*b*) or multiple tanks in a battery arrangement (Fig. 22-3*c*) or a tank of staged compartments either horizontal (Fig. 22-3*d*) or vertical (Fig. 22-3*e*). Extensive use is made in large-scale plants of economy of scale and product quality control.

Semicontinuous Stirred Tank. Semicontinuous operation is performed in a stirred tank with several options: (1) one reactant is charged at the beginning and another fed continuously during the reaction, (2) both reactants are fed continuously without removal until the reaction is completed, and (3) reactants are initially charged to the reactor and one of the products is removed during the reaction.

Mixing Efficiency. Power input per unit volume and impeller tip speeds are used as measures of the mixing efficiency. These measures are based on the vessel dimensions, impeller type, and proper baffling. Some ranges recommended by Walas[1] for various conditions are:

Operation	kW/m³*	Tip speed (m/s)
Blending	0.05–0.1	
Homogeneous reaction	0.1–0.3	2.5–3.3
Reaction with heat transfer	0.3–1.0	3.5–5.0
Gas-liquid, liquid-liquid	1–2	5–6
Slurries	2–5	

*1 kW/m³ = 5.08 HP/1000 gal. Agitator power P_q is correlated through the power number N_p, with the impeller Reynolds number N_{Re}, in Fig. 22-4.

Heat Transfer. Temperature control is one of the major considerations for the design and operation of reactors. Material and energy balances as presented in the tables are needed to determine the required heat transfer. The data needed are heat transfer surface areas, thermal conductivities, and heat transfer coefficients. The correlations for heat transfer which contain these data are the Nusselt, Stanton, Prandtl, and Reynolds dimensionless groups. Figure 22-5 illustrates several options of reactor configuration for heat transfer to stirred tanks. Overall heat transfer coefficients U are given in Table 22-1 for jacketed vessels and in Table 22-2 for immersed coils.

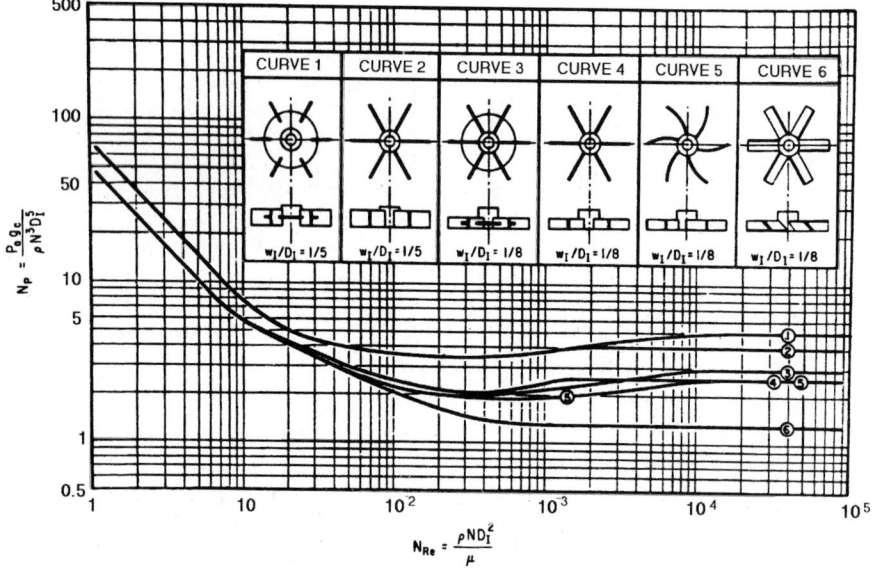

Figure 22-4. Power correlation for baffled turbines. [*From V. W. Uhl and J. B. Gray (eds.), Mixing, vol. 1, Academic Press, New York, 1966; reprinted with permission from H. F. Rase, Chemical Reactor Design for Process Plants, vol. 1, Wiley Interscience, 1977.*]

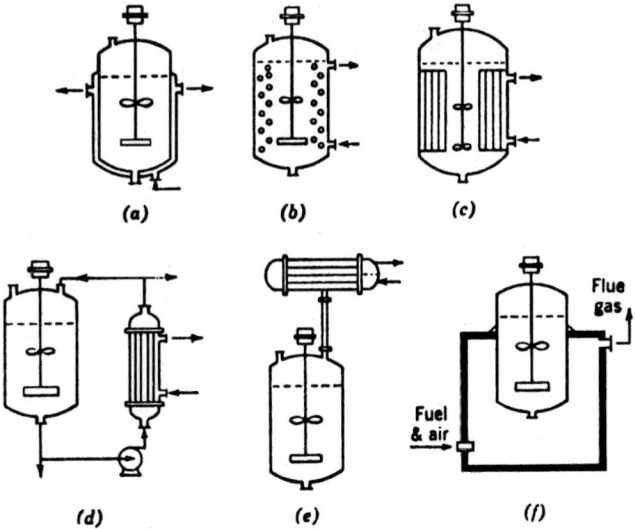

Figure 22-5. Heat transfer to stirred tank reactors. (*From S. M. Walas*, Reaction Kinetics for Chemical Engineers, *McGraw-Hill, New York, 1959; with permission from S. M. Walas*, Chemical Process Equipment, *Butterworths Publishers, Boston, 1988.*)

Homogeneous Tubular Reactors. The ideal tubular flow reactor is the plug flow model (PFR) in which all reacting components flow through the reactor in the same time (residence time). This type of reactor is constructed as several tubes in parallel (Fig. 22-6*a*) or a single continuous tube (Fig. 22.6*b*). The reactants enter and the products leave with a continuous change in composition and temperature with the length of the tube because of reaction. Any backmixing that occurs is incidental and a good approach to plug flow is achieved for gases or liquids in empty tubes when the flow is turbulent ($N_{Re} > 2000$) and the tube length to diameter ratio (L/D) is ≥ 50.[2,3] Tubular-type reactors for homogeneous reactions are favored by reactions requiring plug flow and high heat transfer surface-to-volume ratio obtainable with small-diameter tubes. Major examples of homogeneous tubular reactors are fired-tubular, molten metal, preheated tubular, shell and tube, adiabatic pipeline, adiabatic tower combination heat exchanger, and adiabatic and flame reactors.

22.3.2   Heterogeneous Reactors

Heterogeneous Fixed-Bed Catalytic Reactors. The fixed-bed catalytic reactors have catalyst particles in the range of 2- to 5-mm diameter. Some possible options for loading catalysts include:

- A single large bed
- Several horizontal beds

Table 22-1. Jacketed Vessels Overall Heat Transfer Coefficients

Jacket fluid	Fluid in vessel	Wall material	Overall U^*	
			Btu/(h · ft² · °F)	J/(m² · s · K)
Steam	Water	Stainless steel	150–300	850–1700
Steam	Aqueous solution	Stainless steel	80–200	450–1140
Steam	Organics	Stainless steel	50–150	285–850
Steam	Light oil	Stainless steel	60–160	340–910
Steam	Heavy oil	Stainless steel	10–50	57–285
Brine	Water	Stainless steel	40–180	230–1625
Brine	Aqueous solution	Stainless steel	35–150	200–850
Brine	Organics	Stainless steel	30–120	170–680
Brine	Light oil	Stainless steel	35–130	200–740
Brine	Heavy oil	Stainless steel	10–30	57–170
Heat-transfer oil	Water	Stainless steel	50–200	285–1140
Heat-transfer oil	Aqueous solution	Stainless steel	40–170	230–965
Heat-transfer oil	Organics	Stainless steel	30–120	170–680
Heat-transfer oil	Light oil	Stainless steel	35–130	200–740
Heat transfer oil	Heavy oil	Stainless steel	10–40	57–230
Steam	Water	Glass-lined CS	70–100	400–570
Steam	Aqueous solution	Glass-lined CS	50–85	285–480
Steam	Organics	Glass-lined CS	30–70	170–400
Steam	Light oil	Glass-lined CS	40–75	230–425
Steam	Heavy oil	Glass-lined CS	10–40	57–230
Brine	Water	Glass-lined CS	30–80	170–450
Brine	Aqueous solution	Glass-lined CS	25–70	140–400
Brine	Organics	Glass-lined CS	20–60	115–340
Brine	Light oil	Glass-lined CS	25–65	140–370
Brine	Heavy oil	Glass-lined CS	10–30	57–170
Heat-transfer oil	Water	Glass-lined CS	30–80	170–450
Heat-transfer oil	Aqueous solution	Glass-lined CS	25–70	140–400
Heat-transfer oil	Organics	Glass-lined CS	25–65	140–370
Heat-transfer oil	Light oil	Glass-lined CS	20–70	115–400
Heat-transfer oil	Heavy oil	Glass-lined CS	10–35	57–200

*Values listed are for moderate nonproximity agitation, CS = carbon steel.
SOURCE: *Perry's Chemical Engineers Handbook,* McGraw-Hill, New York, 1984.

- Several packed tubes in a single shell
- A single bed with imbedded tubes
- Beds in separate shells

Heterogeneous Gas/Liquid/Solid Catalytic Reactors. Three-phase reactors can be classified into two main categories: (1) fixed-bed reactors in which the solid catalyst is stationary, and (2) suspended-bed reactors in which the solid catalyst is suspended and is in motion. In gas-liquid-solid catalytic reactors, the direction of flow of gas and liquid provides a number of options of

Table 22-2. Overall Heat Transfer Coefficients with Immersed Coils
[U expressed in Btu/(h · ft^2 · °F)]

Type of coil	Coil spacing, in.*	Fluid in coil	Fluid in vessel	Temp. range, °F.	$U^\dagger$ without cement	U with heat-transfer cement
$^3/_8$-in o.d. copper tubing attached with bands at 24-in spacing	2 3 $^1/_8$ 6 $^1/_4$ 12 $^1/_2$ or greater	5–50 lb/in^2 gage steam	Water under light agitation	158–210 158–210 158–210 158–210	1–5 1–5 1–5 1–5	42–46 50–53 60–64 69–72
$^3/_8$-in o.d. copper tubing attached with bands at 24-in spacing	2 3 $^1/_8$ 6 $^1/_4$ 12 $^1/_2$ or greater	50 lb/in^2 gage steam	No. 6 fuel oil under light agitation	158–258 158–258 158–240 158–238	1–5 1–5 1–5 1–5	20–30 25–38 30–40 35–46
Panel coils		50 lb/in^2 gage steam Water Water	Boiling water Water No. 6 fuel oil Water No. 6 fuel oil	212 158–212 228–278 130–150 130–150	29 8–30 6–15 7 4	48–54 19–48 24–56 15 9–19

*External surface of tubing or side of panel coil facing tank.
†Data courtesy of Thermon Manufacturing Co.

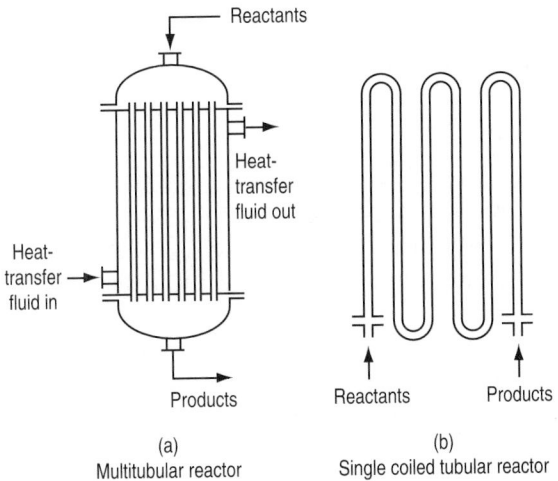

Figure 22-6. Tubular reactors. [*Reprinted with permission from R. H. Perry et al. (eds.),* Chemical Engineers Handbook, *6th ed., McGraw-Hill, 1984.*]

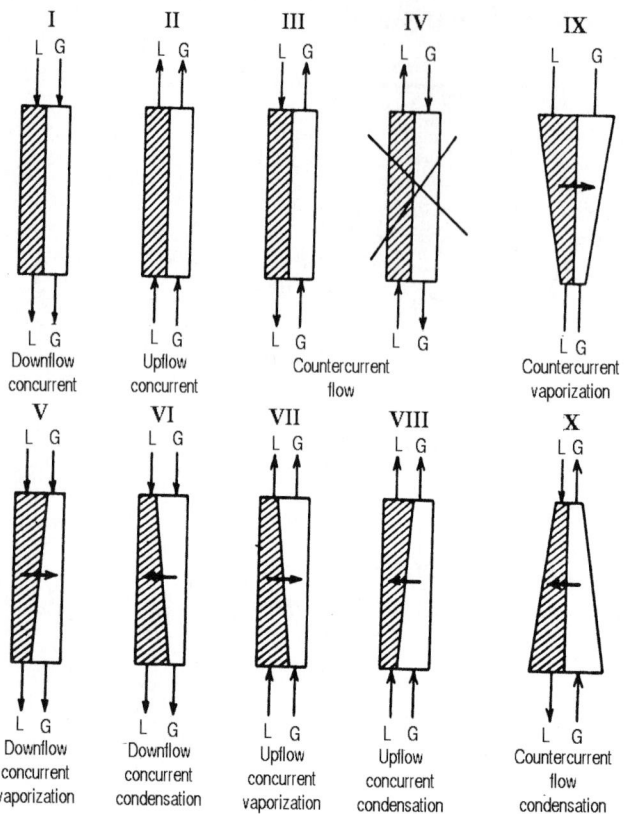

Figure 22-7. Various kinds of gas-liquid-phase fixed-bed reactors. (*Reprinted with permission from M. O. Tarhan,* Catalytic Reactor Design, *McGraw-Hill, New York, 1983.*)

possible reactor types as illustrated in Fig. 22-7. This multiphase reactor has generally been modeled with a gas-phase reactant A(g) and liquid-phase reactant B(1). However, it is also possible to start with gas-phase reactant B(g) and A(g) in a system with liquid. The model must incorporate the resistance for mass transfer in each phase: gas, liquid, and solid.

The overall rate of reaction in three-phase reactors is often limited by the mass transfer processes. The rate of mass transfer processes is generally faster in suspended-bed reactors than in fixed-bed reactors. This is because smaller particles can be used in suspended-bed reactors; the rates of liquid-to-solid mass transfer and intraparticle diffusion are therefore higher, leading to a more efficient utilization of the catalyst. The catalyst loading (amount per unit volume of reactor) is, however, lower in suspended-bed systems. Since the rate of reaction is also proportional to the catalyst loading, the rate per unit volume of reactor is generally higher in a packed-bed reactor except for highly active catalysts, where the mass transfer process dictates the overall rate. Thus, the rate of reaction per unit volume of reactor is lower in suspended-bed systems, while the rate per unit weight of catalyst is likely to be higher. Homogeneous side reactions may contribute more in suspended reactors than in fixed beds due to the relatively large liquid holdup.

In addition to mass transfer, the rate of reaction is influenced by the mixing patterns of the liquid and gas phases. Two extreme situations of mixing are the ideal reactors—plug flow and completely backmixed flow. This axial dispersion model provides for intermediate mixing conditions. Backmixing generally lowers the reactor performance. The liquid phase in a suspended-bed reactor is generally backmixed, while in the fixed-bed reactor the liquid flow pattern approaches the plug flow behavior. Hence, when high conversion of the liquid reactant is desired, fixed-bed operation is preferable to suspended-bed reactors.

Temperature control is easier in slurry reactors because heat transfer is more efficient in suspended-bed reactors than in fixed-bed reactors. This leads to uniform temperature at the active sites of the catalyst and avoids formation of hot spots. Large liquid holdup in slurry reactors also facilitates better temperature control due to the large heat capacity of the liquid phase.

Separation of the catalyst poses difficulties in the continuous operation of suspended-bed reactors mainly due to the problems of filtration, while in fixed-bed reactors this problem is not encountered. In processes where frequent removal and replacement of the catalyst is necessary, fixed-bed reactors may not be suitable because replacement of catalyst generally requires shutdown and dismantling of the reactor column. Also, the problem of bed plugging, experienced in trickle beds due to the formation of nonvolatile residues, is eliminated in suspended-bed systems by use of small, free-moving particles.

Fixed-Bed Gas/Liquid/Solid Catalytic Reactors. In the fixed-bed reactor, the two fluid phases move over a stationary bed of catalyst particles. The various options of flow are (see Fig. 22-7): (1) cocurrent downflow of both gas and liquid (type I), (2) downflow of liquid and countercurrent upflow of gas (type III), and (3) cocurrent upflow of both gas and liquid (type II). These three options are illustrated in Fig. 22-8. Type I has conventionally been called a "trickle-bed" reactor[4] and type II has been called a "fixed-bed bubble" reactor.[5]

Trickle-bed reactors—Cocurrent downflow. The downflow cocurrent reactor illustrated as type I in Fig. 22-7 is the most commonly used of all gas-liquid fixed-bed reactors. It is called the trickle-bed reactor. Both liquid and gas flow downward. The flow rates in the downflow operation are not limited by flooding as is the case with the upflow operation. The only limitation on the flow rates is the available pressure head.

Different flow regimes exist, as illustrated in Fig. 22-9. However, the trickle-bed operation is characterized by low flow rates of gas and liquid. The gas phase is continuous and the liquid phase dispersed.

Both plug flow (PFR) and continuous tank (CSTR) isothermal models for the liquid phase have been developed for the trickle-bed reactor. Reactor modifications such as particle wetting, liquid distribution, and axial dispersion to

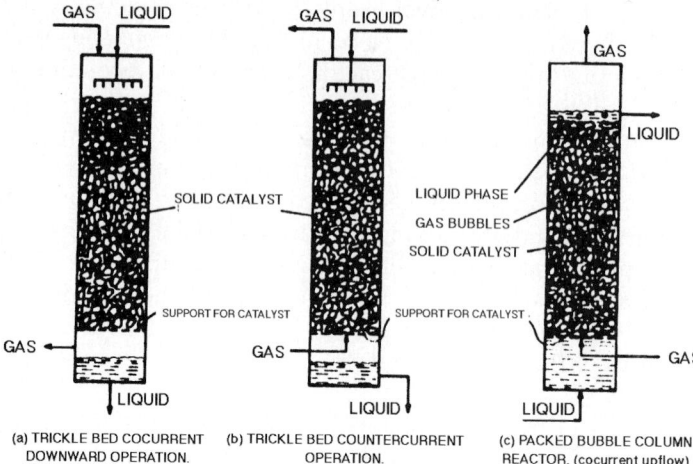

Figure 22-8. Schematic diagrams of three-phase packed-bed reactors. (*Reprinted with permission from P. A. Ramachandran and R. V. Chaudhari*, Three-Phase Catalytic Reactors, *Gordon and Breach Science Publishers, Philadelphia, 1992.*)

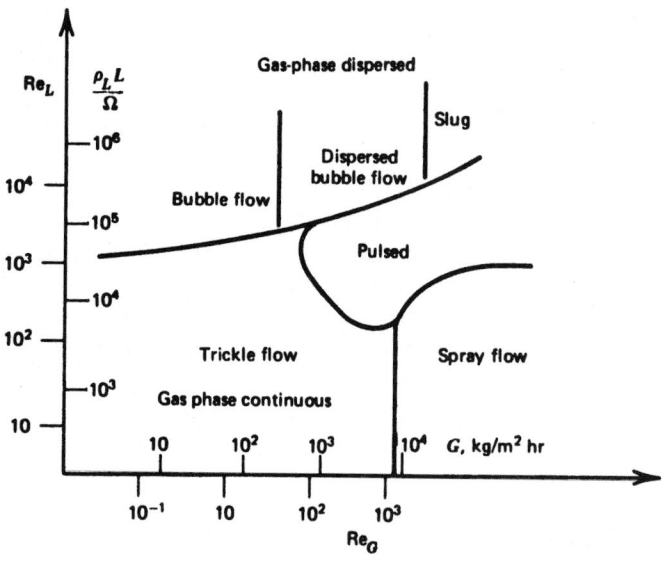

Figure 22-9. Flow regimes in downflow packed beds. (*Reprinted with permission from G. F. Froment and K. B. Bischoff*, Chemical Reactor Analysis and Design, *John Wiley and Sons, New York, 1979.*)

account for deviation from plug flow may be needed when the rate is dependent on the concentration of liquid reactant, B(1), and higher conversion of B is desirable. Operating conditions may also create temperature gradients in the reactor which would require the total energy balance and a knowledge of heat transfer coefficients. Discussion of these modifications has been presented by Ramachandran and Chaudhari.[6]

Fixed-bed bubble reactors—Cocurrent upflow. The typical fixed-bed bubble reactor (FBBR) is an upflow cocurrent gas-liquid reactor, illustrated as type II in Fig. 22-7, with relatively low gas and liquid flow rates [G_G < 100 lb/(hr)(ft²)]. The liquid phase is the continuous phase and the gas phase is the disperse phase. Other flow regimes will occur at higher gas flow rates.[7] When the liquid flow rate is above the fluidization velocity, the fixed bed of catalyst expands into a suspended-bed reactor (ebullated bed).

Comparison of fixed-bed reactors. Trickle-bed reactors operated at low liquid rates may not wet all of the catalyst particle and thus the catalyst may only be exposed to the gas-phase reactant. This may lead to hot-spot formation, temperature run-away, and, in some cases, poorer utilization of the catalyst. In a packed bubble-bed reactor, the catalyst is completely wetted and these complications may be avoided. The pressure drop in downflow operation is less than that in upflow (the packed bubble-bed operation), thus reducing the pumping energy costs. Packed bubble beds have less problem than trickle beds with catalyst deactivation due to deposits of tarry or polymeric materials because of higher liquid rates. Heat transfer efficiency and heat removal are better in packed bubble beds than trickle beds due to larger liquid holdup and higher liquid velocity. Homogeneous liquid-phase side reactions are less in trickle beds than packed bubble beds due to lower liquid holdup.

Heterogeneous Suspended-bed Gas/Liquid/Solid Catalytic Reactors. Suspended-bed reactors are a class of heterogeneous gas-liquid phase reactors in which the catalyst particles are loosely supported in the liquid phase. It is possible to classify these into three categories: (1) the mechanical agitated slurry reactor/continuous stirred tank (CSTR), (2) the bubble column slurry reactor, and (3) the three-phase fluidized-bed reactor (ebullated-bed and three-phase transport). The difference in the three is the method used to suspend the catalyst. Schematic diagrams are shown in Fig. 22-10.

Mechanical agitated slurry reactor/continuous stirred tank. The continuous stirred tank reactor which contains finely divided catalyst maintains the suspension by vigorous mechanical agitation. Several modifications include:

1. Large circulation of liquid through the reactor by spraying the liquid onto the upflowing gas

2. Gas recirculation to increase gas residence time

3. Inert liquid holding vessel to suspend catalyst and improve heat transfer

Bubble column slurry reactors. In bubble column slurry reactors, the particles are suspended by means of a gas-induced agitation by the movement of the gas

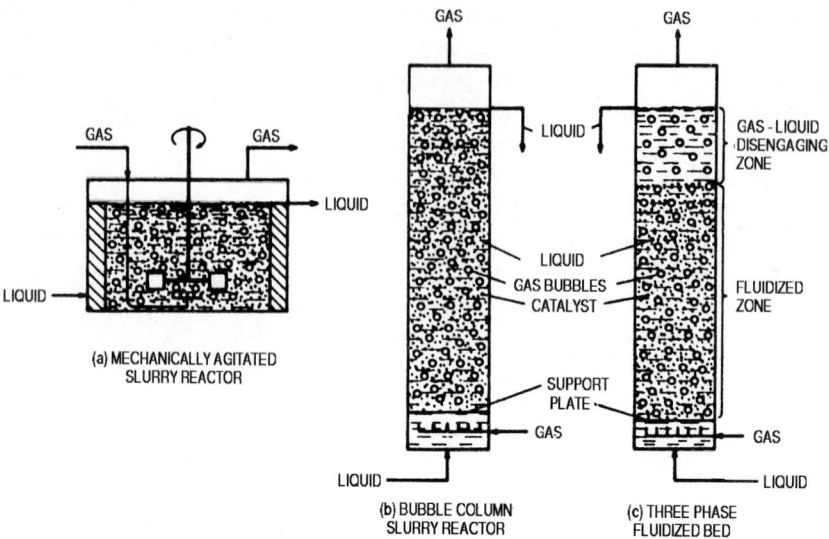

Figure 22-10. Schematic diagram of three-phase slurry reactors. (*Reprinted with permission from P. A. Ramachandran and R. V. Chaudhari,* Three-Phase Catalytic Reactors, *Gordon and Breach Science Publishers, Philadelphia, 1992.*)

bubbles. Finely divided catalyst particles are used in these reactors. The reactor columns have height-to-diameter ratios in the range of 4 to 10. The reactor may be operated in a semibatch mode for conversion of a liquid reactant or in the continuous mode for reaction between gaseous and liquid reactants.

These reactors have the advantage of no moving parts, minimum maintenance, greater heat transfer rates than fixed beds due to high liquid circulation rates, lower power requirements and less floor space than mechanically agitated reactors, and minimum intraparticle diffusion resistance due to small catalyst particles.

These reactors give lower conversion due to backmixing, may require additional energy to overcome column pressure drop, and have rapid decrease in bubble interfacial area because of gas bubble coalescence at column length-to-diameter ratio greater than 10.

Three-phase fluidized-bed reactors. In a three-phase fluidized-bed reactor, the particles are suspended by the combined action of bubble movement and cocurrent liquid flow. However, the major difference between the three-phase fluidized bed and the bubble column is that the suspension of the particles is primarily by the movement of liquid. Therefore, a relatively large particle size can be used in this operation and separation of catalyst is much easier compared to other slurry reactors. Cocurrent flow of gas and liquid is normal, but the countercurrent is possible. The cocurrent upflow of gas and liquid in which the entrainment of the catalyst particles does not carry the particles out of the bed has been termed the *ebullated-bed reactor*.[8] When the liquid velocity is increased, the liquid flow entrains the catalyst particles out of the reactor and this slurry is recycled to the bottom of the reactor. It has been termed a "three-phase transport" reactor.[9] Various conditions of operation of three-phase fluidized-bed reactors have been reviewed by Ostergaard.[10,11]

Three distinct flow regimes are observed in a three-phase fluidized bed: bubble flow, slug flow, and gas-continuous (dispersed liquid flow). The bubble flow regime is the normal mode of operation and is characterized by swarms of discrete gas bubbles flowing in the liquid phase. The flow generally occurs at average superficial gas velocities u_g, less than 10 cm/s and at the ratio of liquid to gas velocities $u_l u_g$, less than 10.[12]

Particle entrainment has been observed by Paige and Harrison[13] to decrease with decrease in bubble size and frequency, increase in liquid velocity and particle size, and baffling in the free board space. Entrainment can be significant for particles with $d_p < 0.1$ cm and $\rho_p < 3$ gm/cm^3. The gas is assumed to move through the bed in plug flow. Liquid backmixing is greater at lower liquid velocity. Complete backmixing is assumed for short columns. Plug flow is assumed for beds of large particles (greater than 6 mm).

Comparison of suspended-bed reactors. The mechanically agitated reactor has the advantage of higher heat and mass transfer efficiencies compared with the bubble column and three-phase fluidized reactors. Catalyst attrition can be significant in mechanically agitated reactors, while this problem is not so serious in the other two types. The mechanical design of bubble columns and fluidized-bed reactors is simpler than that of agitated slurry reactors which require suitable provision for stirrer seals. Catalyst separation is relatively easy in a three-phase fluidized bed as the particle size range used is larger than in the other two types of reactors. The power requirement is highest for the mechanically agitated reactor and is lowest for the bubble column. The catalyst distribution is relatively uniform in an agitated slurry reactor, while in a bubble column and a three-phase fluid bed, a nonuniform distribution of particles can exist.

Acknowledgments

This project has been funded in part with federal funds as part of the program of the Gulf Coast Hazardous Research Center which is supported under cooperative agreement R815197 with the United States Environmental Protection Agency and in part with funds from the state of Texas as part of the program of the Texas Hazardous Waste Research Center.

References

1. S. M. Walas, *Chemical Process Equipment*, Butterworths, Boston, 1988.

2. H. F. Rase, *Chemical Reactor Design of Process Plants*, 2 vols., Wiley, New York, 1977, p. 404.

3. H. F. Rase, *Fixed-Bed Reactor Design and Diagnostics*, Butterworths, Boston, 1990.

4. C. N. Satterfield, "Trickle-bed Reactors," *A.I.Ch.E.J.* **21**:209 (1975).

5. H. Hofmann, "Multiphase Catalytic Packed-bed Reactors," *Catal. Rev. Sci. Eng.* **17**:71 (1978).

6. P. A. Ramachandran and R. V. Chaudhari, *Three-Phase Catalytic Reactors*, Gordon and Breach Science Publishers, Philadelphia, 1992.

7. J. L. Turpin and R. L. Huntington, "Prediction of Pressure Drop for Two-phase, Two-component Cocurrent Flow in Packed Beds," *A.I.Ch.E.J.* **13**:1196 (1967).

8. E. S. Johanson, "Gas-Liquid Contacting Process," U.S. Patent 2,987,465, 1961.

9. B. B. Pruden and M. E. Weber, "Evaluation of the 3-Phase Transport Reactor," *Can. J. Chem. Eng.* **48**:310 (1970).

10. K. Ostergaard, "Gas-Liquid-Particle Operations in Chemical Reaction Engineering," *Adv. Chem. Eng.* **7**:71 (1968).

11. K. Ostergaard, "Three-phase Fluidization," in J. F. Davidson and D. Harrison (eds.), *Fluidization*, Academic Press, New York, 1971, p. 751.

12. R. N. Mukherjee, P. Bhattacharya, and D. K. Taraphdar, "Studies on the Dynamics of Three-phase Fluidization," in H. Angelino et al. (eds.), *Fluidization and Its Applications*, Capedues Editions, Toulouse, 1974, p. 372.

13. R. E. Paige and D. Harrison, "Particle Entrainment from a Three-Phase Fluidized Bed," in H. Angelino et al. (eds.), *Fluidization and Its Applications*, Capedues Editions, Toulouse, 1974, p. 393.

Further Reading

Achwal, S. K., and J. B. Stepanek, "Holdup Profiles in Packed Beds," *Chem. Eng. J.* **12**:69 (1976).

Akita, K., and F. Yoshida, "Gas Holdup and Volumetric Mass Transfer Coefficient in Bubble Columns," *Ind. Eng. Chem. Proc. Des. Dev.* **12**:76 (1973).

Akita, K., and F. Yoshida, "Bubble Size, Interfacial Area and Liquid-Phase Mass Transfer Coefficient in Bubble Column," *Ind. Eng. Chem. Proc. Des. Dev.* **13**:84 (1974).

Alvarez-Cuenca, M., M. A. Bergougnou, and M. A. Nerenberg, "Oxygen Mass Transfer in Three-Phase Fluidized Beds Working at Large Flow Rates," in *Fluidization*, Henniker, N.H., 1980*b*.

Beek, J., "Design of Packed Catalytic Reactors," *Adv. Chem. Eng.* **3**:203–271 (1962).

Bern, L., J. O. Lidefelt, and N. H. Schoon, "Mass Transfer and Scaleup in Fat Hydrogenation," *J. Am. Oil Chem. Soc.* **53**:463 (1976).

Bernard, R. A., and R. H. Wilhelm, *Chem. Eng. Progr.* **46**:233 (1950).

Calderbank, P. H., "Physical Rate Processes in Industrial Fermentation, Part 1: The Interfacial Area in Gas-Liquid Contacting with Mechanical Agitation," *Trans. Instn. Chem. Engrs.* **36**:443 (1958).

Calderbank, P. H., Review series no. 3, "Gas Absorption from Bubbles," *Chem. Eng. J.* **45**:209 (1967).

Calderbank, P. H., and L. A. Pogorsky, *Trans. Inst. Chem. Eng.* (London) **35**:195 (1957).

Campbell, T. M., and R. L. Huntington, *Petrol. Refiner.* **31**:123 (1952).

Coberly, C. A., and W. R. Marshall, *Chem. Eng. Prog.* **47**:141 (1951).

Deckwer, W. D., R. Burchart, and G. Zool, "Mixing and Mass Transfer in Tall Bubble Columns," *Chem. Eng. Sci.* **29**:2177 (1974).

Dhanuka, V. R., and J. B. Stepanek, "Gas-Liquid Mass Transfer in a Three-Phase Fluidized Bed," in J. R. Grace and J. M. Matsen (eds.), *Fluidization,* Plenum Press, New York, 1980*a*, p. 261.

Doraiswamy, L. K., and M. M. Sharma, *Heterogeneous Reactors: Analysis, Examples and Reactor Design,* 2 vols., Wiley, New York, 1984.

Dorweiler, V. P., and R. W. Fahien, "Mass Transfer at Low Rates in a Packed Column," *A.I.Ch.E.J.* **5**:139 (1959).

El-Temtamy, S. A., Y. O. El-Sharnoubi, and M. M. El-Halwagi, "Liquid Dispersion in Gas-Liquid Fluidized Beds," *Chem. Eng. J.* **18**:151, 161 (1979).

Ergun, S., "Fluid Flow through Packed Columns," *Chem. Eng. Prog.* **48**:89 (1952).

Fahien, R. W., and J. M. Smith, "Mass Transfer in Packed Beds," *A.I.Ch.E.J.* **1**:25 (1955).

Fair, J. R., "Designing Gas-Sparged Reactors," *Chem. Eng.* **74**:67 (1967).

Fogler, H. S., *Elements of Chemical Reaction Engineering,* 2d ed., Prentice Hall, Englewood Cliffs, N.J., 1992.

Froment, G. F., and K. B. Bischoff, *Chemical Reactor Analysis and Design,* Wiley, New York, 1979.

Furzer, I. A., and R. W. Michell, "Liquid-Phase Dispersion in Packed Beds with Two-Phase Flow," *A.I.Ch.E.J.* **16**:380 (1970).

Hanratty, T. J., "Nature of Wall Heat Transfer Coefficient in Packed Beds," *Chem. Eng. Sci.* **3**:209 (1954).

Harmathy, T. Z., "Velocity of Large Drops and Bubbles in Media of Infinite or Restricted Extent," *A.I.Ch.E.J.* **6**:281 (1960).

Hiby, J. R., *Interaction between Fluids and Particles,* Institution of Chemical Engineers, London, 1962.

Hochman, J. M., and E. Effron, "Two-Phase Cocurrent Downflow in Packed Beds," *Ind. Eng. Chem. Fundam.* **8**:63 (1969).

Holland, C. D., and R. G. Anthony, *Fundamentals of Chemical Reaction Engineering,* 2d ed., Prentice Hall, Englewood Cliffs, N.J., 1989.

Hopper, J. R., C. L. Yaws, T. C. Ho, and M. Vichailak, "Waste Minimization by Process Modification," *Waste Mgt.* **13**:3 (1993).

Kim, S. D., C. G. J. Baker, and M. A. Bergougnou, "Phase Holdup Characteristics of Three-Phase Fluidized Beds," *Can. J. Chem. Eng.* **53**:134 (1975).

Kobayashi, T., and H. Saito, "Solid-Liquid Mass Transfer in Bubble Columns," *Kagaku Kogaku* **3**:210 (1965).

Kunii, D., and J. M. Smith, "Heat Transfer Characteristics of Porous Rocks," *A.I.Ch.E.J.* **6**:71 (1960).

Kwong, S. S., and J. M. Smith, "Radial Heat Transfer in Packed Beds," *Ind. Eng. Chem.* **49**:894 (1957).

Larkins, R. P., R. R. White, and D. W. Jeffrey, "Two-Phase Cocurrent Flow in Packed Beds," *A.I.Ch.E.J.* **7**:231 (1961).

Leva, M., *Fluidization,* McGraw-Hill, New York, 1959.

Leva, M., and M. Grunner, "Heat Transfer to Gases through Packed Tubes: Effect of Particle Characteristics," *Ind. Eng. Chem.* **40**:415 (1948).

Levenspiel, O., *Chemical Reaction Engineering,* 2d ed., Wiley, New York, 1992.

Levenspiel, O., *The Chemical Reactor Omnibook,* Oregon State University, Corvallis, 1989.

Levins, D. M., and J. R. Glastonbury, "Applications of Kolmogoroff's Theory to Particle-Liquid-Mass Transfer in Agitated Vessels," *Chem. Eng. Sci.* **27**:537 (1972*a*).

Mangartz, K. H., and T. Pilhofer, "Interpretation of Mass Transfer Measurements in

Bubble Columns Considering Dispersion of Both Phases," *Chem. Eng. Sci.* **36**:1069 (1981).

Mears, D. E., "Tests for Transport Limitations in Experimental Catalytic Reactors," *Ind. Eng. Chem. Process Des. Dev.* **10**:541 (1971).

Michel, B. J., and S. A. Miller, "Power Requirements of Gas-Liquid Agitated Systems," *A.I.Ch.E.J.* **8**:262 (1962).

Michell, R. W., and I. A. Furzer, "Mixing in Trickle Flow through Packed Beds," *Chem. Eng. J.* **4**:53 (1972).

Nienow, A., "Suspension of Solid Particles in Turbine Agitated Baffled Vessels," *Chem. Eng. Sci.* **23**:1453 (1968).

Oldshue, J. Y., *Fluid Mixing Technology,* McGraw-Hill, New York, 1983.

Perry, R. H., D. W. Green, and J. O. Maloney (eds.), *Chemical Engineers' Handbook,* 6th ed., McGraw-Hill, New York, 1984.

Plautz, D. A., and H. F. Johnstone, "Heat and Mass Transfer in Packed Beds," *A.I.Ch.E.J.* **1**:193 (1955).

Reiss, L. P., "Cocurrent Gas-Liquid Contacting in Packed Columns," *Ind. Eng. Chem. Proc. Des. Dev.* **6**:486 (1967).

Roy, N. K., D. K. Guha, and M. N. Rao, "Suspension of Solids in a Bubbling Liquid," *Chem. Eng. Sci.* **19**:215 (1964).

Sato, Y., H. Hirose, F. Takahashi, and M. Toda, "Performance of Fixed-Bed Catalytic Reactor with Cocurrent Gas-Liquid Flow," *1st Pacific Chemical Engineering Congress,* 1972, p. 187.

Sato, Y., H. Hirose, F. Takahashi, and M. Toda, "Pressure Loss and Liquid Holdup in Packed-Bed Reactor with Cocurrent Gas-Liquid Downflow," *J. Chem. Eng. Japan* **6**:147 (1973a).

Satterfield, C. N., *Mass Transfer in Heterogeneous Catalysis,* MIT Press, Cambridge, Mass., 1970.

Schlunder, E. U., "Transport Phenomena in Packed Bed Reactors," *Chemical Reactor Engineering Reviews,* Houston, ACS Symposium 72, American Chemical Society, Washington, D.C., 1978.

Shah, Y. T., *Gas-Liquid-Solid Reactor Design,* McGraw-Hill, New York, 1979.

Smith, J. M., *Chemical Engineering Kinetics,* 3d ed., McGraw-Hill, New York, 1981.

Specchia, V., G. Baldi, and A. Gianetto, "Solid-Liquid Mass Transfer in Cocurrent Two Phase Flow through Packed Beds," *Ind. Eng. Chem. Proc. Des. Dev.* **17**:362 (1978).

Stiegel, G. J., and Y. T. Shah, "Backmixing and Liquid Holdup in Gas-Liquid Cocurrent Upflow Packed Column," *Ind. Eng. Chem. Proc. Des. Dev.* **16**:37 (1977b).

Tarhan, M. O., *Catalytic Reactor Design,* McGraw-Hill, New York, 1979.

van Krevelen, D. W., and J. T. C. Krekels, "Rate of Dissolution of Solid Substances. Part I: Physical Dissolution," *Rec. Trav. Chim. Pays Bas.* **67**:512 (1948).

Walas, S. M., *Reaction Kinetics for Chemical Engineers,* McGraw-Hill, New York, 1959.

Yagi, S., and N. Wakao, "Theoretical Temperature Distribution in Packed Beds," *Kagaku Kogaku* **23**: 161–163 (1959).

Zwietering, T. N., "Suspending of Solid Particles in Liquid by Agitators," *Chem. Eng. Sci.* **8**:244 (1958).

23

Separations Technologies

Prakash T. Palepu

Satya P. Chauhan

K. P. Ananth

Battelle Memorial Institute
Columbus, Ohio

Separations is a critical element of any pollution prevention technology portfolio because of its applicability to a broad set of problems. The range of technologies included in separations is diverse and ranges from chemical engineering unit operations such as distillation and extraction to gas separation via adsorption and membranes to solids separations via classifiers and magnetic fields. The discussions in this chapter focus on separations technologies as used in pollution prevention.

23.1 Role of Separations in Pollution Prevention

The term *separations* in pollution prevention usually relates to the removal or isolation of component(s) from process streams to enable in-process recycling or recovery and reuse of the component. As shown in Fig. 23-1, these applications could be found in process streams in all media.

In a liquid matrix or process stream, the technologies of primary interest are tied to removal of

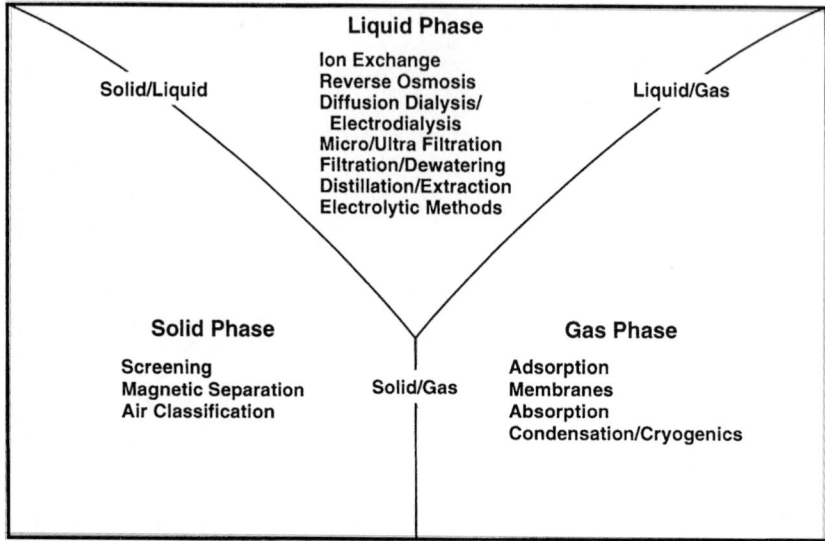

Figure 23-1. Key separation technologies in pollution prevention.

- Suspended solids
- Dissolved solids (e.g., dissolved metals and salts)
- Miscible and immiscible liquids (e.g., wastewater containing organics, and oil-water systems)
- Dissolved gases (e.g., volatile organic compounds, or VOCs)

For gaseous process streams, the technologies relate to

- Suspended-solids removal (e.g., particulates in gas streams)
- Miscible gases removal (e.g., VOCs)
- Suspended liquids (or mists)

In the case of solid wastestreams, the separations technologies typically focus on segregation and reuse of waste components from mixtures of industrial and municipal solid waste. Since these technologies are the traditional solids separations schemes such as screening, magnetic separation, and air classification, the reader is referred to the reference section at the end of the chapter, where several standard texts are listed.

The following sections of this chapter will discuss the *principles* of the separations technologies applicable to the above outlined cases, the specific *applications* currently in use for pollution prevention, and *vendors and suppliers* for these technologies.

23.2 Separation Technologies for Liquid Waste

Separation technologies for pollution prevention where the primary medium being processed is a liquid constitute the widest range of technologies in practice and in research and development. In this category, we include separation of (1) suspended solids (e.g., filtration), (2) dissolved solids (e.g., membrane-based separations), (3) miscible liquids (e.g., distillation), and (4) dissolved gases (e.g., air stripping of volatile organics in liquids).

The primary industries where liquid-based pollutants are likely to be produced are significant in the economy—chemical, petroleum, food, paper, metal-forming, and metal-finishing industries. The wide range of technologies being used and developed is a direct consequence of the need caused by the large volumes being processed and strict regulations on disposal of liquid-based pollutants due to the ease of migration of these pollutants into soils, rivers, and groundwater.

We have classified the separation techniques into two categories: phase (i.e., phase-change-based) separations (distillation, liquid extraction, and air stripping); dissolved-solids and molecular separations (ion exchange, diffusion dialysis, electrodialysis, pressure-based membrane separations, and electrolytic methods). The categorization of various technologies is somewhat arbitrary; for instance, electrolytic methods can be classified as phase-change-based separations or dissolved-solids separation. Ultrafiltration and microfiltration are strictly speaking phase separations but are discussed under pressure-driven membrane separations along with reverse osmosis due to similarities in equipment and use.

23.2.1 Phase-Change-Based Separations

Distillation. *Distillation* is the separation of two or more miscible (volatile) liquids based on the differences in vapor pressure. Distillation is an established and common separation technique in the chemical-processing industry and the reader is referred to any standard text on its principles, operation, and equipment.[1] The focus here is on the applicability of distillation for pollution prevention. It should be noted that distillation is energy-intensive due to the phase change of feed solution and that equipment costs can be high. Hence, in pollution prevention, distillation is attractive where large differences exist in the vapor pressure of the liquids being separated so that one stage (flash evaporation) or a few stages are sufficient for the required separation. Distillation is preferred when the species to be separated are present in significant proportions rather than as minor and major species.

If one of the species is present in trace amounts, other techniques such as carbon adsorption or ion exchange (liquid and resin based) are generally preferable to distillation. If the less volatile species is the minor constituent (e.g., dilute aqueous solution of high-boiling organics), liquid-liquid extraction of the organic followed by distillation of solvent-organic mixture is preferred over distillation. The guiding factor is the cost of energy associated with the phase change of the major species. Despite these considerations, distillation is practiced in pollution prevention because of its versatility; miscible species can be separated as long as there is a difference in the vapor pressures.

Applications for Distillation.   The classic example of distillation in pollution prevention is in solvent recovery. For instance, waste ink generated in newspaper printing contains a mixture of low-boiling solvent (20 percent), water (15 percent), and high-boiling ink (65 percent). Commercially available units perform a flash distillation to separate solvent and water from the ink.[2] The solvent and water are further separated using binary distillation. The water is discharged, the ink residue is centrifuged and filtered to remove suspended solids (paper dust) and reused in the process along with the recovered solvent. Another example is the batch distillation of used antifreeze. Finish-Thompson of Erie, Pennsylvania, sells a batch unit that recovers pure ethylene glycol and pure water as the two separate distillate products, leaving the additives such as sulfates, nitrates, and corrosion inhibitors in the residue. The recovered ethylene glycol and water are remixed in proper proportions along with fresh additives and reused as fresh antifreeze. In the electroplating industry, spent acids from cleaning tanks, etching tanks, and pickling tanks can be recovered by distillation.[3] Since the contaminants are dissolved metals or salts, they are nonvolatile and acid values are recovered using flash distillation. If the acid is a mixture (e.g., HCl/HNO_3), it is further fractionated to recover pure acids. The binary distillation is performed under reduced pressure to maintain low operating temperatures. Results from spent acid distillation show acid recoveries up to 90 percent with trace amounts of metal (<1 ppm) and waste volume reductions of 85 percent. A list of vendors of distillation systems for pollution prevention is given below.

Suppliers of Distillation Systems for Pollution Prevention

Finish-Thompson, Inc.
921 Greengarden Road
Erie, PA 16501

Hastings Engineered Systems
1220-T State Street
Hastings, MI 49058

Giant Distillation and Recovery Company
900 N. Westwood Avenue
Toledo, OH 43607

Hoyt Corporation
251 Forge Road
Westport, MA 02790

Solvent-Kleene
131 Lynnfield Street
Peabody, MA 01960

Evaporation.   *Evaporation* is recovery of a volatile solvent, usually water, from a solution containing dissolved solutes. It differs from distillation in two respects: the dissolved species is nonvolatile and no attempt is made to fractionate the vapor. In pollution prevention, evaporation is used as a means of volume reduction of hazardous solutions. The vapor product, usually pure water, is reused in the process. The concentrate is sent to waste treatment, although there are examples in the plating industry where the concentrate is recycled to the plating tank (metal recovery). The advantage of evaporation is that it can be used for any mixture of nonvolatile solutes and volatile solvents. The disadvantages are that it is highly energy intensive, and that if the concentrate is returned to the process, any nonvolatile contaminants in the feed are also returned to the process.

Evaporators function in a boiling-heat-transfer mode or flash evaporation mode. *Boiling heat transfer* is the evaporation of the solvent on the heat transfer surface in the evaporator. Heat is supplied to the evaporator surface by circulating steam or hot water on the other side of the surface. *Flash evaporation* is the vaporization of the solvent by introduction of the solution into the enclosed evaporator, where the pressure is maintained below the saturation pressure of the solvent. It can be accomplished by compression of the solution or by evacuation of the chamber or both. Commercial evaporators are available in various configurations: climbing-film evaporators, submerged-tube evaporators, and forced-circulation flash evaporators. All evaporators have a vapor-liquid separator above the evaporator. Energy efficiency of an evaporator can be increased by utilizing the heat released by the condensing vapor. This can be accomplished by mechanical recompression of vapor to use it as the heat source in the same evaporator prior to condensate collection or to use it as the heat source in the next evaporator in a chain of evaporators (multiple effect). However, the savings in energy come at the expense of increased capital cost. For pollution prevention efforts, energy-efficient multiple-effect evaporators are not usually practical due to the relatively small feedflow rates.

Applications for Evaporation. In pollution prevention, evaporation is primarily used as a means of volume reduction. In the metal-finishing industry it is used in the volume reduction of spent plating baths, cleaners, strippers, and caustic solutions. Volume reductions of 80 to 95 percent are common. The recovered distilled water is used for rinsewater makeup. The concentrate is sent to waste treatment. Evaporation of chrome rinsewater is an unusual example where the concentrate is reused in the plating bath. A list of vendors for evaporators is given below.

Suppliers of Evaporators for Pollution Prevention

Aqua-chem. Inc.
P.O. Box 420
Milwaukee, WI 53201

Drew Resource Corporation
1717 4th Street
Berkeley, CA 94710

Calfran International, Inc.
P.O. Box 269
Springfield, MA 01101

Environmental Management Technologies, Inc.
16 Hughes Street
Suite C-103
Irvine, CA 92718

Solvent Extraction. In pollution prevention, *solvent extraction* usually means the removal of organics, mostly oil and grease, from aqueous solutions or from oily sludges. For solvent extraction to be effective, the species to be removed should have high solubility in the solvent, the solvent should have high selectivity for the species to be removed relative to other materials in the feed, and furthermore the solvent should have low solubility in the feed material or solution. Once such a solvent is found, it is necessary to be able to easily separate the species from the solvent after the extraction is done. It is difficult to meet all the above criteria in an economical manner. A complete solvent extraction system consists of a multistage extractor (e.g., rotating-disk contactors or mixer-settlers or packed columns), equipment to separate species and solvent from the extract (e.g., distillation column), and sometimes even a raffinate treatment system (e.g., air strippers) to remove trace solvent from the feed solution. Hence the capital costs are high. Consequently, examples of solvent extraction in pollution prevention are few.

Most commercial uses of solvent extraction in pollution prevention are not for recovery and reuse but for decontamination of solutions, soils, and sludges. Examples are the removal of phenols from wastewater effluents from petroleum refineries using methyl isobutyl ketone (MIBK) as the solvent, and recovery of acetic acid from industrial wastewaters using ethyl acetate as the solvent. In these examples, the value of recovered product is insignificant compared to the cost of contaminated-stream disposal—e.g., by deep-well injection. Recently, two commercial systems have been demonstrated that use solvent extraction technology in recovery and recycling of refinery wastewater and sludges.[4,5] The first uses triethylamine, a unique solvent that extracts both water and hydrocarbons. Upon extraction, oil- and water-free solids are separated and dried. The solvent-hydrocarbon-water mixture is heated to 130°F, at which temperature water becomes insoluble, and the water phase is decanted. The solvent-hydrocarbon mixture is distilled to recover the solvent, and hydrocarbons are reused in the refinery. The advantage of the above extraction process is its ability to recover water and organics from a variety of wastes: from soils containing 5 percent organics and 10 percent water to emulsified sludges with 80 percent water and 10 percent oil to filter cakes containing 40 percent water and 20 percent oil. Another advantage is the low-temperature, ambient-pressure operation. Organic recovery is in the 99 percent range.[4]

The second process uses either liquefied CO_2 or liquefied propane as the extractant to recover organics from refinery wastewaters and sludges, oily wastewater from chemical plants, and slop oil emulsions. Many organics are soluble in CO_2 or propane and the separation of extract from the solvent is easily accomplished by vaporizing solvent. The advantage of using CO_2 is that it is nontoxic, nonflammable, and easy to separate from the organics. Liquefied gas extraction has been used for the extraction of acrylonitrile, halogenated hydrocarbons, alcohols, and aromatics from aqueous streams. The concentrations of organics in feed range from a few hundred parts per million to a few percent, and the concentrations in the treated aqueous stream or sludge are in the few

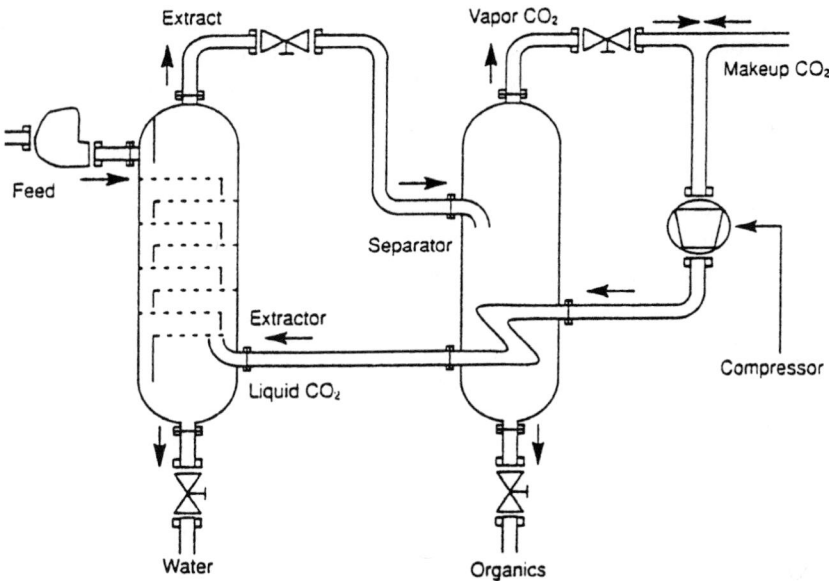

Figure 23-2. Liquefied CO_2 extraction process for aqueous stream containing organics. (*Ref. 5.*)

parts per million range. The recoveries of organics usually are above 99 percent and the treated feed stream is sewerable. A flowchart of the liquefied CO_2 extraction process for organics/water feed is given in Fig. 23-2.[5]

Suppliers of Solvent Extraction Systems for Pollution Prevention

Chem-Pro Equipment Corporation
27 Daniel Road
Fairfield, NJ 07006

CF Systems
3D Gill Street
Woburn, MA 01801

Resources Conservation Company
3006-T Northup Way
Bellevue, WA 98004-1407

Air Stripping and Steam Stripping. VOCs can be removed from water or aqueous wastestreams by air stripping. The driving force is the concentration gradient of the VOCs between the liquid and gas phases. VOCs having Henry's law constants of 10 atm are easily stripped by air at ambient temperatures. Examples are chlorinated hydrocarbons and aromatics. Usually the concentration of VOCs in the feed is in the range of a few hundred parts per million and treated feed has about 10 ppm of VOCs left. The equipment consists of a packed tower with countercurrent feed of the aqueous stream and air from top and bottom, respectively. The advantages of air stripping are low capital and operating costs. The disadvantage is that water-based pollutants become air pollutants unless the off-gas is incinerated.

Steam stripping is more versatile than air stripping. It can handle several percent by weight of VOCs in the aqueous phase, less volatile organics, and even nonaqueous wastestreams. The driving force is the vapor-liquid equilibrium between the liquid and gas phase. The additional advantage of steam stripping is that the overhead vapor is condensed and the VOCs can be recovered from the condenser by phase separation. Steam stripping should be considered as a means of concentrating the VOCs from a dilute aqueous stream to a smaller-volume stream that is more amenable to thermal or chemical treatment. Steam stripping is commonly used in the regeneration of spent activated carbon beds.

23.2.2 Dissolved-Solids or Molecular Separations

The technologies for this class of separations are relatively new and this is also where most research and development activities are concentrated. The technologies are ion exchange (liquid and resin based) and membrane-based technologies such as reverse osmosis, micro- and ultrafiltration, diffusion dialysis, and electrodialysis. The impetus for application and development in these areas is being provided by pollution prevention needs in the plating and surface-finishing industry. Further reading on membrane-based separation technologies can be found in *Membrane Handbook,* edited by W. S. W. Ho and K. K. Sirkar.[6]

Electrodialysis.   In *electrodialysis* (ED) ions in solution are selectively transported across ion exchange membranes under the influence of an applied dc field. The ion exchange membranes are either anion-selective (i.e., permeable to anions) or cation-selective. For example, polystyrene-divinylbenzene membranes when functionalized with sulfonyl groups ($-SO_3^-$) allow the passage of cations and when functionalized with ammonium groups ($-NR_3^+$) allow the passage of anions. The membrane itself acts as a barrier for nonelectrolytes and the solution. A schematic of an ED unit with alternating anion (A) and cation (C) exchange membranes is shown in Fig. 23-3.[7] The feed containing dissolved species is introduced into alternate compartments. Under the imposed polarity, the anions and cations migrate in opposite directions and concentrate in the two adjacent compartments; i.e., ions can pass through one compartment but are blocked by the next one, where they concentrate. ED units are capable of concentration factors of 10 or more. The concentrate and the diluate (feed solution depleted of most of its ions) are collected and sent to their separate uses; for instance, in a metal-plating operation the concentrate is sent to the plating bath and the diluate to the rinse tanks.

A typical ED stack consists of hundreds of alternating anion and cation exchange membranes with spacers between them. The membranes are about 0.5 mm thick and the spacers are 0.75 to 1 mm thick. The advantages of ED are low energy requirements for the degree of concentration and minimal waste generation since both the concentrate and diluate are reused in the process. The

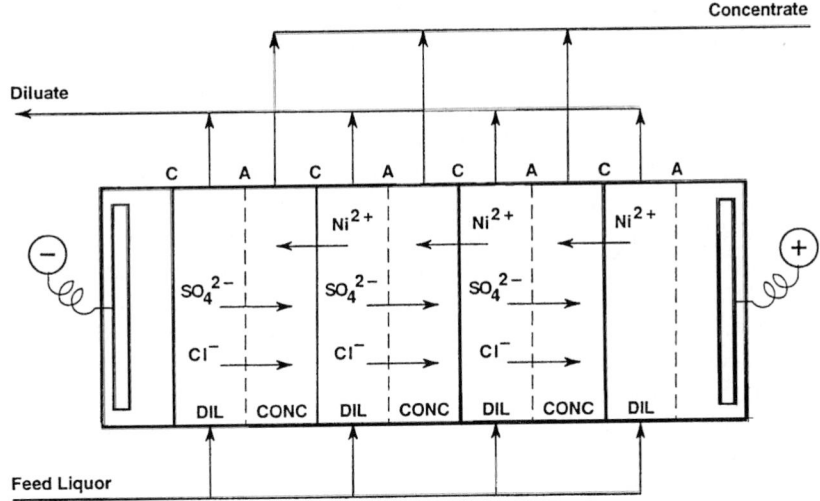

Figure 23-3. Principles of electrodialysis. (*Ref. 7.*)

disadvantage is the fouling of membranes, but this has been improved significantly by the development of antifouling membranes and the use of periodic reversal of electrodialysis.

A variation of traditional ED (which consists of hundreds of alternating anion- and cation-permeable membranes) is the use of ion-selective membrane in an electrolytic cell to separate the anode and cathode compartments. The solutions in the anode and cathode compartments, called the *anolyte* and *catholyte*, respectively, can be different electrolyte solutions (e.g., acid and base). In the recovery of metal-contaminated acids, feed solution is introduced into the anode compartment and a cation-permeable membrane is used. Metal cations are electrotransported across the membrane to the catholyte. Hydrogen ion production at the anode reforms acid with the anions in the feed solution. If the catholyte is maintained under basic conditions, migrated cations form insoluble metal hydroxides and are filtered out. If the catholyte is also an acid, cations are plated out at the cathode and recovered as metal. The reformed acid from the anolyte is reused in the process. This technique has been commercialized in the recovery of spent chromic acid etch solutions and rejuvenation of hard chrome-plating baths.[8] By extending the concept to a three-compartmented cell separated by two cation exchange membranes, alkali hydroxides can be recovered from multivalent metal hydroxides.[9]

Applications of Electrodialysis.   Since ED cannot achieve 100 percent separation (as in evaporation), it is best suited when significant concentration factors are needed at a low cost and the diluate with some dissolved ions and/or the concentrate can be reused in the process. If the concentrate cannot be reused in the process, then it can be treated further by evaporation for additional concentration for metal recovery. Hence the largest use of ED is in the production of concentrated brines for salt production from seawater.

In pollution prevention, it is used in the metal-finishing industry in the recycling of rinsewaters (e.g., nickel galvanization wash waters). Several cases of "closed loop" drag-out rinsewater treatment, especially in the precious-metal-plating industry and chrome plating have been documented.[10] Presently ED is being pilot-tested in the recovery of caustic in alkali cleaners and rust removers.[11] The development and availability of specialized membranes is likely to increase the use of ED in pollution prevention. These specialized membranes include (1) monovalent cation-selective membranes, (2) cation exchange membranes permeable to base (OH^-) diffusion, (3) membranes resistive to organic solvents, and (4) water-splitting bipolar membranes. A list of vendors of ED systems is given at the end of this section.

Diffusion Dialysis.   *Diffusion dialysis* (DD) is another separation process that uses ion exchange membranes. However, the mechanism of separation and the driving force distinguish it from electrodialysis. Since the most prevalent use of diffusion dialysis is in the metal-finishing industry, its principles are best explained with an example from that industry. The steel-finishing industry generates large volumes of spent pickling liquors, which are essentially metal-contaminated acids. DD is used to recover the acid values from spent liquor. The unit consists of a stack of anion exchange membranes with spacers between them. A schematic of the system is shown in Fig. 23-4 (see the portion labeled

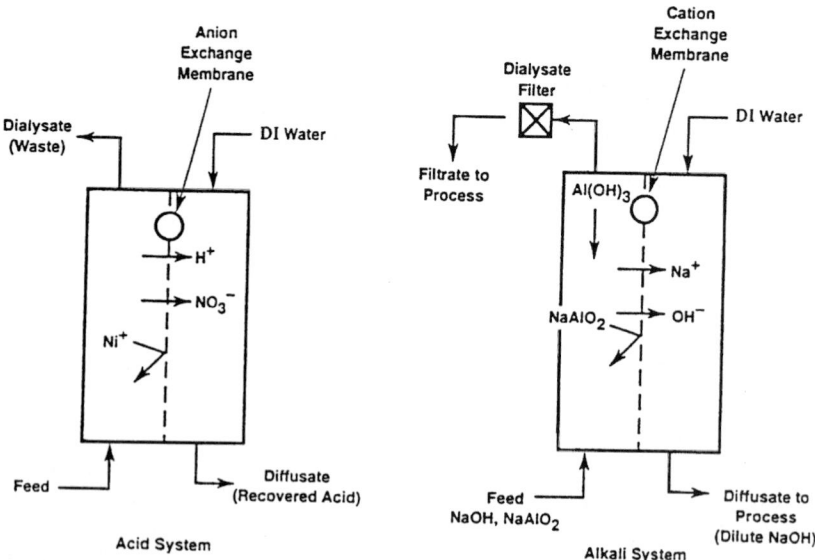

Figure 23-4. Principle of diffusion dialysis for acid and alkali recovery.

"Acid System"). Spent liquor is fed into one compartment and water is fed countercurrently into the other. The anions diffuse across the membrane while the cations remain in the feed stream with the exception of H⁺. Although anion exchange membranes are impervious to cations, the hydrogen ion alone moves across the membrane due to its small size and high mobility. Thus, acid values migrate across the membrane into the water stream. Countercurrent flow of feed liquor and water ensure that the acid concentration in the diffusate is approximately equal to that in the feed liquor. The recovered acid is sent back to the process and the acid-depleted feed with the metal contaminants is sent to waste treatment. Although only two compartments are shown in Fig. 23-4, DD units consist of hundreds of alternating compartments separated by anion-permeable (or cation-permeable) membranes.

Figure 23-4 also shows an alkali recovery system; in a two-compartment system, cation exchange membranes that are impervious to all anions except hydroxyl (OH^-) ions are used to recover alkali values. The example shown is the application of DD in the chemical milling of aluminum. In the process tank, etched aluminum is solubilized as sodium aluminate. As the caustic values migrate to the diffusate, the solubility equilibrium between sodium aluminate and caustic is disturbed, allowing the precipitation of aluminum as aluminum hydroxide, which is filtered out. The aluminum-depleted feed and the diffusate containing caustic are sent back to the process tank. The advantage of diffusion is low cost and simplicity of operation. The disadvantage is that the volume of recovered acid and depleted feed is about twice the volume of the feed liquor so that the separation becomes economical only when the recovered acid stream can be used in the process.

Another variant of dialysis that is similar to DD is *Donnan dialysis*. Here, ion exchange membranes and countercurrent flow are used, but the strip stream is mineral acid (e.g., $0.1M$ H_2SO_4, pH = 1) instead of water. Solution containing dissolved metal salts (e.g., $CuSO_4$, pH = 7) constitutes the feed. Hydrogen ions diffuse through the anion-impermeable membrane, setting up an electrical potential gradient between the two compartments, forcing the metal ions to diffuse into the stripping solution. This flux of metal ions into the stripping solution continues even after the concentration of metal ions in both compartments is equal. The flux of metal ions against their concentration gradient is possible because of the electrical potential caused by the hydrogen ion diffusion. Metal-depleted feed solution is recycled back into the process.

Applications of Diffusion Dialysis. Pollution prevention applications of DD and Donnan dialysis are mostly in the metal-finishing industry. DD is used in alkali recovery from caustic cleaners and chemical milling of aluminum and in acid recovery (HCl, HNO_3, HF, and H_2SO_4) from spent pickling liquors and aluminum anodizing baths.[7,12,13] Its low operating costs and low maintenance requirements make it attractive for use in acid and alkali recovery. The recovery rates of acid and alkali values are between 80 and 90 percent and metal rejection is close to 100 percent.[11]

Suppliers of Electrodialysis and Diffusion Dialysis Systems

HPD Incorporated
1717 North Naper Road
Naperville, IL 60566

The Graver Company
2720 U.S. Highway 22
Union, NJ 07083

Asahi Glass America Inc
1185 Avenue of the Americas
New York, NY 10036

Kinetic Recovery Corp.
7517 Washington Avenue, S.
Edina, MN 55435

Ionsep Corp. Inc.
P.O. Box 258
Rockland, DE 19732

Pressure-Driven Membrane Separations for Liquids. These processes are classified into three categories based on the membrane pore size and consequently the differences in operating pressures and the species being separated. The three categories are (1) reverse osmosis, (2) ultrafiltration, and (3) microfiltration. The differences in operating pressures and retained species sizes (or the cutoff molecular weight) are given in Table 23-1. In all three types of separations the solvent (water) permeates a membrane and is then recycled to the process. In some cases the concentrate containing the dissolved species is also recycled to the process.

Reverse Osmosis. When two solutions containing the same solute are separated by a semipermeable membrane, the solvent permeates from the dilute solutions through the membrane to the concentrated solution until the chemical potential of the solvent is equal on both sides of the membrane. At equilibrium, the pressure difference between the two sides is equal to the osmotic pressure of the solvent. *Reverse osmosis* (RO) is the application of pressures (greater than osmotic pressure) to the concentrated solution side to make the solvent flow from the concentrated side to the dilute side. In this sense, the separation principle of RO is different from that of other pressure-based membrane filtration methods, which are based only on size exclusion. In RO, pressure is used to overcome the natural osmotic pressure to make the solvent flow

Table 23-1. Characteristics of Pressure-Driven Membrane Separation Techniques

Technique	Operating pressures (psi)	Size retained (micrometers)	Typical molecular weight retained	Typical species retained
Reverse osmosis	200–1000	0.0001–0.001	100–1000	Inorganic salts ($NaCl$, $MgCl_2$)
Ultrafiltration	50–200	.001–0.1	1000–100,000	Proteins, polymers
Microfiltration	15–50	0.2–10	—	Emulsified oil, fine suspended solids

against its concentration gradient. Since the solute rejection is close to 100 percent, RO is used to produce clean water from aqueous salt solutions. The disadvantage of RO is that any contaminants in the feed remain in the concentrate.

RO membranes consist of a thin skin, which is supported by a porous substructure; it is the thin skin that acts as the selective layer, allowing the flow and retaining the solute. The thin film is about 0.5 μm thick with 0.0001-μm pores, and the substructure is 50 μm thick with 1-μm pores. The membranes can be made of a single polymer such as cellulose acetate (single-material membranes are called *anisotropic membranes*) or of thin-film composites. Due to the small pore size, RO membranes are susceptible to plugging and it is necessary to pretreat the feed, typically with 1-μm to 5-μm filters. In addition, there are limitations on the allowable pH and temperature of feed due to physical instability of the membrane materials in harsh environments. For instance, cellulose acetate membranes can be used only in the pH range of 3 to 8 and at a maximum temperature of 50°C. However, development of thin-film composites has increased range of allowable pH to between 1 and 13 and temperature limit to 80°C.

Since the osmotic pressure increases with solute concentration in the concentrate, there are limits on the possible concentrations attainable by RO. At solution concentrations of 10 percent, equipment pressure limitations (>1000 psi) preclude the use of RO for further concentration. RO membrane modules are available in four configurations: (1) plate and frame, (2) tubular, (3) hollow fiber, and (4) spiral wound. Figure 23-5 is a schematic representation of the four types of RO membrane modules. The choice of the type of membrane depends on the flow rates, suspended-solids content of the feed, and cost. Tubular membranes are easy to clean when plugged but are expensive and have low membrane surface-to-volume ratio. Hollow-fiber and spiral-wound membranes are compact with high surface-to-volume ratio but are susceptible to plugging and difficult to clean. Plate-and-frame membranes offer simpler operation with average size-to-volume ratio and some susceptibility to plugging.

Applications of Reverse Osmosis. Since RO produces clean water without the high energy cost associated with a phase change, the largest application of RO is in desalination. However, the potential for using RO in treating wastewater is immense and some of those applications are being utilized now. In the electroplating industry, RO is especially suited for obtaining a closed-loop rinsewater system, with the concentrate being sent back to the plating bath. There are over 200 documented cases of RO use in Watts nickel, bright nickel, silver cyanide, copper sulfate, and brass cyanide plating baths.[14] In the paper and pulp industry, where large volumes of wastewater are produced, RO is used for the removal of color, dissolved solids, and trace organics from wastewater.[15] It is also used in the concentration of spent sulfite liquor from 6 percent (in the paper industry) solids to 12 percent solids so that it can be further concentrated by evaporation.[16] In the textile industry RO is used to recover valuable dye stuffs (soluble and colloidal) from wastewater[17] and to reduce the volume of wastewater.[18] A partial list of RO system vendors is given at the end of this section.

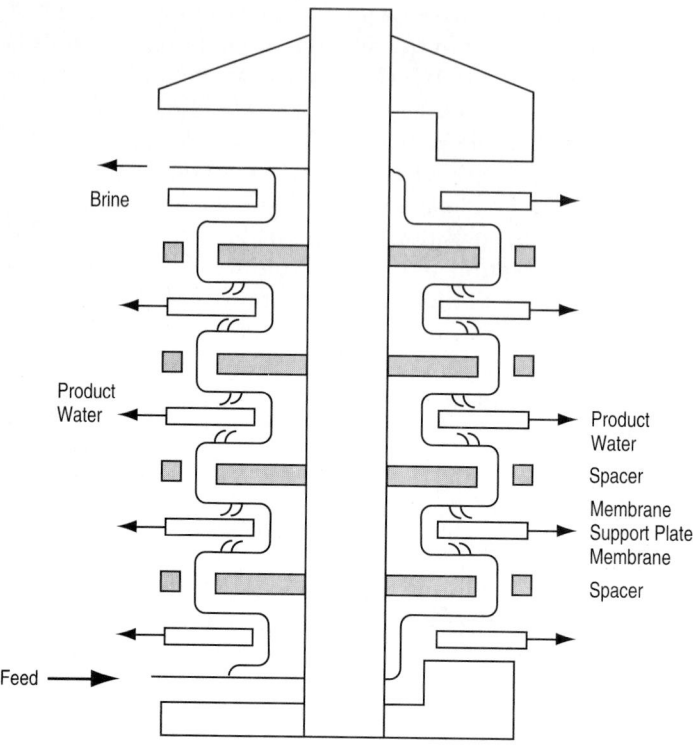

(a) Plate and frame filter module

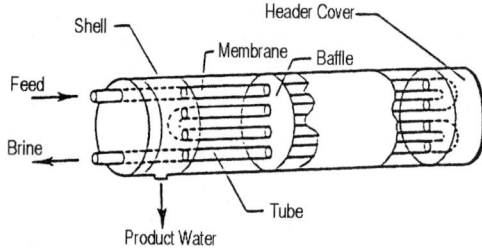

(b) Tubular filter module

Figure 23-5. Schematic of four types of reverse osmosis membrane modules. (*Courtesy of W. S. W. Ho and K. K. Sirkar, 1992, pp. 294—299.*)

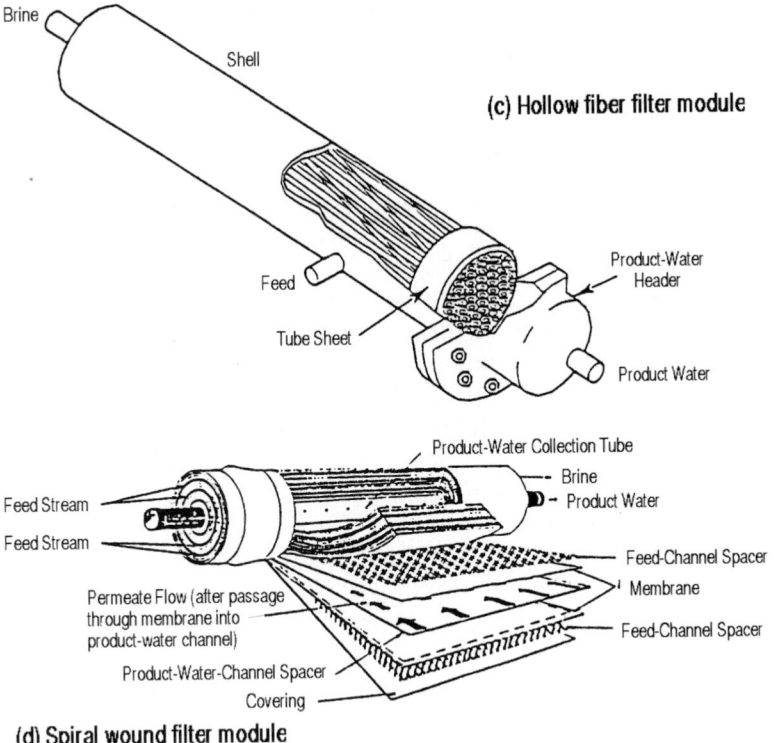

(c) Hollow fiber filter module

(d) Spiral wound filter module

Figure 23-5. (*Continued*)

Ultrafiltration. *Ultrafiltration* (UF) separates species in solution solely based on size. The membrane pore diameters are an order of magnitude bigger ($\approx$0.01 μm) than for RO. The membranes retain proteins, polymers, and complexed heavy metals such as iron complexed with EDTA. Since low-molecular-weight solutes flow through the membrane, osmotic pressure is not an issue. However, since retained large molecules and colloidal particles have low diffusivities in the liquid medium, UF membranes are more susceptible to fouling and concentration polarization than RO membranes. *Fouling* is the deposition of solutes on the membrane surface or within the pores either due to solubility limits (at the surface) or due to the pore geometry / tortuosity or solute–pore wall interaction. *Concentration polarization* is the increase in solute concentration at the membrane surface due to solute retention. Both these effects increase membrane resistance and reduce the primary flow rates. These effects are reduced or countered by various means in UF: (1) the use of hydrophilic membranes for aqueous solutions, (2) the use of cross-flow membranes where the feedflow cleans membrane surface on the retentate side, (3) turbulent flow to displace retained solutes, (4) periodic backflush and removal of retained solutes, and (5) the use of surfactants and cleaning agents including acids and alkalis to solubilize the solute. Even electrically charged membranes have been used (in electrocoating) to repel charged solutes from the membrane surface.

UF membranes are asymmetric in structure like those used in RO; i.e., a thin skin acts as the size-selective barrier and it is supported by a thicker macroporous substructure. Similar to RO, they are available in four configurations: tubular, plate and frame, hollow fiber, and spiral wound (see Fig. 23-5); however, the channel spacing or tube or hollow-fiber diameters are an order of magnitude larger than those used in RO due to lesser pressures being used and large solute sizes and concentration. In plate-and-frame UF modules, the membranes are usually stacked horizontally instead of in a vertical configuration. UF hollow-fiber filters are constructed differently from those used in RO. The feed solution is introduced inside the hollow fiber at one end of the module and it flows through the length of the fiber and the concentrate is collected at the other end of the module. The permeate "weeps" through the membrane to the shell side and is collected there.

Recently, ceramic UF membranes with skin pore sizes of 0.004 µm to 0.1 µm were introduced by Alcoa, Amotec, and others. The substructure (with 15-µm pores) is made of α-alumina and the skin of γ-alumina or zirconia. The advantage of ceramic membranes is their ability to withstand the elevated temperatures and wide pH ranges that polymeric membranes cannot withstand. At present, ceramic UF membranes are more expensive, and the chemical instability of the skin layer (γ-alumina) has limited large-scale usage. Ceramic membranes made solely with α-alumina are being successfully used in microfiltration.

Applications of Ultrafiltration.   In pollution prevention, the first commercial use of UF was in electrodeposition of paint.[19] During electrocoating, undeposited paint (resin, pigment, and organic solvent additives in an aqueous solution) is washed off with water sprays. Paint losses due to this "drag-out" are higher than 50 percent of paint used. UF is used to treat the wastewater, which contains large quantities of paint. The concentrate containing the paint resin, pigment, and also the organic solvents is returned to the process tank and the permeate water is reused in the sprays. This is a classic example of separation technology being used in pollution prevention. Both the retentate and permeate are recycled, waste generation is minimized, and, as a further benefit, UF is used to maintain the water balance in the process tank by controlling the retentate concentration—i.e., process control.

Other applications of UF are in the recovery of polymers such as latex from washwaters generated in cleaning of polymerization reactors and in the recovery of polyvinyl alcohol used in the manufacture of synthetic yarn.[20] In the metal-finishing and metal-forming industries, UF is used in the separation of oil-water emulsions.[21] In metal finishing, the permeate water is reused as fresh rinsewater, and in metal forming the recovered oil is reused as a tool lubricant and coolant. A partial list of UF system vendors is given at the end of this section.

Microfiltration.   *Microfiltration* (MF) consists of retention of fine suspended solids and microemulsions of hydrocarbons. The pore sizes are large enough not to retain molecules and dissolved species. The permeate fluxes are high (≈ 10 gal/ft$^2 \cdot$ h) relative to RO and UF, and transmembrane pressures are low (15 to 50 psi). MF can be done in two configurations: (1) dead-end filtration, as is done in laboratory Buchner funnels, or (2) cross-flow filtration, where feedflows are tangential to the filter surface and liquid permeate flowout is through the filter surface. Filter elements can be (1) membrane-type filters with pore sizes smaller than the size of suspended solids or (2) depth filters that have pore sizes larger than particle size but trap particles in the interstices (e.g., cartridge filters). In the last decade, cross-flow MF with membranes has become predominant because of its self-cleaning ability, lower pressure requirements, and consequently high permeate fluxes. Since the feedflow is tangential to the filter surface, high shear caused by the flow prevents concentration polarization and the associated cake buildup. A steady-state permeate flux is achieved after the initial buildup of a thin cake layer. In a typical operation, the feed is continuously circulated until concentration of retained solids and emulsions becomes high enough to allow removal as sludge, which is then sent to a filter press for further volume reduction. The ratio of feedflow to permeate flow is typically 20:1. The permeate is recycled to the process or discharged to the sewer.

MF technology has developed to the extent that there is a wide choice of membrane materials, their structure, and configurations. Membrane materials can be polymeric [e.g., cellulose acetate, polytetrafluoroethylene (PTFE), polypropylene, nylon, polyvinylidene fluoride (PVDF), and acrylic] or ceramic (α-alumina, zirconia, and silica). Their structure can be symmetric or asymmetric (same polymer), composite or track-etch. Track-etch membranes have pores with a very narrow pore diameter distribution. The others are spongelike in their structure. Membrane configurations can be of the tubular, hollow-fiber, spiral-wound, or pleated-sheet types. Polymeric membranes have service lives of 2 to 4 years, whereas ceramic membranes can last up to 10 years. Recently, due to their ability to withstand high temperature and excellent chemical resistance, ceramic membranes have become popular, although they cost twice as much as polymeric membranes for the same permeate flux. Essentially all MF membranes need to be cleaned periodically to remove the deposits on the surface and the inevitable plugging inside the pores. This is achieved by circulating acid (for inorganic scales), detergents (for colloids emulsions), alkali (for biological materials), or solvents (for organics).

Applications of Microfiltration.   In pollution prevention, MF is used in the recovery of aqueous (caustic) cleaners and rust removers in the metal-plating industry. With use, these caustic-based cleaners accumulate dirt, grime, free and emulsified oil, and metal particulates. Although they have considerable alkali values, they used to be dumped because of reduced cleaning ability caused by suspended solids. MF is used to obtain clean permeate of caustic cleaners, which are then reused. MF has also found an application in the treatment of metal-plating wastewaters after the hydroxide precipitation step. Some metals are amphoteric and are difficult to precipitate. They can be hydrous, with densities close to that of water, and usual methods of clarification or centrifugation are ineffective. MF has been successfully applied to reduce the total suspended solids content from these wastewaters by more than 99 percent so that the permeate can be discharged into the sewer.[22] MF installations are very compact relative to clarifiers. A partial list of MF system vendors is given below.

Suppliers of Reverse Osmosis, Ultrafiltration, and Microfiltration Systems

Ionics, Inc
65 Grove Street
Watertown, MA 02172

Memtek Corp.
28 Cook Street
Billerica, MA 01821

Zenon Environmental Systems, Inc.
845 Harrington Court
Burlington, Ontario, Canada L7N 3P3

Lancy Environmental Systems Inc.
181 Thorn Hill Road
Warrendale, PA 15086

Prosys Corp.
187 Billerica Road
Chemsford, MA 01824

Osmonics, Inc.
5951 Clearwater Drive
Minnetonka, MN 55343

Ion Exchange. In ion exchange (IX), fixed functional groups on a polymeric resin matrix exchange their hydrogen (or hydroxyl) ions with other cations (or anions) in the feed solution. The exchange is reversible, the resin being regenerated with acid or alkali, and the exchanged ions are recovered in a concentrated solution. The polymeric resin is usually polystyrene cross-linked with divinyl benzene with functional group permanently attached to the surface or even impregnated into the resin (in the latter case the matrix is called a *macroporous resin matrix*). IX is either cationic or anionic as shown below, with regeneration being the reverse reaction:

$$2RH + M^{2+} \rightarrow R_2M + 2H^+ \text{ (Cationic exchange)}$$

$$ROH + A^- \rightarrow RA + OH^- \text{ (Anionic exchange)}$$

Ion exchange occurs because the functional group has higher affinity for cation (or anion) in solution relative to hydrogen (or hydroxyl) ion. Ion selectivity is a function of ionic charge (e.g., $M^{3+} > M^{2+} > M^+ > H^+$) and hydration radii of the ions. Regeneration occurs because higher concentration and more than stoichiometrically equivalent amount of acid (or alkali) are used during regeneration.

Cation exchange resins can be further classified into strong-acid cation resins or weak-acid cation resins. Strong-acid cation exchange resins have polysulfonic acid groups which have low affinity for H^+ ions. These resins exchange cations over the entire pH range, but the low affinity for H^+ requires many times the stoichiometric equivalent of acid for regeneration to H^+ form. Weak-acid cation resins have carbocyclic acid groups with higher affinity for H^+. Hence, they are easier to regenerate but function only in the pH range above 4. Similarly, anion exchange resins are also classified as strong-base anion resins and weak-base anion resins. They contain quaternary ammonium (NH_4^+) with low affinity for OH⁻ and tertiary amine groups with higher affinity for OH⁻, respectively. Strong-base anion resins exchange anions in the entire pH range and weak-base anion resins function only under acidic conditions. The above requirements for regeneration also hold for anion resins. In addition to the traditional H^+ and OH^- exchange with cations and anions, IX is also practiced in the exchange of Na^+ ion with cations and Cl^- ions with anions. The intent here is to replace less desirable species in solutions, such as heavy metals, with relatively more benign species, such as sodium. The total dissolved-ion concentration in the processed solution remains the same.

There are resins that exhibit very high selectivity for certain metals such as copper, nickel, lead, and mercury. These resins operate on the chelation of the metal with a specific functional group on the resin; e.g., picolylamine group (DOWEX XFS 4195 resin) selectively chelates copper even at very low concentrations in the presence of large concentrations of other ions. In addition to their anion exchange capabilities, strong-base anions resins have a tendency to *adsorb* whole acid molecules. This ability, known as *acid retardation,* is used in the recovery of nitric, sulfuric, and hydrochloric acids from spent acid etchants and pickling liquors. The adsorbed acid is recovered by eluting with water. Although some applications of IX for organics from water exist, here again the mechanism is adsorption of the organic on the resin rather than true ion exchange.

Since the action of IX resins is based on charge equivalency, they are best suited to remove undesirable ions from large volumes of very dilute solutions, e.g., rinsewaters and wastewaters. The metal-ion-depleted solution is reused in the case of rinsewaters or sent to the sewer in the case of wastewater. When the ion exchange beds are regenerated, the concentrate containing the cations and anions is usually processed further by other techniques such as electrowinning or sent to waste treatment. The advantage of IX is in its versatility (i.e., treatment of mixed wastes), volume reduction, and reuse of solute-depleted stream. Occasionally the regenerated concentrate is reused in the process—e.g., chromic acid recovery from hard-chrome rinses. IX beds are operated in a cyclic fashion, usually duplexed. One bed is in service while the other is being regenerated. Bed exhaustion is usually determined by a conductivity probe on the downstream side and the feed is then switched to the newly regenerated bed. The process feedflow and regenerant flow can be concurrent or countercurrent; concurrent is simpler to operate, countercurrent requires less regenerant.

Applications of Ion Exchange. IX units are now common in most electroplating facilities. They are used in a variety of applications. IX can be an effluent-polishing system that removes the traces of heavy metals from clarifier effluent before final discharge into the sewer. Commercially available IX units are used in the recovery and reuse of rinsewater and metal concentrate in hard-chrome, acidic nickel, and copper plating. Rohm & Haas have demonstrated greater than 99 percent recovery of phenol from wastewater containing up to 5 percent phenol using IX. This company is also developing resins that remove mercury (GT-73 resin) and metal cyanide complexes (IRA-958 resin) from wastewater with relatively good efficiency. Enthone has developed a chelating resin to recover copper from electroless copper solutions. Suppliers of IX systems are given below.

Suppliers of Ion Exchange Systems

Kinetico Engineered Systems, Inc.
10845 Kinsman Road
Newbury, OH 44065

Penfield, Inc.
8 West Street
Plantsville, CT 06479

Eco-Tec Limited
925 Brock Road South
Pickering, Ontario, Canada L1W 2X9

Rohm & Haas
Independence Mall West
Philadelphia, PA 19105

Emulsion Liquid Membranes.   The use of *emulsion liquid membranes* (ELMs) is an emerging technology that depends on the creation of double emulsions, i.e., water/organic/water (W/O/W) or organic/water/organic (O/W/O). The principle of separation employed is explained below, using the example of the W/O/W system. An aqueous phase is emulsified in an organic solvent to form the first emulsion. This aqueous phase is called the *internal phase*. This emulsion is then dispersed in another aqueous phase called the *external phase* to form the double emulsion with the organic solvent separating the two aqueous phases and forming the liquid membrane. A schematic of ELM is shown in Fig. 23-6. The internal-phase droplets are about 2 μm in diameter and the oil globules are 100 to 1000 μm in diameter. Separation occurs in the transport (by diffusion) of solute from the aqueous external continuous phase through the organic liquid membrane to the internal aqueous phase. Once the separation is complete, the external aqueous phase is separated from the organic phase by settling, followed by the breaking of the first emulsion to recover the internal aqueous phase as the extract containing the solute. Figure 23-7 shows a schematic of a continuous ELM extraction process.

The driving force for the transport of solute from the external phase to the internal phase is the concentration gradient between the two phases. The con-

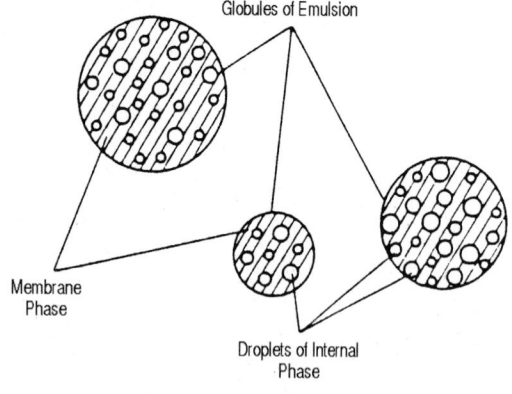

Globules of Emulsion

Membrane
Phase

Droplets of Internal
Phase

External, Continuous Phase

Figure 23-6. Schematic of an ELM system. (*Courtesy of W. S. W. Ho and K. K. Sirkar, 1992, pp. 600–601.*)

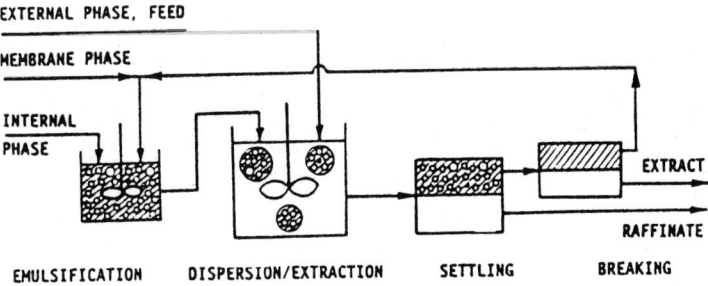

EXTERNAL PHASE, FEED

MEMBRANE PHASE

INTERNAL PHASE

EXTRACT

RAFFINATE

EMULSIFICATION   DISPERSION/EXTRACTION   SETTLING   BREAKING

Figure 23-7. Schematic of a continuous ELM extraction process. (*Courtesy of W. S. W. Ho and K. K. Sirkar, 1992, pp. 600–601.*)

centration gradient can occur by one of two mechanisms. In the first mechanism the solute reacts with a constituent of the internal phase, forming a product that is insoluble in the organic phase and hence is immobilized in the internal phase. This essentially maintains a zero concentration of solute at the internal phase–organic phase interface, thus maintaining a concentration gradient for the diffusion of solutes from the external aqueous phase. In the second mechanism, the organic phase (liquid membrane) contains an extractant that exchanges ions (e.g., protons) for the solute's ions (e.g., metal cations) in the external phase, then diffuses to the internal phase interface where reverse ion exchange takes place. For instance, a mineral acid in the internal phase can strip the metal cations from the organic extractant and exchange them with H^+ ions.

Applications of Emulsion Liquid Membranes. The reaction mechanism is used to extract phenol from wastewater. Aqueous sodium hydroxide is used as the internal phase. The caustic reacts with phenol to form sodium phenolate, which is insoluble in the organic phase. The ion exchange mechanism is used in the recovery of zinc and nickel from wastewaters and rinsewaters. Mineral acids in the internal phase exchange their protons with metal cations in the external phase. Both the above examples have been commercialized.[23,24] Extraction of nickel from rinsewater containing 400 ppm of nickel resulted in effluent nickel concentration of 1 ppm and extract nickel concentrations of 60 g/L.[25]

The advantage of ELM is the extremely large surface areas generated by the emulsions, resulting in high mass transfer rates. Typical internal-phase mass surface areas are 10^6 m^2/m^3 and external-phase mass transfer areas are in the range of 10^4 m^2/m^3. Compared with 10^2 m^2/m^3 for conventional solvent extraction equipment, ELM has considerable efficiency. However, the technology is still evolving and there are no systems vendors. Traditional chemical companies such as AKZO, Henkel, Shell, Mobil, and King Industries supply various organic extractants, but the complete systems are custom designed for a specific application.

Electrolytic Methods. Electrolytic methods fall into two categories: electrowinning, where metal is recovered from solution such as rinsewater; (2) dummy plating, where trace contaminant metals are selectively plated out to purify an electroplating bath. The use of electrolytic methods in pollution prevention is practiced in the electroplating, metal-finishing, and electronic industries. Electrowinning constitutes the major portion of electrolytic methods, with dummy plating limited to decontamination of a few specific baths.

Electrowinning. Sometimes referred to as *electrolytic metal recovery, electrowinning* (EW) uses plating-cell methods to recover metal values from rinsewaters, wastewaters, and in some instances even process streams. The purpose of EW can be (1) to recover a valuable metal such as gold or silver, (2) to remove metal from rinsewaters for water recycling, (3) to remove metal from wastewaters prior to precipitation to reduce sludge, or (4) to meet regulatory limits on metal content in discharge waters. There are also examples of rejuvenation of process solutions using electrolysis.

Electrowinning technology is not new. However, its application to pollution prevention is fairly recent, with modifications made to the traditional electrochemical cell configuration and the modified designs used for pollution prevention. An electrolytic cell is simply a metal anode and cathode connected to an external dc power supply and immersed in a conducting aqueous solution (electrolyte). Metals are deposited at the cathode and electrolysis of water causes evolution of hydrogen at the cathode and oxygen at the anode. Other gases could also evolve based on the electrolyte composition. In traditional electroplating, metal concentration is usually about 5 percent by weight. In pollution prevention cases, the metal concentration is much lower, ranging from a few parts per million to 0.5 percent by weight. This low concentration of metal tremendously aggravates the usual problem in electroplating: cathode polarization. Since metal ions are deposited at the cathode, the solution layer next to the cathode is depleted of metal ions and there is diffusion limitation for the migration of metal ions at the cathode. If this occurs, most of the energy is consumed in the electrolysis of water, resulting in low current efficiency. The problem is alleviated by various means: (1) agitation of solution to bring fresh solution to the cathode, (2) increasing the surface area of the electrodes to maintain low current density, and (3) heating the solution to increase ionic diffusivity. All electrochemical cells used in electrolytic recovery use one or more of these methods.

The advantages of electrolytic recovery are low capital and operating costs and minimal maintenance requirements. It does not require a high skill level, and the same personnel who operate the plating line can operate the electrolytic recovery cells.

Simple modifications to the electrochemical cell include mechanisms to agitate the solution or to flow solution past the cathode and the use of large-surface-area electrodes. Rotating cathodes have been used to overcome the mass transfer limitations and to extract metal from solution concentrations of a few parts per million. The current densities are in the range of 5 to 10 amp/ft^2. Advanced designs in electrowinning cells include the fluidized-bed cell, where electrodes are immersed in a cell containing small glass beads. The glass beads are fluidized by the upward flow of solution through the cell and their motion effectively destroys the diffusion boundary layer.

Other advances in electrolytic recovery are the *extended-surface electrolysis cell* (ESE) and *high-surface-area cell* (HSA). In the ESE, cathode and anode separated by a porous insulator are rolled as a spiral and encased in a pipe. This construction allows for a large electrode area, and flowing solution brings fresh material to the cathode. The simplest example of an HSA electrode is steel wool. In commercial cells, however, carbon filament mats are used. In both ESE and HSA cells, the plated metal cannot be reclaimed as metal because of the cells' intricate construction. Usually, the plated metal is dissolved using acid or other suitable solution. Then the metal concentrate is reused in the plating process tank as metal replenishment or it is plated out in another traditional electrowinning cell and reclaimed.

Applications of Electrolytic Recovery. Conventional electrowinning has been used to recover gold, silver, and tin from their rinsewater.[26] An HSA cell has been used in the recovery of cadmium from the rinsewater used in cadmium cyanide plating.[27] The treated water is returned to the rinse tank. The deposited cadmium is stripped with sodium cyanide and returned to the plating tank as cadmium replenishment. An added benefit is the complete destruction of cyanide in the rinsewater by oxidation at the anode. An unusual example of electrolytic recovery is the rejuvenation of alkali-based permanganate cleaning baths. These baths are used to remove carbon deposits in aircraft engine manifolds. Permanganate oxidizes the carbon to sodium carbonate. With usage, Na_2CO_3 and MnO_2 accumulate in the bath. General Atomics, Inc., devised an electrolytic cell to oxidize manganate to permanganate, thereby extending the bath life significantly. The bath solution is chilled to crystallize the Na_2CO_3, which is filtered out. Electrowinning is also used in the recovery of hexavalent chromium from chrome-plating rinsewaters. The chromium is recovered as chromic acid, which is returned to the plating bath, and rinsewaters are recycled.[28]

Dummy Plating. The principle of dummy plating is that some metals are selectively plated out in the presence of others. The standard potential of the metal in the electrochemical series determines the selectivity. The higher the potential, the higher the selectivity (e.g., $Au^{3+} > Ag^+ > Cu^{2+} > Fe^{3+} > Pb^{2+} > Sn^{2+} > Ni^{2+} > Cr^{3+}$). In addition to the standard potential, stability of a metal complex (such as nickel cyanide complex) also affects the ability to plate out. This principle is applied to removal of trace impurities from plating baths. For instance, metal contaminants in nickel strike bath are copper, iron, and lead. These can be selectively removed by low-current-density electrolysis with minimal loss of nickel even though nickel concentration is three orders of magnitude higher.

23.3 Separation Technologies for Gaseous Waste

The federal Clean Air Act Amendments (CAAA) of 1990 incorporate a preventive approach to air pollution control. The amendments clearly mention "pollution prevention" as a primary goal in dealing with air pollutants. This section describes separation technologies that can be used for reduction or elimination of hazardous air pollutants (HAPs), focusing on those technologies that allow recycling or reuse of HAPs. The HAPs are divided into two categories: particulates and vapors. While the vapor emissions can be both organic and inorganic, the biggest concern for industry is the control of volatile organic compounds (VOCs).

The key technologies for controlling VOCs are oxidation (thermal or catalytic), adsorption, membranes, condensation, and absorption, all except oxidation (combustion) being pollution prevention techniques. These technologies are compared in Table 23-2.[29] The details on principles of operation and applications of the nonoxidation options are provided below.

23.3.1   Adsorption

Adsorption occurs when a molecule is brought to a surface such as activated carbon and held there by physical and/or chemical forces. The quantity of a compound adsorbed depends on the balance between the forces that keep the compound in vapor phase and those that attract the compound to the adsorbing surface. An isotherm test can be used to determine how effectively an organic molecule can be removed from a carrier gas, such as air. Basically, the isotherm shows the relationship between the amount adsorbed and its concentration in the feed at any given temperature. The adsorption capacity is inversely related to both the molecular weights of the VOCs and the temperature of the gas stream.

Table 23-2. VOC Control Technologies

Technology (Recovery/destruction efficiency)	Advantages	Disadvantages
Catalytic incineration	High destruction efficiency Essentially less expensive than thermal oxidation	No recovery of organics Potential catalyst poisoning High capital cost
Adsorption (50–95%)	Effective for solvent recovery Low capital cost	Not selective Moisture constraints with activated carbon
Membranes (90–99%)	Effective for recovery of lower-volatility compounds No need for regeneration	Requires careful pretreatment to achieve long life Lower selectivity
Condensation (50–90%)	Effective for product recovery	Limited applicability; used for pretreatment
Absorption (>90%)	Effective for product recovery, especially for organic vapors	Limited applicability

SOURCE: Adapted from an EPA document entitled "Control Technologies for Hazardous Air Pollutants" (EPA/625/6-91/014), *Chemical Engineering*, June 1993.[29]

Typically, the VOC-laden gas is passed through an adsorption bed. When the bed's adsorption capacity is nearly exhausted, the feed gas is switched to a second adsorption bed. While the second bed is in use, the first one undergoes regeneration by steam or hot air. The dissolved vapors are recovered by condensation as shown in Fig. 23-8.[30]

A variety of adsorbents can be used: activated carbon, molecular sieves, zeolites, and polymeric materials. The use of activated carbon has, however, dominated the marketplace. While the activated carbon typically has high adsorption capacity for most VOCs, it has some disadvantages. A major problem is the difficulty of on-site regeneration, which is related to the difficulty of on-site steam generation and the number of times the carbon can be used before replacement. Another problem with activated carbon is its affinity for moisture, which lowers its capacity to adsorb organic contaminants. The generation of acids when treating halogenated solvents also presents problems.

More recently, there have been efforts to develop lower-cost polymeric materials for VOC adsorption. One such material was developed by Dow Chemical and offered by PURUS, Inc., under the process name PADRE™ (see Fig. 23-9).[31] The adsorption cycle utilizes a combination of temperature and pressure to remove the organics from the bed. The compounds are then condensed and transferred

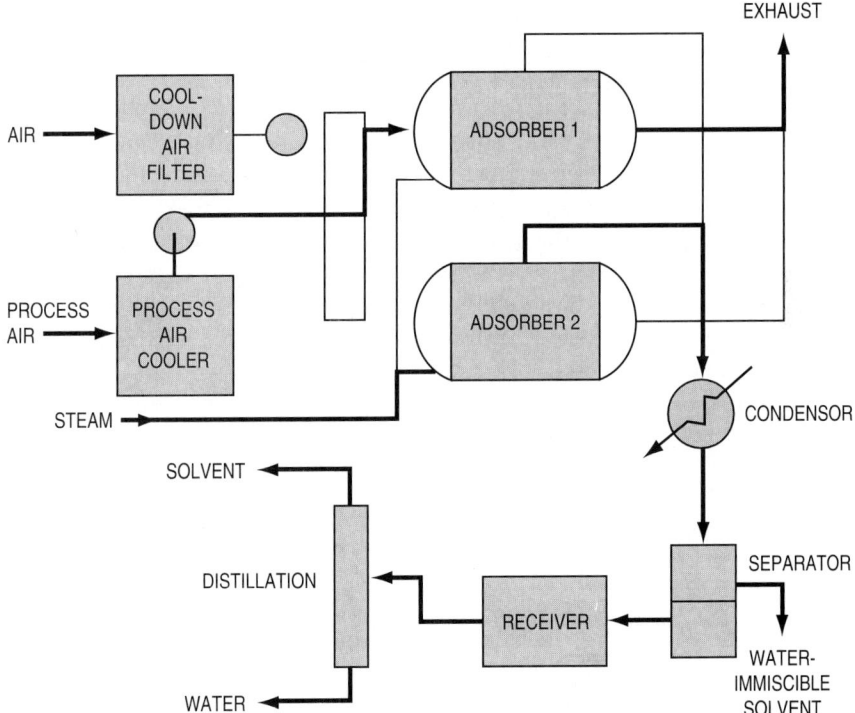

Figure 23-8. Carbon adsorption and steam regeneration. (*Courtesy of Scheihing and Engleman, June 1991.*)

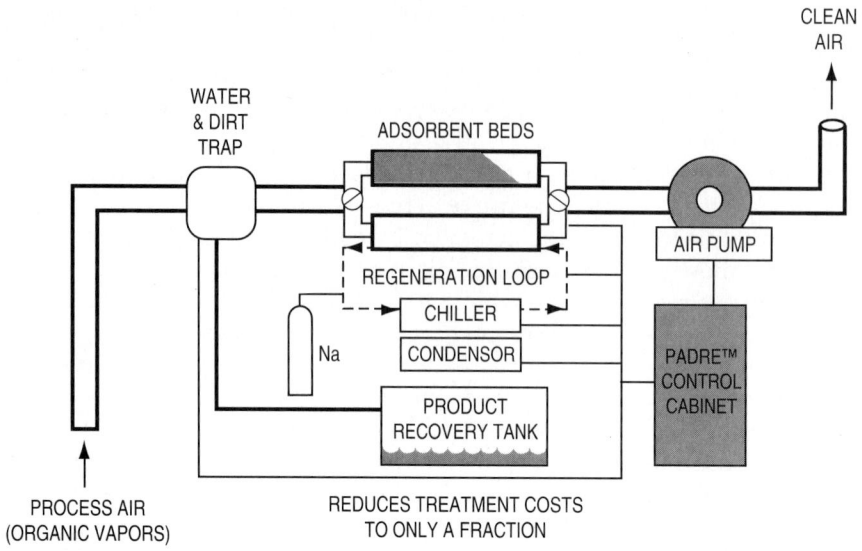

Figure 23-9. Schematic of PADRE™ adsorption system. (*Ref. 31.*)

as a liquid to a storage tank. The condenser system has two stages, one set at 2°C for water collection and the other at −45°C to capture any low-boiling solvents. The recovered organics can be reclaimed by solvent recyclers or fuel blenders. A comparison of two solvents used in the PADRE™ process, namely PurSorb™ 100 and PurSorb™ 200 with activated carbon, is given in Fig. 23-10.[31]

A large variety of organics can be adsorbed on activated carbon and PurSorb™ adsorbents. Examples of the compounds that can be removed from air are alkanes (e.g., hexane), ethers (e.g., tetrahydrofuran), esters (e.g., methyl methacrylate), chlorinated olefins (e.g., vinyl chloride), chlorinated aliphatics (e.g., carbon tetrachloride), aromatic compounds [e.g., benzene-toluene-xylene (BTX)], ketones (e.g., methyl ethyl ketone), and alcohols (e.g., isopropanol).

Regardless of the type of adsorbent used, it is necessary to pretreat the gas to remove liquid droplets (aerosols), solid particles, and polymerizable substances. The most common types of particulate-centered devices are cartridge filters, fabric filters, electrostatic precipitators, venturi scrubbers, and mist eliminators. The principles and applications of these have been extensively published elsewhere.[32]

23.3.2 Membrane Separation

While the principles of gas separation by membranes have long been known, their successful introduction to industrial applications is barely over a decade old. Membranes can be used for gas and vapor separation in a variety of applications, including VOC removal and/or recovery, as shown in Table 23-3.[33]

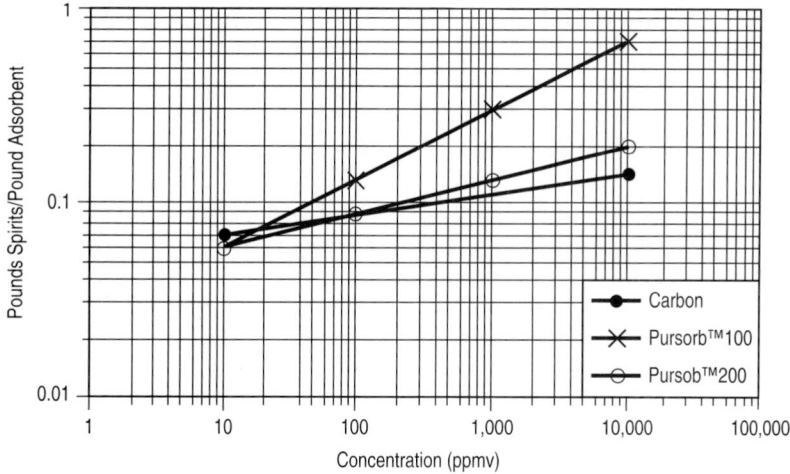

Figure 23-10. Comparison of carbon adsorption with polymeric adsorption. (*Ref. 31.*)

Table 23-3. Gas Membrane Application Areas

Gas separattion	Application(s)
Hydrocarbons/air	Hydrocarbon recovery, pollution control
O_2/N_2	Oxygen enrichment, inert gas generation
H_2/hydrocarbons	Refinery hydrogen recovery
H_2/CO	Syngas ratio adjustment
H_2/N_2	Ammonia purge gas
CO_2/hydrocarbons	Acid-gas treatment, landfill gas upgrading
H_2O/hydrocarbons	Natural gas dehydration
H_2S/hydrocarbons	Sour-gas treating
He/hydrocarbons	Helium separations
He/N_2	Helium recovery
H_2O/air	Air dehumidification

SOURCE: R. W. Spillman, "Economics of Gas Separation Membranes," *Chem. Eng. Prog.*, **85**(1):41–62, Jan. 1989. Used with permission of the American Institute of Chemical Engineers.

Membrane Separation Principles.   A gas permeates a membrane depending on the concentration difference, traveling from the higher-pressure side to the side with the lower pressure. While the primary material for membranes is polymeric, some ceramic, glass, and ametallic membranes also exist. The common polymeric materials are polysufane, polyimides, polyamides, and cellulosic derivatives. Membranes can be *porous* or *nonporous* (also called *porous* or *dense*). In a porous membrane (such as Gortex®), gases travel through by diffusion; in a homogeneous membrane (such as silicone rubber), gases dissolve into the material, diffuse through, and desorb on the other side. In porous membranes, separation is on the basis of molecular size and membrane pore size. This type of a membrane gives a low separation factor, limiting its use. The nonporous membranes, on the other hand, have advanced in the last 15 years to provide practical applications for industrial use. In these membranes, the membrane separation is governed by Fick's law (diffusivity) and Henry's law (solubility).

The key performance attributes that determine the economics of membrane separation are selectivity, flux (permeability), and membrane life. At present, gas separation membranes do not offer a separation solution or product gas that is particularly unique compared to existing processes. Instead, gas separation has to compete on the basis of economics and convenience of operation.

Membrane Process Design and Suppliers.   Economics of gas separation membranes can be greatly affected by the design of the modules and processes. For many applications, a single-stage membrane provides adequate separation and of course the lowest capital investment. But for some applications, involving the recovery of higher-value chemicals, a multistage membrane can be justified.

The primary mode of forming membranes is as flat sheets or as hollow fibers. Both of these can be packaged into modules to give high membrane areas per unit volume. The flat-sheet membranes are typically packaged into spiral-wound modules, shown schematically in Fig. 23-11.[34] A typical hollow-fiber module is similar to that used in reverse osmosis and is shown in Fig. 23-5.

The gas separation membrane was firmly established after Monsanto introduced a hollow-fiber gas membrane, the PRISM® system, in 1929. Today a number of commercial system suppliers exist, as shown in Table 23-4. Most of the early applications were diverted to petrochemical, refinery, and air separation. Recently, some attention has been given to pollution prevention, as discussed below.

Solvent Vapor Recovery by Membranes.   Membranes can be utilized to remove and concentrate VOCs. Nippon Kokan K.K. (NKK) offers a system for use with filling of industrial gasoline storage facilities such as tanks and railroad cars. And Membrane Technology and Research (MTR) markets a system for smaller industrial applications. The NKK system for gasoline recovery is shown in Fig. 23-12.

The MTR System (Vapor-Sep™) combines compression-condensation with membrane separation. The Vapor-Sep™ can be used to treat a wide range of VOC streams. A list of some organic compounds that can be recovered by this process is shown in Table 23-5.[35]

The key parameters that affect recovery performance are membrane area, condenser temperature, and the pressure ratio across the membrane. As an example, a recently installed Vapor-Sep™ system recovers 99.99 percent of the

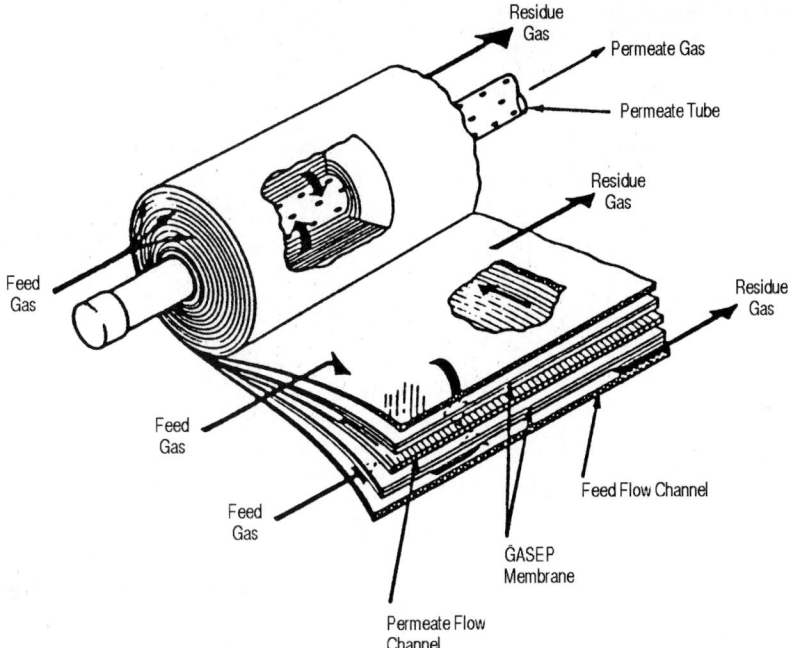

Figure 23-11. Schematic diagram of a spiral-wound membrane module. (*W. H. Mazur and M. C. Chan,* "Membranes for Natural Gas Sweetening and CO2 Enrichment," *Chem. Eng. Prog.,* **78**(10): 38–43, Oct. 1982. Used with permission of the American Institute of Chemical Engineers.)

Table 23-4. Commercial-Scale Membrane Suppliers

Company	CO_2	H_2	Air O_2	Air N_2	Other*
A/G Technology (AVIR)™	X		X	X	
Air Products (Separex)	X	X			X
Asahi Glass (HISEP)			X	X	
Cynara (division of Dow)	X				
Dow (division of Generon)			X	X	
Du Pont		X			
Grace Membrane Systems	X	X			X
International Permeation	X				X
Membrane Technology & Research					X
Monsanto	X	X	X	X	X
Nippon Kokan K.K.					X
Osaka Gas			X		
Oxygen Enrichment Company			X		
Perma Pure					X
Techmashexport (former Soviet Union)			X		
Teijin Ltd.			X		
Toyobo			X		X
Ube Industries		X			
Union Carbide (division of Linde)		X	X	X	
UOP/Union Carbide		X			

*Includes solvent vapor recovery, dehumidification, and/or helium recovery membranes.
SOURCE: R. W. Spillman, "Economics of Gas Separation Membranes," *Chem. Eng. Prog.,* **85**(1):41–62, January 1989. Used with permission of the American Institute of Chemical Engineers.

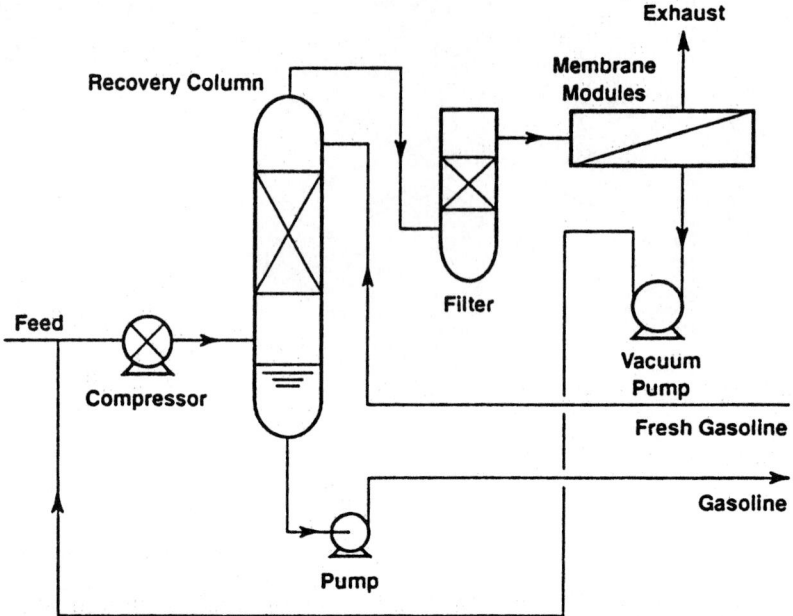

Figure 23-12. Solvent vapor recovery membrane process utilized by Nippon Kokan K.K. (*Ref. 33.*)

Table 23-5. VOCs That Can Be Recovered by the Vapor-Sep™ Process

Acetone	Isopropanol
Benzene	Methanol
Butane	Methylene chloride
Carbon tetrachloride	Methyl ethyl ketone
CFC-11	Octane
CFC-12	Perchloroethylene
CFC-113	1,1,1-trichloroethane
Halon 1301	Trichloroethylene
HCFC-123	Toluene
HFC-134a	Vinyl chloride

(*Courtesy of M. L. Jacobs et al., June 1993.*)

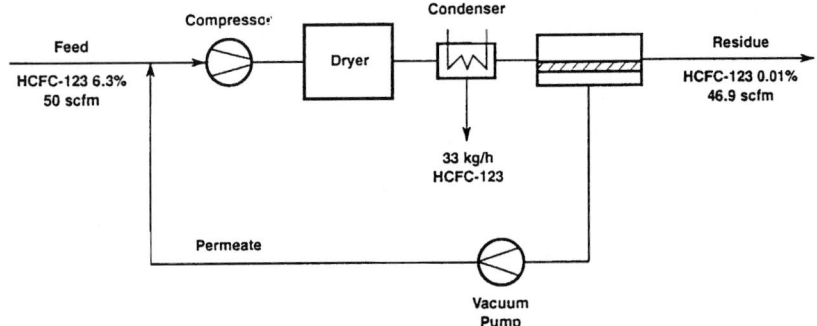

Figure 23-13. Flow diagram and photograph of a Vapor-Sep™ membrane system to recover 99.99 percent of the HCFC-123 lost in a film-drying operation. (*Courtesy of M. L. Jacobs et al., June 1993.*)

HCFC-123 lost in a film-drying operation (see Fig. 23-13).[35] The vapor stream, containing 6.3 percent HCFC-123 is compressed to 125 psig and cooled to −15°C. A dryer removes water vapor from the stream to prevent ice formation in the condenser. The condenser exhaust is channeled into the membrane modules. The unit recovers 70 lb/h of HCFC-123, lowering the VOC exhaust concentration below 100 ppm. It is claimed by MTR that the membrane system has a payback period of 650 hours of operation, if one assumes a price of $5 per pound for the HCFC-123 recovered during each hour of operation.

23.3.3 Condensation

Condensation is generally used as a crude separation step to reduce the VOC lead prior to use of separation equipment such as carbon beds, incinerators, or membrane modules. An example of such an application was given in the subsection on membrane separation. The typical recovery is 50 to 90 percent except in the case of cryogenic cooling, in which case it is well over 90 percent.[35]

Both surface condensaters, involving indirect cooling, and direct-contact condensaters can be used. The surface condensaters are typically shell-and-tube heat exchangers, with the coolant circulating through the tubes and the VOCs condensing on the outer surface of the tubes. The condensed liquids then drain to the bottom of the condenser chamber. The direct-contact condenser method consists of spraying cold liquid directly into the gas stream, which is more efficient than indirect cooling. However, it is difficult to recover valuable organics in the case of direct-contact cooling.

Recovery of VOCs such as methyl ethyl ketone and hexane with the cryogenic condensation technique developed by Liquid Carbonic Industries has been reported.[36] Similarly, the Monsanto Company recovers p-dichlorobenzene (PDCB) used in making mothballs by cooling the waste vapors.[37]

23.3.4 Absorption

Absorption, or *scrubbing,* involves the selective transfer of one or more compounds in the gas phase into a liquid such as mineral oil (for absorbing VOCs) or live solution (for absorbing SO_2). The absorption can be chemical or physical. Chemical absorption involves a reaction between the compound and a solvent, while physical absorption merely involves dissolution of the compound.

A number of vapor-liquid contacting schemes can be used. Packed towers are used for corrosive materials or liquids with foaming tendencies. Tray towers are often used when the liquid flow rates are low. And spray towers are commonly used to contact particles and more soluble gases.

References

1. J. King, *Separation Processes,* McGraw-Hill, New York, 1971.

2. Separation Technology Inc., Anaheim, Calif., product literature.

3. E. O. Jones, "Treating Metal-Bearing Spent Acids with a Transportable Test System," paper presented at the 11th AESF/EPA Conference on Environmental Control for the Metal Finishing Industry, Orlando, Fla., 1990.

4. S. P. Tucker and G. A. Carson, *Environmental Science and Technology,* **19**(3):215–220, 1985.

5. W. E. McGovern and J. M. Moses, "The Treatment of Solvent-Contaminated Waste Using Liquified Gas Extraction," paper presented at the 3d Annual Waste Source Reduction Conference, Worcester, Mass., October 1986.

6. W. S. Winston Ho and K. Sirkar (eds.), *Membrane Handbook,* Chapman & Hall, New York, 1992, pp. 294–299, 600–601.

7. K. Asada, L. Gerdes, and T. Kawahara, "Electrodialysis and Diffusion Dialysis of Effluents from Treatment of Metallic Surfaces," Proceedings of the 79th Annual Technical Conference, Vol. 2, American Electroplaters' and Surface Finishers Society, Winter Park, Fla., 1992, pp. 905–919.

8. W. J. Herdrich, "Recovery of Acid Etchants at Imperial Clevite Inc.," paper presented at the 4th Conference on Advanced Pollution Control for the Metal Finishing Industry," EPA-600/9-82-022, January 1982.

9. G. A. Addison and D. J. Vaughan, "Electrodialytic Separation of Alkali Hydroxides from Multivalent Metal Hydroxides," paper presented at the 1st AIChE Separations Division Conference on Separation Technologies, Miami, Fla., November 1–6, 1992.

10. W. G. Millmam and R. J. Heller, "Some Successful Applications of Electrodialysis," paper presented at the 4th Conference on Advanced Pollution Control for the Metal Finishing Industry, EPA-600/9-82-022, January 1982.

11. M. G. Barth, Jr., "A Prototype Electrodialytic Caustic Recovery System for Aluminum Etchants," paper presented at the 7th Annual Aerospace Hazardous Materials Management Conference, St. Louis, Mo., 1992.

12. M. Jaffari and C. Byszewski, "New Membrane Based Caustic Recycle Process," paper presented at the 7th Annual Aerospace Hazardous Materials Management Conference, St. Louis, Mo., 1992.

13. T. A. Davis, "Recovery of Sodium Hydroxide and Aluminum Hydroxide from Etching Waste," U.S. Patent 5049233, September 1991.

14. P. Werschulz, "New Membrane Technology in the Metal Finishing Industry," in *Toxic and Hazardous Waste*, I. J. Klugman (ed.), Technomic Publishing, Lancaster, Pa., 1985.

15. D. Morris, W. Nelson, and G. Walraver, "Recycle of Papermill Waste Waters and Application of Reverse Osmosis," U.S. EPA Rep. 12040 FUB01/1072, Cincinnati, Oh., 1972.

16. O. Olsen, "Membrane Technology in the Pulp and Paper Industry," *Desalination,* **35**:291–302, 1980.

17. J. Porter and G. Goodman, "Recovery of Hot Water, Dyes and Auxiliary Chemicals from Textile Waste Streams," *Desalination,* **49**:185–192, 1984.

18. C. Brandon, D. A. Jernigan, J. L. Gaddis, and H. G. Spencer "Closed Cycle Textile Dyeing: Full Scale Renovation of Hot Wash Water by Hyperfiltration," *Desalination,* **39**:301–310, 1981.

19. F. Forbes, "Ultrafine Filtration for Electrophoretic Painting," *Product Finishing,* **23**(11) November 1970.

20. I. K. Bansal, "Concentration of Oily and Latex Waste Waters Using Ultrafiltration Inorganic Membranes," *Industrial Water Engineering,* October–November 1976.

21. W. Eyekamp, "Ultrafiltration of Aqueous Dispersions," paper presented at the 79th National Meeting of the AIChE, Houston, Tex., March 1975.

22. H. L. Liu and J. Blacklidge, "Cross Flow Filtration Technology for Metal Finishers," paper presented at the 4th Conference on Advanced Pollution Control for the Metal Finishing Industry, EPA-600/9-82-022, January 1982.

23. X. J. Zhang, J. H. Liu, and T. S. Liu, "Industrial Application of Liquid Membranes for Phenolic Waste Water Treatment," *Water Treatment,* **2**:281, 1988.

24. J. Draxler, W. Furst, and R. J. Marr, "Separation of Metal Species by Emulsion Liquid Membranes," *Journal of Membrane Science,* **38**(3):281–293, Sept. 1988.

25. R. J. Marr, H. Lackner, H. J. Bart, and A. Nickl, European Patent 88/00637.

26. E. F. Hràdil and G. Hradil, "Electrolytic Recovery of Precious and Common Metals," *Metal Finishing,* **82**(11), November 1984.

27. P. Horelik, "Recovery and Electrochemical Technology," paper presented at the 4th Conference on Advanced Pollution Control for the Metal Finishing Industry, EPA-600/9-82-022, January 1982.

28. J. I. Bishara, J. R. Brannan, and R. J. Hovarth, "Recovery of Hexavalent Chromium from Plating Rinse Water," paper presented at the AESF-SUR/FIN Conference, Atlanta, Ga., June 1992.

29. K. S. Kumar, R. L. Pennington, and J. T. Zmuda, "Capture or Destroy Toxic Air Pollutants," *Chemical Engineering* (Environmental Engineering Supplement), **100**(6):12–20, June 1993.

30. P. Scheihing and V. S. Engleman, "National Energy Benefits from Recovery and Recycling of Volatile Organic Compounds and the Evaluation of Control Options," paper no. 91-41.8, presented at the 84th Annual Air and Waste Management Association Meeting, Vancouver, British Columbia, June 16–21, 1991.

31. P. G. Blystone, B. Mass, and W. R. Haag, "Recovery of Volatile Organic Compounds from Air Streams Using Specialized Adsorbents," paper presented at the Emerging Technologies in Hazardous Waste Management Meeting, American Chemical Society, I&EC Division Special Symposium, Atlanta, Ga., Sept. 21–23, 1992.

32. D. W. Green and J. O. Maloney (eds.), *Perry's Chemical Engineers Handbook*, McGraw-Hill, New York, 1978.

33. R. W. Spillman, "Economics of Gas Separation Membranes," *Chemical Engineering Progress*, **85**(1):41–62, January 1989.

34. W. H. Mazur and M. C. Chan, "Membranes for Natural Gas Sweetening and CO^2 Enrichment," *Chemical Engineering Progress*, **78**(10):38–43, October 1982.

35. M. L. Jacobs, R. W. Baker, J. Kaschemekat, and V. L. Simmons "Industrial Applications for Membrane Vapor Recovery Systems," paper presented at the 86th Annual Air and Waste Management Association Meeting, Denver, Colo., June 13–18, 1993.

36. G. T. Sameshima and J. D. Eisenwasser, "Cryogenic Recovery of VOC Emissions," paper presented at the 84th Annual Meeting of the Air and Waste Management Association, Vancouver, British Columbia, June 16–21, 1991.

37. *The Wall Street Journal,* June 11, 1991, p. A-6.

24

Pollution Prevention Through Process Control

Randy D. Down, P.E.

The Sear-Brown Group
Rochester, New York

The ultimate objective of most process plants is to optimize overall production efficiency and quality of the end product. In addition to the obvious benefit of increased profits, enhanced production efficiency often results in emissions reductions, or reduction in unusable solid waste by-products. Rising costs of storage, posttreatment, disposal, and recycling will add further incentive to minimize waste production.

This section discusses instrumentation and control methods that generally result in more efficient process control and reduced generation of pollutants and waste materials.

24.1 The Effect of Measurement Accuracy on Pollution Control

Overall process efficiency, as well as operator safety and product quality, are greatly influenced by control system performance and reliability. These system characteristics are further influenced by measurement accuracy and control methodology, particularly when applied to continuous measurement of air pollutants, which may require accurate measurement and control at a resolution below one part per million (ppm).

Significant improvements in sensor technology over the last twenty years in the areas of microelectronics and material science have resulted in the development of measurement and control devices with greater accuracy, repeatability, and reliability. Furthermore, they are often available at a lower cost, are physically smaller, and require less energy to operate than their predecessors.

Control system efficiency can be attributed to a combination of the following characteristics:

- Measurement accuracy, stability, and repeatability
- Sensor location(s)
- Controller response action (proportional, integral, derivative, cascade, feedforward, stepped)
- Process dynamics
- Final control element (valves, dampers, relays, etc.) characteristics and location
- Overall system reliability

Instrumentation should be selected that enables the control system (or control system operator) to derive a consistently accurate and reliable measurement of the true process conditions. Accurate and reliable measurement is a prerequisite of efficient, precise control of most processes.

Whenever analytical instrumentation and control devices are selected for an application, the following questions should be considered:

How reliable will this device be in terms of human safety, product quality, and loss prevention? Does the risk to human safety, or the potential for product loss or uncontrolled pollutant emissions, outweigh the expense of redundant backup devices?

Will the device be exposed to harsh environments, such as condensation, corrosive fumes, vibration, or temperature extremes, that may degrade or damage the device?

What would the consequences be if this device fails or loses power (failsafe condition)?

Are the following characteristics appropriate to attain optimal control efficiency in your application?

—Full-scale accuracy

—Rangeability

—Repeatability

—Stability

—Materials of construction

What is the availability of replacement devices or replaceable components?

Must the device be intrinsically safe (incapable of igniting a flammable substance)?

How difficult are calibration and routine maintenance of this device?

Amazingly, a great many accurate, and often expensive, analytical instruments are misapplied or improperly installed. As a result, measurement reliability of these devices suffers. The finest instrumentation made cannot overcome an improper installation or incorrect application of the device. Poor control system performance will often result in increased waste production or uncontrolled pollutant emissions.

Generation of waste by-products and pollutant emissions can often be reduced by improving control system performance. Individual processes and their combined effect on overall plant performance must be evaluated to determine the cost/benefit ratio of control system enhancements. Consideration should be given to the cost factors discussed in the following section when potential savings are evaluated (see Table 24-1).

In many cases, the value gained by precise, reliable measurement and control will outweigh the upfront investment in quality instrumentation and system controls. This applies to most industrial and commercial processes, including combustion, dilution, distillation, electroplating, emulsion, evaporation, exhaust, filtration, etc.

24.2 Optimizing Control System Performance to Reduce Pollutants

In many applications, improvements in process control efficiency also result in reduced waste production or pollutant emissions.

Table 24-1. Costs and Potential Savings Resulting from Improved Process Control

Costs	Potential savings
Purchase of instrumentation and control devices	Improved production efficiency
	Improved product quality
Installation, programming, and start-up	Improved safety
Training	Cleaner environment
Maintenance	Reduced waste and pollution by-products: reduced cost of waste storage reduced posttreatment costs reduced waste handling costs reduced landfill cost and liability

Optimizing the performance of control systems requires a detailed analysis of process dynamics. When challenged with a unique, new process, the analysis is replaced by a prediction of how the process will behave under dynamic conditions. Predictions are made based on computer simulations, comparisons to the dynamics of similar processes, or by performance data collected through a pilot operation.

Many pollution control system strategies go well beyond the limitations of basic single-loop control, consisting of a sensor, controller, and final element (see Fig. 24-1). More complex control strategies require the use of cascade control, feedforward control, and the control of multivariable processes. These strategies require networking of multiple controllers to provide control actions that can compensate for long lags (deadtime) in process response time, or very rapid process variations.

24.2.1   Systems Networking

Plantwide efficiency improvements and reductions in waste production may be obtained by networking together individual process control systems throughout a facility into a common distributed control system (DCS), programmable logic controller (PLC) network, supervisory control and data acquisition (SCADA) system, or a hybrid of these networks using data highways, modems, and local area networks (LANs) (see Fig. 24-2). In recent years, these technologies have merged. An old rule of thumb was to use a DCS when many of the input/output points being monitored and controlled were analogue in nature. PLCs were applied where the majority of input/output points were binary (also referred to as discrete, relay, or on/off). As each system's capabilities have evolved, their differences have become less significant. It is not uncommon

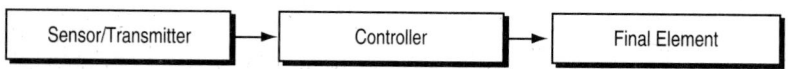

Figure 24-1. Basic control system.

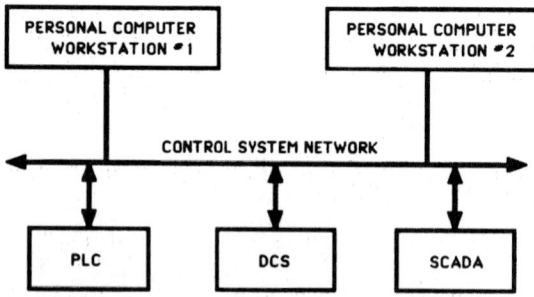

Figure 24-2. Control system networking.

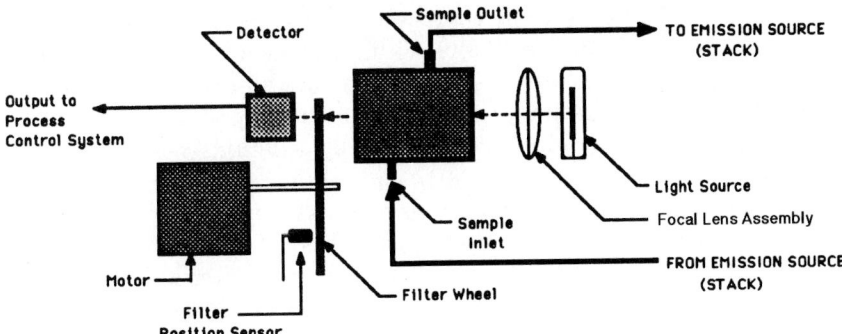

Figure 24-3. UV gas analyzer.

now to see DCS and PLC systems consolidated into a common network that takes advantage of both systems' strengths.

Continuous emissions monitoring systems (CEMs) are also available as a networked system to allow simultaneous monitoring of multiple, remote emissions sources in a facility. Measured emission concentrations (see Fig. 24-3) can be converted to a control signal which, when applied to their respective production processes as a feedback signal, act as a high limit for process activity, preventing emissions from exceeding regulatory limits.

24.2.2 Advantages of Systems Networking

Among the common advantages of a networked measurement and control system are the following capabilities:

- On-line performance monitoring
- Inventory control
- Remote data monitoring and historical recording
- Easier implementation and modification of control strategies
- Trending and graphing of process conditions
- Report generation
- Shared data between subsystems
- Generally easier system upgrades
- Statistical analysis
- Plantwide energy management and pollution prevention strategies
- Interactive process simulations
- "Pinch" technology

These advantages may be used to measure and reduce the generation of waste by-products and limit pollutant emissions within a facility. Results can be measured and recorded for future reference. Software simulations allow a system's initial control strategies and tuning to be developed off line with simulated system dynamics. This avoids on-line trial-and-error variations while the actual process is running, which would result in waste product generated during system commissioning. In addition, simulations software has been developed that allows "pinch" technology to be applied to multiple processes. By analyzing requirements of each process in terms of location, raw materials, and energy requirements, pinch technology identifies beneficial interactions that can be created between the individual processes that will produce the optimum overall benefits in terms of emissions reductions and plant efficiency. Again, the tradeoffs among safety issues, pollution prevention, energy conservation, and capital and operating costs must be weighed. Any decisions must adhere to applicable safety, environmental, and energy regulations.

Emissions monitoring and control required under the Clean Air Act, as well as future changes in federal, state, and local environmental regulations, will be more easily accommodated with a networked control system. Changes in control strategies can be more easily accomplished by downloading new control strategies and tuning parameters to remote-control panels from a central computer workstation than by manual reconfiguration, or replacement, of each controller. Application of "intelligent" transmitters will allow zero and span adjustments (measurement range) of remote-sensing devices to be modified via the system network. This is particularly advantageous on stack analyzers and other devices that are not readily accessible.

24.2.3   Process Control Innovations

Recent innovations in digital controller design and their application have significantly improved the potential performance of process control systems in general, and as they apply to pollution control. Among the innovations are enhanced self-tuning or adaptive controllers, and the application of fuzzy-logic and statistical process control (SPC) strategies.

Self-tuning Controllers.   Self-tuning controllers "learn" repetitive dynamic characteristics of a process and automatically choose their own tuning parameters, based on their analysis, to allow a quicker return to the desired setpoint value. This fine-tuning occurs on a continuous basis, allowing a controller to remain tuned for optimal control without operator intervention. There are a variety of self-tuning controllers on the market, each having a different degree of sophistication and using a different method of learning or analysis. The most beneficial models continuously update their PID tuning parameters by using either heuristic techniques, mathematical modeling, or automatic step tests. Their greatest value is found in their ability to limit the amount of overshoot that occurs in a process after a change in setpoint. As a result, they are commonly found in applications involving temperature control.

Fuzzy Logic.   *Fuzzy logic* is derived from a branch of mathematics known as *fuzzy set theory*. This theory is based on the concept that traditional true/false (discrete) logic cannot deal well with situations or conditions that involve numerous exceptions. In essence, fuzzy logic provides a systematic way to exploit the inaccuracies, using everyday language to describe the condition in "degrees of truth." Examples would be conditions that could be described as "fairly high," "slightly high," or "extremely low." Fuzzy logic has been applied to processes that are considered difficult, if not impossible, to control properly with traditional PID controllers, such as the control of nonlinear systems or systems with very long deadtimes. An additional benefit is the comfort level of control system operators who may not be adept at software programming but who are able to describe the desired control action by using descriptions in plain English.

Fuzzy-logic controllers are cost effective, because they do not require the sophistication of more traditional digital controllers designed for precise calculations using very complex algorithms. Fuzzy-logic controllers are now gaining acceptance in many process control applications. Their use in pollution control applications will increase significantly in the next two or three years.

Statistical Process Control.   Statistical process control determines the optimal control conditions based on the results of statistical data. By making small incremental adjustments to the process and observing the results over time to establish that they are real improvements and not the result of some other temporary factor affecting the process, one can establish and verify true improvements in process conditions. A typical control chart is shown in Fig. 24-4. Through use of control charts, it is easy to establish baseline conditions, measure improvements in process efficiency, and detect any isolated, abnormal variations that are inconsistent with the norm. It is also possible to establish preset control limits that act as an early warning system, notifying the operator that the controlled process performance is degrading or is out of control. Data from these charts may be directly input to a control system as a measurement device, assuring continued optimization of your process.

24.3 Process Control Strategies to Prevent Pollutant Emissions

Numerous control strategies have been created to provide source control of pollution, as required by the Pollution Prevention Act of 1990 and Resource Conservation and Recovery Act (RCRA) hazardous waste regulations. The RCRA regulations require that hazardous waste generators have a program in place to reduce the volume and toxicity of waste generated to the extent that it

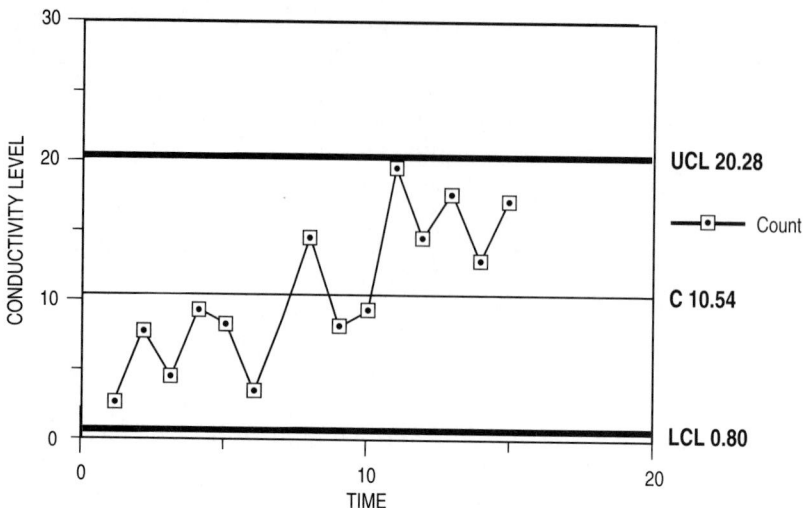

Figure 24-4. Statistical control chart.

is economically practical. Source control can be defined as the pretreatment of waste materials to minimize, or prevent, the release of toxic substances into the natural environment. They typically fall into the general categories of water treatment processes or air emissions control. Solid waste materials are either burned in a waste-burning facility (if it has sufficient heat value), to generate heat and power, or containerized and stored at a licensed storage facility or landfill.

Application of process controls to prevent pollution (source control) can be divided into several categories, including reduction of waste production through process efficiency improvements (as described earlier in this section), pretreatment of pollutant-containing effluents using chemical reactions, capture and recycling of pollutant-containing waste by-products, and collection and storage (containment) of pollutant-containing waste by-products. Examples of wastewater pretreatment technologies include neutralization, chemical precipitation, and filtration. Wastewater may contain any one, or a combination of, toxic materials in the form of organics, metals, and nonmetallic organics.

Analytical instrumentation, such as that shown in Fig. 24-5, is used in wastewater pretreatment to measure such characteristics as pH, turbidity, conductivity, temperature, and color. These measurements can be used as indirect methods to detect and control various contaminants. As an example, turbidity and conductivity are two methods used to determine concentrations of suspended solids in wastewater.

Single-loop process controllers compare measured variables to a desired value (setpoint). Amplitude of deviation of the measured variable from setpoint is entered into a control algorithm which, in turn, establishes the rate and magnitude of the controller's response. A corresponding change in controller

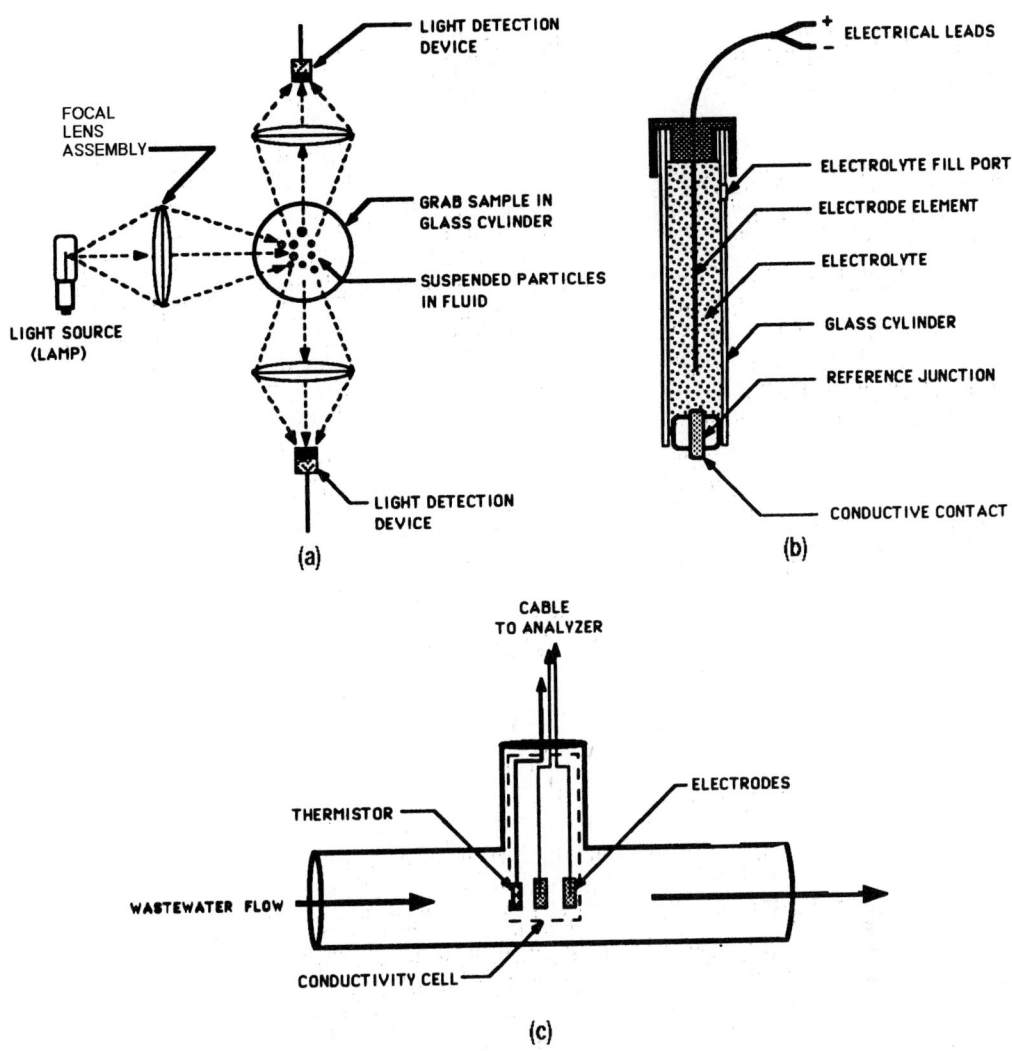

Figure 24-5. (*a*) Turbidity measurement; (*b*) pH referencing electrode; and (*c*) conductivity probe.

output signal causes a change in the physical position of an electromechanical device (relay closure, valve modulation, pump rate, etc.) to correct the process (eliminate any deviation). An example would be the injection of sulfuric acid into a wastestream which exhibited a high measured pH value in order to reduce the pH to an acceptable level before release into the environment, as shown in Fig. 24-6. The control system allows time for the agitated wastewater's pH level to equalize throughout the tank before its release or transfer for further treatment. This is referred to as the *contact* time. PLCs are well

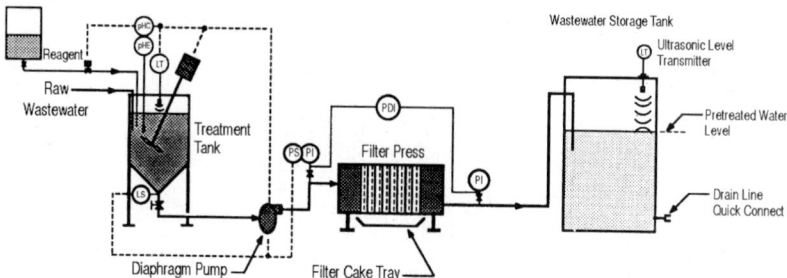

Figure 24-6. Wastewater P&ID.

suited for this type of batch operation due to their sequential control capabilities and built-in software timers.

24.3.1 Organics

Organics are typically treated or removed from a wastewater stream by any one of several methods, including air-stripping, oxidation, carbon absorption, and biological processes. These processes, and the instrumentation and control systems typically associated with them, are described as follows.

24.3.2 Air-stripping

Air-stripping processes convert volatile chemicals from their liquid state to a gaseous state using a stream of moving air. A blower moves air through a tower filled with material designed to maximize wet surface contact with the airstream as a counterflow pattern of wastewater is sprayed into the air. As a result, volatile toxic substances are released into the airstream by a process known as *mass transfer*. Depending upon the concentration of volatile organics released into the air, dispersion of gases from the stripper may become a concern. Further treatment may be necessary in order to limit the emission of organics released into the atmosphere to levels below prevailing regulatory limits. A sampling or in-situ emissions analyzer may be used to monitor and record emission concentrations, as shown in Fig. 24-7.

24.3.3 Oxidation

Oxidation processes apply a chemical to carbon-based pollutants which produces an oxidation reaction, resulting in their conversion to nontoxic by-products such as carbon dioxide. Process controls for this application include control and measurement of such variables as pH, temperature, and reaction times to ensure that full oxidation has occurred with the minimal amount of catalyst used.

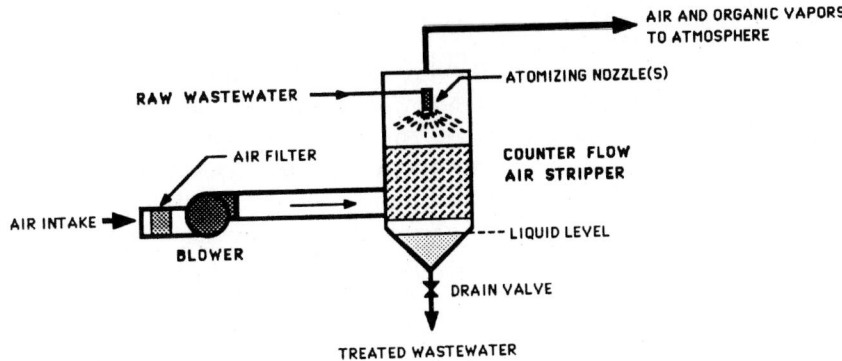

Figure 24-7. Air-stripper system.

24.3.4 Biological Processes

Biological processes use living organisms (bacteria) to feed on and digest (break down) toxic organics into nontoxic components. This process is not as costly or as complex and requires less sophisticated process control techniques than other methods. Instrumentation associated with this type of process typically includes measurement and control systems to regulate flow rates, tank levels, temperature, dissolved oxygen levels (see Fig. 24-8), and chlorination to maintain bacterial growth.

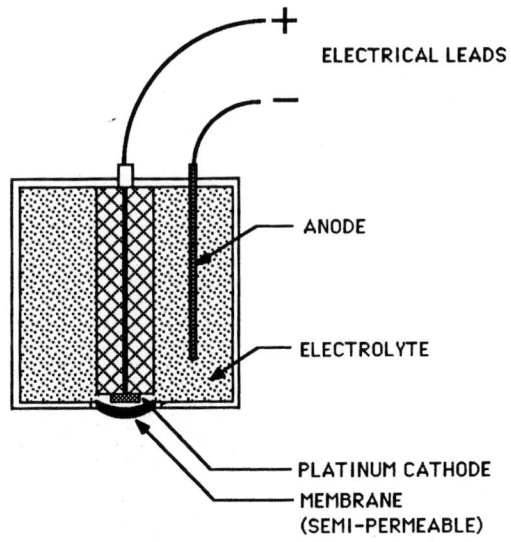

Figure 24-8. Dissolved oxygen monitor.

24.3.5 Carbon Adsorption

Carbon absorption processes pass water through an activated carbon filter to capture toxic organics, which have a tendency to collect at the surface of carbon media through a process called *adsorption*. This process typically uses an activated charcoal that can be removed from the wastewater stream and regenerated by application of heat. Instrumentation and controls associated with this type of process are typically limited to the regulation of water flow rate and differential pressure across the filter to maximize removal of organic vapors, as shown in Fig. 24-9.

Failure of source control methods to limit concentration of toxic materials within the wastewater discharge to levels below applicable regulatory limits will require posttreatment, perhaps in the form of a chemical precipitation or ion exchange process, as shown in Fig. 24-10.

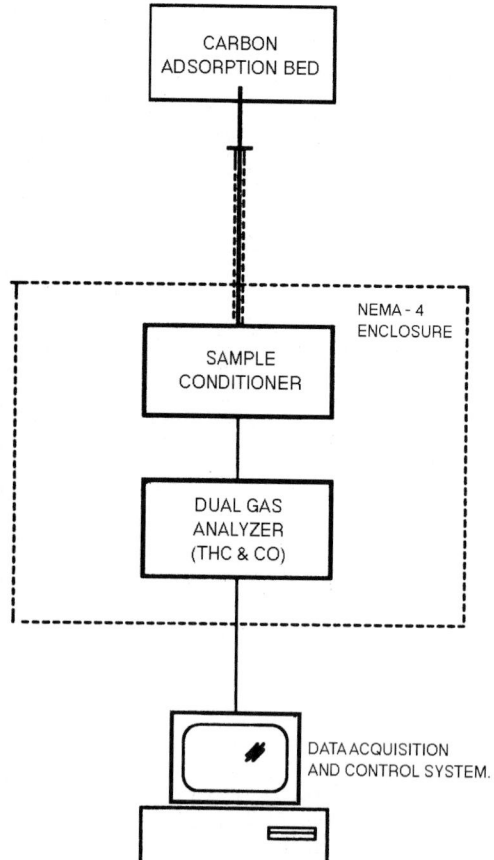

Figure 24-9. Carbon adsorption monitoring.

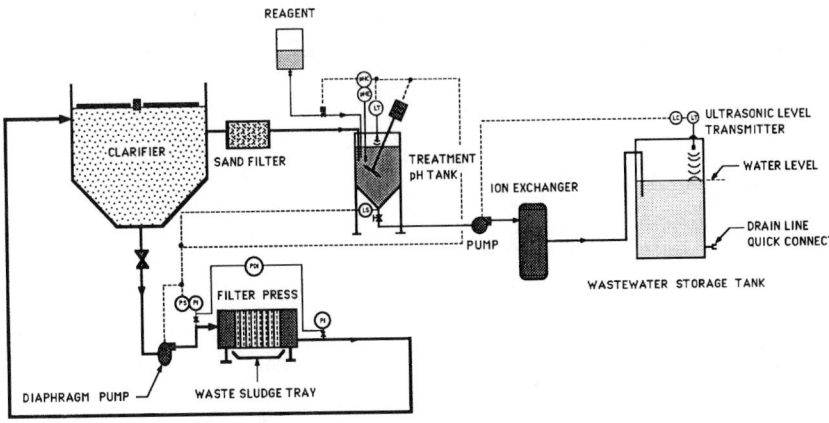

Figure 24-10. Wastewater P&ID.

24.3.6 Continuous Emissions Monitoring

A wide variety of continuous emissions monitoring (CEM) systems are available for analyzing and recording constituents and concentrations of gases and particulate from emissions sources. They range in sophistication from gas chromatographs to opacity monitors. Each is designed for measurement of a particular substance or classification of substances that causes a change in some measurable characteristic when exposed to light, a magnetic field, heat, or other form of radiation.

24.4 Conclusion

The wide variety of process control applications that, when more efficiently controlled, would reduce pollutant emissions and waste generation is too large a topic to fully cover within this section. However, many of the rules that apply to one application can be effectively applied to other applications.

Source control of pollutants is often the best approach, and is encouraged by the U.S. Environmental Protection Agency (EPA). Guidelines for selecting a process control strategy, as described within this section, should start you in the right direction.

Further Reading

Andrew, W. G., and H. B. Williams, *Applied Instrumentation in the Process Industries*, 2d ed., Gulf Publishing Company, Houston, Tx., 1980.
Considine, Douglas M. (ed.), *Handbook of Applied Instrumentation*, McGraw-Hill, New York, 1964.

Murrill, Paul W., *Application Concepts of Process Control*, ISA Publications, Research Triangle Park, N.C., 1988.

Down, Randy D., *Environmental Control Systems*, ISA Publications, Research Triangle Park, N.C., 1992.

Samdani, G., K. Fouhy, and S. Moore, "Fuzzy Logic—More Than a Play on Words," *Chemical Engineering*, February 1993, pp. 30–33.

Samdani, G., and S. Moore, "Pinch Technology: Doing More With Less," *Chemical Engineering*, July 1993, pp. 43–48.

Shaw, John A., "SPC for the Process Industry," *INTECH*, ISA Publications, Research Triangle Park, N.C., December 1989, pp. 34–37.

25

Pollution Prevention Through Process Simulation

Demetri Petrides, Vital Aelion, Subir K. Mallick, and Konstadinos Abeliotis

New Jersey Institute of Technology
Newark, New Jersey

25.1 Introduction

In 1990, 3.6 billion pounds of toxic chemicals were released to the environment, excluding transfers such as to landfills or to off-site treatment facilities. Eighty percent of this pollution came from the chemical, primary metal, paper, transportation, and plastics industry sectors.[1] One way to reduce pollution is to reduce industrial activity, yet most of us will agree that our goal is exactly the opposite because industrial activity results in prosperity and an improved standard of living.

It can be argued that it is impossible to build and operate chemical manufacturing facilities without releasing some harmful chemicals to the environment. Recognizing the physical, chemical, and manufacturing limitations of our processes, we can try to keep pollution to the absolute minimum. Achieving this minimum involves adopting new corporate attitudes, applying new technologies, managing information more effectively, evaluating environmental impact more accurately, and, most importantly, realizing that *pollution prevention* is a better course of action than treatment and disposal.

This realization is the foundation of today's regulations which are based on the experience of the past, when billions of dollars were spent on environmental restoration. The U.S. Environmental Protection Agency (EPA), through its

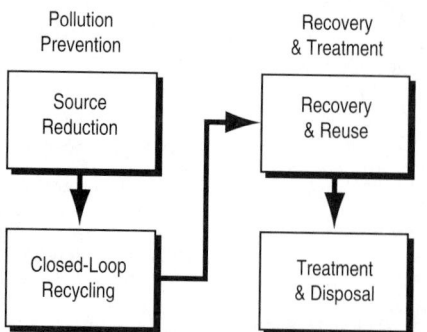

Figure 25-1. Waste management hierarchy.

regulations and its incentive programs, clearly favors pollution prevention, defined as the use of procedures and processes that reduce or eliminate the generation of pollution at the source (see Fig. 25-1). Pollution prevention includes reducing the volume of waste generated as well as its toxicity.

The New Jersey Institute of Technology (NJIT) has undertaken the Integrated Pollution Prevention Initiative (IPPI), an effort to assist in research, application, and dissemination of pollution prevention principles. Under IPPI, a variety of new environmental technologies have been developed and existing technologies have been applied, including work on computer-aided process design applications. In this chapter, we discuss the benefits of computer-aided process design and simulation for pollution prevention. The discussion draws heavily upon the experience gained from our research work at NJIT under IPPI.

25.2   Process Design

The goal of process design is to identify operating alternatives that perform well with respect to a variety of process objectives, such as capital and operating costs, safety, reliability, and environmental impact. The search for these alternatives is limited by the available technologies and the financial and time resources available to the engineer. Information technologies that help perform this search more quickly and/or more effectively play an important role in the realization of improved designs.

Process design follows the loop shown in Fig. 25-2. Typically, we (1) propose a process, (2) analyze it either by simulation or by experimental means, and (3) evaluate it based on whatever objectives we deem important. We then use the information generated during this design loop to modify the proposed process and we go through the loop again. The design process is terminated either because we arrive at a satisfactory process or because we run into time and/or financial constraints.

Process design is almost always hierarchical. The most basic decisions, such as process chemistry, are addressed initially and more detailed issues, such as oper-

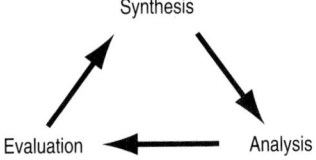

Figure 25-2. Process design loop.

ability and process scheduling, are considered at later stages. This is so because high-level decisions have a profound impact on the overall structure of the process and it would be counterproductive to consider peripheral decisions early.

Process synthesis uses new as well as existing technologies to select process chemistry, solvents, reaction schemes, separation technologies, and so on. Hierarchical synthesis methodologies have been discussed by Rudd, et al.[2] and by Douglas.[3] Synthesis methodologies adapted to address waste management appear in work by Douglas[4] and by Rossiter et al.[5]

The result of process synthesis is a flowsheet for continuous processes or a batch sheet for batch processes. This is the input to the analysis step. If analysis is not done experimentally, then it becomes synonymous with process simulation. The results of process simulation usually include material and energy balances, capital and operating costs, and other measures of system performance. Process simulators used for pollution prevention need to explicitly estimate the environmental impact of integrated processes.

Evaluation concludes the design loop by assigning a measure of overall performance for each process alternative. Evaluation typically considers multiple aspects of performance. Traditional evaluation schemes for chemical and petrochemical processes include capital and operating costs along with safety and flexibility. However, processes also exhibit frequently ignored "hidden" costs, many of which originate in the environmental impact. Two methods which take into account environmental issues are *total cost assessment* (TCA) and *life-cycle analysis* (LCA).

TCA is an accounting methodology which differs from conventional cost evaluation methods because it considers a broader range of items, including probabilistic costs and savings, indirect or hidden costs, liability costs, and less tangible benefits. Examples of indirect or hidden costs include costs associated with (1) compliance, (2) liability, (3) liability insurance, (4) on-site waste management, and (5) operation of pollution control equipment. Liability costs include those incurred from penalties, fines, personal injury, and property damage. Finally, less tangible benefits may include increased revenue from enhanced product and/or corporate image, reduced health insurance costs from improved employee health, and so on.[6]

LCA of a product is a rigorous assessment of the total pollution produced during (1) raw material acquisition, (2) material manufacturing, (3) product manufacturing, (4) product use, and (5) product disposal.[7] It can be used by the private sector as a basis for product substitution, as well as by the government as a policy-making tool. LCA differs from other evaluation techniques in that it evaluates the environmental impact of the *product* in addition to that of the *process*. Typical examples of LCA are the life-cycle inventories of soft drink containers, polystyrene versus paper containers, and diaper systems.[7]

The results of analysis and evaluation are used to generate a new process. As we mentioned in the beginning of this section, any methodology or tool which makes any of the design steps quicker and/or more thorough is bound to result in a better process design.

25.3 Computer-aided Process Design for Pollution Prevention

It is widely recognized that the current process simulation and design tools lack many essential features required for explicitly assessing the environmental impact of new and existing processes. To address this need, the EPA, the Department of Energy (DOE), the Center for Waste Reduction Technologies (CWRT), and the American Institute of Chemical Engineers (AIChE) brought together 50 experts in the field to identify how environmental factors should be incorporated into process synthesis and simulation tools for the chemical processing industries.[8] The goal of the workshop was to identify and prioritize research and development needed in this area. Some of the recognized needs are following:

- Characterize the environmental properties of process wastestreams
- Develop integrated design tools that combine simulation with process synthesis and take into account environmental aspects of design, such as raw material and solvent selection, assembly of reaction and separation steps, and consideration of in-process recycling alternatives
- Implement methodologies and tools that identify and evaluate alternative environmentally benign reaction pathways
- Develop tools that can integrate the design of manufacturing with that of end-of-pipe treatment processes and accurately estimate the real costs of waste treatment and disposal, including intangible costs, such as liability, public relations, legal, etc.
- Invent algorithms for accurately tracking trace components in process and wastestreams

Our group at NJIT has been conducting research in computer-aided process design for pollution prevention motivated by the recognized needs of the EPA workshop. The current application emphasis lies with pharmaceutical and specialty chemical processes, which operate frequently in batch mode. Both specialty chemical and pharmaceutical process industries constitute interesting test cases for environmental design because they make extensive use and release large amounts of volatile organic compounds (VOCs). Also, these industries tend to be less sophisticated in process engineering and automation compared to producers of commodity chemicals, so development of computer-aided process design tools has an opportunity for a greater impact.

We have developed a user-interactive computing environment, called BatchPro Designer, which performs simulation and evaluation, and provides advice for retrofit design of environmentally benign chemical processes. BatchPro Designer is based on BioPro Designer, a process simulator for bio-chemical processes whose development was initiated at the Massachusetts Institute of Technology as a part of Dr. Petrides' Ph.D. thesis.[9]

25.3.1 Description of BatchPro Designer

Scope and System Architecture.   The scope of BatchPro Designer is based on the realization that some pollution is unavoidable because it is dictated by the process chemistry, therefore there is a need for (1) a manufacturing module, where the designer considers alternative processes for pollution prevention during the production of the target molecule, and (2) an end-of-pipe treatment module, where there is an attempt to reduce the environmental impact of the unavoidable waste. The conceptual framework of BatchPro Designer is depicted in Fig. 25-3.

Each module in Fig. 25-3 consists of three components: (1) process simulation, (2) economic evaluation, and (3) a knowledge-based waste minimization advisory system for process modifications. The tasks performed in each component are shown in Fig. 25-4.

User Interface.   BatchPro Designer makes use of a graphical interface to enhance the human/computer communication and reduce the learning period, resulting in a tool that is simple to use and easy to learn, even for occasional users with limited process design and environmental experience. All input-output information is provided or displayed through dialogue windows. Figure 25-5 shows how information about a flowsheet is displayed on the main window. Figure 25-6 shows a typical input dialog window for initializing unit operation models. BatchPro Designer also provides on-line help.

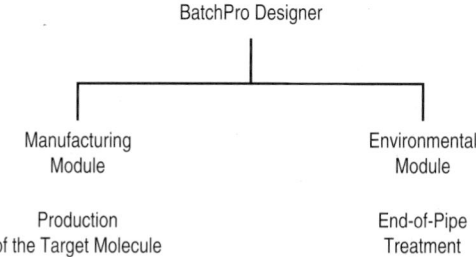

Figure 25-3. Scope of BatchPro Designer.

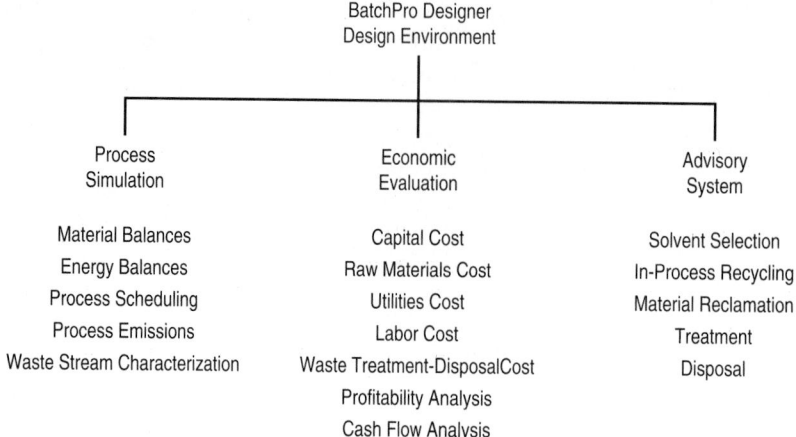

Figure 25-4. Computing components of BatchPro Designer.

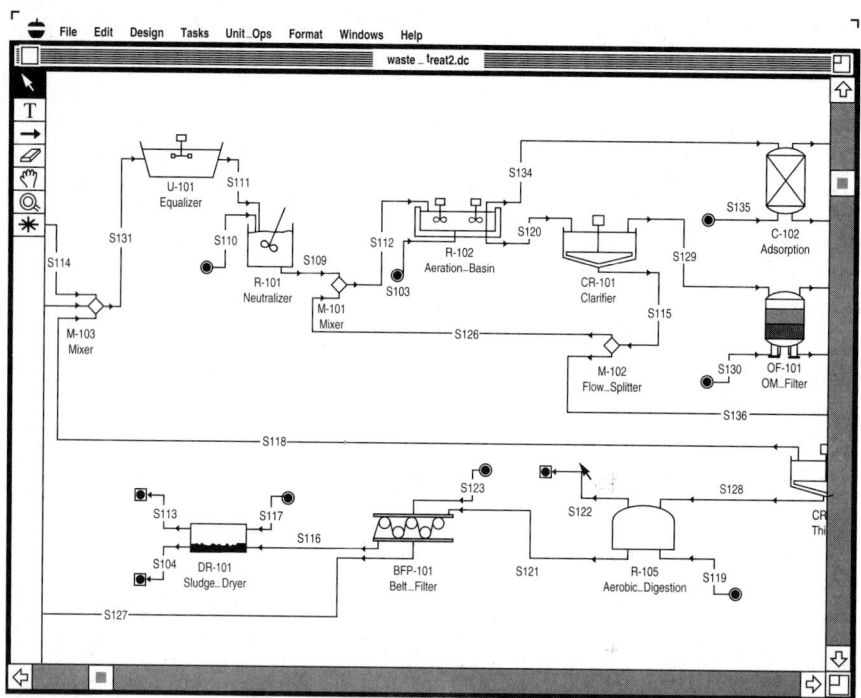

Figure 25-5. Advanced graphics interface facilitates interaction of the user with the computer.

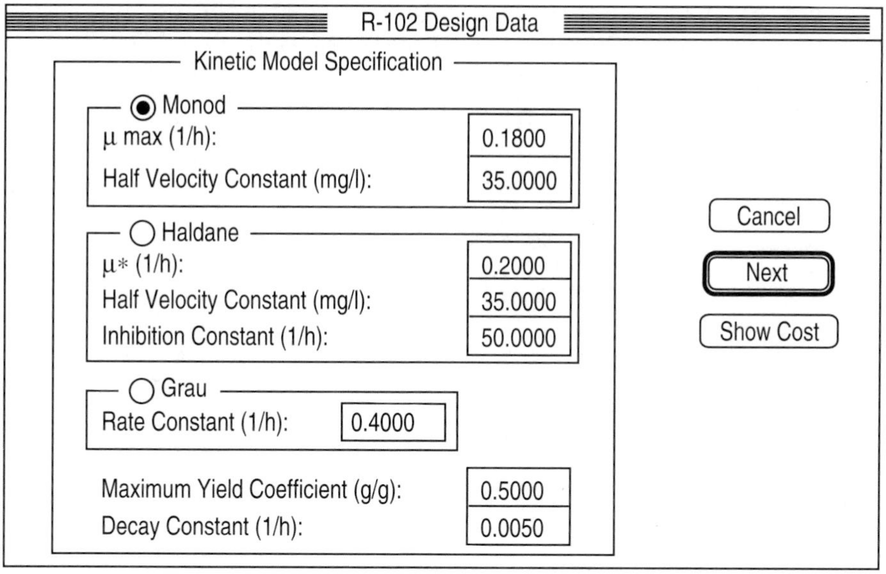

Figure 25-6. First input/output dialog window of an aeration basin.

Process Analysis.   The simulation and evaluation modules are tightly linked into an interactive environment that enables engineers to quickly develop, analyze, and evaluate manufacturing facilities integrated with waste recycling, recovery, treatment, and disposal processes. Using BatchPro Designer, the user can readily investigate the effect of operating choices, such as the replacement of an extraction solvent, the addition of a recycle stream, or the change in operating conditions of a unit operation on the environment, on the economics of the process, and on product quality. Both dynamic and steady-state unit operation models are included in the program. Table 25-1 shows a list of unit operation models included in the current version of BatchPro Designer.

The analysis component of BatchPro Designer facilitates pollution prevention by helping engineers to gain a better insight and strengthen their understanding of an industrial process. Also, it enables engineers to adopt a systems approach and examine the effect of decisions on the entire process as well as on the environment.

Environmental Impact Assessment.   BatchPro Designer estimates and characterizes wastes from a manufacturing facility. The stream report displays the flow rates of all substances, including those with concentrations down to sub-ppm levels. Based on stream composition, BatchPro Designer calculates and displays a number of environmental stream properties (see Fig. 25-7). These properties apply to liquid wastestreams and are indicators of environmental impact on living systems in aqueous environments.

Table 25-1. Unit Operation Models in BatchPro Designer

Reaction
- Batch/continuous fermentor
- Air-lift fermentor
- Well mixed reactor
- Plug flow reactor
- Fluidized-bed reactor

Cell disruption
- High-pressure homogenizer
- Bead mill

Mechanical separators
- Membrane microfilter
- Membrane ultrafilter
- Diafilter
- Reverse osmosis
- Dead-end filter
- Air filter
- Plate and frame filter
- Rotary vacuum filter
- Basket centrifuge
- Disc-stack centrifuge
- Decanter centrifuge
- Bowl centrifuge
- Cyclone/hydrocyclone

Chromatography
- Gel filtration
- Ion exchange
- Reverse phase
- Affinity

Phase separation
- Centrifugal extractor
- Differential extractor
- Mixer-settler extractor
- Short-cut distillation
- Flash drum
- Scrubber/stripper
- Crystallizer
- Decanter tank

General Unit_ops
- Compressors
- Fan/blower
- Pumps
- Heat exchangers
- Heat sterilizer
- Flow mixers
- Flow/component splitter
- Storage tanks
- Blending tanks

Drying/evaporation
- Freeze dryer
- Tray dryer
- Fluid-bed dryer
- Rotary dryer
- Evaporator
- Rotary evaporator

Environmental units
- Equalizer
- Neutralizer
- Aeration basin
- Trickling filter
- Aerobic digester
- Anaerobic digester
- Clarifier
- Thickener
- Granular media filter
- Sludge belt filter
- Sludge dryer
- Incineration
- Flaring
- Baghouse filter
- Electrostatic precipitator
- Activated carbon

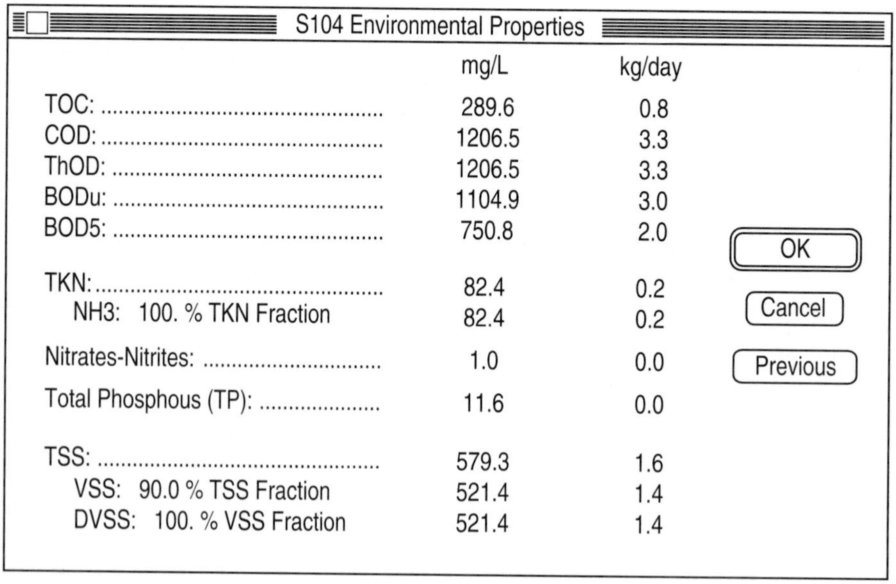

S104 Environmental Properties	mg/L	kg/day	
TOC:	289.6	0.8	
COD:	1206.5	3.3	
ThOD:	1206.5	3.3	
BODu:	1104.9	3.0	
BOD5:	750.8	2.0	OK
TKN:	82.4	0.2	
NH3: 100. % TKN Fraction	82.4	0.2	Cancel
Nitrates-Nitrites:	1.0	0.0	
Total Phosphous (TP):	11.6	0.0	Previous
TSS:	579.3	1.6	
VSS: 90.0 % TSS Fraction	521.4	1.4	
DVSS: 100. % VSS Fraction	521.4	1.4	

Figure 25-7. Output dialog window for stream environmental properties.

To estimate these environmental stream properties, BatchPro Designer retrieves information from a database on the contribution of various chemicals to these properties. The values for some of these factors (e.g., ThOD) come from theoretical calculations based on fundamental principles, while other factors (e.g., BOD/COD ratio) come from experimental data. Figure 25-8 shows such contribution factors for the top five (TOC, COD, ThOD, BODu, and BOD_5) environmental stream properties. The current database includes such contribution factors for a few hundred chemicals used in the pharmaceutical and specialty chemical industries. The user can edit the factors of the existing components and add values for new components.

The stream report of BatchPro Designer identifies all components that are solvents, solids, acids, bases, and biocides. This information is retrieved from its component database. Reports are generated with all substances that are included in the SARA Title III and CERCLA lists.

The widespread use of organic solvents in the pharmaceutical and specialty chemical industries results in substantial emissions of VOCs to the atmosphere. VOC emissions can be classified into four major categories: *storage and handling* (associated with storage and transfer of solvents), *process* (emissions from processing equipment), *fugitive* (unintentional releases from pumps, valves, flanges and other connectors, open-ended lines, and sampling connections), and *secondary* (emissions from waste treatment and other supporting facilities).

Process emissions represent the major component of a plant's total emissions. Sources for these emissions include process reactors, condensers, adsorption and absorption columns, as well as product separation and purification equip-

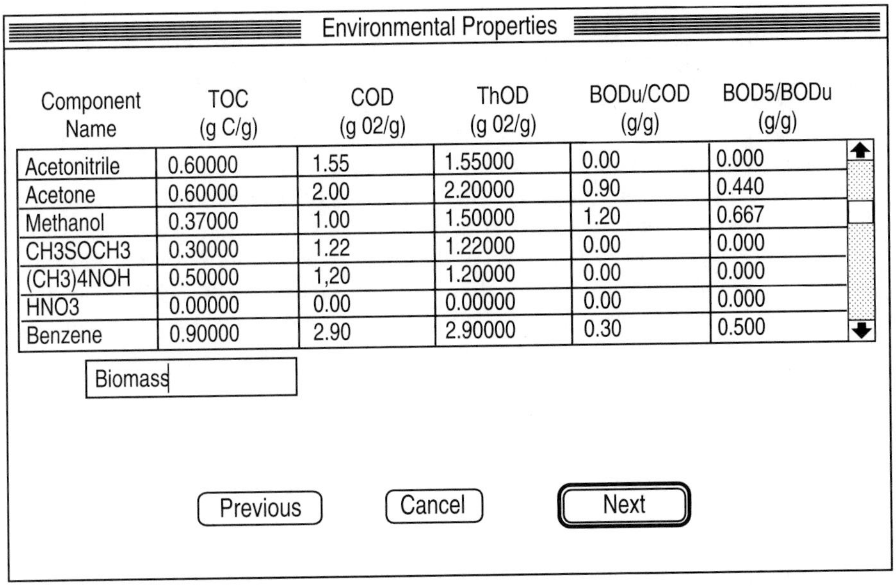

Environmental Properties

Component Name	TOC (g C/g)	COD (g 02/g)	ThOD (g 02/g)	BODu/COD (g/g)	BOD5/BODu (g/g)
Acetonitrile	0.60000	1.55	1.55000	0.00	0.000
Acetone	0.60000	2.00	2.20000	0.90	0.440
Methanol	0.37000	1.00	1.50000	1.20	0.667
CH3SOCH3	0.30000	1.22	1.22000	0.00	0.000
(CH3)4NOH	0.50000	1,20	1.20000	0.00	0.000
HNO3	0.00000	0.00	0.00000	0.00	0.000
Benzene	0.90000	2.90	2.90000	0.30	0.500

Biomass

Previous   Cancel   Next

Figure 25-8. Multipliers for environmental stream estimation.

ment such as filters, centrifuges, dryers, chromatography units, and distillation columns. Process emissions in the pharmaceutical industries primarily result from the following process operations: charging, evacuation, nitrogen or air sweep, heating, gas evolution, vacuum distillation, and drying. To estimate process emissions, BatchPro Designer is equipped with a module based on the EPA's guidelines and simplifying assumptions.[10] These guidelines assume ideal vapor and liquid phases and make use of the Antoine equation for all vapor-liquid equilibrium calculations. Also, other assumptions are made to produce conservative emission estimates in most cases.

In the charging operation, for instance, the assumptions include: (1) the volume of gas displaced from the vessel is equal to the volume of liquid charged into the vessel, and (2) the air displaced from the vessel is saturated with the VOC vapor at the exit temperature. Subject to these assumptions, the process emissions are calculated by use of the following algorithm:

- Calculate the rate of air displacement V_r
- Determine the mole fraction X_i of each VOC in the vessel during the charging

$$X_i = \frac{\text{moles of } i \text{ in liquid mixture}}{\text{total moles of liquid mixture}}$$

- Calculate the vapor pressure P_i of each VOC using Antoine's equation

$$\log 10 P_i = a - \frac{b}{c + T_i} \qquad P_i \text{ in mmHg, } T \text{ in } °C$$

- Calculate the amount of each VOC emitted S_e (kg/h)

$$S_e = \frac{P_i X_i V_r MW_i}{R\,T} \qquad MW_i \text{ molecular weight of VOC}$$

To predict the fate of a chemical entering a biological treatment unit (e.g., aeration basin), the current version of BatchPro Designer is equipped with a heuristic algorithm developed by Merck & Co.[11] Figure 25-9 illustrates some elements of the algorithm. Properties such as volatility (expressed by Henry's law constant), solubility in water, octanol to water ratio, BOD/COD ratio, etc., are used to estimate the fraction of a substance that biodegrades, the fraction that volatilizes, and the fraction that is sorbed on biomass. These initial fractions are then adjusted to reconcile the overall material balances around an aeration basin with those predicted by the kinetic model. Three kinetic models for the aeration basin are currently available: a Monod type, one that considers substrate inhibition (Haldane), and a third that assumes first-order kinetics (Grau).

Waste Minimization Advisory System.   The advisory system of BatchPro Designer consists of five expert system modules implemented with Nexpert Object (a commercial expert system shell from Neuron Data). The *solvent selection* module assists the user in making a preliminary solvent selection. It is interfaced to a database of a large number of solvent compounds. For each com-

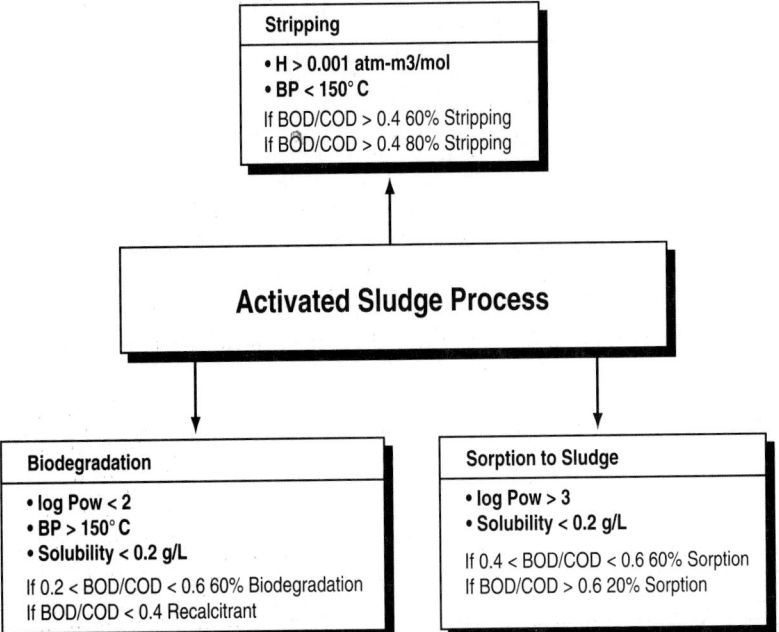

Figure 25-9. Fate prediction heuristic algorithm. (*Adapted from Venkataramani et al., 1992.*)

pound, the system provides physical and environmental properties and warns the user when environmentally unfriendly solvents are selected.

The *source reduction/in-process recycling* module analyzes the wastestreams of a manufacturing facility to identify any opportunities for source reduction and closed-loop recycling. Recommendations for in-process recycling are based on the flowrate and composition (e.g., number of solvents and amount of solvents relative to water and solids) of the various wastestreams.

The *material reclamation* module analyzes the wastestreams of a process plant and recommends alternatives for recovery of valuable components. Decisions are based on the relative amounts and physical or chemical properties of the various compounds. For instance, if a certain solvent is present in a wastestream at a high concentration, distillation may be recommended to recover and reuse it. If, on the other hand, a solvent is present at a low concentration, then steam stripping may be recommended to remove the solvent and reduce secondary VOC emissions.

The *waste treatment* module recommends alternatives for waste treatment for streams from which all valuable components have been recovered. Before a stream is sent to an activated sludge treatment plant, this module checks for the presence of any hazardous or toxic compounds, biocides, and odor-causing compounds. If there are any, it recommends alternatives for their removal.

The *waste disposal* module recommends alternatives for waste disposal for whatever cannot be recovered or treated. Disposal options include various types of incineration as well as landfill and land application of sludge. The results of the advisory system are primitive flowsheets that can be transferred to the simulation component for analysis and evaluation.

Economic Evaluation.   Based on the material and energy balances, as well as the equipment sizing, BatchPro Designer's evaluation component estimates the purchase cost of equipment, the fixed capital investment, and the annual operating cost, and carries out a detailed economic evaluation. The cost estimates fall within a maximum error of ±25 percent, rendering BatchPro Designer suitable for preliminary economic evaluation. Equipment cost is estimated as a function of equipment capacity, material of construction, and other design variables. An estimate of the fixed capital investment is based on the total purchase cost of equipment and is made through use of user-modifiable multipliers.

Two examples follow that illustrate the use of BatchPro Designer for evaluating waste minimization alternatives in an industrial environment.

25.4   Examples

25.4.1   Environmental Impact Analysis of Citric Acid Production

This example demonstrates the use of the process simulation component of BatchPro Designer in assessing the environmental impact of a citric acid production facility. Citric acid is a commodity organic acid produced via bioconversion. It is primarily used as a food preservative.

Conventional Technology—Recovery by Precipitation

Process Description.   Figure 25-10 shows the key process steps of citric acid production with the conventional technology.[12] Inexpensive molasses (a by-

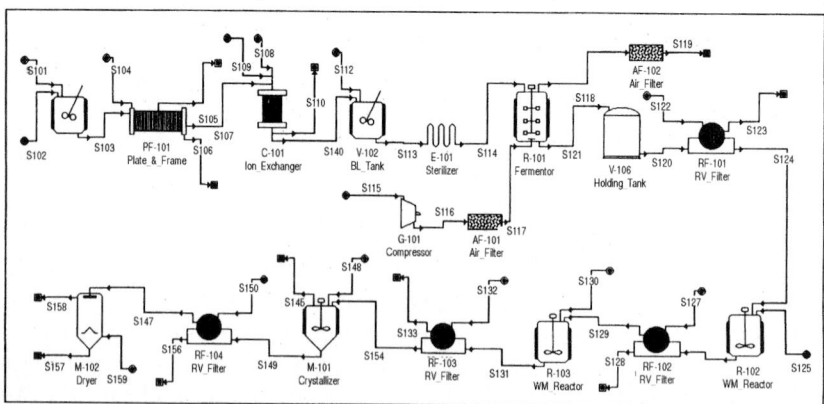

Figure 25-10. Citric acid production—conventional technology.

product of the sugar industry) is used as carbon and energy source. The presence of various impurities in molasses necessitates pretreatment of raw materials. A plate and frame filter (PF-101) is used to remove any undissolved particulate material. Metal ions, particularly iron, that affect product synthesis are removed by an ion exchange column (C-101). Nutrients—i.e., sources of ammonium (NH_3), potassium and phosphorous (KH_2PO_4), magnesium ($MgSO_4$), copper ($CuSO_4$), and zinc ($ZnSO_4$)—are mixed into the mixing tank (V-102). The resulting solution is sent to a continuous sterilizer (E-101). Citric acid is produced through metabolization of molasses in a stirred-tank fermentor (R-101). The seed and inoculum fermentors are not shown in the flowsheet. Air is supplied by a compressor (G-101) at an average rate of 0.5 VVM (volumes of air per volume of medium per minute). Jacket cooling water removes the heat produced by the exothermic process. The fermented broth is transferred into a holding tank (V-106) from which it is fed continuously to the product recovery and purification section of the plant.

Purification starts with the removal of biomass by a rotary vacuum filter (RF-101). Next, the product is precipitated in the form of calcium citrate by use of lime $Ca(OH_2)$ in an agitated reaction vessel (R-102). Calcium citrate is separated by a rotary vacuum filter (RF-102) and the filtrate free of citrate runs into waste. The calcium citrate cake is transferred into the next agitated reaction vessel (R-103) where it is acidified with dilute sulfuric acid (H_2SO_4) to form a precipitate of calcium sulfate (gypsum). A third rotary vacuum filter (RF-103) removes the precipitated gypsum and yields an impure citric acid solution. The resulting solution is concentrated by evaporation and crystallized in M-101. The crystals formed are separated by filtration (RF-104) and dried (M-102).

Environmental Impact. Figure 25-11 shows the overall material balance for the key chemical components generated by BatchPro Designer. The design basis represents a rather large plant producing 15,300 tons/yr (≈30 million lbs/yr) of citric acid that can satisfy 20 percent of the United States demand. The following process aspects render the traditional citric acid production technology environmentally problematic:

1. A large amount of gypsum is generated (17,600 tons/yr and larger than the amount of product). The gypsum is contaminated with biomass. It is questionable whether it has any commercial value and its disposal can cost as much as $100/ton depending on the location of the plant.

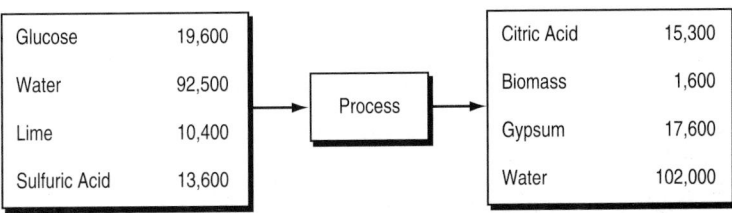

Figure 25-11. Overall material balance for key components (ton/yr).

2. A large amount of wastewater is generated (approximately seven times the amount of product).

3. Biomass waste also presents a disposal problem because it is contaminated with filter aid, which prevents its use as animal feed.

Process Modifications.   Retrofit process design will focus on the three waste issues we identified. The amount of wastewater, most of which is removed by the second rotary vacuum filter, can be substantially reduced via recycling. Recent studies have shown that wastewater recycling has no negative impact on yeast fermentation,[13] which may be an indication that it also can be implemented with *Aspergillus niger* for citric acid production. Investigation of this alternative requires experimentation to establish that wastewater recycling does not compromise product quality. Recycling will also reduce the amount of molasses and other fermentation nutrients since any unreacted raw materials will be recycled along with the wastewater.

The waste biomass can be converted into a by-product if an alternative removal technique is utilized that does not require the use of filter aid. Centrifugation or membrane filtration can be the basis of such alternatives.

The primary waste minimization problem, however, is with gypsum. Using the traditional technology, gypsum generation is unavoidable. An alternative product recovery technology that is based on product extraction was developed in the 1970s and is analyzed in the following section.

Alternative Technology—Recovery by Extraction

Process Description.   An alternative citric acid production flowsheet is depicted in Fig. 25-12.[12] This process makes use of high-dextrose equivalent

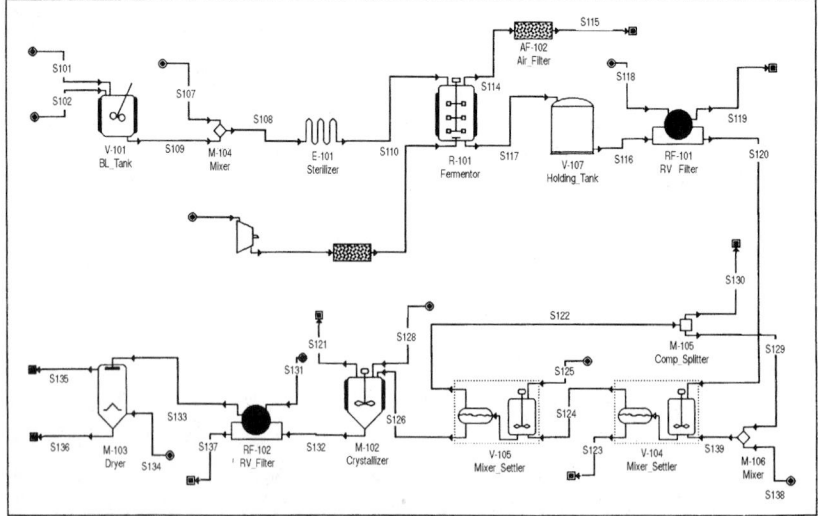

Figure 25-12. Citric acid production—alternative technology.

corn syrup instead of molasses as carbon and energy source. The high purity of the corn syrup eliminates the need for raw materials pretreatment. The fermentation and biomass recovery process steps are similar to those of the conventional technology. The main difference is in product isolation which is accomplished by an extraction and then a back-extraction step.

More specifically, the product is extracted from the filtered broth in a series of mixer-settler units operating countercurrently at 25°C (V-104). The extraction solvent is a mixture of 60% isooctanol and 40% triamine. This solvent is essentially immiscible in water and has high extraction efficiency with respect to citric acid. The organic phase, which contains the solvent and most of the citric acid, is sent to the back-extraction section.

Back-extraction is accomplished by countercurrently stripping the citric acid into a hot aqueous phase in another series of mixer-settler units (V-105) at 80°C. Back-extraction produces an aqueous solution of 35% weight citric acid. The organic phase is cooled and recycled to the extraction mixer-settlers. Final purification of the acid follows the same processing steps as those of the conventional technology. Concentration to 60% citric acid takes place in an evaporator-crystallizer (M-102). Finally, the citric acid slurry is separated (RF-102) and dried (M-103) to produce anhydrous citric acid crystals.

This technology eliminates the generation of gypsum and the use of lime and sulfuric acid as precipitation agents. However, it requires the use of an extraction solvent mixture that contributes to operating cost and to environmental impact. Even though the alternative technology has the potential of being environmentally friendlier, it is not at all obvious which technology is more economical. Based on an economic analysis with BatchPro Designer, it was found that at least a 99.5-percent recycle of extraction solvent is required for the alternative technology to become more attractive than its traditional counterpart. The performance of the extraction-based technology also was found to be sensitive to the partition coefficient of product during extraction and back-extraction.

This example has illustrated the use of process simulation in consideration of the environmental impact of alternative technologies. When applied to existing manufacturing facilities, this analysis can be used to provide targets for retrofit design. During process development, this analysis can be used for setting process performance milestones and for selecting among alternative technologies.

25.4.2 Environmental and Process Analysis of an Industrial Waste Treatment Plant

This example focuses on the process/environmental analysis and retrofit design of an industrial wastewater treatment plant.[14,15] The analysis was done with the simulation component of the environmental module of BatchPro Designer.

Figure 25-13 shows the flowsheet of the base case. A typical industrial wastewater treatment facility consists of two main sections: (1) wastewater treatment, and (2) sludge handling. The wastewater treatment section typically consists of primary, secondary, and tertiary treatment.

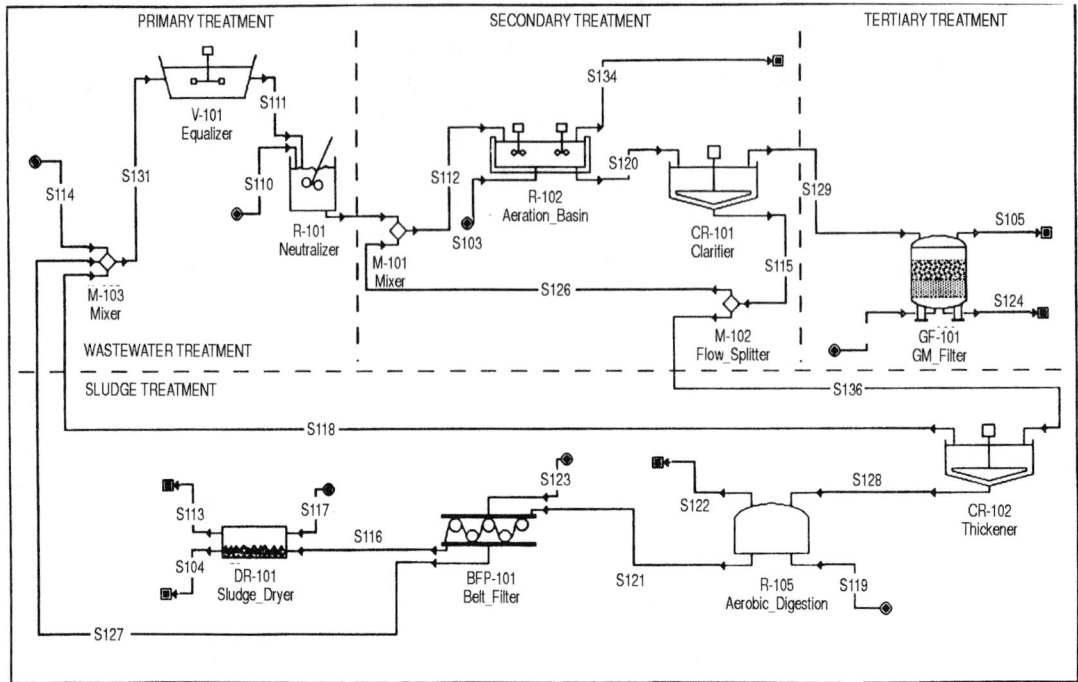

Figure 25-13. Typical industrial wastewater treatment facility.

Process Description. The primary treatment includes a tank with a large capacity (V-101) to equalize the flowrates and concentrations of the incoming streams and a neutralizer (R-101) to neutralize the pH. The secondary treatment includes an aeration basin (R-102) for the biological oxidation of organic materials and a clarifier (CR-101) for the removal of sludge and carried-over solids. A fraction of the sludge (30 percent) is recycled to maintain a biomass concentration in the aeration basin of 3500 mg/liter. The excess sludge (S136) is sent to the sludge treatment section. The overflow from the clarifier (S131) goes to the tertiary treatment section. A granular media filter (GF-101) is used to remove particulate material and colloidal solids escaping the clarifier. The sludge treatment starts with the thickening of the sludge taking place in a thickener (CR-102). The thickened sludge passes through an aerobic digester (R-105), a belt filter press (BFP-101), and a sludge dryer (DR-101).

Table 25-2 shows the composition and flowrate of the feedstream (S114) to the treatment facility. This is a rather small treatment facility (1 MGD), serving few chemical plants. The feedstream to this plant includes a number of regulated chemicals (e.g., benzene, methylene chloride, and heavy metals).

A key question concerning the environmental engineer is the fate of each of these chemicals when they enter such an integrated waste treatment facility. Water-soluble and easily biodegradable substances will be metabolized in the

Table 25-2. Composition of
Entering Wastestream (kg/h)

Water	180,000
Ammonia	40
Acetonitrile	85
Acetone	31
Methanol	112
Nitric acid	8
Benzene	24
Tetrahydrofurane	25
Ethanol	72
Ethyl acetate	18
Methylene chloride	27
Glucose	100
Toluene	29
Heavy Metals	1

aeration basin by the biomass and converted into CO_2 and H_2O. A fraction of recalcitrant and nonbiodegradable compounds with low water solubility will be sorbed on biomass and follow its path through the treatment plant. Finally, a fraction of volatile compounds will be stripped off in the aeration basin and end up in the atmosphere, contributing to VOC emissions.

The amounts of VOCs emitted by some waste treatment facilities may be substantial, resulting in violation of regulations governing air emissions mandated by the 1990 Clean Air Act Amendments.[16,17] Using the heuristic algorithm for the aeration basin depicted in Fig. 25-9, BatchPro Designer predicted the composition shown in Table 25-3 for the gas outlet of the aeration basin (S134). A

Table 25-3. Aeration Basin Gas
Outlet (kg/h)

Carbon dioxide	84.0
Ammonia	6.5
Oxygen	1700.0
Nitrogen	7200.0
Acetone	15.5
Methanol	11.2
Benzene	9.6
Ethanol	7.2
Toluene	5.8
VOCs total	49.3

total of approximately 50 kg/h or 1200 kg/day of VOCs are emitted from this facility. This plant is a major source of VOCs according to the 1990 Clean Air Act Amendments (the limit is 25 tons/yr) and application of Reasonably Available Control Technology (RACT) will be required to reduce the amount. Such technologies include wet scrubbing (packed or mist towers), adsorption (based on activated carbon or zeolites), thermal destruction, and biodegradation (biofilter or bioscrubber).

Process Modifications.   The use of an activated carbon adsorption unit to control VOC emissions was evaluated as a part of this example (Fig. 25-14). It was assumed that the basin is covered with a roof to collect the exiting gases. The carbon column was designed to remove at least 95 percent of all VOCs. The addition of the carbon column and the roof on the aeration basin increase the capital and operating cost of the facility. For the rather small treatment facility of the example (7500 kg BOD removal per day), the treatment cost increased from 1.22 (without VOC control) to $1.44/kg of BOD removed (with VOC control).* This represents an increase of 20 percent and is the price society has to

---

*To account for the cost of the roof, it was assumed that the capital cost of the aeration basin is 30 percent higher.

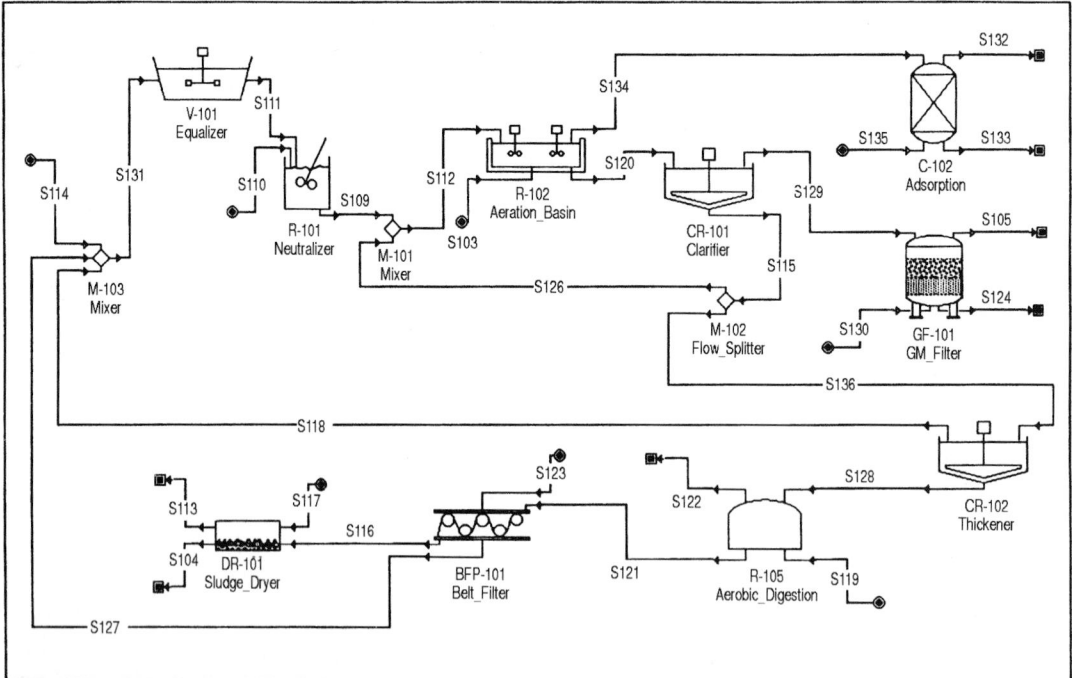

Figure 25-14. Modified flow diagram of waste treatment facility with VOC control.

pay for breathing cleaner air. If, through the use of pollution prevention efforts, these compounds were retained and reused within the manufacturing processes, there would be no need for the extra pollution control cost.

Table 25-4 shows the overall material balance based on the environmental properties that BatchPro Designer calculates. More than 90 percent of TOC and BOD are removed. Note also that approximately 80 percent of heavy metals associate with biomass, classifying it as hazardous solid waste. Such biomass cannot be land-filled and, if incinerated, will require stabilization treatment for the ash.

Tables 25-5a,b,c show excerpts of the economic evaluation reports generated by BatchPro Designer for this example. The sludge dryer is the most expensive piece of equipment, followed by the aeration basin. The total fixed capital investment for a plant of this capacity (7500 kg/day BOD_5 removal) is approximately $14 million. The total annual operating cost is $4.4 million and $3.1 million (or $1.62 and $1.13 per kg of BOD_5 removed), including and excluding depreciation, respectively.

This example illustrates how environmental simulators can be used to track the fate of VOCs and hazardous chemicals in integrated waste treatment facilities. The fate of VOCs is of great importance to industry and to Publicly Owned Treatment Works (POTWs) that are being confronted with increasingly strict regulations. This example also illustrates how environmental simulators can be used to evaluate process modifications and extensions necessitated by new regulations. This can be used by regulatory agencies and industry to estimate the economic burden of stricter environmental regulations. The results of such analyses also can act as incentives for pollution prevention strategies that reduce waste generation at the source and minimize the need for investment in pollution control.

Table 25-4. Overall Material Balance (kg/day)

	Influent	Effluent	
Environmental stream property	(S114)	Water (S124)	Sludge (S104)
Total organic carbon (TOC)	6,920	590	2,710
Chemical oxygen demand (COD)	21,050	1,560	10,100
Biochemical oxygen demand (BOD)			
Ultimate	12,870	275	9,300
5-day	7,660	172	6,320
Nitrogen			
Total Kjeldahl (TKN)	1,480	595	63
Free ammonia	870	535	63
Nitrites/nitrates	42	265	9
Total suspended solids (TSS)	744	7	5,573
Heavy metals	1	0.2	0.8

Table 25-5a. Major Equipment Specification and FOB Cost (1993 prices)

Quantity/standby		Description	Unit cost ($)	Cost ($)
1/0	R-102	Aeration basin volume = 4130.69 m^3 loading = 54,112.88 kg BOD5/day	396,000	396,000
2/0	C-102	Adsorption column diam. = 0.61 m, length = 1.23 m	21,000	42,000
1/0	CR-101	Clarifier vol. = 378.06 m^3, depth = 5.00 m	104,000	104,000
5/0	GF-101	Granular media filter diam. = 2.84 m, height = 0.60 m	73,000	365,000
1/0	CR-102	Thickener area = 1.25 m^2	68,000	68,000
1/0	R-105	Aerobic digestion vol. = 156.84 m^3 loading = 168,647.47 kg BOD5/day	258,000	258,000
1/0	BFP-101	Belt filter width = 0.47 m	99,000	99,000
1/0	DR-101	Sludge dryer concrete, duty = 305,481 kcal/h	715,000	715,000
1/0	V-101	Equalizer concrete, basin vol. = 1893.00 m^3	99,000	99,000
1/0	R-101	Neutralizer CS, vol. = 580.00 m^3	95,000	95,000
		Cost of unlisted equipment 20.0% of total		560,000
Total Equipment Purchase Cost				**$2,801,000**

Table 25-5b. Fixed Capital Estimate Summary (1993 prices)

A. Total Plant Direct Cost (TPDC) (physical cost)

1. Equipment purchase cost	(PC)	$2,801,000
2. Installation	(0.38 × PC)	1,077,000
3. Process piping	(0.35 × PC)	980,000
4. Instrumentation	(0.30 × PC)	840,000
5. Insulation	(0.01 × PC)	28,000
6. Electrical	(0.10 × PC)	280,000
7. Buildings	(0.05 × PC)	140,000
8. Yard improvement	(0.15 × PC)	420,000
9. Auxiliary facilities	(0.40 × PC)	1,120,000
		TPDC = $7,686,000

B. Total Plant Indirect Cost (TPIC)

10. Engineering	(0.25 × TPDC)	$1,921,000
11. Construction	(0.35 × TPDC)	2,690,000
		TPIC = $4,611,000

C. Total Plant Cost (TPDC+TPIC) — TPC = $12,297,000

12. Contractor's fee	(0.05 × TPC)	615,000
13. Contingency	(0.10 × TPC)	1,230,000
		Σ(12 + 13) = $1,845,000

D. Direct Fixed Capital (DFC) TPC + 12 + 13 — $14,142,000

Table 25-5c. Annual Operating Cost (1993 prices)

1. DFC-dependent Items	(DFC = $14,142,000)	
Depreciation		$ 1,343,000
Maintenance material	(0.03 × DFC)	424,000
Insurance	(0.01 × DFC)	141,000
Local taxes	(0.02 × DFC)	283,000
Factory expense	(0.05 × DFC)	707,000
		$ 2,898,000
2. Labor-dependent Items		
a. Operating labor	(7861 h × 12.0 $/h)	$ 94,000
b. Maintenance labor	(0.01 × DFC)	141,000
c. Fringe benefits	(0.40 × (a + b))	94,000
d. Supervision	(0.20 × (a + b))	47,000
e. Operating supplies	(0.10 × a)	9,000
f. Laboratory	(0.15 × a)	14,000
		$ 399,000
3. Administration and Overhead Expense (0.6 × (a + b + c))		$ 169,000
4. Raw Materials		$ 7,000
5. Other Consumables		$ 0
6. Utilities		$ 960,000
Total Annual Operating Cost		
Including depreciation		$ 4,433,000
Excluding depreciation		3,090,000

25.5 Other Environmental Simulation Tools

The stricter environmental regulations that have increased the need for better design of environmental processes have motivated the development of commercial process simulators by software companies. Two such simulators are described here.

ENPRO[18] is a process simulator used in the design and operation of water and wastewater treatment plants. It has the capabilities to model many of the physical, chemical, and biological processes used in a water treatment facility. In ENPRO, using a building block concept, the user has to define a treatment facility. The user first defines the water quality and flow. Then in a sequential manner, each treatment step in the plant is defined. The concept of unit operations is used to allow the user to select and define each step in the treatment process.

ENPRO may be used in both rating and design modes of calculation. In the rating mode, it is used to model existing operations and predict plant performance under changing conditions. In the design mode, plant layout and equipment alternatives are evaluated and operating conditions set to achieve specified effluent requirements.

Within ENPRO, two fundamental methods are available to characterize the streams. These are: (1) compositional method and (2) noncompositional method. In the noncompositional method, the properties of the feedstream are entered by the user. These properties describe the whole stream rather than its components. Currently, there is a limitation in ENPRO concerning which of the two stream types can be fed to a particular unit operation model. For flow-sheets involving both compositional and noncompositional stream analysis, a stream conversion utility is used to interchange the stream types.

Environmental Simulation Program (ESP) is a set of modeling and simulation tools for environmental and other applications.[19] ESP includes models for biological treatment, clarification, neutralization, precipitation, crystallization, generic separation, absorption, stripping, flash, distillation, incineration, and others.

ESP is superb in modeling reaction and equilibrium phenomena in aqueous systems involving molecular as well as ionic species. It is equipped with a component physical property databank that includes data for nearly all the elements of the periodic table. Based on the molecular species present in an aqueous system, the program automatically generates all ionic species that may be present and develops a set of equilibrium reactions that describe the system.

ESP can be used for rigorous calculations of precipitation of heavy metals from an aqueous solution. Also, all vapor-liquid equilibrium calculations involving an aqueous or an aqueous and an organic liquid phase are very rigorous.

Other universities and companies are also involved in the formulation and development of simulation and design tools for environmental applications. This list is in no way intended to be complete.

25.6   Conclusion

Even though the use of simulation and design tools is quite widespread in the chemical process industries for process design and optimization, limited work has been done so far on tools that systematically and explicitly consider the environmental impact of integrated processes. BatchPro Designer, ENPRO, and ESP constitute an attempt in that direction.

Simulation and design tools have been widely used in the chemical and petrochemical industries in the last four decades, resulting in significant cost savings from improved utilization of raw materials and energy. This technology can play a major role in cleaning up the environment and designing industrial processes that are environmentally friendlier.

The use of tools like BatchPro Designer in industry will enable engineers to consider the environmental issues during the early stages of process development when it is easy to make process modifications. Further, the estimation of a dollar value for the cost of waste recovery, treatment, or disposal early on will raise flags and warn scientists and managers that process modifications to reduce waste generation may make more sense than end-of-pipe treatment.

Last but not least, such software packages have the potential to become successful educational tools that will train students and engineers how to consider environmental constraints during the design of new processes and the revamping of existing ones.

Acknowledgment

Financial support for our work on BatchPro Designer from the Emissions Reduction Research Center at the New Jersey Institute of Technology, Eli Lilly, Envirogen, Pfizer, Schering Plough, and SmithKline Beecham is gratefully acknowledged.

References

1. W. W. Doerr, "Plan for the Future with Pollution Prevention," *Chem. Eng. Progress* **89**(1):26–29 (1993).

2. D. F. Rudd, G. J. Powers, and J. J. Siirola, *Process Synthesis*, Prentice Hall, Englewood Cliffs, N.J., 1973.

3. J. M. Douglas, *Conceptual Design of Chemical Processes*, McGraw-Hill, New York, 1988.

4. J. M. Douglas, "Process Synthesis for Waste Minimization," *Ind. Eng. Chem. Research* **31**(1):238–243 (1992).

5. A. P. Rossiter, H. D. Spriggs, and H. Klee, "Apply Process Integration to Waste Minimization," *Chem. Eng. Progress* **89**(1):30–37 (1993).

6. U.S. Environmental Protection Agency, *Total Cost Assessment: Accelerating Industrial Pollution Prevention through Innovative Project Financial Analysis*, EPA/741/R-92/002, May 1992.

7. D. T. Allen, N. Bakshani, and K. S. Rosselot, *Pollution Prevention: Homework & Design Problems for Engineering Curricula*, Center for Waste Reduction Technologies, New York, 1992.

8. J. Eisenhauer and S. McQueen, *Environmental Considerations in Process Design and Simulation*, CWRT-AIChE, ISBN 0-8169-0614-9, 1993.

9. D. P. Petrides, "Computer-Aided Design of Integrated Biochemical Processes; Development of BioDesigner," Ph.D. thesis, Dept. of Chemical Engineering, Massachusetts Institute of Technology, Cambridge, Mass., 1990.

10. U.S. Environmental Protection Agency, *OAQPS Guideline Series: Control of Volatile Organic Emissions from Manufacture of Synthesized Pharmaceutical Products*, EPA-450/2-78-029, 1978.

11. E. S. Venkataramani, M. J. House, and S. Bacher, *An Expert System Based Environmental Assessment System (Easy)*, Merck & Co., Inc., P.O. Box 2000, Rahway, NJ 07065-0900, 1990.

12. L. R. Roberts, in *Encyclopedia of Chemical Processing and Design*, Vol. 8, Marcel Dekker Inc., New York, 1979, pp. 324–333.

13. T.-Y. Hsiao and C. E. Glatz, "Water Reuse in a Yeast Fermentation," *Annual AIChE Meeting*, St. Louis, 1993.

14. Metcalf and Eddy, Inc., revised by George Tchobanoglous and Franklin L. Burton, *Wastewater Engineering—Treatment, Disposal, and Reuse*, 3d ed., McGraw-Hill, New York, 1991.

15. W. Wesley Eckenfelder, Jr., *Industrial Water Pollution Control*, 2d ed., McGraw-Hill, New York, 1989.

16. G. P. Van Durme, "Capping Air Emissions from Wastewater," *Pollution Engineering,* September 1993, pp. 66–71.

17. R. McInnes, "Emissions Monitoring System Helps Wastewater Treatment Plants Comply with Air, Water Regs," *HazMat World,* September 1993, pp. 52–53.

18. Simulation Sciences, Inc., *ENPRO User Manual,* Simulation Sciences, Inc., Aurora, Colo., 1992.

19. OLI Systems, Inc., *ESP User Manual,* OLI Systems, Inc., Morris Plains, N.J. 1993.

26

Pollution Prevention through Chemistry

Joseph J. Breen

Paul T. Anastas

Steven M. Hassur

Paul S. Tobin

Office of Pollution Prevention and Toxics
U.S. Environmental Protection Agency
Washington, D.C.

Prosperity without pollution, and the consideration of how we will achieve this economic and environmental imperative, has become the fundamental environmental theme of the 1990s.[1,2] The new strategy—pollution prevention—will serve as the keystone of federal, state, and local environmental policy. Support for the new approach is broad-based and includes environmentalists, industrialists, lawmakers, academicians, government regulators and policy makers, and the general public. The challenge is to switch from two decades of environmental policy based on pollution controls and government-mandated regulations to a future environmental policy based on pollution prevention, source reduction, recycling, and waste minimization. To make this change will require a new social compact among environmental, industrial, and regulatory interests. The roles and contributions of the chemical engineer, synthetic organic and inorganic chemist, and the process analytical chemist will be integral to the full articulation and implementation of the new vision.[3,4] This chapter focuses on the role of chemistry and the contributions of synthetic and process analytical chemists.[5,6] It also describes the implementation of pollution prevention concepts into the premanufacturing notice review process mandated by Section

5 of the Toxic Substances Control Act (TSCA) and discusses the implications of pollution prevention for chemical safety.

26.1   Alternative Synthetic Pathways for Pollution Prevention

In moving toward a society that is geared toward instituting pollution prevention principles on a national level, it is imperative that there be a consideration of how chemicals are made. This must include a consideration of the individual synthetic and overall synthetic sequence in the production of a chemical substance. Since the beginning of modern approaches to chemistry, chemical synthesis transformations, and their mechanisms, have been the primary focus of chemists. Historically, the most important criterion for selection of one synthetic step in the formation of a chemical over an alternative has been which one had the better "yield." Using this criterion, chemists have studied thousands of synthetic chemical reactions and reported about them in the scientific literature. This collection of synthetic transformations is what the synthetic chemist uses when designing a particular compound or considering how to add a particular functional group onto a compound. From a scientific standpoint, the use of yield as a criterion was kinetically and thermodynamically sound, and in most instances it was favorable from an economic standpoint as well.

In view of the new emphasis on pollution prevention by regulatory agencies and the chemical industry, and with the skyrocketing costs of waste disposal, waste treatment, and regulatory compliance, the practice of evaluating a particular synthetic method based on maximum yield alone is no longer acceptable. It is now recognized that there is a different way, an environmentally benign way, to view the manner in which individual synthetic transformations or overall syntheses are selected to make chemical compounds.[7]

Pursuing this new approach will provide the chemist with more "tools," or synthetic methodologies, to select from when instituting pollution prevention in a manufacturing process. This area of research meets both the chemical industry's and society's needs in developing the concept of pollution prevention and the academic community's need to focus on basic research. The next generation of synthetic chemists will certainly be focusing on how to build new chemical structures, but they will also be incorporating all of its impacts—scientific, economic, and environmental—into their selection of synthetic method for the chemical product.

There have been efforts to promote this type of research in both academia and industry-academic partnerships.[8-12] Notably, the Office of Pollution Prevention and Toxics (OPPT) of the U.S. Environmental Protection Agency (EPA) has funded six university chemistry departments to initiate basic chemistry research in pollution prevention. Emphasis has been placed on alternative synthetic reactions or synthetic pathways for a chemical or chemical class already in commerce and known to generate toxic wastes.

Conceptually, the moment a chemist puts pencil to paper to design how a chemical product will be made, an intrinsic decision is made about

- What hazardous wastes will be generated
- What toxic substances will need to be handled by the workers making the product
- What toxic contaminants might be in the product
- What regulatory compliance issues must be faced in making the product
- What liability concerns there are from the manufacture of the product
- What waste treatment costs will be incurred

By putting forethought into the synthesis of a chemical product such that scientific, environmental, and economic impacts of a particular process are considered, the synthetic chemist becomes a major force in achieving pollution prevention.

In its efforts to promote pollution prevention, OPPT recognizes the fundamental role of the chemist in designing ways to produce chemicals that minimize the use or generation of toxic substances. The Design for the Environment (DfE) Program of the OPPT has several initiatives to support the role of chemists in pollution prevention in industry, academia, and government.

Recent research developments using chemical design to achieve pollution prevention have resulted in totally new synthetic routes which eliminate standard feedstocks. The work of John Frost at Purdue University provides a synthesis of chemicals such as hydroquinone, benzoquinone, catechol, and adipic acid without using benzene as a feedstock.[13] Frost's use of genetically engineered organisms demonstrates how the natural pathways that the organisms use to produce amino acids can be adjusted to produce the important industrial and medicinal compounds mentioned above.

Well-known reactions, such as the Friedel-Crafts reaction, are being investigated to find more environmentally benign alternatives as replacements. George Kraus, at Iowa State University, is developing methods to carry out acylation reactions using aldehydes in place of acid chlorides.[14] The traditional Lewis acid catalysts are replaced by visible light.

Sunlight is also being used as a reagent to replace toxic materials by Gary Epling at the University of Connecticut at Storrs.[15] The cleavage of benzyl ethers, dithiane protecting groups, and oxathianes is accomplished through the use of a dye and irradiation with a sunlamp. This process replaces the traditional alkali-metal halides or catalytic hydrogenation for reductive cleavage, or mercuric chloride in acetic acid for oxidative deprotection.

New synthetic methodologies are being developed to avoid particular toxic feedstocks in the production of large-volume commodity chemicals. The work of Orville Chapman at the University of California at Los Angeles includes use of an intramolecular rearrangement to synthesize styrene without the use of benzene as a feedstock.[16] Because of regulatory controls on benzene as a carcinogen, there is considerable industrial interest in minimizing its use.

The elimination of organic solvents from manufacturing processes is a goal being pursued by environmentalists and the chemical industry as well. Through elimination of volatile organic compounds (VOCs) from industrial processes, environmental releases and associated costs of controlling those releases are eliminated. Research by James Tanko at Virginia Polytechnic Institute and State University has demonstrated how supercritical carbon dioxide can be used as a reaction medium for free-radical reactions.[17]

The OPPT actively promotes alternative synthetic design to achieve pollution prevention. OPPT's DfE Program has other initiatives to promote pollution prevention in the 1990s:

- Collaborating with the National Science Foundation to support environmentally benign chemical synthesis and processing through research grants, education, and outreach

- Developing technical guidance for the chemical industry on how to review processes to assess the pollution prevention potential of alternate synthetic pathways

- Reviewing potential hazards of new chemicals for the human health and environmental impact of the total manufacturing process, and investigating alternative chemistry to mitigate the hazards (see Sec. 26.3)

- Developing software to allow chemists to review the theoretical ways of making a chemical product and to evaluate which approach is most environmentally benign

These initiatives are striving to change the way chemists have traditionally judged whether or not a chemical reaction is "good" or "bad." Much basic research is needed to produce the chemical methods needed to achieve the most profound advances in pollution prevention.

26.2 Process Analytical Chemistry

Process analytical chemistry (PAC) provides the technological means to monitor, in real time, the redesigned industrial processes to verify productivity while documenting reduced or minimized levels of unwanted by-products and pollutants.[2] Global competition and the need for improved quality, improved worker safety, higher productivity, and lower operating costs all couple with the DfE[6] paradigm to drive the demand within the chemical industry for more effective analytical techniques. This is graphically illustrated by the 1993 Dow Chemical–Perkin Elmer Strategic Alliance to develop, promote, and market advanced analytical technology for environmental and industrial monitoring.[18]

For success to be achieved, instruments must be designed from basic principles to meet the particular process requirements. Continuous on-line and at-line analyzers must be rugged, maintain stability over time, and operate reliably and simply, with minimal time for operation and maintenance.

Real-world industrial applications are in place at 3M,[19] Du Pont,[20] Dow,[21] and Amoco.[22] Private-sector development of surface acoustic wave (SAW),[23] chemical array,[24] and electrochemical[25] sensors; membrane interfaces;[26,27] and fiber-optic immunosensors[28] is progressing on a fast track. A highly successful government-industry-university cooperative research center has been established as the National Science Foundation–University of Washington Center for Process Analytical Chemistry.[29]

Modern PAC from industry, as well as developments in sensor, membrane interface, and fiber-optic immunosensor technology are illustrated below.

26.2.1   Real-World Process Analytical Chemistry

Du Pont's on-line analyzer for chlorocarbons in wastewater prevents or minimizes the inadvertent release of contaminants.[20] The apparatus shown in Fig. 26-1 is a new type of sparging system that continuously, rapidly, and repeatedly extracts an equilibrium sample of volatiles from the process stream. The wastewater-sparging Fourier Transform Infrared (FTIR) analyzer allows the

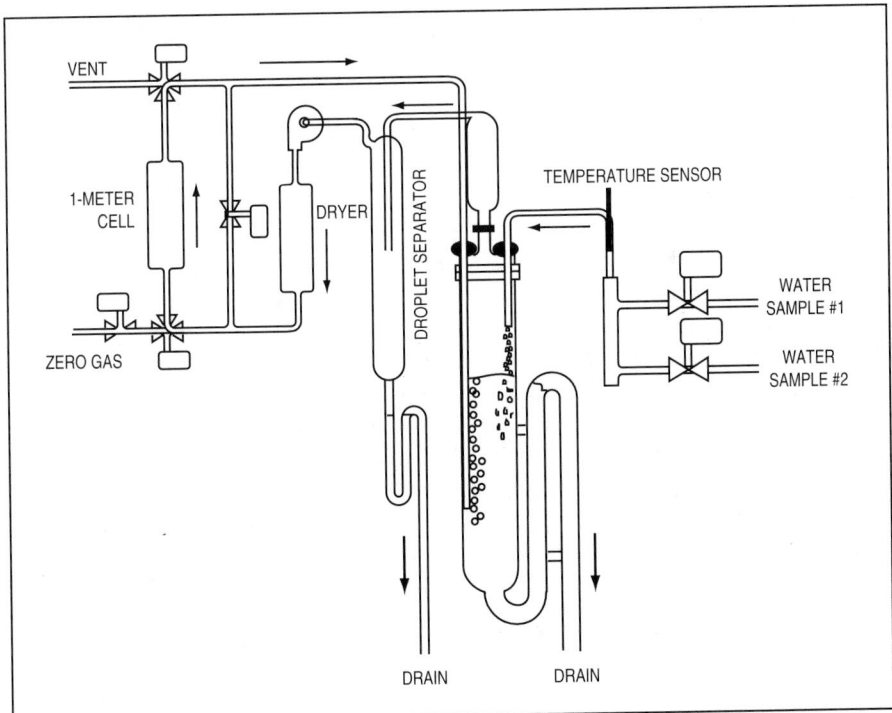

Figure 26-1. Du Pont on-line analyzer for chlorocarbons in process water. System schematic diagram. (*Reprinted with permission from J. J. Breen and M. J. Dellarco (eds.), Pollution Prevention in Industrial Processes: The Role of Process Analytical Chemistry, ACS Symposium Series 508, Washington, D.C., 1992, p. 55.*)

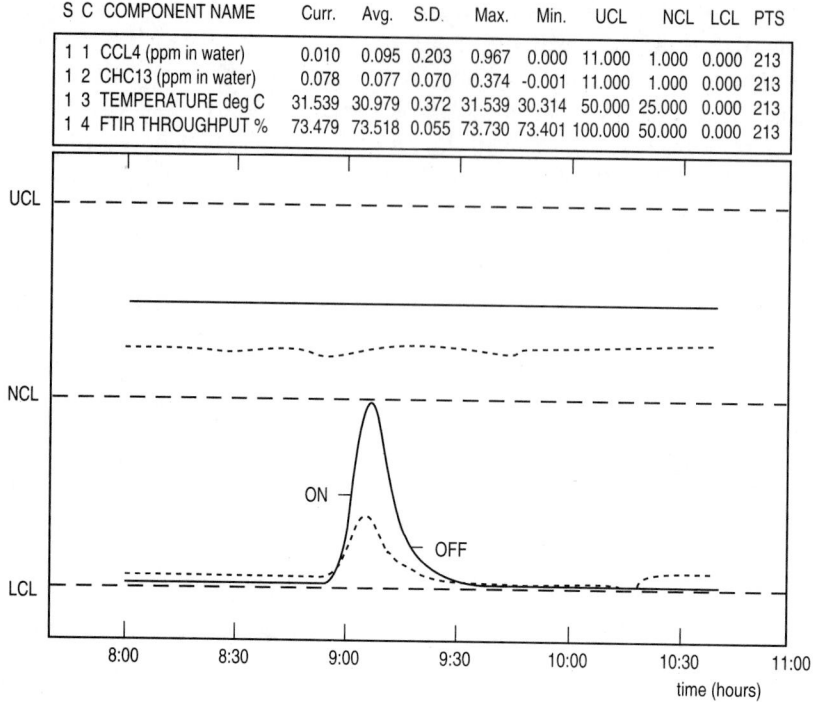

S	C	COMPONENT NAME	Curr.	Avg.	S.D.	Max.	Min.	UCL	NCL	LCL	PTS
1	1	CCL4 (ppm in water)	0.010	0.095	0.203	0.967	0.000	11.000	1.000	0.000	213
1	2	CHC13 (ppm in water)	0.078	0.077	0.070	0.374	-0.001	11.000	1.000	0.000	213
1	3	TEMPERATURE deg C	31.539	30.979	0.372	31.539	30.314	50.000	25.000	0.000	213
1	4	FTIR THROUGHPUT %	73.479	73.518	0.055	73.730	73.401	100.000	50.000	0.000	213

Figure 26-2. Du Pont on-line analyzer for chlorocarbons in process water. On-line data record. (*Reprinted with permission from J. J. Breen and M. J. Dellarco (eds.), Pollution Prevention in Industrial Processes: The Role of Process Analytical Chemistry, ACS Symposium Series 508, Washington, D.C., 1992, p. 60.*)

diversion of out-of-specification flows for subsequent reprocessing (see Fig. 26-2). This strategy had been impossible with previous purge-and-trap methods.

Dow's on-line total organic carbon (TOC) and flow injection analysis (FIA) provides for the detection of amines in wastestreams.[21] The determination of soluble organic content of wastewater streams is important for the operation of biotreatment facilities, for regulatory permit compliance, and for process upset detection. New commercially available TOC analyzers perform acceptably well, as measured by accuracy, precision, sensitivity, and stability parameters. TOC cycle times have necessitated the integration of FIA into the process analytical chemistry (see Fig. 26-3) to ensure early detection of process upsets or spills. The TOC-FIA system ensures compliance, provides data for process modeling, and allows plant personnel to operate plants in a smooth and efficient manner.

26.2.2 Sensors

Surface acoustic wave (SAW) microsensors are ideal for monitoring gases and vapors in industrial chemical processes.[23] SAW devices are very small, rugged,

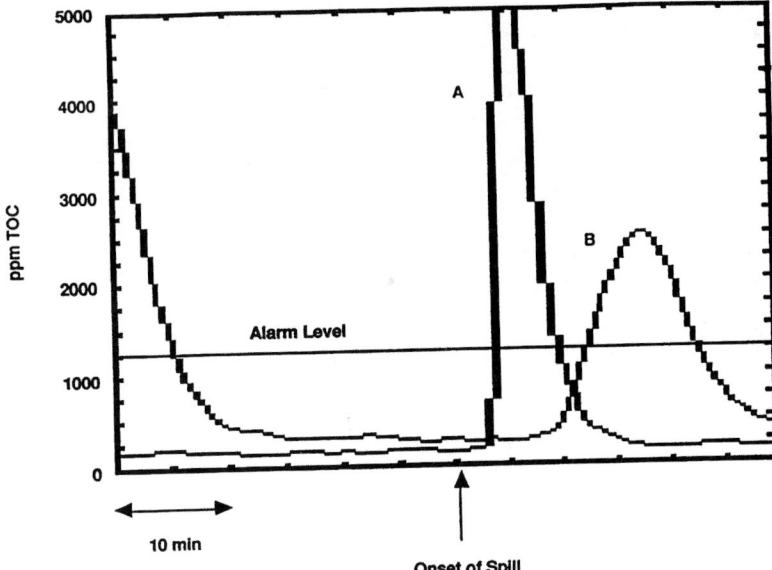

Figure 26-3. Dow Chemical on-line total organic carbon and flow injection analysis. Responses of FIA and TOC analyzers to amine spill. A = output from flow injection analyzer. B = output from total organic carbon analyzer. (*Reprinted with permission from J. J. Breen and M. J. Dellarco (eds.), Pollution Prevention in Industrial Processes: The Role of Process Analytical Chemistry, ACS Symposium Series 508, Washington, D.C., 1992, p. 75.*)

inexpensive, and reliable. They operate over 50°C temperature ranges and provide electronic signals compatible with PC networks for real-time process control. They are sensitive to chemical vapors present in concentrations of a few parts per million, respond in seconds to vapor concentration changes, and can be compound-specific or compound-class-specific. SAW sensors are most effectively used in arrays for industrial monitoring (see Fig. 26-4).

Electrochemical sensors are playing an increasing role in pollution prevention. They provide real-time, on-site information about the presence and concentration of chemicals in a given environment or process stream.[25] Though responsive to a limited number of chemicals with differing sensitivity, they can be arranged to improve the sensor's capabilities. The chemical parameter spectrometer (CPS-100), with four electrochemical sensors, provides a fingerprint of a sample gas, as shown in Fig. 26-5. Microsensors mounted with a housing to provide a gas exposure chamber are also being developed.

26.2.3 Membrane Interfaces

Pervaporation, the process whereby an analyte is transferred from solution phase on one side of a semipermeable membrane into the vapor phase on the other side,

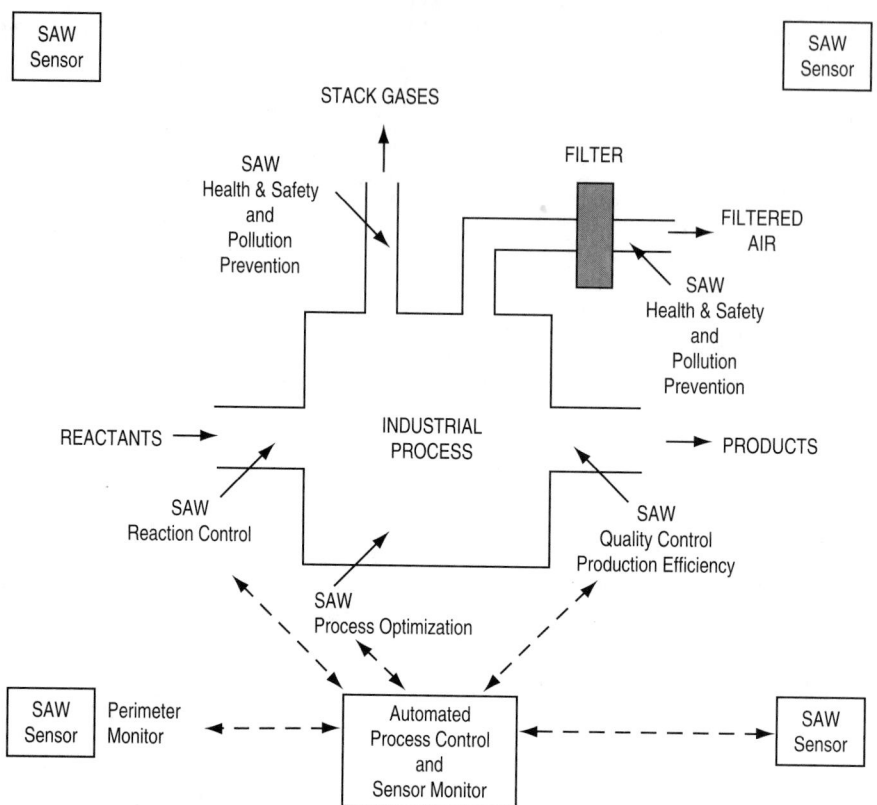

MONITORING VAPOR PHASE INDUSTRIAL PROCESSES
REACTIONS, PRODUCTS AND BY-PRODUCTS

Figure 26-4. Surface acoustic wave chemical microsensors and sensor arrays for pollution prevention. Vapor monitoring of industrial processes. (*Reprinted with permission from J. J. Breen and M. J. Dellarco (eds.),* Pollution Prevention in Industrial Processes: The Role of Process Analytical Chemistry, *ACS Symposium Series 508, Washington, D.C., 1992, p. 101.*)

occurs at widely varying rates for different compounds and therefore can be used as a selective means of sample transfer (see Fig. 26-6).[26]

Membrane introduction mass spectrometry (MIMS) shows great promise for the analysis of environmentally significant compounds from condensed-phase samples at levels of parts per billion without isolation or derivitization steps.[27] Together with FIA, MIMS provides the high-speed analysis for high sample throughput. The direct analysis capability of MIMS has been demonstrated for the detection of toxic and labile compounds such as acrolein, acrylonitrile, and inorganic and organic chloramines. FIA-MIMS is expected to play a significant role in environmental and industrial on-line reaction and process monitoring.

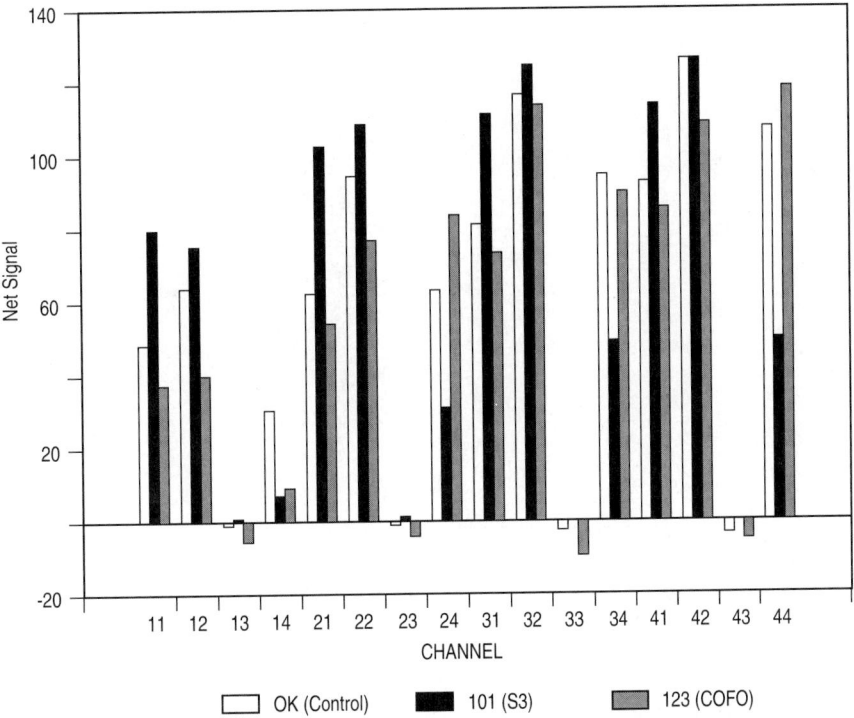

Figure 26-5. The electrochemical sensor and pollution prevention. Average responses for "good," "sour," and "insect" wheat samples. Graph is normalized response patterns. (*Reprinted with permission from J. J. Breen and M. J. Dellarco (eds.),* Pollution Prevention in Industrial Processes: The Role of Process Analytical Chemistry, *ACS Symposium Series 508, Washington, D.C., 1992, p. 112.*)

26.2.4 Fiber-Optic Immunosensors

The combination of fiber-optic technology and advanced optical sensors opens new horizons in medical, clinical, environmental, and industrial monitoring applications. The use of antibody-based fiber-optic fluoroimmunosensors to detect carcinogenic polycyclic aromatic compounds and aflatoxins has been reported.[28] (See Fig. 26-7.)

26.3 Pollution Prevention in PMN Review

Section 5 of TSCA, which is implemented by the EPA's Office of Pollution Prevention and Toxics (OPPT), requires manufacturers of new chemical substances to submit premanufacture notifications (PMNs) prior to their commercialization, allowing the EPA to determine if the chemical may present an

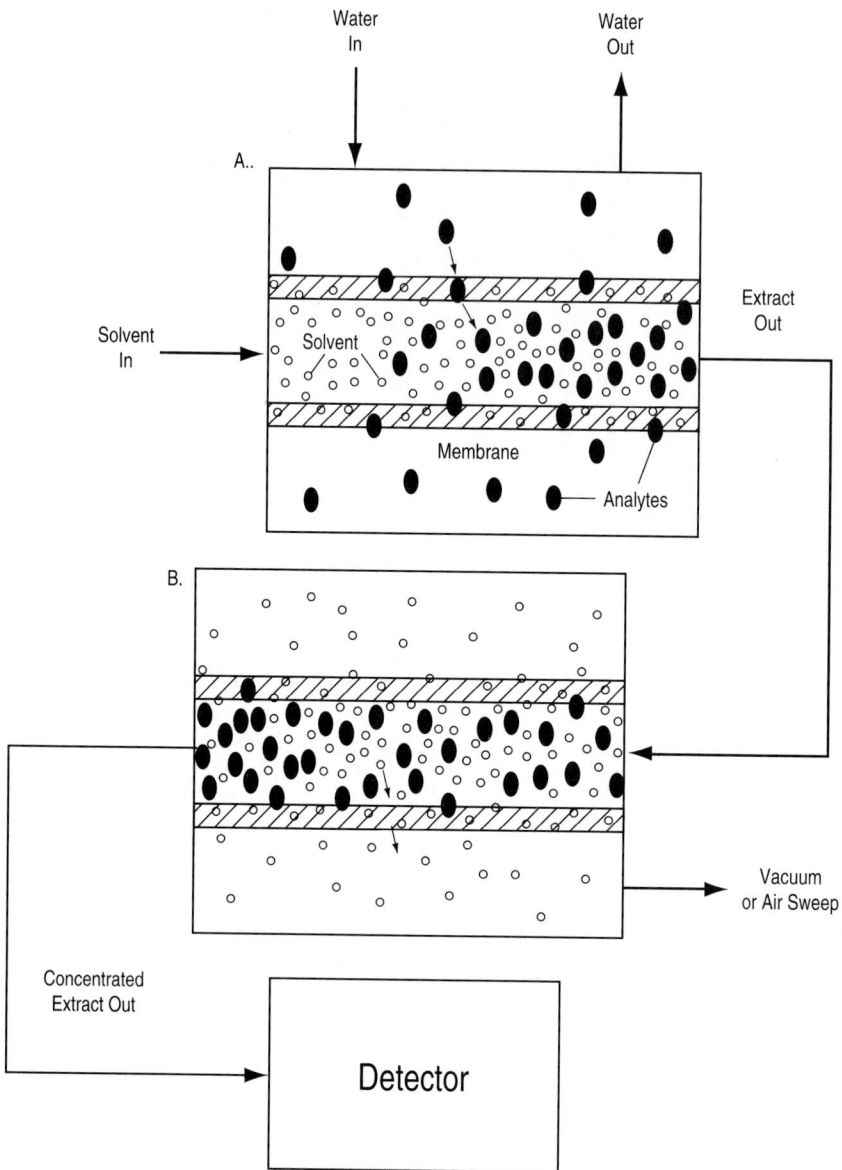

Figure 26-6. Hollow-fiber membrane for continuous extraction and concentration in pollution prevention analysis. *A.* Hollow-fiber extraction process. *B.* Hollow-fiber pervaporation process. (*Reprinted with permission from J. J. Breen and M. J. Dellarco (eds.),* Pollution Prevention in Industrial Processes: The Role of Process Analytical Chemistry, *ACS Symposium Series 508, Washington, D.C., 1992, p. 157.*)

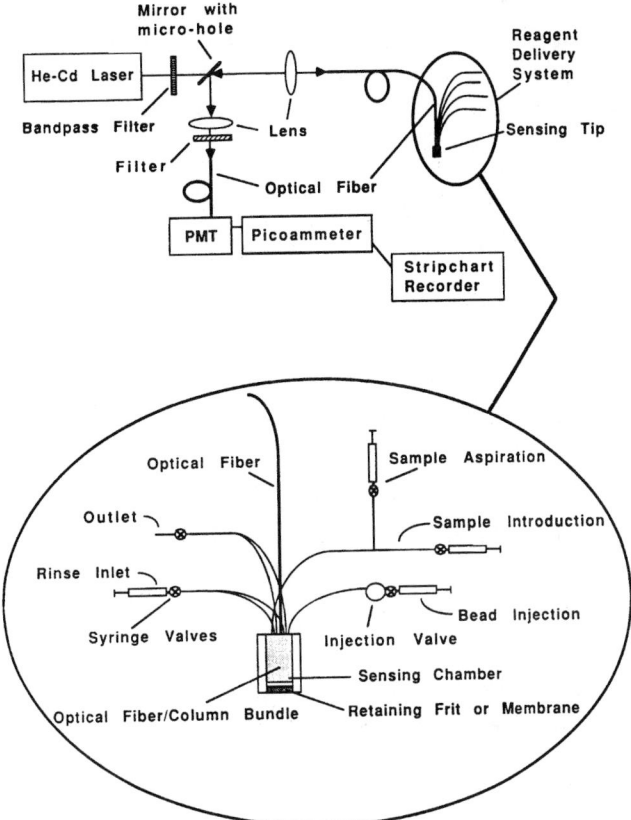

Figure 26-7. Fiber-optic immunosensors for pollution prevention. Schematic of signal acquisition and sensor reagent delivery system for direct fluoroimmunoassay using regenerable fiber-optic-based sensor. (*Reprinted with permission from J. J. Breen and M. J. Dellarco (eds.),* Pollution Prevention in Industrial Processes: The Role of Process Analytical Chemistry, *ACS Symposium Series 508, Washington, D.C., 1992, p. 278.*)

unreasonable risk to human health or the environment. In early 1993, OPPT implemented a new Synthetic Method Assessment for Reduction Techniques (SMART) Review program to assess pollution prevention opportunities during the initial review of PMN substances.

26.3.1   Goal of the New Program

Traditionally, the New Chemical Review Program focused almost exclusively on the PMN substance itself, and the impurities it contained, rather than on other chemicals used in the manufacturing process or generated as by-products, coproducts, or wastes. The Pollution Prevention Act of 1990, however, specifically mandates that EPA facilitate the adoption of source reduction techniques by businesses. To promote toxics use reduction, OPPT is also identifying process inefficiencies and synthetic design variables which may lead to the generation of excessive amounts of toxic materials *before* waste treatment, other control technology, or disposal. The goal is to identify areas of concern for which the PMN submitter may consider voluntary modifications of the proposed process to reduce toxic chemical exposure and/or releases to the environment, and, where possible, to indicate specific opportunities for the implementation of suggested pollution prevention methodologies.

26.3.2   Scope of SMART Review

The objective of toxics use reduction is to minimize the use of toxic chemicals and the generation of toxic wastes or to modify the types of chemical pollutants which are produced in a process. Rather than adjust and control process variables to marginally reduce the waste or emissions produced by a chemical process, a chemical assessment of the chosen synthetic pathway (with its required feedstocks, reagents, solvents, catalysts, etc.) and the associated process parameters (stoichiometry, temperature, pressure, etc.) is undertaken. This can result in the development of alternate approaches, including alternative synthetic pathways, to achieve the source reduction objective. The methodologies used to achieve source reduction are varied and encompass multiple disciplines including process chemistry, chemical technology, chemical engineering, and process management considerations and activities. Detailed reactor design information related to specific unit process equipment can be used to modify processes to control the generation of toxic wastes and increase the efficiency of reactions. However, such data are seldom available to manufacturers at the time of submitting a PMN and few report it. Since more detailed chemistry information is generally provided in the PMN, it is principally the analysis of process chemistry and chemical technology that are the focus of SMART Review.

26.3.3   Cases Subject to Review

SMART Review is a limited assessment applicable to PMN submissions with a production volume $\geq 25,000$ kg/yr (i.e., cases representing both a significant potential for pollution and the opportunity for improvement) and which include sufficient information to conduct the analysis (e.g., a process description or flow diagram). OPPT is not presently reviewing polymers (whose syntheses generally tend to be fairly efficient with few side reactions); reaction mixtures too complex to permit rapid assessment; and test market, low-volume, or polymer exemption notices (since these cases require very rapid reviews which prohibit this type of analysis and are predicated on low exposure or low risk).

26.3.4 Review Process

The review process is divided into two tiers to first screen potential cases and then focus on the development of more detailed information for those that may warrant contacting the submitter. To facilitate these assessments, OPPT encourages submitters of PMNs to use the "Optional P2 Information" page* to discuss the overall net reductions in toxicity or environmental releases and exposures in the proposed manufacturing and use of the new chemical substance.

Preliminary Assessment.   The proposed synthetic method is assessed to identify significant process inefficiencies and their potential causes (e.g., the use of a large excess of a feedstock beyond that required by the stoichiometry of the reaction, low conversions of the feedstocks, competing side reactions producing large amounts of waste, excessively large quantities of reagents and/or solvents used in a synthesis or in product recovery) and to determine the sources, identities, and amounts of generated wastes (based on the estimated third-year production volume provided in the PMN). The assessment includes not only the PMN substance and the separation and purification procedures employed but also chemicals "associated" with the manufacture of the new chemical. A relative toxicity ranking for each associated chemical or generated waste is made based on several previously established lists of hazardous chemicals[30–32] to formulate a level of concern.

Detailed Assessment.   A case meeting any of the following five criteria is given additional review:

1. The process streams contain an extremely toxic substance in any amount.

2. A potentially hazardous (or higher concern level) solvent is not recycled.

3. A potentially hazardous (or higher concern level) coproduct is not commercialized or used captively.

4. A Toxic Release Inventory (TRI) chemical present in a process stream after reaction equals or exceeds its reporting threshold under Section 313 of the Emergency Planning and Community Right-to-Know Act (EPCRA).

5. The estimated quantity (either as a weight percentage of the production volume or an absolute amount) of a waste or wastes exceeds certain "trigger levels" that have been established for specific levels of concern based on considerations of process yield and production volume.[33]

The focus of this more extensive review is on developing solutions for problems identified in the preliminary assessment which could eliminate or minimize the use or generation of toxic chemicals. To this end, more detailed calcu-

---

*Guidance regarding the "Optional P2 Information" page is incorporated in the PMN form instructions available through the TSCA Hotline: (202) 554-1404.

lations are made, including the calculation of a material balance for the synthetic pathway. OPPT plans to incorporate modified software for computer-assisted chemical synthetic design into future chemical assessments to facilitate these analyses.

Notifying the Submitter.   Depending upon the level of concern and the types of wastes involved, as well as how well the process lends itself to reasonable alternatives, the EPA will notify the submitter by letter of specific concerns related to the proposed manufacturing process. If plausible, OPPT chemists will offer suggestions regarding source reduction alternatives which the manufacturer can consider for voluntary action.

OPPT also endeavors to provide specific guidance on pollution prevention where appropriate and identifies those sources of information that will aid the submitter in assessing and modifying the proposed process (e.g., the Electronic Information Exchange System database of the Pollution Prevention Information Clearinghouse,[34] the EPA's *Facility Pollution Prevention Guide*,[35] relevant contacts for Waste Minimization or Technical Assistance Centers[36]).

26.3.5 Potential Impacts of the Program

The SMART Review program will provide selected manufacturers with specific information and ideas to aid them in exploring pollution prevention opportunities prior to the commercialization of new chemical substances. OPPT expects that this program will serve to encourage all manufacturers to more generally consider and implement source reduction in the design of new chemicals—hopefully, while chemists are still engaged in R&D at the bench. Wastes that are generated require treatment and/or disposal, whereas progress in toxics use reduction can result in financial rewards as well as decreased liability and enhanced process and product safety. OPPT intends to develop computerized databases of the information generated by this program (suitably protected for confidential business information). Identifying those chemical processes with the greatest potential for toxics use reduction may help guide future academic and industrial research in developing alternative synthetic pathways (see Sec. 26.1).

26.4 Chemicals and Safety

Industrial processes will change as pollution prevention programs are instituted. A major goal is to maintain safety for workers and surrounding communities as these changes occur. Industry's focus on safe operating conditions is often directed to particular areas because of the need to satisfy regulatory requirements. The overall challenge will be to balance the available options for change and not to focus on eliminating individual hazards at the risk of introducing others. To support this goal, a wide variety of regulations, voluntary programs, and guidance information now exists. The best solutions will be those that introduce pollution prevention and enhance safe working conditions at the same time.

26.4.1 Regulatory Requirements for Safety

As pollution prevention initiatives are adopted, certain regulatory requirements for safety of workers and surrounding communities must be met. Regulatory requirements often try to balance a minimum level of safety with the establishment of simple and cost-effective rules. As a result, a considerable degree of flexibility is usually presented to the regulated community to meet certain "standards" of safety as opposed to specific step-by-step requirements. Some of the requirements and brief descriptions are listed below.

Superfund Amendments and Reauthorization Act.[37] Following the tragedy at Bhopal, India, SARA Title III Sections 301–303 were enacted to set emergency planning requirements at the local level. State Emergency Response Commissions (SERCs) and Local Emergency Planning Committees (LEPCs) were established under the law to carry out the new requirements. Any complying facility that has present, above the threshold planning quantity (TPQ), one or more of 360 listed substances must notify the SERC within 60 days that they are subject to the emergency planning rule. A facility representative must be made available to the LEPC to provide information necessary for the LEPC to develop an emergency contingency plan. Thus, any changes in processes that result in the introduction of listed "extremely hazardous substances" will require new information reporting to the SERC and LEPC.

Clean Air Act Amendments of 1990.[38] The Clean Air Act Amendments Title III(112r), "Prevention of Accidental Releases," lists requirements for a "Risk Management Program (RMP)" for any complying facility that possesses a listed substance above the threshold. The RMP is to consist of three major components: (1) a hazard assessment to estimate potential effects of accidental releases on surrounding communities, (2) a program for preventing accidental releases through maintenance, monitoring, and employee training, and (3) a response program consisting of community notification of accidents and specific response actions. The regulation to list substances and thresholds that will require RMP development has been proposed, and the final list and specific RMP requirements and guidance are expected to be published in 1995. Compliance with these accident prevention requirements of the act is set for three years after the final rule.

OSHA Hazard Communication Standard and Process Safety.[39,40] The Hazard Communication Standard published by the Occupational Safety and

Health Administration (OSHA) poses requirements for the classification of chemicals into hazard categories and also for the preparation of related labeling and Material Safety Data Sheets (MSDS). In addition, OSHA's process safety management rule outlines risk management requirements to control accidents that pose threats in the workplace. Both regulations must be considered under "management of change" practices, which include an evaluation of worker safety and health, time necessary for change, authorization for change, informing of workers, and updating of process information. These requirements will be invaluable in situations where a pollution hazard is eliminated by the introduction of alternative chemicals and processes, some of which, for example, may pose toxicity, flammability, or reactivity hazards. Workers must be informed and be given time and training needed to adjust to the alternative processes.

26.4.2 Nonregulatory Guidance on Hazards

In addition to regulatory requirements and guidance, many nonregulatory guidance programs exist to help ensure safe working and community conditions. For example, the Chemical Manufacturers Association (CMA) has developed a *Community Awareness and Emergency Response Program Handbook*[41] to encourage community outreach programs and emergency planning. Other similar efforts may be available through state and local programs, as well as through trade associations. Many excellent texts have been published on the subjects of emergency planning, accident prevention, and emergency response.[42–47]

26.4.3 Hazards Analysis and Resources

Many resources now exist to assist risk managers in assessing the hazards of toxic, flammable, and reactive chemical substances. For each hazard, an assessment needs to be made of both the severity and probability of an accident. This process is known as *hazards analysis* and should be instituted into the pollution prevention paradigm. On a general level, one may prepare an overview of various hazards for "screening" or prioritization analyses and then reevaluate the highest-priority processes by applying site-specific details. The reevaluation also focuses more on specific scenarios. Because the screening and reevaluation processes rely on different degrees of site-specific information, many different hazard analysis models have been created, incorporating varying degrees of site-specific detail. Thus, toxic cloud release and dispersion, flammable substance heat generation, and vapor cloud and condensed-explosion models vary greatly, depending on the quantitation of the analysis desired. For example, one may assume any toxic substance could be totally released over a specified period of time (e.g., 10 minutes) as a screening assumption, but carry out a detailed valve leak analysis as a reevaluation assessment.

A sampling of the existing hazards analysis guides are listed below along with brief descriptions. Some of these and many other helpful materials (along with information on how to obtain them) are listed in Appendix L of *Technical Guidance for Hazards Analysis,* which is referenced below, and which is available by calling the EPA Emergency Planning and Community Right-to-Know Hotline at (800) 535-0202.

1. *Hazardous Materials Emergency Planning Guide (NRT-1).*[48] A guide to the development of an emergency plan and overview of hazards analysis.

2. *Technical Guidance for Hazards Analysis (Emergency Planning for Extremely Hazardous Substances).*[49] A technical supplement to NRT-1 with detailed hazards analysis screening and reevaluation methods.

3. *Computer-Aided Management of Emergency Operations (CAMEO).*[50] A computer-assisted hazards analysis system for fire departments and emergency planners—includes vulnerable zone calculation and mapping abilities. CAMEO is available by calling (800) 621-7619 or writing National Safety Council, CAMEO Order Department, 444 N. Michigan Avenue, Chicago, IL 60611.

4. Hazardous Materials Information Exchange (H-MIX).[51] A computerized bulletin board system, which provides information on hazardous materials emergency management, training, resources, technical assistance, and regulations. Commercial access number via modem settings 8 and 1 is (708) 972-3275; technical assistance number is (800) 752-6367; contains also the Environmental Protection Agency Planner's Library in User-Friendly Software (PLUS) System, a bibliographic search system.

5. The National Academy of Sciences publishes *Prudent Practices for Handling Hazardous Chemicals in Laboratories*[52] and *Prudent Practices for Disposal of Chemicals in Laboratories,*[53] both of which are excellent sources of lab safety information. Revised editions are now being developed and should be available within a couple of years.

6. *Guidelines for Developing Community Emergency Exposure Levels for Hazardous Substances,*[54] published by National Academy Press [available by calling (800) 535-0202].

7. *Handbook of Chemical Hazard Analysis Procedures.*[55] This publication presents information similar to *Technical Guidance for Hazards Analysis,* but it provides more detailed and site-specific hazards analysis methods as well as information on probability of accidents.

These and many other programs and references to safe handling of chemicals will help to make possible the safe conversion to pollution prevention alternatives. As with many situations, human error is always a major component of accidents, and so it will be up to the responsible individual to maintain a safe environment as well as a nonpolluting one. To achieve this goal will require familiarity with the regulatory and voluntary safety programs now in place and staying informed of the different hazards that chemicals pose through ongoing training and safety exercise programs.

References

1. Joel S. Hirschhorn and K. U. Oldenburg, *Prosperity without Pollution,* Van Nostrand Reinhold, New York, 1991.

2. Joseph J. Breen and Michael J. Dellarco, "Pollution Prevention: The New Environmental Ethic," in J. J. Breen and M. J. Dellarco (eds.), *Pollution Prevention in Industrial Processes: The Role of Process Analytical Chemistry,* American Chemical Society Symposium Series 508, Washington, D.C., 1992, chap. 1, pp. 2–12.

3. Ivan Amato, "The Slow Birth of Green Chemistry," *Science,* **259**:1538–1541, March 12, 1993.

4. Debbie Illman, "NSF and EPA Support Pollution Prevention Research," *Chemical and Engineering News,* March 29, 1993, pp. 5–6.

5. Lois Ember, "Strategies for Reducing Pollution at the Source Are Gaining Ground," *Chemical and Engineering News,* July 8, 1991, pp. 7–16; Lois Ember, "Designing for the Environment," *ChemTech,* June 1993, vol. 23, no. 6, p. 3.

6. Randall Wedin, "Environmental Chemistry: From Paradigms to Para-Xylene," *Today's Chemist at Work,* July–August 1993, pp. 16–19.

7. *Dallas Morning News,* "Learning Green," Sept. 13, 1993, p. 1-D.

8. Deborah L. Illman, "Green Technologies Present Challenge to Chemists," *Chemical and Engineering News,* Sept. 6, 1993, pp. 26–30.

9. Mike Woods, *Sacramento Bee,* "Green Chemistry Developing Methods Safe for Environment," Aug. 26, 1993.

10. "A Spoonful of Sugar Helps the Chemicals Go Down," *Business Week,* Aug. 30, 1993, p. 65.

11. D. Rotman, "Chemists Map Greener Synthesis Pathways," *Chemical Week,* Sept. 22, 1993, pp. 56–57.

12. Gary Stix, "Turning Green: Can Industrial Chemistry Trade Benzene for Sugar?", *Scientific American,* November 1993, vol. 269, no. 5, pp. 104–106.

13. J. W. Frost et al., "Designing Microbes to Be Synthetic Catalysts," American Chemical Society 206th National Meeting, August 22–27, 1993, Division of Environmental Chemistry Abstracts, vol. 33, no. 2, Chicago, Ill., p. 310.

14. G. A. Kraus et al., "A Photochemical Alternative to the Friedel-Crafts Reaction," American Chemical Society 206th National Meeting, August 22–27, 1993, *Division of Environmental Chemistry Abstracts,* vol. 33, no. 2, p. 330.

15. G. A. Epling and Q. Wang, "Preparative Reactions Using Visible Light—High Yields from Pseudoelectrochemical Transformations," American Chemical Society 206th National Meeting, August 22–27, 1993, *Division of Environmental Chemistry Abstracts,* vol. 33, no. 2, p. 328.

16. O. L. Chapman and E. Tsou, "The UCLA Styrene Process," American Chemical Society 206th National Meeting, August 22–27, 1993, *Division of Environmental Chemistry Abstracts,* vol. 33, no. 2, p. 308.

17. J. M. Tanko and J. F. Blackert, "Supercritical Carbon Dioxide as a Medium for Conducting Free-Radical Reactions," American Chemical Society 206th National Meeting, August 22–27, 1993, *Division of Environmental Chemistry Abstracts,* vol. 33, no. 2, p. 313.

18. "Dow-Perkin Elmer Strategic Alliance for Advanced Analytical Technology," *Advanced Instrumentation News,* 1993, p. 1.

19. Jess S. Eldridge, K. M. Hoffman, J. W. Stock, and Y. T. Shih, "Chemical and Biochemical Sensors for Pollution Prevention," in J. J. Breen and M. J. Dellarco (eds.), op. cit., chap. 5, pp. 40–47.

20. Sydney W. Fleming, B. B. Baker, Jr., and B. C. McIntosh, "On-Line Analyzer for Chlorocarbons in Wastewater," in J. J. Breen and M. J. Dellarco (eds.), op. cit., chap. 6, pp. 48–61.

21. W. W. Henslee, S. Vien, and P. D. Swaim, "Determination of Organic Compounds in Aqueous Waste Streams On-Line Total Organic Carbon and Flow Injection Analysis," in J. J. Breen and M. J. Dellarco (eds.), op. cit., chap. 7, pp. 62–78.

22. E. H. Baughman, "Sulfur Recovery to Reduce SO_2 Pollution," in J. J. Breen and M. J. Dellarco (eds.), op. cit., chap. 8, pp. 79–85.

23. H. Wohltjen, N. L. Jarvis, and J. R. Lint, "Surface Acoustic Wave Chemical Microsensors and Sensor Arrays for Industrial Pollution Control and Pollution Prevention," in J. J. Breen and M. J. Dellarco (eds.), op. cit., chap. 9, pp. 86–102.

24. W. Patrick Carey, "Multicomponent Vapor Monitoring Using Arrays of Chemical Sensors," in J. J. Breen and M. J. Dellarco (eds.), op. cit., chap. 21, pp. 258–269.

25. J. R. Stetter, W. R. Penrose, G. J. Maclay, W. D. Buttner, M. W. Findley, Z. Cao, L. J. Luskus, and J. D. Mulik, "The Electrochemical Sensor: One Solution for Pollution," in J. J. Breen and M. J. Dellarco (eds.), op. cit., chap. 10, pp. 103–117.

26. C. L. Fish, I. S. McEachren, and J. P. Hassett, "Hollow Fiber Membrane and Concentration of Organic Compounds from Water," in J. J. Breen and M. J. Dellarco (eds.), op. cit., chap. 13, pp. 155–168.

27. R. G. Cooks and T. Kotiaho, "Membrane Introduction Mass Spectrometry in Environmental Analysis, in J. J. Breen and M. J. Dellarco (eds.), op. cit., chap. 12, pp. 126–154.

28. T. Vo-Dinh, G. D. Griffin, J. P. Alarie, M. J. Sepaniak, and J. R. Bowyer, "Development of Fiber-Optic Immunosensors for Environmental Analysis," in J. J. Breen and M. J. Dellarco (eds.), op. cit., chap. 22, pp. 270–283.

29. McGrath, Elizabeth A., D. L. Illman, and R. Kowalski, "The Role of Process Analytical Chemistry in Pollution Prevention," in J. J. Breen and M. J. Dellarco (eds.), op. cit., chap. 4, pp. 33–39.

30. U.S. Environmental Protection Agency, "Title III List of Lists; Consolidated List of Chemicals Subject to Reporting under *The Emergency Planning and Community Right-to-Know Act (Title III of The Superfund Amendments and Reauthorization Act of 1986),*" EPA/560/4-92-011, January 1992. (with an errata sheet on February 1993).

31. U.S. Environmental Protection Agency, "Priority List of Substances Which May Require Regulation under the Safe Drinking Water Act," *Federal Register*, vol. 56, pp. 1470–1474, Jan. 14, 1991.

32. U.S. Environmental Protection Agency, "New Chemicals Program Chemical Categories," TSCA Assistance Information Service, TSCA Hotline: (202) 554-1404.

33. H. E. Podall and P. T. Anastas, "Process Chemistry Source Pollution Assessments in the U.S. Environmental Protection Agency's Premanufacturing Notice Review Process," *Journal of Cleaner Production*, Butterworth Heinemann, London, in press.

34. U.S. Environmental Protection Agency, "Pollution Prevention Information Clearinghouse," Reference and Referral service: (202) 260-1023.

35. U.S. Environmental Protection Agency, *Facility Pollution Prevention Guide*, EPA/600/R-92/088, Center for Environmental Research Information (CERI), Cincinnati, Oh., May 1992. CERI telephone: (513) 569-7562.

36. U.S. Environmental Protection Agency, *1993 Reference Guide to Pollution Prevention Resources*, EPA/742/B-93-001, February 1993.

37. U.S. Environmental Protection Agency, *The Superfund Amendments and Reauthorization Act of 1986*, P.L. 99-499, December 1986; U.S. EPA, "Emergency Planning and Notification," 40 *CFR* Part 355, 1987.

38. U.S. Environmental Protection Agency, *The Clean Air Act Amendments of 1990*, P.L. 101-549, Nov. 15, 1990; U.S. EPA, "List of Regulated Substances and Thresholds for Accidental Release Prevention; Requirements for Petitions; Final Rule," 59 *Federal Register*, 4478, Jan. 31, 1994.

39. U.S. Occupational Safety and Health Administration, "Hazard Communication Standard," 29 *CFR* Part 1910.1200, *Federal Register*, Aug. 6, 1979.

40. U.S. Occupational Safety and Health Administration, "Process Safety Management of Highly Hazardous Chemicals; Explosives and Blasting Agents," 29 *CFR* Part 1910.119, *Federal Register*, G356, Feb. 24, 1992.

41. Chemical Manufacturers Association, *Community Awareness and Emergency Response Program Handbook*, Washington, D.C., 1985; Chemical Manufacturers Association, *Resource Guide for the CAER Code of Management Practices*, Washington, D.C., 1992.

42. Vasilis M. Fthenakis, *Prevention and Control of Accidental Releases of Hazardous Gases*, Van Nostrand Reinhold, New York, 1993.

43. Robert B. Kelly, *Industrial Emergency Preparedness*, Van Nostrand Reinhold, New York, 1989.

44. Rudolf Meyer, *Explosives*, 3d ed. (revised and extended), sponsored by WASAG-CHEMIE, Essen, Germany, 1987.

45. Tadeo Yoshida, *Safety of Reactive Chemicals*, Elsevier, New York, 1987.

46. Louis A. Medard, *Accidental Explosions*, vol. 1: *Physical and Chemical Properties*; vol. 2: *Types of Explosive Substances*, Wiley, New York, 1989.

47. Daniel A. Crowl and Joseph F. Louvar, *Chemical Process Safety: Fundamentals with Applications*, Prentice Hall, Englewood Cliffs, N.J., 1990.

48. National Response Team, *Hazardous Materials Emergency Planning Guide (NRT-1)*, Washington, D.C., 1987.

49. U.S. Environmental Protection Agency, Federal Emergency Management Agency, and U.S. Department of Transportation, *Technical Guidance for Hazards Analysis (Emergency Planning for Extremely Hazardous Substances)*, Washington, D.C., December 1987.

50. National Oceanographic and Atmospheric Administration and U.S. Environmental Protection Agency, *Computer-Aided Management of Emergency Operations (CAMEO)*, NOS OR CA-65, Seattle, Wash., August 1992.

51. *Hazardous Materials Information Exchange (H-MIX) User's Guide*, sponsored by Federal Emergency Management Agency and U.S. Department of Transportation, RSTA/OHMIT/92-03, September 1992.

52. *Prudent Practices for Handling Hazardous Chemicals in Laboratories*, National Academy Press, Washington, D.C., 1981.

53. *Prudent Practices for Disposal of Chemicals from Laboratories*, National Academy Press, Washington, D.C., 1983.

54. *Guidelines for Developing Community Emergency Exposure Levels for Hazardous Substances*, National Academy Press, Washington, D.C., 1993.

55. Federal Emergency Management Agency, U.S. Department of Transportation, and U.S. Environmental Protection Agency, *Handbook of Chemical Hazard Analysis Procedures*, Washington, D.C., 1989.

27

Mixing as a Tool for Pollution Prevention in Reactive Systems

Fernando J. Muzzio

Rutgers University
Piscataway, New Jersey

Edward L. Paul

Merck Sharp and Dohme Research
Laboratories
Rahway, New Jersey

27.1   Introduction

The industrial importance of mixing can hardly be exaggerated. Chemical, petrochemical, and pharmaceutical processes usually require bringing reactants into close contact by imposing a mixing flow. For fast reactions or viscous fluids, mixing is often slow compared to the rate of reaction, resulting in several important effects: desired reactions are slowed down and stopped before reaching completion, undesired reactions are enhanced, and product selectivity is decreased. Large mixing effects have been recognized in many applications, including reactive turbulent flows,[1] combustion,[2,3] and polymerizations.[4] Mixing is also an essential component of many processes in the environment. Convective transport of pollutants both in the ocean and in the atmosphere can generate complex spatial structures, displaying regions of high and low concentration of the pollutants intermeshed in a highly nontrivial way.[5] Understanding the evolution of these concentration fields is essential for predicting aspects of practical importance, including exposure levels, the distance from the source required for a given dilution, optimal abatement and cleanup techniques, and so on.

The extent of mixing at a given position and time can be quantified in terms of the local scale of segregation,[6] which can be defined in physical terms as the typical distance *across* regions in which an individual component predominates, and the intensity of segregation, which is proportional to the standard deviation of the concentration of a given component across the system. Mixing of soluble fluids can be regarded as a combination of three coupled processes: convection, stretching, and diffusion. These processes are sketched in Fig. 27-1. Convection moves portions of material from one location to another, promoting global uniformity by redistributing in space initially segregated portions of the system. Stretching transforms the initially rounded blob of material into an elongated filament, reducing the scale of segregation. Diffusion, which is generated by the thermal energy of individual molecules, generates uniformity at the molecular scale and reduces the intensity of the segregation. Diffusion and stretching are intimately linked; as segregated portions of fluid are stretched by the flow, the rate of diffusional transport is augmented due to the reduction of the diffusional length scale and the increase of the contact area between components. The situation is considerably more complex in multiphase reactive operations, because processes such as precipitation, dissolution, drop breakup, and coalescence must be considered as well.

This complex combination of processes generates structures in which the local scale of segregation often displays strong spatial fluctuations. Such a situation is sketched in Fig. 27-2, which shows a partially mixed binary system in which the components A (black) and B (white) are intermingled in a complicated spatial pattern qualitatively similar to those observed for many miscible fluids. The upper box displays a region in which A is in local excess; the lower box shows a local excess of B. Often the mixing process is incomplete or inefficient, and unmixed regions persist even after long times. Such inhomogeneities in the spatial distribution of reactants produce local stoichiometric imbalances that are the main causes for selectivity losses and excessive waste production in systems with multiple reactions. Consider, for example, the parallel-competitive reactions A + B → R; A + C → S, where A is the limiting reactant, C is an impurity that enters in the same stream as B, R is the desired product, and S is waste. In order to minimize production of S, the system must be operated under condi-

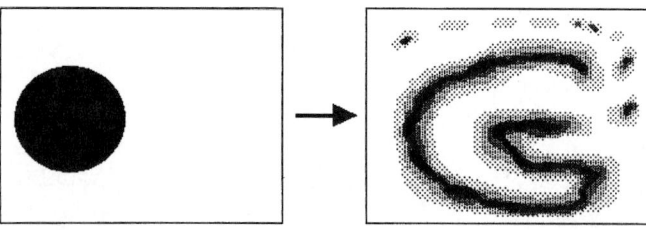

Figure 27-1. Scheme of a mixing process.

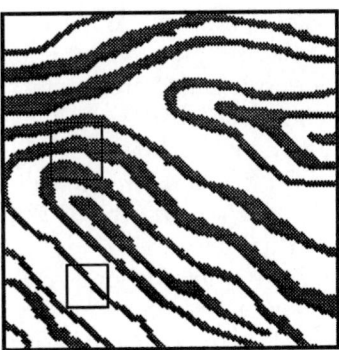

Figure 27-2. Scheme of a partially mixed system.

tions such that the first reaction is considerably faster than the second one. However, due to inhomogeneities, there are regions of the reactor where A is in local excess with respect to B (region inside upper box in Fig. 27-2). At some point in time B is locally depleted in those regions, and large amounts of C react subsequently, increasing the production of S and decreasing the yield of R. Clearly, production of S could be reduced if the mixing process were designed in such a way as to minimize local excesses of A, leading to selectivity enhancements and reduced waste. Similar situations are encountered in a wide variety of industrial systems.

Next, we present a brief review of mixing technology. Several industrial examples in which improved mixing resulted in waste prevention and performance improvements are discussed afterwards.

27.2 Mixing Equipment

Reactive mixing processes are conducted in an enormous variety of configurations. An exhaustive review of design options is beyond the scope of this section; only the most common configurations are discussed here. Processing options for reactor geometry depend on a number of parameters: (1) Does the reactive system involve multiple phases? (2) Are gases involved? (3) Are there suspended solids? (4) What amounts of materials need to be mixed? (5) What are the relative intrinsic rates of the desired/undesired reactions? (6) Is the process batch continuous or semicontinuous? (7) What is the viscosity (and other rheological properties) of the materials to be mixed? and so on.

The most common type of system is the stirred tank reactor, which is commercially available in an enormous variety of sizes, impeller geometries, and baffle configurations. The popularity of stirred-tanks is due to their extreme versatility. Stirred tanks can be implemented for both continuous and batch processes, they can handle reactant volumes from a few gallons to many tons per hour, and they have been successfully used to handle systems involving not only single liquid phases, but also liquid-liquid, gas-liquid, and solid-liquid dispersions. However, stirred tanks have some limitations. Large stirred tanks often have characteristic mixing times in the order of minutes, and are therefore poorly suited for systems with fast competing reactions, because in such systems the product distribution is strongly dependent on mixing and often requires that homogeneity be achieved very quickly. Stirred tanks are also disadvantageous for highly viscous fluids because in such cases the energy demands to achieve efficient stirring are very high. Difficulties can arise as well if the fluids are strongly non-Newtonian, because large unmixed regions commonly occur. Problems are also common for systems that contain large amounts of suspended solids, particularly if the particles are large and heavy, because deposits of solids at the bottom of the tank are often observed. Although many of these difficulties can be alleviated to some extent by one's choosing the proper tank design, it is often the case that performance can be improved considerably more by choosing an altogether different system.

Configurations involving jets are also very common. The two most popular arrangements are based on (1) a jet of one fluid injected within a stream of a second fluid, or (2) two jets, each made of a different component, impinging onto each other inside a small mixing chamber. These configurations are specially well suited for situations in which very fast mixing is required. Depending on the amount of materials being mixed, homogeneity can often be achieved in a fraction of a second. Jet mixers are typically used for relatively small amounts of liquids with moderate to low viscosity and are effective for generating fine liquid-liquid or gas-liquid dispersions. On the other hand, jets are often impractical if large amounts of materials need to be mixed, if the streams contain suspended solids, or if the viscosity is high.

For very viscous materials, the best alternatives often involve either static mixers or extruders. The simplest and most common types of static mixers are baffled pipes. Materials circulating through these systems are mixed through laminar flow mechanisms involving repetitive stretching, cutting, and folding of fluid elements. Static mixers have also been used successfully in many applications involving multiphase mixtures. Extruders typically consist of a shell and one or two rotating screws. The materials being mixed circulate along the shell through channels in the screws. Effective mixing can be achieved by proper design of the screw channel, which can change pitch, width, and depth along the flow direction. The performance of these systems is strongly sensitive on rheological properties of the materials that are mixed.

Reactive mixing processes are also often conducted inside packed beds. Typically, these systems are large diameter pipes filled with packing such as crushed glass or steel rings. The essential function of the packing is to reduce the scale of the system to be mixed. As the material flows through the bed, it is divided into many small streams that merge and split repeatedly along the flow direction. This mixing mechanism is effective in achieving intimate mixing of large volumes of low-viscosity materials such as gases, and it also provides a large amount of contact area for gas-liquid systems. Sometimes the packing plays an essential additional role in facilitating the reaction, such as in reactors filled with a heterogeneous catalyst.

It must be stressed here that reactive mixing is also conducted in many other configurations, which are not discussed here due to the limited scope of this chapter. Additional systems that find industrial applications include two- and three-phase fluidized beds, reactive distillation columns, centrifugal pumps, jet-tank configurations, and so on. Systems with large concentrations of solids are processed in rotating kilns, pans, V-blenders, and many other systems.

27.3 Industrial Examples

The following examples are cited from industrial processes whose wastestreams have been significantly reduced by the utilization of appropriate mixing technology to improve reaction selectivity and/or to modify a process operation to allow elimination of a wastestream or a solvent. The key point in each example is that rapid micromixing at the molecular level is essential to accomplish the process objectives and that failure to recognize the sensitivity of the mixing requirements would have resulted in decreased reaction selectivity with increased waste load or, in two cases, inability to use the most efficient reaction scheme.

Example 1: Line Mixer for a Rapid, Exothermic, Multiple Reaction

Conversion of a mercapto substituted thiazole to methylthiazole, a key intermediate for a major animal health product made in high volume, was being carried out by iron reduction under acidic conditions, a process that was introduced in the early 1960s. Although high yield and quality were being achieved, the environmental issues of hydrogen sulfide generation and disposal of the acidic iron residue prompted development of an alternate route. Alkaline oxidation to replace iron reduction appeared chemically feasible but required solving several reaction system design problems to achieve the necessary control of conditions for plant operation. The primary concerns were competitive and consecutive reactions, optimum mol ratio (both local and overall), control of a large exotherm, and very rapid reaction.

The chemistry of the alkaline oxidation process is as follows:

$$RS + H_2O_2 \rightarrow RSO + H_2O$$
$$RSO + H_2O_2 \rightarrow RSO_2 + H_2O$$
$$RSO_2 + H_2O_2 \rightarrow RSO_3 + H_2O \qquad (27.1)$$
$$RSO_2 + RSO \rightarrow RSO_3 + RS$$

where RSO_2 is the desired product. Common methods for maximizing yield by removal of the product (by distillation or precipitation) are precluded by restrictions in time, temperature, concentration, and pH dictated by the primary reaction. Since the reaction time is about 0.5 to 1.0 s, extremely rapid and effective mixing is essential to prevent local high concentrations of the oxidizing agent from causing the indicated overreactions. Such performance is achieved by selecting a line mixer for plant operation. All of the reactor feeds except the hydrogen peroxide are preblended. This stream is then mixed with hydrogen peroxide in the line mixer with a residence time of 1 s. The temperature is allowed to rise adiabatically to about 95°C and the mixture is fed to the top of a fractionation column for separation of volatiles and reduction in basicity to convert RSO_2 to the product RH which then distills azeotropically. In this example, implementation of properly designed mixing technology made it possible to modify the reaction scheme so that SH_2 and acidic iron residue were completely eliminated from the effluent streams.

Example 2: Selectivity in the Consecutive-Competitive Chlorination of Acetone

The chlorination of acetone is a consecutive-competitive reaction of the type:

$$A + B \xrightarrow{k_1} R \qquad R + B \xrightarrow{k_2} S \qquad (27.2)$$

where A is acetone, B is chlorine, R is monochloroacetone, the desired product, and S represents all overchlorinated undesired products (see Fig. 27-3). The key factor in scale-up of this reaction to plant operation is the gas-liquid mixing.[7] The critical nature of this operation became known after the

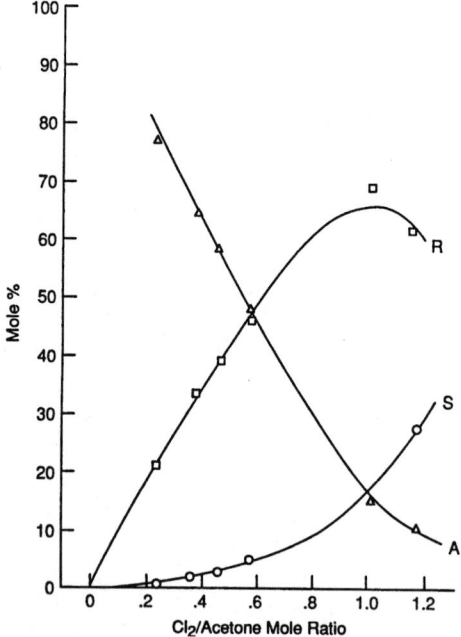

Figure 27-3. Product distribution in the chlorination of acetone.

plant started in operation and the yield of monochloroacetone was 8 to 10 percent lower than expected from laboratory data and from the yield predicted from the rate constant ratio k_1/k_2. The overchlorinated products were correspondingly greater than expected and their presence seriously reduced the yield of subsequent reactions in this process.

The yield loss was corrected by modification of the mixing device for the contacting of acetone and chlorine such that the reaction was complete (chlorine completely consumed) before chlorine vapor could react with monochloroacetone (inside the drops) and produce more overchlorinated by-products than would be expected at the mol ratio of the charged reagents. This reaction at the surface of a liquid drop is shown schematically in Fig. 27-4. The revised mixing system shown in Fig. 27-5 achieved complete consumption of chlorine in the well-mixed zone. The savings achieved were very significant in terms of reduced consumption of chlorine and acetone as a result of the yield increases in this and subsequent reactions. Waste treatment requirements are also significantly reduced by a factor of 3 by the decrease in overchlorinated by-products.

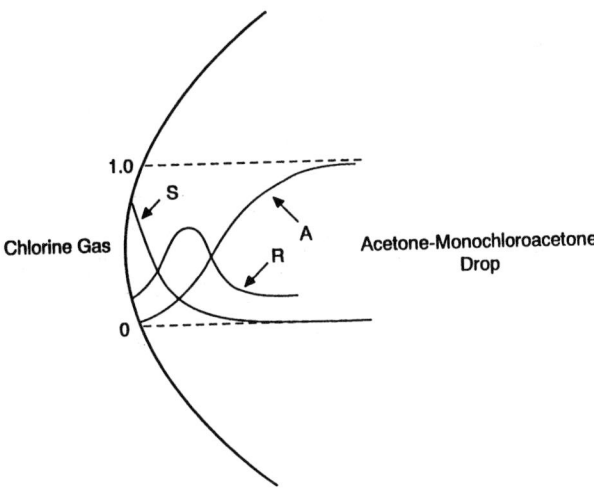

Figure 27-4. Schematic representation of the gas-liquid reaction system.

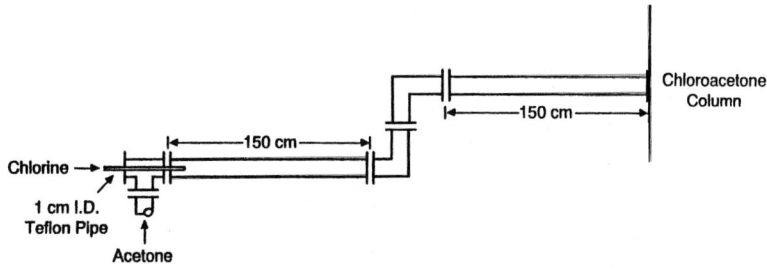

Figure 27-5. Sketch of the gas-liquid mixing device implemented for the chlorination of acetone.

Example 3:   Extractive Alkaline Hydrolysis

A major multistep process includes a hydrolysis step under acidic conditions.[8] The reactant to be hydrolyzed is prepared in methylene chloride in a previous step and is not isolated. The methylene chloride solution is mixed with aqueous HCl to achieve the desired hydrolysis. However, the low solubility of the intermediate in the aqueous phase results in a prohibitively low hydrolysis rate. To accelerate the rate, methanol is added to provide mutual solubility of the two phases. A substantial amount of methanol is needed, requiring extensive downstream recovery facilities.

An alternative reaction system to achieve the hydrolysis under basic conditions was considered. The chemistry is shown in Fig. 27-6. The key issue in this reaction scheme is that both the reactant and product of the hydrolysis can undergo rapid decomposition under the conditions required for the main reaction. Therefore, an extremely efficient mixing and separation device is required to make the alkaline hydrolysis scheme feasible on a manufacturing scale. This is also true for the laboratory scale system since a conventional reaction vessel for hydrolysis followed by separation leads to extensive decomposition even in the laboratory.

A line mixer including a static mixer followed directly by a Podbielniak extractor was selected both for laboratory trials and for manufacturing operations. The system is shown schematically in Fig. 27-7. The transfer of the reactant from the methylene chloride phase to the aqueous phase is achieved in the line mixer by the enolation with base followed by the desired hydrolysis in 2 to 3 seconds. This mixture is then fed to the extractor where phase separation is completed in 5 to 7 seconds. The resulting aqueous phase is then fed directly into an acidic medium, in which the product is stabilized and crystallized. Utilization of this reaction-separation scheme, which is only feasible with rapid and efficient mixing, made it possible to increase yield, increase purity, eliminate methanol from the process, increase productivity, and reduce capital cost.

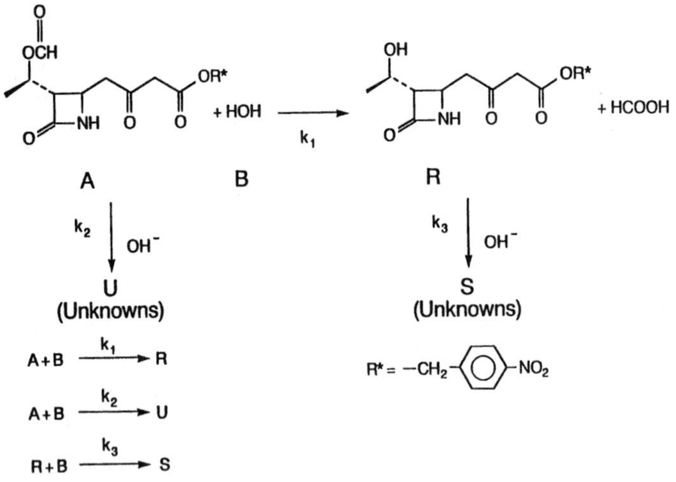

Figure 27-6. Hydrolysis reaction system under basic conditions.

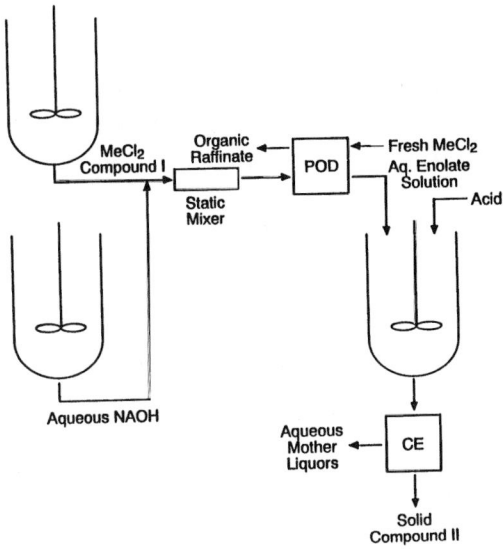

Figure 27-7. Scheme of the mixing/reaction setup for extractive alkaline hydrolysis.

27.4   Reactive Mixing Models

The examples discussed in Sec. 27.3 demonstrate that effective design of industrial mixing operations can lead to the prevention of large amounts of waste. Ideally, a detailed quantitative analysis of the reactive mixing process could be achieved through simultaneous numerical solution of the mass, energy, and momentum differential balances, leading to analytical determination of conditions that optimize performance and minimize by-product formation. Unfortunately, such an approach is impractical in most real systems. The main obstacle is that, for most applications, the fundamental equations that describe reactive mixing cannot be solved explicitly, and one must resort to models. A variety of approaches have been proposed that adopt various levels of approximation. A representative (although admittedly incomplete) list of previous modeling efforts, grouped according to the methods used to describe the reactive system, includes:

- Models that use combinations of plug flow and continuous stirred-tank reactors, bypass, backflow, and recirculation to fit a given (usually measured) residence time distribution of particles in the reactor.[9]

- Molecular simulations of particles undergoing a given type of motion and reaction.[10]

- Mechanistic models, usually expressed in terms of correlation functions, such as the coalescence-redispersion model,[11] the droplet diffusion model,[12] the interaction by exchange with the mean approach,[13] and other probabilistic approaches (several papers in Ref. 3; see also Ref. 14).

■ Stretched lamina models, which describe partially mixed reactive systems as collections of striations either with identical[15–17] or distributed[18] thicknesses. This description of partially mixed structures is based on experimental mixing observations; many laminar flows,[19] as well as some turbulent flows,[20] generate partially mixed layered structures with wide distributions of thicknesses.

Unfortunately, none of these models has been developed to the point that it could be used to investigate the general reactive mixing problem. Several of them are not well suited for the specific purpose of investigating mixing effects because they can incorporate only a limited amount of mixing information. In the first category, it is difficult to assign direct physical meaning to model parameters such as number of CSTRs and fraction of recycle. In the second category, it is often impossible to consider a large enough number of particles to simulate detailed spatial distributions of reactants. In the third category, much information is lost when the system is represented in statistical terms that are difficult to relate to measurable quantities. In particular, probabilistic models assume closures that are difficult to verify. The main drawback of models in the last category is that the striation thickness distribution generated by real mixing processes is unknown. Moreover, lamellar models assume that the layered structure is created first, and then diffusion, reaction, and further mixing take place. This might be a good assumption for some systems with the proper combination of characteristic times for mixing, diffusion, and reaction, but its general validity is clearly questionable.

27.5   Summary

Developing techniques for rational design and optimal control of reactive mixing processes and using them to minimize waste formation in industrial operations is one of the most important current goals in reaction engineering. Mixing can be an effective and relatively inexpensive route to pollution prevention. Potential gains are by no means limited to the handful of examples mentioned here; in fact, optimization of mixing procedures could lead to productivity gains and by-product reductions in a wide range of industrial reactive processes. However, as it should be clear from the foregoing discussion, much work is necessary before this goal can be even partially accomplished. The interactions between mixing and reaction dynamics are not well understood; conditions that lead to optimal performance for a given reactive system could be disastrous for a slightly different one. The state of the art only permits a case-by-case analysis, and a successful assault of the general case must wait until a better understanding emerges.

Nevertheless, recent years have witnessed dramatic progress in understanding mixing in laminar flows. A major factor in this evolution has been the dramatic increase in computer power that took place over the past decade, making it possible to quantitatively analyze flow problems that were considered intractable just a few years ago. These developments suggest that further advances are within reach. Realistic, accurate reactive mixing models are a likely outcome of ongoing research and will be a valuable tool for the purpose of preventing pollution in reactive systems. Another promising approach, not reviewed in this section for the sake of brevity, is direct numerical simulation of reactive flows, both laminar and turbulent. This approach still needs considerable development and is currently restricted to limiting cases (such as laminar flows in the Stokes regime and homogeneous isotropic turbulent flows), but the recent rate of progress has been extraordinary. An exciting challenge in this area is to generate computational frameworks that would enable researchers to describe the spatial distribution of degrees of mixing and extents of reaction. Achieving this goal would make it possible to perform experiments and simulations back-to-back, possibly leading to a practical understanding of reactive flow processes.

Acknowledgments

FJM gratefully recognizes financial support from the Exxon Education Foundation, from the Merck Foundation, and from ARPA/ONR URI Grant N00014-92-J-1888 during the preparation of this section.

References

1. E. L. Paul and R. E. Treybal, "Mixing and Product Distribution for a Liquid-Phase, Second-Order, Competitive-Consecutive Reaction," *A.I.Ch.E.J.* **17**:718 (1971).

2. D. B. Spalding, "Chemical Reactions in Turbulent Flows," in *Physico-Chemical Hydrodynamics*, Proc. 60th Levich Birthday Conf., 321, Advance, London, 1978.

3. P. A. Libby and F. A. Williams (eds.), "Turbulent Reacting Flows," *Topics Appl. Phys.* **44** (1980).

4. E. B. Nauman, "Mixing in Polymer Reactors," *J. Macromol. Chem.—Revs. Macromol. Chem.* **C10**:75 (1974).

5. J. S. Turner, "Convection and Mixing in the Oceans and the Earth," *Phys. Fluids* **A3**:1218 (1991).

6. P. V. Danckwerts, "The Definition and Measurement of Some Characteristics of Mixtures," *Appl. Sci. Res.* **3**:279 (1952).

7. E. L. Paul, W. A. Sklarc, and J. Aiena, "The Effect of Gas-Liquid Mixing on the Selectivity and Product Distribution in a Continuous Industrial Reactor," *A.I.Ch.E. Natl. Meeting*, New Orleans, November 1981.

8. M. L. King, A. L. Forman, C. Orella, and S. H. Pines, "Extractive Hydrolysis for Pharmaceuticals," *Chem. Eng. Progress* 36 (May 1985).

9. J. C. Mecklenburgh and S. Hartland, *The Theory of Backmixing*, Wiley, London, 1975.

10. A comprehensive review is provided by P. Meakin, "Fractal Aggregates and their Fractal Measures," in C. Domb and J. L. Lebowitz (eds.), *Phase Transitions and Critical Phenomena,* vol. 12, Academic, New York, 1988.

11. L. L. Tavlarides, "Mixing Effects on Reactor Modelling and Scaleup. Part IV: Modelling and Scaleup of Dispersed Phase Liquid-Liquid Reactors," *Chem. Eng. Commun.* **8**:133 (1981).

12. E. B. Nauman, "The Droplet Diffusion Model for Micromixing," *Chem. Eng. Sci.* **30**:1135 (1975).

13. J. Villermaux, "Mixing in Chemical Reactors," *ACS Symp. Ser.* **226**:135 (1983).

14. S. B. Pope, "PDF Methods for Turbulent Reactive Flows," *Prog. Eng. Comb. Sci.* **11**:119 (1985).

15. W. E. Ranz, "Fluid Mechanical Mixing: Lamellar Description," in J. J. Ulbrecht and G. K. Patterson (eds.), *Mixing of Liquids by Mechanical Agitation,* Gordon Breach, New York, 1977.

16. J. R. Bourne, and S. Rohani, "Micro-mixing and the Selective Iodination of l-Tyrosine," *Chem. Eng. Res. Dev.* **61**:297 (1983).

17. R. Chella and J. M. Ottino, "Conversion and Selectivity Modification Due to Mixing in Unpremixed Reactors," *Chem. Eng. Sci.* **39**:551 (1984).

18. F. J. Muzzio and J. M. Ottino, "Diffusion and Reaction in a Lamellar System: Scaling with Finite Rates of Reaction," *Phys. Rev. A* **42**:5873 (1990).

19. P. Kolodziej, C. W. Macosko, and W. E. Ranz, "The Influence of Impingement Mixing on Striation Thickness Distribution and Properties in Fast Polyurethane Polymerization," *Polym. Eng. Sci.* **22**:388 (1982).

20. W. Dahm, K. B. Southerland, and K. A. Buch, "Direct, High Resolution Measurements of the Fine Scale Structure of $Sc \gg 1$ Molecular Mixing in Turbulent Flows," *Phys. Fluids A* **3**:1115 (1991).

28

Process Equipment for Cleaning and Degreasing

Arun R. Gavaskar

*Battelle Memorial Institute
Columbus, Ohio*

28.1 Introduction

Cleaning and degreasing processes are applied in industries to remove dirt, oil, and grease (together referred to as *soil*) from manufactured parts. In the metal-finishing industry, cleaning usually follows machining and precedes other surface-finishing steps such as rust inhibition or electroplating. During machining, a variety of oil emulsions or synthetic fluids are sprayed on the workpiece (parts being fabricated) for lubrication and cooling. These fluids must be removed before final finishing. The slightest amount of oil or other residue can render the final surface finish ineffective. Solvents have traditionally been used for cleaning and are widely used during processing in the metal-finishing, dry-cleaning, and electronics industries, as well as in a variety of other industries during maintenance. Solvent cleaning leads to air emissions and occupational exposure during use and environmental hazards during disposal of the spent solvent. Pollution prevention efforts focus on reducing or eliminating use of these solvents.

28.2 Traditional Processes and Add-On Improvements

In the metal-finishing industry, where most parts cleaning is done, common cleaning processes are vapor degreasing, alkaline tumbling, or hand-aqueous washing. Vapor degreasing involves hazardous chlorinated solvents, such as trichloroethylene (TCE) or perchloroethylene (PCE), that are prime targets for pollution prevention. Vapor degreasing is common because it is easy to use, fast, and cleans well. Unlike other cleaning processes involving water, vapor degreasing does not require downstream drying because the solvent vaporizes from the workpiece over time. This eventual vaporization results in significant air emissions and solvent losses.

Conventional *open-top vapor cleaners* (OTVCs) use an open tank where solvent vapor is maintained. Degreasing solvent vapors are heavier than air and occupy the lower volume of the tank. The workload, usually a perforated basket containing the soiled parts, is retained in this vapor layer for a few minutes and rotated to expose all part surfaces to the vapor. As vapor condenses on the parts, the soil is dissolved or mechanically carried away by the condensate, which drips into the tank bottom. When the parts reach the temperature of the vapor, no more condensate is generated and the workpiece is removed. The condensate is redistilled in the same tank or in a separate still to recover the solvent. A residue containing oil and grease with some solvent is evacuated periodically from the bottom of the tank or still.

The height of the tank walls above the vapor layer is called the *freeboard height* and is generally maintained at a *freeboard ratio* (FBR) of 0.75 times the width of the tank opening. To further prevent vapors from escaping, the OTVC has water-cooled coils on its freeboard. Despite these devices, more than 90 percent of the solvent is lost due to air emissions. These emissions result from workload entry and removal (the largest source), idling, and shutdown. As the workload is put in and removed, the air-vapor interface in the OTVC is disturbed and the turbulence results in convective losses to the atmosphere. Also, when the workload is removed, some solvent condensate is dragged out on the cleaned parts and gradually evaporates, causing more air emissions. During idling and shutdown, some solvent escapes through diffusion from the vapor layer to the ambient. Air emissions are a concern for metal finishers because many of the solvents used in vapor degreasing have been targeted for early reduction by the U.S. Environmental Protection Agency (EPA) in the 33/50 Program.

Certain equipment-related and operational factors can reduce emissions from traditional OTVCs.[1] Placing a cover on the OTVC opening during idling and shutdown reduces emissions significantly. Simply increasing the freeboard ratio from 0.5 to 0.75 or 1.0 can cut vapor loss by as much as half. Many metal finishers have converted the water-cooled coils to refrigerated coils to better condense out the escaping vapors. Lowering and raising the workload gently (with a hoist) reduces convective losses by minimizing turbulence. Retaining the workload over the tank opening for a short time allows solvent condensate to drip back into the tank instead of being dragged out.

Aqueous processing using various detergents is an alternative to vapor degreasing. The conventional aqueous processes used in metal finishing are alkaline tumbling and hand-aqueous washing. In alkaline tumbling, the soiled parts are placed in an open, tilted vessel and an aqueous cleaning solution is introduced. As the vessel rotates, the parts tumble over each other. The cleaning solution overflows and clean tap water is added to rinse the parts. The overflow creates large volumes of wastewater requiring treatment and causes considerable loss of cleaner chemicals. In the hand-aqueous process, the workload (perforated basket of parts) is dipped into a series of tanks containing (successively) surfactant solutions and rinsewater. A continuous clean water flow must be maintained through the final rinse tanks, but the surfactant and drag-out tank contents can be used for an entire day without changing. Water and chemical consumption is lower compared with alkaline tumbling. Both aqueous processes require drying at the end before further surface-finishing treatment.

Many metal-finishing plants use both vapor degreasing and aqueous cleaning because each has its advantages. Pollution prevention tends to favor aqueous processes, but many metal finishers persist with vapor degreasing because of its economics and suitability for certain types of parts—e.g., parts that slide into each other to form a close fit, preventing some surfaces from being exposed. Advancing technologies have made both processes more effective and environmentally friendly.

Another cleaning application, used-parts cleaning for overhaul or repair, is usually carried out in large immersion tanks containing liquid solvent. Alternatively, solvent is recirculated continuously in a sink-and-drum assembly and the user rinses the part in the solvent flowing into the sink through a nozzle. Many plants using these methods have arrangements with solvent suppliers who periodically replace used solvent. The supplier then processes the used solvent at a central location, usually by distillation. Alternatively, one supplier installs a patented continuous distillation unit in the plant next to the immersion tank or sink-and-drum assembly to extend the solvent life almost indefinitely. The supplier periodically returns to the plant to evacuate the small amount of residue that collects in the distillation unit and to add make-up solvent.

In the dry-cleaning business, solvent (usually PCE) is emitted at the end of the cleaning cycle when the chamber door is opened. Most dry cleaners now have some means of evacuating the solvent vapors before the door is opened and recovering them by refrigeration or adsorption onto activated carbon.

28.3 Aqueous Processing

Alkaline tumbling and hand-aqueous washing have been used for many years. Recently, new equipment addressing some of the disadvantages has become commercially available. Figure 28-1 shows an automated aqueous washer used at Quality Rolling & Deburring Co. (QRD), a medium-sized metal-finishing plant. Washers with more or less complexity are provided commercially by various manufacturers, and the user should be able to obtain a design suitable for a specific application.

The automated washer in Fig. 28-1 uses a helical screw to transport soiled parts through the five compartments. The parts are sprayed successively with solutions from the holding tanks. The helical screw agitates the parts as it carries them forward. Wastewater is generated continuously in the rinse compartment,

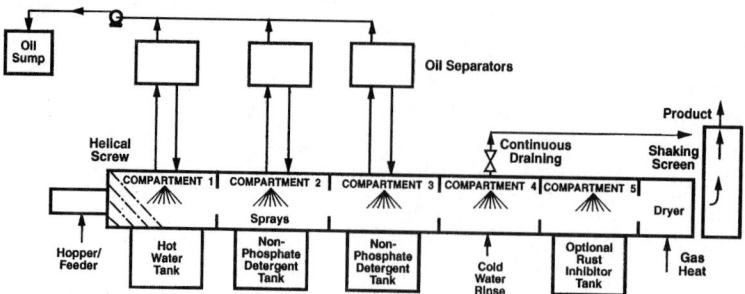

Figure 28-1. Automated aqueous rotary-washing process.

which receives a continuous input of fresh water. In all other compartments, the overspray is captured and taken to oil separator tanks where the oil floats to the top and is skimmed off with a small pump. Other debris settles to the bottom of the tank and can be cleaned periodically. The bulk of the relatively soil-free solution is reused. Each solution can be used up to 1 week. Some make-up detergent may have to be added. At the end of the week, the tanks are emptied and refilled with fresh solutions. Water consumption and wastewater generation are much lower than for either alkaline tumbling or hand-aqueous washing.

Testing at QRD has shown this process to be at least as effective as traditional processes on most types of parts.[2] Table 28-1 shows the reduced wastestreams with automated washing. It should be noted that the waste generated by vapor degreasing, although lower in volume, is much more hazardous than the aqueous waste. At QRD, nearly 6200 pounds per year of PCE emissions are prevented by using aqueous washing. The operating cost of automated washing was lower than that of alkaline tumbling or hand-aqueous washing, mainly

Table 28-1. Waste Volumes When Traditional Processes Are Replaced with Automated Washing

Conventional cleaning		Automated washing	
Wastestream	Volume generated (gal/yr)	Wastestream	Volume generated (gal/yr)
Vapor degreasing*		Automated washing*	
Wastewater in separator	200	Wastewater	143,000
Still-bottom sludge	1,440	Oily liquid	962
Alkaline tumbling†		Automated washing†	
Wastewater	1,010,880	Wastewater	85,800
		Oily liquid	577
Hand-aqueous washing‡		Automated washing‡	
Wastewater	296,400	Wastewater	57,200
		Oily liquid	385

*Based on 5200 barrels per year run on automated washer instead of vapor degreaser.
†Based on 3120 barrels per year run on automated washer instead of alkaline tumbler.
‡Based on 2080 barrels per year run on automated washer instead of hand-aqueous washer.

because of reduced chemical usage and lower wastewater treatment costs. However the automated washer had higher operating costs than the vapor degreaser for similar capacity and slightly higher energy requirements. The automated washer uses energy for a 5-hp motor that drives the helical screw, four 3-hp water circulation pumps, one 1.5-hp oil-skimming pump, and 150 ft^3/h of gas usage in the drier. The drying requirements in aqueous washing more than offset the energy required in the vapor degreaser to keep the solvent vaporized. This example demonstrates the difficulty in assessing a true pollution prevention impact that takes into account secondary pollution sources such as energy generation, chemicals manufacture, and so forth. For sources related to the immediate process, pollution prevention practice does favor aqueous washing by eliminating the need for solvent use.

The automated washer is used mostly for parts small enough to be conveyed by the helical screw. For larger parts, such as engines and transmissions, power washers as shown in Fig. 28-2 can be used. The part(s) are placed on a turntable for the automatically timed cleaning cycle. High-pressure, high-temperature water, usually containing a detergent, blasts the parts clean. Rotation on the table and the number and angle of the sprays enable the water to reach all surfaces. The enclosed design ensures operator safety. The water is captured, cleaned in an auxiliary cleaning device (containing a hydrocyclone or filter, aerator, and oil skimmer), and reused in the next cycle. Some power washers have a hot-air recirculation system for drying.

A power washer with dryer was tested at the Municipality of Metropolitan Seattle's Atlantic Base garage, where three such washers are used to clean wheel hubs, brake drums, and transaxle parts.[3] Previously, the parts had been cleaned in a solvent bath or with a high-pressure water spray in the open. The power washers enabled elimination of seven solvent cleaning stations, and recapture and reuse of the water resulted in some wastewater reduction. The

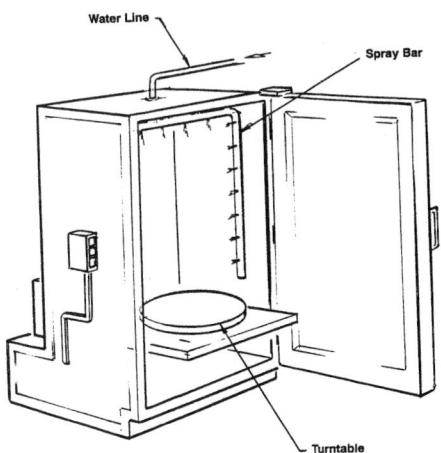

Figure 28-2. Power washer for aqueous cleaning.

quality of the cleaned product in the new washers was equivalent to that from traditional processes. Both operator time and detergent costs were reduced. Based on these savings, the estimated payback period was 4 to 5 years.

For a rebuilding and maintenance shop like the Seattle garage, the cleaned-product quality was acceptable and sometimes better than with the traditional processes. However, at facilities where the slightest oil film cannot be tolerated, this process may not be suitable. Also, the parts cleaned must be able to withstand the high-pressure spray.

28.4   Ultrasonic Cleaning

Ultrasonic cleaning uses high-frequency sound waves to improve the cleaning efficiency of aqueous and semi-aqueous cleaners. By generating zones of high and low pressures in the liquid, the sound waves create microscopic vacuum bubbles that implode when the sound wave moves and the zone changes from negative to positive pressure. This process, called *cavitation,* exerts enormous localized pressures ($\sim$10,000 psi) and temperatures ($\sim$20,000°F on a microscopic scale) that loosen contaminants and actually scrub the workpiece.

Ultrasonic cleaning uses conventional equipment available from a wide variety of vendors. Ultrasonic energy usually is applied to a solution by means of a transducer, which converts electrical energy into mechanical energy. The positioning of the transducers in the cleaning tank is a critical variable. Whether bonded to the tank or mounted in stainless steel housings for immersion in the tank, the number and position of the transducers are determined by part size and configuration, batch size, and tank size. If the transducers are placed with the radiating face parallel to the plane of the rack, the ultrasonic energy is directed at the workpieces. Two types of ultrasonic equipment are available. *Electrostrictive ultrasonics* uses a ceramic crystal to produce sound vibrations, whereas *magnetostrictive ultrasonics* uses metallic elements.

Ultrasonic cleaning proceeds in three stages: presoak, where the part is placed in the heated cleaning solution to remove all chemically soluble soil and gross contaminants; cavitation through ultrasonics; and rinsing. Ultrasonics can be reapplied during the rinse for increased efficiency.

The three-component ultrasonic cleaning system has a liquid solution tank, an ultrasonic generator as the power source, and a transducer to convert electrical to mechanical energy. Most generators convert ac input at 60 Hz to dc. The optimum transducer frequency has been found to be 20 Hz for most applications. Sizes range from 200-W tabletop units to large 1000-W units. Immersible transducers work best when retrofitting existing tanks for ultrasonic cleaning.

Ultrasonic cleaning can be used on almost any parts. Ceramics, aluminum, plastic, and glass—as well as electronic parts, wire, cables, rods, and detailed items that may be difficult to clean by other processes—are ideal candidates. Printed circuit boards and other electronic components also can be cleaned with the more gentle 40-kHz ultrasonic equipment. The higher frequency creates fewer and smaller bubbles, resulting in lower cavitation intensity and gentler cleaning, and it reduces the noise level.

Although most ultrasonic cleaning equipment is designed for batch tanks, cylindrical equipment has been developed that uses peripheral transducers. The resonant-tuned circuit focuses energy along the in-line centerline to allow noncontact cleaning except for the cleaning solution. The concentrated high power reduces cleaning times. The cylindrical form generally is used to clean wire, strip, tube, cable, and rod configurations that must be fed through without bending, require varying customer line speeds, and are immersible in a liquid solution. For best results, testing must be done with each set of parts to obtain the optimum combination of solution concentration and cavitation level.

Temperature has the most effect on the cleaning process.[4] Increased temperature gives higher cavitation intensity and better cleaning, provided that the boiling point of the chemical is not too closely approached.

The simplicity of the equipment and reduced cleaning time result in savings in labor costs when ultrasonics is used. This savings, along with the eliminated solvent purchase and disposal cost, offsets the capital cost of the equipment in a relatively short time. Smaller units can be bought, and existing tanks often can be used if a transducer is added. The ultrasonic technology offers *many* advantages:

- Ultrasonics can be used on a variety of cleaning solutions.
- Ultrasonic cleaning can reach into crevices and small holes.
- Ultrasonics removes inorganic particles as well as oils.
- Processing speed can be increased.
- Health hazards are reduced.
- A lower concentration of cleaning solution can be used, and often neutral or biodegradable detergents can be used.
- Although capital costs may be higher, reduced solvent expense can often quickly pay for a system.

The advantages must be balanced with these limitations:

- Wastewater generated has to be treated and discharged.
- Ultrasonic cleaning requires that the part be immersible.
- Dryers may be needed.
- Testing must be performed to optimize the combination of cleaning-solution concentration and cavitation level for the specific application.
- The electrical power required for large tanks generally limits part sizes that can be cleaned economically.
- The tendency for thick oils and greases to absorb ultrasonic energy may limit their removal.
- Operating parameters have to be more closely monitored.

The Ross Gear Division of TRW Inc.—a manufacturer of hydraulic motors, hydrostatic steering units, and manual steering gears—has used an ultrasonic cleaner since December 1987.[5] TRW uses an intensive machining process known as *lapping* to improve the surface finish. The abrasive material used in lapping must be completely removed after finishing. TRW previously used a solvent (TCE) vapor degreasing system to remove the compound. In 1987, this resulted in approximately 14,090 lb of TCE still bottoms, 3740 lb of filtration powder, and 50,300 lb of fugitive and stack emissions. The three-stage ultrasonic system washer has eliminated TCE use, reduced the quantity of hazardous waste generated at the plant by 50 percent, and significantly decreased disposal costs. The alkaline solution is sent to an ultrafiltration unit to remove oils and then is discharged to the sanitary sewer. Ultrasonic cleaning also has eliminated the potential health hazards associated with TCE.

28.5 Low-Emission Vapor Degreasing

The *low-emission vapor degreaser* (LEVD) is widely used in Europe, where vapor degreasers are regulated as a point source. As seen in Sec. 28.2, much of the solvent (more than 90 percent in some cases) in a conventional OTVC is lost through air emissions. LEVDs are completely enclosed, airtight units. Units such as the one shown in Fig. 28-3 are now commercially available in the United States.

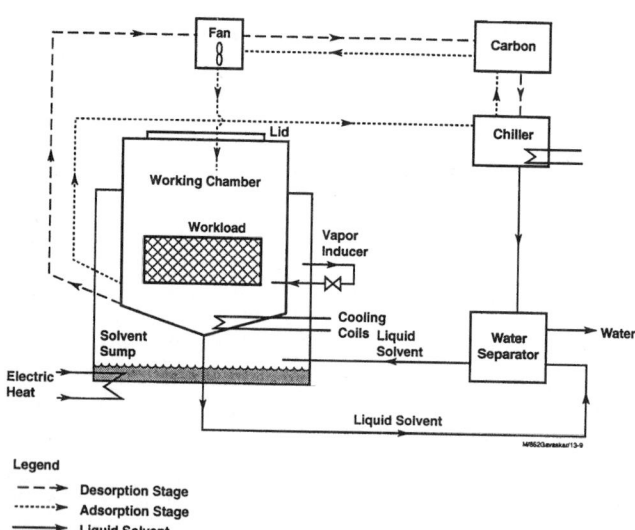

Figure 28-3. Low-emission vapor degreaser. (*Adapted from information provided by Durr Automation, Inc.*)

Approximately 1 hour before the shift begins, a timer on the LEVD unit switches on the heat to the sump. When the solvent in the sump reaches vapor temperature, the vapor is still confined to an enclosed jacket around the working chamber. The parts to be cleaned (workload) are placed in a galvanized basket and lowered by hoist from an opening in the top into the working chamber. The lid is shut, the unit is switched on, and compressed air from an external source hermetically seals the lid shut throughout the entire cleaning cycle. Table 28-2 shows the cleaning-cycle stages. First, solvent vapors enter the enclosed cleaning chamber and condense on the parts. The condensate and the removed oil and grease are collected through an opening in the chamber floor. When the parts reach the temperature of the vapor, no more condensation is possible. At this point, fresh-vapor entry is stopped and the air in the chamber is circulated over a cooling coil to condense out the solvent. Next the carbon is heated up to a temperature where most of the solvent captured in the previous cleaning cycle can be desorbed. The desorbed solvent is condensed out with a chiller. The carbon adsorbs the residual solvent vapors from the air in the cleaning chamber. As shown in Fig. 28-4, the adsorption stage continues until the concentration in the chamber is detected by a sensor to fall below a preset level (usually around 1 g/m^3). When the concentration goes below this level, the seal on the lid is released and the lid can be retracted to remove the workload. Upon retraction, a tiny amount of residual solvent vapor escapes to the atmosphere, the only emission in the entire cycle. EPA studies have shown that the LEVD reduces solvent emission by more than 99 percent compared with an OTVC.[6]

Unlike with a conventional degreaser, with a LEVD there are no significant idling losses between loads or downtime losses during shutdown. The LEVD can

Table 28-2. LEVD Cleaning Cycle

Stage	Vendor-recommended time setting (s)
Solvent heat-up (once a day)	Variable*
Solvent spray (optional)	10–180
Vapor fill	Variable†
Degreasing	20–180
Condensation	120
Air recirculation	120
Carbon heat-up	Variable‡
Desorption	60
Adsorption	60–240§

*Normally requires approximately 1 hour on days following overnight shutdown when sump solvent temperature drops to 70°C. After weekend shutdowns, when sump solvent temperature drops to 20°C, it may take 1 1/2 hours for solvent to reach vapor temperature. Time on unit allows automatic heat-up prior to beginning of shift.

†Varies according to mass of workload and type of metal.

‡Carbon heat-up took approximately 22.5 min during testing.

§At 60 s, if monitor shows that chamber concentration is above 1 g/m^3, then the adsorption stage proceeds to the full 240-s stage. This sequence repeats if necessary.

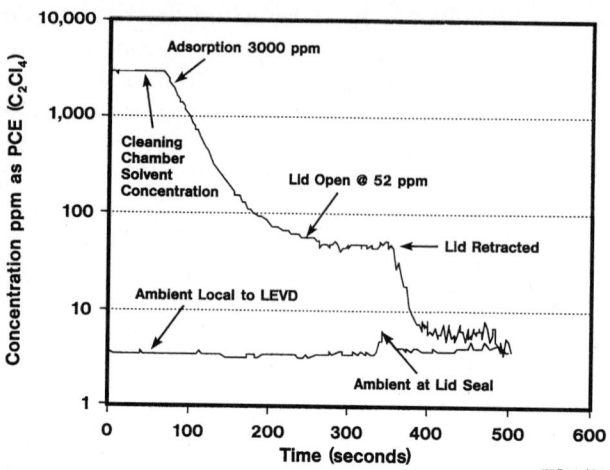

Figure 28-4. Final stage of the LEVD cleaning cycle.

be operated as a distillation unit to clean the liquid solvent in the sump. To distill, the unit is switched on without any workload in the chamber. After most of the solvent is converted to vapor, the residue in the sump is drained out and the vapors in the chamber are condensed in the chiller to recover the solvent. LEVD thus provides a good alternative for meeting pollution prevention objectives.

Given the longer vapor fill time, carbon heat-up, desorption, and adsorption stages of the LEVD, a much larger cleaning-chamber capacity (or batch volume) is required to maintain the same processing rate as with a conventional degreaser. This and other emission control features make the capital cost of the LEVD significantly higher than that of an OTVC, but savings in solvent consumption offset the additional investment. As with a conventional degreaser, for a given shape and size of the parts in the workload, LEVD cycle times are highly dependent on the total mass of parts in a batch and the type of metal being cleaned, as shown in Fig. 28-5.

The economic evaluation in the EPA study[6] compared the operating costs of the LEVD with those of a conventional OTVC with the same production capacity (560 lb of steel parts per hour). The savings for the LEVD were $25,000 per year from reduced labor costs (due to larger batch size) and lower solvent requirement (due to solvent recovery). The LEVD pays for itself in approximately 10 years. The LEVD does not require much of the auxiliary equipment that may be required for a standard conventional vapor degreaser, if the user is aiming to reduce workplace emissions to meet or anticipate increasingly stringent environmental and worker safety regulations. Additional control devices for standard conventional degreasers (e.g., increased freeboard ratio, refrigerated coils, lip exhausts, room ventilation) will add considerably to capital and operating costs. In contrast, the LEVD is a self-contained unit that will require no additional facility modifications to achieve significant emission reduction.

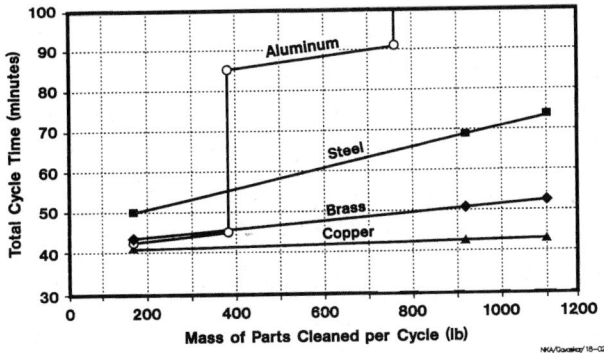

Figure 28-5. Variation of LEVD cycle time for various metals.

A slight variation on the LEVD unit described above[7] is a technology that incorporates an air-lock chamber over the opening of either a conventional vapor degreaser or a dry-cleaning unit to isolate the main cleaning chamber from the outside during cleaning. The workload enters and leaves the cleaning chamber through the air lock. When the cleaned workload enters the air lock, the air is drawn out over an adsorbent bed, where the solvent vapors are captured. The solvent is later desorbed by adjustments to the temperature of the adsorbent bed and returned to the cleaning chamber.

28.6 Alternative Soldering Techniques

In the electronics industry, the cleaning requirement results from soldering. To produce a good solder joint, a fluxing material (such as rosin) is used to promote wettability at the surface and prevent air from being trapped during paste deposition by providing resistance to the oxidation and reoxidation of metals during liquefaction. The flux leaves residues that can affect the reliability of solder joints which may interfere with subsequent testing or coating, or that may be aesthetically undesirable. The residue is cleaned with a solvent, but new processes are seeking to eliminate this solvent usage.

Chlorinated solvents and chlorofluorocarbons (CFCs) traditionally have been used to remove the flux residue during soldering in the electronics industry. One alternative is to use organic fluxes that can be washed off with water (aqueous cleaning). However, a highly acidic wash may be required to avoid tin and lead hydroxide deposition, and the wash can lead to corrosion. Unlike conventional rosin fluxes, water-soluble fluxes do not encapsulate impurities resulting in electromigration. Alternatively, water fortified with surfactants can be used to clean conventional rosin fluxes, but the surfactant may not be able to remove all the flux. Aqueous cleaning and ultrasonics can be used together to clean surface-mount boards that have very small spacings, but as in all aqueous cleaning, the wastewater must be treated before discharge.

A new approach gaining acceptance is the *low-solids flux* (LSF) that leaves very little residue, thus eliminating the need for a cleaning step.[8] Conventional fluxes contain 20 to 35 percent solids, whereas the LSF contains 1 to 5 percent solids. It is very easy to convert an existing system by simply replacing the conventional flux with an LSF.

One limitation is that the activity of an LSF usually is limited to a short dwell time. Lack of adhesion caused by the washing effect of a jet wave sometimes leaves too little active flux at the exit point in the wave to achieve good results. These problems can be mitigated by using a spray fluxer. The AT&T plant in Columbus, Ohio, has developed a new LSF spray applicator and, since switching to this technology in 1988, has eliminated the use of almost 30,000 gallons of PCE solvent annually.[9] This has resulted in a savings of $145,000 per year because the need to purchase, track, and dispose of solvent has been eliminated. The new applicator has an ultrasonic spray nozzle[10] that can be adjusted with great precision to obtain a controlled, uniform flux cover. The operator adjusts the air pressure to control flux deposition over a wide range. The LSF met AT&T's reliability requirements and maintained low defect levels compared with conventional fluxes.

Another limitation of LSF is that even minuscule amounts of residue may be unacceptable in certain demanding applications, as in the military or aerospace.[11] Also, LSFs are volatile, and reoxidation of exposed surfaces takes place during reflow in an oxygen-containing atmosphere. For some applications, soldering can be done in an inert atmosphere to eliminate oxygen where there is no need to keep the solder wave oxide-free. Low-cost, generally available nitrogen often is used to provide an inert atmosphere. Nitrogen is consumed at approximately 10 m^3/h, depending on processing rate. Eliminating both the use of flux and cleaning simplifies the process and results in savings in operating time. Sometimes a combination of LSF and inert atmosphere is used.[12]

28.7   Emerging Technologies

Some of the new cleaning technologies being developed with pollution prevention as an objective are discussed in this section.

2.8.7.1   Vacuum De-oiling

A vacuum furnace uses heat and vacuum to vaporize oils from parts. *Vacuum furnace de-oiling* can be applied where vapor degreasing typically is used to clean metal parts. It also can remove oil from nonmetallic parts. In a typical vacuum de-oiling system, a load of parts is heated in a vacuum to vaporize all oils present. In newer designs, the vapors are condensed and collected for later removal to be reprocessed or recycled. The cycle time depends on the mass of the load and the vapor pressure of the oil being removed. Time and heat determinations depend on emissivity, cross section, and mass. Most equipment is closed to eliminate emissions and to facilitate backfilling the chamber with nitrogen and/or air to cool the parts prior to removal, but the recently designed "hot-wall" eliminates oil deposits on furnace walls by keeping walls at a temperature too high for condensation.

Although capital costs for vacuum de-oiling are higher, the operating costs are claimed to be lower.[13] Unlike other clean technologies, vacuum de-oiling does not leave the cleaned parts water-soaked, so they do not need to be dried. Because the time and temperature of the de-oiling process depend on the material to be cleaned and the oil to be removed, adjustments may be needed for each new material, oil, or combination. Also, the parts must be able to withstand the required temperature and vacuum pressure.

28.7.2 Laser Ablation

The use of laser ablation to clean material surfaces is being explored by Sandia National Laboratories (SNL) in Albuquerque, New Mexico.[14] Short pulses of high-peak-power laser radiation are used to rapidly heat and vaporize thin layers of material surfaces that form a dense cloud of hot vapors that will condense and recontaminate the surface if not removed immediately. To prevent recontamination, the vapors are removed by entrainment into a flowing gas stream. Ablation must be carried out in an inert gas environment to avoid further contamination.

Laser ablation can do localized cleaning in a small area without affecting the entire part. Laser ablation meets waste minimization goals with no solvents or even aqueous solutions needed. The only waste is the small amount of material removed from the surface of the item being cleaned.

28.7.3 Fluxless Soldering

In the electronics industry, conventional soldering requires fluxing to promote wetting and to remove oxidation from surfaces to be soldered. Then halogenated solvents must be used to remove the flux residues. SNL is exploring methods of fluxless soldering[15] with no flux residue created and the cleaning step eliminated, so that no solvent is needed.

Under contract to the U.S. Department of Energy (DOE), SNL is developing the following four technologies to reduce or prevent surface oxidation prior to or during soldering, which is the main function of flux, and to eliminate the need for flux:[15]

1. *Controlled-atmosphere soldering* uses a special atmosphere where soldering is carried out without a flux, such as inert-atmosphere soldering discussed above. SNL also is exploring reactive plasma and a dilute acid vapor–inert gas mixture that functions as either a protective or a reducing cover during pro-

cessing. However, reactive plasma is somewhat destructive and more expensive because it requires sophisticated equipment. Atomic hydrogen atmosphere also is being explored as an alternative.[16]

2. *Thermomechanical surface-activation soldering* uses kinetic or directed thermomechanical energy to spall or ablate the surface oxide and facilitate wetting. Laser, solid-state diffusion, and ultrasonic soldering done in air or in a controlled atmosphere are typical ways to accomplish this.

3. *Metallization technology* uses silver as a nonoxidizing, readily wettable surface. Silver must be controlled precisely in a thick layer to guarantee complete coverage and wettability without too much silver to produce a brittle solder joint. The approach is to apply a thinner layer of silver. The proper balance between a thinner layer and enough to avoid porosity is being investigated.

4. *Inhibitor technology* protects porous metallizations through the application of inhibitors. SNL, with the State University of New York at Stony Brook, is studying the bonding of organic inhibitors on metallic surfaces and their effect on subsequent solder wetting.

28.7.4 Replacement of Tin-Lead Solder Joints

The electronics industry is developing a method to replace tin-lead for soldering with a combination of organic polymers (epoxies, thermoplastics, or silicones) and conductive fillers (carbon, copper, or silver) to eliminate the need for fluxes. Because the electronics industry is the largest user of Freon™-based cleaning solvents, fluxes, and possibly lead, this approach would have significant pollution prevention benefits.

At Interconnect Systems Inc., a large electronics-manufacturing plant, approximately 5 million solder joints are made annually.[17] With tin-lead soldering, this would require several thousand grams of solder paste, several dozen gallons of flux, and more than 50 gallons of Freon™ a year. The use of organic polymers and conductive fibers instead of conventional tin-lead solder would eliminate the use of flux and Freon™. The plant is examining the feasibility of this replacement in terms of electrical resistance in the joint, the mechanical and electrical integrity of the joint over time, the physical form of the adhesives, and the cost effectiveness of the replacement.

References

1. U.S. Environmental Protection Agency, *Halogenated Solvent Cleaners: Alternative Technology Control Documents*, EPA/450/389030, August 1989. Prepared by Radian Corporation.

2. A. R. Gavaskar, R. F. Olfenbuttel, and J. A. Jones, *An Automated Aqueous Rotary Washer for the Metal Finishing Industry*, U.S. EPA Project Summary, EPA/600/SR-92/188, 1992.

3. D. P. Evers and R. F. Olfenbuttel, *Power Washer with Wastewater Recycling Unit,* U.S. EPA Project Summary (expected 1994).

4. John Fuchs, "Ultrasonics Cleaning," in *Metal Finishing Guidebook and Directory,* Metals and Plastics Publications, Hackensack, N.J., pp. 135–140.

5. U.S. Environmental Protection Agency, "TRW, Ross Gear Division, Greenville, Tennessee," in *Achievements in Source Reduction and Recycling for Ten Industries in the United States,* Risk Reduction Engineering Laboratory, Office of Research and Development, Cincinnati, Ohio, September 1991.

6. A. R. Gavaskar, R. F. Olfenbuttel, and J. A. Jones, *On-Site Solvent Recovery,* U.S. EPA Project Summary (expected 1994).

7. J. C. Hickman and H. R. Goltz, "Temporary Vapor Storage Technology," in *Proceedings of the International CFC and Halon Alternatives Conference,* sponsored by the Alliance for Responsible CFC Policy, Baltimore Convention Center, Baltimore, Maryland, December 3–5, 1991.

8. J. Tuck, "A New No-Clean World?" *Circuits Assembly,* vol. 2, no. 7, July 1991, pp. 18–19.

9. U.S. Environmental Protection Agency, "AT&T Bell Laboratories/AT&T Network Systems, Princeton, New Jersey/Columbus, Ohio," in *Achievements in Source Reduction and Recycling for Ten Industries in the United States,* Risk Reduction Engineering Laboratory, Office of Research and Development, Cincinnati, Ohio, September 1991.

10. J. B. Brinton, "Ultrasonics Ease No-Clean Fluxing," *Circuits Assembly,* vol. 2, no. 3, March 1991, pp. 20–22.

11. Roger A. Trovato, Jr., "Inerting the Soldering Environment," *Circuits Assembly,* vol. 2, no. 4, April 1991, pp. 48–52.

12. D. Brammer, "No-Clean Processes Yield Top-Quality SMAs," *Circuits Assembly,* vol. 2, no. 6, June 1991, pp. 37–45.

13. Wayne Mitten, "Vacuum Deoiling for Environmentally Safe Parts Cleaning," *Metal Finishing,* vol. 89, no. 9, September 1991, pp. 29–31.

14. H. C. Peebles, N. A. Creager, and D. E. Peebles, "Surface Cleaning by Laser Ablation," in *Solvent Substitution: A Proceedings/Compendium of Papers,* U.S. DOE, Office of Technology Development, Environmental Restoration and Waste Management, and U.S. Air Force, Engineering and Services Center, Sandia National Laboratories, Albuquerque, N.M., 1990, pp. 1–9.

15. F. M. Hosking, "Reduction of Solvent Use through Fluxless Soldering," report by Sandia National Laboratories to the U.S. Dept. of Energy, Albuquerque, N.M., 1992.

16. Sandia National Laboratories, *Energy and Environment,* a Sandia Technology Bulletin, June 1993, pp. 1–3.

17. Interconnect Systems, Inc. (Bill Werther), "Definition of Electrically Conductive Adhesives for the Replacement of Tin-Lead (Solder) Joints in Electronic Systems," Pollution Prevention by and for Small Business Application Submittal to U.S. EPA, Simi Valley, Calif., 1990.

29

Pollution Prevention in Coating Application and Removal

Michael S. Callahan, P.E.

Jacobs Engineering Group Inc.
Pasadena, California

29.1 Industry Overview

More than 1 billion gallons of paint and coatings are produced annually in the United States. Major categories include architectural coatings, product coatings for original equipment manufacturers (OEM), and special purpose coatings. In addition, a large volume of products involved with paint and coatings removal is produced and sold. Such products include abrasives as well as solvent and aqueous-based stripping solutions.

29.2 Description of Main Processes

Very few substrates can be painted without suitable surface preparation. Proper preparation is essential to ensure adequate adhesion, durability, and dependability of the coating. Contaminants commonly encountered include dirt, soot, grease, mill scale, corrosion, and old paint. Inadequate cleaning before painting is a primary cause of rejects and premature coating failure.

In addition to cleaning, surface preparation often includes a surface treatment step. Iron or zinc phosphate solutions may be used to apply a phosphate coating to steel. The phosphate coating provides an added level of corrosion protection and improves the coating to substrate bond. For aluminum, chromic or sulfuric acid anodizing is practiced.

After cleaning and surface preparation, many different paint application methods may be used. Methods include use of brushes and rollers, dip tanks, flow coaters, curtain coaters, and various spray processes. Methods similar to electrolytic (electrodeposition) and nonelectrolytic plating (autodeposition) have been developed and are in use. The selection of a given method is dependent on the size and shape of the parts, the type of coating applied, the required finish quality, workload, and the availability of utilities such as compressed air and power.

The most versatile and widely used method is the spray process. Spray painting consists of two basic steps: paint atomization and propulsion. Atomization can be accomplished by jetting compressed air into the paint stream (air atomization), by a sudden pressure drop of the paint stream through a nozzle (airless atomization), or by imparting a high-speed centrifugal force to the paint (spinning disk or bell atomization). Propulsion can be accomplished by carrying the atomized paint in a high-volume airstream, by giving the paint a high exit velocity, by creating an electrostatic field between the spray gun and the substrate, or by using a combination of methods.

Conventional air spray painting uses compressed air to atomize and propel the paint. One of the major reasons air-atomized guns are so widely used is that they allow operator control of many variables that affect finish quality. Such variables include fluid pressure and orifice diameter, air atomization pressure, air-fluid pressure ratio, fan shape, and trigger operation.

In conventional airless spraying, paint is supplied to the spray gun at very high pressure. When the paint exits the gun, the sudden pressure drop causes the paint to accelerate, atomize, and move away from the gun. Airless spray is often used for highly pigmented or highly viscous paints where appearance is not critical (zinc-rich primers and sound deadeners).

To control overspray, many painting operations are conducted in spray booths. Both dry filter and water-wash spray booths are common. Water-wash booths employ a water curtain or wall to knock down and trap the overspray. The paint particles may either settle to the bottom of the water sump or float on the water where they are periodically removed as a sludge. To eliminate the generation of a wet sludge, some facilities have converted to dry filters.

The final stage of any painting operation is the cleaning of reusable equipment such as spray guns, pressure pots, and hoses. To clean this equipment, operators might fill the system with solvent and then operate the spray gun to flush out the paint. To minimize solvent emissions, many local air quality agencies mandate the use of enclosed cleaning systems. Cleaning of paint-coated hooks and racks may be performed in thermal burnoff ovens, by high-pressure water sprays, or by chemical-based strippers.

29.3   Typical Wastestreams

The wastestreams from surface preparation may include spent abrasives, solvents, and/or aqueous cleaning and surface treatment baths; air emissions from abrasives and solvents; rinsewaters following aqueous processing steps;

and solvent-soaked rags used for wiping parts before painting. Depending on the complexity of the operation and the nature of chemicals used, the volume and toxicity of wastes generated can vary widely.

Wastes generated by the painting process may include air emissions of volatile organic compounds (VOCs) and particulates, leftover paints and coatings, empty paint cans, disposable brushes and rollers, dirty solvents and thinners used for equipment cleanup, dirty filters from dry filter paint booths, and paint sludge from water-wash paint booths. The toxicity and hazard of the wastes generated is dependent on the concentration of solvent remaining in the waste and the presence of heavy metals such as lead and chromium compounds used in the paint formulations.

In paint-stripping operations, wastes may include spent abrasives; air emissions of dust, combustion products, or solvents; spent stripping baths; and contaminated rinsewaters. With dry-stripping techniques, the major concern is dust emissions and potential lead and chromium compounds in the stripped paint. The major concern with wet techniques is the methylene chloride and phenolic compounds used in cold strippers and the difficulty in handling and treating contaminated rinsewater.

29.4 Pollution Prevention Options

Pollution prevention options recommended for reducing wastes associated with coating application and removal operations are discussed here. Discussion of options suitable for avoiding or reducing the generation of waste from cleaning and surface preparation activities may be found in other sections of this handbook. Readers requiring specific product information should refer to the *Organic Finishing Guidebook*[1] for a detailed listing of equipment and material suppliers.

29.4.1 Paint Application

1. Eliminate the Need to Paint. Paints and coatings perform an important function of protecting materials from serious degradation due to environmental exposure. Without paints, steel structures would rust away and timber would decay. Material repair and replacement costs would be prohibitive. Paints and coatings also provide an aesthetic function, adding color and variety to the local environment.

However, the need for painting can sometimes be avoided by selecting materials that combine both function and aesthetics. Use of injection-molded plastic shells in place of painted metal cabinets is widely practiced in the electronics industry. Building construction employing the use of vinyl siding, PVC and FRP plastics, precolored concrete, and metal trim materials such as stainless steel, copper, bronze, and aluminum are known. To promote the development of surface-coating-free materials, the U.S. Environmental Protection Agency has conducted several workshops.

While much of the government and industry focus has been on the development of low- or non-VOC coating materials, the topic of surface-coating-free materials has received little attention until recently. Some of this may be due to the highly diverse nature of the topic which involves such issues as product design, manufacturing, product usage, and maintenance. Since the trade-off of painting versus surface-coating-free material usage determines actual environmental benefit, this subject may be a natural for the developing science of product life-cycle assessment.

2. Use Low-VOC Coating Materials.   Under the Clean Air Act, limitations are being set as to the amount of VOCs that will be allowed in paints and coatings. While limits are based on specific industry group and application, the VOC limit of 340 g/liter (2.8 lbs/gal) may be viewed as an unofficial national standard. While some limits are set at 3.5 lbs/gal and above, regulations will continue to push allowable levels to lower and lower values.

Substitution of solvent-based paints with water-based or low-VOC-based paints can significantly reduce air emissions. With water-based paints, solvent content may range from 5 to 15 percent compared to 40 to 50 percent for a solvent-based paint. In addition to reduced solvent content, water-based paints do not require solvent for thinning or equipment cleanup. The solvent content for many water-based paints ranges from 1.2 to 2.0 lbs/gal (minus the water and other exempt solvents).

Water-based coatings have exhibited a wide range of performances when used for industrial applications. Performance is not only dependent on paint composition, but on the extent of surface preparation performed and application procedure employed. Some water-based coatings may have different application requirements than solvent-based paints. One major difference with water-based paints is the longer drying time that may be required.

High-solids coatings are another way of reducing the solvent content of paint. High-solids coatings have the consistency of an ink or paste and require the use of high-pressure airless guns. Since the paint is highly viscous, the supply lines to the gun may be heated to reduce its viscosity. One major limitation of high-solids coatings is that they tend to provide a pebbled or textured finish and that the range of available formulations is limited.

A new technology using supercritical carbon dioxide as a replacement for midrange solvents seeks to overcome some of the application problems encountered with spraying high-solids coatings.[2] The fluid delivery system meters and mixes liquid carbon dioxide (supplied from gas cylinders) with the coating concentrate and then heats and pressurizes the mixture into the supercritical range. Use of the supercritical CO_2 is reported to improve paint atomization which results in a high-quality finish. The technology can also eliminate the need for halogenated solvents used in some VOC-compliant paints.

Moving completely out of solvent-based technology are the ultraviolet- (UV-) cured resins and powder coatings. The UV-cured resins consist of specially formulated organic resins that polymerize and cure when exposed to UV light. Powder coatings consist of dry resin and pigment ground to a fine powder. The coating is applied electrostatically and the parts are then heated to effect flow-out and curing. In the maintenance arena, field application of powder coating by flame spraying is sometimes practiced.[3]

3. Use of High Transfer Efficiency Equipment. Assuming that the required amount of coating to be deposited on the painted object is known, total coating usage is a direct function of transfer efficiency. By definition, transfer efficiency is the amount of paint solids deposited on an object, divided by the amount of paint solids sprayed at the object, multiplied by 100 percent.[4] By using spray guns with high transfer efficiency, paint usage (and hence VOC emissions) should be minimized.

Conventional air spray equipment has a transfer efficiency of approximately 40 percent, depending on the shape of the object being painted. As a result, only 40 percent of the paint sprayed lands on the object and the remaining 60 percent is lost. Use of high-volume low-pressure (HVLP) spray equipment, with a minimum transfer efficiency of 65 percent, substantially reduces overspray. HVLP guns use a high volume of air delivered at low pressure to atomize the paint material.

Electrostatic spray equipment directs the paint mist by electrically charging the spray particles. The resulting mist of electrically charged particles is then attracted to, and will adhere to, the nearest grounded object. All surfaces of the grounded object, including the back side of small parts, can be coated by the mist directed at one side of the object (wraparound effect). The transfer efficiency of electrostatic spray equipment often exceeds 90 percent.

4. Set Application Standards. Film thickness may be easily measured by means of wet film gauges or dry film meters. The monitoring of applied film thickness is critical to ensure that a uniform and consistent coating of paint is being applied. Too thin a coat will result in premature failure in the field, while too thick a coat represents excess cost and waste.

Other standards that should be established include crosshatch adhesion, film hardness, and solvent resistance. During curing, the monitoring of the time and temperature profile can ensure that the paint is not being over- or undercured. Next to the painting of dirty parts, improper curing is the second most common reason for coating failure. Specification and adherence to standards can do much to minimize the level of rejects and ease troubleshooting when problems arise.

Specification of desired film thickness should also be included when the transfer efficiency of different processes are compared (even though many air quality agencies do not acknowledge this). As previously stated, efficiency is defined as the amount of paint solids deposited on an object divided by the amount of paint solids sprayed. Because there is a required minimum efficiency for the equipment used, it is assumed that paint usage and VOC emissions will be reduced. This is not always true.

Electrostatic spray guns have very high efficiencies because all of the paint is attracted to the part. Even the overspray will wrap around and coat the backside. This yields a high calculated efficiency. However, the painting of the backside may not be necessary or even desirable. Some shops have discontinued the use of electrostatic sprays because this wraparound effect creates a pebbled finish on the previously coated backside. To smooth out the rough finish, the backsides had to be sanded and repainted. Similar problems have been noted with other spray techniques, such as having to apply an excessive film build to correct for "orange peeling" and other finish defects.

5. Train Operators to Use Equipment Properly. Transfer efficiency is dependent on a large number of parameters such as application technique, target shape and size, spray booth arrangement, paint characteristics, paint and air flow rates, spray gun-to-target distance, spray gun condition, and operator error. Some of these parameters are under the control of the operator, while others are not.

Spray guns are designed to operate at optimum flow rates. Setting flow rates to high can reduce efficiency by increasing the amount of bounceback (paint bounces off target) and overshoot. Excessive air pressure can also lead to premature drying of the paint before it reaches the target (paint fog). When allowed control, many operators tend to operate their air supply at too high a pressure. Proper operation and maintenance of a spray gun can result in a 50 percent increase in efficiency.[4] This effect was noted for airless and electrostatic systems.

6. Standardize Paints and Colors. By standardizing the number of paints and colors employed, the generation of cleanup wastes can be reduced. The frequent changing of colors results in the need to purge supply lines and strip down the paint booths often so as to avoid color contamination. Standardization may also make it feasible for the purchase of paint in returnable bulk containers. These containers may be owned by the coating formulator who picks up, cleans, refills, and returns the empty containers as needed. The use of bulk containers can be very cost effective in addition to eliminating the generation of empty paint cans and leftover paint.

For facilities that cannot standardize the number of colors required but want to benefit from a bulk container arrangement with their supplier, use of non-pigmented paint may be possible. With this arrangement, smaller size lots of the stock paint are drawn off and pigmented as needed. Since the paint is mixed as it is needed, waste due to overpurchase of a color that is no longer used is avoided. To maintain color consistency from batch to batch, the use of a color matching system is often necessary. Such systems may be purchased for $6,000 to $20,000.[5]

7. Implement Strict Material Tracking and Control. Many ways exist in which the indiscriminate use of thinners and cleanup solvents can be reduced. Making employees sign out for cleaning supplies often results in the reduction of material use. Studies in the automotive refinishing industry have shown that rigid inventory control can reduce solvent waste generation rates by 50 to 75 percent.[6] To track solvent use, some companies have issued solvent debit cards to their painters.

The practice of low-volume cleaning techniques is another way to reduce solvent use. Mixing containers and paint pots should first be drained and then scraped free of residual paint by a soft wood or plastic spatula. The containers are then rinsed with a small amount of solvent and wiped clean with a lint-free rag. To promote the reuse of cleaning solvent, the paint sludge initially removed from the containers and paint pots should be stored separately from the dirty rinse solvent. By not loading the solvent with solids, one can decant and reuse it several times. Some shops even use this rinse solvent as thinner in the next compatible batch of paint.

When paint pots are used, disposable polyethylene bags or liners can be utilized. Instead of pouring out the waste paint and cleaning the pot with solvent, one can use a bag or liner which allows the leftover paint to be lifted out of the pot and placed in a collection drum for disposal. Bags and liners typically cost $1 each, which is much less than the avoided cost of solvent purchase and disposal.

8. Use of Enclosed Gun Cleaners.   To reduce air emissions and promote recycling, the use of enclosed cleaning stations is required by many local air quality agencies. Some cleaning stations contain an air-driven solvent-condensing unit and are designed to be mounted on top of an open-top 55-gallon drum. Provided that operators are careful not to introduce too much paint sludge into the drum, reusable cleanup solvent can be decanted from the drum. Enclosed cleaning stations are available for less than $1000. Several companies lease this equipment along with the cleaning solvent.

9. Return Expired Materials to Supplier.   Painting materials often have limited shelf life and, when this is exceeded, may have to be disposed of as hazardous waste. Quite often, the shelf life is based on some minimum time during which the formulator guarantees the quality and performance of the paint. Once the shelf life has been exceeded, there is greater potential that the paint will not be suitable for use. The most likely problems are physical separation of the compounds and clumping of the pigments and resins.

If painting materials are properly stored and routinely shaken, then shelf life may be extended within reason. Some paint suppliers offer a fixed per-gallon fee service whereby they will pick up, remix, recertify, and return the paint. Batches of unused paint can sometimes be mixed together and be retinted to provide a useful product. Retinting can only be performed if the surplus paints are of similar type and color. The availability of such a service should be known before it is needed and the agreed-upon reprocessing fees should be specified in the purchase contract.

29.4.2 Paint Stripping

1. Assess the Need to Strip.   In many equipment-maintenance painting operations, the routine practice is to use chemical strippers and remove all paint down to bare metal. When the reason for repainting is only cosmetic (i.e., the top coat has faded but the base coat and primer are still intact), selective removal of the top coat with abrasives will reduce waste generation. At some facilities, this policy change has resulted in a 90-percent reduction in the number of parts being fully stripped.

2. Use the Most Durable Abrasive Practical.   Abrasives commonly used for parts preparation include steel grit, alumina, garnet, and glass beads. Steel grit creates a rough surface profile on the substrate which aids coating adhesion. Because it is so hard and durable, steel grit can be reused repeatedly and generates the least amount of waste per unit of surface area stripped. To maximize the reuse of steel grit, it must be kept dry to avoid rusting.

Alumina is considered to be a multipurpose material that is less aggressive and less durable than steel grit. Its use results in a smoother surface profile and less removal of substrate material. Garnet and glass beads are the least aggressive and are often used in a single-pass operation (i.e., the abrasive is not recycled). Use of garnet and glass beads is most suitable for preparation of soft materials that are easily damaged and for maintenance of the dimensional tolerance of the part.

3. Use Blasting Surface Standards.   Abrasive blasting a surface longer than necessary creates excess waste and reduces productivity. Blasting standards or measuring devices should be used to define the level of surface scratching or "profile" desired. Most standards use Structural Steel Painting Council (SSPC) classifications for surface cleanliness. Two styles of standards are available: visual disk and photographic. A surface profiler instrument may also be used.

4. Use Abrasives in Place of Chemical Strippers.   Chemical-based paint strippers may be referred to as hot (i.e., heated) or cold. Many hot strippers employ the use of sodium hydroxide and other organic additives. Most cold strippers are formulated with methylene chloride along with other additives such as phenolic acids, cosolvents, water-soluble solvents, thickeners, and sealants. Handling and disposal of spent baths and rinses is a major problem for facilities employing these strippers.

Many facilities have reduced their reliance on chemical-based strippers by converting to abrasive blasting. Manual blast cabinets and automated blasting chambers can be used to remove paint from parts while controlling dust emissions. The abrasive is fed into the cabinet or chamber and directed against the part being stripped. Used abrasive and removed paint are then pneumatically conveyed to the reclaimer. Reusable abrasive is separated from the waste and fines (broken-down abrasives and paint chips) are collected in a dust collector.

Field stripping may be performed in an open area or in an enclosed blast booth. To protect operators from dust, they must wear self-contained breathing equipment. After blasting, the used abrasive may be shoveled or vacuumed from the area and processed through the reclaimer. Some systems combine dust control and abrasive recovery by including a vacuum collection pickup device with the blasting nozzle.

Since the main advantage of chemical-based strippers is that they do not scratch or damage the substrate, most of the abrasives looked at as viable substitutes are relatively soft materials. Glass bead blasting has become popular because it is the least aggressive of the commonly used abrasives. New alternatives include plastic media, wheat starch, ice crystals, carbon dioxide pellets, sodium bicarbonate slurry, and high-pressure water. Some of these techniques are well developed while others are still in the developmental stage.

5. Use of Cryogenic Methods. Cryogenic paint removal uses liquid nitrogen immersion at approximately −200°F, which causes the paint to contract and thus breaks the adhesive bond with the substrate. For small components, a tumbler design is usually employed, where the parts can impact and abrade each other to assist in removing the paint. If the parts have complex shapes, tumbling media may be added.

6. Use of High-Temperature Thermal Technologies. Thermal methods such as flame burnoff are sometimes used to strip paint from large metal structures such as bridges. The method is highly labor-intensive, results in emissions of burned paint that cannot be easily controlled, and typically results in a metal surface fouled by heat scale. This heat scale must subsequently be removed by abrasive methods such as sanding or wire brushing. Most thermal methods are limited to heavy metal parts that will not warp due to thermal expansion and distortion.

Burnoff ovens and molten salt baths are often used to remove paint overspray from hooks, racks, grates, and body carriers used in automotive plants. Stripped parts are left with a residue of ash, which can be removed by rinsing. In addition to burnoff ovens and salt baths, heated fluidized sand is also used.

Use of Less Hazardous Strippers. Because of environmental concerns over the use of methylene chloride and phenolic-based strippers, many new stripping formulations have been researched and developed. Strippers based on formulations of N-methyl-2-pyrollidone (NMP) and dibasic esters (DBE) have found use in the areas of paint application equipment cleanup and the consumer market. They have not found universal acceptance or use in industrial stripping operations because their effectiveness varies from paint to paint. Many of the substitutes suffer from one or more of the following disadvantages compared to the methylene chloride and phenolic standard:

- Effectiveness varies with type of paint
- Effectiveness varies with extent of cure
- Elevated temperature required
- Increased stripping time required
- Formulation may attack substrate

- Formulation is flammable or combustible
- Formulation is photochemically reactive

To resolve some of these issues, the Idaho National Engineering Laboratory (INEL) is currently investigating the use of alternative paint strippers for the U.S. Departments of Defense and Energy.[7] Forty-eight commercial strippers have been evaluated and eleven have passed immersion corrosion tests. Of the eleven tested, the most effective were formulations of ethanolamine/amine (Hotstrip 420, CM-3707, FO 606, and Cee Bee A-477), benzyl alcohol (Cee Bee A-245), and NMP (M-Pyrol).

29.5 Pollution Prevention Successes

The following success story was reported in *Pollution Prevention in California*[8] and it highlights one of the successes of the Southern California Edison (SCE) Clean Air Coatings Technology Program. SCE has an active assistance program directed at retaining customers in the wood finishing, metal coating, automotive refinishing, and printing industries by helping them meet existing and projected air quality requirements. SCE's Customer Technology Application Center allows customers to experiment with powder, high-solids, waterborne, UV, and IR systems. Educational seminars are routinely conducted by authorities in the field of regulatory issues, product coatings, and delivery systems.

Safetrans Systems, Inc. manufactures signal warning systems for railroad grade crossings. As part of the manufacturing process, printed circuit boards are sprayed with a protective coating. Several years ago, the company switched from varnish to a water-based urethane coating to comply with SCAQMD regulations. The new coating material had a lower VOC content but its use resulted in marginal reductions of VOCs overall. This was due to the need to apply more of the coating to the board to achieve the required degree of protection.

The new coating material also created other problems. It was incompatible with the varnish previously used. Hence, the use of varnish was still required for repair work performed on existing boards. A second problem was that the water-based coating required a 24-hour drying time before the boards could be tested. This presented a major obstacle to the adoption of "just-in-time" production.

To overcome these problems, Safetrans decided to switch from a water-based to a UV-cure urethane coating. The UV-cure urethane contains no VOCs and total cure time is two minutes. Since the composition of the sprayed coating does not change until exposed to UV light, overspray may be recovered and reused. The need for cleaning clogged spray equipment has also been eliminated.

The cost for the UV-cure system was approximately $21,700. Annual material costs for the UV-cure coating is $5650 compared to $13,580 for the water-based coating. Savings in annual maintenance labor for the reduced need for cleaning the spray equipment amount to $4870. This equates to an overall annual savings of $12,800 for a payback period of 21 months. Intangible savings for improved process flow and handling have been estimated to be $5000. Given the ability to increase production, reduce board inventory before testing, and remain in compliance, the estimated intangible savings are most likely conservative.

References

1. *Organic Finishing Guidebook and Directory Issue for '93: Metal Finishing* **91**(5A), May 1993.
2. D. C. Busby et al., *Supercritical Fluid Spray Application Technology: A Pollution Prevention Technology for the Future,* Union Carbide Chemicals & Plastics. South Charleston, W.V., 1990.
3. Anonymous, "Case History: Field Applied Powder Coating Weathers the Ups and Downs of Canadian Climate," *Powder Coating,* August 1991.
4. U.S. Environmental Protection Agency, *Transfer Efficiency of Improperly Maintained or Operated Spray Painting Equipment Sensitivity Studies,* EPA/600/2-65-107 (PB86-108 271/AS), September 1985.
5. M. S. Reisch, "Paints & Coatings," *Chemical and Engineering News,* October 18, 1993.
6. California Department of Health Services, *Waste Audit Study Automotive Paint Shops,* January 1987.
7. A. Propp, "Alternative Chemical Paint Strippers." Paper presented at the *3rd International Workshop on Solvent Substitution,* Phoenix, Ariz., December 8–11, 1992.
8. California Department of Toxic Substances Control, *Pollution Prevention in California,* Alternative Technology Division, July 1992.

Further Reading

American Society of Metals, *Surface Cleaning, Finishing, & Coating, Metals Handbook,* vol. 5, 9th ed., Cleveland, Ohio, 1982.
Arnex, "Announcing a Significant Advance in Aircraft Depainting Efficacy and Safety," brochure regarding sodium carbonate paint stripping, Church & Dwight Co., Inc., Princeton, N.J., 1989.
Brantley, M., "How to Reduce VOCs," *The Finishing Line, SME/AFP's Finishing and Coating Quarterly* **6**(2), Second Quarter, 1990.
California Air Resources Board, "Determination of Reasonably Available Control Technology for Metal Parts and Products Coating Operations," ARB/SSD-93-003, CARB, Sacramento, Calif., December 10, 1992.
The Collected SSPC Applicator Training Bulletins from JPCL: 1988–1992, SSPC 92-03, JPCL, Pittsburgh, Pa.
DeVilbiss, *The ABCs of Industrial Spray Finishing,* DeVilbiss Ransburg Industrial Coating Equipment, Maumee, Ohio, 1993.
Duhnkrack, G. B., "Plastic Blast Media, an Alternative to Chemical Stripping," *Pollution Engineering,* December 1987.

Finzel, W. A., "Use Low-VOC Coatings," *Chemical Engineering Progress,* **87**(11), November 1991.

Gashlin, K., and D. J. Watts, *Waste Reduction Activities and Options for a Fabricator and Finisher of Steel Computer Cabinets,* EPA/600/S-92/044, Cincinnati, Ohio, October 1992.

Industrial Finishing, *Cryogenic Stripping of Paint from Hangers,* Industrial Finishing, Wheaton, Ill., 1986.

Kirsch, F. W., and G. P. Looby, *Waste Minimization Assessment for a Metal Parts Coating Plant,* EPA/600/M-91/015, July 1991.

Kirsch, F. W., and G. P. Looby, *Waste Minimization Assessment for a Manufacturer of Military Furniture,* EPA/600/S-92/017, June 1992.

Lee, R., "Keeping Abrasive Dry: A Review of Recent Technology," *JPCL* **10**(9), September 1993.

Modern Metals, *Eliminating VOCs from Paint Booth Emissions,* Modern Metals, March 1989.

Northeim, C. M., et al., *Surface-coating-free Materials Workshop Summary Report,* EPA/600/SR-92/159 (PB93-101 160/AS), December 1992.

Parrish, R. L., "The Art, Science, and Politics of Paint Removal," *Business and Commercial Aviation,* November 1987.

Uhrmacher, C., *Evaluation of the Problems Associated with Application of Low Solvent Coatings to Wood Furniture,* EPA/600/S2-007 (PB87-168 746/AS), May 1987.

U.S. Environmental Protection Agency, *Guides to Pollution Prevention: The Fabricated Metal Products Industry,* EPA/625/7-90/006, July 1990.

U.S. Environmental Protection Agency, *Reducing Risk in Paint Stripping, Proceedings of an International Conference,* PB91-224303, June 1991.

Wolbach, C. D., and C. McDonald, "Reduction of Total Toxic Organic Discharges and VOC Emissions from Paint Stripping Operations Using Plastic Media Blasting," *Journal of Hazardous Materials* **17**(1987).

30

Pollution Prevention in Office Operations

John Houlahan

*Science Applications International
Corporation
Hampton, Virginia*

Carole Bell

*Science Applications International
Corporation
Newport, Rhode Island*

30.1   Introduction

Office operations provide excellent, and often overlooked, opportunities to prevent pollution. At most facilities, attention is directed toward reducing or eliminating hazardous materials and wastes from industrial or manufacturing operations. As a result, offices are often neglected as potential opportunities to reduce or recycle wastes.

This chapter is appropriate for persons interested in pollution prevention for both large and small office operations. It provides suggestions that can be implemented in an isolated office located in a larger industrial facility or at a multistory office complex. The chapter begins by introducing the benefits of practicing pollution prevention in the office. The types of wastes produced from office operations are briefly described, followed by presentation of specific source reduction and recycling options. A discussion of using the purchasing process to reduce waste is also presented. Tips on how to establish a recycling program in the office conclude the chapter.

30.2   Benefits of Practicing
Pollution Prevention in the Office

The most obvious benefits of an office pollution prevention program are reduced waste and cost savings. Substantial amounts of waste can be generated from office operations; common office wastes include paper and paper products, cardboard, metal, wood, and organic materials such as food wastes. Paper accounts for the bulk of the waste from offices; studies have shown that each individual office worker generates between 0.5 and 1.5 pounds of paper per day. Most office waste is not hazardous, although some office products may contain hazardous constituents. Common waste materials from office operations are listed in Table 30-1.

The cost to manage and dispose of office wastes can be high, running into the tens of thousands of dollars for a high-rise office building or large business. The annual waste-handling charges for an office include the cost of internal collection (janitorial services), rental of a dumpster or other container, weekly or biweekly collection, and the per ton disposal fee at a landfill or incinerator. Pollution prevention in the office can save money. AT&T replaced copiers with duplex copiers, cutting paper use by 77 million sheets per year and saving $385,000 in companywide procurement per year. An AT&T employee suggested printing bills on fewer pages, an idea that saved $1 million per year. Mattel Inc. cut computer paper use by almost 7 million pages (from 32 million) and saved $200,000 annually. After a computer printout use audit revealed employees were receiving printouts containing much information they did not

Table 30-1. Common Wastes Generated from Office Operations

Activity	Waste
General	Paper (various grades), cardboard, laser jet printer toner cartridges, photocopying toner cartridges, printer ribbons, typewriter ribbons, pens, pencils
Mailroom and loading dock	Junk mail, envelopes, packaging waste (cardboard, plastic, shrink-wrap, metal bands, polystyrene "peanuts," bubble-wrap, pallets)
Employees' personal consumption	Newspapers, food, aluminum cans, glass and plastic beverage containers, paper and polystyrene cups, paper bags, fast-food packaging
Maintenance	Chemical cleaning products, empty containers, rags or paper towels, light bulbs
Miscellaneous	Surplus or broken equipment, file folders, disposable pens, pencils, paper clips, staples
Miscellaneous (hazardous wastes)	Batteries, correction fluids formulated with solvents (e.g., White Out), cleaning chemicals, mercury ballasts from fluorescent lights

need, the company replaced the paper printouts with computer software, allowing people to access the information on-screen. The switch to on-line access also saves time and effort because many employees no longer need to route and handle printouts.

An additional benefit to including office operations in your pollution prevention program is the education of the office staff and the perception that pollution prevention is a companywide effort. Active participation by everyone is key to a program's success. Practicing pollution prevention in the office brings the concepts of environmental responsibility to a whole new group of people. Too often, pollution prevention is the domain of a select few, such as the environmental coordinator or safety personnel. Pollution prevention in the office involves the clerical staff, purchasing department, contracting personnel, and other "paper pushers" whose jobs are not typically associated with waste generation or waste reduction. An added benefit from this increased involvement is that information about pollution prevention reaches a larger number of individuals, who may apply these concepts in their day-to-day lives.

30.3   Pollution Prevention Options for Office Operations

Office wastes are amenable to a range of relatively simple prevention and recycling techniques. Waste prevention and recycling in the office typically does not involve a large capital investment, engineering changes to production lines, or extensive staff training. Conducting a waste reduction assessment will help you identify opportunities to eliminate, reduce, or recycle waste; a detailed description of how to conduct a waste reduction assessment is presented earlier in this handbook. In Chapter 13, waste prevention strategies are discussed first, followed by recycling options.

30.3.1   Source Reduction

Source reduction or *waste prevention* refers to efforts to reduce the quantity and the toxicity of the materials and products used and then discarded from any operation in your facility.

Source Reduction through Procurement Practices.   Procurement practices can be used to reduce waste. Procurement can be thought of as the gateway through which all materials used and ultimately managed as waste must pass. Targeting procurement focuses attention on the critical first step in the waste generation process. Source reduction procurement is often the most effective way of reducing wastes since it can eliminate or reduce the toxicity or amount of a product or package destined to become waste, at the time that the purchasing decision is made. Source reduction procurement introduces contract specifications or purchasing preferences for less toxic products or products that can be reused rather than disposed. It also includes product substitutions, such as purchase of a product containing a less toxic material. Requiring suppliers to use less packaging is another source reduction procurement activity. Thus, source reduction procurement systematically reduces the amount and/or toxicity of materials a facility must manage by recycling or other disposal method. Avoiding the purchase of toxic materials eliminates the transportation, storage, use, and disposal of these materials, as well as the waste that might result from spills, leaks, or wasteful material usage.

Procurement also can be used in inventory control. Modifying procurement practices and procedures can reduce waste resulting from overpurchasing or expired shelf life. Overpurchasing can create an on-shelf surplus. When new items come on the market and are purchased and used, the older materials may be discarded unused, since a newer product has replaced them. This is particularly true of items with rapidly developing technology, such as computer and telecommunications equipment. Common source reduction procurement practices are listed below.

Requiring full utilization of materials

Using more durable goods

Minimizing unnecessary packaging

Eliminating use of disposable or single-use items

Establishing container or packaging "take-back" policies

Material substitution for toxics reduction

Purchasing energy- or water-efficient products

Requiring modular designs

Purchasing long-shelf-life items

Other Source Reduction Options.   Table 30-2 contains simple ideas that may be applicable to your office. Listed are ideas other businesses have used to reduce waste. You may be able to adapt some of these ideas to your business after considering your company's particular needs and local conditions.

30.3.2   Recycling

Unlike other pollution prevention options, recycling is an "end-of-pipe" activity. Recycling recovers materials destined for disposal and converts them into useful products. Recycling benefits are well documented; the most obvious benefit is the conservation of natural resources. Manufacturers who use recovered materials as feedstock require fewer extracted virgin resources to produce their products, avoiding both the costs and the adverse environmental impacts from obtaining and processing virgin materials.

Table 30-2. Waste Reduction Ideas

For writing and printing paper:

  Make double-sided copies, when possible.

  Keep mailing lists current.

  Make scratch pads from used paper.

  Reuse envelopes or use two-way envelopes.

  Circulate memos, documents, periodicals, and reports rather than making individual copies.

  Use outdated letterhead for in-house memos.

  Use voice or electronic mail or put messages on a chalkboard or central bulletin board.

  Print more words on each page.

  Eliminate unnecessary reports.

  Reuse manila envelopes.

  Where appropriate, use nontoxic fluids and art supplies in your graphic arts department and for general use.

  Reduce the amount of advertising mail you receive by writing Direct Marketing Association, Mail Preference Service, POB 3861, NY, NY 10163-3861 and asking that your business be eliminated from mailing lists.

  Save documents on floppy disks instead of making hard copies.

  Use central files for hard copies.

  Proof documents on the computer screen before printing.

  Donate old magazines and journals to hospitals, clinics, or libraries.

For packaging:

  Order merchandise with minimal packaging, in concentrated form, and in bulk.

  Ask suppliers to minimize packaging on orders.

  Request that deliveries be shipped in returnable containers.

  Return, reuse, and repair wooden pallets.

  Reuse newspaper and shredded paper for packaging.

  Reuse foam "peanuts," bubble-wrap, and cardboard boxes, or find someone who can.

  Set up a system for returning cardboard boxes and foam "peanuts" to distributors for reuse.

  Use reusable boxes for shipping to branch offices, stores, and warehouses.

  Where appropriate, consider rebuilding or fixing packaging material (e.g., reels, wooden pallets).

For equipment:

  Consider renting equipment that is used only occasionally.

  Consider using remanufactured office equipment.

  Invest in equipment that facilitates waste reduction, where feasible, such as:

  ▪ High-quality, durable, repairable equipment

  ▪ Copiers that make two-sided copies

  Reclaim usable parts from old equipment.

  Rotate tires on company vehicles on a regular basis to prolong tire life. Keep tires properly inflated.

  Find uses for worn out tires (e.g., landscaping, swings).

Table 30-2. Waste Reduction Ideas (*Continued*)

For equipment:

Institute maintenance practices to prolong the use of copiers, computers, and other equipment.

Consider using rechargeable batteries.

Consider installing reusable furnace and air-conditioner filters.

Recharge or rebuild fax and printer cartridges.

Sell or give old furniture and equipment to employees, or donate it to a local charitable organization.

For purchasing:

Substitute less toxic or nontoxic materials for toxic materials (i.e., some inks, paints, cleaning solvents).

Where appropriate, use products which promote waste reduction (durable, concentrated, recyclable, reusable, high quality).

Order supplies by voice mail or electronic mail.

Consider using optical scanners, which give more details about inventory, allowing more precise ordering.

Where appropriate, order supplies in bulk to reduce excess packaging.

Avoid ordering excess supplies that will never be used.

For overstocked, exchangeable items:

Donate surplus produce and past-pull-date perishables to food banks, if still edible.

Advertise surplus and reusable waste items through a commercial waste exchange.

Set up an area in your business for employees to exchange used items.

For consumer choices:

Teach your customers about the importance of reducing waste. Effective tools for getting the message across include promotional campaigns, brochures and newsletters (remember to use recycled paper), banners, newspaper advertisements, product displays, store signs and labels.

Encourage reuse of shopping bags by offering customers the choice of buying their own bag, complimenting customers who reuse bags, providing a financial incentive for reuse, implementing a promotional campaign.

Offer customers a rebate when they reuse grocery bags, containers, mugs, or cups for refilling.

Provide a choice of products that include recycled and recyclable materials.

Offer customers waste-reducing choices, such as

- Items in bulk or concentrated form
- Solar-powered items, such as watches, calculators, and flashlights
- Rechargeable batteries
- Razors with replaceable blades
- Durable merchandise
- Repairable merchandise
- Returnable bottles

Encourage the return of metal hangers at the dry-cleaner.

SOURCE: Modified from Washington State Department of Ecology, *Reducing Waste in Your Business,* Washington State Department of Ecology, Olympia, Wash., 1990.

Recycling can also be financially beneficial. Some recyclers will pay for recycled materials such as aluminum, while others will accept the materials but pay nothing for them or charge a fee to collect and process the material. The price paid for each material, or the cost to collect them, is dependent on local markets. The major economic savings from recycling is the money saved from not having to pay for the material to be picked up and disposed of as trash. Every pound or cubic yard of material collected for recycling is that much less material disposed of as waste. If enough material is diverted from the wastestream by recycling, the trash can be picked up less frequently or you can rent a smaller trash dumpster. Depending on your waste-hauling contract this can reduce your company's waste bill. Less material sent to the landfill or incinerator may also translate into savings from reduced disposal (tipping) fees. Even if you are charged by a hauler to pick up the recyclable materials, the charge is usually less than the cost to pick up trash. The difference between the two costs can be considered a savings.

While the office workers may separate materials, the recycling process involves many additional players: collectors, processors, manufacturers, and consumers. The cycle has just begun when the separated recyclables are collected and delivered to a processing facility where they are classified by material, color, and other specific variables. For example, employees separate office paper from other waste. The separated paper is picked up by a collector, who delivers it to a paper dealer. The paper dealer sorts and grades the paper by color and type. Once the dealer has accumulated a large enough quantity, it is sold to a paper mill. The mill will use the office paper as a component in the manufacture of toilet tissue and paper towels. Eventually, consumers will have the opportunity to purchase these recycled paper products.

At present, the availability of markets has the greatest impact on a material's recyclability. The collector or processor must be able to locate a market for the recyclable materials—that is, someone who will use the materials in new products. For example, glass containers may become new glass containers, or the glass may be a component in fiberglass insulation or asphalt sidewalks and roadways. If no market is available, the collected recyclables may end up stored in warehouses or may be sent to a landfill or incinerator for disposal. Without markets, there is little value in collecting recyclable material from your office since there will not be anyone interested in reusing the material.

To complete the recycling process, consumers must buy and use products made with recovered resources. Hundreds of products made with recovered materials, from acoustical tile to wastebaskets, are available for purchase. An entire office can be supplied with everything from wallboard made with recycled newspaper to carpet made from recycled plastic bottles. Supplies of recycled cardboard boxes, plastic strapping, paper for every purpose, stick-on notes, desktop accessories, file folders, scissors, rulers, erasable boards, clipboards, pens, pencils, and even push pins are easily obtained.

It may be helpful to establish a companywide procurement policy stating a preference for reusable, refillable, and less toxic products, as well as those with recycled content. Some companies give a price preference to recycled products; others specify that a certain percentage of purchases must contain recovered material.

There are a number of published sources for recycled products, some of which are listed in the "Further Reading" and "Other Resources" sections at the end of this chapter.

30.4 How to Establish a Pollution Prevention Program in the Office

It is important to remember that it is not possible to accomplish everything at once. Start with a couple of projects and phase in additional projects to increase the effectiveness and long-term success of your waste reduction efforts. As a first step, review existing waste reduction efforts. Are they successful? Can they be improved or enhanced? Where possible, build on current activities prior to initiating new projects. For example, add a new item to an existing recycling collection program. Experience gained in modifying existing projects can be applied to new efforts. People are typically willing to modify their work habits and practices to reduce wastes. Use this to your advantage and try to integrate waste reduction techniques into existing operations rather than institute new procedures or tasks.

Constant reinforcement and education are crucial to the success of a waste reduction program. Employees need to know what is expected of them and why. A series of seminars or training sessions should be held to inform all employees about how the pollution prevention program will work. Face-to-face interaction is far more successful than just distribution of an informational memo. Employees should be given the opportunity to discuss their purchasing practices, to see how recyclable materials will be separated, to offer advice about the location of the collection containers, and to ask questions about the program. The employees should be empowered to identify new products with recycled content and to evaluate and improve the recycling program. Companies that allow employees input into every aspect of the design and implementation of the pollution prevention program seldom hear complaints like "It's not my job!" or "We've always done it this way."

30.4.1 Setting Up a Recycling Program

An effective recycling program is one that is simple in design. To maximize results and minimize problems, the program must be easy for employees. For example, each employee must have easy access to recycling collection containers and must understand what materials are recyclable. The following section briefly describes some considerations in establishing a recycling program in the office.

Identify Recycling Opportunities. Get on the telephone! Find out about recycling collection and marketing opportunities in your community. Determine what materials are recycled locally. Determine how and at what cost materials are collected and processed. It is not unusual to look for markets within a 100-mile radius of your facility. Many state and local governments offer free technical assistance to help you establish and implement source reduction and recycling programs.

Analyze Your Waste. Identify which recyclable materials are generated by your office operations and then determine which of these materials you want to separate for recycling. Target specific operations or wastes for reduction or recycling: for example,

- A waste material that is generated in large quantity, such as paper or cardboard
- A material that has strong market demand, such as aluminum

Negotiate a Recycling Contract. Finding reliable markets for recyclable items is essential. Recycling markets are traditionally unstable, with frequent fluctuations in the amount paid per pound or ton of material. Some recyclers will pay for recycled materials, while others will accept the materials but pay nothing for them or charge a fee to collect and process the material. It is important to find a reliable company to collect your recyclable materials. Figure 30-1 is a list of questions to ask a recycler.

You should choose a company based on more than which one will pay you the most (or charge you the least) to collect your recyclable materials. You want to establish a good, long-term working relationship with the recycler, one that will weather the market fluctuations. Look for a company that can accommodate additions to your program, such as adding another material. Some companies offer free waste reduction assessments as part of their service.

Design a Collection Program. To obtain the highest market value, recyclable materials should be source-separated, that is, collected in separate containers and free of any contaminants. For example, source-separated white computer printout is more marketable than mixed office paper; aluminum cans are more marketable than mixed glass, metal, and plastic beverage containers. In some cases, processed recyclable materials are more valuable than unprocessed materials. For example, some recyclers will pay for baled corrugated cardboard but charge to collect loose cardboard. Make it easy for your employees to participate in recycling. Buy attractive recycling containers that employees will be pleased to have in their offices. Locate recycling containers conveniently. For example, a recycling center on the tenth floor is not convenient for employees on the second floor. Copy machines generate significant amounts of paper for recycling, and employees prefer not to carry misprinted copies back to their desks

1. What recyclable or compostable materials does the company collect?

2. Does this company collect only source-separated materials or will they also process commingled materials (mixed office paper? beverage containers?) What are the additional charges for mixed materials?

3. What additional separation restrictions does the company impose? For example, must staples, paper clips and Post-it™ notes be removed from paper? What about caps, lids, or labels on food and beverage containers?

4. What is the company's policy on rejecting material due to contamination? Are there financial penalties?

5. What are the weekly collection costs for the quantity of material you anticipate? Would these costs vary if collection was more frequent? on call?

6. Does the company provide storage containers? Is there an additional charge for this service?

7. For which materials will the company pay you? Which materials will be collected at no cost? For which materials will the company charge a processing or marketing fee?

8. Is the company willing to establish a contractual arrangement?

9. Can the company provide verification that your materials actually were recycled?

10. Can the company provide actual data on how much material you recycle on a regular (monthly, annual) basis?

11. Can you obtain a list of local references that you can contact?

Figure 30-1. Questions to ask a waste hauler or recycler. (*Adapted from President's Commission on Environmental Quality, Solid Waste Task Force, Workplace Waste Reduction Guide, 1993.*)

for recycling. A recycling container for paper should be located next to each printer and copy machine. Label containers clearly so that employees know which materials should be placed there. To avoid contamination of recyclables, place waste containers nearby.

30.5   Conclusions

Reducing the amount of waste that requires disposal should be a priority for every operation in your facility, including the office. Think about what would happen if every piece of paper that entered your office had to stay there. Within a very short time, the stacks of memos, magazines, orders, reports, newspapers, even napkins would cover every surface. Waste prevention is the first step, and your purchasing practices play an important role in reducing the volume of products and packages that become waste waiting for disposal. Then, recycling and composting recover valuable resources for reuse. What other office program can save energy and resources, demonstrate company commitment to the environment, make employees feel good, and set an example for others, while reducing costs?

Further Reading

Keep America Beautiful, Inc., *Waste in the Workplace*, New York, 1991.

National Office Products Association, *Resource Guide to Office Products Manufacturers' Recycling Programs and Products*, National Office Products Assoc., Alexandria, Va.

Pennsylvania Resources Council, *Guide to Recycled Products for Consumers and Small Businesses*, Pennsylvania Resources Council, Media, Pa. Telephone: (800) 468-6772.

President's Commission on Environmental Quality, Solid Waste Task Force, *Workplace Waste Reduction Guide*, U.S. Government Printing Office, Washington, D.C., January 1993.

U.S. Environmental Protection Agency, Office of Solid Waste, Risk Reduction Engineering Laboratory, EPA/600/R-92/088, *Facility Pollution Prevention Guide*, Cincinnati, Ohio, May 1992.

U.S. Postal Service, *Waste Reduction Guide*, Handbook AS S52, U.S. Postal Service, Washington, D.C., February 1992.

Other Resources

The U.S. Environmental Protection Agency's RCRA Hotline provides general information about the guidelines for purchasing recycled products as well as copies of *Federal Register* notices and guideline fact sheets. Hotline number: (800) 424-9346.

The American Institute of Architects' *Environmental Resource Guide* (discussion of environmental design issues and recycled construction products). Available from American Institute of Architects, 1735 New York Avenue, NW, Washington, D.C. 20006.

The McDonald's Corporation publication *McRecycle USA*® lists recycled materials used by the company in construction and other aspects of their operations. McDonald's Corporation, Kroc Drive No. 062, Oak Brook, Ill. 60521.

31

Pollution Prevention in Laboratory Operations

Russell W. Phifer

Environmental Assets, Inc.
West Chester, Pennsylvania

31.1 Introduction

The scale of chemical usage has the most significant impact on pollution prevention efforts in the laboratory. The average chemical laboratory facility, particularly one involved in the academic or industrial research area, uses thousands of chemicals. Instead of a few chemicals in significant quantities, laboratories use many chemicals in small quantities. This results in both unique problems and unique solutions for those involved in any aspect of pollution prevention.

The Occupational Safety and Health Administration (OSHA) defines a laboratory as "a workplace where relatively small quantities of hazardous chemicals are used on a non-production basis." This includes facilities for teaching, quality control, environmental testing, chemical and medical research and development, and clinical testing. As in other facilities that use chemicals, pollution prevention in the laboratory means source reduction, waste minimization, recycling, and reclamation. Nonetheless, how these procedures are implemented varies significantly from other chemical activities.

31.2   Nature of Waste Generated by Laboratories

An evaluation of wastes generated by laboratories is the first step in assessing possible prevention methods. Since the quantities of individual chemicals are so small—generally less than five gallons of each material—air and water emissions are only a minor part of the total volume entering the environment. As a result, the reduction of laboratory waste is the most significant goal in a pollution prevention program.

Regardless of how they are generated, discarded reagents represent a significant problem to the laboratory waste manager. Between 35 and 50 percent of all wastes traditionally generated by laboratories is composed of discarded reagents. This is slowly decreasing, as more and more laboratories recognize that the true cost of a chemical is its purchase price added to its disposal cost. Unused chemicals are disposed of for a variety of reasons, as shown:

- Reagents/pharmaceutical products past their expiration date
- Chemicals out of specification as received, or contaminated during handling
- Unwanted samples
- Surplus due to overpurchase, canceled project, or retired researcher

Routine wastes represent the remaining volume of the total laboratory wastestream. Routine wastes are those generated by a repetitive process; these include solvents generated from cleaning operations, clinical and environmental testing samples and process wastes, research and development preparations, quality control samples, and spill cleanup material. The keys to reducing these wastes are improved operating procedures, product substitution, good housekeeping, and downscaling of processes.

31.3   Pollution Prevention Assessment

Just as with other industries, in laboratory operations the first step in pollution prevention is a detailed assessment of all emissions. This includes the review of purchasing records to determine a realistic mass balance accounting of all chemicals. This includes not only the purchase of new chemicals, but also an accounting of off-site disposal services and on-site recycling and reclamation efforts. A one-year history is generally a good starting point for such a study, which can either be performed by an individual or committee in-house, or with the help of an outside consultant or industry trade source. Both have advantages: while inside staff is generally more familiar with the specific processes and procedures, an outsider can often spot problems and solutions from previous experience with other laboratories. Regardless of who performs the evaluation, it should include the following areas of concern. Each of these areas represents significant opportunity for pollution prevention.

- List of individual chemicals purchased, with quantities
- The department or individual that purchased the chemical
- The intended use
- The percentage of the chemical which eventually is generated as waste
- The scale of routine processes, with consideration of the smallest quantity which will provide satisfactory results
- A description of chemical storage, including decentralized supplies
- Evaluation of possible substitutes for all routine processes which might generate less hazardous by-products
- Survey of purchasing procedures, particularly for those departments and individuals who generate the largest volumes of unused reagent wastes
- Examination of the inventory management system to determine whether it takes into consideration all supplies of any specific chemical
- Consideration of possible reclamation or neutralization of by-products
- A review of municipal waste policies to determine which wastes go into the trash or are discharged to a public sewer
- Evaluation of recycling/reuse procedures
- Examination of cleaning and housekeeping procedures

31.4 Purchasing and Inventory Management

An interesting statistic available from academia is that students tend to use three to five times the amount of chemical necessary for experiments when they are allowed to dispense their own chemicals from supply bottles. These students later become laboratory professionals, but will they change their habits? By extension of this supposition, isn't it natural to buy a little more chemical than needed, just in case? Tracking the purchase and use of just one common chemical in the laboratory can answer this question. The most frequent problem observed is decentralized storage (in individual laboratories or area storerooms) resulting in surplus chemicals in one physical area of a laboratory facility that could be used in another. Chemical inventory systems developed in the last decade, including bar chart labeling, are designed to prevent this by providing precise location for each chemical purchased. The most difficult situation for centralized inventory management solutions is when individual grant researchers have the ability to purchase their own chemicals outside of normal purchasing controls. In many of these situations, particularly in universities and hospitals, the facility handles all waste disposal without any back-charging to the individual or department. Incentive to reduce waste is then lost.

31.5 Downsizing

The concept of minimizing waste by minimizing chemical use is the most basic of all pollution prevention measures. A reduction in scale can have a number of benefits: less chemical is purchased, waste volumes are smaller, process equipment may be less expensive, and energy costs may be reduced. Storage requirements for both equipment and chemicals are also reduced.

One potential problem with downsizing is the perception that a lower per-unit purchase cost automatically translates to a lower overall cost. Unfortunately, the purchase price of chemical reagents today is largely a factor of such costs as packaging, containerization, labeling, and shipping. These costs do not decrease significantly for smaller quantities of chemical, and the cost of the chemical itself is not enough to completely offset these other costs. Thus, the purchase price of a smaller volume is nearly the same as for a much larger container. The way to beat this argument is to point out that the disposal of surplus chemical can far offset the per-unit purchase difference. Based on 1992 dollars, *the average cost for the disposal of a pound of chemical reagent is $6.73.*[1] Adding this to the purchase price will present a more accurate picture of the cost of a chemical. Combining this with the difference in storage costs can help make the decision to downsize even easier (see Fig. 31-1).

31.6 Product Substitution

While there are many laboratory processes that are etched in stone, such as required Environmental Protection Agency methods for various environmental test procedures or clinical testing, there remains a great deal of flexibility on chemical usage in many laboratory operations. Table 31-1 is a chart summarizing a number of successful substitutions.

In addition to substitution of less hazardous chemicals where possible, efforts should be made to eliminate or reduce the quantities of scale for laboratory procedures which use chemicals that include reactive or explosives, halogenated solvents, or heavy metals such as arsenic, barium, cadmium, chromium, lead, mercury, selenium, and silver.

31.7 Computer Simulation

While purists may object to any replacement of wet chemistry techniques in the laboratory, there are strong arguments for the computer simulation of chemical

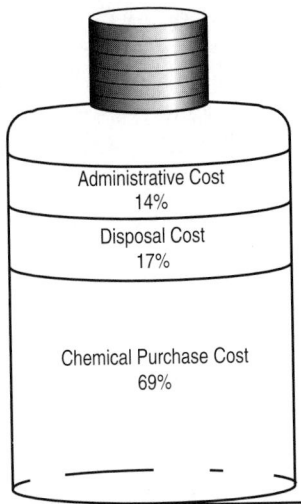

Administrative Cost
14%

Disposal Cost
17%

Chemical Purchase Cost
69%

Estimated costs based on 1993 dollars. Administrative costs include labor for purchasing, inventory control, distribution, and disposal. Costs are based on the purchase of a one pound container at $25.00/lb. which is later discarded as waste.

Figure 31-1. Laboratory reagent purchase/disposal economics.

Table 31-1. Possible Substitutes for Hazardous Chemicals

Procedure	Hazardous chemical	Substitute
Glassware cleaning	Chromic-sulfuric acid solutions	Detergents, enzymatic cleaners
Qual. test for halide ions	Carbon tetrachloride	Cyclohexane
Organic synthesis	Chromate ion	Hypochlorite ion
Qual. test for heavy metals	Sulfide ion	Hydroxide ion
Measurement of vapor pressure-temperature by isotensiscope	Carbon tetrachloride	Isopropyl alcohol
Determination of molecular weight by freezing point lowering methods	Benzene	Cyclohexane
Temperature	Mercury thermometers	Red liquid thermometers
Storage of biological specimens	Formaldehyde	Ethanol or other preservative
Organic synthesis, etc.	Ethyl ether	Methyl t-butyl ether
In-phase change and freezing point depression	Acetamide	Stearic acid

reactions. The opportunities for pollution prevention are the most obvious—no chemicals are used at all. In addition, administrative costs for chemical purchasing, storage, distribution, and waste disposal are all reduced or eliminated. As these cost incentives become more important and the potential advantages in terms of speeding research and development are realized, additional emphasis is placed on computer simulations. It can be expected that improvements in computer memory, speed, and software will encourage more and more workplaces and academic institutions to explore the advantages of simulating experiments. This will undoubtedly contribute to a substantial decline in pollution generation in many laboratories.

31.8 Regulatory Structure

Laboratories represent the proverbial round peg in a square hole as far as regulations are concerned. With few exceptions, laboratories must comply with air, water, and hazardous waste regulations that were clearly designed for industrial facilities dealing with a much larger scale of chemical usage. Regulations that focus on specific volumes of pollutants favor laboratories; those that control concentration without regard to volume represent a potential compliance problem for labs.

31.8.1 Air Discharges

The regulatory structure is clearly advantageous for laboratories in such areas as air discharges. Generally, facility discharges of less than 10 tons per year of volatile organic compounds (VOCs) are not regulated. Few, if any, laboratory facilities emit even 1 ton of VOCs. This, of course, leaves laboratories no clear incentive to reduce or eliminate air emissions except when the material can be effectively and efficiently reclaimed and reused. Nonetheless, there are obvious ways for laboratories to reduce air pollution. The most obvious is to avoid the practice of evaporating solvents in an open system, particularly in fume hoods. Reducing the scale of laboratory processes will also result in a reduction in air discharges.

31.8.2 Water Discharges

Water discharges present a more difficult dilemma for laboratories. While there are few federal limits on the discharge of many chemicals to a publicly owned treatment works (POTW), local regulations may make the discharge of many VOCs and metal-bearing inorganic wastes quite difficult. The discharge of such volatiles as chloroform, methylene chloride, toluene, and benzene—all common laboratory chemicals—may be regulated in the low parts per billion range.

Many POTWs have trouble deciding how to handle laboratory discharges; there are no categorical pretreatment standards for laboratories as there are for other industries. Permit applications must be designed, and analytical costs for the wide range of potential compounds are frequently more stringent than for traditional industrial discharges. Since discharges frequently vary, and may represent only trace quantities of contaminants, the challenge for POTWs is in designing a discharge permit that takes into account both the concentration of contaminants and the overall volume as a percent of wastewater handled.

The challenge for laboratories is the prevention of accidental as well as routine discharges of pollutants. To illustrate this point, assume that the monthly volume of wastewater from a laboratory facility is 2500 gallons (not including domestic sewage). Assuming the allowable discharge concentration limit for benzene is 5 ppb, a total volume of 0.1 ml will exceed the limit. Nonetheless, the POTW will not be able to detect the discharge, and routine aeration processes will result in total volatilization and evaporation of the chemical during treatment at the POTW.

The same scenario would apply for metal wastes. While the discharge permits under NPDES (National Pollution Discharge Elimination System) by a POTW restrict the release of such compounds as zinc and copper to a stream at levels as low as 20 ppm, the total volume that may be present as laboratory discharges would probably not be detectable. Nonetheless, a laboratory could easily exceed discharge limits by concentration.

The solutions to these problems are not easy; the basic tenets of pollution prevention are aimed at the significantly higher volumes generated by traditional manufacturing industries. While laboratories may not feel that the economies of scale dictate a need for pretreatment, this may be far more practical than assumed. Many laboratories perform basic acid-base neutralization, which can precipitate out metals, and there are a variety of possible organic treatment methods. These include carbon filtration, oxidation, and air-stripping procedures.

31.8.3 Hazardous Waste Management

Laboratories have both advantages and disadvantages in complying with hazardous waste regulations. While a large number of different wastestreams may be generated, the quantities are generally small. Off-site disposal can be quite expensive on a per-unit basis; the costs for packaging, labeling, waste handling, and transportation are virtually the same for smaller quantities. There is no economy of scale in this case—the larger the wastestream, the lower the per-unit cost for disposal. Nonetheless, the regulatory structure may allow some flexibility for laboratories, specifically in terms of waste reduction methodology.

One basic tenet of the Resource Conservation and Recovery Act (RCRA) and its amendments is the responsibility of the generator to make waste declarations. Specifically, laboratories may have some flexibility in terms of *when* a waste or process by-product is declared a waste. For example, a routine process which generates a hazardous by-product may be changed (and documented) to include an additional step which results in a less hazardous or nonhazardous by-product. The treatment of hazardous waste is strictly controlled, requiring a

Treatment, Storage, and Disposal Facility (TSDF) permit. The issue of when a waste becomes a waste, then, is crucial. While there are numerous acceptable laboratory methods to render specific compounds nonhazardous, most of these procedures are not legal without a permit. With the exception of acid-base neutralization and treatment in an accumulation container (which may not be legal in some states), treatment of hazardous wastes in the laboratory can result in significant (up to $10,000/day) penalties. While reclamation is generally perceived to be beyond the regulatory structure, at least one state (Florida) has fined a laboratory facility for reclaiming waste solvent without a permit. In other words, treatment of hazardous waste in the laboratory, while logical and achievable in many cases, is a regulatory land mine. The alternative—off-site disposal—is expensive and fraught with its own potential liabilities. This again results in the logical argument that prevention is the best solution.

Table 31-2 charts some laboratory wastes that are amenable to treatment or neutralization. DO NOT ATTEMPT ANY OF THESE PROCEDURES WITHOUT THOROUGHLY INVESTIGATING AVAILABLE DOCUMENTATION OF THE WRITTEN PROCEDURES. SAFETY CONCERNS SHOULD ALWAYS TAKE PRECEDENCE OVER POTENTIAL COST SAVINGS. Thermal treatment procedures are not recommended due to regulatory concerns. Again, any procedures utilized for treatment of by-products should be made part of the overall written procedure for the laboratory process. Again, please note that this list is intended as a guide on possible technologies; do not attempt any of these procedures without complete documentation.

Aside from treatment of wastes under these controlled conditions, reclamation, recycling, and reuse of laboratory wastes is generally acceptable under the RCRA with few restrictions. Any hazardous waste must be managed as such until such time as it undergoes one of these processes, but all are viable in specific instances, and all can represent both effective pollution prevention and economically sound practice. Waste solvents, in particular, may be suitable for reclamation and reuse either in the laboratory or in other areas of the facility. An example might be mixed flammable solvents which can be either distilled and reused or reclaimed and used as a paint thinner or fuel supplement. In-house recycling programs that treat selected surplus reagents as raw material until a user can be found are quite common and usually effective.

31.9   Waste Minimization Requirements in the Laboratory

As with other industries, there are relatively few regulatory mandates for waste minimization. The most visible exception is the signed certification on every hazardous waste manifest. Large (>1000 kg/month) generators of hazardous waste are required to sign a statement on each hazardous waste manifest certifying: "I have a program in place to reduce the volume and toxicity of waste generated to the degree I have determined to be economically practicable and

Table 31-2. Laboratory Waste That Are Amenable to Treatment to Neutralization

Chemical	Treatment/neutralization method
Alkyl halides	Hydrolysis using ethanolic potassium hydroxide
Phenol	Hydrogen peroxide with iron catalyst
Mercaptans, carbon disulfide	Oxidize to a sulfonic acid group with sodium hypochlorite or calcium hypochlorite
Acid halides and anhydrides	Hydrolyze using sodium hydroxide solution
Aldehydes and ketones	Oxidize using permanganate
Aromatic amines	Deamination using hydrochloric acid and sodium nitrite
N-nitroso compounds	Reduction by aluminum-nickel alloy in alkali
Hydroperoxides	Addition to acidified ferrous sulfate solution
Aqueous solutions containing toxic metal ions	Precipitate as insoluble sulfides using sodium sulfide in neutral solution
Oxidizing agents (hypochlorites, chlorates, chromates, etc.)	Reduce using sodium bisulfite
Metal hydrides	Gradual addition of methanol, ethanol, or N-butyl alcohol
Soluble metal fluorides	Treat aqueous solutions with calcium chloride solution
Inorganic cyanides	Oxidation using aqueous sodium hypochlorite
Metal azides	Reaction with nitrous acid
Finely divided metals	Oxidation with water

that I have selected the practicable method of treatment, storage or disposal currently available to me which minimizes the present and future threat to human health and the environment." Small quantity generators must certify: "I have made a good faith effort to minimize my waste generation and select the best waste management method that is available to me and that I can afford."

Other regulatory requirements or industry cooperative projects are aimed specifically at large industry. Laboratories typically represent less than 0.1 percent of air and water discharges of pollutants; regulatory pressures to reduce pollution are rarely directed at laboratories.

31.10 Designing a Laboratory Pollution Prevention Program

Without the support and authority from upper management, any pollution prevention program is destined to fail. Solicitation of management support may be preceded, however, by a thorough assessment of all discharges to the environment, including the off-site disposal of hazardous waste. Management must be able to anticipate some combination of reduced cost, improved public relations, and regulatory compliance.

Once upper management support is received, a written plan to address the specific methodologies is important. Such a plan should be based on the assessment of pollution prevention opportunities. The plan should have a schedule of implementation; most laboratories find that the best way to show progress is to make easy changes first. This might include the development of an internal surplus chemical exchange, modifications to chemical storage procedures, or simple equipment or process changes.

Above all, pollution prevention requires the education and cooperation of all workers (and students) in the laboratory. Without the support of the bench chemist, pollution prevention will fail. With his or her support, great progress can be made.

Reference

1. American Chemical Society Task Force on Laboratory Management, *Less is Better,* Washington, D.C., 1993.

Further Reading

American Chemical Society Task Force on Laboratory Waste Management, *Laboratory Waste Management, A Guidebook,* ACS Books, Washington, D.C., 1993.

Ashbrook, Peter, Cynthia Klein-Banay, and Chuck Maier, *Pollution Prevention in Laboratories; The How-to Guide,* Division of Environmental Health and Safety, University of Illinois, 1992.

National Research Council, *Prudent Practices for Disposal of Chemicals from Laboratories,* National Academy Press, Washington, D.C., 1983.

Potter, Caren D., "Cutting Hazardous Waste Disposal Costs in Lab Research," *The Scientist,* April 19, 1993.

"Waste Minimization: P2 at 3M," *Laboratory Safety & Environmental Management,* June 1993.

"Watch Your Step When Minimizing Your Wastes," *EM Scientist* **2**(3):3 (1993).

32
Solvent Substitutes

Stephen P. Evanoff, P.E., D.E.E.
Lockheed Fort Worth Company
Fort Worth, Texas

32.1 Introduction

Petroleum-based solvents have been used liberally as cleaning agents across most industrial sectors in a variety of applications related to fabrication, manufacturing, assembly, and maintenance activities. Historically, industry has selected low-molecular-weight, high-volatility, petroleum-based solvents in order to achieve maximum removal of soils and contaminants, to ensure compatibility with work surfaces and subsequent process requirements, to minimize the need for user technique and process sophistication, and to minimize associated capital and operating costs. The environmental and industrial hygiene regulations promulgated since 1980, most notably the Superfund Amendments and Reauthorization Act (SARA), the Hazardous and Solid Waste Amendments to the Resources Conservation and Recovery Act (RCRA), and the Clean Air Act Amendments of 1990, have brought about an increased emphasis on user exposure, hazardous waste generation, and air emissions. As a result, industry is performing a fundamental reassessment of cleaning solvents, processes, and procedures. The more progressive organizations have made their goal the elimination of solvents that may pose significant potential human health and environmental hazards. The typical hierarchy when one investigates solvent substitution options is to

1. Eliminate the need for cleaning

2. Substitute nonsolvent materials and processes

3. Substitute alternative solvents and processes with reduced hazards and emissions

4. Implement emissions control procedures and technologies

This chapter discusses solvent cleaning in metal-finishing, metal-manufacturing, and industrial maintenance applications; precision cleaning; and electronics manufacturing. Nonmetallic cleaning, adhesives, coatings, inks, and aerosols also will be addressed, but in a more cursory manner.

32.1.1 Solvent History

Halogenated solvents, high-volatility oxygenated solvents, and high-volatility aromatic and aliphatic hydrocarbons have traditionally served as "general purpose cleaners." Table 32-1[1] summarizes the commonly used industrial solvents and their historical processes. Table 32-2[2] lists several fundamental physical and chemical properties of these solvents. Industry dependence on the five halogenated solvents, particularly 1,1,1-trichloroethane (TCA), evolved due to their reduced toxicity relative to their predecessors, their low flammability, their

Table 32-1. Historical Industrial Solvent Use

Solvent	Process				
	Ambient immersion	Heated immersion*	Vapor degreasing	Spray	Manual
Trichloroethylene (TCE)	X	X	X		X
1,1,1-trichloroethane (TCA)	X		X	X	X
Tetrachloroethene (PERC)	X	X	X	X	X
Dichloromethane (METH)	X	X	X	X	X
1,1,2-trichloro-1,2,2-trifluoroethane (CFC-113)		X	X	X	X
Acetone					X
Methyl ethyl ketone (MEK)	X				X
Methyl isobutyl ketone (MIBK)	X				X
Toluene	X				X
Xylenes	X				X
Low-flashpoint aromatic and aliphatic hydrocarbons and blends†	X				X

*Typically in conjunction with vapor degreasing, i.e., immersion in the boiling-liquid sump.
†Includes PD 680 Type I, PD 680 Type II, and Stoddard solvent.
SOURCE: Derived from Ref. 1, p. 203.

Table 32-2. Relevant Properties of Historical Industrial Solvents

Solvent	Chemical formula	Molecular weight	Specific gravity	Boiling point (°C)	Flashpoint (°C)	Surface tension (dyn/cm)*	Kauri butanol value	Vapor pressure (mmHg)†	TLV (ppm)‡
TCE	$CHClCCl_2$	131.4	1.46	86–88	None	29.3	130	65	50
TCA	CH_3CCl_3	133.5	1.34	72–88	None	25.4	124	100	350
PERC	CCl_2CCl_2	165.9	1.62	120–122	None	31.3	91	21	25
METH	CH_2Cl_2	84.9	1.33	39–40	None	26.5	132	358	500
CFC-113	CCl_2FCClF_2	187.38	1.56	47.6	None	17.3	31	304	1000
Acetone	CH_3COCH_3	58.03	0.79	56–57	−17	23.7	—	220	750
MEK	$CH_3COC_2H_5$	72.1	0.80	79–81	−2.2	24.6	—	84	200
MIBK	$CH_3COCH_2CH(CH_3)_2$	100.16	0.82	112–118	17.6	22.7	—	7	50
Toluene	C_7H_8	92.15	0.87	110.6	4.4	28.5	94–105	26	50
Xylenes	$C_6H_4(CH_3)_2$	106.17	0.86–0.88	138–144	26.7	28.4–30.1	94	8	100

*At 20°C.[1,2]
†At 20°C as interpolated from D-170–D-172, *Handbook of Chemistry and Physics,* 55th ed.[9]
‡OSHA Permissible Exposure Limit, 8-hour time-weighted average, parts per million in air.[1,2]
SOURCE: Refs. 1, 2, and 9.

superior soil and contaminant dissolution capabilities, their compatibility with work surfaces and process equipment, and their relatively low cost. In contrast, toluene, xylene, methyl ethyl ketone (MEK), acetone, and the low-flash-point aromatic and aliphatic naphthas have been restricted to use as ambient-temperature immersion cleaners and manual-wipe cleaners due to their flammability and potential health hazards. Overall, industry dependence on the solvents identified in Table 32-1 resulted from a rational analysis of the key criteria considered at that time. Clearly, environmental and industrial hygiene considerations are causing a shift in these historical applications.

32.1.2 Regulatory Framework

Table 32-3 identifies the principal environment concerns associated with each of the five halogenated hydrocarbon solvents, MEK, MIBK, acetone, toluene, and the xylenes(s). These concerns are being addressed through the following U.S. environmental laws as well as international agreements.

Montreal Protocol on Substances That Deplete the Ozone Layer

Clean Air Act Amendments of 1990

EPA Industrial Toxics Program (voluntary)

Superfund Amendment and Reauthorization Act

Hazardous and Solid Waste Amendments to the Resource Conservation and Recovery Act

The Copenhagen Adjustments to the Montreal Protocol on Substances That Deplete the Ozone Layer, ratified in October 1992, require that developed

Table 32-3. Historical Industrial Solvent Environmental, Health, and Safety Issues

Solvent	Stratospheric ozone depletion	Ozone depletion potential[a]	Volatile organic compound[b]	Hazardous air pollutant[c]	Historical soil and groundwater contamination	Flammable[d]	Carcinogenicity
TCE	No	0	Yes	Yes	Yes	No	See note[e]
TCA	Yes	0.2	No	Yes	Yes	No	No
PERC	No	0	No	Yes	Yes	No	See note[f]
METH	No	0	No	Yes	No	No	See note[g]
CFC-113	Yes	0.8	No	No	No	No	No
Acetone	No	0	Yes	Yes	No	Yes	No
MEK	No	0	Yes	Yes	Yes	Yes	No
MIBK	No	0	Yes	Yes	No	Yes	No
Toluene	No	0	Yes	Yes	Yes	Yes	No
Xylene(s)	No	0	Yes	Yes	Yes	Yes	No

[a]Ozone depletion potential (CFC-11 = 1.0) as listed in Refs. 1 and 2.
[b]Per Title I of the Clean Air Act Amendments of 1990.
[c]Per Title III of the Clean Air Act Amendments of 1990.
[d]Defined as possessing a flashpoint less than 38°C.
[e]The EPA has not formally classified TCE in Category B2 as a "probable human carcinogen," while the International Agency for Research on Cancer (IARC) has classified this solvent in Group 3, a substance not classifiable as to its carcinogenicity in humans.
[f]The EPA has not formally classified PERC in Category B2 as a "probable human carcinogen," while the IARC has classified this solvent in Group 3, a substance not classifiable as to its carcinogenicity in humans.
[g]The EPA has not formally classified METH in Category B2 as a "probable human carcinogen," while the IARC has classified this solvent in Group 3, a substance not classifiable as to its carcinogenicity in humans.

nations eliminate the manufacture of Class I ozone-depleting compounds (ODCs) by January 1, 1996. The Class I ODC cleaning solvents of concern to use are the following: 1,1,2-trichloro-1,2,2-trifluoroethane, or chlorofluorocarbon 113 (CFC-113); and 1,1,1-trichloroethane (TCA), or methyl chloroform. Figure 32-1[2] shows a breakdown of worldwide TCA consumption. In the United States, Du Pont has announced that they will close their CFC production facilities in developed nations by January 1, 1995.[3] Over 50 electronics manufacturers from Canada, Germany, Japan, Sweden, and the United States halted by May 1993 the use of CFC-113 in global operations.[3] Over 60 multinational companies also completed their global phaseout of CFC-113 in 1993.[3] The rapidly escalating cost of ODC solvents, mainly due to federal excise taxes and their decreased availability, is providing incentive to users of TCA and CFC-113 as bulk cleaning agents to eliminate these compounds as soon as technically feasible.

Title VI of the Clean Air Act Amendments of 1990 also addresses ODCs. Title VI contains a list of all Class I and Class II ODCs. Final rules have been promulgated for Title VI, Section 604 and Section 605, which address phase-out of production of both Class I and Class II ODCs, Section 610 on nonessential uses, and Section 611 on products containing or manufactured with Class I ODC. Section 612 establishes a framework for evaluating potential alternatives. The

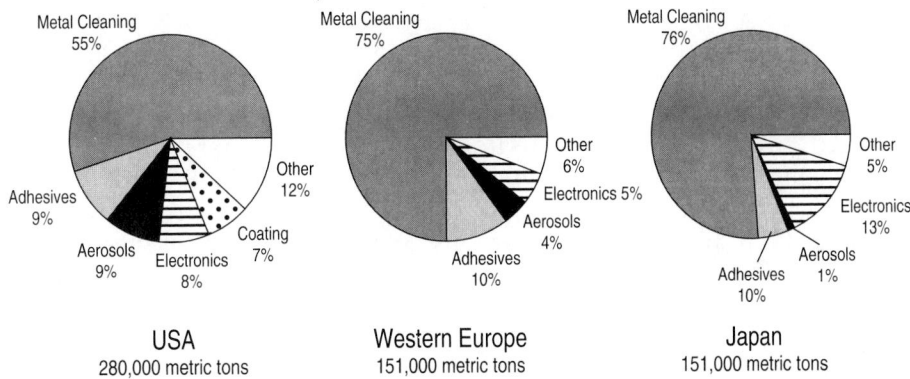

Figure 32-1. Breakdown of worldwide TCA use. (*Courtesy of Chem Systems, 1989.*)

Safe New Alternatives Program (SNAP) of the U.S. Environmental Protection Agency (EPA) will include a list of acceptable substitute categories and a list of chemicals prohibited as substitutes for specific applications. The EPA published the final rule for SNAP in March 1994 in the *Federal Register*.[4] Title III of the CAAA requires regulation of 189 chemicals referred to as *hazardous air pollutants* (HAPs). This list includes all of the solvents discussed above. In addition, Title III will place restrictions on industrial operations located in areas defined as *nonattainment areas* by the EPA. Those solvents defined as *volatile organic compounds* (VOCs) and HAPs (see Table 32-3) are in the process of becoming strictly regulated as to their use and emissions. Industrial users of these solvents will be subject to additional record-keeping and reporting requirements. Significant capital costs may be incurred in cases where these solvents are used in process equipment, due to the need for installation of emission control devices or new process equipment to reduce emissions. The detailed rules for HAPs and VOCs are being promulgated and are anticipated to become effective beginning as early as mid-year 1995. Title III also contains requirements for the prevention of accidental releases of these substances. Industrial facilities will be required to provide a risk management plan for prevention, detection, and response in case of emergency releases.

The EPA Industrial Toxics Program, also called the 33/50 Program, is a voluntary effort to eliminate the emission and off-site transfer of 17 toxic chemicals. The voluntary goal established is a reduction in emissions and transfers of 33 percent by the end of 1992 and of 50 percent by the end of 1995, the 1988 baseline levels. The 17 chemicals include TCA, trichloroethylene (TCE), methylene chloride, perchloroethylene, MEK, methyl isobutyl ketone (MIBK), toluene, and xylenes. Over 1000 major companies had committed to this program as of early 1994.

SARA has required an annual Toxics Release Inventory (TRI) since 1987. The list of off-site chemical transfers and emissions that are to be reported by industrial facilities includes the solvents listed in Table 32-1. In addition, SARA requires contingency and emergency response capabilities for facilities that store and use these chemicals in manufacturing processes.

Lastly, the Hazardous and Solid Waste Amendments to RCRA have placed stringent requirements on disposal of wastes derived from the solvents listed in Table 32-1. These hazardous wastes are restricted from land disposal. Disposal through incineration or as waste-derived fuel-blending stock is increasingly costly and carries potential long-term liabilities for the generator. The chlorinated solvents are the most amenable to recycling. But, under RCRA, recycling of these spent solvents requires management of the spent solvent as hazardous waste and use of the Uniform Hazardous Waste Manifest when these wastes are shipped off-site.

Essentially, the use of any of the solvents listed in Table 32-1 and 32-2 carries a significant environmental regulatory burden that may result in additional capital and operating costs and possible long-term liabilities. Most progressive industries are restricting use of these solvents to those applications in which the use is essential to the product's performance.

The following sections summarize the general technology trends in the elimination of chlorinated and high-volatility petroleum-based solvents use for various applications. The information is organized by industry sector and incorporates information summary tables and a list of observations as to industry trends. A brief list of key references is given at the end of each major section.

32.2 Metal Fabrication and Finishing

In this section, metal-cleaning alternatives are discussed in general terms for this broad industry sector. Relevant categories in the metal-manufacturing industry include automotive, aerospace, rail, ship, appliance, and tooling. Chlorinated-solvent vapor degreasing and ambient-temperature immersion cleaning have been the mainstay of forming-fluid and machining-lubricant removal in metal fabrication and -finishing industries. Wipe cleaning with MEK and high-evaporation-rate solvent blends—e.g., toluene, aromatic naphtha, ethyl acetate, as well as TCA and CFC-113—has been widely used prior to application of coatings to detail parts and during assembly of products. The broad range of metals and alloys and the variety of surface treatments and coatings applied after cleaning make this an industry sector in which alternate materials and processes tend to be numerous and somewhat application-specific.

32.2.1 Summary of Options and Recommendations

Table 32-4[2] summarizes the viable alternative material and process categories for metal cleaning as they pertain to fabrication and finishing. A more detailed summary is given in Table 32-5.[5] Figure 32-2 on page 526 shows the generalized aqueous immersion-cleaning system that is being considered as a replacement for vapor degreasing in most metal-finishing industries. While it is difficult to

Table 32-4. Summary of Metal-Finishing Alternative Materials and Processes

	Alternative process				
Alternate material	Ambient immersion	Heated immersion	Vapor degreasing[a]	High-pressure spray	Manual wipe
Aqueous, alkaline	X	X		X	X[b]
Aqueous, emulsion	X	X		X[c]	X
Low-vapor-pressure solvent blends	X	X[c]		X[c]	X[d]
Aliphatic petroleum hydrocarbons	X[c]				X[d]
Terpene hydrocarbons	X[c]				X[d]
Hydrocarbon-surfactant blends[e]	X[c]				X[d]
Steam				X	
Chlorinated solvents[f]			X		

[a] Specially designed systems which minimize fugitive emissions through enclosed design and vapor recovery devices.

[b] Surfactant-in-water blends under development; requires corrosion-resistant surface and ability to dry.

[c] Appropriate flammability protection required.

[d] Appropriate residue removal technique required.

[e] Includes aliphatic petroleum and terpene hydrocarbons with surfactant chemical additives; these are intended to facilitate rinsing of residues and formation of stable emulsion cleaners when mixed with water.

[f] Includes PCE and TCE, only when used in a specially designed vapor degreaser.

generalize across industrial sectors and for different facilities, the following technical trends would be agreed upon by most scientists and engineers in the metal-manufacturing sector.

A "systems" analysis of the entire manufacturing process frequently reveals opportunities to eliminate redundant or unnecessary cleaning steps and identifies opportunities to reduce the stringency of cleanliness required for a particular operation.

"No clean" technologies are emerging. These technologies will ultimately preclude the need for traditional cleaning methods as they mature; presently they are either not technically feasible or economically unreasonable.

Thermal vacuum de-oiling, solid-film or polymer sheet lubricants, and ultrasonics will emerge as "no clean" alternative metal-forming and -manufacturing techniques.

Aqueous alkaline and emulsion cleaners are dominating the replacement of batch and continuous solvent vapor degreasing and cold cleaning in the United States. Both immersion and spray applications are being employed.

Ultrafiltration can extend aqueous cleaner life.

Hydrocarbons, TCE, and perchloroethylene are being used as well as aqueous cleaners in several European countries as replacements for TCA and CFC-113.

Metals susceptible to oxidation (e.g., D6AC high-strength steel) and metals susceptible to embrittlement and corrosion (e.g., titanium and 7000 series

Table 32-5. Metal-Cleaning Processes

Processes are Listed in Order of Decreasing Preference

Type of production	In-process cleaning	Preparation for painting	Preparation for phosphating	Preparation for plating
Removal of pigmented drawing compounds[a]				
Occasional or intermittent	Hot-emulsion hand slush, spray emulsion in single stage, vapor slush degrease[b]	Boiling alkaline, blow-off, hand wipe Vapor slush degrease, hand wipe Acid clean[c]	Hot-emulsion hand slush, spray emulsion in single stage, hot rinse, hand wipe	Hot alkaline soak, hot rinse (hand wipe, if possible), electrolytic alkaline, cold-water rinse
Continuous high production	Conveyorized spray emulsion washer	Alkaline soak, hot rinse, alkaline spray, hot rinse	Alkaline or acid[d] soak, hot rinse, alkaline or acid[d] spray, hot rinse	Hot-emulsion or alkaline soak, hot rinse, electrolyte alkaline, hot rinse
Removal of unpigmented oil and grease				
Occasional or intermittent	Solvent wipe Emulsion dip or spray Vapor degrease Cold solvent dip Alkaline dip, rinse, dry (or dip in rust preventive)	Solvent wipe Vapor degrease or phosphoric acid clean[d]	Solvent wipe Emulsion dip or spray, rinse Vapor degrease	Solvent wipe Emulsion soak, barrel rinse, electrolytic alkaline rinse, hydrochloric acid dip, rinse
Continuous high production	Automatic vapor degrease Emulsion, tumble, spray, rinse, dry	Automatic vapor degrease	Emulsion power spray, rinse Vapor degrease Acid clean[c]	Automatic vapor degrease, electrolytic alkaline rinse, hydrochloric acid dip, rinse[e]
Removal of chips and cutting fluid				
Occasional or intermittent	Solvent wipe Alkaline dip and emulsion surfactant Stoddard solvent or TCE Steam	Solvent wipe Alkaline dip and emulsion surfactant Solvent or vapor	Solvent wipe Alkaline dip and emulsion surfactant[f] Solvent or vapor	Solvent wipe Alkaline dip, rinse, electrolytic alkaline[g] rinse, acid dip, rinse[h]
Continuous high production	Alkaline (dip or spray) and emulsion surfactant	Alkaline (dip or spray) and emulsion surfactant	Alkaline (dip or spray) and emulsion surfactant	Alkaline soak, rinse, electrolytic alkaline[g] rinse, acid dip and rinse[h]

Table 32-5. Metal-Cleaning Processes (*Continued*)
Processes are Listed in Order of Decreasing Preference

Type of production	In-process cleaning	Preparation for painting	Preparation for phosphating	Preparation for plating
		Removal of polishing and buffing compounds		
Occasional or intermittent	Seldom required	Solvent wipe Surfactant alkaline (agitated soak), rinse Emulsion soak, rinse	Solvent wipe Surfactant alkaline (agitated soak), rinse Emulsion soak, rinse	Solvent wipe Surfactant alkaline (agitated soak), rinse, electroclean[j]
Continuous high production	Seldom required	Surfactant alkaline spray, spray rinse Agitated soak or spray, rinse[k]	Surfactant alkaline spray, spray rinse Emulsion spray, rinse	Surfactant alkaline soak and spray, alkaline soak, spray, and rinse, electrolytic alkaline[j] rinse, mild acid pickle, rinse

[a] For complete removal of pigment, parts should be cleaned immediately after the forming operation, and all rinses should be sprayed where practical.
[b] Used only when pigment residue can be tolerated in subsequent operations.
[c] Phosphoric acid cleaner-coaters are often sprayed on the parts to clean the surface and leave a thin phosphate coating.
[d] Phosphoric acid for cleaning and iron phosphating. Proprietary products for high- and low-temperature application are available.
[e] Some plating processes may require additional cleaning dips.
[f] Neutral emulsion or solvent should be used before manganese phosphating.
[g] Reverse-current cleaning may be necessary to remove chips from parts having deep recesses.
[h] For cyanide plating, acid dip and water rinse are followed by alkaline and water rinses.
[j] Other preferences: stable or diphase emulsion spray or soak, rinse, alkaline spray or soak, rinse, electroclean; or solvent presoak, alkaline soak or spray, electroclean.
[k] Third preference: emulsion spray rinse.

aluminum) may limit the viable substitutes for hydrocarbons or oxygenated solvents for these applications.

Complex part geometries, such as blind holes, tubing, and aircraft honeycomb structures may require special equipment and procedures—including part oscillation, rotation, or vibration in the cleaner bath; ultrasonics; or high-pressure spray—to obtain acceptable cleanliness using aqueous and emulsion cleaners.

Highly carbonaceous soils (e.g., Cosmoline) and high-molecular-weight soils (e.g., paraffin or polyethylene glycol waxes) require the use of terpene hydrocarbons, naphtha blends, or spray inpingement with high-pressure surfactant/steam, solid carbon dioxide, or sodium bicarbonate.

Use of water-soluble or emulsifiable forming, stamping, machining, and cutting fluids will enhance the applicability of aqueous and emulsion cleaners as chlorinated solvent vapor degreasing and cold-cleaning substitutes.

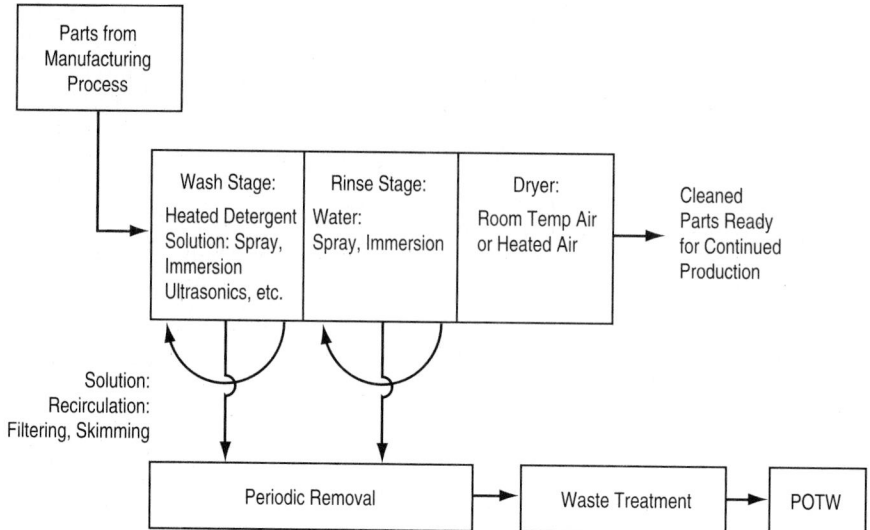

Figure 32-2. Configuration of aqueous cleaning process in the metal-cleaning industry. (*Courtesy of U.S. EPA, 1989a.*)

Manual wiping and cleanup solvents are being replaced with reduced-vapor-pressure solvent blends of 45 mmHg or less composite vapor pressure; many contain fewer HAP constituents.

Alternate wipe solvents and batch degreasers tend to be more application-specific; i.e., "general purpose cleaners" will be replaced by several materials and processes.

Operator and user techniques are more sophisticated and have lower tolerances with the substitute materials and processes as compared to the traditional solvents.

32.2.2   Further Reading and Case Histories

For further reading and case histories relevant to Sec. 32.2 topics, see Refs. 1–9.

32.3   Precision Instruments and Devices

32.3.1   Background[10]

The term *precision cleaning* is difficult to define. The following list provides an indication of the wide and varied applicability of precision cleaning.

Electronics (recording heads, solenoids, micro switches, servo motors)

Medical (heart pacemakers, contact lenses, prosthesis manufacture)

Defense (inertial guidance systems, ball bearings, hydraulic systems)

Aerospace (avionics maintenance, fuel systems, flight controls)

Precision-cleaning applications are characterized by the extremely fine level of cleanliness required to keep delicate instruments and surfaces operating effectively, as illustrated in Fig. 32-3. As Fig. 32-2 implies, precision component cleaning requires the use of more effective cleaning technologies than those traditionally employed in heavy industries, such as those discussed in Sec. 32.2, "Metal Fabrication and Finishing." The key factors that define precision cleaning are those applications in which

- Critical cleanliness standards of particulate and/or organic contaminants need to be satisfied

- Components have sensitive compatibilities

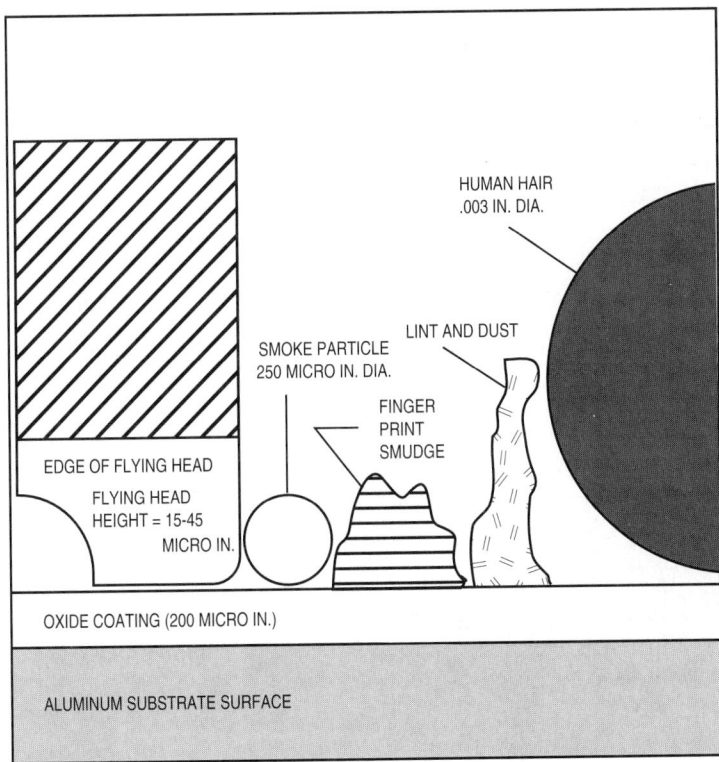

Figure 32-3. Size comparison of computer disk drive head clearance with various contaminants. (*Courtesy of Digital Equipment Corporation.*)

- Components have physical characteristics such as geometry or porosity which make dewatering crucial in the cleaning process
- Components being cleaned are costly

The necessary and desirable solvent properties include[10]

- A volatile solvent with zero nonvolatile residue
- Sufficient solvent power to dissolve organic soils
- Low surface tension to penetrate small spaces between surfaces and components
- High density to assist in displacing and suspending small insoluble particles
- Rapid drying
- Low cost
- Low toxicity
- Nonflammability

Industry has developed a strong dependence on CFC-113, CFC-113 alcohol azeotropes, and TCA as the primary precision instrumentation cleaning solvents.

32.3.2 Summary of Options and Recommendations

A number of options for precision cleaning are available. The following list[10] classifies the options as either alternative cleaning processes or alternative solvents.

Alternative cleaning processes

- Aqueous
- Semiaqueous
- Pressurized gas
- Supercritical fluids
- Plasma cleaning
- Ultraviolet/ozone

Alternate solvents

- Hydrochlorofluorocarbons (HCFCs)
- Alcohols with perfluorocarbons
- N-methyl-2-pyrrolidone
- Perfluoronated alkanes
- Aliphatic hydrocarbons
- Miscellaneous solvents (e.g., ketones, ethers)

Discussion here will be limited to the most widely applicable options. The documents referenced at the end of this section can be consulted to obtain greater detail on each of these materials and processes.

Aqueous. The process design and equipment selection are the key to achieving acceptable aqueous alkaline cleaning in precision applications. Immersion with mechanical agitation or ultrasonics and spray methods are those being most widely incorporated in the substitution for solvent-cleaning methods. Aqueous methods are generally superior to solvent methods in removal of inorganic, polar soils and of particulate matter. However, for high-molecular-weight oils and fluorocarbon greases, multiple cleaning steps or inclusion of cleaning-enhancement methods, such as ultrasonics, may be necessary. Among the significant concerns with aqueous cleaners are soil removal from parts with complex geometries; cleaning-material compatibility with polymers, elastomers, corrosion-sensitive substrate metals, alloys, and porous surfaces; and process control. The rinsing and drying steps are of particular concern. Equipment must be carefully designed and controlled. In addition, energy consumption, the need for high-purity water, and wastewater disposal can have a significant cost impact. Table 32-6 summarizes the advantages and disadvantages associated with each aqueous cleaning method. All of these factors should be considered carefully and within the broader context of the environmental regulatory issues described in Sec. 32.1. Figure 32-4 shows a generalized configuration of an aqueous cleaning process. This general design can also be applied to the general metal fabrication and metal-finishing applications discussed in Sec. 32.2.

Semi-Aqueous. Hydrocarbon/surfactant cleaners can either be emulsified in water and applied in a manner similar to the aqueous cleaners discussed above or they can be applied in concentrated form and rinsed with water. These methods and materials, collectively, are commonly known as *semi-aqueous cleaners*. The general advantages and disadvantages of these materials are similar to those of aqueous cleaners. Semi-aqueous methods can provide greater soil removal and emulsification of higher-molecular-weight hydrocarbon soils and greater compatibility with corrosion-sensitive materials. Semi-aqueous cleaners may pose flammability and odor concerns that must be addressed in the equipment design—e.g., terpene hydrocarbons. Semi-aqueous cleaner stability may also pose particular process control challenges, e.g., auto-oxidation, phase separation, and gelling of organic components. Figure 32-5 is a general conceptualized semi-aqueous cleaning system.

Supercritical Fluids.[6] *Supercritical fluids* (SCFs) are characterized by their existence in a critical temperature and pressure region as follows:

$$0.9 < T_r < 1.2 \qquad \text{and} \qquad 1.0 < P_r < 3.0$$

where T_r and P_r are reduced temperature and pressure, respectively. $T_r = T/T_c$ and $P_r = P/P_c$ where T and P are the actual temperature and pressure, respectively, and T_c and P_c are the critical temperature and pressure, respectively. When elevated to temperatures near or greater than the critical temperature and

Table 32-6. Advantages and Disadvantages of Aqueous Cleaning

Advantages	Disadvantages
Safety—Aqueous systems have few problems with worker safety compared to many solvents. They are not flammable or explosive. It is important, however, to consult the Material Safety Data Sheets for information on health and safety.	Cleaning difficulty—Parts with blind holes and small crevices may be difficult to clean and may require process optimization.
Cleaning—Aqueous systems can be readily designed to clean particles and films better than solvents.	Process control—Aqueous processes require careful engineering and control.
Multiple degrees of freedom—Aqueous systems have multiple degrees of freedom in process design formulation and concentration. This enables aqueous processes to provide superior cleaning for a wider variety of contamination.	Rinsing—Some aqueous cleaner residues can be difficult to rinse from surfaces. Nonionic surfactants are especially difficult to rinse. Trace residues may not be appropriate for some applications and materials. Special precautions should be applied for parts requiring subsequent vacuum deposition, liquid oxygen contact, etc. Rinsing can be improved using DI water or alcohol rinse.
Inorganic or polar soils—Aqueous cleaning is particularly good for cleaning inorganic or polar materials. For environmental and other reasons, many machine shops are using or converting to water-based lubricants and coolants versus oil-based. These are ideally suited to aqueous chemistry.	Floor space—In some instances, aqueous cleaning may require more floor space.
Oil and grease removal—Organic films, oils, and greases can be removed very effectively by aqueous chemistry.	Drying—For certain part geometries with crevices and blind holes drying may be difficult to accomplish. An additional drying section may be required.
Multiple cleaning mechanisms—Aqueous cleaning functions by several mechanisms rather than just one (solvency), including saponification (chemical reaction), displacement, emulsification, dispersion, and others. Particles are effectively removed by surface activity coupled with the application of energy.	Material compatibility—Corrosion of metals or delayed environmental stress cracking of certain polymers may occur.
	Water—In some applications high-purity water is needed. Depending on purity and volume, high-purity water can be expensive.
Ultrasonics applicability—Ultrasonics are much more effective in water-based solvents than in CFC-113 solvents.	Energy consumption—Energy consumption may be higher than for solvent cleaning in applications that require heated rinse and drying stages.
Chemical cost—Low consumption and inexpensive.	Wastewater disposal—In some instances use of aqueous cleaning may require wastewater treatment prior to discharge.

at high pressures, these compounds, which are gases at ambient temperatures, become fluids with high soil dissolution properties, high diffusivity, low viscosity, and low density. The advantages of this technology are its flexibility of application, rapidity of cleaning, and reasonable operating costs. The disadvantages include the high capital cost of designing and operating a system that

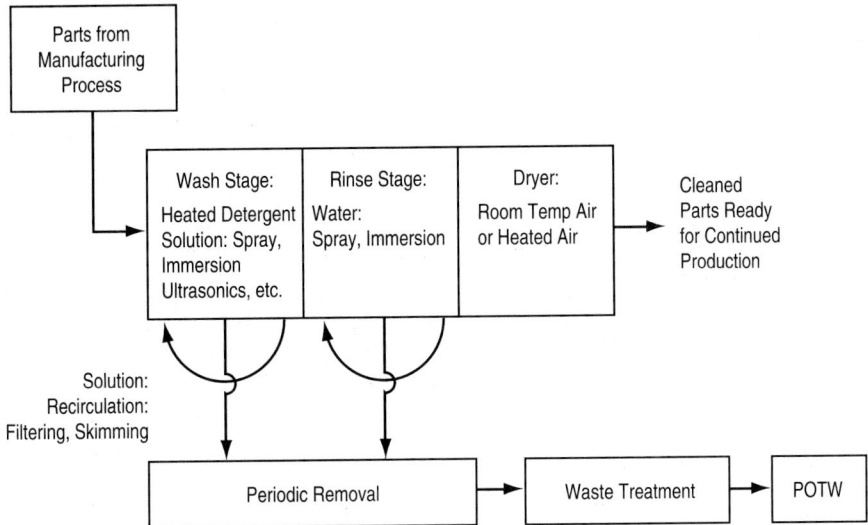

Figure 32-4. Configuration of aqueous cleaning process. (*Courtesy of U.S. EPA, 1989.*)

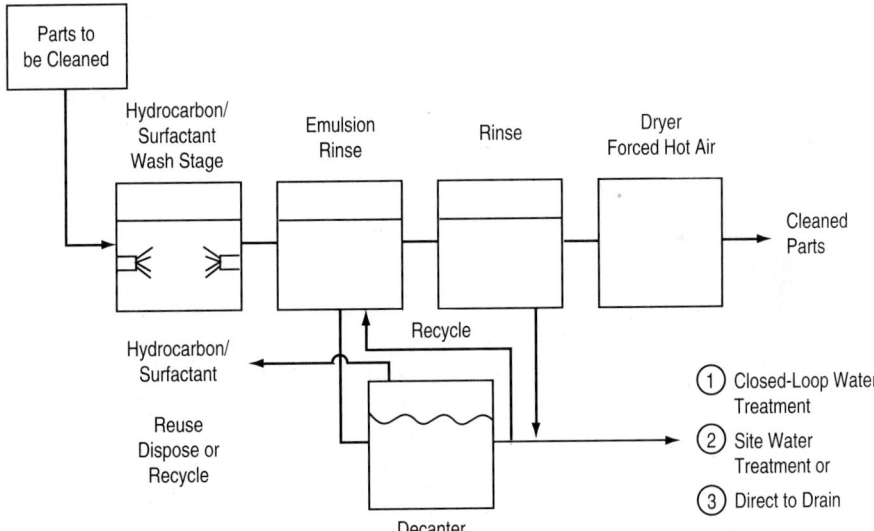

Figure 32-5. Semi-aqueous process for immiscible hydrocarbon solvent.

operates at high temperatures and pressures. These conditions may also be detrimental to the item being cleaned. Water, carbon dioxide, methane, ethane, ammonia, and ethanol are among the compounds being considered for SCF applications. Carbon dioxide has received the most investigation to date. Figure 32-6 shows a conceptualized SCF system that uses carbon dioxide.

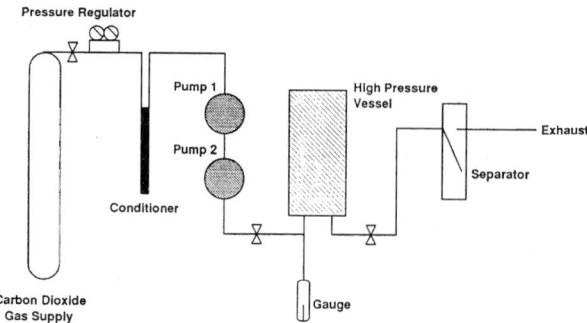

Figure 32-6. Basic model design for carbon dioxide supercritical cleaning system. (*Courtesy of Jackson, 1987.*)

Hydrochlorofluorocarbons.[10]   Use of HCFCs, particularly (1) HCFC-141b and (2) HCFC-225ca/cb, have been proposed as a means of replacing CFC-113 and TCA due to their similar physical and chemical properties. In these instances, similar equipment and procedures could be employed. HCFC-141b is not approved as a vapor degreaser replacement material under SNAP and is more corrosive than CFC-113. HCFC-225ca/cb is under SNAP review. HCFC-225 ca/cb is quite expensive as compared to CFC-113 and will only be available in limited quantities. Although the soil dissolution properties and system requirements make HCFC attractive, these compounds are listed as Class II ozone-depleting compounds in the Clean Air Act Amendments of 1990. As such, they are subject to production limitations and will likely be phased out no later than the year 2030. Figure 32-7 shows an advanced vapor degreaser designed to minimize atmospheric emissions even with low-boiling-point solvents.

Alcohols with Perfluoroalkanes.[10]   Alcohols have excellent soil dissolution properties for polar, polar organic, and hydrocarbon soils, particularly when heated or as a boiling liquid. However, the extreme flammability hazards associated with heated-alcohol cleaning have made this approach impractical in the past. A recently introduced technique, the use of a perfluorocarbon vapor "blanket" above heated alcohol in a cleaning system, renders the alcohol vapor non-flammable. Further, the perfluorocarbons and alcohol are relatively immiscible. Therefore, they can be used together and recovered for reuse with minimal mixing. This technology may prove useful with materials sensitive to aqueous or other solvents or with materials with complex parts geometries. However, such systems retain significant design and operating requirements. Perfluorocarbons are also expensive. For these two reasons capital and operating costs will be high.

Perfluorocarbons.[10]   This group of compounds have all hydrogen atoms substituted with a fluorine atom. Thus, they are highly stable molecules. They have low toxicity and are nonflammable. However, they are very costly. Also, because they are both infrared absorbers and very stable, they have a high

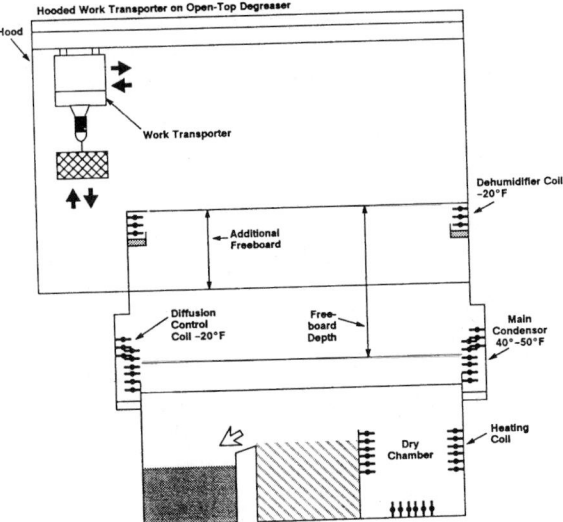

Figure 32-7. Advanced-design degreaser for use with low-boiling-point solvents. (*Courtesy of Du Pont.*)

global warming potential. Perfluoroalkanes' greatest potential application is in cleaning gyroscopes and guidance equipment that contain fluorinated lubricants and where elastomers and polymeric materials require cleaning.

32.3.3 Further Reading and Case Histories

For further readings and case histories relevant to Sec. 32.3 topics, see Refs. 1, 4, and 10.

32.4 Printed Circuit Board Manufacture and Assembly

32.4.1 Background[1]

The assembly of electronics components historically has created the need for the removal of residues from assemblies after soldering. Solder flux residues are characterized as a mixture of raw flux, flux decomposition products, reaction products between the flux components and metal oxides, and residues from soldering oils and contaminants from previous operations left on the boards.[1] CFC-113 became the prevalent defluxing and printed circuit board cleaning solvent by the late 1970s. The use of TCA and TCA-alcohol azeotropes in electronics cleaning has been less, but significant, nevertheless. No other single industry sector has been so completely affected due to the international phase-out of its two most widely used cleaning solvents. Both chemicals have been fundamental elements contributing to product performance and reliability. The electronics industry involves a variety of substrate materials, including numerous polymeric and elastomeric materials, complex geometries, and low-tolerance configurations. The complexities of electronics cleaning will not be discussed in detail here.

32.4.2 Summary of Options and Recommendations[1]

The six primary methods for replacing the highly volatile, ozone-depleting solvents and their applicability are summarized in Table 32-7, with CFC-113 and TCA serving as benchmarks. Table 32-8 provides information on the applicability, costs, and relative strengths and weaknesses of the eight material and process combinations prevalent in the substitution for CFC-113 and TCA. The Technology and Economic Assessment Panel under the Montreal Protocol has noted the following technical trends[3] in replacing ozone-depleting solvents in electronics manufacturing:

- No clean soldering is the first choice to eliminate CFC-113 and TCA. It can decrease costs, avoid hazardous solvent waste disposal, and if operated under controlled atmosphere, minimize dross containing lead oxide. Some manufacturers report increased production efficiency and improved reliability.

- For those applications where cleaning remains necessary, aqueous methods predominate.

Table 32-9 provides a comparison of aqueous-cleaning systems ranging from batch dishwasher types that are preferred by small specialty shops to convey-

Table 32-7. Cleaning Options to Replace CFC-113 and Methyl Chloroform

	No clean	Aqueous saponified underbrush	Semi-aqueous underbrush	Aqueous spray	Aqueous saponified spray	Semi-aqueous spray	CFC-113 and methyl chloroform
Rosin flux—PTH*	X	X	X		X	X	X
Water-soluble flux—PTH				X	X	X	
Synthetic activated flux—PTH						X	X
Rosin flux—SM†	X				X	X	X
Water-soluble paste—SM					X	X	
Rosin flux—mix top and bottom SM and PTH	X				X	X	X
Water-soluble flux—mix top and bottom SM and PTH				X	X	X	
Low-solids flux	X						
Controlled-atmosphere soldering	X						

*PTH=plated through hole.
†=Surface mount.
SOURCE: Ref. 1.

Table 32-8. Summary Comparison of Cleaning Processes

Issue	Hydrocarbon surfactant			Saponified aqueous			Aqueous	
	Underbrush	In-line spray	Batch	Underbrush	In-line spray	Batch	In-line spray	Batch
Equipment cost (mid-range)	$50,000	$160,000	$40,000	$50,000	$100,000	$40,000	$100,000	$40,000
Floor space requirement (minimum) (ft^2)	100	500	40	100	200	40	200	40
Applicable to surface mount technology (SMT)	No	Yes	Yes	No	Yes	Yes	Limited	Limited
Component compatibility issues	Low	Moderate	Moderate	Low	Moderate	Moderate	High	High
Defect rate	Average	Average	Average	Average	Average	Average	Lower	Lower
Throughput rate (average) (ft/min)	5	9	Variable	5	5	Variable	5	Variable
Complexity of installation	Low	High	Low	Low	Medium	Low	Medium	Low
Wastestream handling	Necessary	Necessary	Necessary	Necessary	Necessary	Necessary	Necessary	Necessary
Health and safety issues	Must be evaluated	Must be evaluated	Must be evaluated	Must be evaluated	Must be evaluated	Must be evaluated	Must be evaluated	Must be evaluated
Process flexibility	Low	High	High	Low	Medium	Medium	Low	Low
Idle-time cost	Low	Medium	Low	Low	High	Low	High	Low
Processing cost per square foot compared to CFC-113	5×	4×	3×	4×	2×	1×	1×	1×

Table 32-9. Comparison of Aqueous Process Machine Types*

	Brush	Dishwasher	High-throughput	Tank	Conveyorized
Principle	Underbrush	Spray	Spray	Immersion	Spray
Water soluble (W/S) flux	No	Yes	Yes	Yes	Yes
Saponifier	Possibly	Yes	Yes	Yes	Yes
Hydrocarbon/surfactant (HC/S) wash phase	Possibly	Yes	Yes	Yes	Yes
Surface mounted device (SMD)	No	Possibly	Yes	Yes	Yes
Throughput (m$^2 \cdot$ h^{-1})	10–40	1–2	5–25	2–25	5–50
Water conservation (L $\cdot$ m^{-2})	20–50	25–40	8–15	10–50	10–50
Electricity conservation	0.5–1	1–1.5	0.5–1	0.5–1.5	1–2
Ultrasonics	No	No	No	Possibly	No
Cleaning	Mediocre	Very good	Very good	Good to very good	Mediocre to very good
Drying available	Mediocre	Mediocre	Very good	Variable	Variable
Floor space (m^2)	2–3	2–3	6–10	10–20	10–20
Handling conveyor	Conveyor	Manual	Manual/robot	Hoist	Conveyor
Flexibility	Poor	Excellent	Excellent	Excellent	Poor

*The information given in this table is typical of machines available on the international market. The performance of individual machines may vary considerably and potential users are advised to check with their supplier as to the accuracy of this information under their specific conditions of use.

orized systems that are better suited for major circuit-manufacturing centers. Semi-aqueous cleaners, particularly terpene hydrocarbons, were among the substitutes initially considered and implemented by many companies wishing to make rapid transition from ozone-depleting solvents. Semi-aqueous cleaners appear to be more limited in application than aqueous methods due to cost associated with explosionproof equipment, waste disposal, and wastewater treatment.

32.4.3 Further Reading and Case Histories

For further reading and case histories on Sec. 32.4 topics, see Refs. 1, 3, and 11.

32.5 Maintenance Activities

32.5.1 Background

Maintenance activities at industrial and manufacturing facilities are defined here as the cleaning of products during rework or refurbishment and cleaning of tooling, equipment, work areas, and work surfaces associated with general factory housekeeping. Historically, many facilities used "general purpose cleaners" for both production and maintenance activities. Thus, the solvents listed in Table 32-1 serve as the historical baseline in this portion of the discussion as well.

32.5.2 Summary of Options and Recommendations

The general recommendations made in Sec. 32.2,, "Metal Fabrication and Finishing," apply to factory maintenance activities and rework or refurbishment of parts. However, the soils encountered during maintenance and repair of mechanical devices in products and equipment (motors, compressors, engines, vehicle wheels, etc.) tend to be oxidized, solidified, or reacted; are more carbonaceous in nature; and contain more insoluble particulate matter and salts than encountered during the manufacturing process. In these situations, mechanical impingement, such as plastic media blasting; high-pressure aqueous alkaline cleaner spray; high-pressure surfactant/steam spray; terpene hydrocarbon (particularly D-limonene) wipe and immersion; C7–C12 hydrocarbon wipe and immersion; C4–C6 acetate wipe; and blends of the aforementioned organic solvents as wipe and immersion cleaners offer the greatest degree of soil displacement, emulsification, or dissolution. Special attention must be paid to coated parts and composite materials to ensure that coatings and other sensitive components—e.g., gaskets, seals, and plastic devices—are not adversely affected by the cleaning material or process. Parts with complex geometries may have special drying considerations. In most cases, additional labor time and increased user diligence will be required when any of the above materials and processes are substituted for high-volatility solvents listed in Table 32-1.

32.6   Nonmetallic Materials

These materials include composites (e.g., graphite-epoxy), elastomeric materials, fiberglass, and thermoplastic and thermoset polymers.

32.6.1   Summary of Options and Recommendations

Composite materials typically may be cleaned using the alternatives discussed in Sec. 32.2 and 32.3. However, special care must be taken to perform prolonged, sensitive-exposure bench tests of composites in the candidate cleaning materials in order to identify any subtle embrittlement and corrosion effects. Elastomeric materials vary in their resistance to aqueous solutions. The literature should be consulted to identify situations involving sensitive elastomers. Low-vapor-pressure solvents may cause significant swelling to elastomers, as the slow-evaporating solvent is absorbed into the surface and requires substantial amounts of time for residual solvent to desorb and evaporate. This also increases the potential for irreversible damage to the elastomer. Certain polymeric materials, such as polycarbonate and methacrylate and other acrylics, tend to craze or crack when exposed to low-volatility ester, acetate, and ketone solvent liquids. In cases where sensitive parts are encountered, the use of a higher-volatility, less aggressive solvent is unavoidable. The most benign alternates include methanol, ethanol, isopropanol, acetone, and aqueous solutions of these solvents.

32.7   Adhesives, Coatings, Inks, and Aerosols

32.7.1   Background

Table 32-10[1] lists the common adhesive solvents. It is estimated that 48 percent of the U.S. coatings and inks market in 1986 was solvent-based formulations. TCA has been a popular aerosol propellant, with flammable hydrocarbons such as propane, butane, and pentane also used.

32.7.2   Summary of Options and Recommendations[1]

Alternates to solvent-based adhesives include powder coatings, hot-melt adhesives, radiation-cured adhesives, moisture-cured adhesives, water-based solu-

Table 32-10. Physical Properties of Common Adhesive Solvents

Solvent	Flashpoint, open cup (°C)	Flammable limit in air (volume % at 25 °C)		Water soluble in 100 g solvent (g)	Density (g/L at 20°C)
		Upper	Lower		
TCA	NF*	12.5	7.5	0.05	1.314
Methylene chloride	NF	22.0	14.0	0.17	1.316
Toluene	7.22	7.0	1.3	0.05	0.870
N-hexane	−27.8	6.9	1.25	0.01	0.678
MEK	−5.6	11.5	1.81	11.80	0.804
Ethyl acetate	−2.2	11.0	2.25	3.3	0.900

*NF=no flashpoint.
SOURCE: Ref. 1.

Table 32-11. Summary of Substitute Solvents for TCA in Aerosols

	Major product applications			Performance factors	
	Automotive and industrial products	Pesticides	Household products	Flammability	Density
TCA	X	X	X	None	1.32
Substitute solvents					
Petroleum distillates	X	X	X	High	0.75
Aromatic hydrocarbons	X		X	High	0.87
Alcohols	X	X		High	0.80
Ketones	X		X	High	0.81
Water systems	X	X	X	Low to none	1.00
Dimethyl ether (DME)		X		High	0.66
HCFC propellants*	X	X	X	None	1–1.21

*Includes HCFC-22 and HCFC-142b.
SOURCE: Ref. 1.

tions (a latex, or emulsion), high-solids adhesives, and other solvent-based adhesives, such as acetone, ethyl acetate, heptane, and toluene. Alternative coatings and inks becoming commercially available include powder coatings, water-based coatings, and high-solids formulations. Table 32-11 summarizes substitute solvents for aerosols.

32.8 Material and Process Change Considerations

The most vulnerable point in the process of changing any fundamental manufacturing activity, such as cleaning, is the transition into the factory setting. This activity is the least predictable, the least rational, and cannot be totally simulated in bench or pilot testing. Parameters to investigate, when considering a cleaning material and process change, include equipment and operating parameter variability during scale-up; agitation and foaming in aqueous cleaning processes; floor space, throughput limitations, rinsing, and residues with aqueous and low-volatility solvents in batch or conveyorized systems; drying; energy consumption; labor requirements per unit product; operator skills; and capital and operating costs. The following general conclusions are derived from experience gained in a major aerospace manufacturing facility and may be worthy of consideration:

- A "systems approach" is essential for a successful technology.

- An integrated product team of representatives from all affected functions should be formed at the beginning of the project.

- The users and operators of the material and/or system should lead the transition.

- Much time, energy, innovation, and perseverance is required during the transition.

- Material and process change requires a thorough understanding of the system performance requirements and of the history of the existing materials and processes.

- It is difficult to generalize across industry sectors.

References

1. S. O. Andersen et al., *1991 UNEP Solvents, Coatings, and Adhesives Technical Options Report*, Nairobi, Kenya, United Nations Environmental Programme, December 1991.

2. E. Groshart et al., *Alternatives for CFC-113 and Methyl Chloroform in Metal Cleaning*, U.S. EPA, EPA/400/1-91/019, June 1991.

3. S. O. Anderson et al., *1993 Report of the Technology and Economic Assessment Panel*, UNEP, Nairobi, Kenya, July 1993.

4. U.S. Environmental Protection Agency, "Safe New Alternatives Program (Proposed Rule)," *Federal Register*, vol. 58, no. 90, pp. 28094–28192, May 12, 1993.

5. W. G. Wood et al., *Metals Handbook*, 9th ed., vol. 5: *Surface Cleaning, Finishing, and Coating*, American Society for Metals.

6. H. J. Weltman and S. P. Evanoff, "Replacement of Halogenated Solvent Degreasing with Regenerable Aqueous Cleaners," in *Proceedings of the Purdue Industrial Waste Conference*, Lewis Publishers, Chelsea, Mich., 1992, pp. 851–871.

7. H. J. Weltman and T. L. Phillips, "Environmentally Compliant Wipe-Solvent Development," SAE Tech. Pap. Ser. 921957, presented at Aerotech '92, Society of Automotive Engineers, Anaheim, Calif., October 1992.

8. T. L. Phillips et al., "Development and Implementation of CFC-Free Manual Cleaning Solvents at Air Force Plant No. 4, 93-RP-152.04," presented at Air and

Waste Management Association 86th Annual Meeting and Exhibition, Denver, Colo., June 3–8, 1993.

9. *Handbook of Chemistry and Physics,* 55th ed., CRC Press, Boca Raton, Fla., 1975.

10. Bryan Baxter et al., *Eliminating CFC-113 and Methyl Chloroform in Precision Cleaning Operations,* U.S. EPA, EPA/400/1-91/018, June 1991.

11. Stephen Greene et al., *Aqueous and Semi-Aqueous Alternatives for CFC-113 and Methyl Chloroform Cleaning of Printed Circuit Board Assemblies,* U.S. EPA, EPA/400/1-91/016, June 1991.

33

Maintenance Operations and Pollution Prevention

Ronald L. Berglund

The M. W. Kellogg Company
Houston, Texas

Continuing publicity surrounding environmental concerns suggests that the problems resulting from the generation of waste are approaching crisis proportions with the general public and have reached the highest levels of government. The initial reaction to these concerns has been the development of pollution prevention programs within most manufacturing plants. The long-term response must be an emphasis on pollution prevention in all activities associated with manufacturing—from design and engineering through construction and operation. One area in which a pollution prevention orientation is crucial is maintenance, the vital link assuring that operation of the unit is able to achieve a performance level consistent with design and engineering.[1]

Many activities associated with maintenance, such as good housekeeping, tank cleaning, inventory control, and waste segregation, are among the first areas addressed in a pollution prevention program. Yet, maintenance departments have long been treated as orphans, left alone to fare as best they could,[2] or begrudgingly tolerated as an unpredictable, unmanageable expense. Now, however, as maintenance is emerging as a competitive tool,[3] it needs to be recognized as vital to any sustained pollution prevention activity.

33.1 Definition of Maintenance

A longstanding function like maintenance can be defined in a number of ways. One standard definition is *activities required to keep a facility in as-built condition.*[4]

Included within this definition would be work undertaken in order to restore every piece of equipment to the standards adopted when the equipment was installed, as well as the process of fault discovery, repair, and recommissioning of equipment. More practical definitions might be "controlled degradation" or "organized common sense."[5]

While maintenance is a subsidiary activity to production, in the United Kingdom it is recognized as the largest aspect of a subject known as *terotechnology.*[6,7] Terotechnology represents "a combination of management, financial, engineering, building and other practices applied to physical assets in pursuit of economic life cycle costs (defined as the total costs of an item throughout its life including initial, maintenance, and support costs). Its practice is concerned with the specification and design for reliability and maintainability of plant, machinery, equipment, buildings and structure."[6] Clearly, with the increasing recognition of environmental costs in life-cycle cost assessments, the interrelationship of maintenance and environmental performance cannot be ignored.[8]

All physical equipment in a chemical processing facility is susceptible to failure through breakdown, deterioration in performance through age and use, and to obsolescence due to improvements in technology.[6] Each of these features can affect pollution in the following ways:

Failure results in unplanned losses in output of products or services, generation of wastes, and potential loss of equipment.

Deterioration usually results in an increase in instances of failure, unacceptable levels of quality, and corresponding increases in waste generation.

Obsolescence results in a situation where competitors can achieve a lower unit process cost, lower waste disposal costs, or better environmental performance.

Achieving a balance between expenditures and investment for maintenance, and the added costs of failure and obsolescence, is a continuing challenge to the plant manager and the maintenance supervisor.[3] Maintenance and repairs typically account for more than one-third of the fixed costs at major chemical facilities.[9,10] With this level of investment in maintenance equipment and labor, numerous pollution prevention opportunities should become available.

Deciding what level and when maintenance is needed has always been a complex effort involving trade-offs between committing available financial resources, labor, and time (especially if the unit needs to be shut down) against the risk of equipment breakdown and failure, process efficiency deterioration, and the hazard associated with the potential release.[11]

From an environmental standpoint, the alternative maintenance approaches represent a transition from those that are wholly reactive to those that are mostly proactive (see Fig. 33-1). Reactive maintenance is often unplanned; it may occur on an emergency basis following unit shutdown or a significant environmental release. It literally may include troubleshooting and firefighting. Proactive maintenance, on the other hand, is planned and initiated before there are any adverse consequences in the unit. Reactive maintenance includes

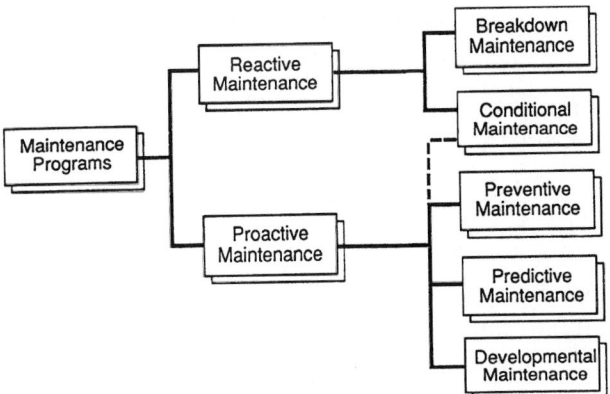

Figure 33-1. Alternative maintenance approaches.

- *Breakdown maintenance.*   Fix it when it breaks; crisis management.
- *Conditional or indicative maintenance.*   Can rely on application of modern inspection techniques to detect problems in their early stages; often relies upon intuitive recognition of potential concern by experienced workers, and effective apprentice program.

Proactive maintenance includes:

- *Preventive maintenance.*   Conduct maintenance on a regular schedule. This would include routine maintenance, comprehensive cleaning, and recalibration. It can also include scheduled inspection of plant equipment to discover and remedy conditions that could lead to premature failures, lost production, and equipment damage.[12] This type of maintenance might be particularly important on analyzers, alarms, and warning systems, which may operate only occasionally and might misfunction at the most inconvenient times.
- *Predictive or reliability maintenance.*   Unlike preventive maintenance, which assumes that equipment and machine behavior follows some sort of statistical average, predictive maintenance closely monitors the performance of each machine in its actual operating environment and directs maintenance to a time predicted by prior performance of the equipment in the unit. Using inspection data to predict when equipment will fail would allow timely corrective action to be taken so that a more serious problem can be avoided. It would also allow the root cause of failure to be determined.[2,13,14]

Because these terms are somewhat qualitative, many programs may be integrated to include all these maintenance activities, even when only one of these terms is used. One additional category of maintenance is

- *Developmental maintenance.*   Differs from the common definition of maintenance, in that the goal is not to bring the unit up to the as-built level, but to bring the equipment to a level that exceeds the original standards of the unit when built.[7]

Developmental maintenance might include trying to reduce fugitive emissions leakage (when this was not considered in the original design), increasing reliability of a population of equipment, or debottlenecking an entire operation during a turnaround.[15,16] Developmental maintenance can play a significant role in a pollution prevention activity, as the results from predictive maintenance programs lead to design improvements and more durable equipment that is easier to maintain.

These maintenance activities also differ in the time frames over which they operate. Most activities of a preventive nature will focus largely on the short term (each shift, day, week, month, etc.), during which time the plant maintenance personnel will be fully occupied with tasks of a commonsense nature. As might be expected, these are also among those areas in which pollution prevention can be most quickly addressed. Even urgent corrective maintenance activities required to restore a critical piece of equipment to production tend to be accomplished in consideration only of current or next operating periods rather than intervals after that. The right balance between immediate corrective maintenance and long-term predictive-preventive or developmental maintenance is a question faced by all maintenance specialists. There must be a compromise between the resources used to keep equipment at a satisfactory level of availability and the resources lost through equipment failures. From a pollution prevention perspective, the maintenance activity should focus not on the failure by itself, but on the consequences of the failure. Depending on the service and impact of a failure of a piece of equipment, the maintenance activity may justifiably sit anywhere along the continuum from predictive and developmental maintenance to a run-till-it-fails philosophy.[14]

Nothing concerning maintenance occurs in isolation. A well-publicized release at a facility on the Gulf Coast occurred as an indirect result of a design improvement on a small filter. Previously, this filter had to be replaced annually. With the design change, the filter could be left in service for a longer period of time. Unfortunately, after the release occurred, it was traced to a gasket upstream of the filter, which previously had been replaced along with the filter and which now was left in service until it exceeded its effective operating life.

33.2   Maintenance Tasks

When specifying a maintenance task, there must be a good reason for why it should be done, and explicit guidelines on how it must be done. "Hard time" (overhaul) tasks in particular should be specified with great caution since they tend to be very costly[17] (both financially and in terms of waste generation) and may, in fact, be unnecessary. Generally, maintenance-related tasks will fall into four basic categories:[17]

1. *Time-directed tasks* (TDT).   Hard time tasks. Task invokes actions that are *directly* aimed at preventing or retarding a failure. An appropriate TDT has the following characteristics:

   The task and its periodicity are preset and the action occurs without any further input information to decide what will be done.

   The action is shown to directly prevent or retard failure.

2. *Condition-directed tasks* (CDT).   When it is not known how or when to directly prevent or retard a failure, or it is impossible to do so, the next best alternative is to detect its onset and predict the point where failure is likely to occur. An appropriate CDT has the following characteristics:

   A measurable parameter can be found which corresponds with the failure onset (e.g., a screening value when looking for equipment leaks).

   A parameter value is known where action should be taken before full failure occurs.

In some cases, the parameter value where action needs to be taken is given in a regulation (fugitive emissions) rather than by potential equipment breakdown. In this case, the control would be based upon the anticipation and minimization of continuous releases to the environment.

3. *Failure-finding task* (FFT).   In some cases, during the normal course of operation, it is not easily known that the failure has occurred. This would be the case for equipment not regularly used such as a relief valve or a spare pump, or equipment not directly associated with the process operation (an analyzer on an emission stream), for which no alternative measures are available. A failure-finding task is one for which a time is prescheduled when the equipment is checked to determine if it works or not. This type of task is often associated with start-up of a unit, when it is checked for leaks and equipment failure, or after maintenance on a piece of equipment has occurred, or just before a standby piece of equipment is to come into service.

4. *Run-to-failure* (RTF).   According to Nicholas and Smith,[17] there are at least three situations where the appropriate maintenance task is simply to run to failure. Such would be the case when:

   There is no identifiable TDT or CDT capable of avoiding the failure.

   There are a number of candidate-applicable tasks, but none is effective. Here it is more effective to wait until the equipment has failed rather than to spend resources attempting to prevent or detect the failure.

   The impact of the failure mode is too low to warrant attention with the allocated resources.

For the first two situations under run-to-failure, an attempt should be made to design the failure mode out of the system.

33.3   The Bathtub Curve

Often, the time-related reliability of equipment, appearance of equipment leaks, and deterioration of process performance can be described by a life characteristic curve, or what has more commonly come to be known as the "bathtub curve" (see Fig. 33-2). The basis of this curve is that there are three types of failure or deterioration in a system, each of which can result in an increase in the generation of waste. These three causes are frequently referred to as (1) quality failures, (2) stress-related failures, and (3) wear-out failures. The sum of these failures gives the overall failure rate for a population of equipment in a process unit. The bathtub curve is characterized by three distinct regions, each generally dominated by one of these failure mechanisms. Understanding each of these regions is important for selecting the appropriate maintenance strategy in the pollution prevention activity.

33.3.1   Decreasing Failure Rate Region (Break-in)

Region I, which begins immediately after start-up of the unit, is typically the portion of the curve that has the largest percent contribution from manufacturing reliability and is most impacted by the effectiveness of the start-up procedures and by the efficiency of the maintenance turnaround. For new equipment, this region of the curve is where most production and warranty failures occur, and it is often influenced by failures caused by mishandling or improper installation of equipment. Some of the causes of failures during this break-in period are summarized in Table 33-1.[18] A unit will generally experience waste generation during the break-in period because:

- There may be no well-defined manufacturing processes and installation procedures that will minimize waste generation.

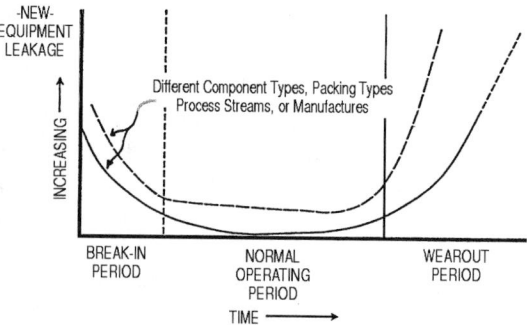

Figure 33-2. Typical life characteristic curve for a population of components (the bathtub curve).

Table 33-1. Summary of Causes of Equipment Failure*

Break-in (Region I) Failure Causes

1. Inadequate test specifications—components or engineering
2. Inadequate quality control
3. Inadequate manufacturing processes or tooling
4. Inadequate materials
5. Improper handling
6. Marginal components
7. Overstressed components
8. Improper setup or installation
9. Improper use procedures
10. Inadequate understanding of installation environment
11. Inadequate training
12. Incomplete final test
13. Subsystem interaction

Region II Useful-life Failures

1. Insufficient design margins
2. Misapplication—overstress
3. Use in wrong environment
4. Inadequate design margins
5. Cause unknown factors
6. Predictable failure levels for design
7. Inherent manufacturing leakage

Region III Wear-out Failures

1. Scratching
2. Material wear, corrosion, friction
3. Aging
4. Misalignment
5. Loose hardware
6. Inadequate or improper preventive maintenance
7. Assembly interference fits
8. Incipient stresses

Note: This table was adapted from the *Handbook of Reliability Engineering and Management*, W. Grant Ireson and Clyde F. Coombs, Jr. (eds.), 1988.

- There may not be a system in place to enforce conformance to procedures that do exist.

- There may not be a process to detect and eliminate the causes of these failures.

In general, this break-in portion of the curve represents the early failure of sub-standard equipment, which may be due to latent material defects, poor installation and assembly methods, poor quality control, or incomplete or inadequate maintenance during the turnaround. Often, an inspection program of equipment after delivery and a short period of in-plant testing can eliminate these potential waste sources from the system. Similarly, screening the entire population of equipment for leakage would be most useful when the population is operating in Region I of the bathtub curve (start-up). In one application, it was reported that over 90 percent of the bearings failed, not because of wear, but due to misapplication, overloading, and misalignment. The first requirement in a pollution prevention program focusing on maintenance is a more careful failure analysis to improve equipment design and installation.

33.3.2 Constant Failure Rate Region (Useful Life)

The flat middle portion of the bathtub curve represents the design failure rate for the specific unit as present during its useful operating life. This region has also been called the *constant failure rate region*, which implies that during this useful-life period, the failure rate is relatively constant. But it might be decreased by restricting usage or by redesign. Region II will include nonrandom as well as random failures. One example of a manufacturing-caused, non-random failure involves materials which will wear with use (i.e., whose critical dimensions do not conform to the necessary specifications). Under some conditions, these could be valve stems or pump seals. Also included might be recurrent leakage from a repaired component.

As can be noted from the causes of unreliability in Region II, the relative contribution of problems which are the responsibility of the design process far outweigh those which are controlled by maintenance or operations. A key assumption that can be used in a pollution prevention program is that, by the time the system has been in operation long enough to be within the useful-life period, the equipment that will have an inherent potential for ultimately failing will already be giving some indication of problems (e.g., noise, oil use, temperature rise, or increased leakage). The probability that a piece of equipment will fail over an operating cycle with a constant failure rate can be characterized by the relationship:

$$F = 1 - e^{-rt}$$

where F is the probability a failure will occur during operating time t
r is the failure rate

33.3.3 Increasing Failure Rate Region (Wear-out)

Finally, as equipment ages, it will reach a wear-out phase characterized by an increasing failure rate (Region III). The increasing failure rate, or wear-out, region is caused by failure mechanisms which slowly change the component in an irreversible way, such as corrosion, polymeric deposits, friction, fatigue, chemical reaction, or catalyst poisoning. Generally, materials selection and tolerance of design for their use determine the useful life of the component. Yet, from a more pragmatic viewpoint, if manufacturing defects were present, they can be expected to show up before the end of design life is reached. Table 33-1 also illustrates some of the causes of failures during the wear-out phase. Again, many of the categories are still a function of the equipment, system, or process design. But, it is within this region that preventive maintenance becomes important. If an impending failure had been properly diagnosed, a system could have been redesigned, repaired, or replaced prior to final failure.

It is fundamental to an effective pollution prevention program to determine where the system is on the life-cycle curve and whether the causes of equipment failure and waste generation are the results of installation, hidden or latent defects, or eventual wear-out of the system, or if they are caused by outside forces not related to the equipment maintenance. In a recent study on United States pipeline failures over the last 10 years that resulted in releases greater than 10,000 gallons each, it was reported that about half were not directly related to the maintenance activity but were attributable to outside forces (damage by others, operators, or natural forces) or operator error. Thirty-seven percent were attributable to equipment corrosion (wear-out) and 13 percent were related to equipment defects or weld defects (start-up).[19] In a British study on in-plant pipe failures which had the potential for causing death or injury, 39 percent of the root causes were associated with maintenance and 27 percent were associated with equipment design. Nearly one-third of the direct causes of failure were attributable to operator error.[20]

33.3.4 Example: A Pump

Consider the maintenance options for a pump operating in hydrocarbon service. An average pump might have leakage at a rate of over 2000 lbs/yr in a refinery and almost 1000 lbs/yr in a chemical plant. A leaking pump (i.e., one with emissions above the regulatory leak definition) might have emissions eight times this amount. A failed pump might release this much material in a day, plus potentially contaminate ground around the pump, overload handling facilities, require cleanup sorbents and protective equipment, expose operating personnel, or increase fire and explosion potential.

- Under *breakdown* maintenance, the pump would be addressed when it stops operating or when a seal blows out.

- Under *indicative* maintenance, the pump would be inspected regularly and addressed when there are indications of leaking (odors, VOC measurements, drips), when the noise or vibration level increases, or when its capacity begins to decrease.

- Under *preventive* maintenance, the pump would be addressed (lubricated, etc.) on a regular schedule, and perhaps repacked on a longer schedule determined by the general turnaround plan for the unit, regardless of whether or not there are indicators of leakage or other problems.

- Under *predictive* maintenance, the history of previous pump failures and leakage in this particular service category would be studied and a program developed that would address the pump before critical problems are allowed to develop.

- Under *developmental* maintenance, alternative packing configurations, pump seal types (dual mechanical seals), or pumps (leakless) would be evaluated for eventual replacement of the existing system should the cost of maintenance and equipment failure prove excessive.

At Ethyl Corp., it was reported that better-designed pumps have reduced the need for repairs due to seal failures by half between 1989 and 1992.[14]

Some well-intended proactive maintenance programs are often doomed to failure because of the manner in which they are originated, developed, structured, implemented, or supported.[12] To be successful, such a program must gain the participation of the maintenance personnel. They will more often support a program if they and their peers have participated in the program's development. Furthermore, maintenance activities are often conducted under rigid time constraints. Adequate planning is essential, especially if pollution prevention is to be considered.

CPI companies typically rate the success of their maintenance programs by the amount of unscheduled downtime,[21] the number and duration of breakdowns, total production, spoilage, or the mean time between failures.[3] From a pollution prevention perspective, waste generation and waste reduction can also be considered.

33.4 Maintenance Wastes

33.4.1 Three Types of Maintenance Wastes

It has been noted[22] that there are ten general sources of waste in a chemical process unit. Almost all of these wastes are in some way related to maintenance. Maintenance-related wastes can be grouped into three categories, which must be addressed in a pollution prevention program (see Fig. 33-3):

1. Wastes avoided through maintenance

2. Process wastes removed during maintenance

3. Wastes generated during maintenance

Avoided Wastes. These are wastes that can be reduced or eliminated through maintenance. In an ideal system, there will be a consistent relationship between production and waste generation that is determined by the design of the unit (see Fig. 33-4*a*). Yet, in reality, such is usually not the case. Once the unit begins operation, wastes will begin to increase. These include increases in

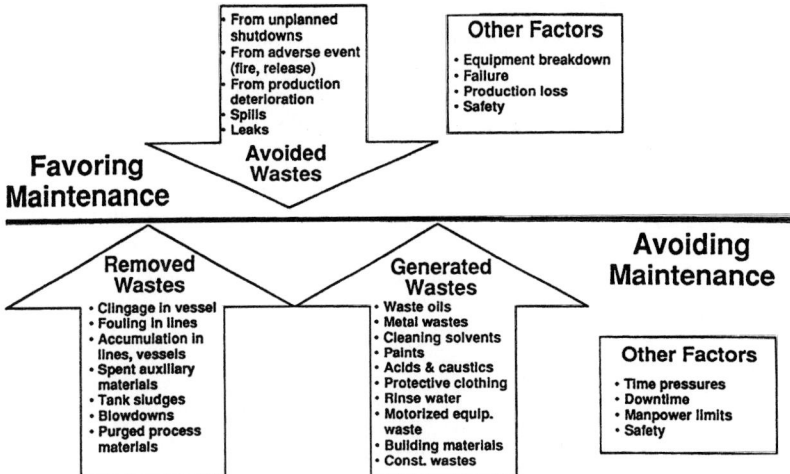

Figure 33-3. A maintenance pollution prevention program needs to balance wastes avoided by maintenance with those generated or removed.

unreacted raw materials, impurities, or by-products generated in process because of reduction in catalyst efficiencies or operational deterioration; process materials purged from vessels and lines during emergency shutdown and start-up; fugitive emissions from equipment leaks; and wastes generated from adverse events (explosions, fires, equipment breakdown, spills). Wastes associated with a recent explosion in a petrochemical plant included damaged drums and cylinders of chemicals, asbestos siding, oil from electrical transformers, spilled process fluids, as well as damaged and contaminated equipment and contaminated soil.[23] These wastes are particularly difficult to evaluate because they are often probabilistic, occurring in some predictable but random pattern, or because they may increase gradually (as increases in by-product formation). Nakajima, the developer of TPM (total productive maintenance), focuses the entire pollution prevention activity on what he calls the six major loss areas in a facility.[24] These are:

Downtime Losses

Losses due to equipment failure

Losses during start-up and process adjustment

Production Losses

Losses from idling and minor stoppages

Losses due to reduced speed

Defect Losses

Losses from process defects

Losses from reduced yield

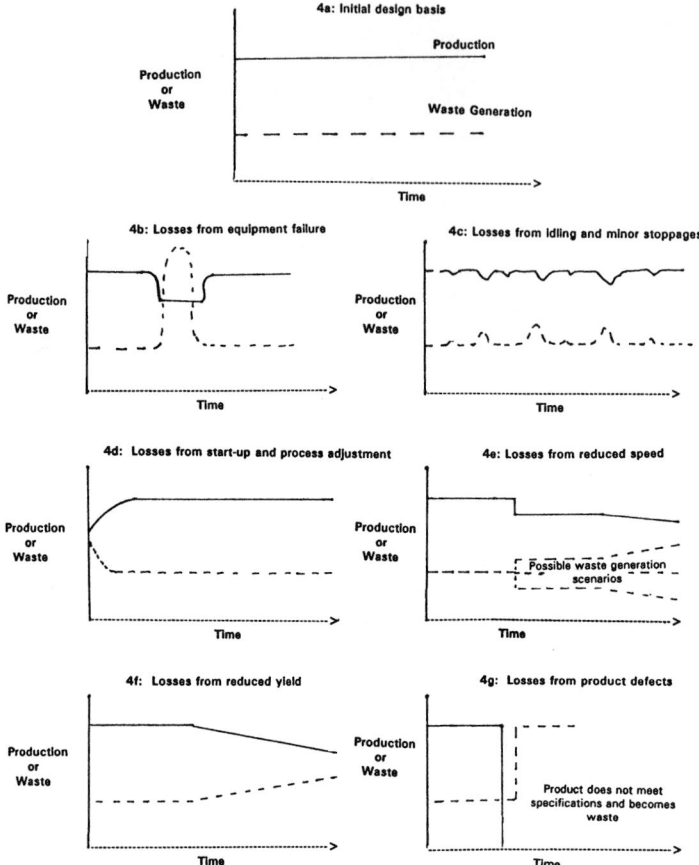

Figure 33-4. Conceptual changes in production and waste generation over time for six major maintenance-related loss areas in a facility.

These six losses affect both waste generation and production, as shown in Fig. 33-4b–g. In each case, waste generation increases with time from what was expected in the original process design.

Process Wastes Removed During Maintenance.   These include unreacted raw materials, impurities, or by-products accumulated in the lines, tanks, or equipment; spent auxiliary materials (catalysts, oils, solvents, etc.); and process materials purged from vessels and lines during planned shutdown and start-up.

Wastes Generated Through Maintenance.   These include cleaning agents (acids, solvents), paint and coating removal materials, spent gaskets, valve packing, equipment packaging, protective clothing, residual paints, solvents, old piping, equipment, and insulation.

A more detailed list of these maintenance wastes is presented in Fig. 33-5.

Avoided Wastes

- Increases in unreacted raw materials, impurities, or by-products generated in process because of reduction process efficiencies due to catalyst or operational deterioration
- Process materials purged from vessels and lines during unplanned shutdown and start-up
- Spills, absorbent, contaminated soil, socks, protective equipment
- Wastes generated from adverse events (explosions, fires, exposure, contaminated soil, washing or firefighting fluids, equipment breakdown) or from release of such materials into the environment
- Wastes generated from process upsets and spills
- Fugitive emissions from equipment leaks

Process Wastes

- *Fouling in lines* from crystallization, sedimentation, chemical reactions, polymerization, coking, corrosion, or bacterial growth
- *Clingage* of process material in reactors, vessels, storage tanks, lines, pipes, pumps
- *Unreacted* raw materials, impurities, or by-products accumulated in the lines, tanks, or equipment
- *Spent auxiliary materials* such as catalysts, oils, solvents, filters, filter media, backwash
- *Process materials* purged from vessels and lines during planned shutdown and start-up
- *Sludges* from feed, waste, and product storage tanks
- *Blowdowns*—exchangers, filters, sightglass

Maintenance Wastes

- Waste oils—hydraulic, compressor, crankcase, coolant
- Waste cleaning solvents—acetone, alcohols, methyl ethyl ketone, part cleaners, turpentine, chlorinated compounds
- Waste paints—enamel, epoxy, lacquers, urethanes, acrylic, water-based
- Paint removal materials—used blasting media and materials, stripping agents
- Construction wastes—demolition waste, pipes, wood, concrete, asphaltic concrete, debris, insulation
- Waste electrical equipment—capacitors, transformers, lines, wiring

Figure 33-5. Wastes associated with maintenance.

- Metal wastes—scrap steel, alloys, aluminum, welding slag, metal filings, copper tubing
- Acid and caustic wastes—metal conditioning, etching, cleaning, stripping
- General maintenance trash—packaging materials, rubbish, paper, pallets, plastic wraps, glass, metals, debris, empty drums
- Automotive and motorized equipment—greases, transmission, power steering, coolant, brake fluid, waste batteries
- Rinsewater, wastewater—from washing, cleaning out vessels, surface preparation, drum rinsate
- Waste building products—grout, sealer, caulking, tars
- Protective clothing—suits, gloves, masks, filters

Figure 33-5. (*Continued*)

33.4.2 Characteristics of Maintenance Wastes

As might be observed from the preceding list and from Fig. 33-5, wastes associated with maintenance have a number of important characteristics that need to be recognized in the pollution prevention program. These wastes tend to be one of the following types.

Intermittent.   Maintenance wastes will be generated on an irregular schedule. When generated, as in a turnaround, for example, process material might be purged from a reactor during a shutdown at a rate of 500 percent or more higher than that during normal operation. These discharges might tax treatment, storage, reuse, or disposal operations.

Variable.   A number of factors would affect the volume and composition of maintenance-related wastes. These include the age of the unit, the time since last maintenance, or the level of maintenance required. In some cases (e.g., fugitive emissions or sandblasting wastes), the estimation procedures for these releases are limited or nonexistent.

Hazardous.   Many solvents, acids, and high-strength or high-temperature materials are used during maintenance or removed from the process during the maintenance shutdown. These often are generated during a turnaround, when a large portion of a plant is taken down and large numbers of people are in the unit for only a few weeks. During such a period, the potential for exposure or hazard conditions is increased.

These wastes are even more difficult to manage if they are unplanned—resulting from spills, upsets, leaks, equipment breakdown, or unplanned shutdowns. Under these conditions, additional pollution resulting from fires, explosions, exposures, or contamination must be considered.

33.5   Environmental Impact

Environmentally related alternatives will follow some hierarchy from the least desirable to the most desirable. Such a hierarchy might be like that shown in Fig. 33-6, which represents a conceptual qualitative comparison of outcomes for a variety of environmental events. These range from the most positive—elimination of the waste at the source—to the most negative—an explosion in the unit. While the list is somewhat arbitrary, it is important to recognize that an effective maintenance program can prevent many of the negative occurrences on one hand (focused on loss prevention), but would generate wastes that need to be reduced or managed on the other. Furthermore, some of the negative occurrences can happen during maintenance, start-up, or shutdown of a unit.

33.6   Selecting Priorities

One of the difficult choices of the process engineer or environmental manager is to distribute a limited pool of financial and technical resources over a myriad of pollution prevention, loss prevention, and waste management opportunities. Criteria often used to prioritize pollution prevention opportunities include: (1)

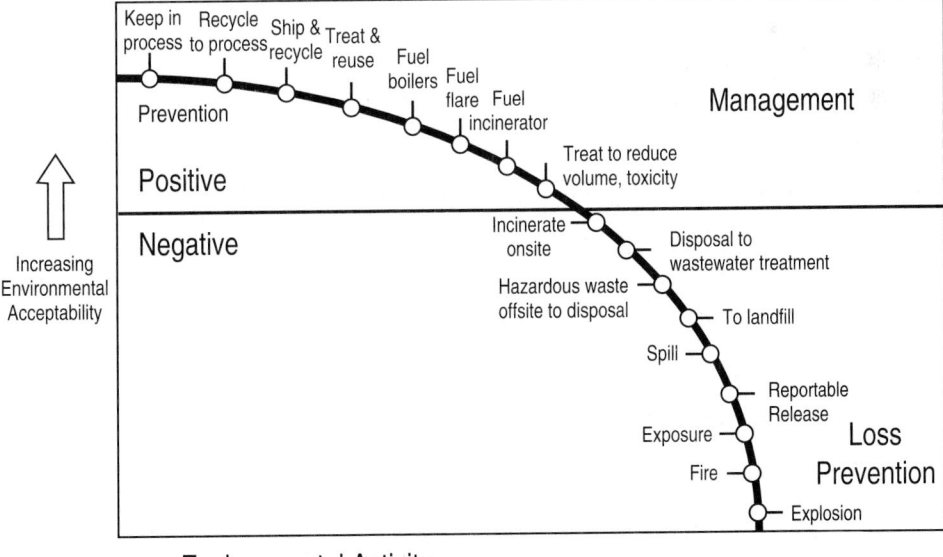

Figure 33-6. Comparison of outcomes of a variety of environmental activities.

the mass of waste being eliminated, (2) its present disposal option or mode of possible entry into the environment, (3) the cost effectiveness of the existing control strategy compared to the pollution prevention option, and (4) the time required for its implementation. Particularly difficult to evaluate are those pollution prevention options associated with irregular wastes resulting from a turnaround that occurs only once a year or once every five or six years, and probabilistic wastes that are generated from spills, leaks, and upsets, whose exact appearance is unpredictable but might follow a statistical pattern.

33.7 Maintenance Trends

There are a number of changes occurring that affect the future role of maintenance in the CPI and the understanding of the relationship between pollution prevention and maintenance. Some coming trends include the move from low-tech to high-tech maintenance, increasing regulatory requirements, and the adaptation of total quality management philosophy.

33.7.1 The Transition from Low-Tech to High-Tech Maintenance[25]

Hyland[26] points out that the traditional "run, fail, fix" attitude toward maintenance will no longer meet the needs of today's process industries. Hyland notes that, historically, low-tech maintenance has been characterized by the following:

- Governed by the clock
- Little or no advance planning
- Short equipment run times
- Long downtimes
- High materials consumption
- Labor intensive
- Seat-of-the-pants diagnostics
- Uneven work quality
- High percentage of breakdown and emergency maintenance
- Reactive response to hazards

While the trend toward high-tech maintenance would be characterized by:

- Governed by equipment condition
- Strong preplanning and scheduling function
- Long and increasing run times
- Shorter downtimes

- Reduced labor commitments
- Stronger technical support
- Statistical approach to failure diagnostics
- Proactive response to hazards
- Increasing documentation and computerized management

The trend toward better diagnostic tools, advanced data-gathering capabilities,[27] and an increasing reliance upon a computerized maintenance management system (CMMS),[28,29] provides an excellent opportunity to begin characterizing wastes and evaluating pollution prevention alternatives during the maintenance activity. Many of the devices on the market for diagnosing a potential problem in the unit are more sophisticated than the machinery they are designed to monitor. For personnel in many plants, there is a gap in knowledge created by this difference in sophistication that must be recognized and addressed in the maintenance and pollution prevention activities, either through increasing use of computerized data management systems or increased training of the operating staff.

33.7.2 Increasing Regulatory Requirements

Whether avoided, removed, or generated, maintenance wastes must be managed in accordance with the requirements of a variety of environmental regulations.[26,30,31] Particularly important will be the RCRA, CERCLA, OSHA, Clean Air Act, and Clean Water Act. In the RCRA, for example, emphasis is given to releases, which are defined as any "spilling, leaking, pumping, pouring, emptying, dumping, emitting, discharging, escaping, leaching, or disposing from an above ground storage tank into the groundwater, surface water or soil." As maintenance staffs receive increased training in safety and environmental concerns, they can become more sensitive to pollution prevention opportunities.

Increasing regulatory requirements are causing changes in maintenance activities in a number of areas, such as process equipment cleaning,[32–34] solvent usage,[34] inspection procedures,[35–37] surface preparation,[38] waste characterization, and training. Many environmentally related regulations (such as OSHA's 29CFR 1910.119 standard for process safety management of highly hazardous chemicals) specifically recognize the important role of maintenance activities (written procedures, training, inspection, testing) in ensuring mechanical integrity and preventing accidental releases.[39] Other rules, such as fugitive emissions regulations, require that covered process equipment deficiencies (e.g., leakage) be corrected when their performance levels are outside acceptable limits (e.g., above the leakage definition).

While hazardous wastes or air toxics are often identified and emphasized in these programs, no wastes should be neglected in the pollution prevention activity.

33.7.3   Adaptation of Total Quality Management Philosophy

In the past, plant and corporate functions, like maintenance, tended to be neatly compartmentalized, a process that optimized parts of the organization at the expense of the whole.[3] Now, however, maintenance personnel are beginning to work more closely with design, engineering, and operations personnel to increase productivity in the unit. Furthermore, a program of total productive maintenance (TPM), which addresses all maintenance activities from a total quality management (TQM) perspective, is already beginning to be adopted in some maintenance activities. Under TPM, maintenance considerations involve and interact with all of the other company functions—from the product and production decisions a CEO makes to the way an equipment operator is involved with his own machine.[3,24] TPM encourages people to work together in different ways using an appropriate maintenance technology and system as tools for change. TPM has been defined as "changing the corporate culture to form a partnership with engineering, maintenance, and production focused on improving equipment effectiveness, improving product quality, and reducing waste, while continually refining teamwork among labor, management and individual work groups." Whereas, in the past, operations staff (concerned with using the equipment to achieve production goals) may have worked independently of maintenance staff (concerned with getting access to equipment to service it) and environmental staff (concerned with meeting regulatory requirements), now it is recognized that they can best meet their individual concerns by working together. As in other areas of the operations, maintenance personnel, when allowed to see the total picture, can be significant contributors to a holistic pollution prevention program. However, increasing the use of programs focusing on "design for maintenance" will enable organizations to address approaches for optimizing maintenance productivity during the design phase. Such activities can provide substantial value to a pollution prevention program. Similarly, since time on many turnarounds is limited, priorities on equipment repair need to be established. By consideration of waste generation and management needs during the maintenance decision-making process, pollution prevention opportunities can be optimized.

33.8   Addressing Pollution Prevention in Plant Maintenance

One approach to assuring that pollution prevention issues are adequately included in maintenance activities is to use a hazard analysis approach such as FMECA (failure modes effects and consequence analyses). Generally, the procedure for including such an approach in the pollution prevention program could include the following steps.[40]

1. List each equipment function and the associated functional failure, which has the potential to generate waste (e.g., a valve leaks vapors, a pump has a reportable release).

2. List the equipment components with failure modes for each functional failure listed in number 1 (e.g., the valve leaks through the stem packing, the pump seal failed; other causes of the functional failure are also possible, and need to be addressed).

3. Develop or estimate failure statistics for each failure mode from historical data or industry databases (e.g., mean time between valve stem leaks: five years; mean time between pump seal failures: two years).

4. Estimate the possible consequences and probabilities (including further accidents and waste generation) associated with each failure mode from historical or industrywide data (e.g., average leaking valve may release 2 lbs/day, pump release may be 100 lbs/day). It is important to consider system failure when the failure mode results in shutdown or failure in a unit or part of the total system. Here it is important to consider the criticality of the event (both probability of occurrence and consequences of occurrence).

5. Identify repair scenarios and probabilities for each component failure mode (e.g., valve packing may need to be replaced but can wait until unit shutdown; pump may need to be replaced, requiring a shutdown as well as cleanup of spill). Determine cost and additional waste generation associated with repair scenario.

6. Summarize the types of tasks that can prevent, detect, or mitigate the equipment failures identified, with frequencies and accompanying waste generation. (Alternatives may include design changes on the pump, instituting a leak detection and repair program, or more frequent maintenance and seal replacement.) It is important to recognize and account for the risk associated with the maintenance task. In one study on a power plant, it was reported that 56 percent of the forced outages in a unit occurred within one week following a planned outage. Unless the maintenance activity is thoroughly evaluated with respect to its potential for reducing equipment failure and correctly implemented, it can amount to little more than tampering and can make the situation worse.

7. Evaluate the mitigation effectiveness of each applicable prevention/detection/mitigation task in terms of waste reduction and cost.

8. Compare the cost and waste reduction achieved with each prevention/detection/mitigation alternative.

9. Develop alternative maintenance strategies from combinations of the different task alternatives for each equipment item under consideration.

10. Create an integrated maintenance plan from a combination of the alternative maintenance strategies.

In one nuclear power plant study, out of some 12,000 individual components identified, only 30 critical components were identified.[5]

33.9 Future Activities

To be successful, a systematic approach to pollution prevention during maintenance activity will require commitment and involvement throughout the organization.

Planning. Consider waste management and pollution prevention issues when developing maintenance programs.

Waste characterization. Quantify and characterize all wastes associated with maintenance.

Statistical analysis. Use reliability engineering for anticipating and avoiding probabilistic waste generation caused by upsets, leaks, and spills.

Information sharing. Transfer pollution prevention opportunities developed in the maintenance arena to programs within other units at the plant or other plants in the corporation.

Linking pollution prevention to modern high-tech maintenance programs can be expensive up front when those programs are staffed by well-trained, highly motivated, skilled people using excellent materials, equipment, and services. But the long-term payoff is a cleaner, safer, smoother operating plant that will run longer between turnarounds and produce less waste.[1]

References

1. C. H. Vervalin, "H P Insight," *Hydrocarbon Processing,* June 1991, p. 17.

2. W. E. Hickman and W. D. Moore, "Managing the Maintenance Dollar," *Chemical Engineering,* April 14, 1986, pp. 68–77.

3. J. Teresko, "The Factory Floor: Time Bomb or Profit Center," *Industry Week,* March 2, 1992, pp. 52.

4. L. Mann, *Maintenance Management,* rev. ed., D. C. Heath & Company, Lexington, Mass., 1983.

5. S. E. Kuehn, "Reliability-centered Maintenance Trims Nuclear Plant Costs," *Chemical Processing,* August 1992, pp. 23–28.

6. T. Hill, "Maintenance," chap. 13 in *Production/Operations Management, Texts and Cases,* Prentice Hall, Englewood Cliffs, N.J., 1991.

7. D. R. Snaddon, "Equipment Maintenance—A Management Policy Guide," chap. 35 in R. Wild (ed.), *International Handbook of Production and Operations Management,* Cassel Ed. Ltd, 1989.

8. H. F. Finley, "Maintenance Management's Importance Is Increasing," *Hydrocarbon Processing,* December 1991, pp. 55–56.

9. Union Carbide Corporation, *1991 Annual Report*, 1992.

10. "Hydrocarbon Processing," *HPI Spending—1992*, Gulf Publishing Company, 1991.

11. R. E. Sanders and J. H. Wood, "Don't Leave Plant Safety to Chance," *Chemical Engineering*, February 1991, pp. 110–118.

12. F. J. Meitz, "Determining Equipment to Be Included in a Preventive Maintenance Program," *Plant Services*, February 1988, p. 18.

13. H. P. Bloch and J. R. Carroll, "Preventive Maintenance Can Be More Effective than Predictive Programs," *Oil & Gas Journal*, July 30, 1990, p. 81.

14. J. J. Rog, "Practical Condition Monitoring," *Maintenance Technology*, May 1993.

15. F. Marschner and H. G. Moertel, "Revamps Increase Efficiency," *Hydrocarbon Processing*, January 1986, pp. 63–66.

16. R. I. Tropp, "More Efficient Turnarounds," *Hydrocarbon Processing*, January 1986, p. 55.

17. J. R. Nicholas and A. M. Smith, "Applying Predictive Maintenance Technologies in a Reliability Centered Maintenance Program," paper presented at *17th Inter-Ram Conference for the Electric Power Industry*

18. D. Elkings, "Reliability in Production," chap. 7 in W. G. Ireson and C. F. Coombs, Jr. (eds.), *Handbook of Reliability Engineering and Management*, McGraw-Hill, New York, 1988.

19. D. J. Hovey and E. J. Farmer, "Pipeline Accident Failure Probability Determined from Historical Data," *Oil and Gas Journal*, July 12, 1993, pp. 104–107.

20. T. A. W. Geyer, L. J. Bellamy, J. A. Astley, and N. W. Hurst, "Prevent Pipe Failures Due to Human Errors," *Chemical Engineering Progress*, November 1990, pp. 66–69.

21. G. Parkinson, "Holistic Maintenance," *Chemical Engineering*, April 1991, pp. 30–35.

22. R. L. Berglund and C. T. Lawson, "Pollution Prevention in the CPI," *Chemical Engineering*, September 18, 1991.

23. T. Wett, "Preplanning Coordinated Team Efforts Help Soften the Blow at Phillips," *Chemical Processing*, August 1990, pp. 68–70.

24. S. Nakajima, *Introduction to TPM: Total Productive Maintenance*, Productivity Press, Cambridge, Mass., 1988.

25. N. Basta and G. Morris, "Managers Tackle Maintenance Problems," *Chemical Engineering*, December 19, 1988, pp. 30–33.

26. R. P. Hyland, "Environmental and Maintenance: Strategies for the 1990's," *Hydrocarbon Processing*, May 1991, pp. 113–116.

27. H. P. Bloch, "Equipment Reliability in the HPI: Getting it All Together," *Hydrocarbon Processing*, January 1992, p. 25 (27).

28. T. R. Welter, "Move the Wrench Over and Pass Me the Computer," *Industry Week*, February 5, 1990, pp. 52–54.

29. G. Gamdani, G. Ondrey, and S. Moore, "Predictive Maintenance," *Chemical Engineering*, November 1992, pp. 30–33.

30. A. K. Rhodes, "Recent and Pending Regulations Push Refiners to the Limit," *Oil and Gas Journal*, December 16, 1991, pp. 39–43.

31. G. Moritis, "Producers Modify Activities as Threat of New Rules Looms," *Oil and Gas Journal*, December 16, 1991, pp. 52–56.

32. "Process Equipment Cleaning," app. B-22 in *Waste Minimization—Issues and Options, Volume II,* U.S. EPA, Report No. EPA-530-SW-86-042, October 1986.

33. C. H. Fromm and S. Budaraju, "Reducing Equipment-Cleaning Wastes," *Chemical Engineering,* July 18, 1991, pp. 117–122.

34. R. L. Price, "Stopping Waste at the Source," *Civil Engineering,* April 1990, pp. 67–69.

35. R. A. Holm, "Fulfilling the New Role of Inspection," *Manufacturing Engineering,* May 1988, pp. 44–47.

36. J. McAlister, "Surveillance—the Forgotten Maintenance Tool?" *Chemical Engineering,* October 13, 1986.

37. "Refiners Focus on Modern Maintenance Techniques," *Oil & Gas Journal,* January 1, 1990.

38. H. E. Howler, "Innovations in Surface Preparation: A Review of Equipment," *Journal of Protective Coatings and Linings,* May 1991, pp. 38–45.

39. D. M. Rhyne, "Beyond Compliance," paper presented at *National Plant Engineering and Maintenance Conference,* Chicago, Ill., March 11, 1993.

40. J. B. Humphries, "Economic Optimization of Maintenance," paper presented at *National Plant Engineering and Maintenance Conference,* Chicago, Ill., March 1993.

34

Measuring the Performance of Environmentally Friendly Cleaning Solvents*

Michael Meltzer

J. D. Shoemaker

Lawrence Livermore National Laboratory
Livermore, California

An important decision factor in the replacement of environmentally risky solvents with more benign materials is performance: that is, can the alternative cleaners equal or surpass the performance of the traditional ones? The purpose

*This document was prepared as an account of work sponsored by an agency of the United States Government. Neither the United States Government nor the University of California nor any of their employees makes any warranty, express or implied, or assumes any legal liability or responsibility for the accuracy, completeness, or usefulness of any information, apparatus, product, or process disclosed, or represents that its use would not infringe privately owned rights. Reference herein to any specific commercial product, process, or service by trade name, trademark, manufacture, or otherwise, does not necessarily constitute or imply its endorsement, recommendation, or favoring by the United States Government or the University of California. The views and opinions of authors expressed herein do not necessarily state or reflect those of the United States Government or the University of California, and shall not be used for advertising or product endorsement purposes.

This work was done under auspices of the U.S. Department of Energy by Lawrence Livermore National Laboratory under Contract no. W-7405-ENG-48.

of this chapter is to discuss some of the measurement techniques that have proved valuable in the evaluation of new cleaners.

34.1   Analytical Tools

Over the last two years, Lawrence Livermore National Laboratory (LLNL) has tested quantitative performance measures that can be used to compare traditional and alternative solvents in a variety of precision-cleaning applications, such as for optical glass and lenses, printed circuit board assemblies, machined metal parts, and vacuum systems. Two fundamentally different analytical approaches have been examined: destructive and nondestructive tests. *Destructive tests* require removing residue from the sample after cleaning in order to perform an analysis. Complete removal is often very difficult to achieve. *Nondestructive tests* have the advantage of measuring residue in situ. Samples subjected to nondestructive testing can also undergo subsequent testing.

LLNL has examined the following techniques to determine residue contaminants remaining on the cleaned surface: optical scanning, ionography, x-ray fluorescence, Fourier transform infrared (FTIR) spectrometry, and gas chromatography/mass spectrometry (GC/MS). Also discussed in this chapter are measurement techniques being developed at other facilities. Table 34-1 summarizes the advantages and disadvantages of the techniques.

Complete details of the LLNL alternative cleaner testing project have been published in a report to the U.S. Department of Energy (DOE) (Shoemaker et al., 1993).

34.1.1   Optical Scanning

Optical-range scanning is a nondestructive technique that measures relative amounts of contamination on smooth substrates. LLNL found this method to be most useful on transparent substrates such as glass, and thus especially good for optics-cleaning applications. It can be used on other smooth surfaces as well, as long as contamination can be differentiated, with only light in the visible spectrum, from surface imperfections such as scratches or craters. The method illuminates a magnified section of the sample surface, and passes the image through a surface scanner, similar to the scanners commonly used to digitize a printed page. The digitized pixels are measured for their level of "grayness," and in this way contamination is distinguished from clean sections of the surface. The output is the percent of the surface containing detectable contamination. While this does not yield an absolute measurement of the amount of residue remaining on the surface, it does provide a very useful way of comparing the levels of cleanliness attained with different solvents.

Approximately five measurements per minute can be performed on a sample. For each contaminant, it is necessary to try several different magnifications to determine which will do the job. Fewer measurements are needed at lower than at higher magnification in order to observe a representative fraction of the

Table 34-1. Advantages and Disadvantages of Analytical Methods

Analytical method	Advantages	Disadvantages
Optical scanning	Nondestructive test Very good for particulates	Measures area of contamination only Limited to transparent substrate Operator dependent Gives only relative quantity of residue
x-Ray fluorescence	Nondestructive test High precision High accuracy	Limited to contaminants that fluoresce strongly enough for detection Not sensitive enough to compare the best cleaners
Fourier transform infrared (FTIR) spectrometry	Reflectance FTIR is a non-destructive test Can analyze films 1–2 nm thick	Calibration is difficult, and requires sophisticated equipment such as ESCA Sample surface must be reflective
Gas chromatography/mass spectrometry (GC/MS)	Equipment is widely available	Destructive test
Electron spectroscopy chemical analysis (ESCA)	Nondestructive test Measures all elements in residue except H and He	Time-consuming process High-vacuum operation (10^{-9} torr) may be difficult to maintain
Ionography	Nondestructive test MIL-spec standard procedure Measures performance by conductivity	Destructive test Limited to ionic contaminants of electronic components that are soluble in isopropyl alcohol solution
Ellipsometry	Nondestructive test Relatively quick analysis Measures film thickness Usable on metal surfaces	Surface to be analyzed must be smooth and free of gross contamination
MESERAN	Nondestructive test Direct measurement of residual surface contaminants	Measures relative amount of organic films Radioactive tracer required

part surface (at least 10 percent of the surface should be measured). For instance, 50-power (50×) magnification appears sufficient to detect most all of the Trimsol* oil (a common machining fluid) residue on a surface. At this magnification, areas of contamination as small as 8 μm in diameter are detectable. For parts originally contaminated with Apiezon vacuum grease, only a fraction

*Trimsol is a trademark of Master Chemical Company.

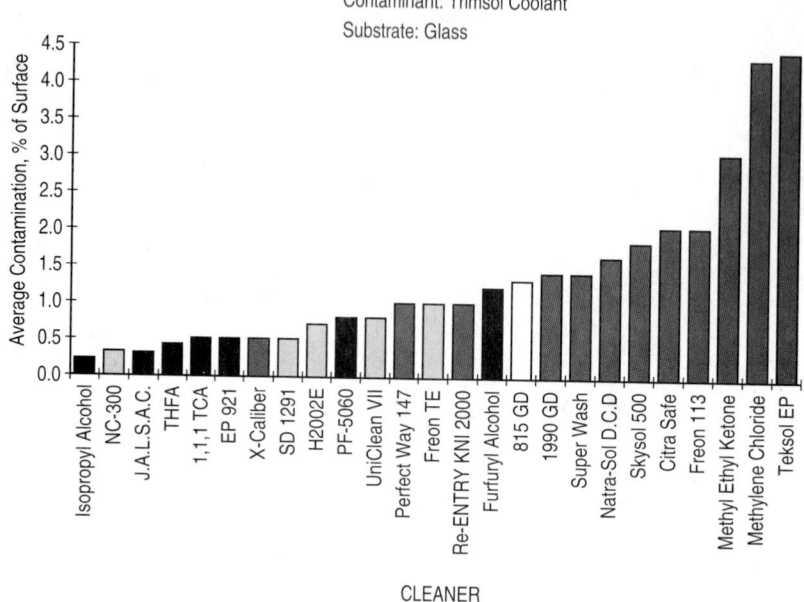

Figure 34-1. Typical optical image analysis results.

of the residue was detected at 50×. At higher magnification, a fine, peppery structure appeared that accounted for a large percentage of the total residue. For Apiezon grease, a magnification of 200× was sufficient. This required 16 times more measurements than at 50× to attain a total measurement of 10 percent of the part surface. Figure 34-1 shows the optical-scanning results for cleaning glass slides contaminated with Trimsol.

34.1.2 x-Ray Fluorescence

Certain contaminants common in hydrocarbon residues, such as chlorine and sulfur, can be detected quite well through their x-ray fluorescence spectra. Thus, their presence can serve as "signatures" of the compounds that contain them. Machining oil, for instance, frequently contains chlorine and sulfur. The relative amount of residue remaining on a surface after cleaning is proportional to the number of counts per second detected when the surface is subjected to x rays.

   In our tests, machined aluminum coupons were contaminated with a chlorine-bearing oil, then cleaned using several of the most promising cleaners we had identified at that time, as well as Freon* TE (a chlorofluorocarbon-ethanol azeotrope) and isopropyl alcohol (IPA). These provided baseline measure-

*Freon is a trademark of Du Pont.

Table 34-2. Results of *x*-Ray Fluorescence Test

Cleaner (manufacturer)	Counts per second*
Uncleaned	301
Isoprophyl alcohol	4
Freon TE	ND
815GD (Brulin)	ND
1990GD (Brulin)	23
Citra Safe (Inland Technology)	2
Natra Sol (Sunshine Chemical Specialties)	ND
Perfect Way 147 (Polychemical Corp)	4
Skysol 500 (Inland Technology)	ND
Super Wash (SWI International)	6
Uniclean VII (Uniclean Products)	7

*ND indicates no detectable contamination.

ments; it was important that an alternative cleaner being considered as a replacement for them be able to give a comparable cleaning performance. While *x*-ray fluorescence was able to clearly identify the most powerful cleaners (all the ones with no detectable contamination remaining), it does not appear sensitive enough to distinguish between the best cleaners. The results of the analysis are listed in Table 34-2.

34.1.3 Fourier Transform Infrared Spectrometry

Fourier transform infrared (FTIR) spectrometry techniques are based on absorbance of infrared signals by the chemicals that are being analyzed. Molecular bonds and groups of bonds resonate and absorb infrared frequencies that match their vibrational modes. Thus, each organic compound exhibits a unique "fingerprint" of absorbed infrared frequencies in the form of a plot of absorbance versus wave number. This method can be used to identify the types and/or amounts of contamination on a part. The more residue present, the more infrared is absorbed and the higher are the absorbance peaks on the plot. Residue thicknesses can be determined by measurement of either the heights of the peaks above background or the areas under them. The two approaches appear to produce very similar results, and both can be measured automatically on FTIR equipment. Figure 34-2 shows the spectrum around wave number 3000 for a sample cleaned with a chlorofluorocarbon (CFC) solvent. Hydrocarbon contaminants produce absorbance peaks with wave numbers near 3000.

Reflectance FTIR spectrometry is a nondestructive type of FTIR methodology. It does not require removal of the contaminant from the substrate, but its use is limited to reflective substrates. This technique has been employed to analyze thin coatings on lenses (Compton and Stout, 1993). Using calibration data

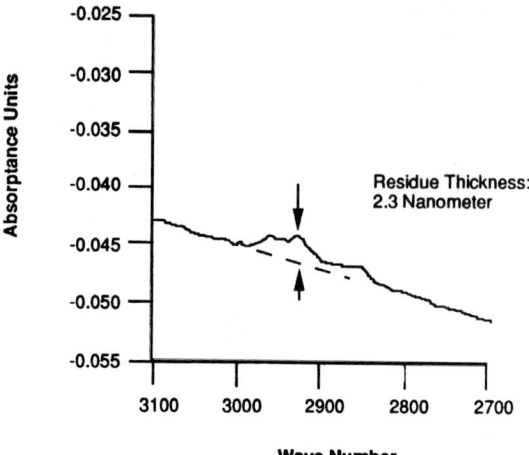

Figure 34-2. FTIR spectra for wave numbers 2700 to 3100.

developed at Oak Ridge National Laboratory, LLNL is measuring hydrocarbon residue layer thicknesses less than a nanometer. This allows precise comparison of the residues remaining after a cleaning with the strongest solvents. The residue thicknesses for these solvents are typically between one and three nanometers; the differences between the performances of different cleaners are frequently only several tenths of a nanometer.

LLNL uses FTIR with a low-angle specular reflection accessory to measure the effectiveness of alternative cleaners at removing organic contamination from aluminum samples. A gold-coated mirror was used as the reference spectrum. The calibration of layer thickness with absorbance was obtained at Oak Ridge National Laboratory from two measurements. The lower value was measured by electron spectroscopy chemical analysis (ESCA), and the higher value was obtained by weighing the thin metal foil substrate used in this test before and after it was contaminated with the hydrocarbon.

In the FTIR method employed at LLNL, the entire sample surface is scanned, and an average thickness is generated. In measurements of over 40 different cleaners, three samples cleaned with each solvent were analyzed, and these results averaged. Figure 34-3 shows the results of these tests. On the right side of the histogram are the residues remaining after using "traditional" hazardous solvents such as Freon TE and 1,1,1-trichloroethane (TCA). The results for ultrasonic and vapor degreasing with the traditional solvents are designated by "U" and "V" respectively. Also included is the temperature at which the cleaning was conducted. Thus, 26U indicates ultrasonic immersion cleaning at 26°C. Cleaning temperatures were limited by such factors as flashpoint, vapor pressure, and boiling point. On the left side of the histogram are the results for ultrasonic cleaning at 40°C with "environmentally friendly" cleaners.

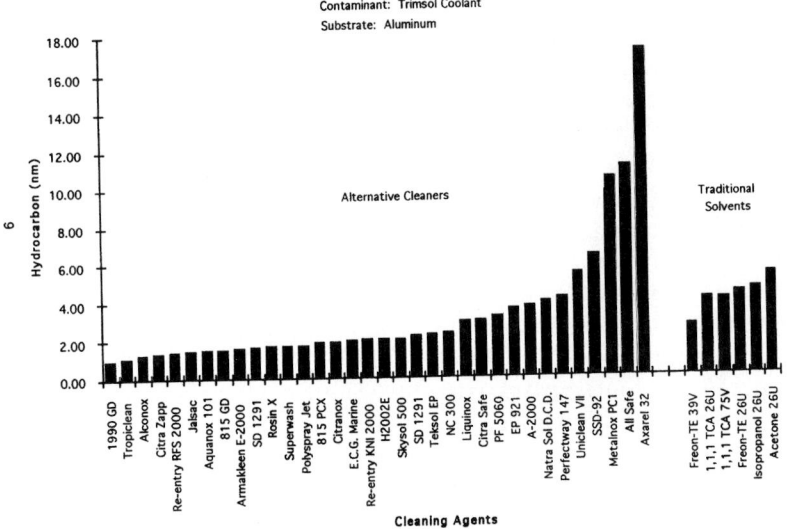

Figure 34-3. Hydrocarbon contamination on cleaned aluminum.

34.1.4 Gas Chromatography/Mass Spectrometry

Although LLNL tried GC/MS to measure the amount and type of contaminant remaining on the cleaned surface, the results were generally unsatisfactory. GC/MS cannot analyze a solid surface, so residual contamination on the surface must be removed with a suitable solvent such as methylene chloride.* The solvent, carrying the residual contaminants, is then analyzed by GC/MS. One reason the technique gave poor results is that many alternative cleaners performed at least as well as methylene chloride. Thus, methylene chloride was unable to remove all of the residual contamination on the surface and the GC/MS could not detect this contamination.

34.1.5 Ionography

Ionography uses electrical resistance to measure the amount of ionic contamination on a surface. It requires the removal of that contamination before the measurement is made. This method is a standard in the electronics industry to determine cleanliness of printed wiring assemblies,[†] and it satisfies the requirements of the U.S. military's specifications MIL-P-28809 and MIL-P-55110.

---

*A 1:1 mixture of methylene chloride and isopropanol is sometimes used for surface extraction and can be much more effective.

[†]ANSI/IPC-RB-276, "Qualification and Performance Specification for Rigid Printed Boards"; ANSI/J-STD-001, "Requirements for Soldered Electrical Electronic Assemblies," ANSI.

In tests at LLNL, printed wiring assemblies were cleaned of solder flux using alternatives to Freon TE cleaning. Residues remaining after cleaning were removed using IPA. Product cleanliness was tested by comparison of conductivities of the residues. MIL-P-28809 requires that a conductivity no greater than 20 micrograms of NACL equivalents per square inch remain on the surface after cleaning. As a point of comparison, an uncleaned board had a measurement of 29.1 micrograms of NACL equivalents per square inch. The instrument employed was an Alpha Metals, Inc., Model 500 Ionograph.

Ionography alone may not be a sufficient test of cleanliness. Researchers at DOE's Allied-Signal Kansas City Division extracted solder flux residue from cleaned circuit boards using a solution of 75 percent IPA and 25 percent deionized water. They employed an Alpha Metals Model 200 Omega Meter to measure the amount of ionic contaminant extracted. These researchers also used the Alpha Metals Model 600 SMD Omega Meter, which heats the IPA solution of 120°F and uses submerged spray jets to improve dissolution of the contaminant. In both measurements, some specimens that showed low ionograph readings (corresponding to low amounts of ionic contamination) actually showed significant contamination when subjected to another type of analysis known as MESERAN—for *measurement and evaluation of surfaces by evaporative rate analysis*—indicating that not all of the residue had been removed by the IPA (Benkovich, 1993). MESERAN is described in Sec. 34.1.6, "Other Analytical Tools," below.

34.1.6 Other Analytical Tools

Electron spectroscopy chemical analysis (ESCA) is a very precise method of measuring type and amount of residue remaining on a surface. It is also known as *x-ray photoelectron spectroscopy* because it uses an *x*-ray beam to excite the contaminant atoms on a sample, causing electrons to be emitted. This technique analyzes areas greater than or equal to 150 μm in diameter, to depths up to 10 nm. ESCA is sensitive to all elements except hydrogen and helium, and permits recognition of different species of a particular compound. Quantitative data (Thompson et al., 1991) can be obtained from a survey scan, while high-resolution scans over narrow energy ranges define stoichiometric relationships (Gill, 1986).

While ESCA gives quite precise measurements of contaminant levels, it is quite time consuming to generate each measurement. The equipment needed is expensive, and operates at a high vacuum (10_9 torr), which may be difficult to maintain. LLNL found ESCA to be most beneficial for calibrating the results from FTIR, which can generate a variety of useful and accurate data more quickly.

Los Alamos National Laboratory (LANL) chose *ellipsometry* to evaluate surface cleanliness for diamond-machined optics because it is capable of submonolayer sensitivity and gives a quantitative value for film thickness on the surface of the sample. This technique measures the state of polarization of a light wave, which changes abruptly as it is reflected at the interface of two optically dissimilar media. Ellipsometry appears to be a viable analytical method for relatively clean substrates that have very smooth surfaces. The lower the ellipsometry value, the cleaner the surface (Noskin, 1992). A single-point measurement can be made in about five minutes. It is necessary to know the identities and physical properties of contaminants to get absolute values of film thickness, but relative cleanliness can be determined even for unknown contaminants with this method.

Allied-Signal Aerospace Company, Kansas City Division, used MESERAN to measure organic contamination of surfaces before and after cleaning, both as a stand-alone technique (Meier, 1993) and as a comparison test for ionograph measurement (Benkovich, 1993). In this method, the surface is coated with a solvent containing the radioactive isotope carbon 14. Organic contaminants absorb the solvent, reducing the evaporation rate and increasing the radiation count, which can be measured by a Geiger counter. This gives a *MESERAN number* (number of counts per 112 seconds). The lower the count, the cleaner the surface (Meier, 1993).

Other analytical techniques used include surface secondary ion mass spectroscopy (Greene et al., 1992), surface carbon analysis, evaporative rate analysis and electrostatic charge decay (Cohen, 1987), and a test method that employs stable (nonradioactive) isotopes (Chauhan et al., 1992).

34.2 Materials Compatibility and Environmental Testing

34.2.1 Effects on Weight, Dimension, Texture, or Visual Appearance

Sandia National Laboratory tested the effects of various alternative cleaners on more than 70 different substrates (primarily organic chemical polymers). The researchers specifically looked for changes in weight, dimension, texture appearance, and glass transition temperature.* Dimensional changes were measured using a linear digital gauge having an accuracy of ±0.001 mm. Scanning electron microscopy was employed to determine texture and appearance, and differential scanning calorimetry was used to measure glass transition temperature (Hoier and Nigrey, 1992). The study concluded that changes in visual appearance and weight were the most useful indicators of cleaners attacking organic materials.

34.2.2 Environmental Stress Testing

Electronic assemblies such as printed circuit boards that are intended for aerospace and military applications may be subjected to extreme environmental changes. These conditions can be simulated in the laboratory, to test whether

---

Glass transition temperature is the temperature at which a polymeric material changes from glassy to rubbery, indicating a loss of mechanical strength.

high temperature or high humidity may reveal a flaw in the cleaning process. Growth of contaminants, deterioration of insulation, and alterations in the surface are examined after environmental stress testing. LLNL used environmental stress chamber analysis to determine the long-term performance of various cleaners on precision electronics assemblies. One month of testing under "85/85" conditions (85°C and 85 percent relative humidity) simulates approximately 15 years of aging under normal temperature and humidity conditions (Hersey, 1993).

Bibliography

Benkovich, M. G., *Ionic Contamination Detection,* prepared by Allied-Signal Aerospace Company, Kansas City Division, for U.S. DOE, Report no. KCP-613-5251, Kansas City, Mo., July 1993.

Chauhan, S. P., P. Schumacher, and J. C. Chuang, *A Method for Cleaning Performance Evaluation Using Stable Isotopes,* Report prepared by Battelle Memorial Institute, Columbus, Oh., for The Aerospace Guidance and Metrology Center, Newark Air Force Base, Aug. 31, 1992.

Cohen, L. E., "How Clean is Your 'Clean' Metal Surface?" *Plating and Surface Finishing,* vol. 25, no. 11, November 1987, pp. 58–61.

Compton, S. V., and P. Stout, "Qualitative Analysis of Thin Coatings on Lenses," *American Laboratory,* vol. 25, no. 8, May 1993, pp. 36N–36R.

Gill, M. E., Surface Analytical Techniques Applied to Printed Board's (*sic*) and Printed Board Assembly's (*sic*)," presented at the 1986 Institute for Interconnecting and Packaging Electronic Circuits Fall Meeting, San Diego, Calif.

Greene, A. C., R. D. Cormia, and Q. T. Phillips, "Evaluating Cleaning Efficiencies of CFC-Replacement Systems in the Disk-Drive Industry Using Surface Analytical Techniques," *Microcontamination,* vol. 10, no. 3, March 1992, p. 37–40, 63–65.

Hersey, R., Lawrence Livermore National Laboratory, Livermore, Calif., personal communication, Oct. 8, 1993.

Meier, G. J., *MC4069 Firing Set Chlorinated, Fluorinated Solvent Substitution,* prepared by Allied-Signal Aerospace Company, Kansas City Division, for U.S. DOE, Topical Report 705914, Kansas City, Mo., February 1993.

Noskin, H., *Los Alamos National Laboratory Solvent Substitution Report,* Los Alamos National Laboratory, Los Alamos, N.M., November 1992.

Sanborn, R., "Evaluation of Cleaners by Low Angle Specular Reflection from Aluminum Surfaces," internal memo, Lawrence Livermore National Laboratory, Livermore, Calif., Apr. 16, 1993.

Shoemaker, J. D., M. Meltzer, D. Miscovich, D. Montoya, P. Goodrich, and G. Blycker, *Cleaning Up Our Act: Alternatives for Hazardous Solvents Used in Cleaning,* Lawrence Livermore National Laboratory, Livermore, Calif., December 1993.

Thompson, L. M., R. F. Simandl, and H. L. Richards, *Solutions for the Chlorinated Solvent Debacle,* prepared by the Oak Ridge Y-12 Plant for the U.S. DOE, Report Y/DV-904/R1, Oak Ridge, Tenn., April 1991.

35

Materials Management

June Bolstridge

GAIA Corporation
Silver Spring, Maryland

Within the realm of pollution prevention, few approaches offer the diversity in cost and range of opportunities provided by materials management. Improving management of materials generates economic, safety, and environmental benefits within individual departments as well as entire corporations. Efficient and effective materials management includes controlling purchasing, managing materials requirements, minimizing on-site storage, and improving use and reuse of materials—approaches that represent the spirit and intent of pollution prevention.

35.1 Business Benefits and Environmental Potentials

Materials management represents the heart of any effective pollution prevention program since reducing the amount of toxic or hazardous chemicals in waste requires knowledge of

Materials purchases (amounts, forms)

Activities and processes—including recycling (use rate, schedule)

Products, by-products, and wastes generated and stored (generation rates, amounts, forms, content)

Materials management information provides a basis for all environmental, safety, and health programs. For example:

■ Inventory data can be compared to thresholds and reporting limits to identify applicable requirements.

- Materials-handling information can be applied to an estimate of a facility's reportable chemical releases.

- Materials to be covered in employee training and emergency response plans can be easily identified.

Effective materials management is more than sound environmental policy, however; it represents good business practices because

- Stockpiles represent either sunk costs or unrealized potential sales to most operations.

- On-site storage is a potential source of releases, exposures, and liabilities.

- Knowledge of on-site materials is essential to meet unanticipated requirements.

- Management of materials required for maintenance assures that delayed repairs will not be the source of releases or spills.

Materials management fits classical philosophies of business management, and can, therefore, be a more easily accepted pollution prevention approach than more esoteric methods.

35.2   Information Sources for Facility Materials Management

Evaluating and improving materials management requires some in-depth information collection concerning where materials are used, stored, and stockpiled on-site, and also where wastes are generated. Sources of existing information that should not be overlooked in the development of an effective materials management program can be assembled under the following headings:

What materials are present on-site?
- Production schedules
- Process flow diagrams
- Purchase and delivery records
- Material balances
- Inventory data for raw materials and products
- Emergency Planning and Community Right-to-Know Act (EPCRA) Section 312 hazardous materials inventory

What are the components of those materials?
- Purchase specifications
- Material Safety Data Sheets (MSDS)
- Container labels
- Vendor literature

- Analytical results
- EPCRA Section 313 supplier notification

What activities require toxic and hazardous chemicals?

- Facility maintenance schedules
- Manuals (start-up, operation, close-out)
- Equipment lists

Which processes generate toxic and hazardous wastes?

- Clean Air Act permit limits and monitoring data
- Clean Water Act permit limits and monitoring data
- Resource Conservation and Recovery Act (RCRA) treatment, storage, and disposal permits
- Solid waste shipment manifests
- RCRA waste shipment manifests
- EPCRA Toxic Chemical Release Inventory (TRI) Reports
- Biennial hazardous waste reports
- Waste disposal invoices

35.3   Inventories Do Count in Pollution Prevention

Knowledge of the types of materials present on-site is a key component to any pollution prevention program. Developing an inventory requires a relatively straightforward process of identifying the amount of the chemical present in all activity and storage areas of the facility.

35.3.1   Working Inventory for Materials Management

Inventories can take many forms for numerous different purposes; however, inventories focused on materials management must consider the current status of the materials present on site. While inventories that track the components of materials and document the average or maximum amount of each chemical or mixture are essential to meet environmental regulatory requirements, such inventories are not effective tools for materials management.

A number of classical texts on materials management for manufacturing operations provide anecdotal examples and graphical presentations of inventory and materials tracking. Table 35-1 shows a simple example of how the necessary data can be structured. Functional categories are established for the total inventory on site ("Stock on hand"), quantities committed to filling existing demand ("Use volume"), planned production or scheduled restocking ("Replenishment"), and net amounts available to promise ("Available").[1]

In the example in Table 35-1, although the apparent inventory (i.e., stock on hand) fluctuates from 80 to 390 units, the actual excess inventory not commit-

Table 35-1. Sample Materials Management Inventory

	Week of operation			
	1	2	3	4
Stock on hand	80	280	180	390
Use volume		100	90	280
Replenishment		200		300
Available	80	90	90	110

SOURCE: Adapted from Ref. 1.

ted to anticipated production activities (i.e., available) remains relatively consistent. Note too that the replenishment through restocking or production that occurs in week 2 is committed to cover use volumes for both weeks 2 and 3. Therefore, the available inventory for week 2 is effectively reduced in anticipation of production to occur in week 3.

35.3.2   Some Concerns Related to Inventory Volumes and Management

The presence of large stores of a material being eliminated through other pollution prevention approaches should receive careful consideration. One U.S. Air Force base was very successful in implementing a bead-blasting procedure for paint removal, only to find that substantial amounts of methylene chloride (previously stockpiled to perform the task) had since passed its shelf life, and so required extremely costly disposal.

Development of any inventory is complicated by the need to uniquely identify the chemicals or materials being tracked. Chemicals are known by a variety of different trade and common names. The Chemical Abstract Service (CAS) numbers are generally a useful code to chemicals known by various synonyms and trade names; however, some caution is required because several chemicals have multiple CAS numbers.

35.3.3   Computer Systems for Inventory Management

A facility that has a computerized centralized purchasing system that logs and identifies all materials can prepare a summary report to identify the materials brought on site during the year. Facilities that lack a centralized record-keeping system for purchasing will need to consider individual purchasing or delivery records unless the amount of the various chemicals present on site can be determined from a review of storage and stockpile areas.

Tracking of on-site locations and amounts of materials can be facilitated by use of a system of bar codes applied to containers and to shelves or cabinets

Chemical Inventory and Tracking System, Chesapeake Software, Inc., Chadds Ford, Pennsylvania, for DEC VAX, $10,000–$75,000

FLOW GEMINI, General Research Corporation, Vienna, Virginia, for DEC VAX and IBM mainframes, $80,000–$295,000

Hazardous Materials Manager, John Systems, Inc., Cedar Rapids, Iowa, for IBM-compatible microcomputers, $4800

HazKNOW, HazMat Control Systems, Inc., Long Beach, California, for IBM-compatible microcomputers, $1750

Hazmat Inventory Tracking System, BSI Systems, Inc., San Diego, California, for IBM-compatible microcomputers, $2500

LogiTrac, Logical Technology, Inc., Peoria, Illinois, for DEC VAX and IBM-compatible microcomputers, $299+ (PC), $9800 (VAX)

Materials Inventory Report System, AV Systems, Inc., Ann Arbor, Michigan, for IBM-compatible microcomputers, $950–10,000

Figure 35-1. Sample environmental chemical inventory software. (*Elizabeth M. Donley, Donley Technology, 1992–1993.*)

containing the materials. Computerized systems can cross reference the information, giving exact location and quantity information on materials.

Many commercially available computer systems are designed to assist in materials management and chemical inventories. Figure 35-1[2] lists a sampling of environmentally oriented inventory management systems. While this listing is not complete, it demonstrates the range of costs for inventory software, which is influenced by the type of inventory and the computer needed to run the program.

35.4 Managing Purchasing, Distribution, and Demand

Materials management requires more than knowledge of the materials already present on site. Ongoing operations mandate continued supply of materials that must be addressed at each stage in the replenishment process.

35.4.1 Purchasing, Distribution, and Material Requirements

Purchasing and distribution are essential elements of effective materials management. The traditional materials management practiced in many organizations involves a series of distributed supply or stock rooms which are designed to provide any and every material requested by the operational personnel of the corporation or facility.

Increasingly, however, corporations are establishing tighter controls over the kinds and types of materials that can be purchased to avoid worker and environmental liabilities. Some of the corporate mandates include

Banning or reducing purchases of chemicals or materials that exhibit certain health or environmental effects, for example, ozone-depleting chemicals or known carcinogens

Mandating detailed information from suppliers concerning the chemical content of mixtures and the potential hazards they represent

Requiring that all purchases go through a central control point where the need for the material can be reviewed and possible substitutes for hazardous or toxic chemicals identified

Occasionally, such material control techniques create additional problems. For example, some companies require suppliers to provide a Material Safety Data Sheet (MSDS) for each item purchased, regardless of its status under the Occupational Safety and Health Act (OSHA) Hazard Communication Standard that establishes regulatory requirements for the MSDS. This results in development of MSDS-type documentation for nonhazardous items, including distilled water and wood shipping pallets, causing additional paperwork and confusion concerning regulatory requirements.

Similarly, a central purchasing approach can only be effective if it is sufficiently responsive to the operation's needs that facility personnel don't resort to purchasing the materials locally using a corporate credit card.

35.4.2 Well-Stocked Lockers and Warm Fuzzy Feelings

Many work areas maintain independent stockpiles for the chemicals and materials involved in their operational tasks. The need for such stockpiles is rarely questioned, because

- Stockpiling is an accepted practice and little (if any) oversight is given to storage areas to recognize the volume of materials they contain.

- Time and effort required to obtain materials is considered wasted, so trips to the central supply are used as a chance to "stock up on a few essentials."

- There are no methods of returning partial containers of materials for others' use.

Such local area stockpiles can represent amazingly large volumes of materials that represent unrecognized sunk costs and potential liabilities to the facility. When the U.S. Navy instituted a central control point for hazardous materials at the Naval Air Weapons Station at Point Mugu, California, materials collected from work area stockpiles totaled 130 percent of annual hazardous material purchases, and were valued at $180,000.[3]

Materials distributed throughout the work areas are required to be included in facilitywide inventories mandated by Section 312 of EPCRA. The effort to document and track such user stockpiles provides incentive to implement more effective approaches to materials management.

Work area materials management can be improved by

- Establishing policies stating that only a minimal amount of material can be removed from the central stock area at any time

- Providing a simple and convenient solution to the problem of return of unused portions of materials to the central area

- Reducing the amount of storage area available at the work sites to discourage stockpiling

35.4.3 Managing Demand

While managing materials produces benefits, stabilizing the demand for materials can also generate extensive and far-reaching benefits. Once an organization's overall material requirements are well understood, the amount of materials maintained on site can be substantially reduced.

Preventive maintenance of chemical-handling equipment is one type of managed demand that can directly impact prevent pollution. By instituting a controlled maintenance program, a facility can assure that the materials required for the scheduled maintenance are available. And the regular maintenance mandated by such a program can reduce accidental releases and waste generation.

While companies cannot always exercise such direct control over demand, most can influence demand, or at least plan for it in an effective manner. Although forecasts are never completely correct, demand management actually improves crisis response because more resources are freed up and recognized as available.

An Approach to Demand Management. Demand management involves the following steps:

1. Forecast production and other activities based on operating schedules and assumptions.

2. Divide each activity into identifiable tasks.

3. Evaluate each task to determine the material requirements, use levels, and rates of consumption.

4. Combine material requirements from individual tasks and activities to determine overall operational requirements.

5. Determine required restocking rates based on manufacture and transportation requirements.

6. Identify the central supply needed to meet the operational demand.

Implementing materials management by managing demand relies on the use of a central supply and distribution network to assure that the operations are provided with the materials on an as-needed basis. A real-time inventory is essential to determine the available stock on hand, the materials committed to filling repetitive needs, and the expected restock dates.

Demand management is most easily applied to consistent operations. Any change in demand that occurs due to product changes, operational levels, or seasonal requirements must be considered in time to allow the supply and distribution system to accommodate the additional material management requirements.

Demand management can be phased in over time as operating units develop confidence in the ability of the central stores to meet their needs. Additional improvements can be achieved by tracking the actual amounts of material used by operations, as compared to the amounts requested. Whenever excess or insufficient materials are being chronically requested, additional efforts to refine predictions can improve overall material management

35.4.4 U.S. Air Force and U.S. Navy Hazardous Materials Pharmacy Concept

One of the most extensive approaches to facilitywide materials purchasing and distribution control is being tested by the U.S. Air Force and U.S. Navy.[3] The Hazardous Materials (HAZMAT) Pharmacy establishes a single point of control for requisitioning, receipt, repackaging, and issue of hazardous materials. The HAZMAT Pharmacy maintains an available inventory of materials to supply anticipated activities and serves as a replacement for existing stockpiles in work areas. Unused materials are returned to the HAZMAT Pharmacy, and any materials still in accordance with military specifications are returned to the inventory for reissue. Off-spec materials are made available for other uses, recycled, or disposed as a last resort.

Model HAZMAT Pharmacy operations at Point Mugu Naval Air Weapons Station reduced annual hazardous material purchases from $132,000 to $43,000 despite program expansions. Similarly, Hill Air Force Base, Utah, reduced hazardous material acquisition costs from $14 million in 1990 to $4 million in 1992.

35.5 Managing Materials Handling to Minimize Waste

A number of low-cost approaches exist within the realm of better handling of materials and overall housekeeping which not only reduce waste but also improve business efficiency. Such approaches include the following:

Inspect containers for damage before accepting shipments.

Order bulk materials when large volumes are used.

Purchase premeasured containers to reduce spillage and waste if weighing is required.

Employ reusable containers.

Train employees and provide needed handling equipment.

Store materials to protect containers from physical damage (falling, abrasion, corrosion).

Protect materials from precipitation and stormwater damage.

Conduct regular maintenance on systems containing or handling materials.

Properly and promptly clean up leaks and spills.

Attempt reuse of spilled material.

Return obsolete materials to supplier.

35.6 Benefits and Effects of Materials Substitution

Replacing toxic or hazardous chemicals with less environmentally damaging alternatives has received a great deal of attention as an appropriate and effective method of pollution prevention. Additional benefits that result from materials substitution are toxic and hazardous chemical use reduction as well as reduced chemical exposure in the workplace. For example, the printing industry has successfully switched from solvent-based to water-based inks for many types of applications.

Replacement of solvents such as TCA used for cleaning and degreasing with citrus-based biodegradable (terpene) cleaners is one of the most widely known success stories of material substitution. APS Materials, Inc., a small metal-finishing company in Dayton, Ohio, achieved net annual savings of $4800 after an investment of $1800 to implement a terpene-based cleaning system that required equipment modifications, including additional heaters and a deionized water system.

In the coatings industry, materials substitution is being relied on to help achieve the reductions in hazardous air pollutants (HAPs) being mandated under the air toxics provisions of the Clean Air Act Amendments of 1990. Waterborne and high-solids substitutes often contain less than half the volatile organic compounds (VOCs) of conventional coatings. Further VOC reductions are possible for some applications by use of powder coatings; however, specialized equipment is required for curing and drying such coatings.

Materials substitution as a means of pollution prevention must include consideration of

Availability of proven substitutes

Customer acceptance and marketability—can be enhanced by explaining environmental benefits

New storage, operational, or environmental hazards of alternative materials

Requirement to retrofit or replace equipment

Required operator training

Ability of existing waste treatment operations to properly handle the new material in wastes

Stock on hand of the material being replaced

35.7 Material and Waste Exchanges

While not a material management technique that will apply to all circumstances, exchange of materials with other companies can net environmental benefits, and reduce costs associated with disposal, even if the material cannot be sold. Off-spec, overordered, or otherwise unusable materials may be sold or transferred to another industry that can still make use of the material. A list of waste exchanges in North America is available from the U.S. Environmental Protection Agency's (EPA's) Pollution Prevention Information Exchange System (PIES) [telephone: (703) 506-1025]. Also see Ref. 4.

35.8 Checklist for Materials Management Success

_____ Can layout changes reduce transfer losses?

_____ Is equipment available for material transfers?

_____ Are materials available to support maintenance schedules to avoid delaying maintenance?

_____ Will new technologies impact materials management?

_____ Are wastes segregated to maximize reuse potential?

_____ What are the impacts of production schedules?

_____ Is inventory control effective and are inventory records accurate and complete?

_____ Have employee training programs addressed materials management and pollution prevention?

_____ Can excess, off-spec, or waste materials be sold?

_____ Have employee incentive programs been considered to identify materials management opportunities?

References

1. Darryl V. Landvater, *World Class Production and Inventory Management,* Oliver Wright Publications, Essex Junction, Vt., 1993, p. 81.

2. "Hazardous Substance Management," in Elizabeth M. Donley (ed.), *Environmental Software Directory,* Donley Technology, Garrisonville, Va., 1992–1993, pp. 39–82.

3. Air Force Center for Environmental Excellence, *Hazardous Materials Pharmacy Pro-Act Fact Sheet,* AFCEE, San Antonio, Tex., 2 pp.

4. U.S. Environmental Protection Agency, *Waste Exchange Information,* Pollution Prevention Information Exchange System, Office of Environmental Engineering and Technology Demonstration, Bulletin 11, Washington, D.C., 1991.

Further Reading

Landvater, Darryl V., *World Class Production and Inventory Management,* Oliver Wright Publications, Essex Junction, Vt., 1993.

Oden, Howard W., Gary A. Langenwalter, and Raymond A. Lucier, *Handbook of Material and Capacity Requirements Planning,* McGraw-Hill, New York, 1993.

36

Generic Pollution Prevention

Water Management Techniques for Pollution Prevention

Richard A. Osantowski

Joseph C. Liello

Charles S. Applegate
Radian Corporation
Milwaukee, Wisconsin

36.1 Introduction

This section establishes an overall strategy for implementing a generic water management pollution prevention program. Approaches are presented for both reducing water usage and minimizing wastewater generation. Projects of this nature are typically performed by a pollution prevention (P2) team that has experience in water management and includes representatives with backgrounds in plant production, economics, and facilities. Company management involvement and commitment are also important for the pollution prevention program to be successful.

36.1.1 Background

Past concern for our environment is well documented in historical references. For example, the Ancient Romans constructed aqueducts to provide a source of fresh water, while sanitary wastes were removed in cloacae or sewers. Water treatment technologies began to surface in Scotland and England in the early 1800s.

In the United States, Congress has recognized that the integrity of our nation's water must be maintained. Concerned that the generation and discharge of large quantities of wastewater pose potential environmental threats to our lakes and rivers, Congress enacted the Clean Water Act of 1972 to achieve a goal of "fishable and swimmable" surface waters. The Water Quality Act of 1987 further strengthened the Clean Water Act, and amendments to the Safe Drinking Water Act required numerous treatment facility upgrades. New regulations included the reduction of viruses and cyst-forming parasites while minimizing disinfection by-products such as chloroform. Although all these acts are dramatic in their protection of our citizens against waterborne diseases and the improvement of our water quality, they placed little emphasis on source reduction or elimination of the root cause of pollution. To address this issue, the Pollution Prevention Act of 1990 was passed. This act crosses media boundaries by establishing a national policy on pollution prevention, including programs that emphasize source reduction, reuse, recycling, and training (see Fig. 36-1). All these areas are key to the successful implementation of a P2 water management program.

36.1.2 Overview

This water management section is a stand-alone document that presents a variety of pollution prevention techniques to reduce potable water usage and wastewater generation. The suggested water management strategy consists of the following wastewater pollution prevention hierarchy, which is outlined in the recommended order of program implementation.

1. Water and wastewater flow reduction
2. Water recycling
3. By-product recovery
4. Water reuse

36.1.3 Section Content

The following section on water management techniques for pollution prevention is organized into six separate subsections to facilitate its use. Section 36.2 presents a general approach to implementing a P2 wastewater reduction program. Section 36.3 discusses in-plant techniques to be considered when evaluating options for wastewater flow reduction. Section 36.4 contains useful information to consider when implementing a wastewater recycling program. In Sec. 36.5, by-product recovery methods are explored. Wastewater reuse

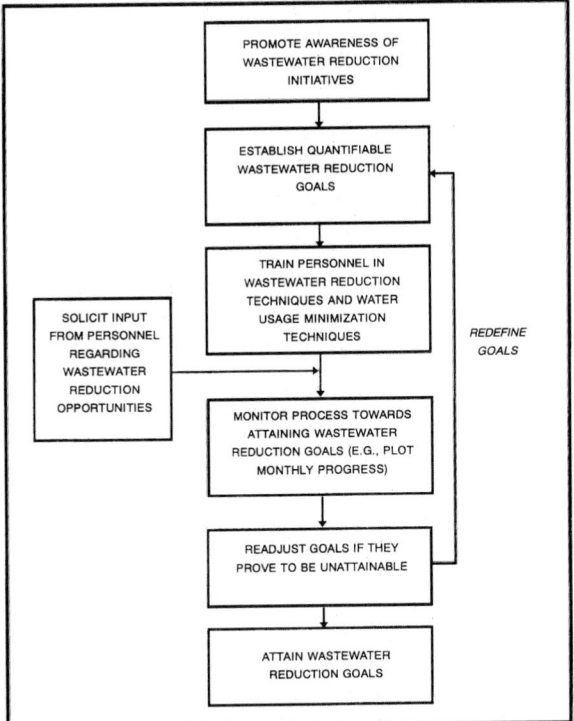

Figure 36-1. Water management training program flowchart.

options, including a brief overview of applicable treatment technologies, are presented in Sec. 36.6.

36.2 General Approach to Initiating a P2 Water Management Program

This section presents an approach that should be helpful in implementing a successful water management program. Maximum benefits will be derived if program objectives consider both the water supply and wastewater generation aspects. If implemented properly, the water management P2 program should provide a number of additional benefits to the facility. Conversely, a variety of engineering constraints may impact how and whether the final project is initiated. Table 36-1 identifies some examples of potential benefits and constraints that plant management must evaluate before proceeding with the program. Each plant must identify its own specific list of detailed benefits and constraints.

Table 36-1. Examples of Potential Benefits/Constraints of a P2 Water Management Program

Issue	Potential benefit	Potential constraint
Technical	Recovery of saleable by-products	Additional controls or treatment needs to meet reuse requirements
	Beneficial use of water (especially in water shortage areas)	Location of reusable water
	Minimizing impact on the environment	Possible equipment warranty voids
Economic	Improved public image	Competition with other projects for capital funding approval
	Employee health and safety	Increased operation and maintenance cost
	Reduction in future capital water projects	Residuals disposal
Regulatory	Possible simplification of the permitting process (e.g., fewer outfalls)	Permitting issues (e.g., Clean Air Act impact on cooling tower emissions; RCRA waste generation)
	Reduced exposure to future regulatory issues (e.g., minimal discharges)	Liability issues (e.g., corrosion)

Once the decision has been made to proceed with the P2 water management program, the key planning steps, outlined here, should be considered.

Step I: Perform an Assessment of Plantwide Water Usage and Wastewater Generation.   This initial assessment of the facility's water and wastewater systems will provide the basis for initiating the P2 water management program. During this phase of the project, historical records should be reviewed to establish an information baseline to prepare the facility water balance. Data gaps must be filled in via a representative sampling and analysis program. Table 36-2 highlights some of the activities associated with this effort.

Step II: Evaluate Potential Areas for Water Conservation, Recycling, and Reuse.   Once the facility's major sources of water usage have been identified, the P2 team should review possible candidate processes for water conservation, recycling, and reuse. This will include sanitary as well as process water uses. Figure 36-2 summarizes factors to be considered.

Step III: Define Supply Water and Wastewater Reduction Goals.   If the water management P2 program is to reach its full potential, goal setting is imperative. Established goals must be specific, quantifiable, assignable, realistic, and time oriented. Although it is impossible to define facility-specific program goals in this section, Fig. 36-3 lists some generic examples.

Table 36-2. Initial Steps in a Facility Water Management Assessment

Review available information from past wastewater surveys	NPDES and POTW permits In-plant surveys Current treatment practices Number of wastewater streams, locations, characteristics, and volume treated
Perform plant water balance	
Water supply data	Source of supply Volume used (average and maximum) Future requirements (plant expansions, changes in types of production, additional shifts, etc.)
Wastewater data	Volume treated (average and maximum) Seasonal and hourly flow variations Impact of infiltration/inflow Evaporation losses Location of wastewater streams

Obtain water chemistry data on water supply and candidate wastewater streams	Water quality criteria
	  pH
	  Temperature
	  Total dissolved solids
	  Organic concentration
	  Hardness, etc.

	Variations in water quality Sampling of candidate wastewater reuse/recycle streams
Prepare plant water balance	The water balance will be used as the foundation of the water management program. From the balance, existing streams will be quality-matched based on flow, chemistry, and location to candidate recycle/reuse processes once they have been identified. The need for additional treatment prior to recycle/reuse will also be established.

Step IV: Implement the P2 Water Management Program. Once you have assessed the facility's water and wastewater systems, evaluated potential areas of promise for pollution prevention, reviewed the system economics, and established the program's objectives, the program can be initiated. The sections that follow summarize techniques for streamlining the water management pollution prevention effort.

36.3 In-Plant Wastewater Reduction Techniques

Numerous in-plant wastewater reduction techniques can be employed to reduce the amount of wastewater generated at a plant. Generally, these techniques entail minimizing the amount of water used in various applications,

1. Identify potential sources for water conservation, recycling, and reuse.

2. Determine water quality requirements for selected reuse/recycling candidate processes (scrubber systems, cooling towers, rinse tanks, cooling systems, etc.).

3. Compare water quality requirements to values that are currently available or potentially available from varying degrees of treatment.

4. Consider locations and volumes of existing wastestreams in relation to potential recycling/reuse sources.

5. Evaluate health requirements and user objections for use of recycled/reused water.

6. Review regulatory requirements regarding the use of recycled/reused water.

7. Develop estimates of current water and treated wastewater costs, as well as projected costs.

Figure 36-2. Some key factors to consider when evaluating potential areas for water conservation, recycling, and reuse.

1. Provide training to all appropriate staff for pollution prevention and environmental awareness by (enter date).

2. Characterize all facility wastestreams by (enter date).

3. Reduce water usage by (enter percentage) by (enter date).

4. Recycle/reuse a minimum of (enter percentage) of wastewater by (enter date).

5. Reduce wastewater flow by a minimum of (enter percentage) by (enter date).

Figure 36-3. Example of water management goals for pollution prevention.

thereby reducing the amount of wastewater generated. These techniques are associated with the following areas.

Personnel water management training. Training plant personnel in water usage minimization practices can reduce wastewater generation.

Minimization of plant production water usage. Process modifications and changes in plant operations can minimize water usage in plant production activities.

Minimization of sanitary water usage. Installing flow restriction and automatic shutoff devices can reduce the generation of wastewater from sanitary water usage.

A combination of techniques from each of these areas can maximize the effectiveness of in-plant wastewater reduction efforts. The following subsections discuss various techniques that are associated with each area and can reduce plantwide wastewater generation.

36.3.1 Personnel Water Management Training

To be effective, P2 wastewater reduction programs should be employed throughout a plant. This requires training personnel not only to be aware of wastewater reduction initiatives, but also to employ various water usage minimization and wastewater reduction techniques. Such training can be provided by developing a personnel water management training program. A water management training program should accomplish the following:

- Promote awareness of wastewater reduction initiatives
- Establish quantifiable wastewater reduction goals
- Train personnel in wastewater reduction techniques and water usage minimization practices
- Implement plantwide wastewater reduction techniques and water usage minimization practices
- Monitor progress toward attaining wastewater reduction goals and readjust objectives if they prove to be unattainable
- Attain wastewater reduction goals

A personnel water management training program should include seminars to promote awareness of wastewater reduction initiatives. Based upon the assessment of the quantity and characteristics of water and wastewater flows through the plant (see Sec. 36.2), water usage and wastewater reduction goals should be established. These goals should include quantifiable reductions in water usage and/or wastewater flows that are emphasized in the personnel water management training program.

The success of wastewater reduction initiatives relies upon active participation by plant personnel. Because personnel are generally the most knowledgeable about various processes, they should be encouraged to propose new techniques for minimizing water usage and/or wastewater generation. Various incentives, such as incorporating wastewater reduction goals into annual job performance evaluations, can promote personnel participation.

After implementing water usage minimization practices and wastewater reduction techniques throughout a plant, you should monitor progress toward attaining wastewater reduction goals (e.g., plot monthly progress against objectives). If the goals prove to be unattainable, the program objectives should be reevaluated, and a revised set of goals should be identified.

36.3.2 Minimization of Plant Production Water Usage

Minimizing water usage through process modifications and/or changes in plant operations can reduce the amount of plant production wastewaters. Each technique is discussed as follows:

Process Modifications.   Where practicable and economical, water usage can be minimized through various process modifications—specifically, changes in equipment. For example, the following equipment changes elicit a reduction in water usage for particular processes:

1. Using high-pressure spray rinses instead of dip tanks
2. Replacing a water curtain spray paint booth with an electrostatic powder painting system
3. Replacing a water-cooled heat exchange system with an air-cooled system
4. Replacing a cooling tower with a refrigerant-based cooling system
5. Replacing a wet scrubber with a bag house

In developing a wastewater reduction program that includes equipment changes, you must weigh the cost/benefits against effectiveness when evaluating potential alternative process equipment.

Changes in Operations.   Changing plant operations can reduce water usage. The applicability of various operational changes differs from industry to industry; however, the following areas are candidates for changes in operations in many industries.

Enhancing equipment cleaning operations.   Wastewater from equipment cleaning operations constitutes a significant amount of wastewater for many industries.

Maximizing the effective life of production water.   Disposal of production water prior to reaching its effective life increases the amount of wastewater generated from a plant.

Optimizing water usage.   Faulty seals on pipes and equipment, as well as unregulated flow rates, contribute to increased wastewater flows.

Table 36-3 summarizes the techniques that can be used to minimize water usage for each area. As described in the following, the amount of wastewater

Table 36-3. Water Usage Minimization Techniques

Area for water usage minimization	Water usage minimization technique(s)
Enhancing equipment cleaning operations	Mechanical cleaning devices
	High-pressure spray heads
	Tank lining
	Countercurrent rinsing sequence
	Cleaning schedule coordination
	Production schedule coordination
	Pipe cleaning
	Dry cleanup practices
	Air curtain
Maximizing the effective life of production water	Countercurrent rinsing sequence
	Conductivity measurement
	Use of deionized water makeup
Optimizing water usage	Improved seals on pumps, pipes, and valves
	Automatic flow control equipment
	Flow restrictors and flow monitoring
	Water level control
	Splash guards
	Lids or silhouettes on tanks
	Heat tracing of pipes

generated from equipment cleaning operations can be reduced through various changes in operations.

Mechanical cleaning devices.   Tanks can be cleaned initially with mechanical cleaning devices such as rubber wipers. This decreases the amount of residual material on various surfaces (e.g., tank walls), thereby reducing the amount of water needed to clean the walls of the tank.

High-pressure spray heads.   By replacing regular spray heads with high-pressure heads, the amount of water needed in some cleaning operations (e.g., cleaning water-based paint from mixing tanks) can be reduced, often with an increase in cleaning efficiency—high-pressure wash systems can reduce water use by 80 to 90 percent.

Tank lining.   Tanks can be lined with Teflon™ to reduce adhesion and improve drainage. Generally, this method only applies to small batch tanks amenable to manual cleaning.

Countercurrent rinsing sequence.   Recycled "dirty" solution is used initially, or after preliminary cleaning with mechanical devices, to clean equipment. Recycled "clean" solution is then used to clean residual dirty material from the equipment. This greatly increases solution life since the level of contamination builds up more slowly in the recycled clean solution than in the recycled dirty solution.

Cleaning schedule coordination.   Cleaning should be scheduled immediately following the use of equipment to prevent residual material from drying onto surfaces.

Production schedule coordination.   Production schedules should be arranged to limit repetitious cleaning of equipment. This can be achieved by scheduling long production runs and/or by scheduling compatible production back-to-back (e.g., scheduling paint production to cycle from light to dark colors).

Pipe cleaning.   Plastic or foam "pigs" (slugs) can be used to clean some materials from pipes (e.g., paint). An inert gas propellant is used to push the pig through a pipe, thereby forcing residual material in front of the pig out of the pipe. This reduces the degree of pipe cleaning that might otherwise generate significantly more wastewater.

Dry cleanup practices.   Whenever possible, dry cleanup practices such as wiping, sweeping, or vacuuming should be used either in place of or prior to wet cleanup. This allows for the collection of solids prior to washing down various surfaces.

Air curtain.   Air curtains can be installed in paint spray booths to collect overspray. This reduces the level of effort and amount of water needed to clean the booths; therefore, the amount of wastewater generated is decreased.

Water usage can also be minimized by maximizing the effective life of production water through the following operational changes:

Countercurrent rinsing sequence.   Similar to cleaning equipment operations, a countercurrent rinsing sequence can be used in particular operations. For example, in parts-cleaning operations, the parts can initially be cleaned using a recycled dirty solution and sequentially cleaned further using progressively cleaner solutions. This increases the life of the solution, thereby reducing the amount of water needed.

Conductivity measurement.   In certain applications, the conductivity of a solution should be measured prior to disposal as wastewater to ensure that the solution has served its maximum useful life. This minimizes the frequency with which a solution is replaced by maximizing the effective life of the solution, thereby reducing the amount of wastewater generated.

Use of deionized water makeup.   Deionized water has a longer effective life than tap water in applications such as rinse tanks; therefore, by replacing tap water with deionized water, the amount of water used and the amount of wastewater generated is decreased. The ion exchange resin will, however, have to be periodically regenerated. This will produce a stream high in dissolved solids.

Water usage can be further reduced by optimizing its use throughout plant operations. Equipment should be maintained to minimize contributions to wastewater flows due to leakage, and water flow rates should be regulated in various operations to the minimum required rates and amounts. The following changes in operation can be employed to optimize water usage:

Improved seals on pumps, pipes, and valves. Water loss from leaks can be minimized by improving the seals on pumps, pipes, and valves and by conducting periodic inspections of such equipment.

Automatic flow control equipment. Automatic flow controls can be installed to regulate water usage to the minimum amount required to meet product specifications.

Flow restrictors and flow monitoring. Flow restricting devices, such as limiting flow valves, can be installed. Subsequent monitoring of flows with meters allows for redundancy to ensure that water usage rates are optimized with respect to wastewater reduction goals.

Water level controls. Continuous flow systems generate a lot of wastewater. Water level controls can be used as a substitute to regulate the flow so that it is not continuous.

Splash guards. The amount of wash water can be reduced by installing splash guards, such as curbing, to prevent overflow and/or overspray into nontarget areas.

Lids or silhouettes on tanks. Makeup water necessary because of evaporation from open tanks can be minimized by using lids or silhouettes on the tanks.

Heat tracing on pipes. Some plants continuously run water through process lines to prevent pipes from freezing, which can produce great amounts of wastewater. However, freezing can be averted by either heat tracing the pipes or by burying pipes below the frost line.

36.3.3 Minimization of Sanitary Water Usage

Wastewater flows from sanitary sources can be decreased by minimizing the amount of potable water used. This can be achieved by using various water usage reduction devices and systems in conjunction with wastewater-generating sources such as clothes washers, drinking fountains, showers, sink faucets, and toilets. Table 36-4 describes various water usage reduction devices and systems that can be employed for each of these sanitary wastewater sources.

Table 36-4. Sanitary Water Use Reduction Devices and Systems*

Service/System	Description
Clothes washers	
Level controller	Adjusts the amount of water used to match the amount of clothes to be washed
Drinking fountains	
Automatic shutoff	Automatically shuts off drinking fountain
Showers	
Limiting flow valve	Restricts flow to a fixed rate
Limiting flow shower head	Restricts and concentrates water flow by means of orifices that limit and divert flows for optimum use
Sink faucets	
Faucet aerator	Increases the rinsing power of water by adding air and concentrating flow, thus reducing the amount of water used
Limiting flow valve	Restricts water flow to a fixed rate
Infrared shutoff device	Uses an infrared detector to initiate the flow of water for a fixed time while a user is present
Temporary flow faucets	Restricts water flow to a fixed period of time once the faucet is activated
Toilets	
Brick in toilet tank	Reduces the volume of water in the toilet tank, thus decreasing the amount of water used in a cycle
Batch flush valve	Delivers a fixed amount of flow for urinals and toilets: 0.5 gal/cycle for urinals and 4.0 gal/cycle for toilets
Dual-cycle toilet	Delivers a fixed amount of flow, depending on the type of waste: 1.25 gal/cycle for liquid wastes and 2.5 gal/cycle for solid wastes
Dual-cycle tank insert	Converts conventional toilet to dual-cycle operation by inserting the device in the toilet tank
Reduced flush device	Prevents a portion of the tank contents from being dumped during the flush cycle or occupies a portion of the tank volume so that less water is available per cycle by inserting the device in the toilet tank
Wastewater recycling system for toilet flushing	Recycles bath and laundry wastewaters for use in toilet flushing
Vacuum-flush toilet system	Uses air as a waste-transporting medium and requires about 0.5 gal/cycle
Recirculating mineral oil toilet system	Uses mineral oil as a water-transporting medium and requires no water. Operates in a closed loop in which toilet wastes are collected separately from other plant wastes and are stored for later pickup by a vacuum truck. Wastes separate from the mineral oil by gravity in the storage tank. The mineral oil is drawn off by pump, coalesced, and filtered before being recycled to the toilet tank
Pressure-reducing valve	Maintains water pressure at a lower level than that of the water distribution system

*Adapted from George Tchobangolus, *Wastewater Engineering: Collection and Pumping of Wastewater*, Metcalf & Eddy, Inc., New York, 1981.

Sanitary water usage can also be reduced by restricting the installation and use of appliances that tend to increase water consumption (e.g., garbage disposals and automatic dishwashers).

36.4   Techniques for Recycling Untreated Wastewater

Recycling can minimize water usage with a corresponding reduction in wastewater flow. By definition, recycling entails using untreated wastewater in an application that is compatible with the wastewater's quality. It is important to understand the primary difference between wastewater recycling and wastewater reuse. Specifically, wastewater recycling efforts forgo treatment, using the untreated wastewater directly in another application, whereas wastewater reuse entails treating the wastewater prior to use in another application. Where viable, recycling effectively extends the useful life of water in industrial as well as sanitary applications, as discussed next.

36.4.1   Industrial Wastewater Recycling

Opportunities for recycling industrial wastewaters vary considerably from industry to industry, depending on the water quality requirements of the receiving process. Also, because wastewater recycling does not involve treatment, the number of opportunities is somewhat less than that involving wastewater reuse. Table 36-5 summarizes the main sources of industrial wastewater and their corresponding applications for recycling.

Cooling towers serve as heat exchangers and, at the same time, recycle cooling water. Therefore, where economically justifiable, cooling towers can be used in cooling applications to maximize the life of cooling water while reducing wastewater flows. During cooling tower operation, water is lost to the atmosphere by evaporation and windage, thus concentrating contaminants in the cooling tower water. Therefore, it is necessary to regularly remove some concentrated water from the tower and replace it with fresh makeup water to control particular contaminant levels, such as total dissolved solids (TDS), that can damage the equipment (e.g., via scaling, corrosion). The concentrated water removed from the tower is known as *blowdown*. Due to its high TDS and contaminant levels, cooling tower blowdown water can only be recycled in applications that do not require high-quality water, such as wash water, pump coolant, scrubber water makeup, utility water, drum seal water, and tank field waste.

Boilers also expel blowdown water that can be recycled; however, like cooling tower blowdown, it also has high concentrations of contaminants such as TDS. Makeup water quality requirements for boilers depend on the pressure

Table 36-5. Potential Industrial Wastewater Recycling Applications

Source of wastewater	Recycling application
Cooling tower blowdown	Wash water
	Utility water
	Flare drum seal water
	Pump coolant
	Tank field waste
	Scrubber water makeup
Boiler blowdown	Lower-pressure boiler makeup water
Rinsewater	Countercurrent rinsing
Once-through cooling water	Cooling pond
	Compressor cooling water
	Noncontact cooling water
	Process water
Condensate from tanks or processes	Boiler makeup water

capacity of the boiler (see Table 36-10). The water quality requirements decrease with a corresponding decrease in pressure capacity. Therefore, blowdown water from high-pressure boilers can sometimes be recycled for use in lower-pressure boilers.

Rinsewater can also be recycled by employing countercurrent rinsing. Countercurrent rinsing can be set up in series or parallel. In series, fresh water (i.e., deionized or tap) is cascaded down a rinse line, and in so doing, the parts are progressively cleaned with cleaner water as they proceed to the end of the line (see Fig. 36-4). Conversely, the rinsewater becomes progressively dirtier as it reaches the parts at the beginning of the line.

A parallel rinsing system employs a similar rationale; however, instead of a single rinsewater, several rinsewaters are used to achieve progressive cleaning (see Fig. 36-5). Specifically, dirty water is recycled at the beginning of the line and is reused as a preliminary cleaning. Likewise, clean water is recycled at the end of line and is reused as a final rinse. This approach prolongs the effective life of rinsewaters for each stage of cleaning, thereby reducing water usage and wastewater generation.

Once-through cooling water (noncontact) is normally discharged; however, this water can be recycled in various processes. The cooling water could, for example, be recirculated into a collecting pond (used as a heat exchanger and for water storage). The water can then be drawn off as needed for various noncontact cooling processes. This effectively creates a closed-loop system in

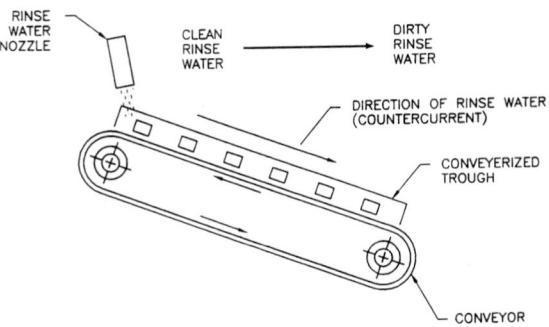

Figure 36-4. Countercurrent rinse wastewater: series configuration.

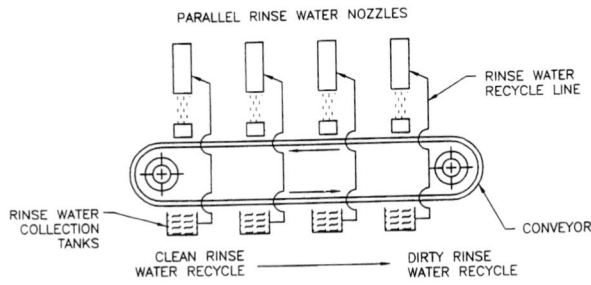

Figure 36-5. Countercurrent rinsewater: parallel configuration.

which there is virtually zero discharge and zero inflow (except for offsetting losses due to evaporation).

Another potential source of wastewater for recycling is condensate from tanks or processes. Condensate water typically contains low concentrations of organics and TDS, and it can often be recycled as makeup water for boilers. The amount of condensate is usually relatively small so it may not be economical to collect for recycle.

36.4.2 Sanitary Wastewater Recycling

The potential applications for recycling sanitary wastewater are somewhat limited. The quality of sanitary "gray" wastewater, composed of wash water (e.g., bathing, washing machine, dishwater), is often conducive to recycling in toilet systems or as irrigation water. A wash water recycle system for toilet flushing uses bath and laundry wastewaters to supplement fresh water inflow into the system, which reduces water usage and wastewater generation. The bath and laundry wastewaters can often also be used for irrigation, depending on the soil characteristics.

36.5 By-Product Recovery

High priority should be given to the recovery of potential pollutants from wastewater and in-process water. After flow reduction and recycling measures, the P2 team should consider by-product recovery to control the release of pollutants to the environment. By-product recovery is unique because it emphasizes the recovery of contaminants instead of destruction or end-of-pipe effluent quality. By-product recovery should be considered when the substance has value or when disposal of the wastewater (or residuals from treating the wastewater) will be costly.

As in all water management and pollution prevention efforts, planning is a key to success. However, since by-product recovery must be integrated with the mainline production, the P2 team must include production management and engineering representatives. The planning should include consideration of the criteria used to determine the feasibility of by-product recovery as discussed in Sec. 36.5.1. Section 36.5.2 discusses technology options available to implement by-product recovery, and Sec. 36.5.3 provides examples of industrial processes that include by-product recovery.

36.5.1 Selection Criteria

Selection of by-product recovery as a pollution prevention measure should be based on an objective review of criteria defined by the P2 team. Table 36-6 provides examples of relevant selection criteria that could be used during this evaluation, including:

- The value of the recovered by-product compared to the cost of its disposal
- The feasibility of recovery at a reasonable cost
- The selection of appropriate recovery technology
- The definition of internal and external institutional concerns

A substance can have a value due to its market cost or because the cost of disposal exceeds the cost of recovery from the wastewater. In some instances, there may be no disposal option regardless of the cost. Internal use of the recovered by-product provides maximum control of production, availability, and end use. However, when a substance is recovered for a second party, additional issues must be considered. The size of the market will determine whether there is a reliable outlet for the recovered by-product. A large market relative to the produced quantity of by-product is desirable. The market should provide multiple users of the by-product to ensure that there is always an outlet. This is necessary so that the facility can proceed with production plans without concern that the outlet for the by-product will be unavailable in the future. In addition, the by-product should be produced in lot sizes and of a quality that the user finds attractive. It may be difficult to find an outlet if production is too small to

Table 36-6. By-Product Recovery Selection Criteria

Selection criteria	Issues
Value of recovered by-product	In-house use
	Market availability
	Market reliability
	Market flexibility
Cost of by-product recovery	Capital
	Operation and maintenance
By-product recovery technology	Availability
	Reliability
	Health and safety concerns
Institutional	Regulatory
	Potential interference with production
	Public acceptance

make up a reasonable order size. In these instances, distribution through a commodities broker may be attractive.

The cost of by-product recovery may be competitive with other alternatives; however, other factors might need consideration. Both capital cost and operating and maintenance cost must be determined. These costs should be balanced against other alternatives and overall plan objectives. For example, there may be a nonintrinsic value on total self-sufficiency and control of waste flows.

The technology for by-product recovery must be reviewed to determine whether the project is feasible. Section 36.5.2 gives examples of candidate technologies. Process reliability is important since by-product recovery is closely tied to the production activities of the facility. Consequences can be costly if the production line is held up due to by-product processing problems. Also, the process must include appropriate safeguards for health and safety and compliance with OSHA requirements.

The P2 plan should address internal and external institutional constraints and the P2 team should consider the impact of air, water, and waste regulations. Contingency should be considered as well so that mainline production is only minimally affected by by-product recovery. When the recovered by-product is used to replace commercially available raw materials, quality standards must be set to limit potential liability associated with the substitution.

36.5.2 Technology Candidates

Technologies that can be used for by-product recovery are highly dependent on the application. Table 36-7 lists some examples of technologies that are applica-

Table 36-7. Examples of By-Product Recovery Technologies

Technology	Description	Applications
Evaporation	Removal of solvent as a vapor from a solution or slurry. The vapor and/or the residue may be the recovered by-product.	Extract proteins Concentrate metals
Selective ion exchange	Uses a solid phase containing bound groups that carry an ionic charge, either positive or negative, with free ions of the opposite charge that can be displaced. Chelating ion exchange resins have high selectivity for certain cations.	Metal recovery from plating solutions Concentration of ammonia
Reverse osmosis	Pressure overcomes the osmotic pressure so that water passes a semipermeable membrane and dissolved salts are rejected.	Recover pesticides Concentrate salts Acid recovery
Ultrafiltration	Filtration using membranes with small pore size. Can be used to filter large soluble molecules or colloidal solids.	Recover pesticides
Biotechnology	Microbes perform specific enzyme mediated reactions.	Methane formation by anaerobic decomposition of organic compounds Concentration of precious metals
Solvent extraction	Two immiscible liquid phases are brought into contact for the transfer of one or more components.	Extract valuable organic compounds from water Concentrate soluble metals
Distillation	The separation of the liquid mixture by partial vaporization and separate recovery of vapor and residue.	Alcohol recovery Solvent purification
Selective precipitation	Desired ions are precipitated by control of pH or counter ions.	Plating solutions
Electrodialysis	An electromotive force is used to transport ionized materials through an ion-selective, semipermeable membrane.	Concentrating brine solutions Recover acids Recover metals
Crystallization	Formation of high-purity crystalline solids from liquid or vapor.	Mineral recovery

ble for the recovery of a range of materials. Many of these technologies are also used for water reuse applications; however, with by-product recovery, more emphasis is placed on the purity of the recovered by-product than on the concentration remaining in the residual.

36.5.3 By-Product Recovery Examples

To illustrate the diversity of by-product recovery applications, this section describes four specific examples representing industries involved in pesticide formulation and packaging, defense, metal finishing, and battery manufacture.

Pesticide Formulation and Packaging. Pesticides are generally formulated and packaged in small batch operations. Between batches, the mixing kettles are washed to prevent carryover of the different formulations. Membrane separation technology has been used to recover residual pesticides from the wash water and to clean the water phase sufficiently for reuse as wash water.

Membrane separation technologies (i.e., membrane filtration) utilize a pressure-driven, semipermeable membrane to achieve selective separations. The pore size of the membrane can be relatively large if precipitates or suspended materials are to be removed from a wastewater (termed *ultrafiltration*) or very small for removal of inorganic salts or organic molecules (termed *reverse osmosis*). During operation, the feed solution flows across the surface of the membrane, clean water permeates the membrane, and the contaminants remain in the rejected concentrate stream. A major advantage of the process is that it can produce a high-quality product waterstream (permeate) and the reject material can potentially be recycled back into the manufacturing process, eliminating the need for disposal.

Recovery of By-Product from Explosives Manufacture. RDX and HMX explosives are manufactured by the Bachmann Combination Process. The process results in the generation of a spent acetic acid stream. This stream is processed by evaporation to recover the acetic acid for recycling. The acetic acid recovery process, in turn, generates a high-nitrate salt by-product, which can be either sodium nitrate- or ammonium nitrate-based. The choice of by-product depends on market conditions. Long-range alternatives are being pursued for both by-products to eliminate market dependency.

Biological denitrification is one long-range alternative considered for the sodium nitrate-based by-product called causticized stripping column bottoms (CSCB). CSCB contains approximately 45% sodium nitrate plus smaller amounts of sodium nitrite, sodium chloride, sodium acetate, sodium formate, and sodium sulfate. Bacteria can use nitrate to oxidize an organic carbon source, such as acetic acid, to carbon dioxide and water. Nitrate is removed by conversion to nitrogen gas.

An alternative use for the high nitrate concentration stream is to use it as an alternative to oxygen in a biological wastewater treatment facility. Under anoxic conditions, the nitrate acts as the electron acceptor in the biologically mediated oxidation of organic pollutants. The use of the nitrate by-product from explosives manufacturing can reduce the energy needed to supply oxygen.

Metal-Finishing Spent Pickle Liquor for Phosphorus Removal. Preparation operations for metal finishing generate spent pickle liquor. This solution of ferrous chloride or ferrous sulfate, depending on the acid used, is a good reagent for phosphorus removal at Publicly Owned Treatment Works (POTW). The metal finisher typically provides transportation and storage tanks at the POTW. For phosphorus control, it is important to add spent pickle liquor in a well-aerated location, such as the activated sludge aeration basin. Aeration ensures that the reduced iron is oxidized, permitting efficient precipitation of phosphorus.

Lead By-Product Recovery in Battery Manufacture. Lead-acid battery manufacture generates lead-contaminated wastewater and solid waste from various processing operations. Lead-bearing wastewater is treated by conventional chemical coagulation, and the sludge from the wastewater treatment is thickened, dewatered, and combined with other lead wastes. Secondary lead smelters accept the by-product for processing into ingots for the battery manufacturer.

36.6   Techniques for Reusing Treated Wastewater

Two principal factors must be evaluated when options are considered for reusing wastewater. First, the quality of the reused wastewater is dictated by the application for which it will be used. Second, a treatment technology capable of achieving the necessary quality must be selected. Both factors are discussed in the following sections.

36.6.1   Applications for Treated Wastewater

Depending on the required quality, treated wastewater can be reused in various applications, minimizing water usage and reducing wastewater flows. Table 36-8 summarizes some applications and potential constraints for reusing treated wastewater. Before a reuse application is selected, equipment specifications should be checked and the manufacturer contacted to ensure that reusing treated wastewater will not compromise effectiveness or void any manufacturer warranties.

Table 36-8. Application for Wastewater Reuse and Potential Constraints*

Applications for wastewater reuse	Potential constraints

Agricultural irrigation
  Crop irrigation
  Commercial nurseries
	Surface and groundwater pollution if not properly managed

Landscape irrigation
  Park
  School yard
  Freeway median
  Golf course
  Cemetery
  Greenbelt
  Residential
	Marketability of crops and public acceptance
	
	Effect of water quality, particularly salts, on solids and crops
	
	Public health concerns related to pathogens (bacteria, viruses, and parasites)
	
	Controls may be needed for use area, including buffer zones, consequently resulting in high user costs

Industrial reuse
  Cooling tower makeup water
  Once-through cooling water
  Boiler feed water
  Process water
	Constituents in reclaimed wastewater relating to scaling, corrosion, biological growth, and fouling
	
	Public health concerns, particularly aerosol transmission of pathogens in cooling water

Groundwater recharge
  Groundwater replenishment
  Saltwater intrusion control
  Subsidence control
	Organic chemicals in reclaimed wastewater and their toxicological effects; TDS, nitrates, and pathogens in reclaimed wastewater

Recreational/environmental uses
  Lakes and ponds
  Marsh enhancement
  Streamflow augmentation
  Fisheries
  Snowmaking
	Health concerns about bacteria and viruses
	
	Eutrophication due to nitrogen and phosphorus in receiving waters
	
	Toxicity to aquatic life

Nonpotable urban uses
  Fire protection
  Air-conditioning
  Toilet flushing
	Public health concerns about pathogens transmitted by aerosols
	
	Effects of water quality on scaling, corrosion, biological growth, and fouling
	
	Cross-connections

Potable reuse
  Blending in water supply
  Reservoir
  Pipe-to-pipe water supply
	Constituents in reclaimed wastewater, especially trace organic chemicals and their toxicological effects
	
	Aesthetics and public acceptance
	
	Health concerns about pathogen transmission, particularly viruses

*Adapted from Metcalf & Eddy, Inc., *Wastewater Engineering: Treatment, Disposal, and Reuse*, 3d ed., New York, 1991.

Of the applications summarized in Table 36-8, the principal applications for water reuse include agricultural and landscape irrigation and industrial reuse. Both are discussed in the following sections, with particular emphasis on industrial wastewater reuse.

Agricultural and Landscape Irrigation. Water quality is the primary constraint associated with reusing wastewater for agricultural and landscape irrigation. In evaluating irrigation as an application for wastewater reuse, one must consider the relationship between desired crop yields (agriculture) and soil properties and the quality of the reused wastewater. Wastewater quality affects the viability of irrigation reuse in four key areas.

1. *Salinity.* Because the presence of salts can affect plant growth, the salinity or TDS of the treated wastewater is the most important factor in determining suitability for this application. Generally, high salinity in the root zone restricts plant growth since plants must expend more energy that would otherwise be used to sustain plant growth, adjusting the salt concentration in their tissue. Salinity can be measured by determining the electrical conductivity of the soil. Acceptable levels of soil salinity vary depending on the tolerance of a particular plant to salt, the temperature, and the wind.[1]

2. *Specific ion toxicity.* Excessive concentrations of specific ions (predominantly sodium, chloride, and boron) can cause a decline in crop growth. Table 36-9 presents the maximum suggested trace element concentrations for irrigation waters.

3. *Water infiltration rate.* The water infiltration rate can be greatly reduced by high concentrations of sodium in the reused wastewater, thereby restricting plant growth.

4. *Health and regulatory requirements.* The presence of biological agents such as pathogens, protozoa, and viruses pose the greatest health risks. Applicable regulations governing wastewater reuse for irrigation can be identified by contacting state agencies.

Table 36-9 shows recommended concentrations for trace elements.

Industrial Reuse. As indicated in Table 36-8, there are several applications for reuse of treated industrial wastewater, including cooling tower makeup water, once-through cooling water, low-pressure boiler feed water, and process water.

Cooling tower makeup water accounts for a significant amount of water use for many industries (e.g., electric power generating stations, oil refining, chemical manufacturing, and metal manufacturing). Therefore, reusing treated wastewater as makeup water in cooling towers reduces fresh water usage and reduces wastewater flows. However, the suitability of the wastewater depends on its quality in relation to scaling, metallic corrosion, biological growth, and fouling.

Table 36-9. Recommended Maximum Concentrations
of Trace Elements for Irrigation Waters*

Trace element	Recommended maximum concentration† (mg/L)
Aluminum	5.00
Arsenic	0.10
Beryllium	0.10
Cadmium	0.01
Chromium	0.10
Cobalt	0.05
Copper	0.20
Fluoride	1.00
Iron	5.00
Lead	5.00
Lithium	2.50
Manganese	0.20
Molybdenum	0.01
Nickel	0.20
Selenium	0.02
Tin	†
Titanium	†
Tungsten	†
Vanadium	0.10
Zinc	2.0

*Adapted from National Academy of Science, National Academy of
Engineering, *Water Quality Criteria 1972*, A Report of the Committee on
Water Quality Criteria, Superintendent of Documents, U.S. Government
Printing Office, Washington, D.C., 1972.
†Tin, tungsten, and titanium are effectively excluded by plants.

Scaling. The formation of hard deposits can reduce heat exchange efficiency. Therefore, the concentrations of calcium and magnesium in makeup
water is of great concern. Calcium and magnesium can be removed by treating
the wastewater via ion exchange, but if phosphate is present, calcium phosphate salt may form and potentially cause scaling. This can be averted, however, through precipitation treatment to remove the phosphates.

Metallic Corrosion. Metallic corrosion occurs when there is a difference in
electrical potential between metal surfaces in the cooling tower. The electrical
conductivity of the makeup water can be increased by contaminants such as
TDS; consequently, corrosion can be accelerated. Corrosion is additionally promoted by components with high oxidation potentials, such as dissolved oxygen
and certain metals (e.g., manganese, iron, and aluminum). Reused wastewater
can be treated with chemical corrosion inhibitors to reduce metallic corrosion
of the cooling tower.

Biological Growth. The availability of nutrients in the makeup water in conjunction with the warm, moist conditions in a cooling tower provide ideal con-

ditions for bacteriological growth. Bacteria can attach to heat exchangers and inhibit heat exchange and water flow. Cooling tower makeup water can either be disinfected or treated with biocides to limit bacteriological growth.

Fouling. The heat exchange capacity of a cooling tower can also be inhibited by the attachment and growth of deposits, including biological growth, suspended solids, silt, corrosion products, and inorganic salts. Makeup water can be treated by adding chemical dispersants to prevent particles from aggregating and settling.

Once-through cooling water refers to water used in cooling processes (e.g., pump, compressor, and bearing cooling; turbine exhaust condensing; and direct contact quenching) that do not involve the recirculation of the cooling water. Instead, the cooling water is discharged after accepting process heat loads by sustaining a temperature increase. The primary cooling water quality considerations include potential deposits from suspended solids and biological growth. Table 36-10 summarizes the water quality requirements for once-through cooling water. Applicable treatment processes to achieve these requirements are provided in Table 36-11.

The reuse of wastewater for boiler makeup has limited applications due to special water quality requirements (see Table 36-10). Raw feedwater for low- to medium-pressure boilers (i.e., 650 psi or less) requires softening, deaeration, and, occasionally, silica removal to prevent corrosion or scaling. High-pressure boilers require demineralized, deaerated water and internal deoxidation chemicals to prevent scaling. Dissolved oxygen and carbon dioxide must also be removed from makeup water to prevent corrosion.[2]

Generally, the high-quality water requirements of high-pressure boilers limit their application in wastewater reuse; however, some wastewaters may be reused for low-pressure boiler feedwater. For example, clean steam condensate, generated when tanks or processes are indirectly heated, can be reused as boiler feedwater instead of being disposed of as wastewater.

Quality requirements for water used in industrial processes depend on the type of industry and the process in which the water will be reused. Therefore, generalizing the water quality requirements associated with industrial processes is not possible. However, Table 36-12 summarizes general water quality standards for feed water used in several industries. It should be emphasized that the information in Table 36-12 does not reflect water quality requirements for any particular process.

36.6.2 Applicable Technologies for Reusing Treated Wastewater

As discussed in Sec. 36.6.1, numerous opportunities exist for reusing wastewater. Table 36-13 summarizes information regarding various treatment technologies that can be used to treat wastewaters for reuse in other applications.

Table 36-10. Water Quality Requirements for Point of Use for Steam Generation and Cooling in Heat Exchangers[a]

| Characteristic | Boiler feed water | | | | Cooling water application | | | |
| | Industrial[g] | | | Electric utilities | Once-through | | Cooling tower makeup | |
	Low pressure	Intermediate pressure	High pressure	—	Fresh	Brackish[b]	Fresh	Brackish[b]
Silica (SiO_2)	30	10	0.7	0.01	50	25	50	25
Aluminum (Al)	5	0.1	0.01	0.01	c	c	0.1	0.1
Iron (Fe)	1	0.3	0.05	0.01	c	c	0.5	0.5
Manganese (Mn)	0.3	0.1	0.01	0.01	c	c	0.5	0.02
Calcium (Ca)	c	0.4	0.01	0.01	200	420	50	420
Magnesium (Mg)	c	0.25	0.01	0.01	c	c	c	c
Ammonia (NH_4)	0.1	0.1	0.1	0.07	c	c	c	c
Bicarbonate (HCO_3)	170	120	48	0.5	600	140	24	140
Sulfate (SO_4)	c	c	c	d	680	2,700	200	2,700
Chloride (Cl)	c	c	c	c,d	600	19,000	500	19,000
Dissolved solids (TDS)	700	500	200	0.5	1,000	35,000	500	35,000
Copper (Cu)	0.5	0.05	0.05	0.01	c	c	c	c
Zinc (Zn)	c	0.01	0.01	0.01	c	c	c	c
Hardness ($CaCO_3$)	350	1.0	0.07	0.07	850	6,250	650	6,250
Alkalinity ($CaCO_3$)	350	100	40	1	500	115	350	115
pH, units	7.0–10.0	8.2–10.0	8.2–9.0	8.8–9.4	5.0–8.3	6.0–8.3	c	c
Organics:								
Methylene blue active substance	1	1	0.5	0.1	c	c	1	1
Carbon tetrachloride extract	1	1	0.5	c,e	f	f	1	2
Chemical oxygen demand (COD)	5	5	1.0	1.0	75	75	75	75
Hydrogen sulfide (H_2S)	c	c	c	c	—	—	c	c
Dissolved oxygen (O_2)	2.5	0.007	0.007	0.007	present	present	c	c
Temperature	c	c	c	c	c	c	c	c
Suspended solids	10	5	0.5	0.05	5,000	2,500	100	100

NOTE: Unless otherwise specified, units are mg/L and values that should not be exceeded. No one water will have the maximum values shown.

[a]Adapted from National Academy of Science, National Academy of Engineering, *Water Quality Criteria 1972*, A Report of the Committee on Water Quality Criteria, Superintendent of Documents, U.S. Government Printing Office, Washington, D.C., 1972.
[b]Brackish water—dissolved solids more than 1000 mg/L.
[c]Accepted as received (if meeting other limiting values); has never been a problem at concentrations encountered.
[d]Zero, not detectable by test.
[e]Controlled by treatment for other constituents.
[f]No floating oil.
[g]Quality of water prior to the addition of internal conditioning chemicals.

Table 36-11. Applicable Processes for Treating Cooling Water or Boiler Makeup*

Applicable treatment process	Reuse application		
	Once-through cooling water	Cooling tower makeup water	Boiler water makeup
Suspended solids and colloidal removal:			
Straining	X	X	X
Sedimentation	X	X	X
Coagulation		X	X
Filtration		X	X
Aeration		X	X
Dissolved solids modification softening:			
Cold lime		X	X
Hot lime soda			X
Hot lime zeolite			X
Sodium cation exchange		X	X
Alkalinity reduction cation exchange:			
Hydrogen		X	X
Hydrogen and sodium cation exchange		X	X
Anion exchange			X
Dissolved solids removal:			
Evaporation			X
Demineralization		X	X
Dissolved gases removal:			
Degasification—Mechanical		X	X
Vacuum	X		X
Heat			X
Internal conditioning:			
pH adjustment	X	X	X
Hardness sequestering	X	X	X
Hardness precipitation			X
Corrosion inhibition		X	X
Embrittlement			X
Oxygen reduction			X
Sludge dispersal	X	X	X
Biological control	X	X	

*Adapted from National Academy of Science, National Academy of Engineering, *Water Quality Criteria 1972*, A Report of the Committee on Water Quality Criteria, Superintendent of Documents, U.S. Government Printing Office, Washington, D.C., 1972.

Table 36-12. Process Water Quality Requirements for Selected Industries*

Characteristic	Textile industry	Pulp & paper industry	Chemical industry	Petroleum industry	Iron & steel industry
Iron (Fe)	0.0–0.3	0.1–10	10		
Manganese (Mn)	0.01–0.05	0.05–0.5	2		
Copper (Cu)	0.01–5.00				
Dissolved solids	100–200	200–500	2,500	3,500	0.5
Suspended solids	0–5		10,000	5,000	0.1–10
Hardness (as $CaCO_3$)	0–50	100–200	1,000	900	0.1–100
Color, units	0–5	5–100	500	25	
Turbidity, units	0.3–5.0	10–100			
Sulfate (SO_4)	100		850	900	
Chloride (Cl)	100	75–200	500	1,600	0.1
Alkalinity (as $CaCO_3$)	50–200	75–150	500	500	0.5
Aluminum oxide (as Al_2O_3)	8				
Silica (as SiO_3)	25		†	85	
Organic growths	absent				
Calcium (Ca)			250	220	
Magnesium (Mg)			100	85	
Ammonia (NH_3)			†	40	
Bicarbonate (HCO_3)			600	480	
pH, units			5.5–9.0	6.0–9.0	5.0–9.0
Odor threshold number			†		
BOD (O_2)			†		
COD (O_2)			†	1,000	
Temperature			†		100
DO (O_2)			†		‡
Sodium (Na) and potassium (K)				230	
Fluoride (F)				1.2	
Nitrate (NO_3)				8	
Hydrogen sulfide (H_2S)				20	
Settleable solids					0.1–5.0
Oil					
0.02–1.00					
Residual chlorine (Cl_2)		2.0			
Silica (soluble) (SiO_2)		20–100			
Free CO_2		10			
Calcium hardness (as $CaCO_3$)		50			
Magnesium hardness (as $CaCO_3$)		50			

*Adapted from National Academy of Science, National Academy of Engineering, *Water Quality Criteria 1972*, A Report of the Committee on Water Quality Criteria, Superintendent of Documents, U.S. Government Printing Office, Washington, D.C., 1972.

†Accepted as received (if meeting other limiting values); has never been a problem at concentrations encountered.

‡Minimum to maintain aerobic conditions.

Table 36-13. Applicable Treatment Technologies for Wastewater Reuse

Treatment technology	Description	Applications	Limitations
Reverse osmosis	Uses the principles of osmosis and differences in pressure to separate dissolved salts in a solution from water by filtering the wastewater through a semipermeable membrane.	Removal of BOD, COD, TSS, NH_3-N, TDS, and phosphorus	Cost Scaling pH and temperature sensitivity Pretreatment may be required Concentrate may require treatment or disposal
Electrodialysis	Electrodialysis concentrates or separates ionic species contained in a solution by passing the solution through semipermeable ion-selective membranes.	Removal of TDS and recovery of metal salts	Chemical precipitation on membrane Pretreatment may be required Cost
Ultrafiltration	Similar to reverse osmosis, ultrafiltration uses porous membranes to remove dissolved and colloidal materials from solutions; however, ultrafiltration operates at lower pressures and is generally limited to removing larger molecules.	Removal of TDS, turbidity, and oil	Cost Scaling pH and temperature sensitivity Pretreatment may be required
Ion exchange	Ion exchange removes specific ions from a solution by exchanging them with ions bound to a specially formulated resin. The resin requires backwashing and regeneration once its capacity has been reached.	Removal of TDS and toxic metal ions; and reduction of hardness by removing calcium and magnesium ions	Spent resins and regenerants (i.e., acid, caustic, or brine) must be disposed High concentrations of suspended solids can reduce the efficiency of the resin
Activated carbon	The activated carbon process uses either granular or powdered carbon to treat wastewater by adsorbing many organic and inorganic compounds. Once the capacity of the carbon has been reached, it must be regenerated.	Removal of many organic and inorganic compounds Treats organic wastes (with high boiling points, low solubility, and polarity), chlorinated hydrocarbons, and aromatics Captures volatile organics in gas mixtures	Cost as a function of the frequency of carbon regeneration Contaminant concentrations should be less than 10,000 ppm Suspended solids less than 50 ppm Dissolved inorganics and oil and grease less than 10 ppm

Table 36-13. Applicable Treatment Technologies for Wastewater Reuse
(*Continued*)

Treatment technology	Description	Applications	Limitations
Sedimentation	Sedimentation is a settling process that allows heavier solids to separate from a solution by gravity.	Removal of solids that are more dense than water	Not suitable for wastewaters consisting of emulsified oils
Filtration	Filtration separates and removes suspended solids from a solution by passing the solution through a porous medium (e.g., fabric, screen, granular material)	Dewatering sludges and slurries Removal of suspended solids from liquids Pretreatment to remove solids to prevent clogging of subsequent treatment devices (e.g., ion exchange, reverse osmosis, carbon adsorption)	Not suitable for reducing toxicity of the wastewater Not suitable for gelatinous solids Limitations on suspended solids concentration of liquids
Evaporation	Evaporation is a physical separation process in which energy is applied to volatilize a solution, thereby separating a liquid from dissolved or suspended solids.	Treatment of hazardous wastes Treatment of solvent wastes with non-volatile constituents (e.g., oil, grease, polymeric resins, paint solids) Separation of dissolved and suspended solids	Effectiveness depends on the volatility of the solution Cost
Dewatering	Dewatering includes any of a number of physical processes (e.g., vacuum filtration, sludge drying, belt filter press) used to reduce the moisture content of sludges.	Reducing the moisture content of sludges	Limitations depend on the dewatering process used for a particular type of sludge

References

1. National Academy of Science, National Academy of Engineering, *Water Quality Criteria 1972*, A Report of the Committee on Water Quality Criteria, Superintendent of Documents, U.S. Department Printing Office, Washington, D.C., 1972.
2. Angelo Morresi et al., "Cooling Water and Boiler Possibilities for Wastewater Reuse," *Industrial Wastes,* March/April 1978.

Further Reading

"A Compendium of Technologies Used in the Treatment of Hazardous Wastes," EPA/615/8-87/014, U.S. Environmental Protection Agency, Center for Environmental Research Information, 1987.

Culp, Russell L., "Treatment Processes to Meet Water Reuse Requirements," *Water/Engineering & Management,* April 1981.

Facility Pollution Prevention Guide, U.S. Environmental Protection Agency, Office of Research and Development, Washington, D.C., May 1992.

Freeman, Harry (ed.), *Hazardous Waste Minimization*, U.S. Environmental Protection Agency, Risk Reduction Engineering Laboratory, McGraw-Hill Publishing Company, 1990.

Pollution Prevention Case Studies Compendium, U.S. Environmental Protection Agency, Office of Research and Development, Washington, D.C., April 1992.

37

An Overview of Potential Environmental Impact from Industrial Activity

Kevin Palmer

*Science Applications International
Corporation
Falls Church, Virginia*

When the environmental issues associated with industry are considered, it is common to focus on the wastes generated during the manufacturing process and to classify and manage industrial wastes and the associated environmental problem by the physical state of the waste. Subsequently, industry tends to focus its attention with respect to the environment on complying with regulations that are intended to directly protect each of the environmental receptors. This approach tends to direct efforts towards end-of-pipe wastes and often results in the neglect of various environmental problems that arise from the entire industrial operation. It is important that regulators and industry understand the environmental impact of all steps of a process when attempting to build a comprehensive environmental protection strategy or program.

Industrial contributions to environmental problems can arise from each of the steps of the life cycle of its products. First, industrial facilities use natural resources and thereby contribute to the environmental problems that arise from the collection of these resources. Second, industry can create additional adverse environmental impact while processing these materials into products. Third, industry may contribute to environmental problems in the packaging and distribution of its products. Fourth, depending on the design and function of a product, the use and maintenance (and even disposal) of the product may also contribute to a variety of environmental problems. The purpose of this chapter is to describe the environmental problems that confront industry in each of the steps of the product life cycle.

37.1 Generalized Environmental Impact

Figure 37-1 depicts the simplified life cycle of any product. Each of the four steps identified in this figure may contribute to adverse environmental impact, which may include the release of chemicals to the environment. Releases of chemicals to the air, water, and soils may degrade environmental quality and result in the destruction of ecosystems and their flora or fauna, bioaccumulation of chemicals in food chains, and/or short- or long-term human health effects in the exposed population. Releases of chemicals may be direct, such as air or water emissions, or indirect, such as solid-phase wastes which are subsequently landfilled or incinerated and landfilled. The generation and management of solid-phase wastes may result in the release of chemicals to the environment through leaching of chemicals (landfilling) or emission of chemicals to the environment (incineration). Finally, any life-cycle step may consume materials, natural resources, and energy that, in the production, collection, processing, or use, create an environmental impact. This impact might include the depletion of ecosystem resources such as the destruction of rain forests, the elimination of old-growth forests, or the disruption of marine coastal systems. Additional impact might include the generation of chemical emissions such as sulfur dioxide and carbon monoxide resulting from the burning of coal in energy production.

It is critical to understand each type of impact at each step of the life cycle, because industry defines and controls the nature of their products and the processes used in manufacturing them. Environmental problems should be linked with the activities that generate the problem. For example, it is important to know that painting operations can result in air emissions of paint and solvents during the application and drying of paint. By identification of the problem and the source, an understanding of the environmental impact may be developed and potential solutions that might alleviate or eliminate the environmental problem may begin to be formulated.

To illustrate the overall environmental impact that an industry can have, the remainder of this chapter covers a hypothetical life cycle that describes many

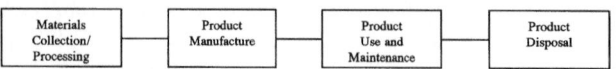

Figure 37-1. A generalized product life cycle.

common industrial processes and related environmental problems. The purpose of this exercise is twofold. First, by using a single life cycle (for a single material), we can illustrate the relationship between separate processes that contribute to overall environmental problems caused by the manufacture and use of the product. Second, we can discuss the environmental problems created by very common types (or classes) of processes. As you will see in our extended example, many of the processes and the related environmental impact discussed are common to a variety of industrial manufacturing operations. This commonality makes such processes likely candidates for pollution prevention research and progress.

To understand the types of environmental problems associated with each step of the life cycle, we will trace the manufacture of a combustion engine vehicle which we will name the ACME Roadster. Specifically, we will trace the life cycle with respect only to steel, a single component of the final product. There are, of course, many other materials, resources, and components in an automobile, but this abbreviated example will demonstrate the magnitude of the impact that might be related to one component of a system. We will use specific examples of processes that illustrate the types of environmental impact that arise from some of the processes used to manufacture and maintain the steel components in the Roadster. We will not, however, provide detailed or analytical discussions of the environmental impact. A detailed exploration of life-cycle analysis (LCA) as an analytical tool is the subject of Chap. 19.

37.2 Generalized Case Study

Figure 37-2 provides a schematic representation of the ACME Roadster life cycle. Each of the example processes and the specific environmental impact of the ACME Roadster life cycle are discussed here. The following simplified life-cycle example is based on the assimilation of a variety of references concerning a multitude of processes and operations and does not reflect a single manufacturing line or an actual life cycle performed for any combustion engine vehicle manufacturer.

37.3 Natural Resource Collection

The ACME Roadster uses a variety of metals and alloys in the frame, body, and engine of the vehicle. One of the largest components is steel. Iron is needed to make the steel parts of the Roadster. To meet this need, an iron producer will

Figure 37-2. Life cycle for steel parts in the ACME Roadster.

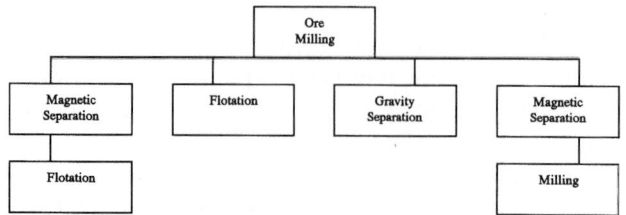

Figure 37-3. General iron beneficiation options.

conduct extraction and beneficiation operations. Extraction of iron is performed almost exclusively through surface operations, including open-pit and open-cut methods.[1] Beneficiation operations, or the concentrating of iron ore, rely on milling and a variety of separation techniques. Figure 37-3 outlines the iron extraction and beneficiation operations commonly used to obtain iron ore.

37.3.1   Environmental Impact

The exact process configuration used to obtain iron to make the steel used in the ACME Roadster is not crucial to the discussion at this point; however, the adverse environmental impact that might result from these processes is the point of this discussion. Examples of some aspects of the impact are discussed here.

Extraction has immediate environmental impact on the localized landscape, flora and fauna, and surrounding aquatic systems that receive runoff (i.e., stormwaters that come into contact with mining operations and residues, or tailings). Even under the best-controlled systems, open-pit and open-cut mining operations will completely disrupt the ecosystem being subjected to the practice. In addition, the beneficiation process will generate fugitive dusts, aqueous wastes, and tailings, at least some of which can be released to the environment. Beneficiation operations can also produce chemical agents that include acids, alcohols, metal chlorides (including mercuric chloride), and organic liquids (including xylene and toluene) which may also be released to the environment.[2]

Over time, the extraction and beneficiation of iron ore will result in large quantities of wastes that may damage the environment. The largest volume of wastes from extraction and beneficiation is the overburden and mine development rock. In addition to the excess mineral matter, extraction, beneficiation, and mine operation result in a variety of mine tailings and surface runoff waters that can transport metals and solids from the mine site into streams, rivers, and lakes. The introduction of such solids and metals can be disastrous to the resident aquatic systems and to all other life supported by these systems.

All of these materials can impact the environment and, therefore, are regulated and managed to minimize their release to the environment. The amounts of waste generated, however, can be significant and thus difficult to control. For example, in 1988, U.S. surface iron ore mines extracted 331,000,000 long tons of material to produce 55,100,000 long tons of ore. In other words, 275,900,000 tons of residues or wastes were produced by iron surface mining operations that year.[3]

Mines such as these and their associated impact and wastes can result in environmental contamination. Specifically, impact results from the migration of chemicals and minerals from the waste management areas that contain overburden, waste rock piles, ore piles, and tailings impoundments to surrounding lands and waters. Since management areas may contain toxic materials, including heavy metals and acid drainage, the damage can be substantial and may involve groundwater, surface water, soils, and air contamination during operation of the mine and after mine closure.

The ACME Roadster does not rely on iron in an unrefined or processed form. In fact, the Roadster includes a variety of metals and alloys that are derived from pure metals obtained from similar extraction and beneficiation processes. In addition, the Roadster contains components that are derived from organic chemicals which are initially derived from petroleum products. Processing of materials from nature (ores, oil, fibers) also produces a variety of wastes which are too extensive to discuss here but should be considered nonetheless.

37.4   Materials Processing

Processing iron ore into steel (or other) alloys can be accomplished through a variety of methods. In general, steel is formed by a two-step process in which the iron ore is first reduced in a blast furnace with a carbon source (coke) to produce pig iron. The pig iron is then processed by one of a variety of oxidation steps into steel (or steel alloy). The resulting steel is then milled into a steel component. A single-step process that eliminates the use of coke by replacing it with natural gas through a direct reduction step is finding more acceptance in today's steel manufacturing industry.[4]

It is important to realize that, regardless of the exact process used, steel production from iron ore is essentially a purification process to remove impurities from the ore in order to produce a more iron-rich blend. Impurities can include carbon, manganese, sulfur, phosphorus, and silicon, which are removed by oxidation. This purification can result in a variety of forms of direct environmental impact and wastestreams.

37.4.1   Environmental Impact

The purification of iron ore can result in types of impact associated primarily with the release of off-gasses from furnaces and coke ovens (employed to purify coke prior to use in the blast furnace) and with the weathering of materials stored prior to processing. In general, steel producers tend to stockpile raw materials such as ore, coal, and the resulting slags for future processing. Due to the volumes stored, these materials are often exposed to the weather, which can lead to contaminated stormwaters, surface waters, and groundwaters. In addition to this direct impact, the purification step is also an energy-intensive process. Energy use results in environmental impact associated with energy production, including gaseous emissions of carbon dioxide, sulfur dioxide, and various metals.

Steelmaking also results in a variety of wastes from each step of the process. In the first step, as mentioned previously, the pig iron production process generates slags that contain metals, carbon, sulfur, phosphorus, and silicon. In general, off-gasses are treated, yielding baghouse dusts that contain similar types of constituents. In addition, scrubbers are employed to clean off-gasses of smoke and fume prior to emission. These scrubber waters are also wastes that must be released (after treatment) to the environment. Cokemaking (from coal) yields a variety of benzene and polynuclear aromatic containing tars, residues, and off-gasses. Compounding these wastes, the second step (oxidation) results in metal-bearing slags which require controlled disposal. These wastes will vary depending on the characteristics of the metal desired. Additives and conditions will determine the exact composition of slags and off-gasses. The grade of iron ore and the composition of coal used for coking will also affect the direct impact and the composition of the wastes.

It is critical to realize that purification occurs not only in the processing of raw materials but also in steps further into the life cycle. For example, numerous purification steps are used to refine organic chemicals and pharmaceuticals. The important point is that when purification occurs, by definition, there will be residual materials that are no longer wanted. These materials—whether generated as gases, liquids, or solids—will be discarded and may have an environmental impact.

37.5 Manufacturing

The ACME Roadster is composed of a variety of metal parts. Some of these are produced at the facility and some are produced externally and then delivered to the plant for use in the assembly. For the purposes of this discussion, an environmental issue is not avoided if a component is manufactured at a different facility or even in a different country. It merely localizes the environmental problem to a different place. It is important to realize that the impact or waste occurs because the product—the Roadster in this case—is manufactured, regardless of location.

Manufacturing generates a variety of wastes in varying quantities. Manufacturing operations often rely on the integration of numerous components that may require the use of a variety of chemicals (some of which are toxic), which frequently result in the generation of toxic or hazardous wastes.

37.5.1   Metal Parts Manufacturing

The ACME Roadster includes hundreds of metal parts, most of which are steel. For steel parts, the manufacturing or milling operation includes all activities used to form parts, including hot forming, cold rolling, forging, and various milling and machining operations. Each of these different processes is used to form metal parts and can include several steps to ensure that the surface remains clean and exhibits the proper characteristics. Specifically, steelmaking includes several steps that prepare the surface for coatings that are applied during finishing.[5]

A hot-forming metal parts manufacturing process is representative of the types of metal manufacturing processes used in the production of the ACME Roadster. Hot forming, used to produce some parts for the Roadster, is the process of shaping hot steel through a series of forming steps to produce finished and semifinished products. It is important to realize that hot-forming lines will include water-based and oil-based treatment steps.

Environmental Impact.   Metal parts manufacturing and forming can result in various forms of adverse environmental impact, most of which are related to the generation of cooling waters and surface cleaning steps. Such waters can contain metals (depending on alloying), dirt, solids, oils, and grease. Furthermore, especially in a process like hot forming, the entire operation consumes power. When released to the environment, metal-bearing wastewaters, fugitive vapors, solids from wastewater treatment, and other discharges can have a serious impact on receiving waters and the associated wildlife.

Such operations that involve water may seem relatively innocuous; however the U.S. Environmental Protection Agency estimates that over 95 percent of all toxic chemicals generated by industry are removed from the facility in water.[6] This means that if water is involved in a process, chemical contaminants related to the process may be exiting the facility through the sewer or by means of direct discharges. Discharge of industrial process wastewaters and cooling waters is common and allowable under the regulations. However, over time, such discharges can reduce the efficiency of municipal wastewater treatment facilities and result in the deposition of metals (or other contaminants) in the environment. Wastewaters from manufacturing (and other) operations are one of the key wastes and sources of environmental impact that must be understood in defining the total impact that a process may have on the environment.

37.5.2   Assembly

At the ACME Roadster assembly line, the steel parts are riveted, welded, and bolted into place. In the ACME facility, parts are coated with oils for protection during storage. Prior to assembly, parts are cleaned using organic solvents. After assembly, the attachment area of a part may require cleaning prior to

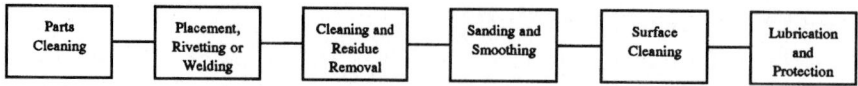

Figure 37-4. ACME Roadster generalized steel parts assembly process.

sanding and painting. The assembly step may also include sanding and metal shaping after initial attachment. Finally, after the last cleaning step, many parts are coated with a variety of petroleum products to protect them from friction and the weather during use. Figure 37-4 provides a simple diagram for a generalized assembly process.

Environmental Impact.   The assembly phase can result in a variety of metal wastes including ruined parts, components, or scrap. Riveting, welding, and soldering can also result in metal-bearing wastes, residues, and emissions. For example, soldering produces lead-bearing residues and fumes from flux. Sanding wastes are predominantly solid-phase but can be transported as fugitive dusts. As mentioned previously, assembly may also require various surface preparation or cleaning steps to insure that the assembly points will hold. Cleaning often results in organic solvents wastes (emissions and residues) as well as waterborne wastes (detergents, metals, organics). All of these wastes can have an adverse impact if released to the environment.

Assembly is common to most manufacturing operations and can result in substantial environmental impact. The nature of the operation will define all of the assembly steps required and which wastes will be generated. For example, assembly facilities that receive components from several sources often have large volumes of packaging materials as well as large cleaning operations to remove oils and protective coatings from parts. High-technology industries (where products are sometimes dirt-sensitive) often require extraordinary cleaning operations to insure that assembled parts result in functional products. Cleaning is often a repeated step in metal manufacturing operations because the surface must be free of contaminants that might interfere with welding, painting, electroplating, or operation of the part.

37.5.3 Painting

Most metal products are painted (or coated) to prohibit rust and protect the surface. In general, painting is the application of a pigment using a solvent carrier medium. Painting for the Roadster includes steps where some parts are painted before assembly and others are painted after assembly. Figure 37-5 illustrates the processes included in the Roadster painting operation. It is important to realize that in Fig. 37-5, the actual painting step may consist of several iterations. Painting may be a dip process or a spray process, depending on the nature of the finish.

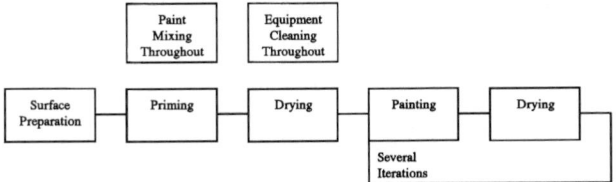

Figure 37-5. Generalized spray-painting operation for the
ACME Roadster.

The painting process is effectively a coating process that relies on a solvent carrier. The paint components will include metal-based pigments, metals flakes (especially for automobile finishes), and a variety of toxic solvents. Use of paints can yield various kinds of direct impact and wastes not only from the painting process but also from the cleanup of equipment. The environmental impact associated with painting is discussed here.[7]

Environmental Impact. The use of any solvent-based system will result in releases of volatile organic compounds to the atmosphere. Such releases can contribute to declines in air quality (smog) and other forms of environmental impact (depletion of ozone from some chlorinated solvents). In the case of the Roadster, the carrier solvent for the primer and the paint contains a number of hazardous constituents, including methyl ethyl ketone and toluene. These carriers can be released directly to the environment. The environmental impact is compounded since the drying process requires the use of energy.

In addition to solvent emissions, painting can result in a variety of wastes. To reduce the air emissions associated with painting, many painting operations include a filtration system to capture particulate matter (dry filters, water falls, or oil falls) and possibly other systems to capture volatile organic compounds (carbon filters). Eventually, these filter media become wastes when they are saturated with filtrants. In addition to such filter wastes, painting operations generate residual paints, thinners used to clean equipment, and various empty containers that held the paints or solvents. The painting operation may also result in a number of parts that are improperly painted. Such parts are either discarded or stripped and repainted. Parts that don't meet quality standards therefore result in additional quantities of paint wastes. The exact quantity of wastes will vary depending on the level of automation, the type of paint system used, the types of spraying and dipping equipment used, the number of cleaning operations performed, and the sequencing of different colors used.

It is important to realize that painting or the application of any solvent-based coating will result in solvent wastes. These solvent wastes (gaseous, liquid, and solid) are generated during spraying and drying operations. Solvents are used to liquefy pigments when placing the pigments on surfaces as well as for removing the pigments from equipment. Thus, solvent wastes are expected as a by-product of the cleaning of equipment.

Solvents are ubiquitous in industry because of their utility. Unfortunately, many of the most commonly used solvents (chlorinated solvents, ketones, and aromatic hydrocarbons) can have an impact on human health, aquatic ecosystems, air quality, and the ozone layer. Whether used in painting or any other industry, solvents present a particular challenge in controlling their use and release to the environment.

37.6   Packaging

The ACME Roadster is now complete and ready for delivery to the dealer for sale. In the past, the Roadster was loaded onto a train or truck and shipped to the dealer. However, this practice has sometimes resulted in damaged finishes. After years of complaints from dealers, ACME Corporation has decided to package the product to ensure safe transport to the dealers. This includes, among other things, coating the exterior finish with an organic coating, which protects the finish from chipping, extreme temperatures, and the weather. This one chemical coating can result in various types of adverse environmental impact.

37.6.1   Environmental Impact

Upon receipt of the new Roadster, the dealer washes the organic coating off the new vehicle. In fact, a hydrocarbon distillate is used to remove the coating. After this gross layer is removed, the vehicle is then washed with detergent and water. In most cases, this cleaning function is performed where all cars are washed— either outside draining to a storm drain, or inside in a mechanics bay draining to the municipal sewer or an outside storm drain. It is unlikely that in these situations such chemicals are collected for more controlled disposal. While the exact impact of the coating on the environment is unknown, the coating and the cleaning solvents can contribute to the general decrease in water quality.

Other manufacturers rely on a shrinkwrap plastic coating to protect finished surfaces, including metals, plastics, fiberglass, and woods. For example, fiberglass boat manufacturers shrinkwrap entire boats for shipping by truck. This type of packaging results in the generation of solid wastes which are landfilled, incinerated, and recycled. The impact associated with these management practices and the use of the materials in the first place should be considered when one documents the environmental effect of manufacturing the product.

This example is important because it demonstrates that even the simplest activities can result in damage to the environment. The initial idea was to apply a protective coating (either an organic coating or plastic) to protect the paint to avoid having to repair a painted surface. However, this coating must be removed before the car is delivered to the customer. Here another waste is generated as a result.

37.7 Use and Maintenance

The Roadster now belongs to an individual. Further generation of wastes and environmental impact is about to begin. Use of the Roadster requires periodic and routine maintenance. Common maintenance activities include replacing lubricants, replacing coolants, tuning the engine, and washing the car. Lubricants are used to protect the engine where moving metal parts come into contact with each other. Coolants are used to dissipate heat generated by combustion and friction. Engine maintenance is performed to reduce engine wear, improve combustion performance, and replace worn or broken components. Washing helps to protect the finish and metal parts and makes the Roadster look good. The schedule for performing these maintenance operations will depend on the vehicle and driving habits of the owner.

37.7.1 Environmental Impact

The maintenance functions mentioned here result in very specific environmental impact from waste oil (oil changes), used antifreeze (radiator flushing), and a variety of components, solvents (engine maintenance), and wash waters (automobile washing). Waste oil can contain lead, cadmium, chromium, and other metals from gasoline and engine wear. In addition to these, oil contains benzene and other organics that are found in petroleum products. When released to the environment, oil is toxic to animals (and humans) and can destroy ecosystems that support various flora and fauna. Antifreeze (which contains ethylene glycol) is extremely toxic to humans and other mammals.

Engine maintenance presents a confusing picture with respect to environmental impact. Regular maintenance lengthens the life of the motor, improves combustion efficiency, and decreases fuel consumption and automotive emissions. Engine maintenance, however, often relies on the use of solvents to clean parts. Solvents make up between 25 and 50 percent of the hazardous waste generated in most maintenance shops.[8] Common solvents include trichloroethylene and methylene chloride, which are volatile and toxic, as well as being suspected carcinogens and depletors of the ozone layer.

Automobile washing is similar to engine maintenance in terms of environmental impact. Regular washing reduces corrosion to the metal parts and protects the finish. Washing also requires the use of water, soaps and solvents, and waxes and polishes. These materials are washed into storm sewers and emitted to the atmosphere. Furthermore, washing removes oils, tars, and dirt that have collected on the vehicle and puts them into the storm sewer. The important idea to remember is that the waste from washing enters directly into the environment. Soaps, oils, solvents, waxes, and polishes all impact the aquatic life that live in and the wildlife that drink from surface waters.

The discussion of maintenance does not seem relevant to a discussion of manufacturing and environmental impact. However, manufacture and use are related through product design. Similarly, manufacture and environmental impact from use are also related through design. By understanding the impact that use and maintenance cause, industries can research and design products that result in minimal environmental impact.

37.8 Disposal

Every twenty minutes, the United States disposes of enough automobiles to form a stack as high as the Empire State Building.[9] These vehicles are often reclaimed for their steel (and other metal) contents. To obtain steel, a recycler will have to segregate metal components prior to reprocessing. The operation used to separate materials from scrap automobiles is presented in Fig. 37-6.

37.8.1 Environmental Impact

The disposal process is critical because it promotes the recycling of steel used in automobile manufacturing (as well as from other steel-containing products). Unfortunately, reprocessing also generates waste that results from all of the nonmetallic components contained in the vehicle. For automobiles, this material, referred to as "fluff," consists of upholstery, foam, plastics, fiberglass, rubber, and vinyls. In addition, fluff may contain crankcase oil, antifreeze, transmission fluid, and other materials from the vehicle. Because it is often landfilled, fluff may create an environmental impact. Further, the water used to separate fluff from the metal components may also contain hazardous chemicals.

It is important to realize that automobile shredding for disposal is a purification step just like the original extraction of iron ore from rock. As such, there are wastes that, in the case of automobiles, happen to contain hazardous materials. The other critical point is that recycling options may also have an environmental impact that should be evaluated when one makes a decision to recycle. This is true for recycling materials like steel, paper, solvents, rubber, and others.

37.9 Conclusions

The simple life cycle for steel in the sample case of the ACME Roadster illustrates that there are a number of types of environmental impact associated with manufacturing. In this hypothetical case, the steel used in the Roadster results in wastes that would be discharged to the air, water, and soils. Furthermore,

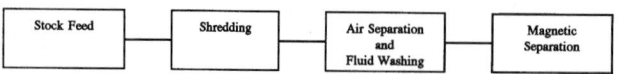

Figure 37-6. Generic automobile-shredding process.

the construction of the Roadster results in a direct impact on the ecosystems in which the materials and resulting vehicles are produced. For example, the lands where iron ore is mined are severely impacted by the mining process. The brief example provided here is for only one material used in the manufactured product. The point of providing this example is to illustrate the complexity of environmental considerations that are associated with industrial activities including the design, materials collection and processing, manufacture, use, and ultimate disposal of a product.

References

1. U.S. Department of the Interior, Bureau of Mines, *Development Guidelines for the Closing of Underground Mines: Michigan Case Histories,* Washington, D.C., 1983.

2. National Institute of Occupational Safety and Health, *National Occupational Health Survey of Mining: Iron Ore Report,* Morgantown, W. Va., (unpublished) 1990.

3. A. Tanner, "Mining and Quarrying Trends in the Metals and Industrial Minerals Industries," in *Minerals Yearbook, Volume 1: Metals and Minerals,* Washington, D.C., U.S. Department of the Interior, Bureau of Mines, 1991.

4. Robert J. King, "Steel," in *Kirk-Othmer Encyclopedia of Chemical Technology,* 3d ed., vol. 21, John Wiley and Sons, New York, 1983.

5. U.S. Environmental Protection Agency, *Development Document for Effluent Limitations Guidelines and Standards for the Iron and Steel Manufacturing Point Source Category,* vol. IV, Washington, D.C., 1980.

6. U.S. Environmental Protection Agency, Office of Solid Waste, *Report to Congress on the Discharge of Hazardous Waste to Publicly Owned Treatment Works,* Washington, D.C., 1986.

7. U.S. Environmental Protection Agency, Office of Research and Development, *Achievements in Source Reduction and Recycling for Ten Industries in the United States,* Washington, D.C., 1991.

8. U.S. Environmental Protection Agency, Office of Solid Waste, "Pit Stops—Be Aware as You Repair," draft, Washington, D.C., 1993.

9. Annette Mills, *Falls Church Recycling Program, Recycling Facts,* Falls Church, Va., 1993.

38

Biotechnology for Pollution Prevention

Krisztina Bordacs-Irwin, Ph.D., P.E.
Newtown, Pennsylvania

38.1 Introduction

Just what exactly is biotechnology? To answer that question, first consider that, for millennia, people have been fermenting sugars and starches to make beers and wines, allowing natural bacteria to sour milk to produce cheeses and yogurt, or warming yeasts to raise breads. All of these processes were used without understanding the underlying science. In our century, we have developed and understood the science, creating the new field that we call *biotechnology*. The scientific "revolution" is not over yet; understanding of genetic mechanisms will open a new door in the near future. But let us go back to the question of definition. Most dictionary definitions describe *biotechnology* as some academic or research activity, but in fact biotechnology passed that stage several decades ago. The best definition is found in the *Webster's 9th New Collegiate Dictionary*: "applied biological science."

Biotechnology uses natural processes; therefore, pollution prevention is almost always incorporated. Using microorganisms, microbial enzymes as catalysts, and natural feedstocks reduces or eliminates the need for pollution control. Many biotechnology processes could be substituted for existing production processes to create "environmentally incorporated manufacturing" or "ecological bioprocessing." Rapid changes and new technical developments characterize the biotechnology field; therefore, only selected aspects and developments can be discussed in a single chapter such as this one.

The most frequently cited argument against using new technology is the cost of implementing the change. Recent advances in biotechnology, along with the adoption of cost calculations that incorporate the ecological impacts of indus-

629

trial processes, will slowly overcome that objection. Using biological processes, one cycles materials back to ecosystems without any dissipative use of material or having any spiraling effect on the global ecosystem. This "natural solution" is accepted by the public and will be widely used by industry.

The other significant opportunity for biotechnology is using it for environmental control. Biotechnology has great advantages, and multiple potential uses of biotechnology for environmental control are described briefly in Sec. 38.8. However, the emphasis in this chapter, as throughout this handbook, will be on preventing pollution, not on mitigating its effects after the fact. Sections 38.2 through 38.7 give examples of biological processes already used or planned to be used in various industries.

38.2   Potential and Advantages of Biotechnology

Biotechnology applications take advantage of natural processes using common and generally indigenous microorganisms. There are several domains of biotechnology where the advantages of "biologicals" compared to "chemicals" are already established. The regeneration of highly selective biocatalysts is easier than regeneration of chemical catalysts. Due to the specificity of the organisms, nonpurified and diluted raw materials can be used under generally less reactive (lower temperature and pressure) conditions. The products of these biological processes are biocompatible and biodegradable. The processes where biotechnology is already in use include

- Classical production of food and beverages
- Production of highly complex organic molecules for health-care needs (antibiotics, proteins)
- Replacement of several sequential reactions for stereospecific and highly selective conversions (chiral compounds, steroids)
- Degradation/treatment of diluted, toxic substances in the environment (industrial wastes in air, groundwater, and soils)

Table 38-1 lists some compounds that are produced by microbial fermentations.

The possible beneficial uses of biological processes are innumerable, and the number of practical applications will increase in the next decade. There are several needs of society that biotechnology can and will fulfill, such as biofuels from agricultural biomass and new plants that survive different climates and are resistant to diseases. The new developments in biotechnology (or "ecotechnology") will lead to the use of the full potential of biodiversity on Earth. They will not be restricted to microorganisms, but will include uses of plants and animals. To present the opportunities, Table 38-2 lists types of chemical reactions that are being or will be replaced by biological reactions in Japan.

Table 38-1. Production of Miscellaneous Compounds by Microbial Fermentation

Compound	Characteristic	Microorganism
Antipain	Protease inhibitor	Various *Streptomyces* spp.
Avermectin	Anthelmintic	*Streptomyces avermitilis*
Carotenoids	Pigments	*Dunaliella bardarvill*
Dopastin	Dopamine-ß-hydroxylase inhibitor	*Pseudomonas* spp.
Elastinal	Elastase inhibitor	Various *Streptomyces* spp.
Gibberellin	Plant hormone	*Gibberella fujikuroi*
Herbicidin	Herbicide	*Streptomyces saganonensis*
Indigo	Pigment	*Escherichia coli*
Inosine	Flavor enhancer	*Bacillus subtilis*
Lysergic acid	Ergot alkaloid derivative	*Clariceps paspali*
Oudenone	Tyrosine hydroxylase inhibitor	*Oudemansiella radiata*
Piericidin	Insecticide	*Streptomyces mobaraensis*
Tetranactin	Miticide	*Streptomyces aureus*
Vitamin B$_{12}$	Vitamin	*Propionobacterium shermanii*

38.3 Pollution Prevention Techniques

Table 38-3 gives the accepted tabulation of pollution prevention techniques. The boldface items on the list are the ones where biotechnology presents opportunities as a pollution prevention method. Sections 38.4 through 38.7 will give examples of biotechnology applications for these pollution prevention categories.

38.4 Product Substitution

38.4.1 Biodegradable Plastics

Our landfills are fast filling up with plastics. Plastic recycling initiated pollution prevention activities in this field; however, the use of recycled plastics is limited since injection-molding processes cannot use more than 30 to 40 percent recycled mixtures. To really solve the problem, research and development efforts had to be directed toward biodegradable plastics.

The first studies were aimed at poly(3-hydroxybutyrate) (PHB) in the early sixties. The development process was speeded up significantly in the seventies when developments in the Middle East sent oil prices skyrocketing. PHB is an energy storage material present in a wide variety of bacteria. The synthesis takes place by a relatively simple biochemical pathway using a wide range of

Table 38-2. Chemical Reactions Replaced by Bioreactions in Japan

Reaction type	Example		
Addition	Olefins	$\rightarrow$ Alcohol	
		$\rightarrow$ Alkyl amines	
		$\rightarrow$ Fatty acids	
Condensation	Carbon monoxide/H_2	$\rightarrow$ Alkanes	
		$\rightarrow$ Olefins	
		$\rightarrow$ Glycols	
Oxidation	Alkanes $\rightarrow$ Epoxides $\rightarrow$ Glycols		
	Aliphatic hydrocarbons $\rightarrow$ Fatty acids		
	Benzene $\rightarrow$ Phenols		
	Butadiene $\rightarrow$ Tetrahydrofuran		
	Cyclohexanol $\rightarrow$ Adipic acid		
Miscellaneous	Aryl chloride $\rightarrow$ Aryl alcohol		
	Biomass $\rightarrow$ Olefins		
		$\rightarrow$ Fatty acids	
		$\rightarrow$ Monosaccharides/polysaccharides	
	Industrial wastes $\rightarrow$ Useful substances		

carbon sources. The biochemical reaction is so effective that PHB purity is too high for crystal formation. In the normal chemical synthesis, remaining impurities serve as nuclei for crystallization. In the case of biotechnologically produced PHB, a nucleating agent is used. The physical properties of PHB are extremely close to that of polypropylene (PP). PHB has lower gas permeability, while PP is substantially tougher. Later development efforts solved that problem. While changing the carbon source and the host organisms, it was discovered that *Alcaligenes eutrophus* could be induced to produce PHB homopolymer as well as poly(3-hydroxybutyrate)/hydroxyvalerate (PHV) copolymer. This copolymer, through the inclusion of PHV units, has improved toughness. An International Chemical Industry (ICI) subsidiary based in England, Marlborough Biopolymers Limited, sells and markets PHB/PHV under the trade name BIOPOL. Increased scale of production caused prices to fall and BIOPOL is widely used in England, Germany, and Italy. In 1992 marketing efforts for BIOPOL were started in the United States.

In the United States, other development efforts have dominated. Starch-based, cellulose-based, and cellulose acetate–based plastics were researched. A Warner-Lambert subsidiary, NOVON Products, developed a starch-based plastic, while Eastman Kodak researched a cellulose-based polymer. While PHB was proven to be melt-processed and therefore has been used in injection and

Table 38-3. Pollution Prevention Techniques

1 Product changes
  - **Product substitution**
  - Product conservation
  - Change in product composition
  - Product redesign
  - **Product recycling**

2. Source control
  - Input material changes
    Material purification
    Material substitution
  - Technology changes/modifications
    Process modification
    Equipment changes
    Automation
    Improved control
  - Good operating practices
    Procedural measures
    Loss prevention
    Management practices
    Wastestream segregation
    Materials-handling improvement
    Scheduling improvement
    Employee training

3. Environmental control techniques

blow molding, starch- and cellulose-based plastics are not ideal candidates for thermoprocesses. Their usability to date has been proven in the production of films and fibers. For these two types, the raw material is also produced by biotechnology.

The newest line of products are poly-alcohol-based. Air Products and Chemicals, Inc., is working on polyvinyl alcohol–based (PVOH) plastics, and other companies are working on pro-degradants to increase the UV and thermal degradation capabilities of the poly-alcohols.

38.4.2 Biopesticides

Pest organisms are susceptible to predation and parasitism as well as diseases caused by fungi, bacteria, and viruses. Biological control seeks to exploit such conditions; however, potential microorganisms are sometimes hard to successfully introduce. The diseases caused by bacteria, fungi, and viruses to pest species are extremely useful for pest control because of their selectivity. The first breakthrough in biological control was the introduction of the vedalia beetle, *Rodolia cardinalis,* to control cottony-cushion scale, and of *Iceryapurchasi* on citrus. *Bacillus thuringiensis* (Thuricide) controls the larval stage of a wide range of lepidopteran species. The Thuricide market itself is about $60 million annually, representing about 0.8 percent of the whole market for the pest control market. An American company, Ecogen, sells and markets a Thuricide under the trade name Cutlass.

Another microorganism that appears to be useful as a microbial insecticide is *Streptomyces avermitilis,* a soil organism. This group secretes microbial products termed *avermectins.* These compounds have potent activity against many agricultural insect pests, phytophagous mites, and parasitic nematodes.

Various species of the fungus *Verticullum lecanii* control whitefly and the peach potato aphid so narrowly that there is no harm to other potentially useful insects like *Encarsia formosa.*

These examples show that several biotechnology-generated pesticides have passed the R&D stage. Natural substances are often more readily accepted by the public as being safer than those from chemical synthesis. This is a misconception, as many of them are very toxic. However, their high target specificity makes them more environmentally acceptable, their production is a biological process with no harmful by-products, and their consumer hazard is minimal due to no residues on food crops. Their R&D cost is lower and manufacturing costs could be brought down significantly (see Table 38-4).

38.4.3 Flavors and Fragrances

An area of considerable interest is replacement of synthetically produced, expensive, and rare fragrance and aroma compounds by similar or identical products produced by microorganisms. The goal is to develop more economical sources of natural substances. For example, a vanilla bean contains about 3 percent vanillin by weight, and 1 ton of strawberries contains only 25 percent flavor compound. These aroma and fragrance compounds are typically extracted with solvents, whether from the natural sources, or during the chemical synthesis of an artificial material. If biotechnology is used, not only is the chemical (solvent) eliminated, but no extraction procedure is required. In the case of vanilla beans that eliminates the disposal of "unused" materials in the beans.

The biotechnology opportunity is well documented in flavor production, such as production of different cheeses. One example from the fragrance industry is the production of the traditional perfumery material Ambroxiale, produced by the sperm whale. Due to the significant decline in whale populations and the fact that the conversion has only a 50 percent yield by chemical synthesis, a biotechnology process was researched. The conversion step can be carried out by *Hyphozyme roseoniger* with essentially qualitative yields.

Table 38-4. Advantages and Disadvantages of Microbial versus Chemical Pesticides

Costs and benefits	Microbial	Chemical
R&D	$0.8–1.6 million	$20 million
Market size required for profitability	Less than $1.6 million	$40 million
Toxicological testing	$500,000	$10 million
Patentability	Still developing	Well established
Discovery	Rational selection for target pests	Screen 15,000 compounds to identify 1 product
Efficacy		
Kill	90–95%	ca. 100%
Speed of kill	Can be slow	Rapid
Spectrum of activity	Generally narrow	Generally broad
Resistance	Only one case known	Often develops
Type of action	Generally only curative	Can be both preventive and curative
Safety		
Operator risk	Low operator risk	Chemicals are hazardous
Environmental impact	Few examples with inundative use of indigenous microorganisms	Many examples
Residues	Crop can be harvested immediately after application	Interval before harvest often required

A wide range of enzymes are important in catalyzing the formation of flavor and fragrance materials. To date less than 1 percent of all enzymes have been scientifically examined. Biocatalytic capability on a commercial scale could be enhanced considerably if a greater number of biocatalysts were available, and as more enzymes are studied, more potential biocatalysts will be found.

38.4.4 Chlorofluorocarbons

Refrigerators and air conditioners use chlorofluorocarbons (CFCs) as a circulating heat-transfer medium. CFCs deplete the ozone layer, and intensive research has been conducted to find suitable replacements for CFCs. Hydrochlorofluorocarbons (HCFCs) and hydrofluorocarbons (HFCs) were found to be effective replacements and they have significantly lower ozone depletion capability. HCFCs and HFCs are being phased in as alternatives to CFCs, with a target date of 1996. Researchers at ENVIROGEN, an American environmental biotechnology company, found that 3 out of 5 HCFCs and 1 out of 3 HFCs tested were biodegradable. The company plans to further develop the technology for emission control.

38.5   Product Recycling

Product recycling is the easiest way to control pollution. Aluminum cans, glass containers, newspapers, plastics, and packaging materials are now being recycled in Europe and the United States. These products are typically reused in the same production. However, recycling of household organic waste will produce a new, useful product: compost. In Switzerland, household organic waste is collected in a separate, degradable bag, called the "green bag." These bags then are transported to a composting facility, and a final composted product is sold to farmers. This is a typical example of the waste being a useful product.

38.6   Materials Changes

Chemicals are widely used in industry. Some companies produce chemicals or have chemicals in their products. Other companies only incidentally use chemicals for additives or lubrication. Most companies fall in between, using chemicals for cleaning, application of paint and coatings, and a variety of other purposes. Chemical use reduction or elimination should be the first priority in pollution prevention if the chemical is highly toxic or produces a high volume of relatively toxic waste. Raw material changes can take the form of purification of existing raw materials or substitution of a new feedstock, catalyst, or other material. The substitute is either less hazardous or results in lower hazardous waste generation, but must continue to satisfy end-products specifications.

38.6.1   Dyes

The modern dye industry relies on chemically synthesized dyes derived from fossil feedstocks, mainly petroleum and secondarily coal. Prior to the synthetic methods, dyes were harvested from plant and animal matter. Except for use in food dyes, natural dyes are today of little economic importance.

The indigo dye used in bluejeans accounts for 11 percent of the U.S. dye market and 3 percent of the world market. The dye has been known for centuries and was first produced by agricultural production. Plants such as *Indigofera tinctoria* contain 0.2 to 0.8 percent indican in their leaves. Indigo was produced from indican-containing plant matter by a harvesting of the plants which were then allowed to ferment. This fermentation-induced hydrolysis of indican to indoxyl and its subsequent oxidation to indigo. Indigo is an aromatic compound and its chemical production is closely linked with the development of aromatic chemistry. In the mid-1800s, due to the excess supply of coal tar, most aromatic compounds were produced from coal tar. BASF, a German company, looked at several components of coal tar to utilize it. Napthalene, a major component in coal tars, was found to be economically oxidized to indigo by concentrated sulfuric acid in the presence of mercury catalyst.

Genetically engineered *Escherichia coli* that has a *Pseudomonas* oxygenase–producing gene inserted showed a novel pathway to indigo. *E. coli* has the tryptophanase enzyme, which hydrolyses tryptophan (a natural amino acid) to indole. The product of the inserted oxygenase gene then oxidizes indol to indigo. The use of tryptophan as an indigo precursor has several advantages. The most important is that tryptophan is produced biosynthetically from inexpensive, renewable, and benign raw materials, mainly plant sugars and ammonia. Thus indigo would be generated using sustainable, environmentally sound technologies. The other advantage is that the biochemical route mimics the agricultural route. The safety of biological indigo has never been questioned.

An American biotechnology company, Genencor, scaled up the microbial fermentation process for the biochemical production of indigo. Depending upon the fermentation cost of tryptophan (likely to come down) and assuming corn as a primary carbon source, a 40-fold higher yield of indigo is expected by the biochemical method than by the agricultural.

Another interesting example is newspaper ink. The commonly used newspaper ink was lead-based, causing several pollution problems during production and safety issues in the handling of newspapers. Recently it has been replaced by carbon black ink. Soy-based inks have now been developed by several companies in the United States, eliminating the need for the lead-based and carbon black ink. Soy-based ink also uses a feedstock that is biological in nature.

38.6.2 Miscellaneous

Several examples of use of microorganisms to separate metals from slurries can be found in the literature. *Penicillium chrysogenum* can bioaccumulate radium, thus separating radioactivity from a wastestream. *Aspergillus niger* bioaccumulates aluminum, and a proteinase enzyme separates silver from photographic solutions. In each case, recycling of the metal to an industrial process is possible, following the lysis of the enzyme or cell material. Similarly, *Chlorella vulgaris* accumulates gold in preference to other metals. Yeast production biomass from the brewing industry, containing *Saccharomyces*, is also used for bioaccumulation of cesium. This is another good example of a waste for which we have found a use.

Peaches and other fruits contain a chemical, *N*-butyl butyrate, that is responsible for their flavor. This natural chemical has been found useful as a replacement for the widely used solvent 1,1,1-trichloroethane (TCA), which belongs to the family of compounds responsible for destroying the ozone layer. The peach-derived solvent was discovered by investigators at the AT&T Bell Laboratories who were looking for a new biodegradable compound.

38.7   Process Changes

38.7.1   Biological Pulping

The pulp and paper industry is one of the contributors to environmental problems. One of the major concerns is the production of black liquor and the subsequent bleaching of the pulp. The black liquor in the pulping process is treated with chlorine to cleave the lignin-cellulose bond. The cellulose is the final product, and lignin stays with the wood pulp. The typical chemical process creates by-products that include chlorinated lignin compound, or organic halogenated compounds (AOX) that are toxic to freshwater species. Research efforts have been concentrated on finding a substitute for chlorine in the bleaching process. Chlorine dioxide was tested and found to produce less AOX compounds. Ozone was very effective and eliminated AOX production, but it is very expensive. Enzymic bleaching has been researched, and has been in use since 1985. The use of enzymes, while it is still in the mill-testing stage for the most part, appears to be economically competitive with chlorine bleaching. There are two enzymes used: lignin peroxidases remove lignin directly, while hemicelluloses dissolve hemicellulose, freeing cellulose from lignin.

Sandoz makes a mixture of hemicellulose enzymes called Cartazine. These enzymes, xylanases, when used to pretreat the pulp, reduced the chlorine requirement by as much as 80 percent. Other xylanase products are now available from a Finnish company, Cultor; Biopulp International, a French concern; and a U.S. company, Genencor International. Sandoz also has a patent for a lignin peroxidase from *Phanerochaeta chrysosporium* to bleach pulp. A Swedish company, Sodra Skogskarna Ab., is researching the lignin peroxidase route to develop a chlorine-free biological pulp-bleaching process. The combination of enzyme treatment and chloride dioxide showed good bleaching results with no dioxin formation.

38.7.2   Chiral Compounds

Several pharmaceutical compounds require extremely high purity and stereospecificity. In a typical chemical synthesis a mixture of isomers is formed, and the product has to be purified with some kind of extraction method. Chiral chemistry uses specific biocatalysts that produce only the correct stereoisomer. These biocatalysts are expensive; therefore, chiral biocatalysts are mainly used in pharmaceutical manufacturing. Manufacturing high-value pharmaceuticals using chiral chemistry creates products with higher than 99 percent purity and eliminates the need for product purification.

38.8   Environmental Control

Environmental control is a major component of environmental biotechnology. Except for dioxins and a couple of PCB congeners, almost everything is biodegradable under the right conditions. Biological degradation provides innocuous end products, such as carbon dioxide, water, and salts. Biological processes can be used in any medium: liquid, vapor, or solid. The knowledge of biochemistry and microbiology allows the treatment of hard-to-degrade or recalcitrant compounds and mixtures of them. Bioengineering creates novel engineered systems to overcome biological and mass-transfer limitations. Instead of end-of-pipe treatment, where typically a wide variety of compounds are present, in-process treatment of a single contaminant can prevent pollution. But as we have noted, by definition, environmental control is not pollution prevention; therefore there are no examples of treatment options here. As a summary, biotechnology has an important role in biotreatment for environmental control, especially for toxic and hazardous wastes.

Further Reading

Moser, A., and M. Narodoslawsky, *Ecological Bioprocessing—Final Report of the Task Group of the European Federation of Biotechnology*, Graz, Austria, March 1993.

Moses, W., and R. E. Cape, *Biotechnology, The Science and the Business*, Harwood Academic, New York, 1991.

Swaminathan, M. S., *Contribution of Biotechnology to Sustainable Development within the Framework of the United Nations Environmental Programme*, UNIDO Report IPCT 148 (SPEC), Vienna, 1992.

U.S. Department of Energy, *Industry Monthly Update, Part II: Waste Reduction*, Washington, D.C., February 1992.

39

Pollution Prevention in the Electronics Industry

Azita Yazdani, P.E.

Pollution Prevention International, Inc.
Brea, California

39.1 Industry Overview

The electronics industry manufactures components and electronics packages. The demand for industry products is expected to go above $370 billion in the United States by the mid-90s. The industry is comprised of three major sectors: printed circuit board (PCB) fabrication, PCB assembly, and semiconductor manufacturing. This chapter describes the industrial processes and pollution prevention measures related to PCB assembly, and to a lesser extent the semiconductor manufacturing process.

Growth in the electronics industry has been significant in recent years, leading to higher use of chemicals in the electronics manufacturing processes. The industry is also constantly in a state of flux due to the competitive nature of the business. The semiconductor industry is comprised of small and large facilities that may be involved in research and production of semiconductors. The production of semiconductors requires high cleanliness standards and extensive capital investment in facilities and sophisticated mechanical equipment. Moreover, the manufacturing processes are continuously changing because of innovations that demand updating and modifying of processes.

39.2 Description of Major Processes

39.2.1 Semiconductor Manufacturing Process

Integrated circuits (ICs) are the major products of the semiconductor industry. Unlike discrete semiconductor devices such as resistors, capacitors, or transistors, ICs are combinations of such devices in a single semiconductor crystal. Semiconductor processes employ 200 different materials and 100 proprietary methods (SRRP, 1991). The two primary steps in the production process are wafer fabrication and wafer assembly. Figure 39-1 shows the steps involved in the semiconductor manufacturing process.

Wafer Fabrication.   The first step in the fabrication process is to grow a single crystal, or "ingot," from a seed crystal. Various materials are used for this purpose but the most common is silicon (SRRP, 1991). Once the ingot is grown, it is sliced into wafers, smoothed, and polished—the so-called *lapping process*—and cleaned. Chlorinated solvents are commonly used for this purpose. In the

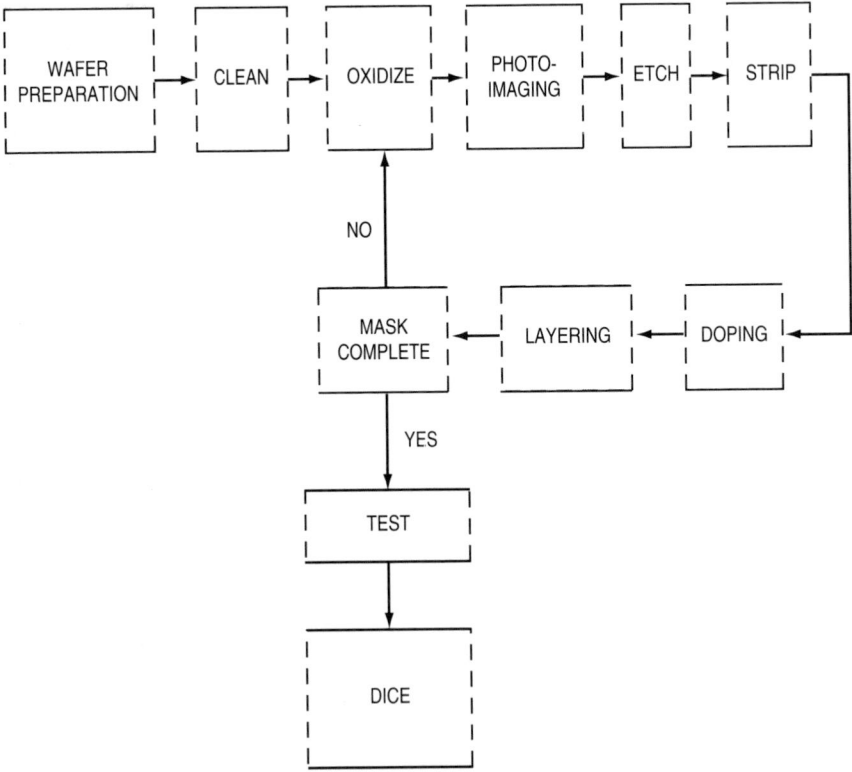

Figure 39-1. Semiconductor production process.

next step, which involves five fundamental steps, semiconductor devices are processed on top of the wafers. The wafers are first cleaned with either an acid plus an oxidant (the most common method), with deionized water, with alkaline cleaners, or with a solvent. The solvent method commonly involves immersing the chips in boiling trichloroethylene (TCE) or 1,1,2-trichloro-1,2,2-trifluoroethane (CFC-113). Some believe that not much solvent is used for these processes any longer. The cleaning is usually performed in a clean room.

The first step in the processing stage of silicon wafers is oxidation, when a silicon dioxide layer is grown on the wafer to provide a base for the photolithography process. The surface of the wafer is coated with silicon dioxide by exposing the silicon wafer to oxidation at a high temperature in a furnace. This layer will prevent diffusion of the dopants (see below) to the silicon substrate. After oxidation, the wafer is cleaned with strong acids and bases. In some cases, 1,1,1-trichloroethane (TCA) is used in the furnace as a source of chlorine which apparently improves the properties of silicon.

In the second step, photoresist is spun onto the wafer in a uniform film. It may be a negative photoresist containing a polymer or a positive photoresist containing resin. In the photolithography, a pattern is transferred from a photo mask to the circuit using light sources such as UV light. In the development step, the polymer is removed from the board. In the case of positive photoresist, the area exposed to the light is polymerized and then removed using organic solvents. In negative photoresist, the area not exposed to the light (the unpolymerized area) is removed using chlorinated solvents.

The third step is etching and stripping. Inorganic acids are used to etch the oxide layer, and the remaining resist layer is stripped with liquid or dry strippers. In certain cases, the strippers may be chlorinated solvents or acids, but dry stripping using oxygen plasma is becoming the method of choice. The fourth step in the sequence is to "dope" the lattice with impurities to change the electrical conductivity of silicon. The fifth step is layering: a thin layer of material, which may be a conductor, semiconductor, or insulator, is deposited over the whole surface of the wafer. The oxidation, photolithography, doping, and layering steps may be repeated several times until the desired number of circuit layers is achieved and the desired circuit has been fabricated. The circuits then go to wafer assembly.

Wafer Assembly. In this process, the chips are tested for defects and the wafers are scrubbed into individual chips. TCA and CFC-113 are reportedly used for cleaning the flux applied to the wafers for testing, although isopropyl alcohol is the most commonly used solvent for removing flux. Then the semiconductors are packaged and tested. A mixture of methylene chloride and CFC-113 is used to test the ink nomenclature on the circuit. Laser testing can also be used, but it destroys certain plastics or metals.

Trends. Greater accuracy will be required in the photolithography process, requiring the use of negative photoresists, leading to lower use of solvents.

39.2.2 PCB Assembly

In the assembly process, electrical components are placed on the board, the boards are fluxed, and the components are attached and soldered to the boards. The flux is applied to reduce the surface tension so the solder will flow evenly. In wave, dip, and drag soldering, a molten solder is deposited on the board and it acts as a heat source and supply of solder. In wave soldering, the board is fluxed and passed through a wave-soldering machine that applies solder which adheres to the metal leads, tin, tin-lead alloy, or gold-plated parts of the board. In the hand soldering, a cold solder is deposited on the board and later heated.

The various types of flux and their characteristics are summarized in Table 39-1. Three of the fluxes noted are most commonly used. These are rosin, organic acid, and synthetically activated fluxes. The flux residue, as well as other contaminants, need to be removed from the board after soldering. In general, contaminants include nonpolar contaminants, such as oils, greases, rosin, and waxes; polar contaminants, such as rosin flux activators, sodium chloride, and plating and etching salts; and particulates, such as dust, and machining, drilling, and punching fragments. Fluxes have different characteristics and are used for different applications (Yazdani, 1991).

Table 39-1. Types of Flux

Flux	Common reference	Composition	Approved by military	Solubility*
Rosin	R	Abietic acid and other isomeric acids	Yes	S,WS
Mildly activated rosin	RMA	Abietic acid and other isomeric acids; amine hydrochloride activators	Yes	S,WS
Activated rosin	RA	Abietic acid and other isomeric acids; amine hydrochloride activators	No	S,WS
Superactivated rosin	RSA	Abietic acid and other isomeric acids; amine hydrochloride activators	No	S,WS
Organic acid	OA	Abietic acid and other isomeric acids; halide or non-halide activators	No	W
Synthetically activated	SA	Alkyl acid phosphates; halide activators	No	S
Resin	Resin		No	W
Low solids	Low solids	2 to 5% solids	No	†

*S = chlorinated solvent soluble; W = water soluble; WS = water soluble with saponifier.
†Cleaning may not be required.
SOURCE: SRRP, 1991.

A number of cleaning processes are used to remove residues of fluxes from the board. These include chlorinated solvents (including CFCs), aqueous, and semi-aqueous systems. A more detailed discussion of these processes is provided in Sec. 39.4. A variety of solvents, such as acetone, alcohol, TCA, and CFC-113, are used in various steps of the PCB assembly process for mechanical and manual cleaning applications.

Trends.   On a traditional PCB, devices are connected to the board by drilling holes in the board, inserting and crimping the leads, and then soldering the components to the boards. Today, there is a movement toward *surface mount technology* (SMT). Surface-mounted devices are small and have no connector leads. Because no holes are needed with these devices, the components may be more densely packed on the board. The movement toward SMT means that cleaners used must be able to penetrate the smaller crevices and spacing between the devices and the board. The recent phase-out of ozone-depleting substances such as TCA and CFCs has led to higher use of aqueous and semi-aqueous cleaning systems. These alternative technologies are described in Sec. 39.4 below.

39.3   Typical Wastestreams

As discussed earlier, there are a number of processes utilized in the electronics industry. Table 39-2 summarizes the multi-media pollutants that are generated from each process. As shown, processes can release air pollutants; wastewater streams, both hazardous and nonhazardous; and solid hazardous wastes. The two major environmental concerns in this industry sector are chlorinated solvents used for cleaning and lead in solders.

39.4   Pollution Prevention Options

Pollution prevention measures discussed in this chapter focus on PCB cleaning and soldering. Other processes such as conformal-coating application and removal are not discussed.

39.4.1   Solvent Substitution in PCB Assembly Cleaning

There are a number of solvents and cleaners used in the cleaning of PCBs after assembly. The phase-out of CFC-113 and TCA has led many users to search for suitable alternatives in this area. Among the various types of solvents being used are alcohols, petroleum solvents, terpenes and heavy hydrocarbons, and detergents with saponifiers.

Table 39-2. Typical Wastes Generated in the Electronics Industry

Process	Materials used	Possible types of environmental releases		
		Hazardous waste	Wastewater	Air pollutants
Wafer cleaning	1. Acidic cleaners 2. Chlorinated solvents or CFCs 3. Alkaline cleaning 4. Deionized water	1. Acidic waste 2. Chlorinated solvents or CFC wastes 3. Tank-cleaning waste from alkaline cleaning 4. Deionized water baths cleaning	1. Acidic rinses 2. None 3. Alkaline cleaning rinses 4. Deionized bath discharge	1. None 2. Solvent emissions 3. None 4. None
Photoresist stripping	1. Chlorinated solvents for negative photoresist 2. Organic solvents for positive photoresist	1. Chlorinated solvent waste 2. Organic solvent waste	1. None 2. None	1. Solvent emissions 2. Solvent emissions
Etching	1. Liquid or dry strippers 2. Chlorinated solvents 3. Acids	1. Liquid or dry stripper waste 2. Chlorinated solvent waste 3. Acidic wastes 4. Contaminated application equipment	1. Rinse wastewater 2. None 3. Rinse wastewater	1. None 2. Solvent emissions 3. Acidic emissions
Wafer assembly	1. Chlorinated solvents or CFCs 2. Alcohol 3. Meth and CFC-113 for nomenclature testing	1. Solvent waste 2. Solvent waste 3. Contaminated application equipment and rags	1. None 2. None 3. None	1. Solvent emissions 2. Solvent emissions 3. Solvent emissions
		PCB Assembly		
Component fluxing and soldering (wave, dip and drag)	*Soldering* 1. Molten solder 2. Flux *Cleaning* 3. Chlorinated solvents or CFCs 4. Aqueous and semi-aqueous cleaners 5. Mechanical cleaning	1. Solder waste containing lead 2. None 3. Solvent waste 4. Bath clean up waste and contaminated bath disposal 5. Contaminated application equipment	1. None 2. None 3. None 4. Rinsewater discharge 5. None, unless the application equipment is rinsed with water	1. Lead 2. None, unless it contains volatile organic compounds 3. Solvent emissions 4. None 5. None

Hydrochlorofluorocarbons. The new chemical substitutes for TCA and CFC-113 are the newly proposed hydrochlorofluorocarbons (HCFCs). Three major HCFCs were proposed initially as replacements for CFCs in the electronics industry. However, the HCFCs have been shown to be suitable substitutes only for a limited number of uses. HCFC-123 has shown chronic toxicity in male rats, which eventually led to the removal of the solvent from the electronics and cleaning market by its manufacturer. HCFC-141b has shown ozone depletion potential equal to that of TCA; thus it is not approved at this time for general cleaning uses. HCFC-141B is available only for limited solvent cleaning applications. HCFC-225 is in the process of toxicity testing.

HCFCs can be used in conventional vapor degreasing systems, where high solvent emissions could occur. A number of equipment manufacturers have introduced totally enclosed or vacuum vapor degreasing systems for the use of HCFCs to control and minimize solvent emissions. Operations from initial cleaning to drying occur in enclosed chambers, where the parts are placed. The solvent vapor is drawn to a storage vessel once cleaning using vapor degreasing or ultrasonic agitation is completed. The (sub)assembly is dried using additional vacuum, and the residual vapor is recovered.

Heavy Hydrocarbons. Heavy hydrocarbon solvents such as terpenes and N-methyl-2-pyrrolidone (NMP) have been introduced to the electronics market as substitutes for the CFCs. These solvents are used in semi-aqueous cleaning systems, either in batch or continuous-cleaning machines. Figure 39-2 depicts the semi-aqueous cleaning process. In these systems, the flux is dissolved in a solvent and then removed with water. Figure 39-2 also shows a semi-aqueous cleaning system designed to recover the cleaning agent. In a true semi-aqueous cleaning system, the cleaning agent is not soluble in water. The residue is removed in the solvent bath, and rinsed off in the water rinse bath. Since the solvent is not soluble in water, the solvent and water can be reused and recycled. However, not all designs allow the recovery of the water and the recycling of the solvent used in the process. The solvent being removed by the assemblies can be replaced with fresh solvent. The emulsion bath removes the flux residue. The rinsewater removes the remaining solvent and flux residue and can be recycled by the use of recovery technologies such as membrane systems. These systems can be designed for low-, medium-, and large-volume in-line cleaners.

Perfluorocarbon Compounds. A new process uses perfluorocarbon compounds in conjunction with heavy hydrocarbons in traditional vapor degreasing equipment. The process, called the advanced vapor degreasing (AVD) process, uses terpene-based compounds as solvating agents. The rinsing agent is generally a perfluorocarbon. The equipment used for this purpose is commonly a modified vapor degreaser with 150 percent freeboard, and a $-20°F$ chiller coil above the usual condenser coils in the freeboard area. Perfluoro compounds, however, are quite expensive and are believed to be greenhouse

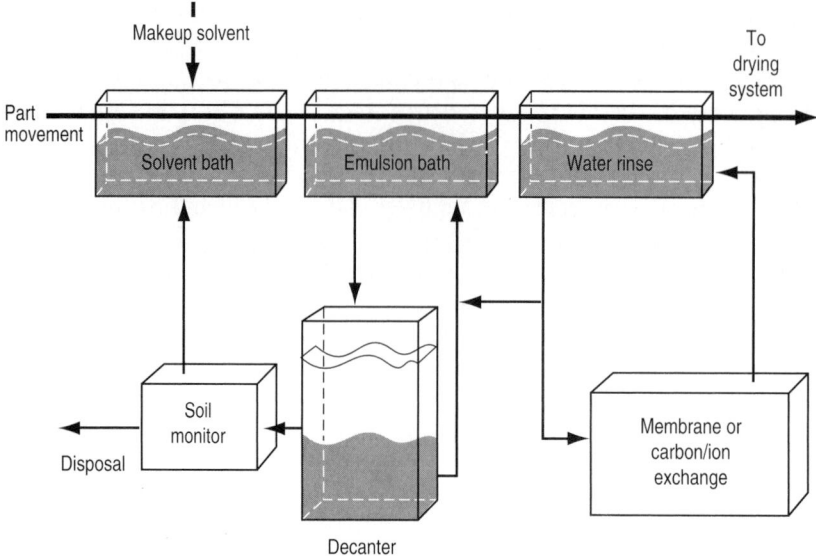

Figure 39-2. The enclosed semi-aqueous cleaning process.

gases, and thus are not approved for use alone by the U.S. Environmental Protection Agency (EPA) in standard cleaning applications. Turnkey and off-the-shelf enclosed semi-aqueous systems are available to users.

39.4.2   Process Substitution

Aqueous Cleaning Process.   Water-based cleaning agents with saponifiers can be used to remove both nonionic contaminants (such as rosin flux, grease, and oil) and ionic contaminants (i.e., flux activators). In fact, aqueous cleaning systems have been used in electronics production for many years. Cleaning with water has its pros and cons. The use of water in cleaning applications will result in a number of wastestreams that must be properly managed. The waste consists of

1. Spent wash solution that contains the materials removed from boards/ assemblies
2. Rinsewater containing impurities dragged over from the wash
3. Filter media (mechanical, carbon, or ion exchange) used for the purification of tap water

Wastewater from aqueous cleaning can be considerable. A significant reduction in wastewater is achievable by employing closed-loop recycling systems where process water is purified continuously and a major portion of the heating energy is recaptured. Turnkey aqueous cleaning systems can be designed that allow recovery and reuse of the wastewater.

No-Flux Soldering. This process utilizes nitrogen-controlled reflow soldering machines with activators that are applied on the surface of the board. The activators are acids of various kinds like formic or abietic acid. Because the oxygen content of the process is controlled, no oxides form on the boards and the residues remaining on the board do not have to be cleaned. There are a number of technical and operational issues that need to be considered in the adoption of this technology; however, it should be noted that this process in a single step eliminates the need for liquid cleaning systems.

39.4.3 Soldering

The presence of lead in solder material has recently become a concern for the electronics industry. Tin-lead solders are widely used throughout the electronics industry, and in recent years a number of European and U.S. suppliers have investigated lead-free solders (Tuck, 1992). Finding a suitable substitute for eutectic solder is an extremely complex proposition. Of the lead-free candidates that have been used, bismuth, a by-product of lead and copper refining, offers an intriguing choice. A number of technical and product issues remain to be resolved before this compound can be considered a viable substitute for lead (Tuck, 1992). Other alloys, e.g., tin alloys with antimony or silver, have also been considered; however, it seems that for the time being they have other technical problems. For example, they liquefy at temperatures above the lead melting point. Also tin-antimony materials form intermetallic needles under certain conditions which can lead to structural problems. Addition of copper can improve the wetting rates of these alloys to rates comparable to measured rates for tin-lead.

A second approach to finding suitable alternatives to lead-containing solders is to replace the metallurgical bond with a connection made by a conductive adhesive. Conductive adhesives are not new; for solder replacement these isotropic adhesives are typically silver-filled epoxies. A number of process variables must be considered before adoption of a suitable adhesive, and all relevant technical hurdles must be overcome before full adoption is possible. It should be pointed out that replacement of lead solders will eliminate the need for cleaning, representing substantial cost savings to the industry.

39.5 Pollution Prevention Successes

Numerous pollution prevention successes can be reported for the electronics industry, mostly involving the move away from the use of chlorinated solvents and CFCs. Two examples follow.

39.5.1   AT&T Laboratories

Scientists from the AT&T Laboratories have been successful in replacing CFC solvents with derivatives of orange peel and cantaloupes. Recently, they have developed solder pastes that eliminate the use of CFCs in some PCB-manufacturing processes. The new pastes are made with natural or synthetic chemicals, including several that occur in common foods and are safer to handle or dispose (*Circuits Assembly*, 1993).

39.5.2   Alcatel Network Systems
Surface Mount Division

Alcatel's Richardson, Texas, plant has been successful in eliminating the use of all ozone-depleting substances (ODSs) in its PCB assembly operations. ODSs were used for removing flux residues, cleaning stencils, wave-soldering fixtures, and other tool-cleaning uses. Terpene-based systems were evaluated and rejected due to flammability issues, and to size and cost of the systems. In-line aqueous cleaning systems were also considered and rejected due to annual operating cost for a large in-line aqueous cleaner. The company finally evaluated, tested, and adopted "no-clean" flux technology. The change to this new technology required upgrading and modification of the existing stencil printing and wave- and reflow-soldering machines (Castenada, 1993).

Bibliography

Castenada, Chris, "Eliminating Ozone-Depleting Chemicals," *Circuits Assembly,* Vol. 4, No. 3, June 1993, p. 34.

Circuits Assembly, "AT&T Eliminates More CFCs with Food Ingredients," Vol. 4, No. 3, June 1993, p. 14.

Tuck, John, "Getting the Lead Out," *Circuits Assembly,* Vol. 3, No. 12, December 1992, p. 24.

———, "A Successor to Solder?" *Circuits Assembly,* Vol. 4, No. 3, June 1993, p. 22.

Source Reduction Research Partnership, *Electronics Products Manufacture,* Chapter 11, California Environmental Protection Agency, Department of Toxic Substances Control, Sacramento, Calif., June 1991.

Yazdani, Azita, "Source Reduction of Chlorinated Solvents in the Electronics Industry," *Plating and Surface Finishing,* April 1991.

Further Reading

Electronic Manufacturing Production Facility (EMPF), Indianapolis, Ind.

Ozonet Database, Institute for Interconnecting and Packaging Electronic Circuits (IPC), Lincolnwood, Ill., (708) 677-2850.

U.S. Environmental Protection Agency, *Manufacture Systems to Produce Semiconductors*, U.S. EPA, EPA/600/S-92/050, June 1990.

———, *Printed Circuit Boards*, U.S. EPA, EPA/600/S-92/033, June 1990.

———, *Printed Circuit Board Industry*, U.S. EPA, EPA/625/7-40/007, June 1990.

Source Reduction Research Partnership (SRRP), *Electronics Products Manufacture*, California EPA, Department of Toxic Substances Control, June 1991.

40

Industrial Waste Recycling at an Automotive Component Manufacturing Facility

John A. Jaffurs

Richard L. Hubler

Denise P. Behaylo
*General Motors Corporation,
AC Rochester Division
Flint, Michigan*

40.1 Introduction

The AC Rochester Division of General Motors Corporation (GM) develops and manufactures automotive components for engine management systems at nine facilities in the United States. Its largest facility is located in Flint, Michigan, and is known as the "Flint East" site. The Flint East site covers nearly two square miles and consists of several plants housing manufacturing operations for spark plugs, glow plugs, oil filters, air filters, air cleaner assemblies, fuel pumps, fuel level sensors, cruise control systems, and other components. The volume and diversity of the scrap and wastes generated from these operations require skillful waste management to provide environmentally safe and cost-effective disposal options.

Over time, a full-scale recycling and waste disposal operation evolved at Flint East. The operation has grown over the past thirty years to handle over 68,000 tons of material annually. It is managed by the By-Products and Waste Management Group, also known as the By-Products Group, which consists of 27 employees. Flint East's program is regarded as a model industrial waste reduction and recycling operation. In 1990, the program was recognized for its long-term success when the Michigan Recycling Coalition honored AC Rochester with the "Industrial Recycler of the Year" award. Elements of the program are presented here as a guide to establishing a successful industrial recycling program.

40.2 Waste Minimization Philosophy

Flint East's recycling program was founded on GM's belief that waste minimization is good business. By reducing and eliminating costly wastes, GM can offer its customers not only a competitively priced product, but also one produced in a manner which is friendlier to the environment. GM's waste minimization techniques include source reduction, recycling, and technology improvements. Priority is given to preventive measures. To utilize these techniques, the By-Products Group is organized along the principles of total quality management with a goal of continuous improvement. Figure 40-1 illustrates the By-Products Group interactions. An organization chart for the By-Products Group is shown in Fig. 40-2 .

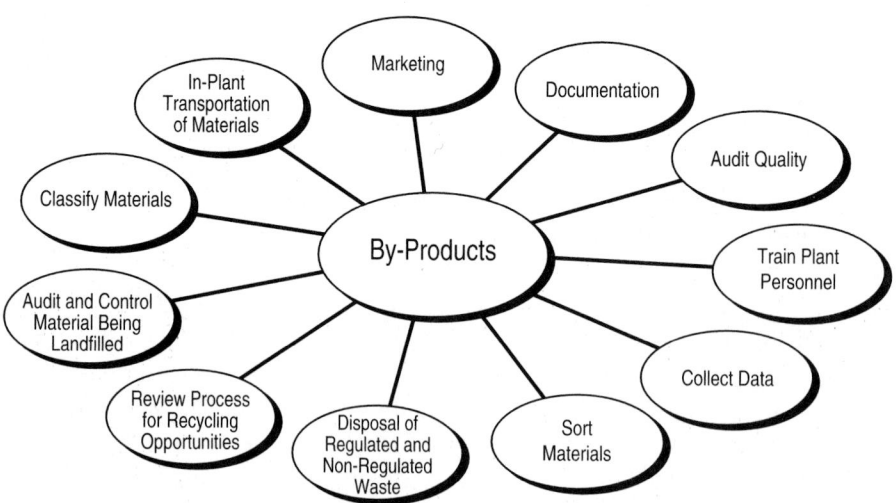

Figure 40-1. By-Products Group interactions.

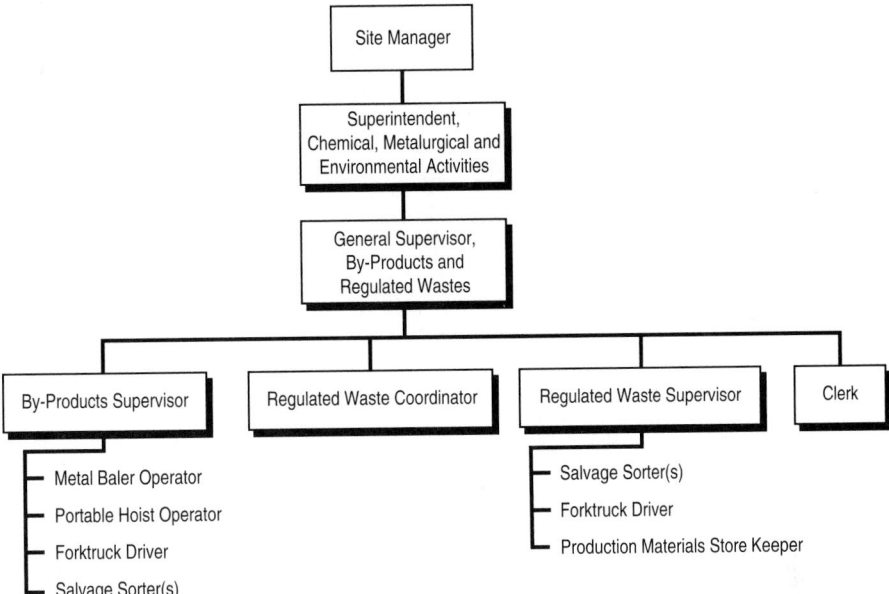

Figure 40-2. Organization chart for the By-Products Group.

40.3 Brief Description of Flint East's Operations

Currently, almost 100 different items are being recycled at the facility, including ferrous and nonferrous metals, plastics, ceramic powders, paper, cardboard, wood, and scrap parts. Materials are collected from dozens of operations, including stamping, pressing, injection-molding, screw-machining, assembly, product-testing, cleaning, research, maintenance, construction, and office activities.

The materials are brought to the By-Products department where they are placed into separate storage containers. The materials are kept in inventory until picked up by their respective buyers. Through coordinated efforts to prevent, reduce, and recycle plant wastes and sell plant scrap materials, the By-Products Group achieves revenues of around $3.5 million each year.

40.4 Recycling Program Functions

40.4.1 Criteria for Recyclable Materials

To ensure that recyclable materials are marketable, a number of criteria must be agreed upon by the By-Products Group and the material buyer. Material quality must be consistent throughout and either meet or exceed a buyer's particular specifications. Buyer specifications for materials can include material sepa-

Table 40-1. Types of Landfillable Materials
Recycled

Blacktop	Ceramic miscellaneous, with steel
Concrete	Ledger paper
Expandable plastic trays	Precious metal (paper rags)
Scrap plastic regrinds	Dunnage styrofoam
Dunnage trays	Unusable uniforms
Auto electrical harnesses	Scrap insulators
Unusable plastic forms	Corrugated containers
Electrical components	Gaylords
Spools	Saggers—broken
Scrap plastic, painted	Spark plug powder

ration, contamination and moisture content limits, oil and lubricant removal, and volume requirements. In general, the criteria change from commodity to commodity and can change over time for the same buyer. Table 40-1 is a list of landfillable materials that are recycled.

40.4.2 Handling and Transportation of Recyclable Material

The By-Products Group is responsible for the handling of spent material from its point of generation at the process until its transfer to a buyer. From the point of generation, the material is transferred to the By-Products department either by forklift or conveyor. It is stored at the scrap dock until the necessary volume is accumulated. Once sufficient volume has accumulated, the material is loaded onto the buyer's vehicle by site personnel using site loading equipment.

The By-Products Group works with buyers to coordinate their shipment pickup times with the availability of site personnel and equipment. The group also works with buyers to reduce transportation costs by evaluating different modes of transportation and different containers for transportation. Transportation practices can change frequently and are influenced by transportation rate changes.

40.4.3 Material Sorting and Classification

Once collected, material is sorted into categories to match buyer specifications. Classification of a material can impact its market value. For example, one material which had been sold as "steel and copper mixed" was found to meet the criteria for "contaminated copper." The material was reclassified and was then able to be sold for a higher price.

40.4.4 Equipment

Many pieces of equipment are used in the By-Products department and in the operations areas for collecting and managing scrap. A partial list of the equipment includes these items: a wood and metal shredder, metal balers, cardboard balers located near the point of generation, a portable hoist, compactors, scrap conveyor systems, a central baling system, scrap containers, and forktrucks with rotating forks.

40.4.5 Material Containers and Storage

Containers for scrap are chosen based on the vendors' needs for specific sizes, shapes, and closures, and can include hoppers, tubs, Gaylord boxes, etc. Other considerations include whether the container is to be returnable, whether a standardized container is available, and which features will help most to maintain the quality of the scrap material. Containers are often sized to be filled in one shift so as to minimize contamination from any change in operation during the next shift. Where possible, the actual shipping container is used at the point of generation to capture material, instead of a transfer container. When this isn't possible, the container and storage systems for each material are chosen to balance the needs of the manufacturing operations, the By-Products Group, and the buyer.

In one operation, small scrap containers used at the point of generation were consolidated into larger ones prior to being transported to the main scrap-handling area for storage. It was expensive to consolidate the containers, but there was a lack of storage room in the originating area. When some floor space opened up adjacent to the point of generation, however, a scrap-handling conveyor was installed to carry material to the storage area. Installing the conveyor reduced the number of times the material was handled by at least three.

In another case, the container required by the buyer made a tremendous improvement in the appearance of the operations department. The operations department had been placing waste ceramic powder in open-topped hoppers which allowed loose powder to spill out. A buyer for the powder required the powder to be shipped in "Supersacks," a more expensive but more secure container. The By-Products Group placed the Supersacks directly at the point of generation to eliminate a step in the transfer of the material, and, consequently, eliminated a housekeeping problem. The improvement in control over the powder was so well received that, when Supersacks were no longer expected to be required by the buyer, the operations department still opted to use them.

40.4.6 Segregation of Scrap at Process Sources

Segregating scrap at the point of generation is one of the simplest yet most cost-efficient procedures that the By-Products Group has helped to establish at many operations. Segregation maintains the purity and identification of the material, which enhances its market value. Segregation at the source also prevents more difficult separation at a later stage, and so reduces handling costs. When one is deciding whether to implement source segregation, the cost of separating the material is compared to the increase in sale price of the separated materials. Caution is taken to assure that any increase in scrap-handling costs does not exceed the gains in revenue obtained for the separated material. The process operators are generally amenable to source separation because "it's just as easy to put the material in the right container as it is to put it in the wrong one," and, they can often label a material more accurately than By-Products Group members can.

Source separation is widespread at the Flint East facility. For example, to make separation of cardboard more convenient, separate balers for cardboard are generally located near the regular trash bins to provide convenient "one-stop" trash disposal. In another example, plastic scrap from plastic injection-molding machines was being placed in a centrally located container. However, this resulted in contamination of the scrap when different machines used different plastics. The contamination was eliminated when each injection-molding machine was provided with its own scrap bin. The same procedure has been instituted in metal operations as well.

40.4.7 Other Waste-Handling Responsibilities

The By-Products Group is also responsible for collecting from operations areas plant trash destined for a landfill. If it is found that an area is not following standard plant procedures for disposal of recyclable materials and is including them in the plant trash instead, the By-Products Group will bring it to the department's attention. Then, the department's operating budget could be backcharged for not properly handling its recyclable materials.

Wastes regulated as hazardous that are recyclable are handled by one segment of the By-Products Group, while nonregulated recyclable wastes are handled by another segment. Regulated-waste handlers receive specialized training, tools, and protective clothing. Regulated materials are transported in approved containers from the operations directly to the regulated-waste accumulation pad, where the operators are also trained in handling techniques for regulated wastes.

40.4.8 Job Titles and Functions

Metal baler operator: Operates a metal baler machine which presses scrap metal pieces into square blocks of material. Also assists in miscellaneous sorting activities.

Portable hoist operator: Loads scrap materials onto vendor-supplied containers for transport. Also loads the shredder with wood pallets and large scrap metal pieces.

Forktruck driver: Transports containers of scrap from point of generation to storage and condenses materials into larger storage containers. Also handles movement and loading of containers within the By-Products area.

Salvage sorters: Sort material into categories either by hand or by consolidating containers. Also responsible for labeling containers and for general housekeeping.

Production materials storekeeper: Assists the By-Products area by reviewing waste containers for proper labeling, completing administrative paperwork, and loading empty drums to be recycled.

Clerk: Provides clerical support, obtains and tracks shipping forms, and prepares bidding process information for the sale of recyclable materials.

By-Products supervisor: Directly supervises the By-Products area and manages metal scrap and trash disposal.

Regulated-waste coordinator: Obtains paperwork approvals and related administrative controls, and tracks regulated solid wastes.

Regulated-waste supervisor: Directly supervises the employees handling regulated wastes and the waste storage area.

General supervisor of By-Products: Responsible for all By-Products, waste disposal, and regulated-waste disposal activities, except for management of the wastewater treatment plant.

Superintendent of chemical, metallurgical, and environmental activities: Manages chemical and metallurgical processes and environmental activities.

40.5 Implementation of the Waste Minimization Philosophy

40.5.1 Waste Minimization Committee

Managers and personnel from different operations are appointed to the committee and act as liaisons between their departments and the committee. Committee members encourage participation in waste minimization and recycling efforts in their areas, and their effort is aided by support of the site newsletter, which often publishes articles about recycling. The committee also facilitates the exchange of waste reduction ideas and ensures consistency in site waste reduction efforts.

The committee sets waste reduction goals to be met by the site. To help the committee track progress toward the waste reduction goals, the By-Products department provides a monthly tally of the volume of recyclable materials sold and the revenues received from them. These tallies are compared with the volume and makeup of trash sent to landfills as part of the continuous search for new waste reduction ideas. Occasionally, outside experts are brought in to solve a waste reduction problem that is beyond the committee members' area

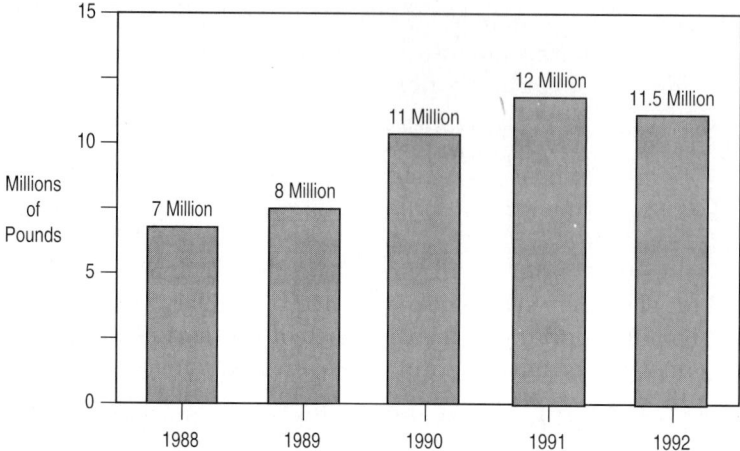

Figure 40-3. Amount of material diverted from landfill due to recycling.

of expertise. The committee serves to instill commitment to waste reduction among the different site departments. Figure 40-3 illustrates an impact of the recycling program.

40.5.2   Waste Reduction and Minimization

Periodically, the By-Products Group and the plant Waste Minimization Committee review wastestreams to determine whether the wastestream can be reduced or recycled, or a less toxic chemical substituted for a toxic chemical. The waste minimization progress is documented and annual goals are established. Each year the actual reduction and targets are compared and resulting improvements documented.

For example, methylene chloride, which is regulated as a toxic chemical and which has low employee exposure limits, was being used by one operation as a purging agent. The By-Products Group informed the operations department about the cost of disposing of the methylene chloride–contaminated waste and related regulatory requirements. As a result, the operations department worked with the manufacturing and process engineering staffs and found a nonhazardous substitute. The switch in materials reduced operational costs and eliminated potential employee exposure issues.

In another case, used solder paste pails were being turned in to the By-Products Group with about 14 percent of the original lead-tin solder still remaining in the pails. The By-Products Group suggested that the operations department remove and then consolidate and reuse the solder. This change reduced the solder remnants in the pails to about 2 percent of the original amount and reduced the operations department's cost for solder purchases.

Through situations like these, the By-Products Group can realize both waste reduction and cost savings opportunities.

40.5.3 Reuse of Material

Reusing scrap material is an effective way of reducing wastes. Several processes at the Flint East site incorporate scrap material into the process. Some plastic injection-molding processes, depending on the products' engineering and durability requirements, mix reground scrap plastic with virgin plastic to form parts. This occurs either (1) in a closed-loop operation, where plastic scrap from the end stage of the operation is reground and reintroduced to the feed-stock, or (2) in an open system, where scrap is reground off site and then re-added to the operation. For the closed-loop operation, the container for the scrap material is covered in order to prevent contamination by foreign plastics or objects. In another case, oil lubricants used in the screw machine operations are recycled within the process.

40.5.4 Developing New Markets and New Businesses

New recycling technology and new markets for spent materials frequently develop. The By-Products department reviews industry publications and jour-nals, and benchmarks against other companies for new and more efficient scrap management practices and ideas.

In one case, the By-Products Group helped to develop a market for a recycled material and subsequently spurred the development of an independent recy-cling company. The By-Products department was searching for a buyer for reground plastic scrap generated by injection-molding operations. Some of the reground plastic was sold to a paper recycler who attempted to market the material. However, the vendor found it difficult to market because different plastics had been mixed together in the regrind. To make the regrind more mar-ketable, the vendor offered to buy the scrap plastic whole from Flint East and to do the regrinding under controlled conditions. After the By-Products depart-ment received approval for the practice, the vendor opened a subsidiary exclu-sively to handle the large volume of scrap plastic. Flint East benefited both from the new market for its scrap plastic and from the quality of the regrind it could buy back from the vendor. The subsidiary is now a separate firm which processes plastics from many companies for regrind and resale.

40.5.6 Documentation and Computerized Aids

The By-Products Group coordinates and manages the shipping paperwork. Total weight, material identity, price, and vendor are verified for each shipment. Each material type is given an identification number which is entered into a computer tracking program for each shipment of the item. The program provides a running total of the amount and value of materials sold that year to date, and it is used to measure progress from year to year. Separately, the By-Products Group also documents how much trash is shipped from each collection point at Flint East and the reasons behind any changes in volume. All of this information is used in reports for the site manager, plant managers, and the Waste Minimization Committee. The Flint East engineering staff has also used the information when determining floor space and materials-handling requirements.

40.6   Management of the By-Products Group

40.6.1   Training of Plant Personnel

Formal training of site personnel in waste management techniques is primarily conducted on an as-required basis. By-Products Group members receive an introduction to the department and then obtain task-specific training while on the job. For the rest of the site, the Waste Minimization Committee facilitates the exchange of waste minimization and recycling ideas and establishes site goals.

Informal training also encourages participation in recycling among site personnel. Employees in the By-Products Group are informally empowered to address the employees in operations. In this manner, arrangements for the collection and separation of scrap material are made, and any problems are corrected.

40.6.2   Managing Material Sales

For recycled metals that can be reused within the corporation, the By-Products Group coordinates its sales with GM scrap management teams. The teams manage the exchange of metals between GM divisions and individual plants. For all other recycled materials, the By-Products Group works with the purchasing department to set up contracts with buyers. Material sale contracts typically last one month. The contracts are put out for bid and awarded to the highest bidder who has an approved credit standing. Buyers of regulated waste material must also meet certain handling requirements and environmental considerations. The sale information is entered into a computer database which generates the shipping forms, and the forms are forwarded to the By-Products Group to be used to document each shipment.

The list of potential bidders is periodically updated and expanded in order to increase the number of potential buyers and the amount of commodities sold. Purchasing and By-Products Group members call current bidders to see if they have developed a new interest in other materials from the site. Contacts are also made with other plants and with industry colleagues to learn of new markets and new buyers. New buyers can also be found through waste exchange services which are located throughout the country and which are run both privately and through government agencies. The Michigan Office of Waste Reduction Services is one such service frequently contacted by the site.

40.6.3 Material Specifications and Working with Buyers

Quality assurance procedures need to be established to assure that the buyer is receiving material within specifications. A reputation with buyers for dependable material quality is important. The higher the quality, the higher the price the material will command.

Maintaining a continuous dialogue with vendors is important in order to keep up with changes in material requirements which may not be immediately communicated. For example, the Flint East site had been separating cold-roll steel, tin plate steel, and painted steel for sale to a mill. By coincidence, a conversation with the mill's truck driver revealed that the three materials were combined at the mill and processed together. As a result, Flint East realized substantial savings in handling costs by discontinuing the material separation.

40.6.4 Budget and Equipment Planning

The By-Products Group is funded through allocations from each operation at the site based on the operation's actual use of the By-Products Group's services. The allocation takes into account the type and amount of disposal activity completed and the number of hours spent by the By-Products Group on behalf of that operation. In this manner, expenses for scrap handling are charged back to the originating department. Scrap revenues are also credited to an operation's piece cost; this in turn encourages the operation's recycling and source reduction efforts. The By-Products Group's budget provides for salaries, equipment, and building expenses such as depreciation and floor space allocation.

Equipment purchases are usually planned in advance by the By-Products Group and are justified through business cases. New equipment acquisitions are evaluated to determine whether it is more cost efficient to purchase the equipment or to lease it.

An example of decision making using a business case is the purchase of a shredder for wood pallets. A large volume of wood pallets were used by the site and then recycled by a local recycler. When the recycler could no longer handle the pallets, a cost-effective and environmentally responsible alternative was needed. Representatives from the By-Products Group, hourly employees, and staff from plant and industrial engineering, safety, operations, purchasing, and finance worked together to brainstorm ideas, evaluate equipment, and conduct financial analyses of the options. Twenty-six options were identified, and the option with the greatest net present value of cash flow was chosen. The shredder was purchased to shred unusable wood pallets, while the reusable pallets were recycled through an outside contractor. The shredded material from the wood pallets was provided to other companies to use in their wood-fired boilers.

40.6.5   Organizational Strength

The recycling operation has maintained success over many years because of its lean organization and because of the resources committed to it. By-Products Group hourly employees have a regular inflow of work. When there is no material that needs to be lifted, loaded, or transported, there is always material to be sorted, labeled, and stored. In turn, By-Products Group managers have responsibility for supervision, material tracking, vendor recommendation, and related administrative functions. It is well known at the site that there is little downtime in the By-Products department.

Despite the work demands, By-Products Group employees are committed to their jobs. There is very low turnover in the department, and, in most cases, employees have chosen to work in the department. One employee views his work in the By-Products Group as compatible with his personal belief in recycling, since he recycles at home. The By-Products Group supervisor has worked in the department for many years because he enjoys the work and is personally challenged to improve each year.

40.6.6   Support from the Top

Without the dedicated support and resources provided by top management at AC Rochester, Flint East's recycling program might not have achieved the level of success that it has. The waste prevention and recycling program received direct support from GM Vice President and AC Rochester General Manager Jan Tannehill, who appointed a member of his staff (the divisional comptroller) to champion the program at the divisional level. At the site level, the superintendent of environmental, chemical, and metallurgical processes was appointed the champion. This manager roused support from the other site staffs by soliciting their representation on the site Waste Minimization Committee. As a result of the active support from all levels of management, the By-Products Group is able to maintain sufficient resources to perform required tasks.

Upper management recognizes the complexities of operating the By-Products area and strongly supports resources for the operation. The department receives support for its ability to maximize the return on its resources. The By-Products Group also draws management support by engaging in long-range planning to secure its monetary and equipment needs. Further support for the goals of the By-Products Group comes from the United Auto Workers (UAW). The UAW supports waste minimization and recycling activities for several reasons, including that these activities decrease the cost of doing business without jeopardizing jobs, reduce employee exposure to chemical materials during handling, and provide an opportunity for employee involvement.

40.7 Recognition of the By-Products Group

40.7.1 Michigan Recycling Coalition "Industrial Recycler of the Year"

The dedication and accomplishments of the Flint East recycling program were recognized by the Michigan Recycling Coalition when AC Rochester was presented the Industrial Recycler of the Year award for 1990. The award honors outstanding recycling operations as identified by members of the recycling industry. Receipt of the award marked a proud moment for AC Rochester. In the following year, AC Rochester personnel made 25 presentations to community and industry groups to pass along recycling techniques and expertise.

40.7.2 Community Outreach

Over the course of a year, AC Rochester gave 25 presentations to schools and community groups on how to apply lessons from the Flint East recycling program to their own recycling programs. Presentations have been given to schools, environmental organizations, economic development groups, government officials, academic forums, and other community groups. In another effort, members of the By-Products department trained community trainers to look for innovative ways to manage and recycle scrap. AC Rochester continues to conduct its community outreach efforts to increase awareness of recycling opportunities.

41

Pollution Prevention and Foundries

Dieter S. Leidel

Tanoak Inc.
Barrie, Ontario, Canada

Foundries have been recycling for decades. Over three-quarters of all castings are produced in green sand. More than 95 percent of all sand in such systems is recycled within the foundry. Foundries also buy back their products at the end of their useful life and recycle this scrap into new products. Despite these accomplishments, gaseous, liquid, and solid wastestreams remain which need to be reduced further.

Gaseous wastestreams are by far the largest, if compared on a weight basis (1000 scfm = 2.25 tons/h). The preferred reduction approach is not clear at this time, with the majority of regulations under the Clean Air Act Amendments of 1990 (CAAA 1990) still in preparation. However, foundry-sponsored research was started a few years ago.

Liquid wastes are a small-stream, mostly noncontact cooling water, and many of the liquid waste problems have been remedied.

Solid waste originates from various points within the foundry such as sand, slag, pollution control dust, spent refractories, etc. In a study conducted by the American Foundrymen's Society (AFS),[1] sand waste has been identified as the most pressing problem to be dealt with. Accordingly, this chapter will concentrate on this multifaceted problem.

Two avenues are open to the industry: reclamation for internal reuse or external beneficial reuse. The latter aspect is presently being investigated and, while some promising results have been achieved, there is still widespread uncertainty, not the least caused by a regulatory framework which is only developing now and which will regulate those conditions under which external reuse is permissible. This leaves reclamation as the most promising alternative.

667

41.1   Definition of Sand Reclamation

Reclamation has been defined by the AFS as follows:

> Sand reclamation is the physical or chemical or thermal treatment of a refractory aggregate to allow its reuse without significantly lowering its original useful properties for the application involved.

This highlights the following point: There is more than one process. Chemical reclamation methods, however, are not in commercial use at this time. There are a multitude of sands, such as silica, zircon, chromite, etc., each of which has its own requirements. And there are numerous requirements imposed by the type of metal poured (from light metals to steel), the casting size (from ounces to many tons), and the binders used (organic or inorganic, etc.). Each of the many possible combinations imposes its own criteria which will dictate what can be considered to constitute "reuse without significantly lowering the original useful properties." Searching for the perfect reclaimer without consideration of the many accompanying requirements would be wasted time.

41.2   General Technical Considerations

41.2.1   Factors to Be Considered

The objective of reclamation is to remove residual binder (and other contaminants) from each individual sand grain to a degree which will permit reuse of this material in the next production cycle without sacrificing product quality.

Attrition.   Exposing the sand to attrition is known technology. However, no attrition system will ever be able to completely remove all residual binder. Figure 41-1 provides a view of a sand grain surface at 500 magnification. Binder that resides in the crevices where no grain-to-grain contact can take place can obviously not be removed by attrition. Even removing the binder from all convex areas on the grain requires long processing times, given the large specific surface areas of commercially used sands (from 2.5 to over 20 acres/ton). If the residual binder is to be removed out of the depression areas, two other methods must be considered.

Wet Reclamation.   Wet reclamation was used in the 1950s and '60s. It is no longer considered a major process for two reasons: (1) most modern chemical binders are hydrophobic and cannot be dissolved in water, and (2) binders which can be dissolved result in contaminated water which now must be

Figure 41-1. Scanning electron micrograph of silica sand, 500×.

cleaned before release to the environment. Additionally, the sand must be dried. Thus, total processing cost is high, both from the aspect of operation as well as in terms of capital requirements.

Thermal Processing.   Thermal processing permits removal of binder out of depression areas, as long as the binder can be combusted. This eliminates inorganic constituents which can be introduced with the binder, the sand, or special additives. This means that, in most cases, reclaimed sand is different from new sand. This was shown by AFS research.[2]

41.2.2   Reclamation Generates Waste

All reclamation processes aim at removing residual binder. Since matter cannot disappear, the process generates waste (solids in attrition, solids and gases in thermal reclamation, and wastewater plus exhaust gases in wet reclamation). The amount of waste varies considerably but is influenced by the sand to be processed. Figure 41-2 provides yields for two different sands, processed in the same reclaimer under identical conditions. While one sand leads to a (sustained) yield of 95 percent, the second sand produces a yield of only 88 percent; i.e., it generates more than double the waste. The largest wastestreams are generated when processing inorganically bonded sands in combined systems, applying thermal processing, followed by postscrubbing (attrition). Systems of this type must deal with the gaseous waste from the thermal process plus the attrition waste, which virtually always exceeds 20 percent and frequently exceeds 30 percent. The high attrition loss in such installations is caused by the present practice of high-temperature calcination which fuses inorganics onto

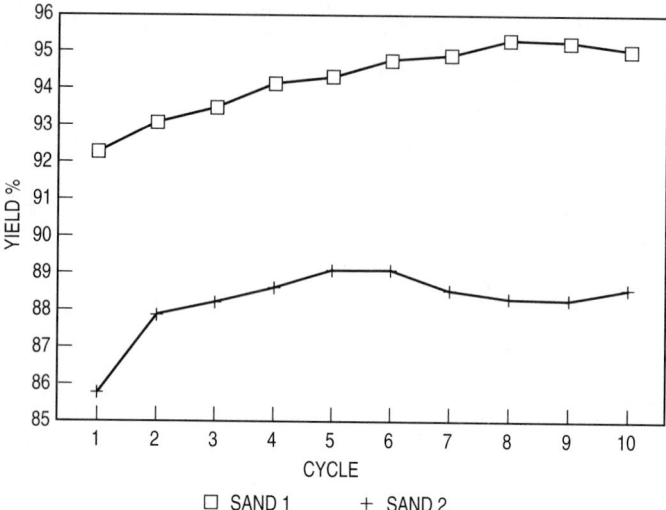

Figure 41-2. Yield of two different sands, processed in the same reclaimer in identical fashion.

the sand grain. This then must be exposed to excessive attrition in order to remove the fused material. The following table highlights this.

Property	Used sand	924°F	1000°F	1200°F	1300°F	1400°F	1500°F	1600°F	New sand
Active clay (%)	8.08	7.25	3.89	1.04	0.0	0.0	0.0	0.0	0.0
Total clay (%)	10.69	10.52	7.91	1.68	1.0	1.09	0.75	0.91	0.47

SOURCE: Adapted from J. M. Svoboda, and M. J. Granlund, "CMP Report #91-7, Electrotechnologies for Sand Reclamation, I. An Overview and Evaluation of Infrared," The EPRI Center for Materials Production, Pittsburgh, Pa., April 1991, p. 6-6.

As clay containing green sand is exposed to increasingly higher temperatures, total clay (which can be removed by a simple wash procedure) diminishes more than 90 percent. Since clay is not combusted, this proves that it is fused to the sand grains where even a wet process can no longer remove it. Such fusion of inorganics can result in lasting damage to the sand.

41.2.3 Problems with Present Testing Methodology

Surface-Related Problems.   Bonding of sand is not a mass-related problem but a surface-related phenomenon. However, virtually all foundry-related test methods are mass-related. The industry expresses binder levels as percent binder weight based on the weight of sand (bos). If sands with high specific weight are covered with the same binder level (bos) as lighter sands (for example, zircon with 4.6 g/cm^3 vs. silica with 2.6 g/cm^3), they display higher strength—not because the binder suddenly has a higher bonding strength but because the system has a thicker binder layer due to the smaller surface area per unit weight for the heavier aggregate. The AFS, with test procedure 109-87-S, offers a method to calculate surface area of sands when the sieve analysis is known. These calculated values are in reasonable agreement with AFS measuring test method 108-87-S. However, when these values are compared with the ones obtained by application of the gas absorption technique (developed by Brunauer, Emmett, and Teller), generally referred to as the BET method (a gas adsorption technique) which is widely used in industry and research, major differences exist. BET values are higher by a factor of 4 for silica, 27 for lake sand, and 130 for olivine. The discrepancy can be explained by the surface morphology as shown in Fig. 41-1. This indicates that effective surface area is dependent not only on the actually measured value but also on binder viscosity. This value differs greatly from binder to binder. Some binders change viscosity during storage, for example, from 100 cp to 500 cp, within a period as short as 30 days. Therefore, there is no undisputed and generally applicable surface measurement technique.

Loss on Ignition (LOI).   This is an often misunderstood parameter. The test places a 50-g sand sample into a muffle furnace where it is heated for two hours at 1800°F. The weight change is then related to the original sample and expressed as percent LOI. The weight change can be a loss or a gain. The presence of one volume percent calcium carbonate ($CaCO_3$) in the sand will cause a weight loss of approximately 0.44 percent. $CaCO_3$ dissociates at approximately 1500°F into CaO and CO_2, with the latter released as gas.

On the other hand, the presence of one volume percent metallic iron, if completely oxidized to Fe_2O_3, causes a weight gain of 1.2 percent. AFS test method 117-87-S does not require the removal of metallic material from the test sample before performing the LOI test. The test result only indicates the sum of all gains and losses. It is therefore wrong to interpret LOI as a measurement of residual organic material ($CaCO_3$ is not an organic substance; neither is the chemically bound water contained in bentonite or in sodium silicate). LOI also cannot be used as a guaranteed indicator for potential gas problems. Figure 41-3 provides gas generation curves for sand samples, originating from one foundry, using two different binders. The test was performed on bonded new sand. The sample bonded with binder II has an LOI value 24 percent higher than sample I. However, the gas generated in the first 0.2 minutes is 500 percent higher than that from sample I. This means that the potential for gas defects is dramatically higher because it is during this period that the metal in the mold is still liquid. The problems associated with erroneous interpretation of LOI were covered in detail in an AFS paper.[3]

Acid Demand Value (ADV).   Similar problems exist with ADV, which is considered to be a major quality parameter. The problem was highlighted by a

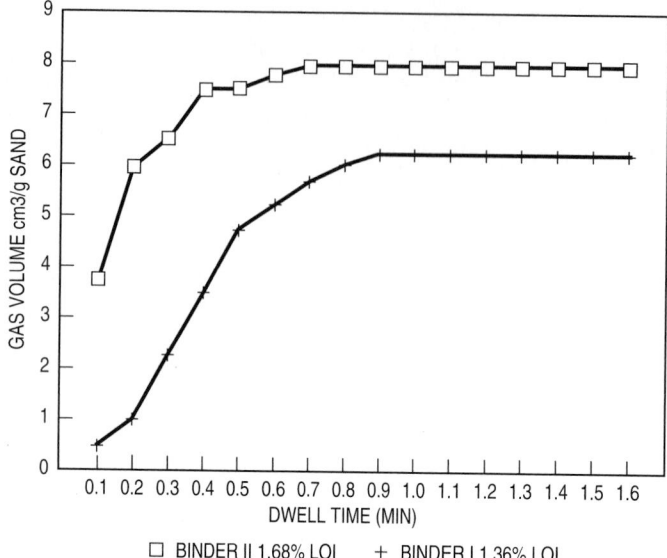

Figure 41-3. Gas evolution curves of new sand, bonded with two different binders.

1982 AFS presentation.[4] New sand showed an ADV value of 21.4; particulated core sand also produced a value of 21.4—i.e., no change. When this particulated core sand was fired at 1800°F (comparable to thermal reclamation) and cooled again, ADV was measured as 36.9. ADV is frequently stated as being a measurement for the alkali contaminant present in the sand. Since simple heating and cooling do not add alkali substances, this interpretation cannot be correct. Hoyt* also documented this with a simple test, the results of which are given in Fig. 41-4. He mixed unadulterated silica sand with varying amounts of alkali $CaCO_3$. As long as $CaCO_3$ was added as 70 mesh material, there was no effect on the setting of the acid-cured furan binder. Strip time remained unchanged at 15 minutes. However, when this test was repeated with contaminant of 270 mesh size, an addition of only 0.1 percent increased strip time by one-third, and higher additions prevented setting of the binder altogether. Thus, it is not so much the quantity of a contaminant that causes bonding problems but its reactivity (and its distribution over the grain surface). These important factors are not recognized by the ADV test. This may lead to problems specifically in thermal reclamation, since the presently applied high-temperature calcination leads to activation of contaminants such as $CaCO_3$ and causes chemical reactions between inorganic contaminants and the base aggregate (mostly silica).

*Daryl F. Hoyt, Vice President of Research & Technology, Wedron Silica Co.

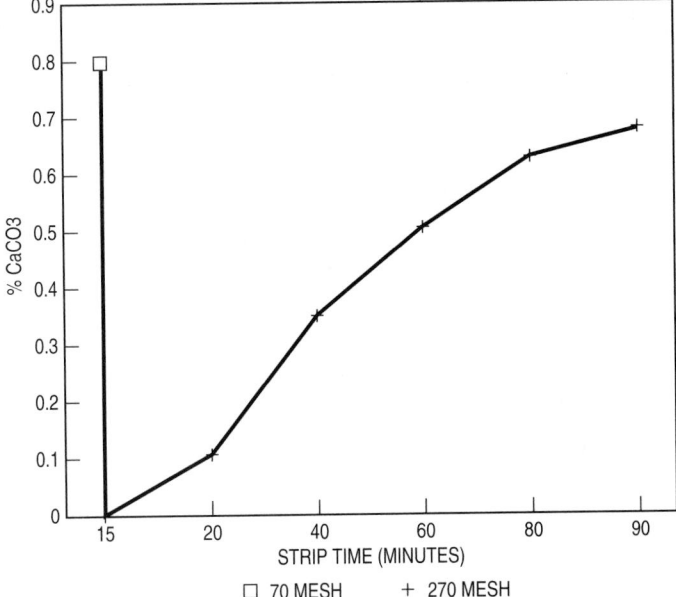

Figure 41-4. The effect of alkali material on stripping time of furan-bonded sand. (*Courtesy of Daryl F. Hoyt, Wedron Silica Co.*)

41.3   Different Reclamation Requirements

There are four different scenarios.

41.3.1   Reclaim Green Sand for Use in Green Sand

This approach, practiced since the 1950s, was limited to steel foundries. Due to the high pouring temperatures in steel (up to 3000°F), the spent sand was generally dry and free flowing. The pneumatic reclaimer developed for this process is unsuitable to processing moist sand. However, specifically in high production iron foundries, the occasional unpoured mold (with moistures over 3 percent) is quite common. This leads to unwanted production interruptions. Also, new sand addition to green sand systems in steel foundries is traditionally much higher than in iron foundries (thermal degradation plus use of facing sand). There was then little incentive for iron foundries to consider this process.

With increasingly expensive disposal cost, the situation changed and a new process was introduced in the 1980s,[5] which also permitted the processing of wet sand. This process was further developed[6] to combine the functions of reclamation, drying, and cooling in order to permit the reclaimed sand to be screened at 20 or 30 mesh (which permits removal of all "coarse" contaminant). This screening is an imperative step if new sand additions to a green sand line are to be completely eliminated.

The process has been commercially used in the United States since the early 1980s. It not only retains sand for reuse but also retains major portions of the clay binder and carbonaceous material, as shown in Fig. 41-5. It is thus the reclamation process with the highest yield (generally 95 percent or more) and simultaneously the lowest cost ($10 to $15 per ton of sand processed for operating plus capital cost).

A further possibility is correction of nitrogen problems. Ammoniacal nitrogen, which is a decomposition product of organic binders and additives, is attached to the sand. If present at excess levels, pinholing results. In the past, this was corrected by increased new sand addition. However, the nitrogen is primarily associated with the clay, which has a much larger (adsorptive) surface than the sand. Increased scrubbing removes more clay (with the nitrogen attached to it) but will retain the sand, thus significantly reducing waste volume.

41.3.2 Reclaim Chemically Bonded Sand for Use in Chemically Bonded Systems

Chemical binders became popular in the 1960s. Contrary to the case of green sand, where 95 percent or more of the sand can be recycled without major processing, chemically bonded sand must be reclaimed completely in each cycle.

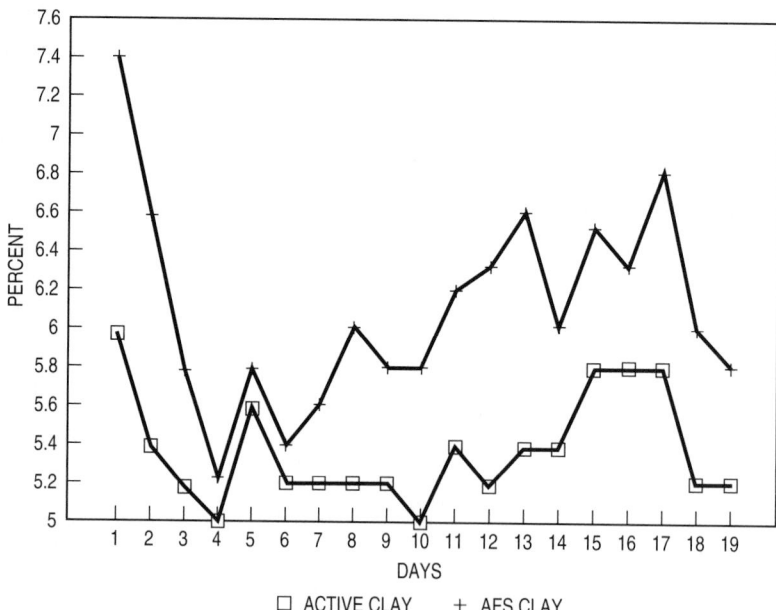

Figure 41-5. Residual clay on reclaimed green sand. (*Reprinted with permission from AFS Transactions p. 18, Fig. 11, 1988.*)

The pneumatic reclaimer, originally designed for green sand in steel foundries, was successfully modified for this application. Limitations remained in cases where noncompatible binders were employed, for example, using acid-cured furan and alkali-cured phenolic urethane. As was shown previously, even the most effective attrition system is unable to reestablish the original grain surface. Thus, residual acid from the furan system interfered with the alkali-curing mechanism. It was expected that thermal reclamation would alleviate this problem. Indeed, as long as the interfering agent can be removed by combustion, thermal reclamation can technically solve this problem. However, we must now also consider the effect of sulfur dioxide (SO_2) emissions, which will be generated by the combustion of sulfur. Given the requirements of the CAAA 1990, the installation of sulfur scrubbers (as they are already applied overseas) may be necessary. This will substantially increase capital cost. But equally important is the fact that thermal reclamation cannot remove inorganic contaminants. There are increasing efforts by binder manufacturers to use more inorganic components (or to switch to totally inorganic binders) in order to meet in-plant emission limits during pouring and shakeout. Zinc, copper, calcium, lead, cobalt, and potassium are found as parts of the binder chemistry. Fe_2O_3 or KBF_4 may be added to a sand system to achieve certain technical results, and new sands themselves carry inorganic contaminants in many cases. Figure 41-6 shows that thermal reclamation of a lead-containing binder rendered the sand leachate toxic, due to accumulation of metallic lead on the sand. A Michigan foundry,[7] using the same binder, solved the problem by abandoning its thermal reclamation process and switching to an advanced attrition system. Its sand, after over 50 cycles, showed a leachate value of 1 ppm; i.e., it was not leachate toxic (with the regulatory threshold value presently set at 5 ppm).

But leachate toxicity is not the only concern. Deteriorating bonding properties as a result of accumulating inorganic contaminants can be equally damaging. The following table documents this.

Sand type	Process temp. °F	Attrition time (min)	LOI (%)	Potassium (%)	24 h tensile (psi)
Single-cycle	Ambient	2	0.383	0.0113	68
Multicycle	Ambient	6	0.490	0.0113	39
Multicycle	250	8	0.450	0.0113	30

SOURCE: Adapted from S. R. Iyer and C. K. Johnson, "Reclamation of Phenolic Ester Cured Nobake Sands," *AFS Transactions*, p. 614, tables 4 and 5 (1988).

Phenolic-ester-cured nobake binder is known to suffer deteriorating bonding strength as a result of accumulating potassium.[8] The first line shows single-cycle sand—i.e., sand that was bonded once, then reclaimed, and rebonded. An attrition time of two minutes led to a value of 0.0113 percent potassium (K) and a tensile value of the rebonded sand of 68 psi. When compared with multicycle sand (used multiple times), a reclamation time of six minutes provided a higher

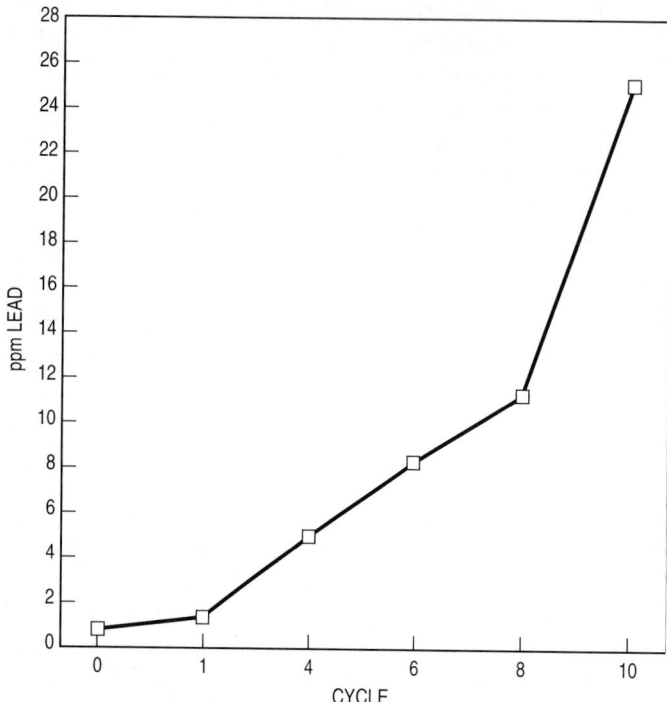

Figure 41-6. Increasing leachate values for lead in thermally reclaimed sand. (*Adapted with permission of Michigan Technological University from* Development of a Steel Foundry Waste Sand Reclamation Process, *September 1992, p. 81.*)

LOI value but the same value for K. This time, however, tensile value was only 39 psi, i.e., 43 percent lower. Frequently, the claim is made that LOI values higher than those for new sand are responsible for poor bonding. This is clearly disproved by the third line of the table. The same multicycle sand, this time processed for eight minutes (an increase of 33 percent), but simultaneously heated to 250°F, provided a slightly lower LOI value (−8.2 percent), the same value for K, but a bonding strength of only 30 psi (−23 percent). This proves that there is no correlation between LOI and bonding strength. There is also no correlation between percent K and bonding strength if percent K is expressed as mass percent. Obviously, the distribution of the contaminant over the grain surface and its reactivity are important. A temperature as low as 250°F can influence this. Despite a longer attrition time, the deleterious effect of thermal treatment is so great that bonding strength falls. This suggests that reclamation of chemically bonded sands must consider the binder composition. The presence of any inorganic element (even if it is part of an organic compound such as an organic lead salt), can seriously affect rebonding if the sand is thermally processed.

Pollution Prevention and Foundries   **677**

Conclusions with respect to bonding cannot be based on a first-cycle test but must take process equilibrium (steady state) into consideration.

41.3.3 Reclaim Chemically Bonded Sand for Use in Green Sand

Chemically bonded sand enters the green sand system automatically in most foundries due to the mixing of sand from collapsed cores with the clay-bonded sand at shakeout. This is normally not associated with a reclamation process. However, it has lately been suggested that instead of adding new sand to such a system, the core scrap should be particulated and used in lieu of new sand. European research[9] points out that such additions not only reduce the stream of discarded sand but actually have beneficial effects due to the closing of the "pores" on the surfaces of sand grains (see Fig. 41-1). This "sealed" sand is easier to reclaim for reuse in the core room if the bentonite layer is not in direct contact with the silica surface but is separated by the layer of residual resin. Reclamation of such waste sands (core scrap) can be accomplished by relatively simple attrition methods. This effect had been investigated in the United States 10 years earlier and an extensive report was provided.[4] The feasibility of this approach has been confirmed, with the following aspects being important:

- Sand must be properly particulated. Complete binder removal is neither necessary nor desirable.

- Sand must be "metered" into the green sand system. The introduction of "slugs" must be avoided.

- A slight extension of the mulling cycle to accomplish proper coating of these resin-coated grains with bentonite may become necessary.

41.3.4 Reclaim Green Sand for Use in Chemically Bonded Systems

This is the most difficult (and expensive) reclamation process. Successful technology was developed in Japan approximately 10 to 15 years ago. The process included drying and preparticulation of the spent sand, calcining at high temperature, and postscrubbing to remove residual inorganic contaminant not removed by the calcining process. As environmental pressures increased in the United States and Europe, this process was adopted. However, some major differences were not considered:

Despite its high cost, the process was economically justified in Japan because most Japanese foundry sands have to be imported and new sand cost is in excess of $100/ton, much higher than the price in North America. The high cost of this process has prevented large-scale introduction in both Europe and the United States.

Japan uses the shell process extensively. This binder system is much less sensitive to residual inorganic contaminant from clay-bonded systems than are the cold box and hot box systems, predominantly used in the United States.

Japanese foundries mixed high-quality imported sands with angular sands of Japanese origin. The intensive attrition applied after high-temperature calcination resulted in substantial attrition losses, necessitating high new-sand additions. Losses of 20 percent and more are common and a European system reports approximately 30 percent new sand requirement. The "sounder" North American sands withstand attrition considerably better, and yields as high as 90 percent or more are possible. However, the remaining contaminant requires additional new sand for dilution. This, then, puts the whole process approach into question.

Low-temperature processing has therefore been suggested. The objective is to remove residual clay from the sand grain surface. As the earlier table from the EPRI study showed, the higher the temperature, the larger the amount of clay fused onto the sand grain. Low-temperature processing prevents this interaction between clay and base sand. Indeed, a low-temperature process has been successfully applied in a United States foundry for over 10 years.[10] More recent research has confirmed the viability of this approach.

Research with respect to the possibility of reclaiming green sand to coreroom quality was performed by the AFS.[2] The report presents results from three different cases, A, B, and C. The rebond data reported in Case A are averages from three different reclamation systems, two applying high-temperature calcination and one applying low-temperature processing. Case B reports results achieved with the low-temperature system only. The properties of the two refuse sands were as follows:

	Case A	Case B
Grain fineness AFS GFN	58.2	61.1
AFS clay (%)	1.66	3.2
Loss on ignition (%)	0.76	0.85
Acid demand value ADV	5.3	3.54
pH	7.4	7.73

SOURCE: Adapted from G. Good et al., *Total Sand Reclamation*, American Foundrymen's Society, Inc., Des Plaines, Ill., July 1987, pp. 42, 47–48.

It is obvious that sand from Case B had a higher level of contamination; hence, it would be expected that it would present greater problems in reclamation and rebonding.

Figure 41-7 shows the results when the reclaimed sand was rebonded with the cold box system. Since the sands processed in Case A and B were different, rebond results are expressed as a percent of the values achieved with new sand to permit comparison. The rebond data from Case B are clearly better than from

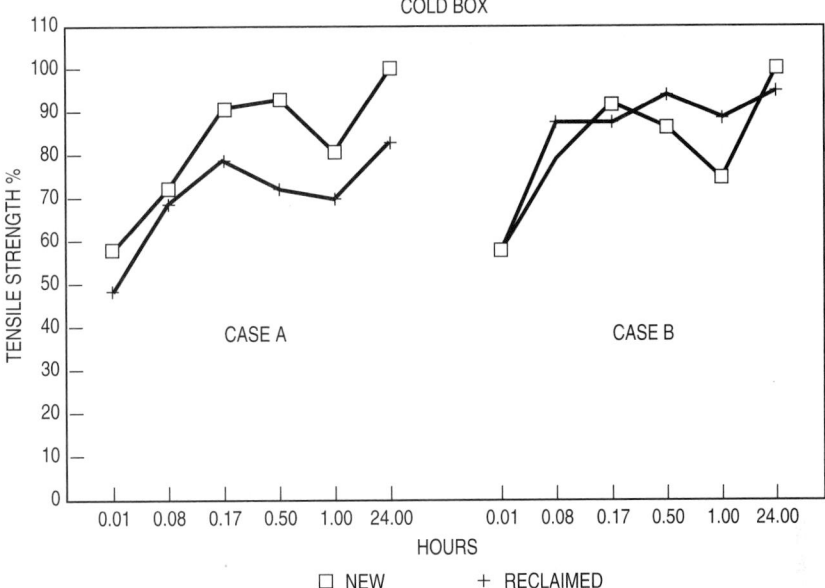

Figure 41-7. Strength data of reclaimed green sand, bonded with cold box. (*Adapted with permission of American Foundrymen's Society from* Total Sand Reclamation, *pp. 47, 53.*)

Case A. This indicates that the low-temperature process must be better than the high-temperature process. The average reported in Case A was depressed by the lower rebond strength obtained with the sand from the high-temperature process.

Figure 41-8 presents a specifically significant aspect—bench life. It has been pointed out that good tensile values can be achieved if the sand mixture is gassed immediately after mixing but that substantial strength loss is experienced if some delay occurs between mixing and gassing (which is unavoidable in most production environments). The data documents the superiority of the low-temperature approach (Case B). However, it also shows that even this process leads to strength loss when compared with new sand. The German Foundry Association VDG issued Recommendation R93 which postulates that reclaimed sands to be used with cold box binder should be gassed one hour after mixing and that the strength value obtained should be at least equal to 80 percent of the value of new sand. This was possible with the low-temperature process applied in Case B; it was not possible with Case A where the average values were heavily depressed by the results of the two high-temperature processes which are included in this average.

Figure 41-9 shows the results when rebonding reclaimed sand with hot box binder and Fig. 41-10 reports results with a nobake binder. In all cases, the advantage of the low-temperature process is confirmed.

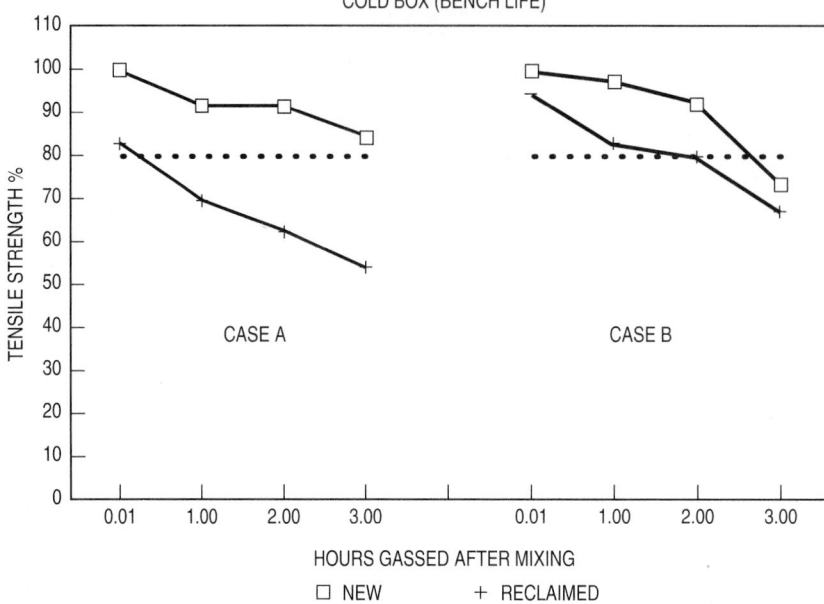

Figure 41-8. Bench life of reclaimed green sand, bonded with cold box.

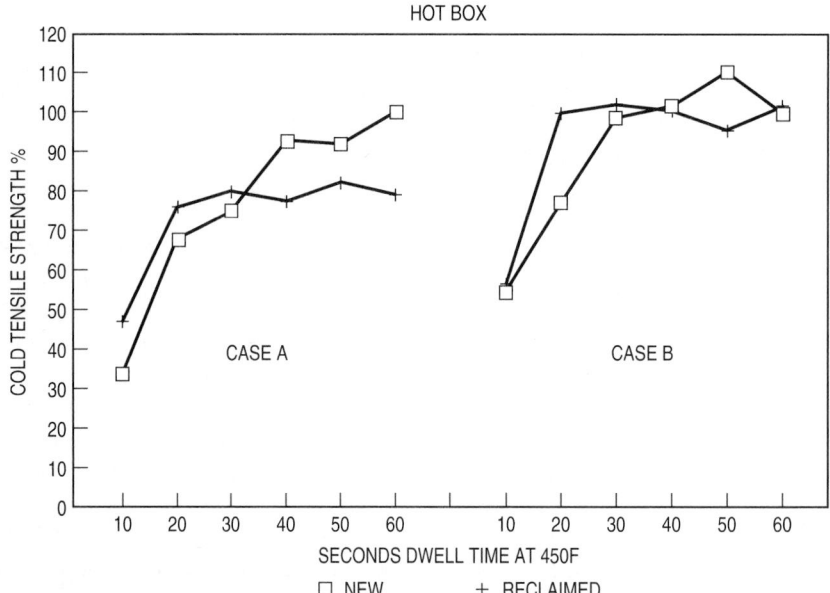

Figure 41-9. Strength data of reclaimed green sand, bonded with hot box. (*Adapted with permission of American Foundrymen's Society from* Total Sand Reclamation, *pp. 46, 52.*)

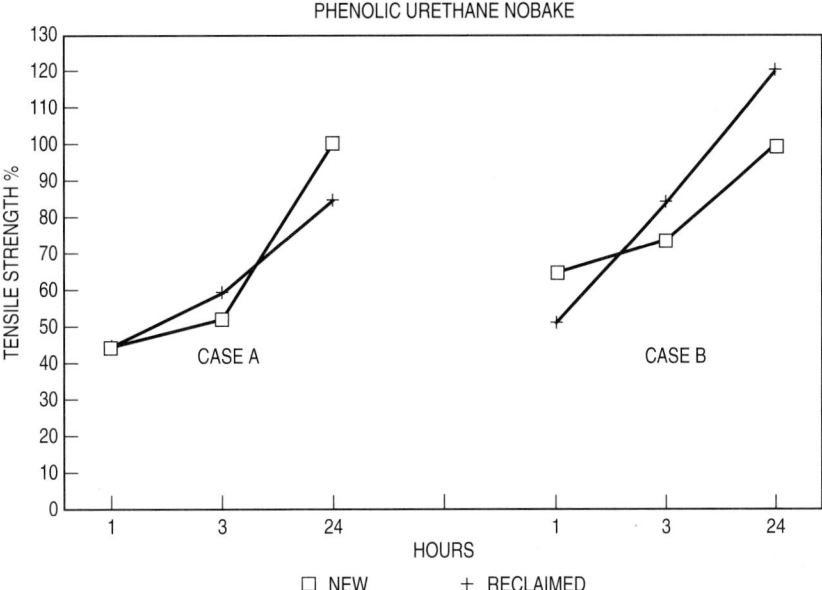

Figure 41-10. Strength data of reclaimed green sand, bonded with phenolic urethane nobake. (*Adapted with permission of American Foundrymen's Society from* Total Sand Reclamation, *pp. 46, 52.*)

To further evaluate this, two members of the AFS Sand Reclamation Committee investigated the effect that high-temperature calcination has on new sand which was never bonded before. The sand selected was a lake sand (which, as new sand, already carries a much higher level of inorganic impurities than does silica sand).

Table 41-1 shows that scrubbing changes the sand properties very slightly when compared to new sand while calcining changes the properties dramatically. Once the sand is heat damaged, even extensive postscrubbing cannot reestablish the original properties.

Table 41-2 shows rebonding results with cold box binder. Sand which was scrubbed only displayed a slight depression of strength values. Sand after calcining, even if scrubbed extensively, became unusable. Table 41-3 confirms the same for rebonding with hot box binder.

There are undoubtedly reclamation requirements which cannot be satisfied with attrition alone. However, operating at lower temperature levels (flash calcining or *thermal embrittlement*) offers a solution which is more appropriate for the situation in North America with heavy emphasis on cold box and hot box binders versus the Japanese technology with shell sand.

To avoid costly errors, each individual foundry will have to carefully evaluate its specific situation since there are so many different combinations with respect to metal poured, casting program, sand used, and binders applied.

Table 41-1. Data from New (Unbonded) "Reclaimed" Lake Sand

	New sand	Scrubbed 3000#/hc*	Scrubbed 1500#/hc*	Calcined 1500°F	Calcined & scrubbed 3000#/hc*	Calcined & scrubbed 1500#/hc*
Active (MB) clay (%)	0.0	0.0	0.0	0.0	0.0	0.0
AFS clay (%)	0.8	0.8	0.9	0.3	0.6	0.6
Loss on ignition (%)	0.38	0.31	0.31	0.11	0.08	0.10
pH	7.9	8.2	8.2	11.6	11.0	10.6
ADV	8.7	7.7	9.4	18.8	3.9	2.8
Grain fineness GFN	49.23	51.22	53.76	48.80	55.92	56.72

*3000#/hc means the sand was scrubbed in a pneumatic reclaimer at a rate of 3000 lbs/h per cell. Accordingly, 1500#/hc represents half the throughput, which means twice the retention time—i.e., more intensive scrubbing.

SOURCE: R. Iyer and M. Granlund, unpublished research results presented to the AFS Sand Reclamation Committee 4-S in September 1992.

Table 41-2. Cold Box Bonding Data of "Reclaimed" New Lake Sand

Time after mixing*	Time after gassing†	New sand	Scrubbed	Calcined	Calcined & scrubbed
Immediate	1 min	162	162	27	122
	1 h	262	225	58	170
	24 h	315	273	33	135
1 h	1 min	177	135	0‡	0
	1 h	230	190	0	0
2 h	1 min	162	128	0	0
	1 h	210	175	0	0
3 h	1 min	155	122	0	0
	1 h	170	167	0	0

*Time after mixing sand and binder together.
†Time after gassing the blown tensile specimen.
‡Value of 0 indicates that sand set up, no core could be blown.

SOURCE: R. Iyer and M. Granlund, unpublished research results presented to the AFS Sand Reclamation Committee 4-S in September 1992.

Table 41-3. Hot Box Bonding Data of "Reclaimed" New Lake Sand

Dwell time (sec)	New sand	Calcined	Scrubbed	Calcined & scrubbed
10	188	0	135	0
20	384	0	226	92
30	478	53	425	124
40	486	93	452	129
Hot tensile from 30-sec core	58	0	53	0
Bench life 1 h from 30-sec core	480	0	419	101

SOURCE: R. Iyer and M. Granlund, unpublished research results presented to the AFS Sand Reclamation Committee 4-S in September 1992.

Consideration must be given to the fact that inorganics always lead to accumulations. This means that steady-state conditions must always be considered and first-cycle tests may be grossly misleading.

References

1. D. L. Twarog et al., *Waste Management Study of Foundries' Major Waste Streams: Phase I*, HWRIC Project RRT-16, The Illinois Hazardous Waste Research and Information Center, Champaign, Ill., November 1992.

2. G. Good et al., *Total Sand Reclamation*, AFS Special Report, American Foundrymen's Society, Des Plaines, Ill., July 1987.

3. J. Bachmann, and D. Baier, "Some Aspects of Gas Evolution from Carbonaceous Materials Used in Foundry Molding Sands," *AFS Transactions*, p. 465 (1982).

4. G. Good et al., "Particulated Core Scrap as a Replacement for New Sharp Sand Additions to a Green Sand Molding Line," *AFS Transactions*, p. 217 (1982).

5. D. Lawson, and D. S. Leidel, "Reclaimed Shakeout Sand as New Sand Substitute in Green Sand Molding Lines," *AFS Transactions*, p. 1 (1984).

6. J. W. Kucharczyk, and D. S. Leidel, "Combined Cooling, Reclamation, and Particulation for Improved Green Sand Performance," *AFS Transactions*, p. 13 (1988).

7. D. L. Smith, and D. S. Leidel, "Changing from Thermal Reclamation to `Attrition Only'—a Step Back to the Future," AFS paper 93-010, *AFS Conference*, Chicago, Ill., 1993.

8. S. R. Iyer, and C. K. Johnson, "Reclamation of Phenolic Ester Cured Nobake Sands," *AFS Transactions*, p. 611 (1988).

9. D. Boenisch, "Reclamation of Spent Sands Containing Bentonite—Guidelines for an Economical Process Leading to Minimized Waste," *Proceedings of International Sand Reclamation Conference Novi/MI*, March 1991, p. 211.

10. *Modern Casting*, issue May 1988, pp. 33–34.

Further Reading

Aoki, T., and J. Takeuchi, "Outline of Reclamation Plant for Used Sand from Foundries," *Proceedings of International Sand Reclamation Conference Novi/MI,* March 1991, p. 193.

Bauch, G., and U. Dieterle, "Production Experience with Reclamation of Mixed Spent Sands and Their Use in Core Production," *Giesserei* **79**(3):102 (1992).

Boenisch, D., "Recycling of Spent Core Sands—Part 1: Cold Reclamation of Resin Bonded Sands," *Giesserei* **78**(21):733 (1991).

Leidel, D. S., "Towards the Closed System by Reclaiming Green Sand," *AFS Transactions,* p. 303 (1983).

———"Sand Reclamation—Its Possibilities and Limitations," paper #21, *51st International Foundry Congress,* Lisbon, Portugal, 1984.

———"Myths, Misconceptions and Mistakes in Sand Reclamation," *AFS Transactions,* p. 21 (1988).

Simmons, C. W., and D. S. Leidel, "The Universal Sand Reclaimer: Eight Years of Foundry Experience in Processing Furan and ECP Bonded Sand," *AFS Transactions,* p. 725 (1992).

Stutzman, W., and V. Godderidge, "Single and Multiple Step Sand Reclamation System in Modular Design—Design Concept and Production Results," *Giesserei* **79**(3):110 (1992).

Wesp, S., and W. Engelhardt, "Thermal Reconditioning of Core Sand in an Aluminum Foundry—a Contribution to Environmental Protection," *AFS Transactions,* p. 227 (1991).

42

Pollution Prevention in the Power Generation Industry

Michael S. Callahan, P.E.

Jacobs Engineering Group Inc.
Pasadena, California

Terry Sciarrotta

Southern California Edison
Los Angeles, California

42.1 Industry Overview

Many different fuel and energy sources are used in the generation of electricity. According to the U.S. Bureau of the Census, the net generation of electricity in the United States in 1990 was 2805 billion kWh. More than 50 percent of this electricity was generated by the burning of coal. The second major source was nuclear, followed by hydro, natural gas, and oil. Changes in the pattern of fuel and energy usage that have occurred over the last 20 years are presented in Fig. 42-1 and the data table.

Power plants designed to convert these fuels or energy sources into electricity vary widely in terms of number, size, and design. More than 10,000 power-generating systems were operating in the United States in 1987. Systems ranged in size from small diesel-fired internal combustion engines up to large-scale nuclear and conventional fuel steam generation systems. Data regarding the type, size, and number of electrical generating units operated in the United States is presented in Table 42-1.

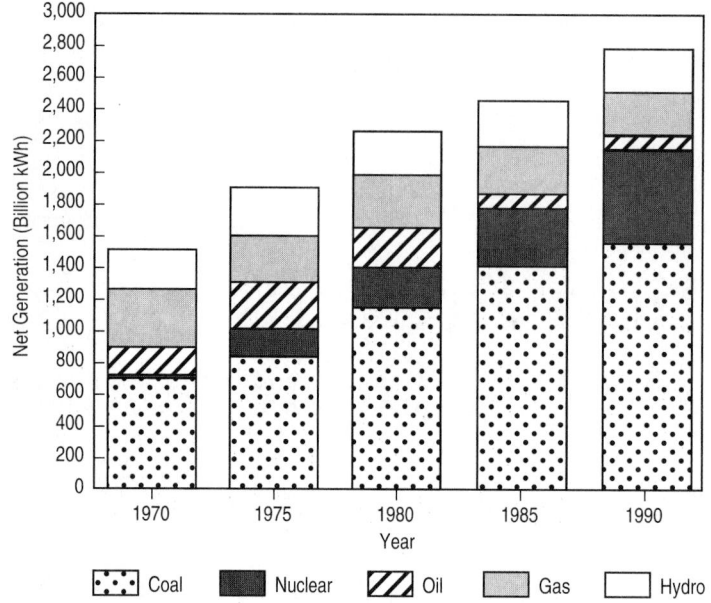

Figure 42-1. Electric power generation in the United States by type of fuel. [*U.S. Bureau of the Census,* Statistical Abstracts of the U.S.: *1991 (111th ed.) and 1992 (112th ed.)*].

Data for Fig. 42-1 Electric Power Generation in the United States by Type of Fuel

	Year				
	1970	1975	1980	1985	1990
Total, billions of kWh	1532	1918	2286	2470	2805
Coal, percent	46.0	44.6	51.0	57.2	55.9
Nuclear, percent	1.4	9.0	11.0	15.5	20.6
Oil, percent	12.0	15.1	10.8	4.1	4.2
Gas, percent	24.3	15.6	15.1	11.8	9.4
Hydro, percent	16.2	15.6	12.1	11.4	10.0

SOURCE: U.S. Bureau of the Census, *Statistical Abstracts of the U.S.:* 1991 (111th ed.) and 1992 (112th ed).

42.2 Description of Major Processes

Most electricity produced in the United States is produced by the conventional steam process. The conventional steam process consists of three stages: generation of heat from fossil fuel combustion (coal, natural gas, or oil), conversion of

Table 42-1. Number of Electrical Generating Units in the United States by Type of Prime Mover

Prime mover	Number of units	Nameplate capacity (millions of kW)	Average capacity (MW/unit)
Steam, conventional			
Less than 100 MW	1198	45.3	38
100–500 MW	926	210.0	227
Greater than 500 MW	313	215.9	690
Total	2437	471.2	193
Steam, nuclear			
Less than 500 MW	6	1.5	250
Greater than 500 MW	103	100.1	972
Total	109	101.6	932
Hydro			
Total	3488	86.0	25
Gas Turbine			
Less than 25 MW	759	14.0	18
Greater than 25 MW	649	38.4	59
Total	1408	52.4	37
Internal combustion			
Less than 5 MW	2783	4.0	1
Greater than 5 MW	134	1.1	8
Total	2917	5.1	2
Other			
Total	47	1.7	36
Overall			
Total	10406	718.0	69

SOURCE: Adapted from U.S. Bureau of the Census, *Statistical Abstracts of the U.S.*, 1991 (111th ed.). Data shown are for the year 1987.

heat into high-pressure steam capable of doing mechanical work, and conversion of mechanical work into electricity.

The heat energy released in the burner from the combustion of fossil fuel is transferred through the walls of the boiler tubes to boil water and generate high-pressure steam. Steam is used as a "carrier" or transfer medium to convert heat energy into mechanical energy in the steam turbine. As the high-pressure steam is allowed to expand, it acquires high velocity and exerts tangential force on the turbine blades, causing the turbine shaft to rotate and power the generator.

To improve the efficiency of the system, the exit steam is condensed as it exits the turbine. Cooling water may be provided by means of once-through or closed-loop systems. Exhaust from the boiler may be used to preheat the combustion air before discharge. Removal of particulate matter (i.e., ash) from the exhaust may be performed by electrostatic precipitators, baghouses, or scrubbers. Particulate controls are not common on plants fired primarily on natural gas.

42.3   Typical Wastestreams

The type and quantities of wastes generated by a power plant depend on plant configuration, size, age, and fuel type. Wastes directly attributable to the power generation process include atmospheric emissions of nitrous oxides, sulfuric oxides, and fly ash; aqueous wastes including boiler blowdown, cooling-water discharges, and feedwater regeneration brines; and solid wastes such as bottom ash and slag. These wastes tend to be generated on a routine basis dependent on the level of power production. Other power-production-related wastes include wet ash from fireside washes and spent acids from chemical cleaning of boilers.

Wastes that are not dependent on the level of power production include coal pile runoff and yard drainage, maintenance wastes such as spent cleaning solutions and solvents, leftover paints and coatings, dirty lube oils and hydraulic fluids from rotating equipment, and sanitary wastes. Nonroutine wastes include contaminated soils due to fuel oil leaks, and wastes from replacement of chromium-treated cooling-tower wood and replacement of asbestos insulation.

Expanding beyond the boundary of the plant, many other types of waste are generated by field operations. Such wastes include spent motor oils, filters, batteries, transmission fluids, and engine coolants from vehicle maintenance. Some of the wastes collected and handled by field maintenance operations include treated wood poles, lead-lined electrical cables, PCB and non-PCB transformers and capacitors, solvents used for electrical equipment cleaning, and dirty rags. Some of these wastes are more hazardous than the wastes routinely generated at the power plant but they also tend to be of lesser volume.

42.4   Pollution Prevention Options

Table 42-2 lists many pollution prevention options applicable to the power generation industry. The major wastestreams addressed include fireside wastes, boiler chemical-cleaning wastes, rotating-equipment maintenance wastes, and facility maintenance wastes. The sections below discuss fireside wastes and boiler chemical-cleaning wastes in more detail.

42.4.1   Fireside Wastes

Fossil fuels, such as coal, oil, and to a lesser extent natural gas, contain impurities which form both gaseous and solid combustion by-products. In addition, incomplete combustion of the fuel and corrosion of the boiler internals can lead to formation of solids. Some of these solids collect within the boiler and are periodically removed as fireside debris. Fireside debris mainly consists of bottom ash and damaged refractory brick which may be contaminated with heavy metals from the ash. Ash removal may be performed manually during boiler

Table 42-2. Pollution Prevention Options for the Power Generation Industry

Options	Comments
Fireside Wastes	
1. Use cleaner fuels.	Natural gas is cleanest-burning fossil fuel, but availability limits widespread use. Cleaner-burning fuel oils and coals are available but may be cost-prohibitive.
2. Use alternative cleaning methods.	Soot blowers and sonic horns may be used to reduce the need for washing. Dry ash has higher potential for reuse. Abrasives may be used but add to waste created.
3. Recycle or reuse fireside wastes.	Lime sludge from treatment may be sold to copper smelters. Vanadium recovery from fuel oil ash may be feasible. Coal ash can be used as a substitute for cement in concrete or as structural fill.
Boiler Chemical Cleaning	
1. Improve boiler water supply.	Regenerate ion exchange resins promptly. Install reverse osmosis equipment ahead of ion exchange system to reduce mineral loading and reduce regeneration frequency.
2. Control boiler water chemistry.	Use hydrazine to control dissolved oxygen and morpholine to control carbon dioxide.
3. Reduce contaminant ingress.	Improve equipment seals to prevent air and cooling-water leaks into the boiler.
4. Base cleaning on fouling.	Use coupons to measure scale buildup, and schedule cleaning accordingly.
5. Use on-line cleaning.	Sodium polyacrylate injection may be used to remove deposits without having to shut down boiler. Further research required.
6. Reuse wastewater.	Wastewater may be used for cooling-tower makeup or as feedwater to ash scrubbers and flue gas desulfurization units. Some pretreatment and/or segregation may be required.
7. Reuse lime sludge.	Sludges from lime treatment of chemical-cleaning wastes may be sold to copper smelters for reuse.
Rotating-Equipment Maintenance	
1. Use high-quality fluids.	While costing more initially, high-quality fluids may last twice as long in service.
2. Routinely monitor fluid condition.	Waste fluid generation can be reduced by switching to a replacement schedule based on fluid condition. Low-cost testing services can provide detailed information.

Table 42-2. Pollution Prevention Options for the Power Generation Industry (*Continued*)

Options	Comments
Rotating-Equipment Maintenance	
3. Use nonleak equipment.	Use dry-disconnect hose couplings, self-sealing lock nuts, and elastomeric flange gaskets to reduce oil leakage. Canned or magnetically driven pumps, bellow valves, and bellow flanges are also effective.
4. Clean and recycle dirty fluids.	Dirty fluids may be cleaned for extended use by small filtration devices. More complex systems may use centrifugation or vacuum distillation.
5. Use waste oils as boiler fuel.	Depends on boiler size, PCB content, and halogen content of the waste oil. Would not apply to synthetic hydraulic fluids.
Facility Maintenance	
1. Eliminate use of hazardous materials.	Major accomplishments have been made in this area, including eliminating the use of PCBs, asbestos insulation, chromium-based cooling-water-treatment chemicals, and leaded paints.
2. Replace TCA and CFCs with non-ODS cleaners.*	Petroleum distillate and D-limonene blends are effective cleaners for electrical equipment. Detergents are good for general purpose cleaning but must be kept out of yard drains and oil-water separators.
3. Use high-transfer-efficiency painting equipment.	Brushes, rollers, and hand mitts are very efficient but labor-intensive. Airless spray is common for field use since a source of clean, dry air is not required.
4. Use an enclosed cleaning station.	Several air districts mandate the use of enclosed gun cleaners and prohibit the spraying of cleanup solvent into the air.
5. Avoid the removal of leaded paint.	Removal of lead-based paint should only be performed when the paint fails to provide adequate protection. Use wet-blasting or vacuum collection devices to prevent the generation of leaded paint dust.
6. Purchase materials in bulk.	Purchase of products in reusable drums and bulk bins can reduce empty drum waste. Mega-drums and bulk bins are returned to the vendor for cleaning and refilling. Use of 55-gal deposit drums is also viable.
7. Require on-site contractors to practice pollution prevention.	Some maintenance wastes generated at the plant are due to activities conducted by on-site contractors. Pollution prevention practices and procedures should be part of the contract.
8. Recycle fuel-oil-contaminated soils.	Depending on soil composition, oil-soaked soils may be recycled as hot-batch or cold-mix asphalt. Thermal desorption and reuse of the clean soil as fill is practiced.

*TCA = 1,1,1-trichloroethane; CFCs = chlorofluorocarbons; ODS = ozone-depleting substance.

overhaul or continuously by means of an ash-handling system. Manual removal is most common at oil- and gas-fired plants, while continuous systems are used at coal-fired plants.

In addition to fireside debris, some ash adheres to the boiler tubes and reduces thermal efficiency. This reduction in efficiency is monitored by measuring the temperature of the stack flue gas. When the flue gas temperature has increased 20 to 30°F or more with respect to a clean boiler baseline value, a fireside wash is scheduled. During the fireside wash, the ash is washed off the boiler surfaces with copious amounts of water.

Wash wastes contain suspended soot, ash, and elevated levels of dissolved iron, nickel, chromium, vanadium, and zinc. In addition to these metals, the wash wastes will contain dissolved alkali (sodium), earth alkali metals (calcium), and associated anions such as sulfates, chlorides, and nitrates. The soot consists of a wide variety of organics including polynuclear aromatics.

The required wash frequency for a natural-gas-fired boiler is generally very low, less than once per year. Oil-fired boilers may require washing two to six times per year. Air preheaters are sometimes washed monthly. For a coal-fired boiler, washing is typically less frequent, unless slagging occurs. Coal ash tends to be less sticky than ash from oil, and hence cleaning methods other than fireside washing may be employed to maintain thermal efficiency.

Processes commonly used for treatment of fireside wastes include coponding with other low-volume wastes, physical-chemical treatment by lime [$Ca(OH)_2$] precipitation, pond evaporation, and sedimentation. Ferrous sulfate is sometimes added before lime precipitation to reduce complexed vanadium compounds and allow subsequent removal during the lime precipitation process.

Combustion ash from fossil-fueled electric power plants is specifically excluded from federal hazardous waste regulations [40 *CFR* 261.4 (b) (4)]. This provision includes boiler-cleaning solutions such as the fireside wash wastes. Since fireside cleaning wastes can often be disposed of as nonhazardous wastes, the direct incentive for pollution prevention may be low. This status may change in the future, however, and the following waste reduction options may be considered.

1. Use of Cleaner Fuels to Minimize NO_x, SO_x, and Ash. This approach may be applicable for utilities that currently use fuel oil or coal. Replacement of these fuels by natural gas will dramatically reduce the amount of deposits formed in the boilers and hence the generation of fireside wastes. Use of natural gas will also reduce the generation of NO_x and SO_x emissions. For facilities faced with the need to install NO_x and SO_x emission controls, conversion to natural gas may be more cost effective than installation of wet or dry flue gas scrubbers. Access to reliable sources of natural gas is highly dependent on plant location and limits the applicability of this option for most facilities.

Burning of cleaner coal is another way to effectively reduce SO_x emissions. The sulfur content of U.S. coals ranges from less than 1 percent to more than 7 percent (Miller, 1986). Low-sulfur coal is mainly located in the western United

States and in the Central Appalachian region. Plants located in the eastern United States switching to low-sulfur coal must not only consider higher transportation costs, but also the different handling and combustion characteristics of the coal. Low-sulfur coal has a lower heating value, higher ash content, and is harder to grind than high-sulfur coal. Conversion could require derating of the boiler, increased fouling and slagging, need to upgrade particulate control systems, and improved coal-handling equipment. Most coal contains 6 to 20 percent ash by weight.

For utilities using fuel oil, specification of low-sulfur fuel oil is a viable way to reduce SO_x, NO_x, and particulate emissions. Fuel oils currently used by utilities in California are hydrotreated at the refinery to reduce sulfur content. Hydrotreatment also reduces the nitrogen and metal content of the fuel oil. A typical distillate oil may contain less than 0.06 percent sulfur and 0.014 percent ash by weight. A heavy fuel oil may contain as high as 4 percent sulfur and 0.5 percent ash by weight. The avoided cost for emission controls must be weighed against the higher fuel costs incurred.

2. Substitution of Alternative Cleaning Techniques for Washing.   As previously discussed, a fireside wash consists of washing the heat-transfer surfaces with copious amounts of water. A number of alternative cleaning techniques are available which may reduce the amount of water used. These options offer the obvious benefit of conserving water, and also avoid the generation of a wet-ash sludge. In addition to being easier to handle, there are more opportunities for the reuse of dry ash (see option 3 below). Alternative fireside cleaning techniques include steam or air soot blowers, sonic horns, manual removal methods, and abrasive cleaning techniques.

Soot blowers using steam or air are widely used to clean fireside fouled heat-transfer surfaces. The effectiveness of soot blowers depends on the location and spacing of the blowers; the number, type, and size of the nozzles; the angle of attack; the frequency of blower operation; and the type of fuel burned. Soot blowers are generally employed for boilers burning fuel with an ash content of 0.08 percent or more. Removed soot and ash deposits may be reintroduced into the combustion process, may be redeposited where they are more accessible for manual removal, or be entrained in the flue gas and removed by the particulate control equipment.

The operation of sonic horns is based on the generation of sound waves, which cause the heat-transfer surfaces to vibrate and dislodge the ash and soot. Sonic horns are most effective at removing light and dry deposits as opposed to sticky and tenacious deposits. Sonic horns are often used on a continuous basis, while soot blowers are typically operated once per shift.

Manual cleaning, such as brushing, sweeping, and vacuuming, is an effective way to minimize the generation of wet ash during fireside washes. On the downside, manual cleaning is limited to readily accessible areas of the boiler and is labor-intensive. This may lengthen the downtime of the boiler considerably. Manually operated abrasive cleaning methods such as sand or walnut shell blasting may be suitable for reducing water use, but each will increase the volume of dry solids generated and may damage the boiler tubes and refractory.

One promising abrasive technique that may be applicable is the use of dry carbon dioxide pellets. Such pellets sublime upon impact and do not add to the volume of material removed from the cleaned surface. While the overall performance of this technique continues to improve over time, equipment costs are much higher and cleaning rates are slower than with conventional abrasive cleaning methods.

3. Recycling and Reuse of Fireside Wastes Off Site. To treat the washwaters generated during a fireside wash, many facilities employ lime (calcium hydroxide). The resulting metal hydroxide sludge often contains an excess of lime, which can be reused by metal smelters. Recyclers often accept lime sludge from the treatment of fireside washes for sale to copper smelters. This option also applies to lime treatment of boiler chemical-cleaning wastes.

Off-site recovery of vanadium contained in the ash from fuel-oil-fired plants may be economically feasible, depending on vanadium content and market value. The market value for vanadium typically fluctuates between $2 and $11 per pound of vanadium pentoxide (V_2O_5). In general, recovery is economical if the vanadium content of the ash exceeds 10 to 12 percent by weight. Some fuel oil ashes, especially slags formed on superheater tubes, may contain as much as 35 percent V_2O_5 or 20 percent vanadium by weight.

For coal ash, which does not contain vanadium, off-site reuse may be practical. Each year, approximately 60 to 70 million tons of coal ash is generated, with 20 to 25 percent being reused in commercial products. Almost 50 percent of the coal ash reused is used as a substitute for cement in concrete. Other major uses include use as a structural fill material and use as road base. The Electric Power Research Institute (EPRI), the American Coal Ash Association (ACAA), and the Edison Electric Institute (EEI) actively promote the reuse of coal ash and research new potential markets.

42.4.2 Boiler Chemical Cleaning

Boiler chemical-cleaning wastes, or waterside cleaning wastes, are generated by the removal of deposits and corrosion products from the waterside of boiler tubes. Even with proper control of the boiler feedwater chemistry, the internal surfaces of the tubes collect deposits over time. These deposits restrict heat transfer and reduce boiler efficiency. Excessive scale and deposit buildup inside the tubes restricts water flow and may cause tube failure due to localized overheating.

One commonly occurring deposit is magnetite (Fe_3O_4). Magnetite deposits are formed by the reaction of elemental iron and steam. These dense deposits form immediately after a chemical cleaning and continue to grow slowly during boiler operation. While magnetite buildup is one of the major causes of reduced thermal efficiency, some buildup is beneficial since it prevents excessive corrosion of the metal. Other sources of contamination include leakage of cooling water into the boiler water system (e.g., from condenser tube leaks) and poor feedwater treatment. Chemical contaminants commonly found in boiler water systems include calcium and magnesium salts, silica, and alumina.

The choice of chemicals used for cleaning depends on the type of deposits to be removed, boiler configuration, and boiler tube metallurgy. Laboratory tests are usually conducted to aid in defining cleaning requirements. Inhibited hydrochloric acid is the most prevalent cleaner for high-pressure drum-type boilers. Ammonium bromate may be used before cleaning with hydrochloric acid if copper deposits are heavy. Other chemicals commonly used include ethylene diamine tetraacetic acid (EDTA) and hydroxy acetic/formic acid (HAF). EDTA is an organic complexing agent used to remove iron and copper deposits. HAF is used to clean once-through boilers and other stainless steel equipment which cannot tolerate exposure to hydrochloric acid.

A survey performed by the Electric Power Research Institute showed a range of two to five years between cleaning of high-pressure boilers typically used in the utility industry (EPRI, 1987). The specific volume of boiler chemical-cleaning waste generated during a cleaning episode depends on boiler size, type of deposits present, type of cleaner used, and number of rinses required for removing the cleaner from the boiler. A 1500-MW coal-fired plant reportedly generates 600,000 gal of cleaning waste and rinsewater during each boiler cleaning (Holcombe et al., 1987).

The composition of cleaning wastes depends on boiler feedwater composition, boiler tube metallurgy, time between chemical-cleaning tasks, task time, and cleaner used. Some of the more prevalent contaminants encountered include iron, copper, aluminum, chromium, manganese, nickel, antimony, vanadium, and zinc. Processes commonly used for treatment of cleaning wastes are similar to those used for fireside wastes and include coponding with other low-volume wastes such as fly ash, physical-chemical treatment, evaporation, and sedimentation.

Treatment of cleaning waste is often required to meet the federal effluent discharge guidelines of 0.2 mg/L for chromium, 1 mg/L for iron, 1 mg/L for copper, 1 mg/L for zinc, and 100 mg/L for total suspended solids. Treatment sludges may be considered hazardous wastes under the criteria of the Resource Conservation and Recovery Act (RCRA), due to the concentration of heavy metals (EPRI, 1987).

Given the increasingly stringent restrictions concerning land disposal of metal-bearing wastes, it is anticipated that the future disposal of these sludges will only become more difficult and expensive. Pollution prevention options which reduce the generation of boiler chemical-cleaning wastes and waste sludges are discussed below.

1. Improve Removal of Minerals from Boiler Feedwater. Boiler feedwater must be treated to remove the harmful impurities that are typically encountered in water supplies. Such impurities include calcium and magnesium salts, silica, sulfates, chlorides, dissolved oxygen, dissolved carbon dioxide, and many other naturally occurring minerals and compounds. Use of improperly treated boiler feedwater can lead to excessive fouling and severe corrosion. Most utilities employ fine filtration to remove suspended matter, followed by chemical treatment and ion exchange (IE).

For optimum performance, the IE resins must be regenerated as soon as their capacity is reached. The need to regenerate the resin beds can be detected by monitoring the electrical conductivity of the treated water (very pure water has low conductivity). A saturated bed will allow some untreated water to break through, which increases the conductivity of the resulting water. During regeneration, both acid and caustic are used, and the resulting waste is a brine solution containing the removed contaminants.

To reduce the frequency of resin regeneration, and reduce the overall volume of acid and caustic required, many facilities have installed reverse osmosis (RO) systems as pretreatment units ahead of their IE systems. The RO system can produce a relatively pure water for feed to the IE system without the use of chemicals. Since the IE system now receives a purer feedstream, the frequency of regeneration required for producing a given volume of treated water decreases. The major drawback to RO systems is that they tend to reject more water than produced. The ratio of treated water produced to feedwater rejected is directly related to the degree of impurity removal required.

2. Improve Control of Boiler Water Chemistry. Boiler water chemistry is critical to safe, economical boiler operation. Improving boiler water chemistry can minimize the amounts of deposits formed in the boiler tubes, thereby reducing the required cleaning frequency and the waste generated. Two common methods include use of demineralized boiler feedwater to prevent the introduction of contaminants into the boiler and use of chemicals to control the level of harmful contaminants.

One such contaminant is dissolved oxygen, which can lead to waterside corrosion. Typically, most dissolved oxygen is removed by the de-aerator, while an oxygen scavenger such as hydrazine is added to remove residual oxygen. Another source of corrosion is carbon dioxide dissolved in the steam condensate. The carbon dioxide forms carbonic acid, leading to acid attack on the metal. In the condensate return system, neutralizing amines such as morpholine may be used to control acid formation. By implementing a rigorous program of contaminant monitoring and control, the need for boiler chemical cleaning can be minimized.

3. Reduce Contaminant Ingress Due to Leaks. Contaminants leaking into the waterside system can result in extensive corrosion and damage. Oxygen ingress in the steam cycle may be a significant cause of corrosion inside the condensers. Measures that can be taken to prevent or combat oxygen ingress include installation of improved seals on steam cycle components, proper material selection and installation to minimize leakage, and improvement in the performance of oxygen de-aerators.

Another major contributor to corrosion in the steam cycle equipment is the infiltration of cooling water, especially river and ocean water used in once-through systems. It is of paramount importance that steam condenser integrity be vigorously maintained to prevent water infiltration. A review of the chemical and mechanical problems that occur with pressurized-water reactors used in nuclear power plants, and how these problems were corrected, was presented by Green (1987).

4. Base Boiler Cleaning on Extent of Fouling. Optimally, the frequency of boiler tube cleaning should be based on the thickness of deposits inside the boiler. Cleaning based on deposit thickness, as opposed to cleaning on a predetermined schedule, tends to reduce the number of cleanings performed over time and thus reduces waste generation. According to a survey performed by EPRI, one California utility monitors both scale thickness and composition by means of small, retrievable test strips placed inside the boiler. Base unit boilers are now cleaned about once every 72 months and recycling units are cleaned once every 48 months. Other California utilities report cleaning schedules as often as once every 24 months (EPRI, undated).

5. Use On-Line Cleaning Techniques. *On-line boiler cleaning* consists of injecting a sodium polyacrylate additive into the boiler feedwater. This additive removes the outer layer of loosely bound magnetite, the cause of most heat-transfer resistance. The dense inner magnetite layer which protects the tube wall from corrosion is not removed, as it is by traditional cleaning methods. Monitoring of a boiler using this technique indicates that the removed iron may remain in suspension since the concentration of iron dissolved in the boiler water did not increase. This may simplify waste management of the blowdown stream, since suspended solids can be physically removed via filtration or sedimentation.

The efficiency of this method for removing magnetite can range from 14 to 50 percent of the magnetite present. The sodium polyacrylate solution decomposes readily in the boiler to form stable carboxylic acids. Boiler blowdown occurs within several hours of chemical injection, and the total cleaning cycle requires one to two days. Advantages of this technique include continued operation of the boiler during cleaning, reduced cleaning duration, use of less hazardous cleaning chemicals, and ease of waste handling. Cost savings for a 300-MW unit were estimated to be $25,000 to $30,000 per year (EPRI, 1987).

There are also several disadvantages to this technique that may limit its use. The process is less rigorous in removing deposits than shutdown cleaning with acids, and hence full efficiency may not be restored. No data are available on copper removal, and there is a potential for introducing contaminants into the steam turbine and other process units.

Another way to perform on-line cleaning is by use of mechanical systems such as brush and baskets or sponge-rubber balls. These systems are quite effective at maintaining heat-transfer efficiency in heat exchangers and condensers. These scrapers are run through the heat exchanger and condenser tubes on a routine basis and dislodge contaminants. System design, as well as the design of the equipment being cleaned, is very important to ensure that the brushes or balls will not become stuck inside the tubing. For a properly designed system, improvement in thermal efficiency and the reduced need for acid cleaning can be substantial. Information regarding system design can be found in Someah (1992).

6. Reuse Boiler Chemical-Cleaning Wastewater. In arid or semi-arid regions, it may be desirable to reuse wastewater streams such as boiler chemical-cleaning wastewaters and fireside washes. Uses for this water may include makeup for cooling towers, fly ash scrubbers, or flue gas desulfurization systems (EPRI, 1987). Reuse of the chemical-cleaning water may require that the final rinsewaters be segregated from the concentrated wastes, and appropriate physical-chemical treatment be employed.

7. Recycle or Reuse Chemical Cleaning Sludges. While the recycling of a chemical-cleaning sludge for its metal content is seldom economical, some copper smelters will accept lime (calcium hydroxide) sludges for reuse. Copper smelters use a large volume of lime in the recovery of copper from ore, and the presence of metals in the sludge does not prohibit its use. Depending on the type of metals present in the sludge, they may be intimately incorporated into the copper matte (the product), the slag, or both.

42.5 Pollution Prevention Successes

Southern California Edison (SCE) has been very successful in reducing the amount of waste going to land disposal. Since 1986, the volume of waste sent to landfill has been reduced by 83 percent at a cost savings of over $3 million per year. This reduction has largely been accomplished via source reduction, recycling, and reuse technologies developed by SCE, a consortium of other California utilities, and EPRI, which is based in Palo Alto, California.

In the area of source reduction, materials substitution is playing a major role. Significant reductions have been achieved by elimination of PCBs from transformer oils and by using nonabestos insulation. Reductions have also been achieved by elimination of the use of lead- and chromium-containing maintenance materials (e.g., paints).

Active programs are also under way to eliminate the use of other hazardous materials. Research is being conducted to identify which products currently in use should be targeted for substitution and to establish guidelines for the selection, testing, and adoption of environmentally preferable products. One major success of the program has been the replacement of 1,1,1-trichloroethane (TCA), an ozone-depleting substance used for cleaning electrical equipment, with a terpene and aliphatic hydrocarbon blend. Additional research has been conducted to determine the biodegradability, toxicity, and recyclability of this replacement solvent. To promote material substitution and the use of safer products throughout the network, a multidepartmental product screening committee has been established.

In addition to materials substitution, materials-tracking and control measures have minimized excess procurement and eventual waste generation. A tracking system based on bar codes has been initiated at a central materials warehouse to reduce the generation of outdated supplies and to promote the shared use of materials among facilities. The bar code system keeps track of a chemical product's fate from procurement to disposal and allows materials managers to locate excess or reusable products.

In addition to source reduction, recycling and reuse technologies have played an important role in SCE's waste management practices. SCE has built a large, central oil-water separator to accept oily wastes from the entire network. Oils and greases from automotive steam rack clarifiers, power generation retention basins, and PCB-free transformer oils are sent to the separator to recover the hydrocarbons for use as cutter stock in the fuel delivery and storage facilities. This reuse activity saves SCE $360,000 per year in disposal fees.

Caustic sludges from the treatment of boiler chemical-cleaning solutions are recycled off site rather than being disposed of in a landfill. In 1986, SCE paid $170,000 for landfill disposal of boiler-cleaning waste. In 1990, SCE paid $31,000 to send 250 tons of lime filter cake to a commercial copper smelter in Arizona. In addition to this 82 percent reduction in disposal fees, state fees and taxes were reduced by an additional $13,000.

The safe reuse of petroleum-contaminated soils is also being practiced. SCE has conducted extensive environmental tests to support the reuse of these soils in cold-mix asphalt. This research has been successful for soils contaminated with fuel oil, mineral oil, and diesel, as well as contaminated soils from a man-ufactured-gas plant. Research is continuing with a variety of metals-contaminated soils. SCE has used this technology to process 1000 to 6000 tons of soil per year since 1989. Cost savings amounted to 80 percent over land disposal costs.

SCE is also developing, in cooperation with the Idaho National Engineering Laboratory and the U.S. Bureau of Mines, a reuse technology that can potentially eliminate the need for land disposal altogether. The technology is called *electrovitrification*, and involves use of an electric arc to burn organics and fuse inorganic wastes into an amorphous mass. The inorganic melt, after cooling and crushing, closely resembles obsidian and can safely be used as construction aggregate. The vitrifier can accept a wide range of feed materials such as asbestos, soil, refractory, catalyst, metals, and cleaning wastes. For some of these wastes, landfilling is the only option currently available. By means of this technology, facilities may one day achieve the ideal of zero hazardous waste discharge.

Bibliography

Electric Power Research Institute (EPRI), *Manual for Management of Low-Volume Wastes from Fossil-Fuel-Fired Power Plants*, EPRI CS-5281, Palo Alto, Calif., July 1987.
———, *Special Waste Management Study*, vol. 2: *Boiler Chemical Cleaning Waste*, EPRI, Palo Alto, Calif., undated.
S. J. Green, "Solving Chemical and Mechanical Problems of PWR Steam Generators," *Chemical Engineering Progress*, vol. 83, no. 7, July 1987, pp. 31–45.
L. J. Holcombe et al., "Methods for Removal of Copper and Iron from Boiler Chemical Cleaning Wastes," *Environmental Progress*, vol. 6, no. 2, May 1987, pp. 74–81.
M. Miller, "Retrofit SO_2 and NO_x Control Technologies for Coal-Fired Power Plants," *Environmental Progress*, vol. 5, no. 3, August 1986, pp. 171–177.
K. Someah, "On-Line Tube Cleaning: The Basics," *Chemical Engineering Progress*, vol. 88, no. 97, July 1992, pp. 39–45.

Further Reading

California Environmental Protection Agency, *Waste Audit Study of the Electric Utility Industry*, Cal-EPA, Department of Toxic Substances Control, December 1991.
Callahan, M. S., and T. Sciarrotta, "Case Study: Pollution Prevention at a Power Facility by Means of Material Management and Product Substitution," *Pollution Prevention Review*, vol. 4, no. 1, Winter 1993–1994, pp. 31–46.
Cavender, M. R., et al., "Optimize Ion Exchange Resins Replacement," *Chemical Engineering Progress*, vol. 88, no. 9, September 1992, pp. 56–59.
Cohan, D., et al., "Beyond Waste Minimization: Life-Cycle Cost Management for Chemicals and Materials," *Pollution Prevention Review*, vol. 2, no. 3, Summer 1992, pp. 259–275.
Electric Power Research Institute (EPRI), *Manual on Chemical Cleaning of Fossil-Fueled Steam Generation Equipment*, EPRI, CS-3289, Palo Alto, Calif., January 1984.
———, *Manual for Management of Low-Volume Wastes from Fossil-Fuel-Fired Power Plants: Spent Solvents Waste Management*, Research Project 2215-1, EPRI, Palo Alto, Calif., March 1989.
Galeucia, G., et al., "Controlling Acid Deposition by Seasonal Gas Substitution in Coal and Oil Fired Power Plants," *Environmental Progress*, vol. 6, no. 3, August 1987, pp. 190–197.
Garret-Price, B. A., et al., *Fouling of Heat Exchangers*, Noyes Publications, Park Ridge, N.J., 1985.
Miller, M., "Pollution Prevention in the Electric Utility Industry," *Pollution Prevention Review*, vol. 2, no. 2, Spring 1992, pp. 153–166.
Pelosi, P. F., and C. J. Cappabianca, "Corrosion Control in Steam and Condensate Lines," *Chemical Engineering*, vol. 92, no. 13, June 24, 1985, pp. 39–45.

Sundin, D., "New Options in Transformer Oils," *Chemical Engineering,* vol. 98, no. 12, December 1991, pp. 125–130.

U.S. Environmental Protection Agency, *Waste Reduction Activities and Options for a Nuclear Powered Electrical Generating Station,* EPA/600/S-92/025. U.S. EPA, Risk Reduction Engineering Laboratory, August 1992.

———, *Waste Reduction Activities and Options for a Fossil Fuel Fired Electrical Generating Station,* U.S. EPA, EPA/600/S-92/061, October 1992.

———, *Waste Reduction Activities and Options for an Electrical Utility Transmission System Monitoring and Maintenance Facility,* U.S. EPA, EPA/600/S-92/063, October 1992.

43

Pollution Prevention in the Fabricated Metals Products Industry

Dale Denny

Brian Frewerd

Tracy Hoefling Pava

Elaine Appley
Concurrent Technologies
Johnstown, Pennsylvania

43.1 Industry Overview

The United States metal fabrication industry is an essential part of both domestic and international economies and plays a key support role in the appliance, automotive, defense, electronics, furniture, and other assembly industries. Identified as Standard Industrial Classification Code (SIC Code) 34, the industry processes and manufactures a wide range of metal components including cans, cutlery, hand tools, general hardware, ordnance, forgings, stampings, and structural metal products.

The industry's environmental compliance problems arise from increasingly restrictive discharge limitations and from the product phaseout of ozone-depleting chemicals (ODCs) as mandated in the 1990 Clean Air Act Amendments. Hazardous raw materials in some metal fabrication operations are regulated under the Occupational Safety and Health Act (OSHA). Some facility discharges are regulated by the Resource Conservation and Recovery Act (RCRA). The major pollutants of concern are volatile organic compounds (VOCs), ozone-depleting compounds (ODCs), hazardous air pollutants, heavy metals, acids, and oils.

43.1.1   Pollutant Characterization Study

The extent and the nature of the metal fabricating industry's environmental discharges were determined from analysis of the U.S. Environmental Protection Agency's (EPA) 1990 Toxic Release Inventory (TRI). TRI is a database containing comprehensive information on ground, air, and water discharges. The following material summarizes the significant information contained in the TRI.

More than 2900 individual metal fabrication facilities identified a total discharge of 195,000,000 lb of reportable chemicals nationwide. There were 106 individual chemical species or groups of compounds included in the TRI listing. The chemicals discharged in quantities greater than 1 million lb are listed in Table 43-1. These 22 compounds account for 94 percent of the total chemicals released. Thirteen are solvents (137 million lb), three are acids (26 million lb), and the remaining six are metals or metal compounds (17 million lb). An asterisk is located next to the ODCs for which production is being discontinued. The industry is pervasive with discharges of greater than 1 million lb in at least 33 states. Ohio, Illinois, Texas, and California are the only states with total discharge values greater than 10 million lb and no state has greater than 11 percent of the national total.

Most discharges go to ambient air. As shown in Fig. 43-1, over 40 percent of the TRI pollutants are discharged from stacks, point air sources, and another 25 percent are released as fugitive air emissions. Eighty percent of the total air emissions have been identified as candidates for regulation under various provisions of the Clean Air Act. The amounts sent to deep well injection, surface waters, land disposal, and publicly owned water treatment facilities (POWT) is small. Approximately 30 percent of the wastes are sent off site for treatment, disposal, or recycle.

Over 30 companies were interviewed by telephone to identify the general class of sources for the pollutants and to determine the approach industry was taking to deal with the regulatory compliance issues. Five common parts-processing operations were identified as major pollutant discharge sources: (1) stripping, (2) cleaning, (3) painting, (4) inorganic surface treatment, and (5) inorganic surface finishing. The first three operations are sources of VOC and ODC discharges, while the last two are sources for acids and metals. Most companies are currently dealing with compliance through end-of-pipe or add-on control systems. Very few companies have established pollution prevention programs.

Table 43-1. Chemical Listing

Substance	Amount (lb/yr)
Glycol ethers	24,730,245
1,1,1-trichloroethane*	24,091,271
Xylene (mixed isomers)	18,732,507
n-Butyl alcohol	13,558,995
Methyl ethyl ketone	12,934,845
Sulfuric acid	12,266,912
Toluene	11,354,278
Trichloroethylene	10,919,027
Hydrochloric acid	9,993,273
Zinc compounds	8,307,487
Acetone	4,684,016
Dichloromethane	4,175,908
Nitric acid	4,155,061
Methyl isobutyl ketone	3,834,677
Nickel	3,693,765
Freon 113*	3,624,162
Tetrachloroethylene	3,404,762
Zinc (fume or dust)	2,573,685
Chromium	2,259,946
Copper	1,497,629
Chromium compounds	1,406,117
Ethylene glycol	1,122,054

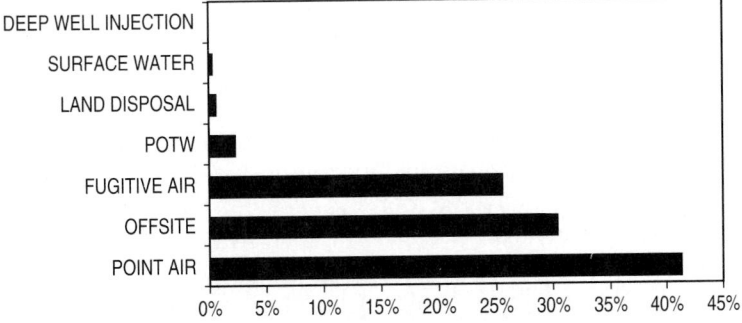

Figure 43-1. Total pollution breakdown to media.

The following material contains a section for each type of process described here. Each section contains a description of the conventional manufacturing operations, the process function, and associated environmental problems. A figure is included in each section that identifies alternative materials and technologies to the conventional processing equipment. Each of the alternative technologies is described briefly and the significant operating characteristics are noted. Each section concludes with an example of successful pollution prevention. Because of the prevalence of end-of-pipe treatment systems in SIC 34, the prevention examples are from other industry sectors. The technologies cited could be used in the metal fabrication industry.

43.2   Stripping

Stripping in SIC 34, small parts manufacture, is used as the first step in reworking out-of-specification products. Defective molded rubber forms and poor quality organic coating applications are removed by mechanical or chemical means. Environmental problems arise from the applied chemicals and removed coatings. Chemicals create air emission problems because they contain VOCs and ODCs. The removed material results in solid waste and may be hazardous, depending on the contents of the paint.

43.2.1   Conventional Technologies

Abrasive blasting is a technology currently used in industry to physically remove paints and other organic coatings from metallic and nonmetallic surfaces. This technology uses mechanical energy to hurl particles at high speed to remove the paint film. The particle media may be composed of sand, glass, plastic, and steel shot. Blasting is generally used on large stationary workpieces or small pieces that cannot withstand heat or chemical stripping. The main environmental problem associated with this process is solid waste disposal. The waste is composed of the media particles and materials removed from the part surface. In some instances, the materials removed from the surface may be RCRA hazardous waste.

Manual grinding processes are used to physically remove very tenaciously adherent organic or inorganic coating with hand or power tools which abrade, cut, or scrape away the film. This process is labor-intensive and is not always practical for large or irregularly shaped parts. The environmental problem associated with grinding is direct worker exposure to the abraded material during the removal process. A second major environmental problem is disposal of the abraded material, which may be classified as hazardous waste.

Chemical stripping processes remove the organic coatings by direct application of chemicals to the coated surface. The solvents soften or dissolve the coatings, and are then usually scraped away or otherwise mechanically removed. Chemical stripping processes are used when the workpiece to be stripped is uneven, has many small crevices, or is not amenable to mechanical removal. The major environmental problems associated with chemical stripping are high emissions of VOCs, ODCs, explosion or fire hazards, and solid or liquid waste disposal.

43.2.2 Pollution Prevention Options

Table 43-2 contains a summary of the significant performance characteristics of replacement technologies relative to those used in conventional practice. Replacements include blast substitutes, high-pressure water, high-energy light, and organic solvents. The developmental status of each is characterized as commercial, pilot scale, or research. Commercial technology is being used in production operations. Pilot scale means large-scale tests have been completed but the process is not in a commercial application. A process in research status is either still a concept or has been subjected to limited small-scale testing. These definitions apply to all the replacement technology descriptions.

Media substitution replaces environmentally unacceptable media used in blasting with environmentally acceptable media such as plastics, wheat starch, CO_2 pellets, and bicarbonate of soda stripping (BOSS). The environmental

Table 43-2. Stripping Replacement Technology Characteristics

Replacement technology	Conventional technology	Replacement status*	Coating removal performance	Operating simplicity	Relative capital cost	Relative operating cost	EHS risk
Blasting substitutes	Abrasive blasting	C	Same	Same	Same	Same	Same/Less
	Chlorinated solvents	C	Same/Less	Better	Less	Less	Less
High-pressure water	Abrasive blasting	P	Better	Same	More	Same	Less
	Grinding	P	Better	Better	More	Same	Less
	Chlorinated solvents	P	Better	Same	More	Less	Less
High-energy light	Abrasive blasting	R	Better	Worse	More	More	Same/Less
	Grinding	R	Better	Worse	More	More	Same/Less
	Chlorinated solvents	R	Better	Worse	More	More	Same/Less
Organic solvents	Abrasive blasting	C	Better	Better	Less	Less	Same/Less
	Chlorinated solvents	C	Same/Less	Same	Same	Same/Less	Less

*C = commercial, P = pilot scale, R = research.

advantages are specific to each media type, but can include biodegradable waste, treatable wastestream, and improved worker health and safety. Most of the replacement media listed here are presently used in commercial practice.

High-pressure water stripping is another replacement technology. High-pressure water stripping removes paint using mechanical energy by spraying a stream of high-pressure water at the surface of the part. The advantages of this process include a readily available media (water), an easily treatable waste-stream, and no fume or hazardous waste production. A disadvantage of this process is that robotics is usually required for application due to the extremely high pressure of the water stream, or where the base material is susceptible to damage.

High-energy light uses optically directed beams of photon energy emitted by lasers or flash lamps (typically xenon lamps) to ablate the paint. High-energy light removes paint with decreased operating cost, minimal wastestream, and low possibility of material damage. Disadvantages of this process are high capital cost and precision robotics requirements.

Organic solvent stripping replaces chlorinated solvents by utilizing completely organic compounds to soften and remove paint. Organic solvents reduce or eliminate the environmental impact of VOCs associated with chlorinated solvents. Disadvantages of this process include determining the specific applicability of the organic solvent to the substrate or coating, longer softening times required, explosion and fire hazards, and treatment of the waste solvent and paint mixture.

43.2.3 Pollution Prevention Success

A shipyard was forced to stop using hydroblasting and solvent stripping as paint removal techniques because liquid wastes generated were prohibited by state environmental law. The facility replaced hydroblasting and chemical solvent stripping techniques with CO_2 pellet blasting. They were able to limit waste production to the material actually removed from the ship hulls, thereby reducing the hazardous wastestream by at least 65 percent. In addition, paint removal rates were higher and material costs went down. Further, there was less undercoat damage and the CO_2 pellet blast system did a more complete job of paint removal.

43.3 Cleaning

Cleaning, as used in SIC 34, requires the removal of contaminants such as oil, grease, grit, and metal chips from metal parts. This process prepares the part for further treatment, such as painting or plating, by affording better adhesive properties as well as increasing the quality of the finish. The technologies currently utilized by industry include solvent, semi-aqueous, and aqueous cleaning.

43.3.1 Conventional Technologies

Solvent cleaning removes contaminants from metal parts with chlorinated solvents via a variety of mechanisms such as spray, wipe, vapor degreasing, or immersion. These processes do not require any rinsing or forced drying. Chlorinated solvent cleaning is applicable for a wide range of substrates and soils. The processes are relatively easy to operate but have high VOC emissions, health, and safety risks.

Semi-aqueous cleaning can be used on a wide variety of substrates and contaminants by utilizing both organic solvents and water in a series of tanks. The part is first cleaned in a pure organic solvent to dissolve the soils. It is then rinsed with water to remove any residual soil and/or solvent. Forced drying is common practice for this process to prevent flash oxidation and to increase production rates. Although typical environmental issues associated with this process include VOC emissions and hazardous wastewater discharges, the resulting environmental threat is less than that of conventional solvent cleaning.

Another conventional cleaning technology is aqueous cleaning, which uses a water-based cleaning solution ranging from pure water to combinations of water, detergents, saponifiers, surfactants, corrosion inhibitors, and other special additives. The solution is usually more than 95 percent water and is applied to parts via spraying or immersion, followed by rinsing and forced drying. Aqueous cleaning requires more scrutiny in selecting a system that will accommodate the more peculiar substrates and soils. Some disadvantages of this cleaning method are its high water consumption rate and its hazardous wastewater discharges. Aqueous processes apply to a wide range of products and are environmentally safer than the chlorinated solvent processes.

The environmental issues arising from these conventional processes are ODC and VOC emissions and hazardous liquid waste generation. Ozone-depleting solvents such as 1,1,1-trichloroethane and CFC-113 will be banned from production under the Montreal Protocol. Fugitive emissions from chlorinated solvent processes such as trichloroethylene, perchloroethylene, and methylene chloride are regulated as hazardous air pollutants. Hazardous wastewaters are treated to meet environmental discharge limits.

43.3.2 Pollution Prevention Options

Table 43-3 contains technology alternatives that can be substituted for conventional cleaning processes. These replacement processes include improved aqueous and semi-aqueous systems, nonchlorinated organic solvents, supercritical fluids, pressurized gases, and plasma cleaning. Also shown are characteristics of these replacement technologies as compared to their respective conventional technologies.

Conventional aqueous systems use immersion or spray techniques for cleaning. Immersion cleaning can be improved with supplementation of a mechanical or ultrasonic agitation. An advantage of improved aqueous cleaning is zero dis-

Table 43-3. Cleaning Replacement Technology Characteristics

Replacement technology	Conventional technology	Replacement status*	Range of substrate applicability†	Operating simplicity	Relative capital cost	Relative operating cost	EHS risk
Improved aqueous systems	Chlorinated solvent	C	Less	Worse	More	More	Much less
	Semi-aqueous	C	Less	Worse	Same	Less	Less
	Aqueous	C	Same	Same	More	More	Less
Improved semi-aqueous systems	Chlorinated solvent	P	Same/Less	Worse	More	More	Much less
	Semi-aqueous	P	Same	Same	More	Same/More	Less
Organic solvents	Chlorinated solvent	C	Same	Same	Same	Same	Less
Supercritical fluids	Chlorinated solvent	P	Less	Same	More	Less	Less
Pressurized gases	Chlorinated solvent	P	Less	Better	Less	Less	Less
Plasma cleaning	Chlorinated solvent	P	Less	Worse	More	More	Less

*C = commercial, P = pilot scale, R = research.
†Refers to the relative number of substrates that the replacement process can accommodate.

charge of wastewater through recycling. Although greater care must be taken when selecting cleaning agents for particular substrates and soils than with conventional solvent cleaning, an improved aqueous system should be used, whenever possible, as a replacement available for all conventional technologies.

Improved semi-aqueous systems remove soils in the same manner as conventional semi-aqueous cleaning. Improvements in this process include better filtration to remove contaminants from wastewater, as well as recycling of the organic solvent. This process is the next best alternative. Whenever possible, improved aqueous cleaning should be chosen first.

Organic solvents are being used to replace chlorinated solvents for cleaning parts. Replacement solvents include terpenes, isopropanol, n-methylpyrrolidone, certain hydrocarbon mixtures, and perfluorocarbons. Favorable properties in choosing a solvent include low volatility, low viscosity, ozone compatibility, high solvency power, biodegradability, and efficient drying characteristics. Replacement solvents are applied in the same fashion as chlorinated solvents by spray, wipe, vapor degreasing, or immersion. Although organic solvents have less of an environmental impact than chlorinated solvents due to reduced or eliminated VOC and ODC emissions, they require greater care in the selection of the solvent to the particular substrate or soil. They can also pose a fire and explosion risk or tropospheric ozone problem. This process should be utilized only if improved aqueous or improved semiaqueous processes cannot be utilized.

Pressurized gas cleaning uses clean, dry, inert gas or air which is fed to a pressurized gas gun to physically remove the contaminant from the substrate surface. Advantages of this process are low capital cost and the fact that non-flammable gases are generally used. However, this technology may not be effective in removing all soils and it may damage the substrate.

The remaining alternatives are not readily available commercially but are being developed for metal-cleaning applications. A brief description of each follows.

Supercritical fluids are applied at temperatures and pressures above their critical point to remove contaminants from parts. CO_2 is the most commonly used fluid in this process due to its environmentally benign nature and wide availability. One advantage of this process is that it is compatible with stainless steel, copper, silver, porous metals, and silica. It also leaves no solvent residue after cleaning and has low operating costs. However, capital costs are very high.

Plasma cleaning is a process that utilizes an electrically charged gas containing ionized atoms, electrons, highly reactive free radicals, and electrically neutral species to remove soils. Plasmas can be used in a wide range of temperatures and pressures. The advantages of this process include low operating costs and lessened disposal costs. However, initial capital costs can be high.

43.3.3   Pollution Prevention Success

A degreasing facility employing vapor degreasers and cold cleaners to clean metal and electronic parts explored the possibility of implementing alternative methods of degreasing due to increasing environmental regulations concerning the release of ODC and chlorinated solvents. This facility implemented an aqueous cleaning alternative to vapor degreasing and cold cleaning. Aqueous defluxing of circuit boards reduced CFC emissions by nearly 500 kg/yr. Replacement of wash rack solvent cleaning with steam cleaning reduced emission of CFC-113 by nearly 1000 kg/yr.

This facility was able to achieve a 59 percent reduction in CFC-113 emission, a 30 percent reduction in 1,1,1-trichloroethane, and a complete reduction of CFC-11 through the use of the aqueous cleaners and equipment modifications.

43.4   Painting

Paint application is a finishing step in the manufacturing process that imparts decoration, corrosion, and oxidation protection to finished items. The overall manufacturing process is illustrated as follows:

strip → clean → inorganic surface treatment → paint application → cure

Currently, environmental problems caused by the industry's painting operations are large amounts of VOC emissions. Solvent carriers in paint formulations contain VOCs that are emitted while the paint is being applied and while it is curing. These VOC emissions give rise to costly vapor recovery or treatment systems. Also, paint application transfer efficiencies are typically poor, less than 50 percent. The resulting paint overspray creates environmental, health, and safety problems for industry and must be collected and disposed of as hazardous waste.

43.4.1 Conventional Technologies

Solvent spray methods of paint application are attractive to industry because of low equipment costs and ease of operation. These methods utilize a hand-held or automated gun to atomize and project the paint onto the part surface. However, a major deficiency with solvent spray systems is their inability to transfer a substantial portion of the paint to the part. Typically, conventional solvent spray transfer efficiencies range from 30 to 40 percent. This inefficiency leads to larger amounts of VOC emitted per part painted. Generally, the overspray is collected by contact with water and the resulting mixture must be disposed of as either solid waste or hazardous waste, depending on the paint components. New spray equipment such as airless and air-assisted airless guns, while better, have not been able to alleviate the overspray problems associated with conventional solvent painting. However, high-volume/low-pressure (HVLP) guns are a potential alternative which increase the transfer efficiency to the 70 to 90 percent range, considerably lowering environmental problems.

Industry uses electrostatic spray methods of paint application because of its ease of operability. In this process, atomized paint particles from a hand-held or automated spray gun are positively charged with static electricity, causing an attraction between the paint and the grounded metal part. The spray guns used are similar to those used in solvent spray systems and include air, airless, and air/airless rotary bells and discs. Electrostatic spray systems have higher initial capital costs than solvent spray, but paint transfer efficiencies are increased to the 60 to 80 percent range. This greater efficiency translates to decreased emissions as compared to solvent spray methods of application. However, overspray is still produced and requires cleanup and disposal of potentially hazardous waste.

Dipping/flow applications are also employed in the painting of metal parts. In dipping, the part is immersed in a tank of a liquid coating, then withdrawn and allowed to dry. During flow coating, the part receives a shower of coating in a paint booth, followed by drying. Another dip/flow coating method is curtain coating which is a high-speed painting process using a revolving applicator roller to coat a flat part. Dipping/flow methods have the same low degree of operating difficulty as solvent spray but transfer efficiencies are greater than 90 percent. Since a great deal of paint is efficiently transferred to the part, VOC emissions come mostly from paint evaporation. The number of applications for dip/flow methods of painting is limited because of the inherent difficulty in controlling the coating thickness.

43.4.2 Pollution Prevention Options

Pollution can be prevented by implementing alternate processes (powder coating system and electrodeposition) or materials (high-solids solvent and water-based paints) which increase transfer efficiencies and eliminate hazardous substances in paint. Table 43-4 contains technology areas that can be substituted for conventional painting processes. Included are characteristics of replacement technologies as compared to conventional technologies. A further explanation of the replacements is also presented.

Water-based paints utilize water as the solvent and can be substituted for most solvent-based paints. Application is nearly identical to solvent spray systems, meaning that only minor equipment changes are required and transfer efficiencies are the same. These similarities result in similar operating and capital costs when compared to solvent spray or dip/flow systems. Another advantage of these paints is that surface operation is about equal to that of both solvent spray and dip/flow systems. Also, water-based paints adhere well to aqueous-cleaned substrates, which could eliminate environmental problems in preparation steps. Cleanup of these paints is easier since they are water soluble, reducing or eliminating the need to dispose of solid wastewater. Shortcomings associated with water-based paints are their limited resistance to wear and corrosion on some steels and aluminum. Also, forced-air drying may be required in humid areas.

High-solids solvent paints contain 50 percent or more solids by weight and have similar properties as solvent-based paints, thus making substitution easy. The paints are applied at a high velocity using conventional spray or electrostatic guns, which requires no equipment changes. Because application equipment remains the same, capital and operating costs are unchanged. Since transfer efficiencies are not much greater than conventional spray methods, wastewater

Table 43-4. Painting Replacement Technology Characteristics

Replacement technology	Conventional technology	Replacement status*	Surface preparation required	Operating simplicity	Relative capital cost	Relative operating cost	EHS risk
Water-based spray	Solvent	C	Same	Same	Same	Same	Less
	Dip/Flow	C	Same	Same	Same	Same	Less
High-solids solvent	Solvent	C	Same	Better	Same	Same	Less
	Electrostatic	P	Same	Same	Same	Same	Less
Powder coat	Solvent	C	More	Same	Higher	Less	Less
	Electrostatic	C	More	Same	Higher	Less	Less
Electrodeposition	Solvent	C	More	Same	Higher	Less	Less
	Electrostatic	C	More	Same	Higher	Less	Less
	Dip/Flow	C	More	Same	Higher	Less	Less

*C = commercial, P = pilot scale, R = research.

discharges and VOC emissions, though reduced, are still a problem. Also, high-solids solvent paints can require long drying times and provide nonuniform coatings which can limit production capabilities.

Powder-coating systems apply coating as a powder onto the parts without the use of solvents. Dry powder particles from a spray gun are given an electric charge, which creates an attraction to the grounded part. This electric potential holds the powder on the part until it is oven-cured. The chief advantage of this process is a transfer efficiency in the 95 to 100 percent range; hence, VOC emissions are virtually nonexistent and no wastewater is discharged. Cleanup and recycling of the minimal overspray is easy since the paint is a solid. This process has a high initial capital cost, but a low operating cost, making it most beneficial for high-volume production lines. However, more surface preparation may be required to achieve a uniform electric potential and a high-quality coating.

Electrodeposition (E-coat) methods of painting immerse the part to be coated in an aqueous bath containing ionized paint materials. A current is run through the part, causing the paint to deposit on the surface. After being withdrawn from the bath, the part is drained and is then cured in a conventional oven. Capital costs are higher because equipment is more complex than that required for conventional painting processes. However, operating costs are lower due to the fact that an aqueous solution is used. Because water is the carrier, air emissions are extremely low, thus eliminating the need for a vapor recovery unit. These high equipment costs and low operating costs make E-coating beneficial for large-volume production lines. If the part is sufficiently cleaned prior to immersion, a uniform coating will form even in highly recessed areas of the part. In addition, E-coat processes are very effective with typical transfer efficiencies ranging from 95 to 100 percent.

Changing the curing process also reduces environmental problems, since considerable amounts of VOCs are emitted during the curing process. Conventional drying techniques use either air or heat. Radiation-curable coatings use IR or UV light to dry the coating. These coatings do not use or contain the organic solvents that cause problems. Instead, reactive monomers are applied to the part surface, then exposed to radiation such as UV or IR light. While the part is under exposure to radiation, reactive cross-linking occurs. Another alternative curing process is the use of thermoplastics. Thermoplastics contain non-cross-linking polymers that flow at elevated temperatures and solidify at an ambient temperature without a change in chemical composition. Since neither type of coating contains organic solvent, VOC emissions are greatly reduced.

43.4.3 Pollution Prevention Success

A metal parts painting facility that was utilizing manual paint application methods was having difficulties complying with emission regulations. Their paint system utilized solvent paints and was manually operated. Problems included inconsistent paint coverage, release of VOCs, generation of paint sludge waste (which represents a time and material loss and had to be disposed of as hazardous waste), and various disposal problems.

This plant installed a 100 percent solids electrostatically applied powder coating system. This system was fully automated, produced a uniform finish, had higher transfer efficiency, and reduced overspray. The little overspray that did occur was easily recycled. VOC emissions were completely eliminated, thereby eliminating environmental problems. Overall, both production capability and transfer efficiency increased. Also, part quality and appearance improved, resulting in fewer rejects, which also reduced the need to rework parts.

43.5   Inorganic Surface Treatment

Inorganic surface treatments fall into two categories: chromating and phosphating. They are utilized by industry to improve the surface properties of a metal for specific needs, including corrosion resistance, better paint adhesion, and overall durability. These processes are batch operations in which the metal workpiece is either immersed in a series of chemical baths or sprayed with the specific material. The environmental problems associated with inorganic surface treatment include the release of heavy metals and acids into the wastestreams.

43.5.1   Conventional Technologies

Chromating is a process which uses hexavalent chromium and proper pH control to deposit a protective film on metal surfaces. This process works well on a variety of metals and alloys. The film is usually deposited through immersion and requires a high degree of preliminary surface cleaning to prevent imperfections in the film. The chromate film is resistant to dissolution in water, thus forming a mechanical barrier which resists corrosion. This property is very important in applications where the metal part is exposed to marine environments. The equipment used in chromating is relatively inexpensive; however, the process releases high levels of hexavalent chromium and acids into the wastestream.

Industry also utilizes phosphating to pretreat metal parts. Zinc phosphate is the predominant phosphate used in commercial applications. The film is deposited on the metal surface through either immersion or spray application. It is very important that the metal part be free of grease, scale, or rust prior to treatment with the phosphate in order to ensure proper coverage and adhesion. Good phosphate films generally provide excellent corrosion resistance; however, phosphating does not provide satisfactory films on stainless steel, Monel, and certain high-alloy steels. Following the phosphate-coating step, the part is rinsed to remove remaining chemicals, then rinsed again with chromic acid. The waste severity of this process is less than with chromating, but heavy metals and acids are still discharged.

43.5.2 Pollution Prevention Options

The replacement technology described in the following section can be substituted for conventional inorganic surface treatment processes. Table 43-5 indicates how the replacement technology compares with each conventional technology.

Material substitution eliminates or reduces heavy metal and acid discharges by using solutions that do not contain chromium. Instead of acid, baths contain molybdate, tungstate, and permanganate solutions. Some alternatives such as permanganate and molybdates impart greater corrosion resistance than traditional phosphating methods. However, some aluminum applications result in poor adhesion.

43.5.3 Pollution Prevention Success

A facility using current chromate coating systems generated many forms of hazardous wastes (airborne and wastewater). These wastes exceeded EPA, DOT, and OSHA regulations; therefore, alternative conversion coatings were sought to eliminate these wastes. The facility replaced the current chromate coating process with a dry-in-place, waterborne emulsion conversion coating. The product is completely chromium free and adaptable to heat spray applications. The permanganate-based product is considered environmentally safe at ambient temperatures. In fact, small residual amounts of the potassium salt that is deposited as the primary coat are desirable in the industrial wastewater treatment because it aids in treatment of other common wastewater contaminants.

Replacement of the chromate conversion coating process resulted in a coating that was strongly bound to the metal surface, thus it provided good corrosion resistance. Uniform coverage was easily achieved, as indicated by a color indicator. The unpainted coating does not rub off on workers hands or emit toxic fumes upon welding. The parts can be painted immediately, which lessens the time required to complete the part finishing.

Table 43-5. Inorganic Surface Treatment Replacement Technology Characteristics

Replacement technology	Conventional technology	Status of replacement*	Surface preparation required	Operating simplicity	Relative capital cost	Relative operating cost	EHS risk
Material	Chromating	C	Same	Same	Same	Same	Less
substitution	Phosphating	C	Same	Same	Same	Same	Same

*C = commercial, P = pilot plant, R = research.

43.6 Inorganic Surface Finishing

Inorganic surface finishing improves surface qualities of metals by decreasing corrosion, oxidation, and wear. In some cases it also provides decoration. Usually the coating is metallic and is applied using one of the following processes: electroplating, cladding, anodizing, case hardening, or dipping.

43.6.1 Conventional Technologies

Although electroplating is most commonly considered a decorative finish, it can be applied to a variety of metals and alloys to provide protection from corrosion. In electroplating, metals are deposited on the parts using either an electric potential (electrolytic) or a controlled chemical reduction (electroless). In electrolytic plating, the metal is designated as the cathode, while the material being deposited is the anode. There is a wide variety of coating materials that are plated on common substrates such as steel, brass, and zinc. Electroless plating replaces the electric potential used in electrolytic plating with a chemical reducing agent which reacts at the surface to allow deposit of metals. Electroplating processes typically require a high degree of surface preparation and emit high levels of hazardous waste.

Cladding is a mechanical plating process in which the coating is metallurgically bonded to the part surface by combining heat and pressure. An example of cladding is a quarter: the copper inside is heated and pressed between two sheets of molten nickel-alloy, thus bonding the materials together. Cladding is used to deposit a thicker coating than electroplating. It requires less preparation and emits less waste than electroplating; however, equipment costs are higher.

Anodizing electrolytically forms a stable film on metal surfaces and is sometimes used as a pretreatment for painting. The metal substrate is designated the anode; it reacts with the electrolyte (commonly a salt or acid) to form an insoluble metal oxide. The reaction continues and forms a thin, nonporous layer that provides good corrosion resistance. Equipment costs for anodizing are lower than the cost for cladding equipment. Wastes severity with this process is greater than with cladding but less than with plating because of the presence of organic acids.

Case hardening is a metallurgical heat treatment which modifies the surface of the metal. The metal is heated and molded, then the temperature is quickly quenched. This process results in a very hard surface on a ductile metal. The case-hardening method is used to produce Samurai swords, because the hardened surface can easily be sharpened but the sword remains pliable. This method has low waste emissions and requires a low degree of preparation. Operating difficulty and equipment costs are about the same as for anodizing, although case hardening imparts improved toughness and wear.

Dip/galvanized coatings are applied primarily to iron and steel to protect these base metals from corrosion. During the dipping process, the part is immersed in a molten metal bath commonly composed of zinc compounds. The metal parts must be free of grease, oil, lubricants, and other surface contaminants prior to the coating process. Operating difficulty and equipment cost are low, which makes dipping an attractive coating process for most industrial applications. However, dipping does not always provide a high-quality finish.

43.6.2 Pollution Prevention Options

Existing replacement technologies, if used properly, can avoid most of the environmental problems associated with conventional coating methods, such as emissions of heavy metals, acids, and caustics. Table 43-6 contains alternatives to conventional processes.

There is not one replacement technology appropriate for every coating need. However, each of the replacement technologies described here is appropriate for many applications and does reduce environmental impacts.

Chemical vapor deposition (CVD) is generally used to coat irregularly shaped parts. This method is used to increase the strength, purity, and density of the coated material. CVD creates a coating by depositing a chemical vapor on a substrate through thermal or chemical reaction. Variations of CVD include

Table 43-6. Inorganic Surface Finishing Replacement Technology Characteristics

Replacement technology	Conventional technology	Replacement status*	Surface preparation required	Operating simplicity	Relative capital cost	Relative operating cost	EHS risk
Chemical vapor deposition	Plating (electrolytic)	P	Same	Better	Higher	Same	Lower
	Plating (electroless)	P	Same	Better	Higher	Same	Lower
	Cladding	P	More	Better	Higher	Higher	Same
	Anodizing	R	Same	Better	Higher	Same	Lower
Ion beam techniques	Plating (electrolytic)	P	Same	Better	Higher	Higher	Lower
	Plating (electroless)	P	Same	Better	Higher	Higher	Lower
	Cladding	R	More	Better	Higher	Higher	Lower
	Case hardening	C	More	Better	Higher	Higher	Lower
	Dip/Galvanizing	R	More	Better	Higher	Higher	Lower
Plasma and spray	Plating (electrolytic)	P	Less	Same	Higher	Same	Lower
	Plating (electroless)	P	Less	Same	Higher	Same	Lower
	Cladding	P	More	Same	Higher	Same	Lower
Evaporation	Plating (electrolytic)	R	More	Better	Same	Same	Lower
	Plating (electroless)	P	More	Better	Same	Same	Lower
	Cladding	C	More	Better	Higher	Same	Lower
	Dip/Galvanizing	R	More	Better	Higher	Higher	Lower
Material substitution	Plating (electrolytic)	C	Same	Same	Same	Same	Lower
	Anodizing	P	Same	Same	Same	Same	Lower
	Case hardening	P	Same	Same	Same	Same	Lower

*C = commercial, P = pilot plant, R = research.

atmosphere, low-pressure, and plasma-assisted methods. In atmospheric vapor deposition, the part must be thoroughly cleaned before chemical deposition. The purity of the finished CVD product leads to extensive use in optical coating applications. Proper material selection results in reduced emissions and high-quality finishes. Plasma-assisted CVD can be used on parts which may be deformed by the high temperatures of atmospheric CVD. Performing CVD under vacuum produces more uniform coatings with higher purity. Waste products originate from the low material utilization and the hazardous components in the pretreatment stages.

The ion beam technique implants metal ions under an inert environment into the surface of metal parts. This type of process includes ion vapor deposition (IVD), ion implantation, and ion-beam-enhanced deposition (IBED). The IVD aluminum coating process works well in replacing cadmium coatings in that it alters conductivity, hardness, wear, and corrosion. However, coating control can be difficult. Ion implantation processes work by altering the chemical composition near the surface of the metal to enhance conductivity, hardness, wear, and corrosion resistance. This option is attractive due to low operating temperatures and minimal waste generation. IBED, used when lattice defects are present, is more effective in coating than direct implantation.

Plasma and spray replacement technologies combine a gas with metallic powder, heat to a molten state, and apply the coating to the part surface. Plasma processes include detonation gun and vacuum plasma. Detonation gun uses a combustion flame spraying process, which uses oxygen and acetylene. There is a limited range of materials that can be used as coatings or substrates. With employment of a vacuum, properties of the coats can be improved. High temperatures limit applications in many cases. The plasma vapor process's main pollutant is overspray; however, generation of ozone via the plasma arc is also of concern. Metal and acid emissions are generally reduced using this method. Thermal spray processes include HVOF (high-velocity oxygen fuel), slurry, and plasma arc. HVOF applies metals and other materials in powder form through flame spraying. A high degree of surface preparation is required in some cases to increase bond strength. Slurry applications are commonly used in the coating of jet engine components. A slurry of aluminum powder, ceramics, binders, and solvents is applied to the metal substrate through spraying or dipping techniques, dried, and fired. This method requires minimum surface preparation. Plasma arc techniques allow a variety of materials to be applied through spraying because high temperatures are obtainable. Porosity of coatings deposited by this method can be varied to suit the desired application. Operating costs are higher due to the need to replace parts that have been thermally damaged more frequently. Care must be taken to protect the workers from fumes and metal dust emitted in this process.

Evaporation methods coat metal through evaporation of the metal from a molten pool using an electron beam. Evaporation methods include electron beam reactive deposition, thermoionically enhanced evaporation, and coaxial electron beam. This method is widely used in the computer industry for coating memory boards. Thermoionically enhanced evaporation methods produce improved films as a result of the increased fraction of ionization. Evaporation requires more surface preparation than conventional techniques do, is generally easier to operate, and reduces emissions; however, operating this system can be difficult because pressure must be closely regulated.

As shown in Table 43-6, material substitution of toxic substances within a process is a viable solution for pollution prevention. Substitutions include trivalent chromium for hexavalent chromium coating, cyanide-free bath solutions in electroplating processes, and zinc-nickel for cadmium in electrolytic processes. The preparation, operational simplicity, and costs are comparable to conventional technologies, but environmental risks are lower. Replacements are suitable in most applications; however, wear and corrosion resistance is sometimes less desirable.

43.6.3 Pollution Prevention Success

A manufacturer of chrome-electroplated farm equipment was seeking a way to lower environmental problems. Hexavalent chromium discharges were of particular concern. The farm parts are subjected to excessive wear. A replacement process would have to provide durability equal to or greater than that of the chrome-plated part.

A firm specializing in ion implantation was contacted as a possible alternative. Metal parts were ion-implanted and given to the manufacturer for testing. The parts were tested and found to be equally resistant to wear and corrosion as a chrome-electroplated part. Environmentally hazardous waste discharges are zero.

Bibliography

Bishop, C., and G. Loar, "Practical Pollution Abatement Methods for Metal Finishing," *Plating and Surface Finishing*, February 1993, pp. 37–39.

Lancaster, Fred, Concurrent Technologies Corp., personal communications with Elaine Appley, MTS Technologies, Inc., August 18, 1993.

"Metal Finishing," *Guidebook and Directory* **91**(1A), January 1993.

"Metal Finishing," *Organic Finishing Guidebook and Directory* **91**(5A), May 1993.

Pirrotta, Dick, Concurrent Technologies Corp., personal communications with Elaine Appley, MTS Technologies, Inc., August 18, 1993.

Roberts, Thomas, Concurrent Technologies Corp., personal communications with Brian Frewerd and Tracy Pava, MTS Technologies, Inc., August 19, 1993.

Spearot, R., "Review of Waste Reduction Technologies for the Metal Finishing Industry," *Finishers' Management*, National Association of Metal Finishers, June/July 1993, pp. 34–37.

Standard Industrial Classification Manual, Executive Office of the President, Office of Management and Budget, 1987.

Stinner, Bob, ISM Technologies, Inc., personal communications with Brian Frewerd, MTS Technologies, Inc., August 18, 1993.

"Surface Cleaning, Finishing, and Coating," *Metals Handbook,* 9th ed., vol. 5, American Society for Metals, Metals Park, Ohio, 1990.

Surface Preparation and Finishes for Metals, Society of Manufacturing Engineers, McGraw-Hill, New York, 1971.

U.S. Environmental Protection Agency, *Guides to Pollution Prevention: The Fabricated Metal Products Industry,* EPA/600/7-90/006, 1990.

U.S. Environmental Protection Agency, *Industrial Pollution Prevention Opportunities for the 1990s,* EPA/600/8-91/052, 1991.

U.S. Environmental Protection Agency, *Guides to Pollution Prevention: The Metal Finishing Industry,* EPA/625/R-92/011, 1992.

U.S. Environmental Protection Agency, *Toxic Release Inventory,* EPA/700/C-92/002, 1992.

Villareel, David, Amber Plating Works, Inc., personal communication with Paul Ritchey, MTS Technologies, Inc., July 14, 1993.

The following companies also provided valuable information:

EPRI
Center for Materials Fabrication
505 King Avenue
Columbus, OH 43201

Pollution Prevention Information Clearinghouse (EPA)
401 M Street, SW, PM-211A
Washington, D.C. 20460

44

Pollution Prevention in the Chemical Industry

Frederick L. Moore

Union Carbide Corporation
Danbury, Connecticut

44.1 Introduction

44.1.1 The Chemical Industry

The chemical industry is often termed a *basic industry*. The term is appropriate because most of its products are essential raw materials or inputs to the production processes of other industries. Only a small portion of the output of the chemical industry ends up in finished consumer goods directly. Most chemicals made depend upon petroleum-based feedstocks somewhere in their life cycle. In addition, the front end of the chemical industry also utilizes a variety of different catalysts, many derived from rare metals (see Fig. 44-1). Currently, the U.S. chemical industry is responsible for the largest U.S. trade surplus of any industry (see Table 44-1), even surpassing agriculture.

44.1.2 Energy Requirements

The chemical industry's energy consumption, both for raw materials and energy feedstocks, amounts to 5.4 quadrillion BTUs. That's about 7 percent of total U.S. energy consumption. The industry's energy use currently equals about 2.5 million barrels of crude oil per day and consists of a wide mix of energy sources:

Crude oil and derivatives	45%
Natural gas	30%

Electricity	17%
Coal	6%
Others	2%
Total	100%

The estimated annual energy bill for the U.S. chemical industry is about $24 billion. This represents about 8% of the value of industry shipments. The industry's pro rata share of $50.8 billion of U.S. oil imports in 1991 was $3.8 billion. This supported not only U.S. production of about $290 billion in shipments, but exports of $43 billion and a trade surplus of $18.8 billion.[1]

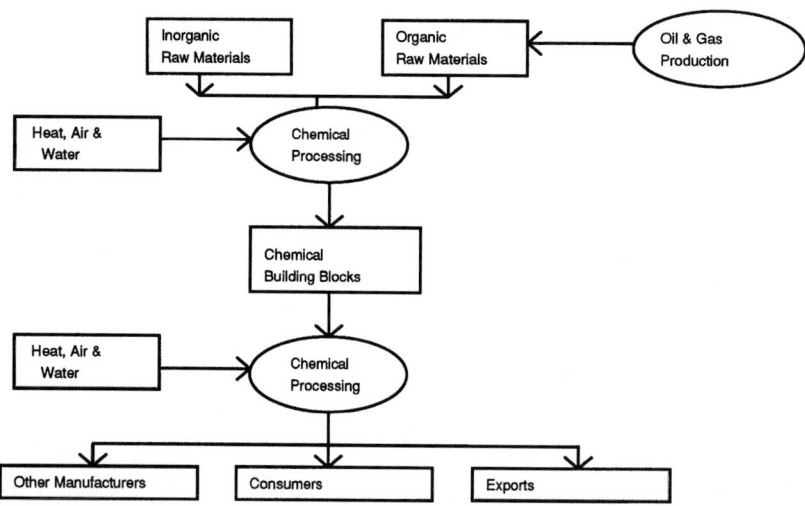

Figure 44-1. Elements of chemical production.

Table 44-1. Key Indicators of Chemical Industry Performance, 1987–1992

	1987	1988	1989	1990	1991	1992
Production index (1987=100)	100.0	105.9	109.2	111.7	111.3	115.0
Chemicals and allied products shipments (billions of $)	229.5	259.7	278.1	288.2	289.0	296.8
Operating rate (%)	81.8	83.9	83.6	82.8	80.3	80.9
Exports (billions of $)	26.0	31.9	37.4	39.0	43.0	44.0
Imports (billions of $)	15.6	20.3	20.7	22.5	24.2	27.7
Trade surplus (billions of $)	9.5	11.6	16.6	16.5	18.8	16.3

SOURCE: U.S. Department of Commerce.

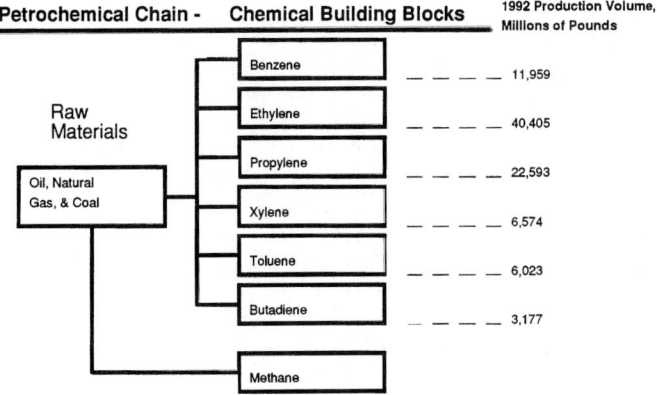

Figure 44-2. Basic building blocks of the chemical industry.

44.1.3   Chemical Chain—The Building Blocks of the Industry

The chemicals shown in Fig. 44-2, derived from petroleum feedstocks, are the basic building blocks of the chemical industry. These basic building blocks are used to produce the vast majority of what the U.S. public has come to accept as part of today's standard of living. For example, one building-block chemical, ethylene, eventually ends up in products such as latex paints, food wrap, garbage bags, plastic bottles, records, video and audio cassette tapes, adhesives, safety glass, pharmaceuticals, cosmetics, fragrances, aspirin, synthetic fibers, detergents and shampoos, brake fluids, heat-transfer and hydraulic fluids, and even the fluid used for storing human hearts awaiting transplant.

44.2   Pollution Prevention—a Review of Chemical Industry Performance

44.2.1   Reduction of Releases to the Environment and Off-Site Transfers

The Chemical Manufacturers Association (CMA) member companies, which represent more than 90 percent of U.S. chemical manufacturing capacity, reduced releases and off-site transfers of chemicals listed in Section 313 of the Superfund Amendments and Reauthorization Act (SARA) in each of the years from 1987 through 1992 (see Fig. 44-3). From 1987 to 1992, releases to the environment decreased 35 percent and off-site transfers decreased 37 percent.[2] (SARA, Title III, Section 313 requires that U.S. chemical plants, among others, provide data to the public that identify specific chemical releases to all environmental media and off-site transfers for treatment and disposal.)

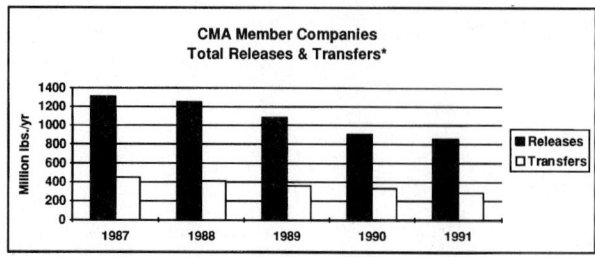

Does not include Energy Recovery in "Transfers" since 1991 was the first reporting year for this new category. Does not include ammonia. because EPA made a major change in 1990 allowing industry to report only the ammonia portion of ammonium sulfate. This chart allows an "apples-to-apples" comparison of the chemical industry.

Figure 44-3. CMA member companies' total releases and transfers. (*Courtesy of Chemical Manufacturers Association, Preventing Pollution in the Chemical Industry: Five Years of Progress, 1993.*)

44.2.2 Pollution Prevention Data, 1991

The Pollution Prevention Act of 1990 required, for the 1991 reporting year and onward, additional data (see Table 44-2) to be collected and reported under SARA Section 313. These new data elements included quantities of waste generated, recycled, and burned for energy recovery. Analysis of these data is complicated by the lack of a standard definition of a *wastestream,* but provides a useful look at the quantities of waste generated at chemical plants. The new data will be useful in long-range thinking as the industry continues to reduce overall risks and has adequate resources to address less risky operations that may include generic waste generation. In addition, the data allow those chemical facilities that were not already doing so to track trends in the generation of waste at their facilities.

For the 1991 reporting year, the pollution prevention wastestream data, as required by the Pollution Prevention Act of 1990, for all Standard Industrial Classification (SIC) Codes reporting under SARA 313 totaled ~37.1 billion pounds as reported by the U. S. EPA. The 1992 data showed an increase in total waste generation of 0.5% or ~200 million pounds total. The U.S. EPA has not

Table 44-2. Chemical Manufacturers Association Pollution Prevention Data, 1991

(In Millions of Pounds)

	On site	Off site	Total
Energy recovery	919.5	180.7	1,100
Recycled	6,293	639.9	6,933
Treated	3,844	208.6	4,053
Released			868
Total	11,056	1,029	12,954

SOURCE: Ref. 2.

yet promulgated a regulatory definition of wastestream and the 0.5% increase should be regarded with caution. Actual chemical releases to the environment for all SIC Codes were down 6.6% between 1991 and 1992.

44.3   Implementing Pollution Prevention

44.3.1   Overview of Opportunities

Figure 44-4 illustrates the life cycle of the chemical chain. The life cycle of chemical manufacturing is important in understanding the driving forces that contribute to voluntary initiatives for reducing risk and preventing pollution. Chemical manufacturing is unique in that the industry spends much of its resources simply converting one chemical into another. The industry relies upon the very reactive nature of chemicals, which also makes them hazardous, to make its end products.

For the chemical industry, where one chemical is converted to another and the chemicals manufactured are a function of market demand, a focus on chemical use cannot provide a reasonable measure of success in preventing pollution. Rather, a focus on releases to the environment and their associated overall risk would provide a better benchmark. As long as a market demand exists for a specific product, the chemicals necessary for its manufacture will be made somewhere, whether in the United States or abroad. Chemical use will be affected only where market demand shifts occur or where actions are taken to

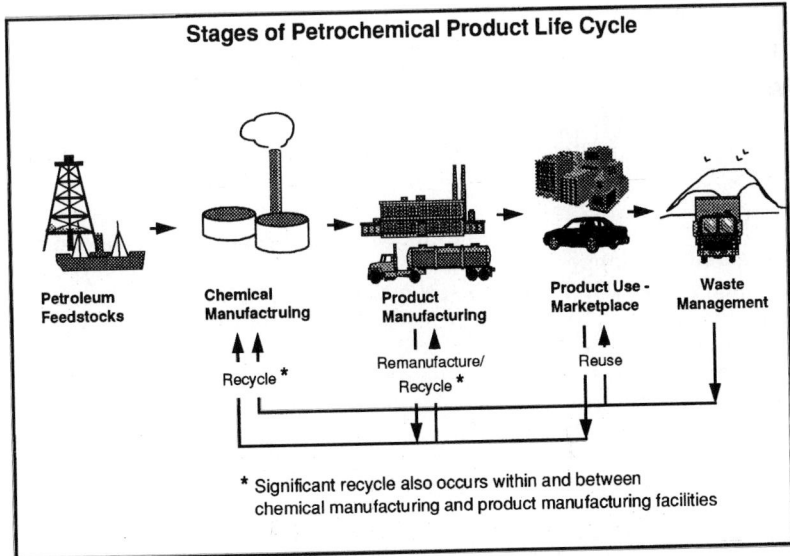

Figure 44-4. Stages of petrochemical product life cycle. (*Adapted from* Green Products by Design: Choices for a Cleaner Environment, *OTA-E-541.*)

regulate or restrict a particular chemical use or product application, as in the case of polychlorinated biphenyls (PCBs) and chlorofluorocarbons (CFCs).

A focus on risk, and on pollution prevention as a tool for reducing risk, will create a market force. The creative tension generated by public sharing of data about releases to the environment, via the EPA's Toxic Release Inventory (TRI), is an example of how the desire to significantly reduce releases in communities where chemical plants operate has become a market force in the eyes of senior management in the chemical industry. They have come to recognize that their ability to operate is a privilege given by the communities in which they operate.

The chemical industry's voluntary initiatives to reduce both releases to the environment and off-site transfers are illustrated by the fact that 100 percent of the CMA member companies originally invited became participants in the EPA's 33/50 Program. The 33/50 Program is a voluntary program designed to reduce the releases of 17 targeted chemicals by 50 percent by the year 1995, from the baseline year of 1988.[3]

44.3.2   A Corporate Strategy for Pollution Prevention

As Frances Cairncross noted,

> What distinguishes the management of the most environmentally adept companies, first and foremost, is that management has drawn up a clear statement of environmental principles and objectives and gotten it blessed by the board of directors. That establishes its high priority credentials. A second earmark is a specific plan and detailed guidelines flowing from the policy statement—management's marching orders, if you will. Finally, the execution of the plan is closely monitored by a senior executive with personal responsibility for the results.[4]

Where to Start?   Dr. Stephan Schmidheiny, a Swiss businessman and the Chairman of UNOTEC, states it well:

> The best people, those who are able to think beyond the end of their day and who worry about the longer-term future, are interested in the ecological behavior of their employer. They will see better chances with a corporation that promises to adopt and to implement sustainable (environmental) growth concepts....Businesses that fail to do so will in the future no longer be in a position to attract and motivate the best people, and the management of such companies will more and more come under pressure from their own employees who are increasingly concerned about environmental issues....Especially in this time of increased reliance on the creative potential of employee empowerment, and when employees are being asked "to do more with less," their feeling toward the company is important.[5]

A key role for senior managers is to provide leadership. The essence here is to "walk the talk." Robert D. Kennedy, CEO and Chairman of Union Carbide, puts it this way: "People who work in organizations know how to read signals....they're going to do only what their bosses indicate is truly important."[6]

In 1988, the CMA's member companies embarked on an initiative called Responsible Care.® The Responsible Care® initiative includes six codes: (product stewardship, employee safety and health, community awareness and emergency response, process safety, distribution, and pollution prevention) and 106 specific management practices under the codes. All member companies must abide by the Code of Management Practices as a condition of membership. The 14 management practices specific to pollution prevention are shown in Fig. 44-5.[7]

Integrating Pollution Prevention. Making environmental management, and especially pollution prevention, a part of a company's existing setup isn't easy. Pollution prevention, in particular, cuts across the traditional management boundaries of R&D and product development, manufacturing, engineering, business management, customer and supplier relationships, and risk management (which looks to episodic risks presented by operations, storage, and transportation, and to chronic risks from the routine releases to the environment associated with the particular manufacturing facility). Many chemical companies have integrated pollution prevention as a tool in their overall risk reduction and risk management strategy. This allows the company to balance environmental risks presented, not only from the generation of waste or release of chemicals into the environment, but from episodic risks presented by chemical processes as well.

In 1990, the U.S. chemical industry's annual costs for all types of pollution abatement were about $3.8 billion. The effect of the Clean Air Act Amendments of 1990 will be to raise total abatement costs to $6.4 billion in 1995 and $11.0 billion in 2005, as expressed in 1990 dollars.[8] It is this extraordinary level of expenditure that makes flexible solutions and provision for regulatory incentives so important.

Pollution prevention is one tool, albeit a very important tool, in reducing risk to employees, communities, and the environment. Risk decisions are often complex and balance one benefit against another. The best answer will often lie in the eye of the beholder, which explains why the Responsible Care® initiative has become so important in establishing a dialogue with communities where chemical plants are located. In 1991, a major chemical-manufacturing facility in Texas was faced with balancing their capital expenditures to reduce risk. While additional opportunities existed for pollution prevention, management decided to reduce the episodic risk presented by a large ammonia inventory necessary for the ammonia-based refrigeration system. The decision was further complicated by the fact that current technology and process requirements called for replacing the ammonia-based system with a CFC (now identified as an ozone-depleting chemical) system. A decision was made to install the new refrigeration system using CFCs, with the knowledge that additional restrictions would likely apply under the Clean Air Act, and that ultimately the system would require replacement by alternate technologies. While it is arguable whether or not this was the best decision, it is clear that the decision makers

1. A clear commitment by senior management, through policy, communications, and resources, to ongoing reductions, at each of the company's facilities, in releases to the air, water, and land and in the generation of wastes.

2. A quantitative inventory at each facility of wastes generated and releases to the air, water, and land, measured or estimated at the point of generation or release.

3. Evaluation, sufficient to assist in establishing reduction priorities, of the potential impact of releases on the environment and health and safety of employees and the public.

4. Education of, and dialogue with, employees and members of the public about the inventory, impact evaluation, and risks to the community.

5. Establishment of priorities, goals, and plans for waste and release reduction, taking into account both community concerns and the potential health, safety, and environmental impacts as determined under practices 3 and 4.

6. Ongoing reductions of wastes and releases, giving preference first to source reduction, second to recycling and reuse, and third to treatment. These techniques may be used separately or in combination with one another.

7. Measurement of progress at each facility in reducing the generation of wastes and in reducing releases to the air, water, and land, by updating the quantitative inventory at least annually.

8. Ongoing dialogue with employees and members of the public regarding waste and release information, progress in achieving reduction, and future plans. This dialogue should be at a personal, face-to-face level, where possible, and should emphasize listening to others and discussing their concerns and ideas.

9. Inclusion of waste and release prevention objectives in research and in design of new or modified facilities, processes, and products.

10. An ongoing program for promotion and support of waste and release reduction by others.

11. Periodic evaluation of waste management practices associated with operations and equipment at each member company facility, taking into account community concerns and health, safety, and environmental impacts and implementation of ongoing improvements.

12. Implementation of a process for selecting, retaining, and reviewing contractors and toll manufacturers, taking into account sound waste management practices that protect the environment and the health and safety of employees and the public.

Figure 44-5. Pollution Prevention Code of Management Practices.

13. Implementation of engineering and operating controls at each member company facility to improve prevention of and early detection of releases that may contaminate groundwater.

14. Implementation of an ongoing program for addressing past operating and waste management practices and for working with others to resolve identified problems at each active or inactive facility owned by a member company, taking into account community concerns and health, safety, and environmental impacts.

Figure 44-5. *Continued*

chose to balance community, employee, and environmental risk in a realistic fashion within their system of prioritizing overall risk reduction.

The concept of risk management is not foreign to the EPA. In May of 1993, the EPA's Office of Pollution Prevention and Toxics stated their organizational "vision"[9] as follows:

I. Promote pollution prevention as a principle of first choice to achieve environmental stewardship throughout society;

II. Promote the design, development, and application of safer chemicals, processes and technologies in the industrial sector of the economy;

III. *Promote risk reduction and responsible risk management practices throughout the life cycle of major chemicals of concern,* and;

IV. *Promote the public understanding of the risks of chemicals and public involvement of information* on toxic chemicals....(*emphasis added*)

The EPA author, in discussing goals to achieve this vision, went on to say,

> The first goal is to provide for an educated public which is cognizant of chemical risks and benefits. The first goal will act as a catalyst for the second: a more environmentally and scientifically literate public which will be an integral part of the regulatory process and help insure that government sets appropriate priorities and makes the best choices on toxics control.[9]

As in the case of the CMA pollution prevention code, dialogue with employees and the community is important in balancing this equation. The sharing of "worst case" hazard scenarios under the accidental release provisions of the 1990 Clean Air Act Amendments will make this even more challenging, as the communities where the industry operates learn even more about chemical plant operations and the potential risks they present. A more informed public will hopefully lead to public policy more in step with the need to balance overall societal risks and available resources.

Another important consideration in developing and implementing a pollution prevention program is finding a way to integrate pollution prevention with the quality process. The quality concept is founded on the philosophy of continuous improvement. If waste is viewed as a "defect," then this concept is applicable to pollution prevention. As part of the quality process, a system of checks or verification may also be needed to ensure the integration of pollution prevention into the corporate and facility infrastructure.

One major chemical company puts the issue of verification this way:

> The objective of the compliance audit program is to provide independent assessment of line management location performance against the following criteria:
>
> - Compliance with applicable internal standards and governmental requirements, and
> - Management systems that assure continued compliance.[10]

This company also announced in 1992 that all the Responsible Care® codes of management practice were being integrated into their corporate health, safety, and environmental standards worldwide.[11] In a case like this, where pollution prevention is integrated into the internal standards of the corporation and these standards are verified by independent assessments, senior management can be assured that it has not only penetrated the organization from a standard-of-performance perspective, but also that management systems are in place to assure its continued application worldwide.

44.4 Looking for Pollution Prevention Opportunities

The chemical industry is perhaps the most difficult industry to "look into" as an outsider. Many chemical plants are large (typically employing 500 people, and sometimes more than 1000), complex (typically involving the manufacture of many different chemicals, and in some cases, literally hundreds of chemicals), and highly integrated (for example, the light gases coming off the front end of an olefins facility are used as fuel for the thermal-cracking furnaces that make the ethylene and propylene used later in the facility; the ethylene can then be used to make dozens of other chemicals, depending upon the processes employed, with recycle streams integrated into various stages of chemical production).

This means that, more often than not, it requires not only an experienced chemical engineer, but one experienced in the particular site's operations, to evaluate complex pollution prevention options. A 1991 EPA publication on pollution prevention examples helps to put it in perspective. Early in the section on the chemical industry, the author notes that

> it proved difficult to obtain process-related pollution prevention information about the [chemical] industry, either because of competitiveness and resulting confidential nature of this information, because of relative lack of candor with an enforcement agency or because options often are specific to a single process, a single piece of equipment, or a single product. Neverthe-

less, the industry has been a leader in the pollution prevention (earlier identified as waste minimization) effort, and has documented innovative if specific changes in a plant, process, or piece of equipment....It appears that more opportunities for reuse exist in complex plants....The chemical industry, on the basis of cost-consciousness, has been very aware of the need to minimize losses of solvents, raw materials, etc.[12]

44.4.1 Designing Pollution Prevention into the Process

Opportunities for cost-effective solutions to pollution prevention problems are most prevalent at the R&D and process engineering stages for new units and plants. After the basic process chemistry and basic system design philosophy are in place, significantly fewer opportunities exist to make cost-effective decisions in the detailed engineering, construction and start-up, operations and maintenance, or dismantling phases of the project. Figure 44-6 illustrates this concept.

Cost effectiveness of instituting pollution prevention measures during the earliest stages of new-facility development is illustrated by the following example:

In 1991/1992, a major chemical plant modified their olefins unit process and not only eliminated the need for methanol, but reduced secondary and fugitive emissions of benzene and methanol by more than 150,000 pounds per year. They redesigned the unit and relocated the furnace gas dryers

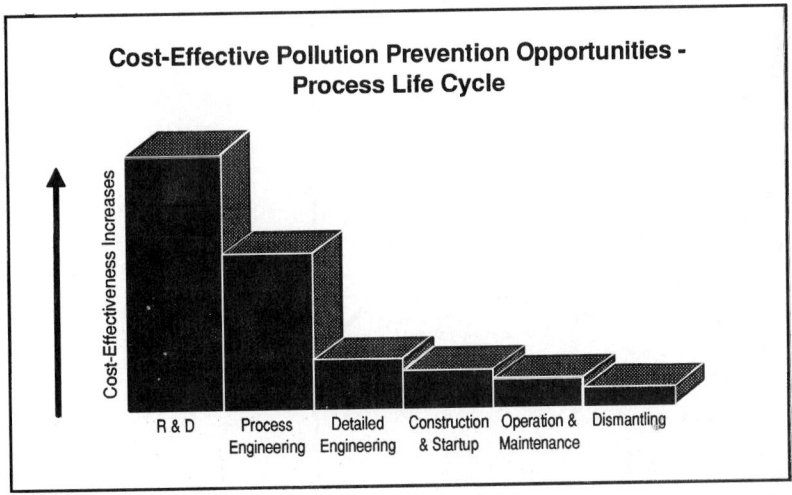

Figure 44-6. Cost-effective pollution prevention opportunities—process life cycle. (*Courtesy of* Designing Pollution Prevention into the Process—Research, Development, and Engineering, *CMA, 1993.*)

upstream of the precooler in a compressed furnace gas process. This process modification eliminated the need for methanol injection at the precooler stage, used to prevent hydrate formation, and allowed the resulting benzene stream to be recycled back to the process (uncontaminated by methanol).[13] This project cost upwards of $7 million with a savings of only about $250,000 per year in methanol purchases. At this rate of return, this particular project would not have been done without the additional benefit of reducing risk to employees and the community from benzene emissions. However, if this design had been built into a "new" unit, there would have been little additional original capital cost and the savings of $250,000 would have been available from day one.

44.4.2 Preventing Pollution— A Checklist of Ideas[14]

Pollution prevention requires that the entire process be assessed for opportunities. Figure 44-7 provides an overview of process elements that should be considered.

While the focus is necessarily on the process, other factors, such as off-spec product, maintenance wastes, leaks, and spills, can also be significant causes of pollution. Table 44-3 outlines several examples of potential sources of pollution and waste. Once the points of waste generation are identified, opportunities for prevention of the pollution must be evaluated. Table 44-4 offers some ideas about where to look for solutions.

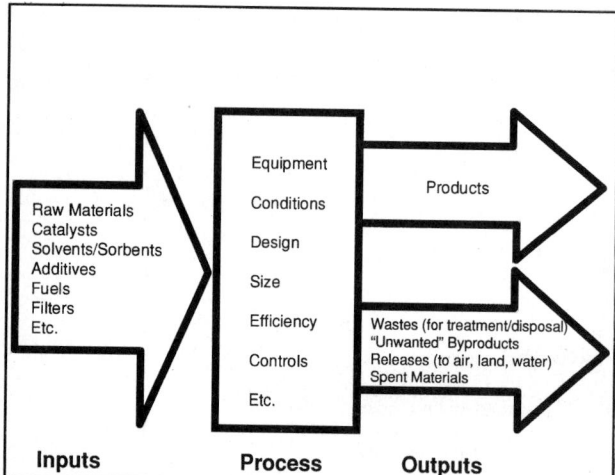

Figure 44-7. Process elements that can be targeted for pollution prevention. (*Adapted from* Designing Pollution Prevention into the Process—Research, Development, and Engineering, *CMA, 1993.*)

Table 44-3. Potential Sources of Pollution and Waste

Potential Sources of Air Emissions

Point-source emissions: stack, vent (e.g., laboratory hood, reactor, storage tank vent), material loading and unloading operations, and others

Fugitive emissions: pumps, valves, flanges, mechanical seals, relief devices, tanks, and others

Secondary emissions: wastewater treatment unit, cooling tower, process sewer, sump, spill and leak areas, and others

Potential Sources of Liquid (Organic or Aqueous) Wastes

Equipment wash solvent and water, lab samples, surplus chemicals, product washes and purifications, seal flushes, scrubber blowdown, cooling water, steam jets and vacuum pumps, leaks, spills, spent or used solvents, housekeeping (pad washdown), waste oils and lubricants from maintenance, and others

Potential Sources of Solid Wastes

Spent catalysts, spent filters, sludges, wastewater treatment biological sludge, contaminated soil, old equipment and insulation, packaging material, reaction by-products, spent carbon and resins, drying aids, and others

Potential Sources of Groundwater Contamination Due to Leaks or Spills

Unlined ditches; process trenches; sumps; pumps, valves, and fittings; wastewater treatment ponds; product storage areas; tanks and tank farms; aboveground and underground piping; loading and unloading areas and racks; manufacturing maintenance facilities.

SOURCE: Adapted from Ref. 14.

After a thorough pollution prevention analysis, there are normally several alternatives. An engineering evaluation of each alternative will include an assessment of technical feasibility and the usual cost factors for an engineering project. However, because of environmental consequences, less obvious environmental cost factors, such as those listed in Table 44-5, should also be considered.

Finally, documentation of pollution prevention results is important. It can be used as the starting point for future studies as well as a measure of progress against management goals.

44.5 Voluntary Pollution Prevention and Regulatory Flexibility

The United States is just now coming to terms with the concepts of how we might promote voluntary initiatives and incorporate regulatory flexibility. One of the most important considerations, from the chemical industry perspective, is how overall risk reduction strategies can become market forces that drive voluntary initiatives. The report of the National Commission on the

Table 44-4. Where to Look for Pollution Prevention Solutions

Inventory Management and Operations

Improved material purchasing, receiving, storage, and handling practices which minimize material that may exceed its shelf life, be left over or not needed, or have the potential for accidental release

Installation of the right equipment and maintaining strict preventive maintenance programs to prevent leaks, spills, or accidental releases

Production scheduling: reducing the number of production changes to minimize waste generated from production transitions or turnover

Equipment

Installation of equipment that produces less waste

Redesign of equipment or production lines to produce less waste and to enhance or permit recovery or recycling options

Improved operating efficiency of equipment

Elimination of sources of leaks and spills

Process

Process changes which optimize reactions and raw material use to reduce waste generation or releases

Input material changes: material purification and material substitution which allow a change in the process to reduce waste or eliminate a hazardous constituent used in the product or during manufacture

Product substitution: developing a new product that is less hazardous in use or ultimate disposal than the original

Additional automation to increase the reliability of the operations and reduce the occurrence of process upsets and production of off-specification material

Recycling

Participation in waste exchanges

Installation of closed-loop systems for in-process recycle

Recycling on-site at other process units or off-site for reuse

Finding new uses for previously unwanted by-products

Segregation of wastes by type to allow for recovery

Reclamation: processing of waste for resource recovery

SOURCE: Adapted from Ref. 14.

Environment of the World Wildlife Fund suggests a broad policy for promoting this goal:

> The government should give high priority to efforts to narrow the gap between public perceptions of risk and expert evaluations of risk. These efforts should include promoting citizen participation....Working with

Table 44-5. Less Obvious Environmental Cost Factors

Direct Costs
Capital for pollution control equipment
Operation and maintenance
Energy
Waste transportation

Indirect Costs (charged to overhead account)
On-site waste management
Insurance
Waste taxes and fees
Increased safety protection

Regulatory Compliance Costs
Permitting
Compliance monitoring and reporting
Waste manifests
Time delays for permits and regulatory approval
Compliance with future regulations

Liability Costs
Penalties and fines
Employee health
Personal injury and property damage
Legal claims and awards
Settlements for remedial action

Less Tangible Costs
Company product image
Increased revenue from enhanced sales
Reduced health maintenance costs from improved employee health
Increased productivity from improved employee relations

SOURCE: Adapted from Ref. 14.

Congress, the Department (Environment) should develop legislation for integrating all pollution control functions as soon as it is practicable.[15]

Today, some of the most relevant impediments exist from society's early attempts "to do things right" rather than "do the right thing." In many cases, the costs of using environmental resources have not been adequately accounted for in our market economy. In the future we must seek to better describe these costs, measure them, and inform the public about the policies that foster correct choices and volunteerism.

Although the TRI has probably been the best attempt at creating an environment for making voluntary and correct choices, it still focuses on a list. As such, it can only show that a potential hazard exists—no relation can be directly drawn to the risk a chemical or group of chemicals presents in a particular setting. Furthermore, without a database that describes all sources of environmental stress (e.g., from transportation, nonpoint sources, and power generation), we have potentially misled the public by providing only a part of the equation that should form their foundation for making decisions about societal risk.

The Clean Air Act Amendments of 1990 are another example of an attempt to promote voluntary pollution prevention actions. However, under the provisions for credits for early reductions, credit was only available back in 1987. If senior management made major voluntary investments in 1986 that now are not available for consideration in this program, they may be inclined in the future to wait until they are sure about the features of legislation that will provide credit before expending resources for pollution prevention.

The World Wildlife Fund report (quoted above) describes the rationale for flexibility: "Comprehensive reform is imperative to re-focus the regulatory system on coherent policies that can bring about sustainable development, encourage environmentally benign technologies, and institute effective incentives for innovation and behavioral change."[16]

References

1. Allen J. Lenz, *The U.S. Chemical Industry Performance in 1992 and Outlook*, CMA, Washington, D.C., January 1993.

2. Chemical Manufacturers Association, *Preventing Pollution in the Chemical Industry: Five Years of Progress*, CMA, Washington, D.C., 1993.

3. Environmental Protection Agency, *The Industrial Toxics Project, The 33/50 Program: Forging an Alliance for Pollution Prevention*, EPA 560-1-91-003, Washington, D.C., March 1991.

4. Frances Cairncross, "How the Best Companies Do It," *ECO Business & the Environment*, June 1993, p. 36.

5. S. Schmidheiny, "The Entrepreneurial Mission in the Quest for Sustainable Development," in *The Greening of Enterprise*, J. O. Williams (ed.), International Chamber of Commerce, Paris, June 1990.

6. R. D. Kennedy, "Achieving Environmental Excellence: Ten Tools for CEO's," in *Prism*, Arthur D. Little, Inc., Cambridge, Mass., 3d Quarter, 1991.

7. Chemical Manufacturers Association, *Pollution Prevention Resource Manual*, CMA, Washington, D.C., 1991.

8. Lenz, *op. cit.*

9. George Bonia, Deputy Director of the U.S. EPA Office of Pollution Prevention and Toxics, letter dated May 27, 1993.

10. Paul D. Coulter, "Union Carbide's Audit Classification Program," in *Corporate Quality/Environmental Management*, Union Carbide Corporation, Danbury, Conn., 1991.

11. Union Carbide Corporation, *1992 Responsible Care® Progress Report*, Danbury, Conn., 1992.

12. Ivars J. Licis, *Industrial Pollution Prevention Opportunities for the 1990's*, U.S. EPA Risk Reduction Engineering Laboratory, Office of Research and Development, EPA/600/8-91/052, Cincinnati, Oh., August 1991, pp. 16–17.

13. Chemical Manufacturers Association, *Preventing Pollution in the Chemical Industry 1987 to 1990*, CMA, Washington, D.C., Spring 1992, p. 40.

14. Chemical Manufacturers Association, *Designing Pollution Prevention into the Process—Research, Development and Engineering*, CMA, Washington, D.C., 1993.

15. World Wildlife Fund, *Choosing a Sustainable Future*, The Report of the National Commission on the Environment, Island Press, Covelo, Calif., 1993.

16. Ibid., p. 15.

45

Pollution Prevention in the Petroleum Refining Industry

Carl H. Fromm

Steven L. White
Jacobs Engineering Group Inc.
Pasadena, California

45.1 Background

The petroleum refining industry is primarily engaged in the manufacture of fuels, lubricants, and petrochemical intermediates using petroleum crude oil as a principal input material. Crude oil is a complex mixture of paraffinic, olefinic, naphthenic, and aromatic hydrocarbons; the carbon content usually varies between 82 and 87 percent by weight and the hydrogen content from 12 to 15 percent by weight. In addition to hydrocarbons, crude oil contains small concentrations of organic and inorganic compounds of sulfur, oxygen, nitrogen, and metallic elements (mainly vanadium and nickel). Dissolved alkaline metal salts found in seawater, such as sodium chloride, are found in the brines which are extracted along with the crude oil from underground deposits.

Petroleum refining is the process that separates the hydrocarbons into fractions through distillation, which upgrades the hydrocarbons through chemical conversion to render them suitable for intended product use and separates unwanted impurities (generally compounds of sulfur, nitrogen, oxygen, and metallic elements) from the hydrocarbons. In general, the sources of waste from the refining process can be categorized as follows:

1. *Losses of oil* (and hydrocarbons in general). Such losses are mainly caused by entrainment, emulsification, and, to a lesser extent, dissolution of hydrocarbons in an aqueous phase in all operations involving contact between hydrocarbons and water, and by inadvertent leakage from equipment and piping.
2. *Consumption of auxiliary input materials,* such as catalysts, solid adsorbents, liquid-absorbing solutions for gas cleanup (such as amines or Stretford solutions), caustic, mineral acids, heat transfer fluids, lubricants, water treatment chemicals, and water.
3. *Irreducible wastes* generated as a result of separation of unwanted impurities in the process of refining; e.g., salts in desalter brine. Note that impurities which originally come with crude oil or are often created during refining are converted to useful and commercially viable products, such as ammonia, sulfur, phenolics, and metals.

Consequently, pollution prevention (P2) as applied to petroleum refining should seek opportunities to reduce waste by preventing oil/hydrocarbon loss, by decreasing consumption of auxiliary input materials, and by improving conversion of incoming impurities into useful products. This chapter will focus on P2 techniques that have found or could find applications in minimizing or eliminating reducible solid waste and wastewater from the petroleum refining process. Air emissions are not covered.

The following commonly encountered wastes are considered in this chapter:

- Oily sludges
- Spent caustics
- Spent catalysts
- Miscellaneous process wastes
- Wastewater
- Maintenance and materials handling wastes

Following a brief description of waste components and sources, specific P2 techniques are presented in tabular form for each of these wastestreams. None of the P2 techniques presented is discussed here in any detail—the intent is to give the reader a menu of potentially effective P2 options to consider, along with the references where a more detailed discussion may be found. Some of the options presented were advanced in the original references merely as suggestions or plans for improvement. No effort was made to verify their efficacy or applicability in this compilation.

45.2   Oily Sludges

Oily sludges are stable emulsions formed through emulsification of oil with water, usually in the presence of suspended fines. The mechanism responsible for sludge formation is thought to be the adsorption of asphaltenes to hydrophilic solid particles at the oil-water interface. Heavy and coarse solids will cause the sludge to settle; fine and light solids will result in a sludge that floats on the water surface. Common oily sludge wastes include

Sludges from API or corrugated plate separators

DAF sludge

Slop oil emulsions

Heat exchanger bundle cleaning sludges

Cooling tower sludges

Tank bottoms

Currently available oily sludge management alternatives are depicted in Fig. 45-1. Raw watery sludges are filtered using pressure belt filter press or are centrifuged to remove solids, which destabilizes emulsion, causing it to separate

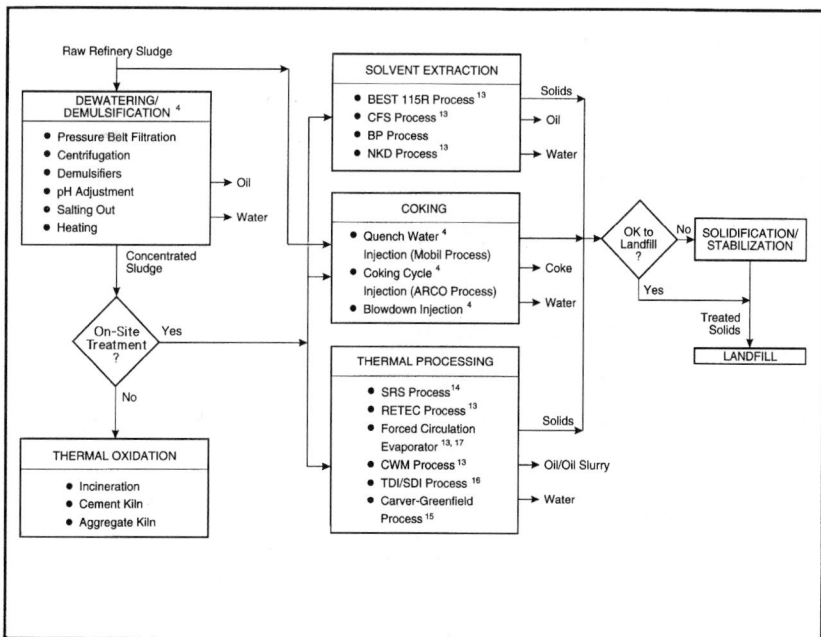

Figure 45-1. Refinery sludge management alternatives.

into three phases: oil, water, and dewatered sludge. The dewatered sludge can then be further treated to recover the oil, using either solvent extraction or thermal processing. Often, off-site disposal is more economical, especially if the sludge volumes are small. If the heat content of dewatered sludge can be increased above 5000 BTU/lb (e.g., using combustible filtration aids), it can be used as a waste fuel in cement or aggregate kilns; otherwise, hazardous waste incineration must be employed.

However, it is generally more preferable to minimize the generation of oily sludges at the source rather than to rely solely on the recycling and treatment techniques briefly presented here. Source reduction techniques generally rely on minimizing the release of solids, oil, and surfactants into the refinery wastewater system. Reduction of the solids' ingress into the oily water sewer is particularly effective since 1 lb of solids can result in generation of 5 to 10 lb of sludge. A list of potentially applicable source reduction techniques is presented in Table 45-1.

45.3   Spent Caustics

Solutions of sodium hydroxide are used primarily to wash the hydrocarbon products or intermediates in order to remove dissolved sulfides, mercaptans, phenolics, and other acidic compounds. The phenolic and sulfitic spent caustics should be segregated in order to enable recycling. Discharge of spent caustics into the process sewer is undesirable—as surfactants, they promote emulsion formation and impose an additional load on wastewater biotreatment facilities.

P2 options for reduction of spent caustics are presented in Table 45-2. Of particular importance is the prevention of phenolics formation in the fluid catalytic cracking (FCC) reactor by limiting the intrusion of oxygen carried with the regenerated catalyst. This can be done through inert gas stripping of the regenerated catalyst.

45.4   Spent Catalysts

With time, all catalysts lose their activity through sintering, poisoning, or buildup of surface deposits. Fluid-bed reactor catalysts, such as FCC zeolites, are also depleted by attrition where catalyst particles break up into fines, which are subsequently elutriated from the fluid bed. The development of more stable and/or regenerable catalysts is obviously the leading pollution prevention measure. Other more easily implemented measures are listed in Table 45-3.

Table 45-1. Source Reduction Techniques for Oily Sludges

Source of waste	Pollution prevention technique	References
Ingress of particulates into the process sewer	Paving/dust control of exposed dirt areas	1,4,6,12,17
	Planting ground cover	4
	Use of street sweeper or vacuum truck to collect dust and catalyst fines	1,4
	Install water seal inserts on sewer drains	9,17
	Install wind walls	17
	Periodic cleanout of drain openings and use of drain covers in process areas	1
	Redesign of sewer boxes to prevent soil entry	2
	Route runoff from unpaved, noncontaminated areas (such as future expansion or tank farms) separately to impoundment downstream of WWTP	4,6
	Use central vacuum systems on areas where fines/dust settlement is frequent (e.g., FCC catalyst loading area)	4
	Avoid introduction of sandblast solids into sewer	4
	Avoid spillage of filter aids (such as diatomaceous earth) during changeouts	4,17
	Replace diatomaceous earth filters with cartridge filters where possible (e.g., amine units)	17
	Install vent filters on FCC catalyst hoppers	17
	Install drain screens or overflow dams around the drains in the heat exchanger cleaning area to keep solids out during heat exchanger cleaning operations	4,17
	Recover coke fines for reuse, thus preventing sewer entry	1,4
	Remove and recycle coke fines by installing filter on coke calciner scrubber effluent prior to sewer discharge	17
	Use dust control on coke piles	4
	Prevent entrainment of coke-cutting fines into sewer by adding an inclined plate separator	12
	Prevent overfilling of coke rail cars by adding a double roll crusher with intermediate silo rather than filling the cars directly with coke cut from coking units	12
	In the oil field, minimize carryover of formation fines and precipitates by balancing injection water with formation water	4

Table 45-1. Source Reduction Techniques for Oily Sludges (*Continued*)

Source of waste	Pollution prevention technique	References
Ingress of oil into the process sewer	Provide secondary containment for intermediate and product transfer tankage pumps	1
	Install tanks or a separator for recovery of oil from desalter washwater effluent prior to sewer discharge	1,11,17
	Reuse oily stripping steam condensate as desalter makeup water	5
	Reuse vacuum tower steam jet condensate as desalter makeup water	5,11
	Replace third-stage steam jets with mechanical vacuum pumps	5,17
	Reuse FCC unit syn tower condensate as desalter makeup water	5,17
	Replace once-through barometric condensers with surface condensers	5
	Upgrade pump seals	17
	Assign a full-time person to continually survey small leaks from pump seals, valve packing, sample lines, and to expedite repairs	4,11,17
	Install area oil-water separators to reduce contact time between oil and water in the sewer	11
	Convert steam strippers to reboiled strippers	3,5
	Use low-shear in-line static mixer instead of high-shear mixing valve for contacting washwater and crude in desalter in order to reduce turbulence and decrease emulsion formation	4,17
	Decrease turbulence in the desalter vessel by using mud rakes instead of water jets to move desalter sediment toward a boot	4,17
	Reduce specific gravity of oil by adding naphtha to desalter brine followed by demulsification using polymer or acid addition and phase separation in corrugated plate separator equipped with electrostatic enhancer	4
	Use kerosene to extract organics from the desalter brine	17
	Reduce tubesheet leaks by designing out vibration, specifying seal-welded tubesheet joints, or double tubesheet designs	17
	Improve oil breakthrough control of tank draw-offs	17
	Allow asphaltenic solids to settle in crude oil storage tanks rather than introducing them to desalter and later into sewer	4
	Drain residual hydrocarbons into the slop oil system, and not into sewer upon equipment shutdown and preparation for steamout	17

Table 45-1. Source Reduction Techniques for Oily Sludges (*Continued*)

Source of waste	Pollution prevention technique	References
Ingress of surfactants into the process sewer	Keep surfactants out of sewer	1
	Minimize detergent use in tank truck washing, unit pad washing, in the lab, and in the maintenance shop	4
	Minimize polymer use in desalter feed, dirty desalter washwater, API separator feed, and in slop oil treatment	4
	Do not discharge spent phenolic or sulfitic caustic into sewer, but collect for reuse/recycle	4,17
	Eliminate inadvertent releases of amine solutions into sewer	17
	Use sour water stripper effluent and other phenol-bearing wastewater as desalter wastewater makeup so as to transfer phenol to crude	4,11,17
	Replace phenol as extraction solvent in dewaxing with less hazardous materials to eliminate troublesome leaks and spills	4
Heat exchanger cleaning	Reduce oversurfacing of heat exchangers during design	3,17
	Reduce film temperature in heaters during turndown operations by increasing fluid velocity in the tubes using a pumparound	3,17
	Reduce fouling rate by use of smooth (electropolished) heat transfer surfaces, use of antifoulants, and removal of fouling precursors	1,3,17
	Collect and filter hydroblast water prior to sewer discharge or reuse	5
Tank sediments	Use side-entry or in-tank mixers to keep solids in suspension	1,3,4,17
	Minimize intermediate tankage	1,17
	Minimize contact between crude oil and air to prevent formation of settleable gummy solids through inert blanketing or floating roofs	3
	Separate oil and water phases of tank bottoms using filters or centrifuges	4
	Add emulsifiers to prevent settling	4
	Use warm oil circulation with dispersant for tank bottom sludge cleanout to recover entrapped oil	8

Table 45-2. Source Reduction Techniques for Spent Caustics

Source of waste	Pollution prevention technique	References
Formation of phenolic compounds	Minimize air intrusion into FCC reactor system, e.g., using inert gas or fuel gas to strip regenerated catalyst and for instrument purges	11,17
Excessive caustic consumption	Use dilute instead of concentrated caustic wash to reduce caustic consumption and carryover of phenolics into aqueous phase	11
	Use hydrotreating to remove H_2S from jet fuels and home heating oils and eliminate caustic use in nonregenerative scrubbing	4
Discharge of spent caustic into process	Segregate sulfitic and phenolic spent caustic streams to enable recycle	4,17
	Send spent phenolic caustic off site for phenolics recovery through extraction and distillation by commercial recycling firms	4
	Recover phenolics from spent caustics on site through acidification and phase separation	4
	Send spent sulfite caustic for off site reuse in mining and pulp industries or for sodium recovery	4

Table 45-3. Spent Catalysts

Source of waste	Pollution prevention technique	References
Catalyst deactivation and/or attrition	Sell spent catalyst for regeneration (e.g., hydrotreating Co-Mo or Ni-Mo catalyst)	4,7,17,18
	Sell spent catalyst for metal recovery (e.g., Pt-based reforming catalyst or Pd-based hydrocracking catalyst)	4,18
	Recycle spent FCC catalyst as cement additive in cement manufacturing for its alumina and silica content	1,4,7,17,18
	Reduce FCC catalyst loss by demetalizing and desulfurizing gas oil feed through mild hydrotreating	3
	Reduce FCC catalyst attrition loss by minimizing aeration and purge steam rates	3
	Reduce FCC catalyst loss by extracting metals from the equilibrium catalyst (Coastal Corp. technology) and recycling it back to the process	17
	Recycle spent alumina (Claus) catalyst as raw material for alumina-based products	7

45.5 Miscellaneous Process Wastes

Table 45-4 presents P2 techniques for the following wastes:

FCC slurry oil sludge

Spent clay

HF alkylation sludge

Amine solution purge

Stretford solution purge

These wastestreams result mainly from application of auxiliary materials or mass separation agents. The P2 techniques listed in Table 45-4 invoke prevention of waste by limiting the formation or ingress of a troublesome impurity, e.g., preventing formation of acid-soluble oils and attendant HF loss by selective hydrogenation of diolefin precursors in the olefin feed to the HF alkylation unit. Other techniques rely on regeneration of scrubbing solutions through selective removal of a target impurity, e.g., regeneration of DEA solution through the crystallization and removal of amine salts.

45.6 Wastewater

Table 45-5 presents pollution prevention options for the following streams:

- Cooling tower (CT) blowdown
- Process water

The main themes are water conservation and prevention of impurities ingress into the wastewater collection and treatment system.

45.7 Maintenance and Material Handling Wastes

These wastes include

Empty drums and containers

Solvents used for maintenance cleaning and paint thinning

Overage supplies

Spent heat transfer fluids

Spent lubricants

Spent hydraulic fluids

Spent sandblasting media and stripped paint

Table 45-4. Miscellaneous Process Wastes

Source of waste	Pollution prevention technique	References
FCC slurry oil/decant sludge	Filter slurry oil to remove catalyst fines using electrofilter (Gulftronic Separator System) and render slurry oil suitable as a carbon black feedstock instead of No. 6 fuel oil additive; recycle fines to reactor.	4,17,18
	Use hydrocyclones, centrifuges, or porous metal filters to remove fines from slurry oil	18
	Maintain cyclones in good operating condition	18
Spent clay	Backwash spent clay from diesel or jet fuel filtration with water or steam to recover hydrocarbon, extend service life of clay, and/or render it nonhazardous	4
	Replace clay filtration with hydrotreating for kerosene and lube oils	3
	Regenerate/recycle spent clay at high temperatures using Thermofor kiln technology following naphtha wash and drying	3
HF alkylation sludge	Convert calcium fluoride slurry to fluorspar for sale as fluxing agent to steel and glass industry	4
	Upgrade alkylation unit feed by selective hydrogenation of diolefins which are acid polymer precursors	3
	Avoid upsets and improper operation of alkylation unit	3
	Reduce sludge volume by decreasing $CaCl_2$ excess and pH of CaF_2 precipitation reaction with spent caustic	17
Amine solution purge	Minimize DEA degradation by continuous slipstream filtration in addition to carbon filtration and inhibitors addition	1
	Minimize film temperature in amine regenerator reboiler	3
	Use amine degradation inhibitors	3,17
	Regenerate DEA by crystallizing out amine salts	6
Stretford solution purge	Minimize thiosulfate formation by replacing sulfur melter with pressure belt filter	10
	Minimize thiosulfate formation by removal of Na_2CO_3 from solution prior to sending it to sulfur melter using centrifuge-wash-centrifuge steps	18
	Oxidize thiosulfate to sulfate using formaldehyde or peroxide addition	18
	Oxidize thiosulfate using sulfuric acid	18
	Precipitate thiosulfate using nickel ethylenediamine	18

Table 45-4. Miscellaneous Process Wastes (*Continued*)

Source of waste	Pollution prevention technique	References
Stretford solution purge (*cont.*)	Use DOW Stretford Chemical Recovery Process to reclaim vanadium and ADA from purge stream using IX and carbon adsorption	4,18
	Coprocess Stretford purge with spent hydrotreating catalyst	18
	Maintain proper concentrations of ADA and vanadium	10
	Use ADA with higher content of ADA-2,7 isomer	10
	Reduce CO_2 concentration in the Claus unit offgas	10
	Add sodium tartarate to prevent vanadium oxysulfide formation	3
	Remove thiosulfates by oxidation to sulfates which are then removed by crystallization (Global Sulfur System)	6
	Replace Stretford Unit with SELECTOX, SCOT, LOCAT, SULFEROX, UNISULF, or other processes which are less waste-intensive	3,4,10,17

These wastes are generated by nearly all industries and are not specific to petroleum refining. Subsequently, they are covered elsewhere in this handbook—the reader is hereby referred to Chap. 29.

The lists of P2 techniques compiled in this work were based on the review of available literature and in-house project experience. These lists are offered with the intent of helping the reader initiate the hunt for suitable P2 solutions. However, they are not a substitute for the creativity, imagination, and hard work needed to work out all application-specific details necessary to make any P2 technique work in a real refinery setting.

References

1. H. G. Hethcoat, "Minimize Refining Waste," *Hydrocarbon Processing*, August 1990, pp. 51–54.

2. H. Klee Jr. and M. K. Podar, "Pollution Prevention in Petroleum Refining," *AIChE Summer Meeting*, August 18–21, 1991, Pittsburgh, Pa.

3. U.S. Environmental Protection Agency, *Waste Minimization Issues and Options, Vol. II*, prepared by Jacobs Engineering Group under contract to Versar Inc. for the Office of Solid Waste and Emergency Response, EPA/530-SW-86-043, October 1986.

4. American Petroleum Institute, *Waste Minimization in the Petroleum Industry—A Compendium of Practices*, API Publication No. 849-30200, Washington, D.C., November 1991.

Table 45-5. Source Reduction Techniques for Refinery Wastewater

Source of waste	Pollution prevention technique	References
Cooling tower blowdown	Maximize use of air cooling to reduce CT thermal load	1,5
	Pretreat CT makeup water by cold lime softening, reverse osmosis, electrodialysis, or zeolite softening	1,3,4,5,20
	Air-cool cooling water return at night or during cold weather	5
	Reuse treated refinery effluent as CT makeup (and also as washwater or fire water)	5
	Use boiler blowdown from low-pressure boilers as CT makeup	5
	Reuse treated refinery effluent for once-through cooling applications (e.g., cooling of pumps, compressors, and surface or barometric condensers) to reduce CT duty	20
	Use sidestream softening to achieve zero CT liquid blowdown discharge	22
	Collect and reuse low TDS rainwater from nonprocess areas as CT makeup water	17
	Reuse demineralizer ion exchange resin rinsewater as CT makeup water	17
	Use ozonation treatment of CT water to reduce scale, inhibit corrosion, and decrease CT blowdown	19,24
Process water	Reuse treated refinery effluent as quench water for heater decoking	17,23
	Minimize steam to side strippers on atmospheric crude tower	17
	Recycle part of desalter effluent back as inlet washwater	21
	Use stripped sour water as desalter washwater	5,17,21
	Reuse oily crude tower stripping steam condensate or vacuum tower steam condensate as desalter washwater or as a washwater injected upstream of overhead condensers	5,11,17,21
	Collect washwater injected into wet gas compressor intercoolers for injection upstream of the main FCC fractionator overhead condenser	21
	For product washing, recirculate part of washwater to a midpoint of a countercurrent packed tower contactor so as to increase contaminant concentration and reduce fresh water requirements	5
	Use stripped sour water as washwater injected upstream of condensers in crude unit and hydrotreaters	21
	Recycle coke cutting water; use treated refinery effluent or stripped sour water as makeup to coke cutting water loop	17,21
	Minimize steam-to-steam jet eductors on vacuum tower	17

5. New Mexico Energy Research and Development Institute, *Water Use, Conservation and Wastewater Treatment Alternatives for Oil Refineries in New Mexico,* prepared by Jacobs Engineering Group, NMERDI 2-72-5628, Sante Fe, N.M., May 1985.

6. California Environmental Protection Agency, Department of Toxic Substances Control, private communication, 1993.

7. P. L. Avery, ARCO Watson Refinery, private communication, 1990.

8. J. W. Barnett, "Better Ways to Clean Crude Storage Tanks and Desalters," *Hydrocarbon Processing,* **60**(1):82–86 (1980).

9. *EPA Benzene Waste Operations,* NESHAP Workshop, U.S. EPA Office of Air Quality and Standards, Research Triangle Park, N.C., February 11, 1993.

10. G. Lorton, C. H. Fromm, and M. Meltzer, *Waste Minimization Assessment Report— Case Study of Minimization of Stretford Solution Purge at an Oil Refinery,* by Jacobs Engineering Group under a contract to Hazardous Waste Engineering Research Laboratory, U.S. EPA, 1988.

11. R. V. Willenbrink, *Waste Minimization in the Petroleum Refining Industry, in Hazardous Waste Minimization,* H. M. Freeman (ed.), McGraw-Hill, New York, 1990.

12. J. A. Balik and S. M. Koraido, "Identifying Pollution Prevention Options for a Petroleum Refinery," *Pollution Prevention Review,* Summer 1991.

13. California Department of Health Services, *Incinerable Hazardous Waste Minimization Workshop Proceedings,* Berkeley and Irvine, Calif., January 14–18, 1991.

14. C. Wanberg, "MX-2500 Thermal Process for the Treatment of Petroleum Refining Wastes and Contaminated Soils," *Environmental Progress,* vol. 12, no. 2, May 1993.

15. Foster Wheeler Corporation, *Carver-Greenfield Process Brochure,* Clinton, N.J., undated.

16. TDI Services Inc., *HT-5 Thermal Distillation Process Brochure,* Houston, Tex., 1989.

17. Jacobs Engineering Group Inc., past project experience, Pasadena, Calif., 1986–1993.

18. P. Allen, A. Jackman, and R. Powell, *Petroleum Refining Industry Waste Audit,* University of California—Davis under the auspices of Toxic Waste Reduction Project for California Department of Health Services, State of California, May 1990.

19. J. T. Echols and S. T. Mayne, "Cooling Water Cleanup by Ozone," *Chemical Engineering,* May 1990.

20. B. S. Langer, "Wastewater Reuse and Recycle in Petroleum Refineries," *Chemical Engineering Progress,* **79**(5):67–76 (1993).

21. K. S. Eble and J. Feathers, "Water Reuse Within a Refinery," presented at the *1992 National Petroleum Refinery Association Meeting,* New Orleans, March 22–24, 1992.

22. P. R. Puckorius and T. Harris, "Zero Cooling Tower Discharge Achieved by Sidestream Softening at a Petroleum Refinery," *International Water Conference—43rd Annual Meeting,* Pittsburgh, Pa., October 25–27, 1982. (Proceedings published by Engineers' Society of Western Pennsylvania, Pittsburgh.)

23. J. H. Tay and P. C. Chui, "Reclaimed Wastewater for Industrial Application," *Water Science and Technology,* **24**(9):153–160 (1991).

24. A. Pryor and M. Bukay, "Water Conservation Through Cooling Tower Ozonation," *Ultrapure Water,* May 1990.

46

Pollution Prevention in Electroplating Industries

Frank Altmayer

Scientific Control Laboratories, Inc.
Chicago, Illinois

46.1 Process Description

Electroplating facilities fall into two major categories: job shops and captives. Job shops function as a service, taking parts manufactured by others and electroplating them with any one or a combination of over a hundred different metallic coatings available. Job shop electroplaters tend to be very small businesses, averaging less than 50 employees and with annual sales less than $5 million, although larger shops exceeding 100 employees and with sales over $10 million do exist. Job shop electroplaters in the United States number in the neighborhood of 3000 companies, most of them located in or near the major metropolitan areas, notably Chicago, Detroit, Cleveland, California, New York, and the New England states. Captive electroplating facilities perform electroplating operations for in-house manufactured parts. Captive electroplating facilities can be found spread out throughout the United States and include numerous large (Fortune 500) manufacturing and service corporations, including major airlines, aerospace firms, computer and electronics manufacturers, hardware manufacturers, and all automotive firms.

Electroplating operations are utilized in the manufacture of strategic and consumer products, including printed circuit boards, and hundreds of automotive parts such as piston rings, dashboards, electronic sensors and controls, air bags, metallic and plastic hardware, fuses, lights, and most engine components.

All aircraft contain hundreds of electroplated components and mechanisms, including the engines, landing gear, and cockpit instruments. Even the uranium that goes into atomic bombs and nuclear generators is electroplated. In general, most any metallic component in a manufactured item is commonly electroplated, to enhance corrosion resistance, appearance, magnetism, non-magnetism, solderability, weldability, lubricity, bearing performance, wear resistance, or current-carrying capability. Electroplating is not limited to application over metallic substrates. Most any plastic, ceramic, or other substrate can be electroplated, using specialized solutions and equipment.

The electroplating process utilizes a wide variety of chemicals, depending on the types of metals that are processed for electroplating and the types of metallic coatings that the facility applies to the processed substrates. Table 46-1 is a partial list of typical chemicals that may be found in an electroplating facility. Since parts are typically processed in a water-based solution containing a combination of these chemicals and are then rinsed, the chemicals purchased by an electroplating facility normally will find their way into one or more wastewater streams, can be emitted into the air through process exhaust systems, and can find their way into the soil through leaching from landfills and from past land disposal practices or poor facility maintenance. A typical job shop electroplating facility will operate a wastewater pretreatment system to comply with federal and local water discharge regulations. They will also operate air-scrubbing systems to comply with the Clean Air Act and local air discharge regulations. The solid waste generated by these facilities is regulated under the Resource Conservation and Recovery Act (RCRA), and most electroplating facilities will generate a relatively low hazard "filter cake" from their wastewater treatment system that is classified by the U.S. Environmental Protection Agency (EPA) as F-006 hazardous waste. Many electroplating facilities also generate concentrated wastes that may contain high concentrations (in the range of ounces per gallon) of toxic metals such as lead, cadmium, or hexavalent chromium, or other toxics such as chlorinated solvents and cyanide. These wastestreams typically originate from stripping operations, unrecoverable contamination of electroplating solutions, or use of processing solutions with a finite utility life. Quantities of waste generated are so variable that a generalization cannot be made. Some facilities generate no more than a few gallons of waste per day, while others generate 40 cubic yards per week. Table 46-2 is a list of typical electroplating processes and the types of wastes generated.[1]

Electroplating involves processing the part to be electroplated through a series of water-based solutions containing one or more chemicals that either clean, deoxidize, or coat the part. Because each processing step utilizes specialized chemicals that would react unfavorably with the subsequent process, most every processing step is followed by a water rinse. The rinsewater thus becomes contaminated with the processing chemicals and needs to be treated for purification prior to discharge. Rinsing is performed in either flowing-water rinse tanks or in counterflow or controlled-flow rinse tanks that conserve water usage. Certain processes are equipped with accessory equipment such as fil-

Table 46-1. Some Chemicals Used in the Metal-Finishing Industry

Chemical	pH adjustment required	Cyanide or chromium treatment required	Waste products*
Aluminum potassium sulfate	X		MS, DS
Aluminum silicate	X		MS, DS
Ammonium acetate	X		DS, NH_3
Ammonium bifluoride	X		MS, DS, NH_3
Ammonium chloride	X		DS, NH_3
Ammonium citrate	X		DS, NH_3, O
Ammonium hydroxide	X		DS, NH_3
Ammonium molybdate	X		MS, DS, NH_3
Ammonium nitrate	X		DS, NH_3
Ammonium sulfate	X		MS, DS, NH_3
Anisic aldehyde	X		O
Antimony potassium tartrate	X		MS, DS, O
Barium carbonate	X		MS
Barium sulfate	X		MS
Benzene (benzol)	X		O
Boric acid	X		DS
Cadmium cyanide		X	MS, DS
Cadmium sulfate	X		MS, DS
Calcium nitrate	X		DS
Chromic acid		X	MS, DS
Citric acid	X		DS, O
Cobalt carbonate	X		MS
Cobalt sulfate	X		MS, DS
Cupric sulfate	X		MS, DS
Diammonium phosphate	X		MS, DS, NH_3
Ferric nitrate	X		MS, DS
Fluoroboric acid	X		MS, DS
Formaldehyde	X		O
Glue	X		O
Glycerine	X		O
Hydrazine sulfate	X		DS
Hydrochloric acid CP	X		DS
Hydrofluosdilcic	X		MS, DS
Hydrogen peroxide	X		
Hydroxyacetic acid	X		DS, O
Hypophosphorus acid	X		MS, DS
Indium sulfate	X		MS, DS
Iron oxide	X		MS
Isopropanol	X		O
Lard oil	X		O
Lead fluoborate	X		MS, DS
Lead oxide	X		MS
Lime (calcium hydroxide)	X		MS
Magnesium sulfate	X		MS, DS
Manganese carbonate	X		MS
Manganese sulfate	X		MS, DS
Methanol	X		O
Monoammonium phosphate	X		MS, DS, NH_3

Table 46-1. Some Chemicals Used in the Metal-Finishing Industry (*Continued*)

Chemical	pH adjustment required	Cyanide or chromium treatment required	Waste products*
Nickel carbonate	X		MS
Nickel chloride	X		MS, DS
Nickel sulfate	X		MS, DS
Nickel sulfamate	X		MS, DS
Nitric acid	X		DS
Oxalic acid	X		MS, DS
Phosphorus acid	X		MS, DS
Potassium bromate	X		DS
Potassium citrate	X		DS, O
Potassium chloride	X		DS
Potassium copper cyanide		X	MS, DS
Potassium cyanide		X	MS, DS
Potassium ferricyanide		X	MS, DS
Potassium hydroxide	X		DS
Potassium phosphate	X		MS, DS
Potassium stannate	X		MS, DS
Potassium thiocyanate	X		DS
Sodium acid pyrophosphate	X		MS, DS
Soda ash (sodium carbonate)	X		DS
Sodium bicarbonate	X		DS
Sodium bisulfite	X		DS
Sodium bifluoride	X		MS, DS
Sodium citrate	X		DS, O
Sodium copper cyanide		X	MS, DS
Sodium cyanide		X	DS
Sodium dichromate		X	MS, DS
Sodium fluoroborate	X		MS, DS
Sodium gluconate	X		DS
Sodium hexametaphosphate	X		MS, DS
Sodium hypophosphite	X		MS, DS
Sodium hydrosulfite	X		DS
Sodium hydroxide (caustic soda)	X		DS
Sodium metasilicate	X		MS, DS
Sodium molybdate	X		MS, DS
Sodium nitrate	X		DS
Sodium orthosilicate	X		MS, DS
Sodium polysulfide	X		MS, DS
Sodium stannate	X		MS, DS
Sodium sulfate	X		DS
Sodium sulfide	X		MS, DS
Sodium sulfite	X		DS
Sodium tripolyphosphate	X		MS, DS
Stannous fluoroborate	X		MS, DS
Stannous sulfate	X		MS, DS
Stearic acid	X		O
Sulfamic acid	X		DS
Sulfur (liquid)	X		MS
Sulfuric acid	X		MS, DS

Table 46-1. Some Chemicals Used in the Metal-Finishing Industry (*Continued*)

Chemical	pH adjustment required	Cyanide or chromium treatment required	Waste products*
Tallow glyceride	X		O
Tartaric acid	X		O
Tetrapotassium pyrophosphate	X		MS, DS
Tetrasodium pyrophosphate	X		MS, DS
Toluene (toluol)	X		O
Trichloroethylene	X		O
Trichloroethane	X		O
Trisodium phosphate	X		MS, DS
Xylene (xylol)	X		O
Zinc chloride	X		MS, DS
Zinc cyanide		X	MS, DS

*MS = metal sludge; NH_3 = ammonia; DS = dissolved solids; O = organic matter.

Table 46-2. Pollution Sources and Characteristics

Processes	Spent process liquors and wastewaters					Solid wastes	Air emissions
	Chromium	Other metals	Cyanide	Oils	Solvents		
Deburring	X	X		X		X	
Polishing	X	X		X		X	X
Solvent cleaning				X	X	X	X
Alkaline cleaning		X	X	X	X	X	X
Pickling	X	X				X	X
Etching	X	X				X	X
Bright dipping	X	X				X	X
Chromating	X	X					
Phosphating	X	X		X		X	X
Passivating	X	X					X
Plastic coating					X	X	X
Paint coating		X			X	X	X
Hot-dip coating		X					X
Electroless plating		X					X
Anodizing	X	X					X
Electropolishing	X	X					X
Electroplating	X	X	X			X	X
Electrocleaning	X	X	X	X		X	X

ters, heat exchangers, and air scrubbers. These accessories can be additional sources of chemical wastes, along with other operations such as process tank maintenance, pickling, derusting and etching of metallic parts, solvent degreasing, and production of de-ionized water for critical rinsing and tank makeup. The equipment used for electroplating operations can be as primitive as a series of tanks made from plastic drums and rectifiers, all the way to automated systems driven by computers, process controllers, and programmed hoists.

46.2   Pollution Prevention Options

The most commonly practiced pollution prevention option in electroplating is the utilization of drag-out rinses (see Fig. 46-1). These involve use of water-bearing rinse tanks that parts are rinsed in after processing but before rinsing in a flowing-water rinse. The water and chemicals collected in the drag-out rinse are returned to the process tank to make up for evaporative losses. The system obviously works only on chemical processes that operate with sufficient evaporative losses to make room for the returned drag-out rinse volume.

The second most popular pollution prevention alternative is substitution. For example, there are several different chemistries available for zinc plating, including some that contain cyanide and others that do not. By successfully making enough changes to allow the use of a noncyanide zinc-plating process, cyanide can be eliminated from the facility. The electroplater making a substitution will almost always be paying a price for this and needs to know as much as possible about the substitute process to make an intelligent decision. Popular substitutes for existing coatings or chemical processes in electroplating facilities are discussed below.

46.2.1   Trivalent for Hexavalent Chromium

Regulations under the Clean Air Act, along with local constraints on chromium emissions to the air and water, plus the stigma attached to hexavalent chromium as a health hazard, are the driving forces behind the push for substituting baths made from trivalent chemistries for hexavalent chromium–plating processes. The trivalent process is currently suitable only for applications that are considered "decorative." There is no commercially viable hard-chromium-plating process using the trivalent chemistry currently on the market. The trivalent processes tend to produce a deposit that is generally considered to be darker in appearance than the hexavalent processes unless impurities and plating conditions are carefully controlled.

Trivalent baths produce a deposit that is microporous. This enhances the corrosion resistance of the plated coatings when applied over nickel plating that is greater than 0.00025 in thick. When the nickel thickness is less than 0.00025 in, the trivalent chromium provides less corrosion resistance than a "crack-free"

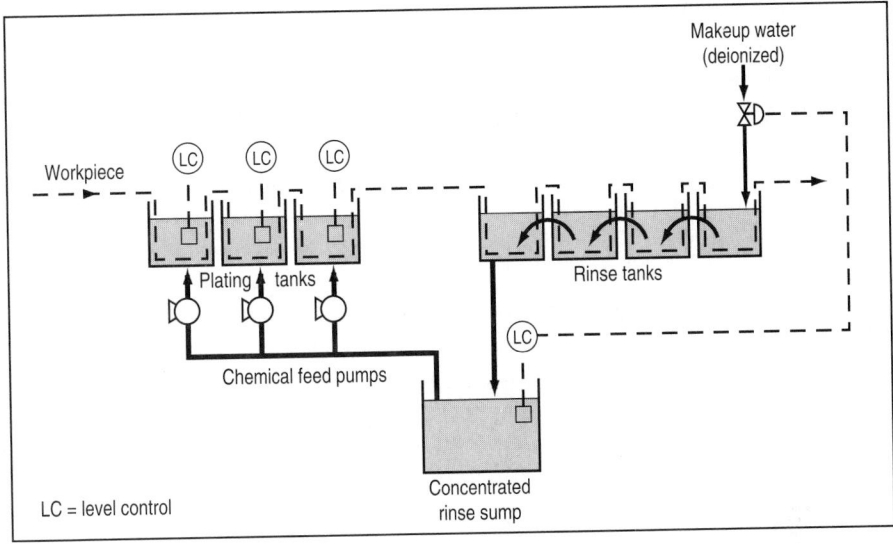

Automated Drag-Out Recovery System

DRAG-IN-DRAG-OUT RINSES

Figure 46-1. Drag-out rinsing systems.

deposit from hexavalent baths. Trivalent deposits in excess of 0.0005 in tend to be stressed and have very low corrosion resistance.

The wear resistance (hardness) of a trivalent chromium deposit is similar to that of a hexavalent deposit. Stripping trivalent deposits with hydrochloric acid is difficult and slow. Reverse current stripping is usually employed.

Trivalent chromium chemistries can be expected to generate between 5 and 10 percent as much solid waste from waste treatment of rinsewater, due to lower chromium content in the drag-out, when compared to hexavalent baths (assuming no recovery of drag-out). Trivalent baths require no reduction step, only pH adjustment. This translates to lower capital costs for waste treatment and no "hexavalent" violations due to malfunctions of the reduction system. For companies that must comply with chemical oxygen demand (COD) regulations, trivalent baths can increase COD loading because they contain a significant amount of organic additives.

No exhaust from the process means no scrubbing equipment, and no treatment and/or disposal of scrubbing solutions.

46.2.2   Alkaline Noncyanide Copper for Cyanide Copper

Since 1990, alkaline noncyanide copper-plating processes have been available as substitutes for the cyanide process. The obvious benefit is the elimination of cyanide from the wastewater stream. Not so obvious benefits include faster barrel-plating speed, no cyanide in the F-006 waste from waste treatment (making delisting of the F-006 a more likely possibility), lower sludge volume generation due to lower metal concentrations, simplified wastewater treatment, no trouble with carbonate buildup, elimination or reduction of TRI (Toxic Release Inventory, a.k.a. "Form R") reporting, and lesser safety concerns under the Occupational Safety and Health Act (OSHA). While it is unlikely to happen on a frequent basis, if the cyanide copper solution ever requires disposal, the cost is astronomical compared to the alkaline noncyanide process.

Balancing the benefits of alkaline noncyanide copper-plating processes are disadvantages such as higher operating costs, inability to use the process on zinc surfaces (die castings and zincated aluminum or magnesium), greater sensitivity to impurities, and a chemistry that is more difficult to control. At least one of the noncyanide processes utilizes a "purification" cell in addition to the normal plating tank.

Both cyanide and noncyanide processes for copper plating yield a fine-grained, dense deposit of equal metallurgical property, with the possible exception of purity, since the noncyanide process incorporates a trace of organic into the deposit from the additives. An ideal application for the non-cyanide process is for thick deposits used as heat-treating (carburizing) stop-off on steel parts. The dense deposit is an excellent diffusion barrier for carbon.

The cyanide and alkaline noncyanide processes typically function at elevated temperature, making recovery through drag-out control and other recovery systems viable. The noncyanide process contains one-half to one-fourth as much copper as a full-strength cyanide copper bath, translating to lower sludge generation. Waste treatment is accomplished by pH adjustment with lime, or magnesium hydroxide, eliminating the two-stage chlorination system from the waste treatment system and eliminating use of more dangerous chemicals such as chlorine or sodium hypochlorite. A potential negative pollution prevention effect would occur if the noncyanide bath frequently became contaminated beyond control (as happened during pilot-scale testing in one case). The bath would then require treatment and disposal. Cyanide-based copper-plating solutions are far more tolerant of impurities and can last many years before requiring treatment and disposal.

The noncyanide process does not create a cyanide-bearing F-006 waste. If copper plating on steel is the only source of cyanide in the F-006 waste, substitution of the noncyanide process may create a waste that qualifies for delisting.

46.2.3 High-pH Nickel for Copper Strike

High-pH nickel-plating solutions have been available for a long time as a substitute for cyanide copper strike on zincated aluminum surfaces. Cyanide copper has fewer cleaning and analytical requirements, but the nickel process allows for the elimination of one more cyanide-bearing process.

Typical formulation and operating conditions for the nickel process are provided in Table 46-3. To obtain optimum results, the plater must balance the ratio between the nickel sulfate and the sodium sulfate. Parts with complex shapes require higher sodium sulfate concentrations. Zinc contamination should be continuously removed through low-current-density dummying in a purification cell.

The higher the sodium content of this nickel-plating bath, the more brittle the deposit becomes. The bath should therefore be used only as a "strike" before conventional nickel plating. Parts that undergo fatigue cycles or extreme temperature changes may experience fatigue failures and less corrosion resistance.

The ammonium ion present in the high-pH nickel formulation may cause waste treatment problems unless its concentration is minimized through drag-out recovery techniques. This bath contains a higher metal content than the cyanide copper process and twice the metal content as the alkaline noncyanide process. Sludge volume from wastewater treatment would be affected accordingly.

46.2.4 Zinc Alloy for Cadmium Plating

When cadmium electroplate is used mostly for enhanced corrosion resistance to salty environments, zinc alloy processes are suitable candidates as substitutes. Even pure zinc is a suitable substitute for cadmium deposits greater than 0.001

Table 46-3. High-pH Nickel-Plating Solution Formulation

Nickel sulfate	10–15 oz/gal
Ammonium chloride	2–5 oz/gal
Sodium sulfate (anhydrous)	10–15 oz/gal
Boric acid	2–3 oz/gal
pH	5.3–5.8
Temperature	70–90°F
Current density	12–36 ASF

in thick. When cadmium is specified for enhanced lubricity, solderability, low electrical contact resistance, ease of disassembly after corrosion has occurred, or to obtain benefits from the toxicity of cadmium (organisms such as fungus or mold won't grow on it), zinc alloy deposits may not be suitable substitutes.

While noncyanide cadmium processes are presently commercially available, the most desirable substitute for a cadmium-plating process would do away with both cyanide and cadmium.

There presently are numerous zinc alloy processes commercially available, including zinc-cobalt, zinc-nickel, zinc-tin, and zinc-iron. For these, analytical control of the alloy-plating solutions is far more critical than it is for cadmium plating. A wide variety of variables must be controlled to obtain the "right" alloy composition, including pH, temperature, chemistry, and agitation level. Zinc-nickel alloys can be plated from a chloride-based process similar to chloride zinc baths or from an alkaline process similar to alkaline noncyanide zinc.

Brightening agents and other additives make the alloy processes more expensive to purchase and operate than cyanide cadmium baths. The alloying metal is usually added as a chemical concentrate that is purchased from the supplier. Zinc alloy deposits are fine grained and generally harder than cadmium or pure zinc. Yellow chromates of the zinc alloys may have a significantly different appearance from pure zinc. For example, zinc-nickel may look purple when chromated. The zinc-nickel and zinc-cobalt deposits can be very bright, while the zinc-iron and zinc-tin are less so. Solderability is marginal for all but the zinc-tin alloy.

Zinc alloy deposits are much harder than cadmium and offer far less lubricity in torque-tension applications, with the possible exception of zinc-tin. Cadmium has a fine-grained structure that offers excellent torque-tension properties (lubricity). Cadmium metal is toxic, so organisms such as mold and fungus will not grow on cadmium-plated surfaces. Corrosion products of cadmium tend to break apart readily, while zinc and zinc-alloy corrosion products tend to hold fast.

Cadmium, ion-vapor-deposited (IVD) aluminum, zinc, and zinc-nickel (7–9 percent nickel) are the most effective corrosion protection of steel in natural marine, industrial, and rural atmospheres. Cadmium, IVD aluminum/topcoat, tin-zinc, zinc, and zinc-nickel plating exhibit the lowest disassembly breaking torques.

Cadmium plating is the least likely coating to promote stress-corrosion cracking, while all currently available substitutes tend to enhance it. Zinc-nickel- and tin-zinc-plated deposits yield the best alternate performance.

Wastewater treatment of the rinsewater from zinc alloy electroplating processes is normally accomplished by simply adjusting pH, thereby eliminating the need for cyanide oxidation. The zinc-cobalt, zinc-tin, and zinc-iron processes do not add any metals to the process that are presently regulated under federal clean water programs. The alloy processes are suitable for use with presently available recovery technologies, although recovery may pose increased chemical control problems.

46.2.5 Substitute Processes for Hard-Chromium Plating

While a viable substitute for conventional hard-chromium deposits is presently not on the market, there is a considerable amount of research being conducted to develop coatings that could function in place of heavy chromium deposits. One such process is an electrodeposited alloy of nickel and tungsten with fine particles of silicon carbide dispersed throughout the deposit. The chemistry was patented by Takata in 1990 and is currently under consideration by several large manufacturing companies.

The plating solution for this process contains nickel sulfate (35 g/L), sodium tungstate (65 g/L), and ammonium citrate (110 g/L). Fine silicon carbide particles (30 g/L) are suspended within the plating bath by continuous agitation. The silicon carbide particles are incorporated into the deposit as it forms, and provide a major portion of the abrasive resistance of the coating, while the nickel-tungsten electroplate provides a hard matrix for the retention of the particles. The bath operates at pH 6–8 and is adjusted with ammonium hydroxide and citric acid. Bath temperature is 150 to 175°F. Plating current density is 100 to 300 ASF.

The alloy process can deposit the same thickness in half the time compared to a conventional hard-chromium process due to a current efficiency of 24 to 35 percent versus 10 to 12 percent for chromium plating. The alloy process also has significantly better throwing power, although complicated shapes still require auxiliary anodes. There presently is no good technique for stripping the alloy process, or performing "spot repairs," while the chromium process is easily stripped and some parts can be spot-plated. The alloy process is sensitive to metallic contamination, while the chromium process is very tolerant. The alloy process utilizes citrates at relatively neutral pH, creating an environment conducive to biological activity such as mold growth.

The alloy process utilizes a tungsten compound that is very expensive and difficult to obtain. Some companies investigating the process have resorted to producing their own sodium tungstate from other tungsten compounds.

The alloy process performs better in abrasive-wear applications than chromium when tested with a Taber abrasion apparatus.

The successful substitution of the alloy/matrix plating process would eliminate hexavalent chromium from the wastewater and air discharges of the metal-finishing facility. However, the alloy process utilizes nickel and tungsten compounds which have also been linked to cancer. The net effect is to replace one potentially carcinogenic material with two that are also carcinogenic suspects. Since the plating efficiency of the alloy process is only 24 to 35 percent, the process requires exhaust and scrubbing to reduce air emissions. High concentrations of ammonium ions in the wastewater stream would cause serious wastewater treatment problems due to chelation of metals such as copper and nickel. On the positive side, none of the ingredients in the alloy bath is an oxidant (such as chromic acid), so storage of the chemicals is simplified and the hazards of unwanted reactions and explosions caused by accidental combination of oxidizer and reducing agent are eliminated.

Other substitute baths based on other nickel alloys such as nickel-tungsten-boron have been investigated and reported in the literature but have not been commercialized.

46.2.6   Carbon for Electroless Copper

The circuit board industry is looking at a proprietary process that produces conductivity in the through-holes with carbon. The process uses conventional plating equipment and reduces the number of processing tanks it takes to produce plated through-holes. The process also would eliminate the use of formaldehyde, which is a suspected carcinogen.

The printed circuit boards are prepared prior to carbon coating in the same manner as for electroless copper, including etch-back. Immediately prior to carbon coating, the boards are processed through a proprietary cleaner and proprietary conditioning solutions which are alkaline and contain weak complexing agents. The carbon-coating solution is also slightly alkaline and contains extremely fine carbon particles on the order of 0.15 to 0.25 µm (6 to 10 millionths of an inch) in diameter. The process has been commercially available since 1989 and is presently in use by many circuit board facilities. Suitability for plating on plastics is unknown.

The U.S. military standard MIL-P-55110D permits the use of this technology as a substitute for electroless copper. Elimination of electroless copper removes a heavily chelated process from the wastewater stream, and, at the same time, eliminates a suspected carcinogenic material (formaldehyde). There are disadvantages, however. The extremely fine, suspended carbon may cause problems in wastewater treatment, by coating over probes, clogging filters, and interfering with the proper operation of a clarifier. The carbon will not be removed by conventional precipitation systems and needs to be controlled at the source. Carbon in the wastewater discharge will significantly increase COD loading to the publicly owned treatment works (POTW). The carbon may also act as an organic "collector," increasing total toxic organic concentrations in the wastewater discharge. Some POTWs have "excessive coloration" regulations. Carbon-containing discharges would probably fail to meet such regulations.

46.2.7   Autophoretic Coatings as Substitutes for Zinc Electroplate

A number of electroapplied organic coatings (E-Coat) and at least one commercially available "autophoretic" coating are available as viable nonmetallic substitutes for electroplated zinc coatings applied primarily for corrosion resistance to steel substrates. Properties and process characteristics of such systems are too diverse to detail here, but more information can be obtained from the sources cited in the "Further Reading" section at the end of this chapter.

46.3 Recovery and Recycling

The electroplating industry utilizes numerous recovery-and-recycle techniques to return a portion or all of the process chemicals to the origin. The typical plater will first evaluate the efficacy of drag-out rinsing; will determine if a viable, less polluting or nonpolluting substitute exists; and will make those changes before investing in recovery-and-recycle equipment.

Recovery-and-recycle equipment is generally expensive, requires reduction of water usage to be economically feasible, and increases the maintenance workload and operational complexity of the electroplating facility. However, such systems can reduce the amount of solid waste generated by the electro-plater and can often yield net savings in chemical costs that can often pay for the equipment in a matter of a few months.

Direct reuse involves the reuse of a waste material without processing it either as a feedstock in a production process or as a substitute for a commercial process. Recycled chemicals are used or reused in other industrial processes or are used as substitutes for other chemical products. Electroplating operations where reuse and/or recycling is commonly employed include

- Reuse of concentrated rinsewater in the electroplating process
- Nonrecycle recovery of metal or concentrates
- Sale of by-product sludges
- Regeneration and reuse of process solutions
- Recycling of treated wastewater

Some of the more commonly used technologies for reuse and recycling in the electroplating industry are

- Evaporative recovery
- Electrolytic recovery
- Ion exchange
- Reverse osmosis

Uncommonly used reuse and recycling technologies include

- Membrane systems for purification of process solutions
- Resin systems for regeneration of contaminated acids
- Electrodialysis
- Crystallization
- Carbon adsorption

A typical recovery-and-recycle system for a metal-finishing process tank involves either a drag-out tank, a counterflow rinse, or a combination of both.

The water used in the rinses should be de-ionized to protect the process solution from buildup of minerals that are present in tap water. Such buildup can slowly make a process solution contaminated to such an extent that the solution is totally lost and must be transported to a commercial disposal facility, often at very high cost. The recovery "unit" (see Fig. 46-2) can operate off the drag-out tank, off the first rinse of a counterflow rinse system, or off the process tank itself, depending on the design of the unit and its recovery efficiency. The recovery unit can recover all or only a portion of the chemical dragged out of the process. Rinsewater and unrecovered chemicals are typically discharged to a wastewater treatment system. The following is a general discussion of commonly used recovery-and-recycle equipment in electroplating.

46.3.1 Evaporative Recovery

Evaporation of water from processes that operate at elevated temperature can be very significant, and increases dramatically as the solution temperature goes beyond 140°F, or if the solution is air-agitated. The evaporation rate is also dependent upon the relative humidity in the air. For example, at 140°F and 50 percent relative humidity, an air-agitated tank, with 100 ft² of surface area may evaporate 10 gal/h of water, while a non-air-agitated tank can evaporate only 4 gal/h. At 160°F, the same numbers are estimated at 20 and 10 gal/h, respectively. This evaporative loss creates "room" in the process tank to return a portion of the drag-out loss. This return can be accomplished in several ways, including simply manually returning the drag-out liquid to the process tank. When the drag-out rate is high, however, there will not be enough room in the process tank to efficiently return the dilute drag-out solution. An evaporator can then be used either to create the additional room in the process tank, or to concentrate the rinsewater so less volume goes back to the process tank. Evaporators (see Fig. 46-3) are commonly employed on decorative chromium-plating solutions, nickel-plating solutions, and cyanide copper-plating opera-

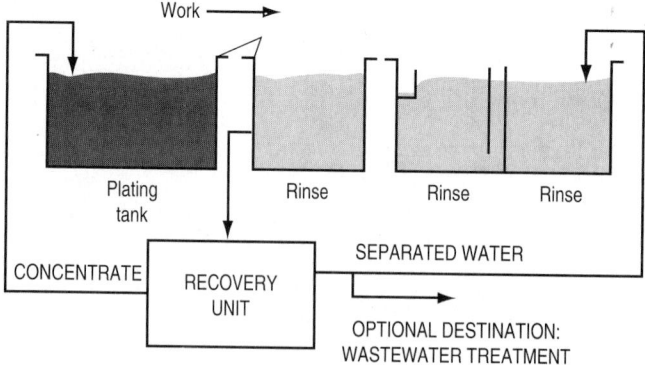

Figure 46-2. General schematic for recovery-and-recycle systems.

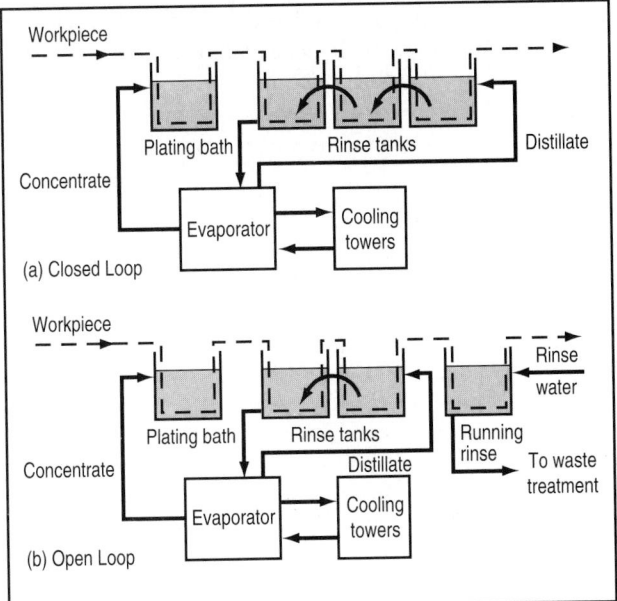

Figure 46-3. Schematic for evaporative recovery systems.

tions, although they are not necessarily limited to these. Evaporators fall into two general categories: vacuum and atmospheric.

Atmospheric Evaporators. Atmospheric evaporators operate by spraying the dilute wastestream over packing media, grid, or plates, and then blowing plant air over the packing, to effect evaporation. Advantages include low cost and simple maintenance. A major drawback is the inability to evaporate on days when air humidity levels approach 80 to 90 percent. In areas of general high humidity, the solution to be evaporated will need to be heated to some degree. Atmospheric evaporators generally have low capacities, ranging from 10 to 40 gal/h, although a number of them can be installed in parallel to attain whatever evaporation rate is desired (as long as the humidity in the air is not too high). On cyanide-based plating solutions, use of any evaporative system can increase the rate of carbonate buildup in the plating bath due to carbon dioxide adsorption from entrained air and thermal breakdown of cyanide.

Vacuum Evaporators. There are a number of different vacuum evaporator designs on the market, normally differing in how the vacuum is achieved (eductor, vacuum pump), or how much energy is employed (single effect, double effect), or how much vacuum is achieved. The higher the vacuum, the lower the temperature at which water will boil. By boiling the water off at lower temperatures (110 to 130°F), the vacuum evaporators protect some delicate ingredients of processing solutions that might decompose at higher temperatures.

Advantages of these systems include recovery of both a concentrate and condensed water, and operation in all weather conditions. Disadvantages include high energy and maintenance costs, and foaming of some process solutions (one evaporator manufacturer has a design that can handle foam).

Single-Effect Evaporators.   A single-effect unit usually uses steam or high-temperature hot water to heat the liquid to its boiling temperature. The steam is passed through a steam coil or jacket, and the vapors produced by the boiling liquid are drawn off and condensed. The concentrated liquid is then pumped from the bottom of the vessel. The process requires about 1200 BTUs per pound of water evaporated.

Multiple-Effect Evaporators.   A multiple-effect unit consists of a series of single-effect evaporators. Vapor from the first evaporator is used as the heat source to boil liquid in the second evaporator. Boiling is accomplished by operating the second evaporator lower than the first. The process can continue for several evaporators (effects). Depending on the number of effects, multiple effect units can require as little as 200 BTUs per pound of water evaporated.

Vapor Recompression Units.   The vapor recompression evaporators use steam to initially boil the liquid. The vapor produced is compressed to a higher pressure and temperature. The compressed vapor is then directed to the jacketed side of the evaporator instead of more steam being used, and it is thus used as a heat source to vaporize more liquid. These units require as little as 40 BTUs per pound of water evaporated.

46.3.2   Electrolytic Recovery

Electrolytic recovery (see Fig. 46-4) is a technology that uses special electroplating equipment to lower the concentration of dissolved metals in drag-out rinses

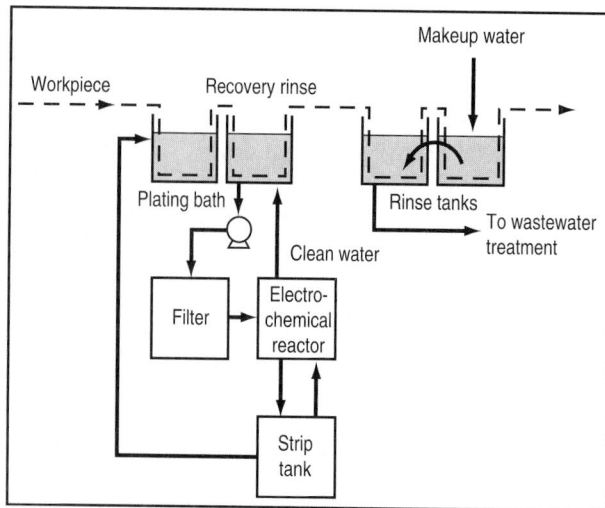

Figure 46-4. Schematic of electrolytic recovery system.

and concentrated-rinse tanks. Benefits include reduction of sludge generation, some electrolytic destruction of cyanide, and reuse or sale of scrap metal plated out. Disadvantages include incomplete recovery (some waste is generated), tendency to spontaneous combustion of plated metal, and energy costs. Electrolytic systems fall into two primary categories:

1. Equipment for primarily removing metal from the wastewater stream, in a form that has little potential for recycling

2. Recovery of metals with recycling of some other ingredient potential

The major difference between these systems is the type of cathode used to plate the metal out of the wastestream. A "conventional" system might be "home made," and simply use corrugated steel or stainless steel electrodes. A high-surface-area cathode system that improves the efficiency of metal removal is available from one supplier. Another supplier offered an "HSA" (high-surface-area) reactor that plated out cadmium from a drag-out rinse and simultaneously destroyed cyanide by electrolytic oxidation. The manufacturer is no longer in business, but the systems are still available on the used equipment market. Typical applications of these systems include after acid copper plating, cyanide cadmium plating, cyanide zinc plating, and cyanide copper plating. While attempts have been made to utilize these systems on nickel-plating processes, the lack of a suitable insoluble anode has held back most progress.

46.3.3   Ion Exchange

Ion exchange (see Fig. 46-5) has been in use in the metal-finishing industry for decades. One "centralized recovery" facility in Minnesota utilizes ion exchange cylinders to remove dissolved metals and cyanide from rinsewaters. The cylinders are rotated and returned to the central facility, where they either are treated to recycle beneficial materials or are waste-treated. A typical metal-finishing ion exchange system has a fixed bed of resin with the ability to exchange or remove cations or anions such as chromates from rinsewaters. In general, divalent and trivalent ions are easier to remove than monovalent ions using ion exchange.

When the useful capacity of the ion exchange column is exhausted, it is regenerated using dilute acid (sulfuric or hydrochloric) for cationic resins and sodium hydroxide solution for anionic resins. The hardware of prepackaged units consists of pressure vessels ranging from 2 to 6 ft or more in diameter, handling flow rates up to 300 gpm or more. Custom-made units, with 12-ft-diameter columns, have been built to handle flows as high as 1150 gpm. The loading typically is 10 gpm per square foot of resin cross section.

The chief advantage of ion exchange is that it can be selective in what it removes. In a recycling application, this means that undesirable impurities need not be recovered along with the desired materials. Close control of the pH of the feedstream is important, with lower pH reducing the capacity of the resin, and higher pH tending to clog the resin with solids (metal hydroxides). The systems also have the drawback of not having suitable instrumentation to

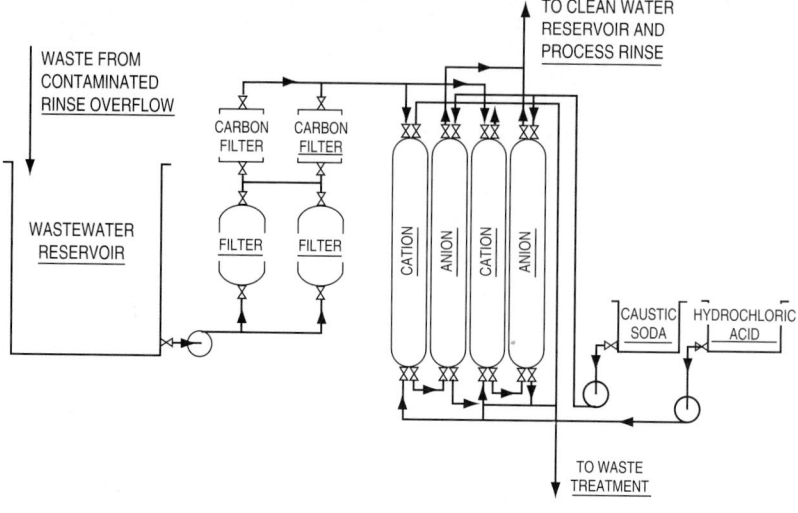

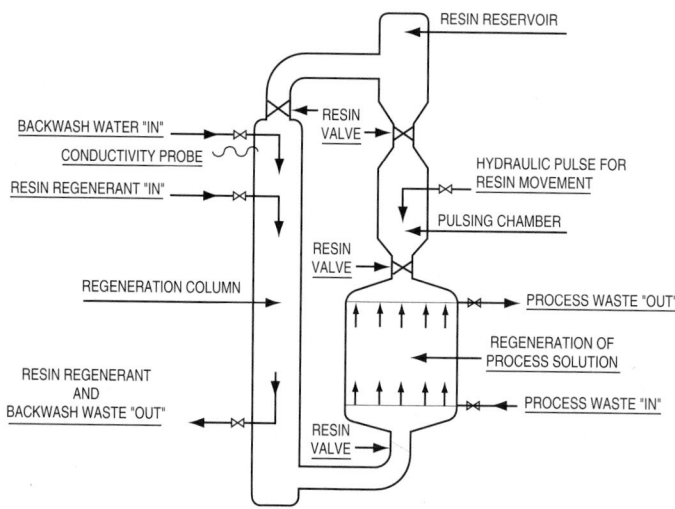

Figure 46-5. Schematics of ion exchange systems, conventional (top) and moving bed (bottom).

tell when the resin is saturated. Saturated resin will discharge an effluent containing high concentrations of the pollutant it is supposed to remove. Operators typically compensate for this drawback by taking ion exchange cylinders off line and replacing them on a tight time-based schedule, the schedule being based on a study of the resin loading rate.

Additional disadvantages of ion exchange recovery systems include the requirement of additional treatment equipment for modifying the regenerant stream to a chemistry suitable for reuse in some selected cases. An example is a nickel recovery system, which requires de-acidification of the regenerant before it can be used in the plating bath. The regenerant stream is often not highly concentrated (2 to 4 oz/gal is typical), and often evaporative systems are needed to preconcentrate the recovered stream to a usable level. The technique is considered to be an expensive recovery method that requires a lot of care and knowledge of operation.

Ion exchange also can be used to remove specific ions from rinsewater at the rinse tank. A typical application may be to remove nickel, copper (from an acid bath), or chromium from their respective rinses and then to return the regenerated metal ions to the plating bath as concentrates.

Ion exchange is also frequently utilized to remove tramp heavy metals from hard-chromium-plating solutions. The presence of heavy metals such as iron, copper, and nickel can significantly reduce the efficiency of the chromium-plating solution, creating excessive misting and solution loss through the exhaust stack. Such systems do generate a regeneration waste that needs to be treated to remove the heavy metals prior to sewer discharge.

46.3.4 Reverse Osmosis

Reverse osmosis (RO) (see Fig. 46-6) is used to separate water from inorganic salts, through the use of a "membrane" that allows transfer of water and "rejects" the salts. The system utilizes pressures of 400 to 800 psi, generated by pumps to force the water through the membrane ("permeate"), leaving a concentrated

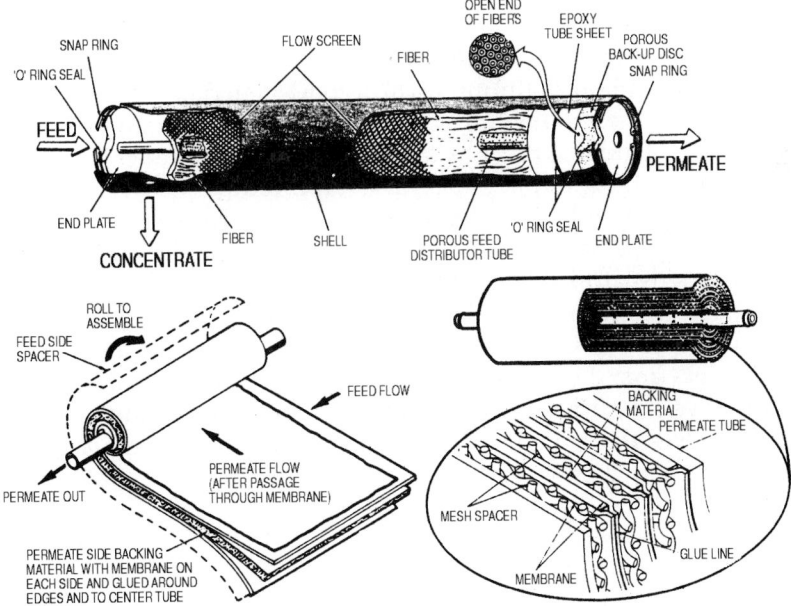

Figure 46-6. Hollow-fiber RO membrane (top) and spiral-wound RO membrane (bottom).

residual liquid ("rejectate") behind. To prevent fouling of the membrane, feed solutions must be pretreated to remove materials such as magnesium, calcium, lead, iron, carbonates, particulates, oils, and other fouling materials.

The membranes are made out of cellulose acetate (similar to the clear plastic covering cigarette packages), aromatic polyamides, and cross-linked polyamides, and can often represent 50 percent or more of the cost of the equipment. They come in various configurations, including spiral wound, tubular, and hollow fiber. The tubular membrane is inserted onto or into the surface of a porous tube. This type of RO is used for low-volume (low-pressure) applications. The spiral-wound membrane is a flat sheet separated by a mesh spacer that is spirally wound around a perforated plastic tube that acts to channel the permeate flow. The hollow-fiber membrane consists of millions of membrane fibers. The most common (99 percent of all) application of RO would be on the rinsewater of a nickel-plating tank. The rejectate is often too dilute to directly return to the tank, so evaporation is often added to the system. RO systems have been developed to be used on alkaline solutions such as cyanide copper and brass, but the rinsewater must be filtered extremely well and must have a low carbonate level, and the water used in the plating and rinse tanks must be de-ionized. Aside from the fouling problems, some RO systems are very temperature sensitive. For example, a cellulose acetate membrane cannot withstand a temperature greater than 96°F. The membrane also rapidly deteriorates if the pH of the feedstream is less than 3 or higher than 10. The most popular process in the electroplating industry utilizing RO recovery equipment is nickel plating.

46.3.5 Electrodialysis

Electrodialysis (ED) (see Fig. 46-7) is a process which uses a "stack" of closely spaced ion-selective membranes through which ionic materials are selectively transferred or rejected. The driving force causing the ions to migrate within the stack is an electrical potential applied by a rectifier to two electrodes. In a plating operation, ED is typically applied to a rinse tank, and the ED system separates the dissolved metal and associated anions from the rinsewater. Because each pass through the system does not create a very concentrated product stream, the system is usually connected to a recirculating drag-out rinse tank. Typical applications are on drag-out tanks after plating tanks for gold, silver, nickel, and acid tin plating. On nickel-plating systems, ED does not recover organic contaminants from the rinsewater, thereby eliminating organic contaminant buildup that would exist with evaporative systems. ED equipment is very expensive and involves the same level of membrane fouling and replacement cost problems as RO.

46.3.6 Others

There are a number of other recovery-and-recycle schemes available to a metal finisher. One unique proprietary process is a resin-based system for purification of acid contaminated with heavy metals. A typical application would be on

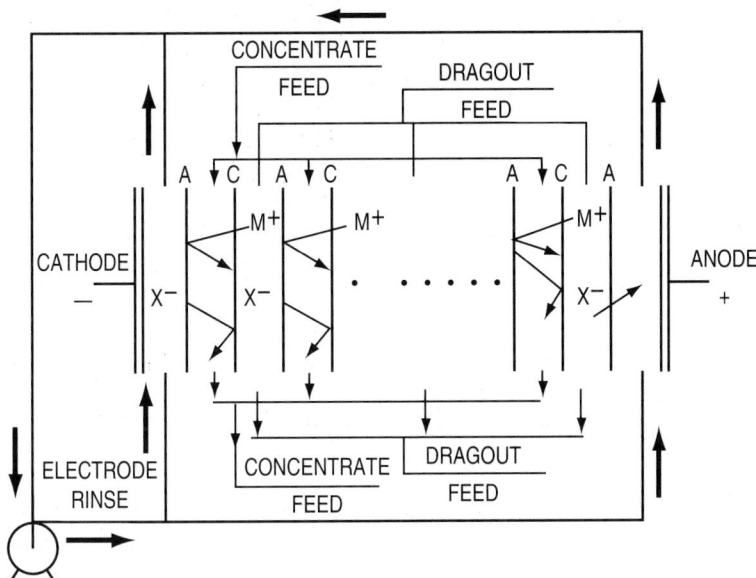

Figure 46-7. Schematic of electrodialysis stack.

aluminum-anodizing solutions based on sulfuric acid and stripping solutions based on nitric acid. The resin retains the acid and allows the heavy metal to pass through. Upon backwashing with water, the acid (which may be as concentrated as 50 percent prior to purification) is released by the resin and is returned to the process.

Crystallization, or "freeze-drying," systems are available and are used in a number of large manufacturing firms that have electroplating facilities. Crystallizers are commonly used in printed circuit board manufacturing facilities to recover copper sulfate crystals from the etching and plating wastes. Freeze-drying systems have been pilot-plant-tested on nickel-plating solutions, yielding mixtures of crystals of nickel sulfate and nickel chloride. These systems are typically too expensive for most all job shops to justify installing.

Carbon adsorption systems for recovering organic solvent lost from vapor degreasing and solvent-spraying operations are also available (at high cost).

The U.S. Bureau of Mines in Rolla, Missouri, developed a process for the recovery and recycling of heavy metals and chromic acid from chromic acid–based copper-stripping solutions. Several home-made systems have been constructed and have worked reasonably well (the main cost is replacement of the membranes on a periodic basis). A license to construct a system costs $1.00.

46.4 Pollution Prevention Successes

There are far too many pollution prevention success stories in electroplating to detail in this handbook. Reference 2 has specific case studies.

While we have covered the equipment and chemical processes available for a pollution prevention program in electroplating, it should be noted that no pollution prevention program has a chance to succeed without the firm and unwavering commitment of management to solve those problems that invariably arise as a result of instituting such a program and making changes in processes and equipment. There are hundreds of pollution prevention tips, techniques, and hints available through a large number of pollution prevention guides published by the EPA, and by various individual states, universities, and societies. Some of these are also listed in the "Further Reading" section below.

References

1. U.S. Environmental Protection Agency, "In-Process Pollution Abatement: Upgrading Metal Finishing Facilities to Reduce Pollution," EPA Technology Transfer Seminar Publication.
2. U.S. Environmental Protection Agency, *Pollution Prevention Case Studies Compendium*, U.S. EPA, Industrial Environmental Research Laboratory, Cincinnati, Ohio, EPA/600/R-92/046, 1992.

Further Reading

Altmayer, Frank, "Pollution Prevention: Both Sides," *Plating and Surface Finishing*, August 1992, p. 23.

California Department of Health Services, *Hazardous Waste Reduction Checklist and Assessment Manual for the Metal Finishing Industry*, California Department of Health Services, Toxic Substances Control Program, Alternative Technology Division, Technology Clearing House, February 1992.

Control and Treatment Technology for the Metal Finishing Industry, Evaporators, U.S. EPA, Industrial Environmental Research Laboratory, Cincinnati, Ohio, EPA/625/8-79/002, 1979.

Control and Treatment Technology for the Metal Finishing Industry, In-Plant Changes, U.S. EPA, Industrial Environmental Research Laboratory, Cincinnati, Ohio, EPA/625/8-82/008, 1982.

Control and Treatment Technology for the Metal Finishing Industry, Ion Exchange, U.S. EPA, Industrial Environmental Research Laboratory, Cincinnati, Ohio, EPA/625/8-81/007, 1981.

Economics of Wastewater Treatment Alternatives for the Electroplating Industry, U.S. EPA, Industrial Environmental Research Laboratory, Cincinnati, Ohio, EPA/625/5-79/016, 1979.

Environment Canada, *Pollution Prevention and Control in the Metal Finishing Industry*, Environment Canada, Canadian Water and Wastewater Assistance, Ottawa, Ontario K1N 5P3. Telephone: (613) 238-5692.

Facility Pollution Prevention Guide, U.S. EPA, Industrial Environmental Research Laboratory, Cincinnati, Ohio, EPA/600/R-92/088, 1992.

Guide to Clean Technology, Alternative Metal Finishes, U.S. EPA (unpublished).

Guides to Pollution Prevention: The Fabricated Metal Products Industry, U.S. EPA, Industrial Environmental Research Laboratory, Cincinnati, Ohio, EPA/625/7-90/006, 1990.

Guides to Pollution Prevention: The Metal Finishing Industry, U.S. EPA, Industrial Environmental Research Laboratory, Cincinnati, Ohio, EPA/625/R-92/011, 1992.

Guides to Pollution Prevention: The Printed Circuit Board Manufacturing Industry, U.S. EPA, Industrial Environmental Research Laboratory, Cincinnati, Ohio, EPA/625/7-90/007, 1990.

Ingle, Mark, and James Ault, "Corrosion Control Performance Evaluation of Environmentally Acceptable Alternatives for Cadmium Plating," in *Proceedings, SUR/FIN '92*, American Electroplaters and Surface Finishers Society, Orlando, Fla., 1992.

Swalheim, D. A., *Recovery and Re-Use of Chemicals in Plating Rinses*, American Electroplaters and Surface Finishers Society, Orlando, Fla., 1988.

Swalheim, D. A., *Rinsing, Recycle and Recovery of Plating Effluents*, American Electroplaters and Surface Finishers Society, Orlando, Fla., 1988.

Treatment of Electroplating Wastes by Reverse Osmosis, U.S. EPA, Industrial Environmental Research Laboratory, Cincinnati, Ohio, EPA/600/2-76/261, 1976.

U.S. Army Corps of Engineers, *Evaluation of Aluminum Ion Vapor Deposition as a Replacement for Cadmium Electroplating at Anniston Army Depot*, Contract DAAA15-88-D-001, April 1992.

Waste Minimization Opportunity Assessment Manual, U.S. EPA, Industrial Environmental Research Laboratory, Cincinnati, Ohio, EPA/625/7-88/003, 1988.

47

Pollution Prevention at General Motors

Todd A. Williams

Sandra S. Brewer

P. N. Mishra, Ph.D.

O. Warren Underwood
General Motors Corporation
Detroit, Michigan

47.1 Industry Overview

Automotive manufacturing is a complex industry involving thousands of processes to complete the final car or truck. The motor vehicle industry contributes about 4.5 percent of the GNP of the United States and is a major economic presence in every industrialized country. The industry is a major consumer of iron, steel, plastics, rubber, glass, chemicals, zinc, platinum, aluminum, and electronics. General Motors (GM) is a highly vertically integrated car and truck manufacturer and the largest manufacturer in the world. GM also produces components for its own vehicles, as well as for other domestic and foreign automobile manufacturers, and replacement/service parts markets.

The most common vision of automotive manufacturing is that of the final vehicle assembly line. This vision is somewhat outdated and oversimplifies the assembly process. The assembly plant has the challenge of receiving the thousands of parts necessary for today's vehicles, coating and painting the body, and assembling the components into the final vehicle. A typical assembly plant can produce more than 200,000 vehicles annually on a two-shift operation.

The power train of a vehicle, basically the engine and transmission, are manufactured in similar facilities. These facilities have extensive metal machining operations, followed by the assembly of the final engine or transmission. These plants also receive many parts, such as castings, fasteners, belts, hoses, fuel systems, and ignition systems, from other GM divisions or outside suppliers.

GM and outside suppliers manufacture the many other components necessary to produce today's automobiles. The processes are numerous and include ferrous and aluminum casting, forging, machining, plating, welding, metal stamping, and plastic molding. Some of the major components include fuel management systems, spark plugs and ignition systems, batteries, heat transfer systems, chassis, brakes, axles, wire harnesses, electronics, instrument panels, seats, body trim, and steering systems.

47.1.1 Wastestreams

Wastewater. Automotive assembly plants typically generate wastewater containing paint, sludge, and dissolved metals, mainly zinc and lead. Paint particles in the wastewater are detackified (killed) and removed for disposal. The next step in wastewater treatment is hydroxide precipitation of the metals, followed by clarification, before the wastewater is discharged to a sanitary sewer system. The many different machining operations of power train and components plants generate oily wastewater. This type of wastewater contains both emulsified and free oils, as well as cleaning solutions, and is generally treated on-site with physical and chemical processes. The oily sludge may be recovered on site for reuse or sent off site for re-refining or energy recovery. Component operations that involve certain processes, such as plating, produce a metal-bearing wastewater. Metals removed during treatment (usually hydroxide precipitation) may be recovered or properly disposed.

Solid Waste. The largest solid wastestreams generated by GM include scrap metal, plant trash, wastewater treatment sludges, and waste oils. Scrap metals include mainly steel, cast iron, aluminum, and copper from stamping press offal, machinings, scrap parts, and machinery. GM recycles virtually 100 percent of the scrap metal produced, due mainly to the available markets and economic value. All cast iron poured at GM's foundries is melted from scrap iron from internal and external sources. Plant trash is mainly composed of packaging waste such as fiberboard (corrugated or "cardboard"), wood, and plastic. Plant trash also contains office waste, cafeteria wastes, and other general refuse. Many facilities segregate and recycle corrugated, wood, plastic, and office paper, which greatly reduces the amount that must be landfilled. Some facilities have on-site waste-to-energy units to supplement fuels used to produce steam for space heating and plant processes. Wastewater treatment sludges vary in composition and some may be recycled or landfilled. Waste oils, such as spent hydraulic, quench, and machining oils, may be recovered at the source or recovered during wastewater treatment. Virtually all waste oils generated at GM facilities are recycled or burned for energy recovery.

Air Emissions. Automotive manufacturing air emissions are associated mainly with automotive coating (painting), powerhouse operations (boilers), and metal casting. These sources meet strict emissions standards and are abated with state-of-the-art treatment when appropriate. Emissions from coating operations have long been a focus of source reduction due to the high cost of automotive coatings and strict emission standards for volatile organic compounds (VOCs). Paint technology has advanced greatly in the last two decades, greatly improving transfer efficiencies and reducing VOC emissions. These improvements have been achieved by using high solids, waterborne, and powder coatings in conjunction with high transfer efficiency equipment and processes. Many GM facilities have converted boilers from coal to natural gas for both economic and environmental benefits.

47.2 GM's Pollution Prevention Program

General Motors' corporate pollution prevention initiative is called "WE CARE." This is an acronym for the phrase **W**aste **E**limination and **C**ost **A**wareness **R**eward **E**veryone. WE CARE's foundation lies in the GM WE CARE mission which states:

> To minimize the impact of our operations, we will reduce emissions to the air, water, and land by putting priority on waste prevention at the source, elimination or reduction of wasteful practices, and the utilization of recycling opportunities whenever available. The responsibility for achievement of this goal is primarily dependent on both management's support and actions of every employee to modify existing methods, procedures, and processes and to incorporate waste prevention into all new endeavors.

This program was initially piloted in 1990 at selected facilities within GM before being introduced throughout GM's United States and Canadian operations in 1991, followed by GM's Mexican facilities in 1992.

The WE CARE program guides each individual facility in setting up a multi-discipline committee to direct pollution prevention efforts. The committee should include representatives from such departments as maintenance, quality control, materials management, production, engineering, purchasing, and environmental, as well as representative(s) from the local union. This approach makes pollution prevention a part of everyone's job—not just that of the environmental engineer. These committees coordinate local pollution prevention efforts, including brainstorming, prioritizing efforts, investigating and evaluating different pollution prevention options, implementing projects, tracking results, and communicating and rewarding outstanding performance. WE

CARE supports the Environmental Protection Agency's environmental management hierarchy, which places highest priority on source reduction, followed by recycling, treatment, and, finally, environmentally sound disposal.

The corporate WE CARE Committee is responsible for promoting facility efforts, tracking overall performance, facilitating the exchange of information within the corporation and reporting progress to government agencies and the public. This committee also develops and coordinates corporatewide pollution prevention strategies and initiatives, such as the Automotive Pollution Prevention Project, the EPA's 33/50 program, a used-oil recycling initiative, and pollution prevention training.

General Motors facilities are engaged in numerous pollution prevention activities and projects. As mentioned earlier, painting technologies have been advancing significantly in the last two decades. These changes include high-solids paints, waterborne paints, powder coatings, high-transfer efficiency equipment, and many other pollution prevention advancements. GM has instituted pollution prevention, recyclability, and other manufacturing considerations into the design process of the vehicle. This approach is effective in reducing the total life-cycle impact of GM vehicles. For the purposes of this handbook, two broad-based initiatives that have applications outside of the automotive industry will be discussed. First, *chemicals management* is a new supplier relationship that reduces the overall chemical cost while providing incentives for the chemical supplier to find ways to reduce chemical usage. The second initiative, *packaging reduction and recycling,* has been a great success at many of GM's facilities. The goal is to reduce the amount of packaging coming into a plant and to make sure that the packaging that is received is easily recycled or returned.

47.3 Chemicals Management

As competition has grown in the automotive industry, innovative techniques have been developed to improve processes and stay competitive. With shrinking work forces, one technique to improve processes is to leverage resources and expertise from other sources. Another way is to reshape the relationship between supplier and customer. Chemicals management utilizes both of these techniques to reduce waste and save money. This approach is quite different from the traditional way of doing business. Automotive companies are large users of chemicals that aid manufacturing processes. These chemicals include cleaners, machining fluids, hydraulic fluids, quenching fluids, water treatment chemicals, and solvents. These chemicals are often called indirect chemicals because they are not directly incorporated into the final product (vehicle). Direct chemicals include automotive paints, vehicle lubricants, and fluids that are incorporated directly into the final product.

Generally, a manufacturing company's focus is on producing quality products and not on the use of chemicals. Therefore, knowledge of chemicals within the company will tend to be limited and diffuse, and the management of chemicals at a facility will tend to be more reactive than proactive. Instead of using the limited resources to become chemical experts, it is often better to form a partnership with a chemical supplier who can bring chemical expertise to the plant floor. In this way, the supplier plays an integral role in the management of the chemical services in the plant.

47.3.1 Objective

The objective of chemicals management is to reduce manufacturing costs associated with indirect chemical services. By developing and implementing an effective chemicals management system in the facility, we reduce the usage of chemicals at the source and thus reduce waste treatment and disposal costs. While minimizing the use and overall cost of chemicals, it is essential to meet or exceed production and quality goals. The company must increase the supplier's responsibility and share the risks and rewards while maintaining or improving the competitive position.

47.3.2 The Old Way

A single facility could have many different chemical suppliers—perhaps more than a dozen. Often there are several suppliers supplying cutting fluids, several suppliers for lubricating oils, and several for cleaners. This situation will result in a greater number of chemicals used at a facility than is actually needed. Furthermore, interacting with this many suppliers requires a great deal of time.

With this system, most chemicals are purchased by volume. Therefore, the more chemical that is used, the more money that the chemical supplier makes. This way of doing business does not encourage suppliers to be more frugal or to come up with innovative ways to reduce usage at the source. Furthermore, once the chemical reaches the facility, the facility assumes responsibility for that chemical.

The facility must manage the inventory of all the chemicals. This is difficult when there are so many. Furthermore, carrying a large inventory of chemicals is a poor use of capital. Also, if a problem arises with a chemical system, the facility personnel are usually the ones to do the investigation and troubleshooting. Since they are not in the chemical business, they must rely on somewhat limited knowledge obtained from experience at that facility. The best solution may not be reached in the most timely manner.

This system does not encourage suppliers to introduce new technologies unless it also means more profit. Plants may also be encouraged to use higher-priced formulations in areas where they are not needed. Each department within the facility often has autonomy over the chemicals that are used there. Individual departments will tend to its specific needs and not view the facility's chemicals management as a system. Each department will also tend to resist change if it is comfortable with the status quo. Departments are likely to resist giving up the control of their systems.

In this customer relationship, the plant purchases chemicals by volume. A close look makes it apparent that the goals of the customer and supplier are not in alignment. There is no incentive for the supplier to help the plant reduce its usage of chemicals. This traditional system breeds inefficient use of time, effort, money, and chemicals.

47.3.3   The New Approach

General Motors has now moved to a new chemicals management system. The goal is to have *one* supplier for all indirect chemicals at each facility. Obviously, a single supplier would not be able to supply every chemical for a facility. Thus, the "tier-1" supplier is responsible for obtaining chemicals from other suppliers (tier 2) if they cannot supply chemicals to meet certain needs. This tier-1 supplier provides a chemicals manager for the facility. Under this new system, the facility no longer purchases chemicals. The facility now purchases chemicals service. In addition to supplying chemicals, the supplier provides management, analysis, inventory control, and information management services. The supplier becomes part of the production team and no longer is paid by volume. Under the chemicals management system, the supplier is paid based on production or a fixed amount. A financial cap is placed on the amount the supplier can make. Because the supplier is paid based on production and not the amount of chemical used, there is no longer an incentive for the supplier to sell more chemical. The supplier's incentive is actually to reduce the amount of chemicals that are needed to do the job. In this way, the customer and supplier have the same goals.

The supplier now owns the inventory and is more likely to achieve just-in-time delivery. Further, now that profits are no longer based on the amount of chemical sold, continuous improvement forms a way to achieve joint goals, greater profitability, and better service. The chemicals manager must have up-front involvement in relevant plant projects so that proactive steps can be made and both parties can benefit.

47.3.4   Benefits

The benefits include immediate cost savings through the reduced number of suppliers, types and volumes of chemicals, and chemical inventories. Additional benefits are improved information management, environmental control (waste treatment and disposal), chemical technology application, reduced purchase order processing, and reduced freight. Because of the consolidation of similar chemicals, more chemicals can be handled in bulk or bins in lieu of drums. This reduces drum handling and deposits.

A single supplier who supplies the majority of chemicals at a facility can now afford to have an on-site chemicals manager. This chemicals manager will generally work full-time at large facilities, but is not paid by the facility. The chemicals manager has an incentive to convert as many systems to the tier-1 supplier's products as possible. The chemicals manager also has an incentive to reduce the number of different chemicals used by the facility to reduce the inventory that is now owned by the supplier. With the chemicals manager as a single focal point, chemical information management is more easily coordinated. This makes SARA 313 reporting and other environmental reporting much easier.

47.3.5 Implementation

To ensure the success of a chemicals management program, facilities should follow a proven implementation process. The first step is to introduce this concept to plant personnel. Building an understanding of the initial concept is important in order to get acceptance of this process. The facility should form an implementation team consisting of all the major stakeholders in this process. This team should include representation from the following groups: production, maintenance, plant engineering, process engineering, environmental, health and safety, purchasing, local union(s), wastewater treatment, powerhouse, material management, and chemistry. This committee approach ensures local development and implementation and speeds up the program's acceptance throughout the facility. The following table outlines the entire 16-step process used by General Motors to implement this program.

16-Step Implementation Process

1. Introduce the concept to plant personnel
2. Form a plant committee
3. Involve union members
4. Determine the present status of chemicals in the plant
5. Collect existing cost and production data
6. Obtain committee acceptance
7. Obtain plant manager acceptance and approval
8. Determine the group of chemicals best suited for chemicals management
9. Set up performance requirements for the supplier
10. Develop and issue prequalification questionnaire
    - Company organization
    - Legal forms
    - Size of company and staff
    - Financial qualifications
    - Insurance
    - Types of services
    - Experience
    - Project approach

11. Review responses and develop a "short list"

12. Develop system specifications

13. Develop and issue proposal requests

14. Review responses and interview suppliers

15. Select supplier best suited for plant needs

16. Implement program jointly with supplier

This process takes time and commitment by the entire team, but the rewards definitely make this effort worthwhile. Some reluctance is likely to be encountered both at the plant and by the supplier community, since this is a new way of conducting business. Some may not want to give up control. Once these fears are overcome, however, departments realize that the chemical service provided is usually of a much higher quality and the facility sees overall benefits through the single-system approach of this process.

47.3.6 Success Stories

One of the first assembly plants at GM to implement this program went from 35 different suppliers supplying 348 chemicals to 12 suppliers (including 11 tier-2 suppliers) and 200 chemicals. This is a 66 percent reduction in the number of suppliers and a 43 percent reduction in the number of chemicals. Total savings were well over $750,000 per year, and there were many additional benefits realized which are not as easily quantified.

Because chemicals management includes wastewater treatment, substantial savings and operational improvements are realized. One assembly plant reported going from three treatment chemicals to detackify paint sludge to one, with a 95 percent reduction in polymer usage. This common systems approach ensures that chemicals are compatible through a better coordinated process.

One facility's chemicals manager helped identify and correct many oil leaks that previously ended up in wastewater treatment. At the North American Operations (NAO) Technical Center, the chemicals manager was instrumental in saving 50 percent of the water used, along with the treatment chemicals that would have been needed for treatment. At another facility, a die lubricant was successfully replaced with plain tap water. These types of solutions and this level of service are not typical of the traditional system.

47.4 Packaging Reduction and Recycling

Packaging is one of the largest wastestreams that an automotive assembly plant must handle. Packaging protects the many parts being shipped from suppliers to a facility. Packaging (or container) design for the shipment of parts from manufacturing locations to the assembly plants has come a long way since the days of manually loading and unloading parts into and out of railroad boxcars and truck trailers. Large parts were once loaded by hand and blocked, piece by piece, with wooden bars and spacers to hold the parts in place in the trailer or boxcar. In the early 1950s, the railroads began using their own containers because of the large number of parts being shipped. Since the containers were the property of the railroad, they still had to be unloaded by hand at their destination.

In the mid-1950s, GM started designing and purchasing standardized returnable metal racks and baskets to use as shipping containers. While many of these are still in use today, more containers are now designed specifically for shipping a particular part or family of parts. This trend has significantly increased the number of containers in use. GM uses millions of returnable containers in hundreds of configurations. Today, returnable containers may be made of plastic or paper products, as well as metal.

47.4.1 Packaging Criteria

It takes many individuals to engineer, test, and implement packaging devices for thousands of parts being shipped daily between GM facilities or from outside suppliers to GM facilities. Packaging engineers design containers to meet the following criteria:

1. Protect the quality of parts being shipped to the facility
2. Minimize the labor required to stock assembly lines or to load and unload trailers and boxcars
3. Consider the ergonomics of the workplace and the employees
4. Reduce the amount of expendable packaging
5. Minimize the overall packaging costs
6. Optimize freight utilization

To effectively meet these criteria, packaging engineers must work with product engineers, manufacturing engineers, tool designers, and plant personnel.

Today, GM has more than 100 engineers dedicated to packaging (containerization) throughout NAO. GM's Packaging Action Committee, formed in 1984, brings representatives from all of the different vehicle engineering groups and divisions together in a cooperative effort to share ideas and new technology, including those that protect the environment. Through this committee, packaging methods are standardized, allowing a balance among manufacturing, transportation, and assembly needs, while the quality of the parts is protected at the lowest overall cost. GM's packaging specifications were recently changed to require incoming packaging to be reduced and made more readily recyclable. These specifications reduced the costs by:

- Reducing the number of containers or the volume of packaging materials being used

- Reducing the amount of manual labor required to break down and segregate expendable packaging materials for recycling
- Increasing the amount of recoverable materials
- Reducing packaging material sent to landfill

47.4.2 Returnable vs. Expendable

The obvious answer to returnable versus expendable packaging seems to be that returnable containers are always better. However, in practice, this is not necessarily the case. Looking at total cost, environmental impacts and site-specific considerations will usually lead to some combination of the two.

Returnable Containers.   Returnable containers (returnables) are containers that are shipped back to the original supplier when empty. Examples include metal racks, rigid plastic racks and dunnage, metal skids, returnable bins, and totes. Returnables often contain expendable dunnage to protect the parts. Returnables have a high initial investment; however, in time, that cost can be recovered.

Returnables usually have a greater tare weight which may increase transportation costs and have negative ergonomic impacts if the containers are manually handled. If a part is changed significantly, the returnable containers may have to be altered or completely changed at a relatively high cost. With a large number of suppliers, plants may find that returnables create a logistical problem requiring detailed tracking and at-plant storage. Costs to return these containers may be large, especially if the containers do not break down or nest and if the distance traveled is great. Returnables need periodic cleaning, repair, and maintenance, which is an ongoing expense. Returnables often have features such as "feet," ribs, or other protrusions that may inhibit plant material handling systems. Obviously, these containers do not require disposal and, therefore, use of returnables avoids this ever-increasing expense. Despite some of the drawbacks mentioned, returnables are an attractive option for many situations, especially if the suppliers are relatively close.

Expendable Packaging.   Expendable packaging (expendables) is used once and then discarded or the material content is recycled. These include corrugated boxes, shrink-wrap, styrofoam peanuts, wood skids, plastic, and metal banding. Expendable packaging has virtually no initial investment but has an ongoing cost to the supplier, who will pass this on to the customer. Expendables do not have to be returned to the original supplier, which avoids some logistical challenges. Its light weight may reduce shipping costs. It can be easily modified if parts are changed, which provides greater flexibility. However, simply throwing out used packaging is not only environmentally detrimental, but the cost of disposal continues to rise. Recycling of expendable packaging is a good option, but involves internal labor to sort and handle. If the packaging is well thought out and the proper systems are in place in the plant, sorting and handling costs can be greatly reduced. The markets for recycled materials must be located and prices vary from location to location and from year to year. Expendable packaging may not be suitable near operations sensitive to fiber contamination (paint shops) or near ignition sources (welding).

47.4.3 GM's Strategy

GM is made up of many divisions and business units that are very different from each other and have different needs. One single packaging strategy would not fit the entire corporation and, therefore, each division is responsible for setting its own goals and strategies. Deciding on returnable versus expendable packaging is a complex decision that is dependent on a number of factors. Simply using all returnable containers does not solve all the problems and creates some others. Looking at the total life-cycle cost associated with the packaging systems is essential to making the best choice.

Both Saturn and many of the Cadillac Luxury Car Division (CLCD) plants considered costs, suppliers' locations, facility infrastructure, and floor space considerations before deciding on strategies that rely heavily on returnable packaging. Other divisions of GM have programs in place to improve existing packaging practices by using returnable containers where appropriate, improving the design of returnable and expendable packaging, eliminating the use of unnecessary or excessive packaging, improving the recyclability of packaging, and using recycled content in packaging.

In the summer of 1991, the packaging engineers of the Midsize Car Division, formerly Chevrolet-Pontiac-Canada (C-P-C) Group, piloted an effort in support of the WE CARE program. A zero landfill goal for all packaging waste by January 1, 1994, was set for the division's seven assembly plants. These plants are located all over the United States and Canada, which would make the full use of returnable containers uneconomical for many applications. The following packaging guidelines and requirements were developed and communicated to parts suppliers, both internal and external. These guidelines were so successful that they are now used throughout GM and have been shared with other automotive companies through the Automotive Industry Action Group (AIAG). In December of 1992, the AIAG published the "Solid Waste Management Packaging Material Guidelines," which essentially utilizes the GM standards.

Guidelines

Eliminate—Where possible, eliminate the packaging altogether without putting products at risk of damage

Reduce—Minimize the amount of material used in packaging

Reuse—Where practical, use packages that are either returnable or refillable/reusable

Recyclable/recycled content—Use packaging that is recyclable and that uses recycled material

Supplier Requirements for Expendable Packaging

1. Use corrugated pallets for all loads 500 pounds or less, considering performance and economics

2. Construct pallet cartons with a breakaway feature or other method to easily allow the separation of the wood pallet from the corrugated box

3. Color code nonmetallic strapping based on AIAG standards, and the closure must be by friction sealing (i.e., not metal)

4. Do not staple wood to the corners, walls, or top of a corrugated carton

5. Do not bond dissimilar packaging materials together such as foam glued to corrugated

6. Do not use expanded polystyrene (EPS) as an expendable packaging material

7. Use unitizing adhesives ("lock-and-pop") in place of strapping or stretch-wrap

8. If stretch-wrap films are used, they should be uncolored, either high- or low-density polyethylene, and correctly identified with the SPI code and chasing arrows symbol

9. Do not use wasteful, excessive, and nonrecycled/nonreturnable packaging

10. Identify plastics with the proper SPI code and chasing arrows symbol

11. Do not use lead or cadmium in packaging or labeling material

47.4.4 Success Stories

By working closely with suppliers, many common-sense solutions to packaging problems have been achieved, often resulting in a savings for the supplier that was shared with GM. In this highly competitive market, finding cost-effective solutions to environmental challenges is essential.

The Mid-Size Car Division has made great progress since setting the "zero-packaging-to-landfill" goal. At the onset, the average quantity of packaging waste going to landfill was 82 pounds per vehicle. As of September 1993, the average was down to 18 pounds. This represents more than a 75 percent reduction in just two years! One assembly plant has done an exceptional job and, as of September 1993, "landfills" less than one pound of packaging per vehicle. This facility is in a rather remote location and depends a great deal on expendable packaging. Putting the proper systems in place, along with commitment from all employees, has turned into a win for the environment and for the plant. The "zero-packaging-to-landfill" goal is currently expanding to other divisions, while aggressive goals for packaging reduction are already in place at others.

Since 1987, CLCD has purchased 600,000 returnable containers to support its strategy. As of June of 1993, 1233 tons of plastic material from damaged or obsolete containers has been sold back to the original container manufacturers to be reground and recycled into new containers. Landfill avoidance costs between 1987 and 1992, attributable to the use of returnables, are estimated at over $12 million.

The Delco Remy Division replaced large wire baskets with plastic tray packs to transport starter motors and other components. These new packs weigh 147 pounds, while the old wire baskets, which required plastic or corrugated dunnage, weigh 449 pounds. These new packs nest together when empty, taking up only 15 percent of the space of the same number of wire baskets. The plastic used in these packs is made from 50 percent recycled content. When damaged or obsolete, the packs are sold back to the manufacturer to be reground and recycled into new packs.

48

Wood Furniture Finishing*

Joshua M. Heltzer

Virginia Department of Environmental Quality
Office of Pollution Prevention
Richmond, Virginia

48.1 Industry Overview

Wood furniture finishing (WFF) is performed under the wood furniture manufacturing (WFM) process, as indicated by Fig. 48-1. Finishing involves the
application of protective and/or decorative coatings to products. Coatings,
such as stains and topcoats, are typically delivered to assembled wood items in
spray booths by means of carrier solvent blends. Solvents used in finishing
have been designated as hazardous air pollutants (HAPs) under the Clean Air
Act, contribute to ground-level ozone (the principal component of smog), and
may present health and safety problems for those working closely with the
materials.

The function of furniture finishing is to provide a specific final appearance
and protection which meets the tastes and needs of the end user or customer.
Furniture can be painted, stained, printed, upholstered, and/or overlaid with
plastic, vinyl, or wood veneer. Wood furniture can be classified as high,
medium, or low quality, which traditionally has been "closely correlated with
the number of finishing operations performed on the piece. A low-end piece
might undergo 6 to 12 finishing operations, while a high-end piece could

---

*Although based upon a report the author prepared for the Virginia Department of
Environmental Quality, this chapter was written by the author in private capacity. No official support or endorsement by the Virginia Department of Environmental Quality is intended or should
be inferred.

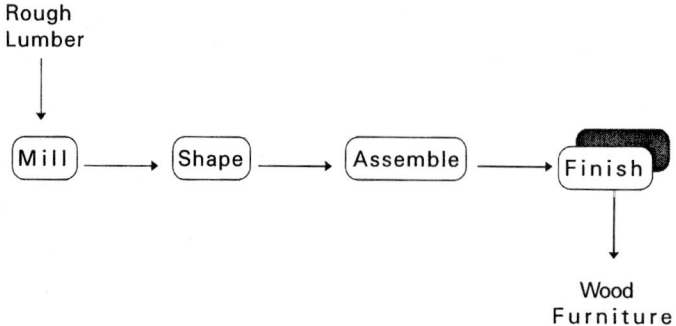

Figure 48-1. Wood furniture manufacturing typical process flow. (*Courtesy of* A Guide to Pollution Prevention for Wood Furniture Finishing, *Dambek et al., 1992.*)

require [30 or more]."[1] Cost of coatings, labor, and equipment are key factors in any finishing function.

High-quality and -priced wood furniture consists of very fine and expensive wood whose natural appearance has been carefully enhanced in a traditional, labor-intensive manner. In contrast, progressive, lower-end, mass-produced furniture may consist of particleboard or fiberboard whose surface has been smoothed by an ultraviolet curable filler and printed to resemble natural wood grain. The latter finishing method represents technological innovation through the refinement of embossing, engraving, and printing techniques, which evolved from the desire to reduce veneer costs and produce low-cost furniture in the 1960s.[2]

One industry expert suggests that manufacturing furniture at moderate prices and in large volume "with most of the aesthetics and durability of very expensive furniture" is "a testimony to the ingenuity of our major furniture manufacturers." This has been accomplished "through the remarkable rate of development of new materials, methods, and machines over recent years."[3]

This chapter will focus on pollution prevention opportunities and options associated with the manual air-spray finishing of wood furniture with conventional, organic solvent-based lacquers. This finishing system represents a level of "no control" with regard to air emissions, other than particulates. All other coating systems can be viewed as improvements over this "no control" system. In short, new levels or norms of control have been the result of innovative manufacturing, made possible by coatings research and improved application technology.

Capturing the essence of spray finishing, one author notes:

> Furniture finishing, as it relates to high quality residential furniture, involves a process which may consist of 30–35 steps. The process may include application of various kinds of stains, fillers, glazes, sealers, wash coats, and top coats which are used to achieve a specific and distinctive final appearance. *It is this appearance that the furniture maker is selling, and the process used to achieve this look is crucial.* [Italics added for emphasis.][4]

48.2 Description of Main Processes

WFF operations are of varying length and complexity, depending on the desired quality and characteristics of the finish. Operation setup also varies, with small operations using one spray booth for all applications and larger operations having a conveyor system to transport the furniture through a series of booths and ovens.

Table 48-1 presents a conventional WFF schedule in terms of two essential concerns in manufacturing: production time and labor. The process is relatively simple: there is one motion mode of loading and unloading a piece [on and off a conveyor]; there are three motion or application modes of spraying, sanding, and wiping/brushing; and, there is one nonmotion mode of drying, which is accelerated or controlled by ovens. Note that out of the 399 minutes allowed in

Table 48-1. Typical Wood Furniture Finishing Schedule[5]

Operation	Time allowed (min)	Number of persons per operation
Load	5	1
Spray uniform stain	1.5	2
Dry	20	0
Spray washcoat	1.5	2
Dry	20	0
Sand lightly	1.5	4
Spray filler	1.5	2
Flashoff filler	2	0
Wipe filler	4	8
Dry	45	0
Spray sealer	1.5	2
Dry	30	0
Sand	3	7
Spray sealer	1.5	2
Dry	30	0
Sand	3	7
Spray glaze	1.5	2
Wipe and brush	5	13
Dry	60	0
Distress	2	4
Spray lacquer	1.5	2
Dry	45	0
Spray lacquer	1.5	2
Dry	75	0
Unload	5	1
Return to load	15	0
TOTAL	399	63

the schedule, 342 minutes, or 86 percent of the time allowed, is accounted for under a drying mode.

As indicated in Table 48-1, typical steps in a finishing process include staining, washcoating, filling, sealing, glazing, and topcoating. The number of steps required to finish a piece of furniture has traditionally been an indication of quality and price (i.e., the greater the number of steps, the higher the quality and price). Characteristics that drive the use of additional steps are associated with uniformity of shading and "depth" of grain.

Referring to Table 48-1, the first layer of stain applied can include a sap or equalizer coating, followed by a body stain. The first stain adds initial color, while sap and equalizers accent the natural wood grain and even out color.

Next, a washcoat is applied to provide additional color uniformity, partially seal the wood from subsequent color addition, and aid in adhesion of the following coatings.

After the washcoat step, a filler is often required. This is usually a high-solids, pigmented wiping stain applied to smooth the surface of open-pore wood. A sealer is then applied as a means of enhancing adhesion of subsequent coats, enabling sanding, increasing build (to create a higher-quality finish), and sealing the wood.

After the sealer step, a glaze coat is sprayed. This is a color coating which highlights and adds substance to the existing finish. Glaze can be in the form of stain applied over a decorative coating and aids in achieving a specific look. Other similar coatings are shading, padding stains, and spatter.

Finally, one or more topcoats are applied to provide a clear coat, protect the color coats, enhance the gloss of the furniture, and provide a durable final finish.

Table 48-2, taken from a 1978 study, provides a summary of the functions of liquid coatings and approximate solids contents.[6]

Table 48-2. Liquid Coating Functionality

Finish	Purpose	% Solids
Stains (sap body, shading, padding, spatter)	Gives color uniformity; develops wood grains and character	3–5
Washcoat	Seals wood surface; prevents subsequent unwanted staining from penetrating filler coat	6
Wiping stain	Gives color uniformity and texture	40
Filler	Fills large pores in wood	60
Sealer	Seals the wood for application of subsequent coats	15–18
Glaze	Penetrates and adheres to sealer	30–40
Topcoats	Provides deep, clear, durable final finish	21–27

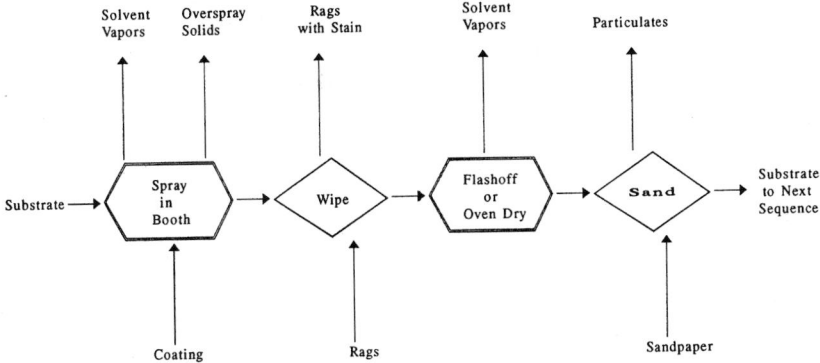

Wiping, sanding, and oven drying are optional.
Flashoff begins as soon as coating is applied and continues until it is dry.
Spray booth filters become wastes, when they are replaced due to solids buildup.
Spent sandpaper is also a waste from finishing.
Energy sinks are spray-gun air compressors and pumps, spray booth fans, sanding units,
HVAC systems, lighting, and ovens. Wood wastes are often utilized to fire boilers for oven steam.

Figure 48-2. Basic process flow of conventional spray finishing. (*Courtesy of Pollution Prevention in the Finishing of Wood Furniture: A Resource Manual and Guide, Virginia Department of Environmental Quality, Office of Pollution Prevention, October 1993.*)

Figure 48-2 distills the schedule presented in Table 48-1 into a process schematic depicting the WFF process in its simplest form. This schematic illustrates a general materials accounting for WFF.

48.3   Typical Wastestreams

Wastestreams from finishing are the result of coatings application and ancillary operations, such as overspray arrestor replacement and spray booth cleaning, rag management, and furniture stripping.

After a coating is sprayed onto a furniture piece, drying is accomplished through volatilization of solvents from the coating base. Essentially, all solvent that is sprayed is lost to the environment. However, not all the finish that is sprayed ends up on the piece being coated. Some film-building solids bounce off the target piece at impact due to their relatively high velocities; others never reach the target due to turbulence from high-pressure air exiting the spray gun nozzle. The amount of solids lost during application is inversely related to transfer efficiency (TE) and depends both on the shape of the furniture, application equipment, and technique being used. Transfer efficiency indicates the ratio of solids adhesion to a target substrate versus the total amount of solids sprayed.

Of all finishing losses, air toxics emissions predominate, accounting for approximately 90 to 95 percent of all losses most facilities report in Toxics Release Inventory reports required under Section 313 of the Emergency Planning and Community Right-to-Know Act. Air losses are concentrated in four particular areas: spray booths, flashoff (natural drying), ovens, and cleanup or ancillary operations.[7] Examples of ancillary operations are line flushing, gun cleaning, spray booth cleaning, and coating preparation. Air emissions may also result from faulty pumps, seals, and unsealed containers. Approximately 70 percent of VOC losses occur in spray booths.[8] Conventional topcoats and stains account for about 65 percent of all VOC losses.[9]

Finishing also results in both hazardous and nonhazardous solid wastes. Empty containers are common wastes. Rags used to wipe stains from and into furniture are sources of emissions and may enter the wastestream carrying hazardous constituents. In stain booths, sorbent material (often clay or sawdust) may be placed on the floor of the booth to adsorb excess material as well as to prevent liquid flow beyond the confines of the spray booth. In addition, spent solvents result from line flushing, equipment and spray booth cleaning, and furniture stripping.

Paper or polystyrene spray booth filters encrusted with overspray are another wastestream associated with finishing. In addition, overspray which never was entrained to the booth filters is a wastestream. Overspray and filters encrusted with overspray may be hazardous wastes, depending on ignitability or reactivity.

Water losses (i.e., discharges to sewerage or water treatment works) occur at the few facilities which still utilize water-wall spray booths. Solids sludge from pretreatment results in waste. Depending on the coating, some solids may be reused.

48.3.1   Coating Losses

The major input of materials in WFF is the coatings themselves. As previously noted, the application of coatings results in an output of air emissions composed of organic vapors and overspray particulates. Air emissions are prevalent at each spraying and drying stage.

The United States Paint Industry Data Base indicates that wood furniture and fixtures manufacturers consumed 270 million pounds of solvent in 1989. Table 48-3 presents a breakdown of solvents by weight. Providing further detail, Table 48-4 presents the composition of coatings utilized in a typical finishing process for residential furniture.

48.3.2   Ancillary Operations

Ancillary operations include but are not limited to filter replacement and booth cleaning, rag management, and furniture stripping.

Filter Replacement and Booth Cleaning.   Maintenance of spray filter systems is essential for proper ventilation. Filters in conventional spray booths are typically constructed of paper, polystyrene, or metal. Overspray solids adhere to the filter material, thus requiring periodic filter change-out or cleaning.

Table 48-3. Wood Furniture Industry Solvent Usage[10]—1989

Solvent	Millions of pounds	Percentage of total
Aliphatic hydrocarbons	8.6	3
Toluene*	71.5	26
Xylenes*	40.4	15
Other aromatics	9.2	3
Butyl alcohol	27.5	10
Ethyl alcohol	27.7	10
Isopropyl alcohol	15.4	6
Other alcohols	2.0	1
Acetone	3.0	1
Methyl ethyl ketone*	15.0	6
Methyl isobutyl ketone*	19.8	7
Ethyl acetate	7.8	3
Butyl acetate	14.3	5
Other ketones and esters	2.8	1
Glycol ethers and ether esters	4.7	2
Total	270	100

*Hazardous Air Pollutant (HAP) and 33/50 Chemical.

Filters constructed of paper or polystyrene may be brushed or shaken to remove initial solids accumulation and then reused a number of times prior to disposal.

Spray operations also deposit solids on booth walls and floors. Although overspray is generally removed from booth floors on a daily basis, a thorough cleaning operation normally occurs during a facility shutdown, generally on an annual basis. Depending on the capture efficiency of the filter media, the interior of the stack/exhaust may need cleaning. The walls, floors, and ceiling are typically scraped of all solids buildup. An organic sovent may be utilized.*

Rag Handling. Rag handling typically involves the recycling of rags saturated with finish materials from wiping. To achieve the desired finish, shop rags are used to "work" the stain into the wood. As a result of this operation, rags sorb excess stain and must be either laundered and reused or discarded. Some states and EPA regions consider used shop rags to be "contaminated wipers" under RCRA, while others allow such rags to be handled in a manner that avoids hazardous waste requirements. In some states, rags are sent off site to an industrial laundry/cleaning service.

*A new EPA study on booth cleaning addresses mechanical methods, masking agents, covers, water-based or low VOC cleaners, and changes in work practices. See *Automobile Paint Spray Booth Cleaning Emission Reduction Technology Review* (in final processing), U.S. EPA Office of Air Quality, Planning and Standards, Control Technology Center, Research Triangle Park, N.C., (919) 541-0800.

Table 48-4. Composition of Finishing Materials[11]

Coating	Constituents	% Weight
Stain (sap)	Methyl alcohol	95
	Diethylene glycol monoethyl ether	1
	Ethylene glycol	1
	Solids (other additives)	3
Toner	Lactol spirits	35
	Toluene	4
	Methyl alcohol	14
	Ethyl alcohol	2
	Butyl acetate	2
	Isobutyl acetate	4
	Ethyl acetate	8
	Isobutyl isobutyrate	4
	Acetone	12
	Methyl ethyl ketone	12
	Methyl isobutyl ketone	2
	Solids (other additives)	1
Washcoat	Aliphatic naphtha	9.6
	Toluol	13.1
	Xylene	5.1
	Methanol	4.7
	Isopropanol	19.4
	Isobutyl alcohol	9.6
	Isobutyl acetate	20.7
	Isobutyl isobutyrate	5.0
	Methyl ethyl ketone	4.9
	Solids (other additives)	7.9
Stain (wipe)	Mineral spirits	30
	VM & P naphtha	16
	Xylene	2
	Petroleum hydrocarbons	31
	Solids (other additives)	2
Stain (shade)	Methyl alcohol	32
	Ethyl alcohol	51
	Ethyl alcohol (PM 4083 ANH)	12
	Diethyl glycol monobutyl ether	3
	Solids (other additives)	2
Sealer	Aliphatic naphtha	18.5
	Toluol	5.2
	Xylene	7.3

Table 48-4 (*Continued*)

Coating	Constituents	% Weight
Sealer (*Cont.*)	Ethanol	9.2
	Isopropanol	3.4
	Isobutyl alcohol	6.4
	Isobutyl acetate	2.1
	Isobutyl isobutyrate	10.9
	Methyl ethyl ketone	18.0
	Solids (other additives)	19.0
Lacquer	VM & P naphtha	6
	Toluol	9
	Xylene	7
	Isopropyl alcohol	4
	Butyl alcohol	7
	Isobutyl alcohol	7
	Butyl acetate	29
	Methyl amyl ketone	11
	Dioctyl phthalate	2
	Solids (other additives)	18

Furniture Stripping. Furniture stripping is performed when a nonconforming finish is identified and quality control measures dictate that the defect should be repaired. Depending on the flaw, either discrete segments or entire surfaces are stripped and refinished. Stripping solvents are applied to remove the layers of coating that have been applied. After most of the finish has been removed from the defective area, this area may be sanded down to the bare wood. A washcoat might then be applied to remove any discoloration remaining in the grain of the wood. The piece may be further sanded before being refinished.

Losses associated with this operation include wasted finish and stripping solvent, soiled rags, sandpaper, washcoat, and associated VOC emissions.

48.4 Pollution Prevention Options

Beyond good operating practices—which involve scheduling, decreased cleanup frequency and solvent usage, and inventory tracking and control—pollution prevention options for finishing fall into two main categories which are not mutually exclusive: (1) the employment of application techniques which increase transfer efficiency and (2) coating reformulation. The capital cost and degree of process changes required varies, depending upon the particular combination of options selected. For example, improving spray operator efficiency, utilizing higher-solids conventional coatings, and purchasing spray guns offering higher transfer efficiencies require no significant process change and is generally a low-cost, quick-payback project. On the other hand, the purchase of flat-line application equipment, use of polyester coatings along with ultraviolet curing equipment, and installation of clean room environment controls amounts to a significant process change and involves relatively high capital costs.

Following is a list of principal options:

- Increased transfer efficiency

    Airless spraying
    Air-assisted airless spraying
    High-volume low-pressure spraying
    Electrostatic spraying
    Dipping
    Roll coating
    Curtain coating
    Automated sensing and spraying
    Operator training

- Coating reformulation

    Higher-solids coatings
    Waterborne coatings
    Reactive or conversion coatings
    Ultraviolet radiation or electron-beam-cured coatings
    Unicarb™ coating system

In addition to these options, there are two additional paths to pollution prevention in WFF. First, reducing the number of coating steps required to achieve a specific finish is a pollution prevention path. In fact, some alternative coatings and techniques achieve an acceptable film build in fewer steps than the process replaced. A second path to pollution prevention is to ascertain the level of quality factors necessary for specific products in specific market segments and to modify finishing requirements accordingly. Both of these paths are related to competitiveness.

48.4.1   Application Techniques and Transfer Efficiency

The application of protective and decorative coatings to furniture has evolved since World War II, when furniture was typically hand-finished. Application techniques have been influenced by developments in coating materials, manufacturing methods, and regulations requiring reduced solvent emissions.

Coatings can be applied by the following actions: brushing, wiping, spraying, rolling, curtain coating, or dipping. A manufacturer's selection of any particular method depends, among other things, on production rate, application efficiency, quality, and equipment price. Each application technique has advantages and disadvantages. In the scheme of manufacturing, the search for the most efficient application method is a key to low-cost production of any particular coated product. Spray-gun application requires the lowest up-front capital costs and is, not surprisingly, the principal finishing method employed by most establishments.

Many types of spray gun systems have been developed over the years. Increased transfer efficiency and associated loss reductions, along with acceptable film builds and gun speeds, are the primary advancements offered by newer-generation gun systems. Transfer efficiency is a term used to describe the amount of sprayed coating solids that remain on a furniture piece. As transfer efficiency increases, solvent losses and overspray solids decrease. Therefore, increasing transfer efficiency is a source reduction measure from both air emissions and solid waste perspectives. Spray systems offering increased transfer efficiency over conventional air spraying are airless, air-assisted airless, high-volume low-pressure (HVLP), and electrostatic systems.

Although there is controversy over which spray gun system gives the highest transfer efficiency, the U.S. EPA assumes a transfer efficiency of 25 percent for conventional air spray and 40 percent for both airless and air-assisted airless spray. Although there is no consensus within the EPA on HVLP, EPA Region IX (San Francisco) considers transfer efficiency for HVLP to be 65 percent or greater.[12]

Significant loss reduction has been achieved by improving upon spraying techniques and equipment. A recent study conducted by the Pacific Northwest Pollution Prevention Research Center reaffirmed that properly trained operators practicing efficient spray techniques keep VOC emissions and materials consumption at optimum levels.[13] Introductory and ongoing training is important. Areas of focus should be spray techniques, coating content, and optimal equipment settings (including the use of variable fluid tips and air caps). Some spray equipment manufacturers provide training information in the form of seminars, literature, and videos. For example, one vendor makes available, free of charge, over forty rental videos which discuss training aspects of application techniques ranging from HVLP to electrostatics.[14]

Given proper operation, HVLP spray systems provide substantial transfer efficiency improvements over conventional air spray systems. In fact, HVLP systems are the "low-hanging fruit" in pollution prevention equipment for the industry.

Not all HVLP systems are alike, however. HVLP guns most commonly utilized by wood finishers are conventional air spray guns which have been modified with a venturi to spray low-pressure compressed air. There is no doubt that these guns provide substantial savings over conventional air spray. However, one manufacturer has developed an industrial-turbine-powered HVLP system, which not only has a significantly higher TE than conventional compressed-air-powered HVLP systems but also operates on about one-third of the electrical power. Materials savings reportedly range between 20 to 40 percent over compressed-air-powered HVLP systems. In addition, the industrial turbine HVLP system reportedly can spray high-solids coatings at rates comparable to airless systems.[15]

48.4.2 Coating Reformulation

If there is one general category of pollution prevention options with significant source reduction potential for the WFF industry, beyond good housekeeping and other quick payback options such as spray gun conversion, it is coating reformulation. Finishers and coating formulators must take into account the following factors: appearance, productivity, environmental regulations, cost-effectiveness, and durability.[16]

Coating systems include the coating, the application method, and the drying or curing process. Coating chemistry, along with desired build and appearance, is the primary determinant of the technical feasibility of any particular application method. The drying or curing process is dependent on the coating constituents.

Coatings can be divided into those that dry by solvent loss (lacquers or nonconversion coatings) and those that cure and chemically cross-link to form a reaction product (conversion coatings). Curing is effected by a catalyst or, in the case of UV-cured coatings, a photoinitiator. The catalyst is either present "in the can as delivered" for slow-curing precatalyzed coatings or added near the application time for faster-curing coatings.

Nonconversion coatings are said to be thermoplastic, because they flow above a given transition temperature but retain their chemical composition when cooled and solidified. Conversion coatings, on the other hand, are often much harder and said to be thermosetting. When some thermosetting resins are exposed to heat in the presence of a catalyst, they tend to chemically cross-link. After solidification, the reaction product is relatively stable when exposed to heat. Lacquers have a relatively high degree of plasticity and are generally easier to rework or repair than conversion coatings.

The majority of the wood furniture liquid coatings used in the United States today are alkyd-modified, nitrocellulose-based lacquers used by the residential wood furniture segment. Acid-catalyzed conversion coatings (also based on alkyd resins) are used predominantly by kitchen cabinet and office furniture manufacturers (formaldehyde emissions are a disadvantage to the use of these coatings). An even smaller volume of polyurethane coatings and unsaturated polyester/unsaturated polyacrylate (UPE/UPA) coatings is utilized. Polyurethane and UPE/UPA coatings are used much more frequently in Europe and Japan than nitrocellulose lacquers.[17] This is largely attributed to the furniture design and the high-volume, capital-intensive finishing processes employed by Western European and Japanese furniture manufacturers.

Organic-solvent, nitrocellulose-based lacquers are highly utilized in the United States wood furniture industry for one main reason: they provide finishes which meet manufacturing needs (ease of application and repairability) and customer demands. Solvents serve two functions: they act to dissolve and dilute resin systems, and, based on their evaporation rates, allow for control of drying times. Solvents used in conventional coatings typically have been alcohols, ketones, ester solvents, glycol ethers, aliphatic solvents, and aromatic solvents.

Efforts were made as early as 1978 to reduce hydrocarbon emissions by the use of waterborne finishes in WFF. A cooperative test program involving coating manufacturers and furniture finishers generated data suggesting that the use of such finishes could result in hydrocarbon emission reductions of between 26 to 94 percent over the use of conventional finishes. However, the program study concluded that

> ...none of the reduced hydrocarbon finish system products evaluated were commercially acceptable to the furniture manufacturers because of grain raising, hardness, lack of depth or sheen, and inadequate smoothness. Furthermore, none of the waterborne systems were as resistant as conventional finishes to household chemicals or printing (loss or transfer of finish materials by direct contact with another object such as a packing carton).[18]

The study recommended that efforts be made to improve the performance of waterborne and low-solvent finishes to the point of commercial acceptability.

Since 1980, the acceptability of waterborne coatings has been improved markedly.[19] Waterborne coatings have long been thought to be acceptable for low-end products. Waterborne coatings have been used successfully on wood furniture for children and on futon frames. Their use in higher-end, medium- to high-quality products is expanding slowly. Office and institutional furniture, including chairs and cases, can be finished with full-waterborne systems, although the technology is still developing. More frequently, waterborne coatings have been selectively targeted at those coatings/steps with the greatest emissions: stains, sealers, and topcoats. A system which utilizes conventional coatings on some steps and waterborne coatings on others is referred to as a hybrid waterborne system.

Although water-based coatings cost between 25 to 50 percent more than conventional coatings on a per gallon basis, overall savings would seem to outweigh the increased cost. Advantages to these coatings are higher solids, reduced VOC emissions, reduced employee exposure to VOC vapors, and a decrease in fire hazard. In addition, savings are attributed to reduction or elimination of solvent for line flushing and cleaning. Further, costs associated with spent solvent disposal are reduced or eliminated. One major disadvantage is that the finish is often less soluble due to its high molecular weight and as a result is more difficult to repair than conventional finishes. In addition, to provide acceptable print resistance, water-based finishing generally requires line reconfiguration or adjustment, often with oven additions, to ensure proper drying. Management of water-based coating wastes, such as line flush, also has to be taken into consideration. For example, surfactants and emulsifiers in waterborne wastes can disrupt the operation of publicly owned treatment works (POTWs) by killing bacteria necessary for wastewater treatment.

Table 48-5 presents a summary of the VOC and solids content of commercially available coatings. Note that the coatings offering the most substantial

Table 48-5. Approximate VOC and Solids Content of Commercial Coatings[20]

Formulation	VOC content (lbs/gal less water)	Solids content percent by volume
Nitrocellulose	6	16
Acid-catalyzed	5.1–5.8	18–26
Waterborne	1.3–2.3	26–30
Polyurethane	3.4	30–60
Polyester (acrylic)	3.0	30–50
Polyester (styrene)	—	100
UV- or EB-cured	0–3.1	56–100

VOC reductions appear to be waterborne and radiation- (UV or EB) cured coatings. The latter coatings will most likely see greater use, although health and safety concerns dictate that some must be applied by automated spray techniques.

When the benefits of alternative coating systems are assessed, it is important to consider trade-offs between environmental losses and toxicities across all media, along with potential impact to worker health and safety. Epoxy resins, styrene, acrylates, and polyisocyanates are substances associated with alternative or low-VOC coatings. The relative human health and environmental risk of using these substances compared to conventional solvents is unknown. Many of the radiation-cured, reactive coatings cure very rapidly and emit very little or no chemicals at and after application.

48.5 Pollution Prevention Successes

In June 1993, an advisory committee was chartered under the Federal Advisory Committee Act for the development of Control Techniques Guidelines (CTGs) and National Emissions Standards for Hazardous Air Pollutants (NESHAPs), under the Clean Air Act, for wood furniture finishing. Regulatory negotiations are being managed through the U.S. EPA Consensus and Dispute Resolution Program and a facilitator. At the table are representatives for various stakeholders: wood furniture manufacturers, coatings suppliers, resins producers, regulatory agencies, and nongovernmental and environmental organizations. Standards are now scheduled to be finalized in November 1994.

Because the EPA is in a standard setting mode and uncertainty over future regulatory requirements abounds, few furniture manufacturers are publicizing any substantial air emissions reduction activities for existing facilities unless a healthy and rapid payback is certain. Nevertheless, some furniture manufacturers have implemented finishing techniques exemplifying pollution prevention.

New England Woodcraft (Forest Dale, Vt.) finishes institutional furniture with a full waterborne system.[21] Sun Tui (St. Paul, Minn.) finishes futon frames with waterborne finishes and electrostatic spray equipment.[22] Loewenstein, Inc. (Pompano Beach, Fla.) produces chairs with high-solids, UV-cured sealers and topcoats in electrostatic disk booths.[23] The South Coast Air Quality Management District, Southern California Edison Customer Technology Application Center, and members of the California Furniture Manufacturers Association have been performing extensive research regarding water-based, high-solids, and UV-curable coating systems.[24]

Several furniture manufacturers in Virginia have successfully produced furniture with waterborne coatings. A manufacturer in Martinsville is finishing chairs with a hybrid waterborne system. A manufacturer in Galax produced children's furniture with a full waterborne system. Another manufacturer, in addition to having successfully produced furniture in trial runs with the Unicarb™ system, is utilizing waterborne stains in some instances for hybrid finishing. Virginia Correctional Enterprises (Jarratt, Va.) finishes institutional wood furniture with a full waterborne system.[25]

In sum, following the establishment of CTGs and NESHAPs, many loss reduction techniques consistent with the practice of pollution prevention will emerge. Subject to increasingly stringent emissions limitations through air program "racheting," furniture manufacturers will be using finishing methods which should demonstrate the feasibility and commercial viability of alternative coating systems.

References

1. H. Van Noordwyk, *Reducing Emissions from the Wood Furniture Industry with Waterborne Coatings,* EPA 600/2-80-160, July 1980, p. 4.

2. Andrew W. Riedell, Facility Manager, PPG Industries, Greensboro, N.C., unpublished paper.

3. Ibid.

4. Vincent Ross (Ross Associates, Asheville, N.C.), "Waste Reduction—Pollution Prevention in the Furniture Industry," presented at *Pollution Prevention: Waste Reduction for Industrial Air Toxic Emissions Conference,* April 24–25, 1989, Greensboro, N.C.

5. H. Van Noordwyk, op. cit., from Technical Paper, Society of Manufacturing Engineers, MS75-251, p. 6.

6. H. Van Noordwyk, op. cit., from "Surface Coating in the Wood Furniture Industry," 20 October 1978, Foster D. Sneel Division, Booz, Allen and Hamilton, Inc., Florham Park, N.J., p. 8.

7. *Control of Volatile Organic Emissions from Wood Furniture Coating Operations,* U.S. EPA, Office of Air and Radiation, Office of Air Quality Planning and Standards, Research Triangle Park, N.C., October 1991.

8. Ibid., p. 4-20.

9. Ibid., p. 4-15.

10. Ibid., p. 2-31; Stanford Research International, *U.S. Paint Industry Data Base* prepared for the National Paint and Coatings Association, Washington, D.C., 1990.

11. Paul Dambek et al., *A Guide to Pollution Prevention for Wood Furniture Finishing,* Tufts University, Department of Civil and Environmental Engineering, Medford, Mass., November 1992. The coating constituents were identified through documentation obtained from the principal case study facility.

12. Ibid., p. 6.

13. *Transfer Efficiency and VOC Emissions of Spray Gun and Coating Technologies in Wood Finishing,* Pacific Northwest Pollution Prevention Research Center, Seattle, Wash., 1992.

14. Binks Manufacturing Company, Training Division, Franklin Park, Ill.

15. J. M. Heltzer, Virginia Department of Environmental Quality, with M. H. Bunnell, President/CEO, CAN-AM Engineered Products, Livonia, Mich. July 21, 1993, telephone conversation. Also see company literature.

16. J. M. Heltzer, Virginia Department of Environmental Quality, with James M. Bohannon, Manager, Eastern Wood Coatings Lab, Valspar Corporation, High Point, N.C., July 15, 1993, telephone conversation.

17. *Control of Volatile Organic Emissions from Wood Furniture Coating Operations,* op. cit., pp. 2–15.

18. H. Van Noordwyk, op. cit., p. 2.

19. Carl Urmacher, *Evaluation of the Problems Associated with Application of Low Solvent Coatings to Wood Furniture,* EPA 600/2-87-007, January 1987.

20. This table is adapted largely from information in *Control of Volatile Organic Emissions from Wood Furniture Coating Operations,* op. cit.

21. "Getting the Most from Water-Based Finishes," *Furniture Design and Manufacturing,* Delta Communications, Chicago, Ill., January 1991.

22. "Futon Maker Plugs into Electrostatic Finishing System," *Wood & Wood Products,* Vance Publishing, Lincolnshire, Ill., January 1993, pp. 52–56.

23. "Loewenstein VOC Dip Continues," *Industrial Finishing,* Hitchcock Publishing, Carol Stream, Ill., May 1993, pp. 14–15.

24. *Evaluation of Low VOC Coatings for Wood Furniture,* South Coast Air Quality Management District, Diamond Bar, California, (909) 396-2000. Study is ongoing.

25. *Pollution Prevention in Wood Furniture Finishing: A Resource Manual and Guide, Overview,* Virginia Department of Environmental Quality, Office of Pollution Prevention, Richmond, Va., October 1993.

Further Reading

Carter, Robert, *Solvents—The Alternatives,* Waste Reduction Resource Center for the Southeast, Raleigh, N.C., April 1993.

An Evaluation of VOC Emissions Control Technologies for the Wood Furniture and Cabinet Industries, American Furniture Manufacturers Association, High Point, N.C., January 1992.

Gardner, Lisa C., and Donald Huisingh, "Alternative Approaches to Waste Reduction in Materials Coating Processes," *Hazardous Waste & Hazardous Materials,* Mary Ann Liebert, Inc., **4**(2):177–191 (1987).

Guide to Cleaner Technologies: Applications for Paint and Coatings Alternatives, U.S. EPA, Cincinnati, Ohio. Contact CERI Distribution Center at (513) 569-7562.

Johnson, Sharon M. (North Carolina Office of Waste Reduction), "Overview of Coating Technologies," *Conference Proceedings, National Roundtable of State Pollution Prevention Programs,* Spring Conference, April 28–30, 1993, San Diego, Calif.

Lawson, Kenneth, et al., "Safe Handling of UV/EB Materials," *RadTech '92 North America Conference Proceedings,* RadTech International North America, Northbrook, Ill., 1992.

Mahon, William F., and Dale L. Nason, "UV Cure Finishing Systems for Wood," *RadTech '92 North America Conference Proceedings,* RadTech International North America, Northbrook, Ill., 1992.

Pollution Prevention Options in Wood Furniture Manufacturing: A Bibliographic Report, U.S. EPA Office of Pollution Prevention and Toxics, Washington, D.C., February 1992.

Proceedings from Pollution Prevention Conference on Low- and No-VOC Coating Technologies (U.S. EPA, Air and Energy Research Laboratory, Research Triangle Park, N.C.), May 25–27, 1993, San Diego, Calif. Specifically, these papers:

> Caldwell, Mary-Jo L., "Lower-VOC Coating System Conversion Costs for the Wood Furniture Industry," Midwest Research Institute, Cary, N.C.

> Huang, Eddy W., et al., "Development of Ultra-Low-VOC Wood Furniture Coatings," Center for Emissions Research & Analysis, City of Industry, Calif.

> Morton, Brian J. (now at Environmental Defense Fund, Research Triangle Park, N.C.), and Bruce Madariaga (U.S. EPA, OAQPS, Research Triangle Park, N.C.), "Economic Incentives to Stimulate the Development and Diffusion of Low- and No-VOC Coating Technologies."

> Yazdani, Azita, "VOC Reduction Study for Furniture Refinishers in the City of Los Angeles," Pollution Prevention International, Inc., Brea, Calif.

Riedell, Andrew W., "Three-Dimensional Curing of Wood Furniture," *RadTech '88 Conference Proceedings,* RadTech International North America, Northbrook, Ill., 1983.

Acknowledgments

This chapter was based primarily on a group project performed by the author in partial fulfillment of an M.S. in Hazardous Materials Management at Tufts University. Excerpts from the project report appear in this chapter. The author wishes to thank for their support and diligent work the other group members: Paul J. Dambek, Kevin D. Kelly, Maria L'Annunziata, and Thomas M. Smith.

49

Encouraging Pollution Prevention through Publicly Owned Treatment Works Activities

Lynn Knight

Eastern Research Group, Inc.
Lexington, Massachusetts

49.1 Background

This chapter covers actions personnel at publicly owned treatment works (POTWs) can take to encourage pollution prevention among their industrial and commercial users. Pollution prevention measures implemented by sewer users will reduce pollutant loadings in their discharges. The objective is for POTW personnel to identify and educate users about the benefits of pollution prevention, while encouraging and assisting them to assess and implement pollution prevention in their own operations. The following text outlines a broadly applicable approach for integrating these pollution prevention concepts into existing POTW pretreatment program activities.

49.2 Why Should POTWs Encourage Pollution Prevention?

POTWs are the recipients of wastewater from an estimated 30,000 significant industrial users, small industrial users, commercial establishments, and domestic sources. These users discharge the full spectrum of heavy metals, volatile organics, and other contaminants that can degrade environmental quality and pose health and safety risks to POTW workers. Although discharges from significant industrial users are regulated by the U.S. Environmental Protection Agency (EPA) categorical pretreatment standards, the EPA estimates that these users, even if in full compliance, would continue to discharge a significant loading of toxic pollutants to POTWs each year. Wastewater from small industrial, commercial, and domestic sources is largely unregulated.

By further reducing the quantity and toxicity of user discharges, pollution prevention can help POTWs meet federal and state environmental quality standards, reduce the transfer of influent contaminants from wastewater to other environmental media (e.g., land, surface waters, groundwater, and air), increase POTW worker safety and reduce collection system hazards from toxic or hazardous gases, and reduce POTWs' high sludge management costs.

49.3   How Does a POTW Promote the Benefits of Pollution Prevention among Businesses?

Industrial and commercial facilities can derive significant benefits from pollution prevention. In many cases, pollution prevention might be the least expensive means of reducing unacceptable toxic discharges. Pretreatment personnel can point out the benefits of pollution prevention to their sewer users. Through pollution prevention, companies can reduce waste monitoring, treatment, and disposal costs; reduce raw materials and feedstock purchases and manufacturing costs; increase productivity and reduce off-specification products; reduce regulatory compliance costs; reduce hazards to employees through exposure to chemicals; reduce costs of environmental impairment insurance; improve public image and employee morale; and reduce potential liability associated with toxic waste.

49.4   Setting Priorities to Focus Pollution Prevention Efforts

POTW personnel should identify the pollutants of greatest concern, identify the sources of those pollutants, and then set priorities to focus pollution prevention efforts (see Fig. 49-1). Pollution prevention provides sewer users with another tool to comply with local limits developed to prevent or remediate problems at the POTW related to specific pollutants in wastewater discharges. For a particular sewer user, the most logical pollutant of concern may be one that is currently over the existing local limits. For a POTW, the most logical targets may be pollutants of concern for their own current or anticipated discharge or sewage sludge use or disposal limits.

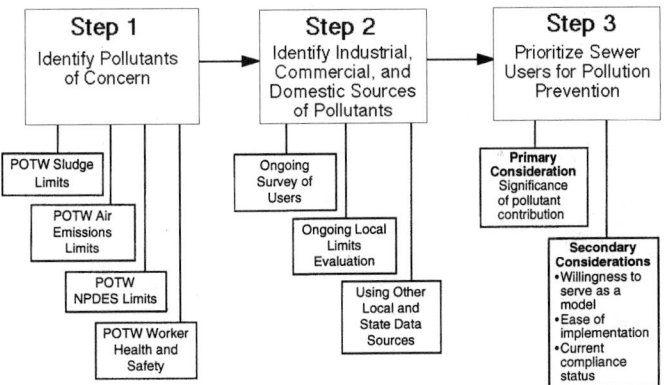

Figure 49-1. Setting pollution prevention priorities.

Once a POTW has targeted particular contaminants for pollution prevention, the POTW must determine which industrial, commercial, and domestic sources discharge those contaminants. Determining which significant industrial users discharge the contaminant of concern should be a relatively simple matter since POTWs routinely collect and receive data on these industrial users. To help locate new or unknown dischargers, pretreatment personnel generally contact local and state agencies to cross-reference records on water users, new utility connections, and building permits. Observation of changes in local businesses while out in the field also provides information about new users.

Many POTWs have discovered that commercial dischargers account for a large percentage of the toxic pollutants in a POTW's influent. Unfortunately, POTWs often have little information about their commercial dischargers since they do not actively inspect them and might not have included them in the initial waste survey. As a first step, the POTWs could develop a comprehensive list of commercial processes that produce the contaminant in question and what types of commercial establishments employ those processes. For example, if mercury is a particular problem, likely commercial contributors might include dental offices and laboratories. Table 49-1 lists some commercial establishments and the types of pollutants they commonly produce.

To define further which commercial establishments produce and discharge the contaminant of concern, the POTWs could survey commercial establishments in the POTW service area that are likely to be discharging that contaminant. Cross-referencing records of businesses with other agencies will help identify previously unknown or new commercial users to include in the survey. The survey can refine the list of potential commercial contributors, estimate average discharge concentrations and flows from each facility, and provide information about the pollution prevention measures the facilities already employ. A well-defined survey instrument can yield enough data on which to base further actions and assess the potential usefulness of pollution prevention in those commercial establishments. The survey instrument can be short and simple.

Table 49-1. Commercial Establishments and Their Potentially Hazardous Discharges

Type of facility	Discharges of concern
Automotive repair and service	Chemical oxygen demand, heavy metals, solvents, paints, surfactants, oil, and grease
Car washes	Chemical oxygen demand, zinc, lead, and copper
Truck cleaners	Chemical oxygen demand, total dissolved solids, cyanide, phosphate, phenol, zinc, aluminum, chromium, lead, and copper
Dry-cleaners	Total dissolved solids, chemical oxygen demand, phosphate, butyl cellosolve, N-butyl benzene sulfonamide, perchloroethylene, iron, zinc, and copper
Laundries	Chemical oxygen demand, ethyl toluene, N-propyl alcohol, isopropyl alcohol, toluene, m-xylene, p-xylene, ethylbenzene, bis(2-ethylhexyl)phthalate, iron, lead, zinc, copper, chromium, phosphate, and sulfide
Hospitals	Total dissolved solids, chemical oxygen demand, phosphate, surfactants, formaldehyde, phenol, fluoride, lead, iron, barium, copper, mercury, silver, and zinc
Photoprocessors	Chemical oxygen demand, ammonia, cyanide, sulfur, phosphates, silver, arsenic, chromium, phenol, and bromide
Laboratories	Chemical oxygen demand, mercury, silver, and toxic organics
Dental offices	Copper, zinc, silver, and mercury

SOURCE: Adapted from U.S. EPA, 1991d.

Households regularly discharge many problem wastes and products, such as used oil, drain cleaners, detergents, paint and paint thinners, and solvents, directly to household drains and storm drains. Although this chapter does not cover techniques for reducing toxic loadings from households, these users should not be overlooked as potential significant sources. As industrial and commercial loadings are reduced, domestic sources can account for an increasingly large portion of toxic loadings.

Prioritizing users of concern is the next step. Generally, the highest priority will be the sources contributing the largest share of a given contaminant of concern to the POTW's influent. Secondary considerations for prioritizing include the willingness of the users to participate, the current compliance status of the users, and the potential ease of implementing pollution prevention measures (i.e., the potential for a successful experience).

Once a POTW has identified problem contaminants and prioritized its users accordingly, it can focus on how pollution prevention can be incorporated into inspection, permitting, and enforcement activities as a full or partial solution to identified problems. In many cases, it might be best to begin with a simple activity and use the experience gained to launch more complex pollution prevention efforts in the future.

49.5   Encouraging Pollution Prevention through On-Site Inspections

One of the most effective ways to identify potential pollution prevention measures and to promote pollution prevention in general is to explore opportunities during routine facility inspections. Because a POTW's staff usually has a close relationship with local industry and commercial establishments, they are in a unique position to educate businesses on the advantages of pollution prevention. Pollution prevention can be incorporated into any facility inspection by the POTW staff asking investigative questions, disseminating basic pollution prevention information, and offering sources of further technical assistance (see Fig. 49-2).

49.5.1   Before the Inspection

Before the inspection, collect facility data, analyze facility operations, and assemble preliminary recommendations and sources of further information. With a solid understanding of the processes, the types of inputs they require, and the wastestreams they generate, POTW personnel can help identify potential problem areas and initiate discussions with facility personnel about implementing pollution prevention measures. Much of the required information to prepare for the inspection is readily available at the POTW. For example, POTW personnel collect process information and wastestream monitoring data on significant industrial users to develop permits and prepare for traditional user inspections. In addition, the revised General Pretreatment Regulations require dischargers to report to wastewater authorities the types and quantities of certified hazardous chemicals they generate and discharge to the sewer (40 *CFR* 403.12[p]).

POTWs interested in inspecting unpermitted industries and commercial facilities might have greater difficulty obtaining current, facility-specific process data. Options in such cases include reviewing industrial waste survey data; contacting other federal, state, and local environmental and public health program offices that might have collected facility-specific information; and requesting process data and information directly from the facility (under the pretreatment program, POTWs have the authority to collect facility-specific information from any discharger). In addition, POTW personnel can gather information about the process in question from general sources, such as EPA guidance documents and other technical manuals. POTW personnel also can contact the Pollution Prevention Information Clearinghouse (PPIC) (see "Other Resources" at the end of the chapter), state technical assistance offices, and trade groups to find out more about specific industrial and commercial processes and applicable pollution prevention techniques.

For permitted facilities, POTW personnel should review information relating to the facility's compliance history. Compliance data can help POTW personnel

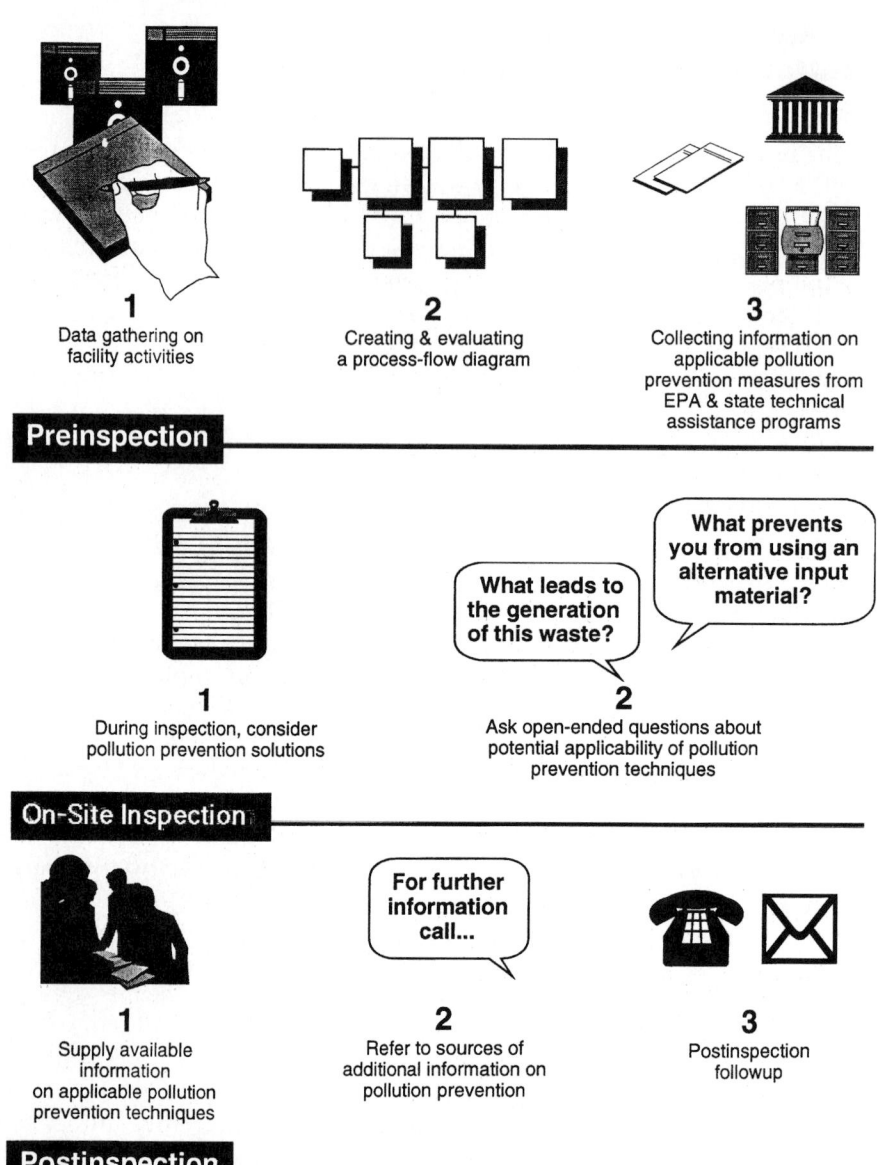

Figure 49-2. Using the on-site inspection to promote the benefits of pollution prevention.

focus preinspection information gathering on pollution prevention options that address the facility's greatest compliance problems. For example, if POTW personnel know that the facility is having or has had problems meeting pretreatment standards for copper, they can make a special effort to investigate pollution prevention measures that have succeeded in reducing copper discharges from similar facilities. POTW personnel also should be aware of any impending pretreatment standards or POTW restrictions that will either require more stringent discharge limits for a particular contaminant or address a previously unregulated contaminant that the facility in question currently discharges. With this knowledge, POTW personnel can advise facilities to start thinking about pollution prevention as a means of meeting future discharge limits.

Drawing on the information gathered from the sources discussed, the following four-step approach will assist in identifying areas of the facility's process where pollution prevention measures could reduce toxic loadings to the sewers:

1. Construct a simple process flow diagram of the operation. Show all inputs and outputs to the process, including raw materials inputs, product outputs, material recovery, and wastestreams.

2. Perform a materials balance assessment to identify significant material losses occurring in the process.

3. Evaluate the sources of identified losses.

4. Identify areas other than process areas, such as storage areas or garages, where losses typically occur.

As the first step, POTW personnel can develop a flow diagram that depicts the sequence and function of all the unit processes and the materials going into and coming out of each unit. This diagram will help POTW personnel define the operation and form the basis for tracking the materials as they go through the process and ultimately end up in the product, recovered materials, or the wastestream. POTW personnel can verify the accuracy of the process flow diagrams during the inspection. Figure 49-3 is an example of a process flow diagram for a photoprocessing operation. (This photoprocessing example will be used to illustrate the application of the four steps outlined above.)

The next step is to account for the majority of the material flows into and out of the process. Based on the process flow diagram, POTW personnel can track the pollutant of concern from its point of origin in the raw material inputs to the resulting products and wastestreams. It is helpful to make a list of all input and output materials. For the photoprocessing example, Table 49-2 itemizes the material inputs and outputs and identifies areas where losses are occurring and wastes are generated. Using raw materials purchasing records, wastestream monitoring and flow data, and product data, POTW personnel can quantify the mass of materials going through the process. For the photoprocessing example, the tracking of silver mass is illustrated in Fig. 49-4. This is similar to, but not as rigorous as, an engineering mass-balance exercise. The mass of input materials

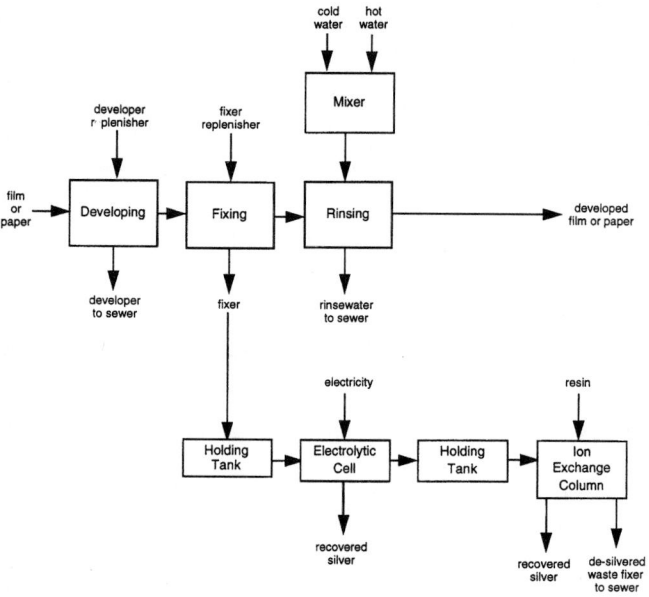

Figure 49-3. Sample flow diagram of photoprocessing operation.

Table 49-2. Sample Materials Accounting List for Photoprocessing Example

Material inputs	Material outputs	Losses/wastes
Photographic film	Developed film and paper	Waste developer
Photographic paper	Recovered silver	Desilvered waste fixer
Developer replenisher		Waste rinsewater
Fixer replenisher		
Stabilizer		
Iron		
Cold water		
Hot water		

should approximate the combined mass of materials output in the product, recovered materials, and the wastestreams. Although the mass balance will be unequal due to the variability in wastestream sampling and flow data and errors in estimating input and output masses, it should be within an acceptable margin of error. The acceptable margin of error varies with the known precision and accuracy of information used to estimate the material mass at each stage. A substantial difference between materials input and output from the

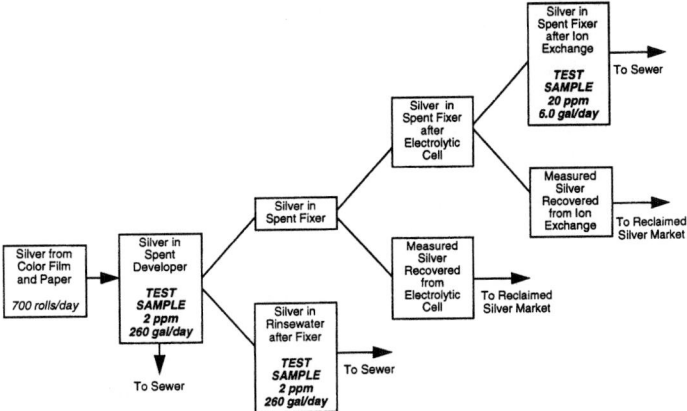

Figure 49-4. Tracking the silver material balance in a color photoprocessing operation.

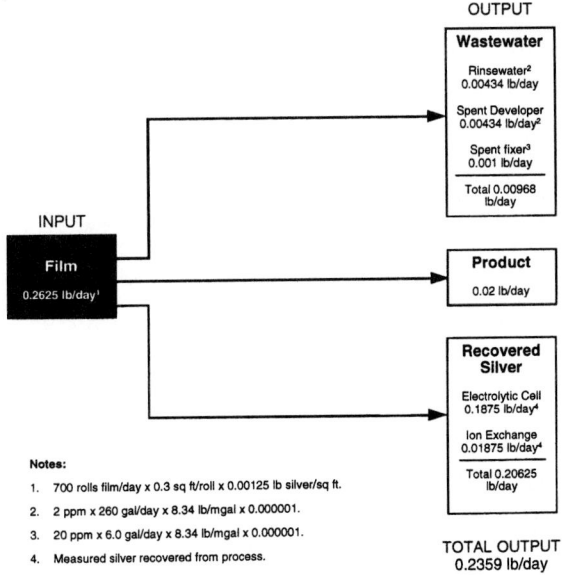

Figure 49-5. Comparing silver input and output in a photoprocessing operation.

process indicates losses of materials that should be investigated. Figure 49-5 illustrates a material-balance calculation tracking the mass of silver going into and out of the photoprocessing example. In this example, the material balance was not exact, but was judged to be within an acceptable margin of error.

Losses can occur during the process for several reasons. They can be related to inefficiencies in the production process itself, maintenance procedures, inventory controls, or internal management of waste residuals. POTW person-nel can speculate about the sources of losses before the inspection; however, at the inspection, through observing operations and questioning facility person-nel, POTW personnel will be better able to draw more informed conclusions regarding the source of—and possible pollution prevention solutions for—the materials losses.

Once a preliminary assessment of materials losses is conducted, POTW per-sonnel should compile a "laundry list" of possible pollution prevention alter-natives that would reduce or eliminate losses. Investigators should focus on collecting as much information as possible about the pollution prevention opportunities available for the industry under investigation, for the purpose of educating facility owners and operators about the usefulness of pollution pre-vention measures, supplying available documents and other materials on pol-lution prevention, and encouraging facility owners to conduct their own pollu-tion prevention assessment of all potentially feasible options. The final decision about the applicability of any pollution prevention measure will be made by the facility based on economic, technical, and feasibility factors.

State and federal pollution prevention technical assistance offices can help POTW personnel with specific pollution prevention questions or information requests. Many of these technical offices sponsor pollution prevention work-shops for industry and state personnel interested in learning about pollution prevention opportunities in a given industry.

Table 49-3 lists some potential pollution prevention measures identified for the photoprocessing example illustrated in this chapter. The options are orga-nized according to the major wastestreams from the process, the developing and fixing steps, and the rinsing unit. There are also some general facility options listed.

49.5.2   At the Inspection

At the inspection, observe, ask questions, and make appropriate recommendations. The inspection provides an opportunity for pretreatment personnel to view facility operations and encourage pollution prevention to the fullest extent. One of the goals of the inspection is to leave an industrial user with a good idea of which areas of the facility can potentially employ pollution prevention measures to help achieve compliance with discharge limits and reduce toxic loadings to the sewer. These goals can be accomplished by (1) setting the appropriate tone, (2) making observations and asking the right questions, (3) giving appropriate advice, and (4) highlighting pollution prevention in the exit meeting.

Most routine facility inspections begin with a meeting. At this time, POTW personnel can inform facility personnel that the POTW is promoting pollution prevention as a means of reducing toxic discharge to the sewers and achieving long-term compliance with pretreatment standards.

During the inspection, POTW personnel can observe the flow of the facility's process, following the train of events that leads to the disposal of contaminants

Table 49-3. Sample Pollution Prevention List for Photoprocessing Example

Process or process step	Opportunity	Option
Developing and fixing steps	Reduce chemical use (to reduce chemical loading to POTW)	Adjust replenishment rates.
		Install silver recovery fixer recirculator.
		Use squeegees to minimize chemical carryover from developer and fixer.
		Evaluate recycling fixer.
		Monitor silver recovery units to assure maximum operating efficiency.
		Use low-silver-content rapid access (RA) chemicals.
		Route fixer overflow drains to silver recovery.
		Segregate high and low silver-bearing streams to enhance silver recovery.
		Check storage areas daily for spills. Chemical storage area could be diked and absorbent pillows could be made available to contain spills.
Rinsing	Reduce water use (to reduce water use, water heating, and silver discharge to POTW)	Install water recirculator.
		Evaluate recycling rinsewater, including recovering silver.
		Check storage areas daily for spills. Chemical storage area could be diked and absorbent pillows could be made available to contain spills.
General facility	Good operating practices	First-in–first-out inventory control.
		Inventory inspection for leaks and spills.
		Use lids or other means to minimize chemical contact with air.

to the sewer and verifying the accuracy of the process flow diagram constructed prior to the inspection. If user or POTW compliance issues were identified prior to the inspection, reducing the sources of problem contaminants very likely will be the primary focus of the inspection. If the materials-balance calculations indicated substantial losses of certain materials, identifying the sources of these losses and reducing them very likely will be another major focus of the inspection.

In addition to the process areas, POTW personnel should investigate the existence of storage areas, pumping stations, laboratories, boiler areas, garages, pollution control equipment, and power-generating facilities. These are areas that should be observed to determine whether good operating practices are being applied to prevent or minimize the discharge of pollutants to the POTW, especially through floor drains, and whether further improvements in existing practices or other pollution prevention options might be appropriate. In addition, based on knowledge of the industry, POTW personnel can identify any periodic maintenance activities, such as equipment or tank cleaning, boiler blowdown, and motor fluid changes, that can periodically generate significant wastestreams potentially discharged to the sewer. Improving operating practices for these activities should be encouraged, and applying specific pollution prevention measures may also be appropriate.

The key to getting facility owners and operators thinking about pollution prevention and how it might work in their facility is to ask open-ended questions about why they use a certain process or input, or why some current practice could or could not be changed. POTW personnel should formulate open-ended questions that solicit thoughtful answers and stimulate further discussion. Ultimately, such discussions might lead to the discovery of a feasible pollution prevention opportunity. Open-ended questions prompt users to think about why they have chosen a given process or input and what prevents them from changing to another process or input. Close-ended, or "yes/no," questions tend to be more accusatory and solicit one-word answers that can effectively end the discussion and might close a potentially promising pollution prevention angle entirely.

POTW personnel must be careful about giving pollution prevention advice. In general, investigators should refrain from specifying products or suggesting that if the firm implements a certain pollution prevention measure, it will achieve compliance with pretreatment standards. POTW personnel should give limited, basic advice in an informal manner and provide examples of other companies that have experimented with a given pollution prevention measure. Also, POTW personnel should be careful about revealing the identity of firms that have implemented pollution prevention measures that seem applicable to other similar facilities. Some of this information might be considered confidential, and therefore POTW personnel should check with facility managers before giving out company names for illustration purposes.

As part of the usual exit meeting, POTW personnel can summarize preliminary findings with respect to compliance and pollution prevention and receive the facility's initial response to those findings and any comments they might have about the inspection process. At this meeting, POTW personnel might wish to disseminate any applicable published pollution prevention information (e.g., EPA or state industry-specific pollution prevention handbooks or fact sheets) and inform owners and operators about state technical assistance offices and other pollution prevention resources.

49.5.3 After the Inspection

After the inspection, put pollution prevention observations in writing and provide additional sources of technical assistance and other information. As part of the normal inspection report, the investigator should include observations about pollution prevention measures for the facility to consider, and should forward more detailed information about measures that seem particularly promising and suggest some additional contacts and references for more information. POTW personnel also might wish to contact the facility after an appropriate amount of time to see if the facility has given any further consideration to the identified pollution prevention opportunities and to discover what problems or successes, if any, the company has had. This information could be very useful in future inspections.

49.6 Encouraging Pollution Prevention through Permitting Activities

POTWs have authority to require users to meet discharge limits and other requirements to prevent pass-through of toxic contaminants and disruptions of normal wastewater treatment operations. In general, the setting of local limits covering a wide range of contaminants and industrial and commercial sources provides a strong incentive for implementing pollution prevention measures. The cost of treatment generally rises with the stringency of local limits; as this occurs, pollution prevention becomes a more desirable means to assist industrial and commercial users in meeting local limits.

POTWs with the appropriate authority, usually established in sewer use ordinances, can use the permitting process as an effective mechanism for instituting pollution prevention as a local requirement for industrial and commercial users. This section will discuss three permitting strategies that either directly require facilities to adopt certain pollution prevention practices or create incentive structures that indirectly promote pollution prevention. These approaches are

- Requiring pollution prevention plans and implementation of *best management practices* (BMPs)
- Controlling discharges from small industrial and commercial users
- Encouraging pollution prevention when responding to user noncompliance

A local pretreatment program could explore the possibility of incorporating a pollution prevention planning provision into the permitting process. Such a provision could require that a facility interested in renewing an existing permit, or obtaining a new permit, must submit a detailed pollution prevention plan.

Enactment of sewer use ordinances where they do not already exist, or amendment of existing ordinances, may be necessary to provide POTW staff with the authority to require submission of pollution prevention plans.

Many industrial users are already subject to pollution prevention planning requirements. Pretreatment personnel should contact appropriate state and local agencies to determine whether any of the POTW's users have filed pollution prevention plans to meet existing federal or state regulatory requirements.

A pollution prevention assessment or audit conducted by facility owners and operators can be the single most effective means of identifying technically and economically feasible pollution prevention opportunities capable of achieving long-term reductions in the generation of toxic wastestreams. Pollution prevention plans should consist of the following elements:

- A process flow diagram showing where toxic constituents enter and exit the manufacturing process
- An estimate of the amount of regulated waste generated by each process
- An assessment of current and past pollution prevention activities, including an estimate of the reduction in amount and toxicity of regulated waste achieved by the identified actions
- A review of pollution prevention opportunities applicable to the facility's operations
- Identification of technically and economically feasible pollution prevention opportunities, including an assessment of the cost, benefits, and cross-media impacts of the identified opportunities
- An implementation timetable

POTWs also can require their dischargers to adopt BMPs, including inventory controls, employee training, and basic maintenance and inspection activities. BMPs generally can be implemented at little or no cost and often can achieve significant reductions in toxic discharges. Most industries have implemented some level of BMPs in an effort to run more efficient operations. Small industrial and commercial facilities, however, may not be aware of these simple steps to cleaner, more efficient operations and could benefit from the POTW's guidance. The most direct means for achieving widespread implementation of BMPs is to require pollution prevention planning as a precondition for obtaining or renewing a discharge permit.

Commercial and small industrial dischargers, such as laundries, dental offices, laboratories, hospitals, printing and publishing operations, photoprocessing facilities, wood refinishers, and motor vehicle operations, are sometimes not required to obtain discharge permits. These facilities, however, may represent a significant portion of the total loading of a toxic pollutant entering a POTW. In this situation, a POTW could benefit greatly from imposing local limits on, and promoting pollution prevention at, commercial and small industrial users. In some cases, a sewer use ordinance alone can provide the necessary control over small industrial and commercial users; however, an ordinance does not allow a POTW to set user-specific requirements that can be incorporated into individual discharge permits.

POTWs can encourage pollution prevention by taking full advantage of their authority to deal with users in noncompliance with pretreatment requirements. As part of the normal program activities of issuing permits and conducting inspections, POTWs can encourage pollution prevention, but they cannot require specific measures beyond those considered BMPs. In response to user noncompliance, however, a POTW can require specific pollution prevention measures as part of a mutually agreed upon compliance schedule with the user.

In requiring the development of a corrective action plan, POTWs can require facilities in noncompliance to conduct pollution prevention planning, to identify cost-effective pollution prevention measures, and to develop an implementation schedule with interim and final milestones. The implementation schedule can then be incorporated into a binding compliance schedule. The user in noncompliance can be required to evaluate pollution prevention options, but should be allowed the flexibility to develop a corrective action plan that includes the most effective mix of pollution prevention measures and traditional treatment options. An example of a compliance schedule that includes pollution prevention and recycling requirements is provided in Fig. 49-6.

49.7 Examples of POTW Pollution Prevention Initiatives

49.7.1 Palo Alto, California, POTW

The Palo Alto, California, POTW's Silver Reduction Pilot Program is an excellent example of using pollution prevention to drastically reduce commercial discharges of a specific contaminant. This POTW discharges to South San Francisco Bay (South Bay), which, over many decades, has become severely polluted by heavy metals. The Palo Alto POTW received permission from the Regional Water Quality Board to conduct a source reduction pilot program targeted at silver, a particular problem in South Bay. At the outset of the program, the Palo Alto POTW discharge concentrations of silver were more than 3.5 times the proposed South Bay limits, and silver concentrations in South Bay clams were many times higher than levels observed in other areas of the bay.

Initial sampling and mass-balance audits conducted by the Palo Alto POTW revealed that small businesses contributed up to 70 percent of the POTW's influent silver loading, regulated industries contributed 25 percent, and residential users contributed 5 percent. POTW personnel already had a solid understanding of the nature of the industrial silver discharges and concluded that commercial dischargers deserved their focus. They surveyed 650 businesses in the service area that might process x-rays and photographic films and negatives. More than 50 percent of the establishments that returned the survey indicated that they produced silver-bearing photographic wastes. The affirma-

---

Compliance Schedule for User in Noncompliance with Permit Limits

A. By July 1, 1992

The user shall submit a preliminary report on corrective action measures to be taken to maintain consistent compliance with permit conditions. At a minimum, the report shall include

- A detailed process flow diagram that identifies and characterizes the input of raw materials, the outflow of products, and the generation of wastes
- Any steps taken to reduce the concentrations and/or mass of regulated pollutants in the user's discharge to the sewer
- Preliminary findings of corrective action planning, including the identification of any pollution prevention, recycle/reuse, and treatment measures that are being considered for implementation

B. By August 15, 1992

The user shall submit to the POTW a corrective action plan for its discharge to the sewer system. The plan shall present an implementation schedule that outlines the steps to be taken to bring the user's discharge into consistent compliance with permit conditions by December 31, 1992. In developing the corrective action plan, the user shall evaluate and identify, for implementation, all cost-effective pollution prevention measures. Once developed, and if deemed technically sound by the POTW, the implementation schedule shall be incorporated into this compliance schedule.

Figure 49-6. Example of compliance schedule that incorporates pollution prevention.

tive responses were received from many small graphic artists, photoprocessors, printers and publishers, medical facilities, and dental offices. About 80 percent of these facilities indicated that they produced less than 5 gallons per day of silver-bearing photoprocessing wastes.

The survey data provided the basis for calculating local limits for commercial photoprocessors. Along with the new local limits, permitted facilities must also comply with various pollution prevention provisions designed to reduce the use and discharge of silver. For example, affected facilities must now conduct studies identifying pollution prevention opportunities for reducing silver discharges as part of the permitting process. Through on-site inspections and workshops, Palo Alto encourages photoprocessors to adopt pollution prevention methods wherever practicable to achieve compliance with local limits.

The program has been immensely successful. The average silver concentration of POTW effluent had decreased by about 75 percent within two years of when the local limits were imposed, and is now well below the National Pollutant Discharge Elimination System (NPDES) permit limit of 2.3 µg/L. Palo Alto estimates the cost of the source reduction project to the POTW at about $320 per pound of silver. This is extremely cost effective when compared to the $2700 per pound cost Palo Alto estimated for an end-of-the-pipe reverse osmosis treatment unit at the POTW.

In an effort to reduce metal and organic contamination in South San Francisco Bay, the Palo Alto POTW recently passed an ordinance requiring BMPs for automotive-related industries (i.e., facilities that repair automobiles, trucks, buses, airplanes, boats, etc.; or that perform services such as parts cleaning; body work; vehicle washing; fuel dispensing; or radiator, muffler, or transmission repair). Palo Alto offered these facilities the option of either sealing floor drains and implementing BMPs or installing treatment systems and meeting local limits. Palo Alto drafted the ordinance with the belief that automotive facilities can virtually eliminate toxic waste discharges by implementing inexpensive BMPs, thereby eliminating the need to apply for permits and install costly treatment systems.

The ordinance stipulates that automotive facilities meet the following requirements:

1. No person shall directly or indirectly dispose of vehicle fluids, hazardous materials, or rinsewater to storm drains.
2. Spilled rinsewater, hazardous waste, and vehicle fluids must immediately be cleaned up.
3. Vehicle fluid removal must take place where spilled fluid will be in an area of secondary containment.
4. No person shall leave unattended drip pans or other open containers containing vehicle fluids.
5. Vehicle service areas shall be cleaned using methods that ensure that no materials are discharged to sanitary or storm drains except in accordance with pretreatment standards. Facilities that use the following three-step process for cleaning floors will not require a permit:
   a. Clean up spills with rags or other absorbent materials.
   b. Sweep floor using dry absorbent materials.
   c. Discharge dirty water from mopping floors to the sanitary sewer via a toilet or sink.
6. Spill prevention and cleanup equipment and absorbent materials shall be kept on hand at all times.
7. Owners and operators shall ensure that all employees are trained regarding BMPs upon hiring and annually thereafter.

The Palo Alto POTW took several steps to ensure that automotive facilities were aware of the new ordinance and that facilities had access to technical assistance prior to the effective date of the ordinance. For example, they distributed a handbook describing automotive facility BMPs that reduce toxic waste discharges (Santa Clara Valley, 1991). If requested by the user, POTW personnel were available to go to the facility to answer questions about the new ordinance and give guidance on the implementation of BMPs. Palo Alto awarded public recognition to automotive facilities that achieved full compliance with pretreatment standards through use of BMPs and other pollution prevention methods by October 1, 1992.

49.7.2 Western Lake Superior Sanitary District

The Western Lake Superior Sanitary District (WLSSD) anticipated that it would have trouble meeting its future NPDES permit level for mercury. After determining that major industrial facilities do not significantly contribute to mercury loadings, WLSSD focused its mercury abatement efforts on unpermitted commercial establishments and residential users. Investigators determined that discharges from dental offices and laboratories were significant sources of mercury. Also, mercury from certain items in solid waste from commercial and residential sources was being transferred to the POTW through discharges from the air pollution control systems of municipal solid waste incinerators.

The WLSSD formed working groups representing dentists and laboratories, two groups of sewer users believed to be collectively significant generators of mercury waste. The purpose of the working groups is to identify means of reducing mercury discharges through use of BMPs and other measures. Also, a local advisory group is exploring the possibility of implementing a thermostat collection program for local construction and demolition companies to reduce this source of mercury in solid waste that is incinerated.

Bibliography

City of Palo Alto, *Silver Reduction Pilot Program; Palo Alto Regional Water Quality Control Plant*, Palo Alto, Calif., 1992.

Greiner, T. J., and P. H. Richard, *Facility Inspections: Obstacles and Opportunities*, 1992, unpublished, Office of Technical Assistance, Massachusetts Executive Office of Environmental Affairs.

Minnesota Pollution Control Agency and Western Lake Superior Sanitary District, *Lake Superior Partnership; Compliance Assistance Program: A Multimedia and Pollution Prevention Inspection Program. October 1, 1991 through April 30, 1992*, Semiannual Progress Report to U.S. EPA, MPCA and WLSSD, 1992.

Santa Clara Valley Nonpoint Source Pollution Control Program, *Best Management Practices for Automotive-Related Industries*, San Jose, Calif., 1991.

Sherry, S., *Minimizing Hazardous Wastes: Regulatory Options for Local Governments*, Local Government Commission, Sacramento, Calif., 1988*a*.

———, *Reducing Industrial Toxic Wastes and Discharges: The Role of POTWs*, Local Government Commission, Sacramento, Calif., 1988*b*.

U.S. Environmental Protection Agency, *Waste Minimization Opportunity Assessment Manual*, U.S. EPA Hazardous Waste Engineering Research Laboratory, CERI, EPA/625/7-88/003, Cincinnati, Oh., 1988.

————, *Interim Policy on the Inclusion of Pollution Prevention and Recycling in Enforcement Settlements*, memo from James A. Strock, Assistant Administrator, Washington, D.C., 1991*a*.

————, *Environmental Research Brief: Waste Minimization Assessment for a Manufacturer of Printed Circuit Boards*, U.S. EPA Risk Reduction Engineering Laboratory, EPA/600/M-91/002, Cincinnati, Oh., 1991*b*.

————, *Achievements in Source Reduction and Recycling for Ten Industries in the United States*, U.S. EPA Office of Research and Development, EPA/600/2-91/051, Washington, D.C., 1991*c*.

————, *Supplemental Manual on the Development and Implementation of Local Discharge Limits under the Pretreatment Program*, U.S. EPA Office of Water, 21W-4002, Washington, D.C., 1991*d*.

————, *Report to Congress on the National Pretreatment Program*, U.S. EPA Office of Wastewater Enforcement and Compliance, Washington, D.C., 1991*e*.

————, *Facility Pollution Prevention Guide*, U.S. EPA Office of Research and Development, EPA/600/R-92/088, Washington, D.C., 1992.

Other Resources

Pollution Prevention Information Clearinghouse, c/o SAIC, 7600-A Leesburg Drive, Falls Church, Va. 22043. For more information about PPIC or technical assistance, call (703) 821-4800. For on-line information from the Pollution Prevention Information Exchange System (PIES), call (703) 506-1025.

50

Pollution Prevention in the Textile Industries

Lesley J. Snowden-Swan

Battelle Pacific Northwest Laboratories
Richland, Washington

50.1 Industry Overview

From the ancient handweaving of linen mummy cloths for Egyptian tombs and fine silk gowns in China to turning out designer blue jeans in mass quantities to the manufacture of fiber-reinforced composites for the Mars mission, the production of textile products has prevailed throughout the centuries. In today's industrial setting, the basic principles of textiles are much the same, but the practice of crafting textiles has grown to become one of the major industries in the United States and abroad. The United States textile industry, including fiber production, fabric production, and apparel workers employs 1.7 million workers, accounting for one of every 11 jobs.[1]

In its broadest sense, the textile industry includes fiber, fabric, and apparel production, and retail sales. Generally, production includes the following steps:

1. Generation of a usable fiber from either a natural source (i.e., cotton, wool, and silk) or from a manufactured source (polyester, rayon, and nylon)

2. Production of yarn for knitted and woven fabrics

3. Construction of product through knitting, weaving, and nonwoven or other methods

4. Preparation, dyeing, printing, and finishing to produce finished fabric

5. Fabrication to produce final product (i.e., garment, household item, industrial, or specialty product)

The context of this chapter focuses on fabric production, which is most commonly viewed as the heart of the textile industry.

Textile fabrics are made from a very wide and diverse range of products, including a wide variety of fiber types. Table 50-1 lists some common fibers used in fabrics today.[2]

Markets for textiles are most frequently divided into apparel (clothing), domestic (household items such as towels and bed coverings), and industrial (automotive components such as tire cord, filters, carpet, and upholstery, and recreational equipment such as backpacks and parachutes). Of the total textile products, cotton fiber has the highest demand, and the apparel market accounts for the majority of fiber consumption.[3]

Traditionally, the textile industry is very energy-, water-, and chemical-intensive. Within the industry, approximately 60 percent of the total energy consumed is within the wet-processing stages.[4,5] Most wet processing involves treatment with chemical baths, which often require washing, rinsing, and dry-

Table 50-1. Fibers Used in Textile Manufacturing

Fiber type	Resource
Natural:	
Cellulosic (natural)	Cotton, flax (linen), ramie
Animal hair	Wool, mohair, angora, camel's hair, giviut (musk ox), cashmere, llama, alpaca
Silk	"Sericulture": cultivation of Bombyx mori silkworm
Manmade	
Rayon (manmade cellulosic)	Wood pulp
Acetate	Purified cellulose from wood pulp + acetic acid + catalyst
Nylon (the first synthetic)	Polyamide production from various monomers
Polyester	Dicarboxylic acid + dihydric alcohol
Olefin	Polymerization of ethylene; polypropylene
Acrylic and other vinyl fibers	Polymerization of acrylonitrile, Saran® (vinylidene-chloride/vinyl-chloride copolymer)
Spandex	Polyester (preformed) or polyether + di-isocyanate, then polymerization
Special use	Aramid (aromatic polyamide), glass, novoloid (contains cross-linked phenolformaldehyde polymer)

ing steps between key treatment steps. Consequently, large volumes of wastewater are generated with a very diverse range of contaminants that must be treated prior to disposal; much energy is consumed to heat and cool chemical baths and washwater and to dry fabrics or yarns.

Not surprisingly, the industry has faced increasing pressure regarding environmental and waste-related concerns as a result of the quantity and toxicity of generated wastewaters; this was illustrated in 1989 when the industry was listed among the top ten toxic waste generators in the United States Environmental Protection Agency TRI report[6] (for 1987 releases), the majority of toxics (52%) being released to water media. The industry has since made significant reductions in waste generation through equipment changes, recycling, and non-process-related measures such as housekeeping, as well as research and development activities centered on technology for waste minimization. Specific examples of these activities are discussed in Secs. 50.4 and 50.5.

Recent developments in the industry are providing for more recycle and reuse of process water and chemicals. Furthermore, interest in media other than the traditional water-based systems, such as solvents and foams, for chemical application is increasing. Accordingly, the industry is finding that, beyond meeting regulations, the potential economic gain through reductions in treatment costs and wasted resources is enormous.

Waste reduction efforts can involve process-related solids (primarily fiber and yarn spinning waste, fabric cutting waste, fly ash from burning of fossil fuels, sludge from wastewater treatment facilities, and packaging materials) and air pollution (i.e., volatiles such as formaldehyde and ammonia emitted during flame-retardant finishing[7]), as well as waste from non-process-related activities. However, the largest impact, especially with respect to water pollution, may be made in the wet-processing operations, primarily those steps taken after the construction of unfinished fabric (commonly called gray goods), because these operations are the most water- and energy-intensive and potentially the greatest waste-generating part of the textile industry.[4]

The following sections describe preparation, dyeing/printing, and finishing operations; the primary wastestreams generated; feasible pollution prevention strategies that are currently available; and success stories demonstrating selected strategies.

50.2 Process Description

Because there is such a diverse product and application range of textiles today, the type of processing used is highly variable and depends on site-specific manufacturing practices, as well as on the type of fiber used and the final physical and chemical properties desired. These properties include tensile strength, flexibility, uniformity, and luster. Even for a constant product type, no two textile mills use exactly the same methods of production. Figure 50-1 shows the typical sequence for the minimum wet-processing steps taken for cotton-con-

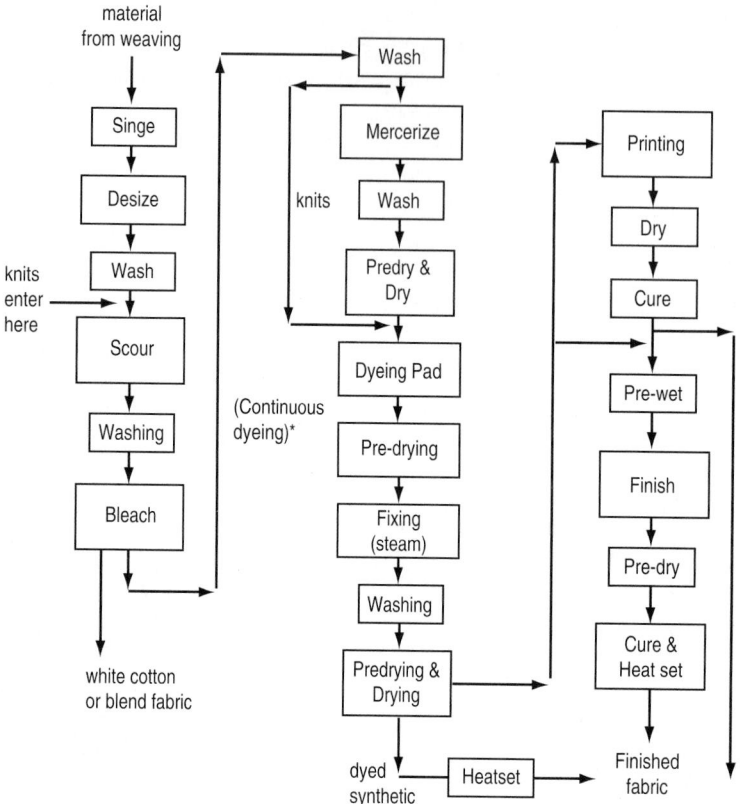

Figure 50-1. Typical process for pretreatment, dyeing, and finishing of cotton and cotton-blended fabrics. (Note that batch, rather than continuous dyeing, is normally used for knits.)

taining (100 percent and cotton-blended) woven fabrics, and will be the focus of the remaining discussion. Variations in this sequence are indicated in Fig. 50-1 for knits, whites, and manufactured fibers. In general, natural fibers receive more processing than manufactured ones to achieve the same product performance.

The common systems used for processes shown in Fig. 50-1 include batch, continuous, and semicontinuous equipment setups. In batch setups, a fixed amount of substrate (cloth, yarn, etc.) is placed in the machine, chemical solution is introduced, and processing proceeds. After the reaction is finished, substrate is removed and chemical solutions are discharged (or reused, if feasible), with any subsequent auxiliary processes (rinsing, washing, etc.) occurring in the same vessel. In continuous systems, chemical solution is placed in the machine and fabric is moved through it without interruption. Because batch systems inherently tend to generate more waste than continuous systems, batch systems are generally used for treating (i.e., dyeing) small quantities of a particular type of product, while continuous systems are more amenable to processing large yardages of product. Further description of the three major phases of wet processing is given to provide information on the wastestreams generated.

50.2.1 Fabric Pretreatment

As shown in Fig. 50-1, the minimum aqueous processing for cotton-containing woven fabric preparation prior to dyeing includes desizing, scouring, bleaching, and mercerizing. Desizing, scouring, and bleaching can be carried out in either batch or continuous modes. In continuous operation, fabric is immersed in solution, moved to a heated reaction chamber, and washed off in a series of wash boxes after reaction completion. In batch operation, the reaction actually occurs while the fabric is immersed in the treatment bath. Typically, the vessel is drained, and neutralization and washing are carried out in the same vessel.

Fabric coming directly from the loom or knitting machine usually contains impurities. For wovens (only), the major impurity is generally a warp-sizing compound, material applied to the yarn during the slashing step to minimize yarn damage from abrasion and to maximize weaving efficiency. The sizing is a film-forming stiffening agent, such as starch, carboxymethyl cellulose (CMC), or polyvinyl alcohol (PVA). PVA is becoming increasingly popular because, unlike starch, it remains intact during desizing and can therefore be recovered and reused. However, current recovery and reuse of PVA is primarily found in large vertically integrated companies where the weaving and preparation operations are geographically collocated. In desizing, fabric is treated with a compound that is complementary to the sizing agent previously used. Typically, enzymes are used as reducing agents, and the system is run at elevated temperatures to increase the speed of the process. The fabric is then washed completely to remove the impurities.

Next, the cotton cloth is treated with hot caustic solution in the scouring process in order to remove substances such as pectin, wax, and other impurities present in natural fiber. Other compounds necessary for removal include yarn spinning and knitting lubricants (for knits only). Impurities are removed from fabric for processing reasons and/or because they cannot be released to the retailer/consumer. For aqueous scouring, there are a variety of chemicals used, including sodium hydroxide, detergents, and occasionally solvents for treating manmade fibers. Fabric is then thoroughly washed to remove scouring solution.

After scouring, essentially all cellulosic-containing fabrics are bleached even if they will eventually be dyed or printed another color. Bleaching provides a uniform white surface for dyeing and/or printing. The majority of bleaching operations (more than 95 percent) use hydrogen peroxide (H_2O_2), while the rest use calcium hypochlorite. Peroxide bleaching is favored by the industry because it is cost-effective and noncorrosive to processing equipment as well as being a safe and environmentally sound alternative to other chemicals. It has also been suggested that less washing is required after peroxide bleaching for removing the reaction solution.[3]

The last step of aqueous processing for cotton woven fabric preparation is mercerizing, which is always carried out in continuous mode. In this process, the fabric is held with or without tension while being saturated with caustic solution. This method results in a desirable change in the physical and chemical properties—in particular, luster, dyeability, strength, and smoothness. Like fabrics, yarns may also be mercerized before fabric construction. Again, the fabric is rinsed thoroughly under tension after mercerizing is finished.

50.2.2 Dyeing/Printing

After preparation, color is applied to fabric through dyeing and/or printing. There are three primary mechanisms for applications of dyes to fiber, yarn, or fabric: (1) chemical reaction with the fiber molecules, (2) attachment to the fiber surface, or (3) absorption into the fiber with no reaction. Dye categories include acid, azoic, basic, direct, disperse, pigment, reactive, solvent, sulfur, and vat dyes. Most commonly in use today are the reactive and direct types for cotton dyeing, and disperse types for polyester dyeing. Direct and fiber reactive dyes have a fixation of 90 to 95 percent and 60 to 90 percent, respectively, while disperse dye is 80 to 90 percent. Individual dyes possess unique chemical characteristics that are suitable for a specific type of fiber, a desired color and quality of the dyed material, the type of equipment to be used, and other considerations.

For some applications, the fiber (stock dyeing) or yarn (package and skein) is dyed prior to fabric construction, or the garment is completed prior to dyeing. However, piece dyeing is most widely used because it is the simplest and least costly method. Piece dyeing also allows manufacturers to color fabrics as ordered, rather than stockpiling and having to risk changes in customer preferences. Methods for piece dyeing include beam, beck, jet, and jib processing, all of which are batch processes, and pad dyeing, which can be either a batch or continuous setup. Typically, in batch (exhaust) dyeing, scouring (see Sec. 50.3) and rinsing are carried out after dyeing to remove all excess dye and chemicals from the fabric. The majority of the dyeing machines in use today are exhaust processes; however, continuous dyeing accounts for about 60 percent of the total yardage of product dyed in the industry.

Of the numerous printing techniques used for commercial production, the most common is rotary screen, but others such as direct, discharge, resist, flat screen (semicontinuous), and roller printing are often seen commercially. Sometimes washing with hot water and detergent is required for "wet printing"; however, for "pigment printing," which comprises approximately 75 to 85 percent of all printing operations, washing is not required. One method developed in the '70s, transfer printing, requires no cleanup and generates little or no waste. Currently, this technology is practical for 100 percent polyester fabric only, and thus is viewed as having limited potential in the future. The application of foam technology to printing has also been demonstrated, the main advantage again being reductions in water and energy consumption.

50.2.3 Finishing

The primary purpose of the finishing process is to apply chemical moieties to the fabric in order to alter properties affecting the care, comfort, durability, environmental resistance, aesthetic value, and human safety associated with the fabric. Finishes include a very large and diverse group of chemicals ranging from antistatic to shrink-resistant to flame-resistant treatments; the most common are those which ease fabric care, specifically the permanent-press, soil-release, and stain-resistant finishes. In wet-finishing, the sequence of steps typically includes chemical finish application, drying, curing, and cooling. Most finishing employs chemical application together with mechanical techniques, the advantages of the latter being improved feel, strength, and abrasion resistance and lower chemical consumption and waste. Developments in low add-on technology and other application methods are in progress.

50.3 Primary Wastestreams

The principal wastestream of concern is water containing natural impurities and processing chemicals (Tables 50-2 to 50-4). Wastewater composition is highly variable due to the wide range of treatments used in batch processing. Generally, wastewater is colored, highly alkaline, high in BOD and COD, and at elevated temperature. Also of particular concern are more specific com-

Table 50-2. Waste Characteristics for Fabric Preparation Operations

Process	Purpose	Species in bath/washwater
Desizing of wovens	Remove size applied for weaving	Enzymes, degraded starch (high BOD), or, alternatively, PVA
Scouring (for cotton only)	Remove natural impurities and handling contaminants	NaOH; chelating agent for Fe; detergents; fats; oils; pectin; wax; cotton seed, stems, and leaves; knitting lubricants; spin finish
Bleaching	Decolorize natural pigments and enhance uniformity of color adsorption	H_2O_2 (most common): sodium silicate or organic stabilizer, sodium hydroxide, surfactants, chelates, sodium carbonate; possibly antichlor (for wool) like sodium peroxide
Mercerizing (for cotton only)	Improve fabric's chemical/physical properties	NaOH (16–24% solution)

Table 50-3. Wastewater Characteristics for Dyeing and Printing Operations

Process	Purpose	Compounds in bath/washwater
Direct and reactive	Cotton dyeing	Color (for blue or green, includes Cu), surfactant, defoamer, sodium salts, (Cl-, S042-), sequestrant, leveling and retarding agents (direct only), diluents, and finish. For postscouring: surfactant, sequestrant (phosphate), acetic acid, fixative (cationic), and NaCl.
Disperse	Polyester dyeing	Color, acetic acid, MSP, EDTA, NTA, phosphates, leveling agents, carrier (methyl benzoate, phenyl benzoate), defoamers, lubricants, dispersants, delustrants, and diluents. For postscouring: NaOH, soda ash, sodium hydrosulfite, and acetic acid.
Printing	Rayon	Urea (up to 20% solution)[8]

Table 50-4. Wastewater Characteristics for Finishing Operations

Finish type	Compounds in bath/washwater
Durable- or permanent press	Dimethyl dihydroxy ethylene urea (DMDHEU), with catalyst (MgCl), softener (fatty compound, polyethylene, silicone), and surfactant*
Soil-release	Acrylic; fluorochemicals
Stain- and soil-resistant	Fluorochemicals; pyridinium compounds
Flame-resistant or retardant	Water solubles (are reapplied after water contact), such as borax, boric acid, or ammonium phosphate
Bacteriostat	Zinc acetate w/H_2O_2 and acetic acid for cotton; methylol melamine w/zinc nitrate; NH_4Cl; zinc chloride

*Most finishing baths contain catalyst, softener, and surfactant.

pounds that are toxic to aquatic life, such as heavy metals, primarily from dyeing and finishing (and water impurities), surfactants (wetting agents), compounds used throughout the wet-processing steps, and other process chemicals. Fabric rinsing and/or washing (using detergent) is usually performed between primary process steps, resulting in large quantities of dilute wastewater in excess of spent chemical baths.

The following sections describe wastewaters from individual steps within fabric pretreatment, dyeing, and finishing.

50.3.1 Fabric Pretreatment Waste

With the numerous process steps that are potentially involved in fabric preparation, wastewaters may contain a complex mixture of chemicals. The desizing step is of particular concern, contributing up to 50 percent of the BOD load in wastewaters from wet processing.[9] Presented in Table 50-2 are typical chemical characteristics of pretreatment effluents.

50.3.2 Dyeing/Printing

Of the 700,000 tons of dyes produced annually worldwide, approximately 10 to 15 percent of the dye is disposed of in effluent from dyeing operations.[10,11] For one dyehouse recently characterized,[12] as much as 50 percent of the dye originally present in the fresh dyebath was discharged after dyeing of synthetic fibers. Wastewater generation from a typical dyeing facility is estimated at 1 to 2 million gallons per day. Including the dyeing, postscouring, and rinsing processes, approximately 12 to 17 gallons of wastewater per pound of product are produced for disperse dyeing, while 15 to 20 gallons per pound is more typical for direct and reactive dyeing. The primary source of wastewater is spent dyebath and washwater, which contain by-products (hydrolyzed dye), some intact dye, and auxiliary chemicals. In addition to process water and chemicals, a major source of toxic pollutants in wastewater is cleaning solvents used in dyeing and printing machine cleaning, such as oxalic acid, hydrochloric acid, and carbon tetrachloride.

With the abundance of individual dyes and the wide range of dyeing equipment in use today, it is difficult to summarize wastestream characteristics. In general, wastewater from batch dyeing is high in volume and pollutant load, and tends to contain heavy metals, aromatics, and halogenated hydrocarbons from the dyebath makeup. All are toxic to aquatic life.[12] In addition to the dyestuff itself, many auxiliary chemicals are used to aid in dye transfer; the majority of these chemicals, including unreacted color, are discharged with the spent bath.

In the most commonly used technique, pigment printing, the main source of waste is from the cleanup, during which unused printing paste is removed from the screen. Consequently, proper planning of paste use and housekeeping are major issues in minimizing waste in printing operations.

Table 50-3 lists chemical characteristics of waste effluent from exhaust dyeing of cotton (direct and fiber reactive) and polyester (disperse) products.

50.3.3 Finishing

As with the dyeing operations, finishing methods are highly variable, due to the broad range of finishes available. Pollutants in wastewaters include natural and synthetic polymers, and a range of other potentially toxic substances. Table 50-4 shows a few common finishes used and serves to illustrate the wide variety of chemicals that may be present in finishing effluent.

50.4   Pollution Prevention Opportunities

Pollution prevention is perhaps most easily conceptualized and followed using the EPA hierarchy of pollution prevention strategies.[13] This model has become the most widely accepted infrastructure for adopting and effecting the pollution prevention philosophy:

> *First and foremost, reduce waste at the origin—through improved housekeeping and maintenance, and modifications in product design, processing, and raw material selection. Next, if you must produce waste, recycle potential wastes back into the process. Finally, if there is no prevention option possible, treat and safely dispose of the waste.*

As in most manufacturing operations, substantial reductions in textile facility waste can be realized through good operating practices, without major investments in new technology. Nonprocess strategies include activities such as improved equipment maintenance (repairing leaking hoses, valves, and pump seals), material handling (inventory management, spills) and production scheduling. For example, with the large number of process steps and chemicals used in fabric preparation, dyeing, and finishing, it is crucial that different effluent types be kept segregated to facilitate effective separation and recovery of process chemicals and water. Generally, the more dilute and complex a stream in composition, the more difficult it is to treat the stream and reclaim valuable components. Ideally, each step (or group of similar streams) would have its own recovery system to maximize separations efficiency and cost effectiveness. For further explanation of general techniques for minimizing waste through improved housekeeping, see Ref. 14, also listed in Further Reading.

Beyond good operational practices, the primary near-term focus for achieving pollution prevention in textiles manufacturing is on reductions in water, chemicals, and energy use, and the utilization of less hazardous, more efficient, and more recoverable compounds through material substitution, process changes, and reuse strategies. Efforts to further identify and characterize potential pollutant sources in commodity (as trace impurities) and specialty chemicals are also underway. Long-term solutions to waste generation may involve increased product reformulation specific to fibers, such as increased reactivity and resistance to abrasion during weaving. However, it is unlikely that drastic changes in the fiber types used will occur. Further discussion on long-term solutions is presented in Sec. 50.4.2.

50.4.1   Available Options

There is a vastly increasing number of available technologies that lend themselves to preventing hazardous waste generation and reducing the environmental impact of textile manufacturing. Prevention strategies for wet processing, along with specific examples of each are presented in Tables 50-5 through 50-7. A more thorough description of each technique is found in the referenced documents and the list of further reading.

As shown in Table 50-5, material change strategies have focused on decreased toxicity, increased recoverability, and improved efficiency of reactions. When the complete textile product life cycle is considered, it is apparent that chemical suppliers (for fibers, dyes, and auxiliaries) have a tremendous influence on the environmental impact of textiles manufacturing. In fact, there is increasing pressure from the textile mills on their suppliers to improve product performance via increased exhaustion and efficiency of dyes, less toxic and fewer auxiliary chemicals needed, and low-volatile and permanent finishes.[12]

Table 50-6 gives process modification strategies to aid in minimizing wastestreams. Process change methods have focused on advanced dyebath

Table 50-5. Material Change Strategies in Wet Processing

Option	Example	Benefit
Alternate desizing agents	1. H_2O_2 rather than enzyme for desizing of starch[14]	Resulting products of CO_2 & H_2O over hydrolyzed starch using enzymes (less BOD in wastewater)
	2. Enzymes that degrade starch to ethanol[14]	Reduced BOD of spent desizing bath and opportunity to recover ethanol as fuel
Bleaching agent		Increased wash efficiency (less water needed)
Alternate dyes	1. Copper-free for producing green shades in 100% cotton fabric[15]	Reduced toxicity (metal content) of spent dye bath and washwater
	2. Improved fixation reactives such as with Remazol™ (95–98% fixation w/pad/batch)[15] and others	Less unreacted and hydrolyzed (degraded) dye in spent bath and washwater improves reuse opportunities
	3. High-temperature reactive (such as Procion™) for simultaneous application of disperse and reactive dyes[4]	Energy reduction, elimination of caustic bath required after disperse dyeing
Substitute auxiliaries (all processes)	1. Substitute for phosphates such as acetic acid (pH control) and EDTA (water conditioner)[14]	Reduction in phosphorus load in wastewater
	2. New washing agents (such as Sandpure RSK™ and others)[15]	Increased wash efficiency, and thus decreased water consumption and improved fastness of reactives

Table 50-6. Process Change Strategies in Wet Processing

Option	Example	Benefit
Pad-batch dyeing	Transfer of dye to cotton, rayon, and blended goods through rollers (continuous method)[14]	Reduction in water (2 gal/lb vs. 20 on becks), energy (2000 vs. 9000 BTU/lb) and chemical use, increased productivity
Low liquor ratio dyeing	Reduction in the weight of water (solution) used to dye a given weight of goods[4,14]	Improved dye fixation, large reduction in energy and water consumption in dyeing (but not necessarily in subsequent washing steps)
Foam technology	Application of dye through foam media (air dispersed in liquid)[4,14] or other solvents for dyeing and printing (finishing and preparation)	Reduced water and energy consumption, chemical waste, and time necessary for drying
Spraying technology	Application of finishes using sprays[4]	Reduced energy and water usage, and chemical waste (most finish remains on fabric)
Washing technology	Countercurrent washing, vibrating-reed jet washers, mechanical means for increased turbulence[4]	Improved washing efficiency, thus decreased water and energy usage
Process consolidation	Single pad-steam-wash sequence using unique combinations of chemicals[14,16]	Decreased water and energy usage, salt in effluent, and process times

techniques for aqueous processing, alternative application media and delivery, and improved washing efficiency. In addition, overall strategies such as combining individual process steps into one step offers perhaps a more systematic and holistic approach to minimizing waste.

Table 50-7 gives specific strategies feasible for reclamation of process water and chemicals. Inherent in water conservation is the added benefit of reduced consumption of secondary resources such as electricity for drying and pumping, fossil fuels used for steam generation, and cooling water. Furthermore, water that has been treated and recovered is generally more valuable than freshwater (less impurities), providing even more economic incentive to recycle options. Recent advances in membrane technology and other treatment schemes have focused on nondestructive treatment to allow recovery, providing improved alternatives to common wastewater treatment methods used in the past (chemical precipitation, trickling filtration, and biological treatment and aeration), which have not necessarily enabled resource recovery.

Along with the specific reuse opportunities listed in Table 50-7, accurate and automated chemical analysis of spent bath and wastewaters is considered

Table 50-7. Recovery and Reuse Strategies for Wet Processing

Option	Description	Benefit
Size* recovery (after desizing)	Separation/concentration of PVA size by ultrafiltration (or reverse osmosis) for reuse[14,16,17]	Reduced BOD load in effluent, and freshwater and chemical consumption
Dyebath reuse†	With low hydrolysis of dye molecules, dyebaths may be recharged with bath chemicals and reused repeatedly[14]	Decreased pollutant load in effluent, freshwater, and chemical consumption
Caustic recovery	Recycling of NaOH mercerizing solution up to 98%[14]	Reduced alkalinity of wastewater from pretreatment and chemical consumption
Metal reuse	Treatment of exhausted dyebath w/biological, chemical coagulation, membrane separation technology to remove and recover metals and water	Reduced freshwater and metals consumption, and toxicity of effluent

*Also involves material substitution (e.g., PVA or similar agent) in the sizing process to enable subsequent recovery.

†Dyebath analysis tends to be difficult with reactive type dyes due to the difficulty of differentiating unhydrolyzed and intact colorant using spectrophotometry.[18]

essential for providing more opportunities for implementation of pollution prevention measures. For example, the feasibility and effectiveness of dyebath reuse increases dramatically if an accurate and timely analysis of used dyebath composition is available for calculating the amounts of makeup chemicals needed to refresh and reuse the bath.

50.4.2 Future Strategies

Until the last year or two, research and development (R&D) efforts addressing environmental matters have been minimal. It is now evident, however, that advanced technology will be essential to the future of the textiles industry. Examples of research and development efforts underway are:

- Sizes able to be removed in cold water for later reuse and sizes that are permanent (no removal necessary, function as dyesites)[16]

- Gaseous reactants for bleaching, including O_3, singlet oxygen, or vapor phase reactions,[4] resulting in reduced aqueous waste and energy and water consumption

- Ultrasonic waves in dyebath to increase dye uptake by threefold

- Electrostatic application of powdered chemicals for printing and finishes,[4,18] resulting in water and energy conservation and reduced effluent volume

- Application of overlapped colors of powdered pigments and binders to fabric, using a process similar to color xerographic methods,[12] resulting in reduced water consumption and effluent volume

- The application of ink-jet printing, a noncontact method of propelling very small droplets of ink onto a surface, to textile products[19]

- Fiber modification through surface-photografted (and other grafting methods) synthetic fibers[20] resulting in enhanced dye absorption and less wasted dyestuff

More overall changes in the production infrastructure may be likely as well. For example, integrating automation into every aspect of production, including everything from dyebath chemical analysis and reuse to garment fabrication, is a major topic of interest and effort within the industry. An essential aspect of automation is *demand activated manufacturing* (quick response), which will hold high importance in the future. An example is the coloring of preassembled items to order rather than carrying a large inventory.[18]

Increased vertical integration of individual functions is also desired, as closely coupled manufacturing processes tend to maximize control and energy and water efficiency, chemical recovery, and production rate. In the future, synergistic relationships will be identified between different industries, and facilities will be built in close geographical location to strive for complete closed-loop systems.

In its efforts to improve operations, the industry has recently reached out into other sectors of the research community. Just recently formed is the American Textiles (AMTEX) Initiative, a cooperative research agreement between the United States Department of Energy (national research labs) and numerous organizations from the textile industry and academic sectors. This agreement has facilitated collaborative research in many aspects of textile manufacturing that are important to enhancing the global competitiveness of United States textiles, incorporating pollution prevention and waste minimization technology.

Case Studies

Chemical Substitution of Sulfides[21]

One textile facility that uses sulfur dyes has investigated possible substitutes for sodium sulfide, which is used to convert water-insoluble dyes to the soluble (lueco) form for application to materials. After several attempts at substitutes, the facility found that they could replace 100 parts sodium sulfide with

65 parts alkaline solution containing 50 percent reducing sugars plus 25 parts caustic soda. As a result, sulfide levels in the effluent from this process have dropped substantially (below 2 ppm).

Recovery and Reuse of Rinse Baths[22]

Through the implementation of recycle schemes for process water used to remove mercerizing solution, a yarn finishing company has drastically reduced pollution load in wastewater, and soda (Na_2CO_3) and caustic consumption. The new process involves reusing the rinse bath three times over rather than using three separate volumes of rinsewater in series. The spent rinsewater is then processed in an evaporator and concentrated caustic is reused in mercerizing. The result of the process changes was an 80 percent reduction in suspended solids, 55 percent reduction in COD, and 70 percent reduction in neutralizing soda in the wastewater. Corresponding reductions in hydrochloric acid used to neutralize the effluent were experienced as well. The investment in new equipment resulted in an annual savings of $189,000, with a payback of under one year.

Reuse of Nonprocess and Process Water and Automated Chemical Addition for Dyeing[23]

In efforts to reuse both noncontact cooling water and contact production water used in the dyeing process, an acrylic yarn producer has realized tremendous savings of water, energy, chemicals, and reduced waste generation. The facility has reduced its water consumption from 320,000 gallons per day to 102,000 gallons per day and simultaneously increased production from 12 to 20 batches per day. Additionally, energy consumption for heating dyebath has decreased substantially. The investment has resulted in a savings of approximately $13,000 a month and paid for itself in 30 days after implementation.

In addition to water reuse, the company has implemented computer technology to automate dyebath flow and temperature in a new facility. With the computer program, addition of auxiliary chemicals (retarders and leveling agents) is precisely controlled, resulting in a clean exhausted dyebath; this eliminates the need for postrinsing and thus reduces amounts of water and chemicals consumed and wasted.

References

1. *People in the U.S. Textile Industry,* employment statistics for 1990, prepared by the American Textile Manufacturers Institute, Washington, D.C.

2. Norma Hollen, Jane Saddler, Anna L. Langford, and Sara J. Kadolph, *Textiles,* 6th ed., Macmillan, New York, 1988.

3. Peyton B. Hudson, Anne C. Clapp, and Darlene Kness, *Joseph's Introductory Textile Science,* Harcourt Brace Jovanovich, Fort Worth, 1993.

4. Joseph S. Badin and Howard E. Lowitt, *The U.S. Textile Industry: An Energy Perspective*, DOE/RL/01830/T-56, prepared by Energetics, Inc. for Pacific Northwest Laboratory, Richland, Wash., 1988.

5. *Energy Conservation in the Textile Industry Phase I & II*, ORO-5099-T1/T2, prepared by the Georgia Institute of Technology, 1977/78.

6. "Textiles Ranks Sixth in Toxic Waste," *Textile World News*, August 1989, pp. 23–25.

7. David Gent, "A Novel Approach to Practical Problems," *Journal of The Society of Dyers and Colourists*, **108**:306–307 (1992).

8. Fred C. Cook, "AATCC Invades Boston with Environmental Ammo," *Textile World*, September 1990, pp. 67–70.

9. W. B. Achwal, "Environmental Aspects of Textile Chemical Processing (Part I)," *Colourage*, October 1990, pp. 40–42.

10. Jack T. Spadaro, Michael H. Gold, and V. Renganathan, "Degradation of Azo Dyes by the Lignin-Degrading Fungus," *Applied and Environmental Microbiology*, **58**(8):2397–2401 (1992).

11. H. Zollinger, *Color Chemistry-Syntheses, Properties and Applications of Organic Dyes and Pigments*, VCH Publishers, New York, 1987.

12. Mervyn C. Goronszy, and H. Tomas, "Characterization and Biological Treatability of a Textile Dyehouse Wastewater," *47th Purdue Industrial Waste Conference Proceedings*, Lewis Publishers, Inc., Chelsea, Mich., 1992.

13. *Facility Pollution Prevention Guide*, EPA/600/R-92/088, May 1992.

14. Brent Smith, *A Workbook for Pollution Prevention by Source Reduction in Textile Wet Processing*, available from the Pollution Prevention Pays Program, North Carolina Department of Environment, Health, and Natural Resources, Raleigh, N.C., 1988.

15. Fred C. Cook, "Environmentally Friendly: More than a Slogan for Dyes," *Textile World*, May 1991, pp. 84–89.

16. Fred C. Cook, "Fabric Processes Beholden to Energy, Environment," *Textile World*, November 1990, pp. 49–54.

17. J. D. Nirmal, V. P. Pandya, N. V. Desai, and R. Rangarajan, "Cellulose Triacetate Membrane for Applications in Plating, Fertilizer, and Textile Dye Industry Wastes," *Separation Science and Technology* **27**(15):2083–2098 (1992).

18. Fred C. Cook, "QR, Environmental Pressure Will Drive Dyeing, Printing," *Textile World*, December 1990, pp. 83–85.

19. Brent Smith and Elizabeth Simonson, "Ink Jet Printing for Textiles," *Textile Chemist and Colorist*, **19**(8):23–29, August 1987.

20. Zhenguo Feng, Magdalena Icherenska, and Bengt Ranby, "Photoinitiated Surface Grafting of Synthetic Fibers, IV: Applications of Textile Dyes onto Surface-Photografted Synthetic Fibers," *Die Angewandte Madromolekulare Chemie*, **199**:33–44 (1992).

21. Case Study for Century Textiles and Industries Limited (Bombay, India), available from Waste Reduction Resource Center, Raleigh, N.C..

22. Michael R. Overcash, *Techniques for Industrial Pollution Prevention*, Lewis Publishers, Chelsea, Mich., 1986, p. 168.

23. Case Study for Amital Spinning Corporation, available from Waste Reduction Resource Center, Raleigh, N.C.

Further Reading

Badin, Joseph S., and Howard E. Lowitt, *The U.S. Textile Industry: An Energy Perspective,* DOE/RL/01830-T56, prepared by Energetics, Inc. for Pacific Northwest Laboratory, Richland, Wash., 1988.

Smith, Brent, *Identification and Reduction of Pollution Sources in Textile Wet Processing,* available from the North Carolina Pollution Prevention Pays Program, Department of Environment, Health, and Natural Resources, Raleigh, N.C., 1986.

Smith, Brent, *A Workbook for Pollution Prevention by Source Reduction in Wet Textile Processing,* available from the North Carolina Pollution Prevention Pays Program, Department of Environment, Health, and Natural Resources, Raleigh, N.C., October 1988.

Acknowledgments

This work was supported by the U.S. Department of Energy under Contract DE-AC06-76RLO 1830. Dr. Brent Smith of the College of Textiles at North Carolina State University provided a technical review and contributed much useful information to this section.

51

Pollution Prevention in the Pulp and Paper Industry

Pamela G. Jenkins, P.E.

Science Applications International Corporation
Olympia, Washington

51.1 History and Process Description

Paper manufacturing is an old industry, dating back to the first century A.D. when the Chinese developed a method for forming a slurry of bamboo fibers into a mat, then pressing and drying it. Paper was made by hand until a continuous paper machine was invented at the end of the eighteenth century. Today, pulp and paper making is a highly sophisticated manufacturing process, carried out in approximately 345 pulp mills and 600 paper and paperboard mills in the United States, with hundreds of mills located in other countries. Over 240,000 people are employed in the pulp and paper industry in the United States, and another 81,000 in Canadian mills, producing a combined total of about 36 percent of the world's paper.[1]

Pulp and paper products play a significant role in nearly every aspect of our lives. We use paper for communicating, recording, and storing a variety of information; paper packaging products (such as cardboard and corrugated cartons) are used more widely than any other type of packaging material; and new applications for pulp and paper are being developed continuously. Nonpaper pulps include dissolving pulp, fluff pulp, and specialty pulps. Dissolving pulp is used in the manufacture of photographic film, cigarette filters, explosives, rayon, and other cellulose chemicals. Fluff pulp is used as the absorbent material in disposable diapers and feminine sanitary products. Filters, laminate materials, and other products are manufactured using specialty pulps.

The primary source of fiber for pulp and paper making is wood. The pulping process separates cellulose fiber from most of the lignin binder. Several methods of pulping are in use today, broadly classified as mechanical, thermal, and chemical methods, and combinations of these techniques.

Originally, the mechanical process involved pressing a block of wood lengthwise against a rotating rough grinding stone. Water washed the loosened fibers away from the stone and the wood. The resulting slurry of fibers and fiber fragments was screened to remove large material, then thickened to form a pulp stock. Current methods of mechanical pulping use a combination of disk refiners, thermal, and chemical processes. A *refiner* is a machine that employs rotating disks to separate fibers from the wood. Thermal and/or chemical softening of the wood chips may be used before the refiner, reducing the energy needed to defiber the wood, and producing a strong pulp that contains fewer contaminants. The resulting pulp, called *thermomechanical pulp*, is stronger than refiner mechanical pulp. The primary use of mechanical pulp is newsprint, which is produced using a combination of mechanical and kraft pulps. Recycled pulp is increasingly used for newspaper production.

The two primary chemical pulping methods are the kraft and sulfite processes. In chemical pulping, the wood chips are processed at elevated pressures and temperatures in aqueous chemical solutions. The lignin is dissolved out of the wood during cooking, leaving most of the cellulose and hemicellulose fibers intact. Chemical pulping typically produces lower yields than mechanical or thermomechanical pulping, usually about 40 to 55 percent of the original wood mass as compared to 85 to 95 percent by mechanical methods.[1] The sulfite process uses an acidic mixture of sulfurous acid (H_2SO_3) and bisulfite ion (HSO_3^-) to dissolve the lignin. The advantages of the sulfite process relative to kraft pulping are production of a brighter pulp than kraft, which facilitates bleaching. Disadvantages include production of a weaker paper, difficulty in using resinous softwoods and hardwoods that contain tannin, and less efficient chemical recovery.

The kraft process involves cooking wood chips in an alkaline solution of sodium hydroxide (NaOH) and sodium sulfite (Na_2S). Under heat and pressure, the alkaline solution breaks down the lignin molecules. A high degree of chemical recovery is achieved in the kraft process, recycling cooking chemicals, extracting useful by-products, and recovering energy from the digestion liquor. The kraft process produces a strong, dark-brown pulp. Unbleached pulps are used for making packaging materials such as paperboard and grocery sacks. Bleached pulp can be made into white pulp or paper products.

A combination of chemical and mechanical pulping methods comprises *semichemical pulping*. The wood chips are initially softened or partially digested with chemicals, then pulped using mechanical methods, usually in disc refiners. The result is a pulp yield within the range of 55 to 85 percent of the wood mass. Some of the processes classified as semichemical are the high-yield kraft, high-yield sulfite, and neutral sulfite processes. Typically, the yield is controlled by the degree of cooking. Less cooking results in a higher yield and more energy required to separate the fibers.

Probably no other industry has made as much progress as the kraft pulp and paper industry in reclaiming waste products. About half of the wood used in making pulp is cellulose; the reclamation of the other ingredients in the wood constitutes a continuing evolution of pollution prevention and economic success. The by-products of chemical pulping include turpentine used in the paint industry, lignosulfonates used as surfactants and dispersants, "tall oil" used in chemical manufacturing, yeast, vanillin, acetic acid, activated carbon, and alcohol. Sulfamic turpentine recovered in the kraft process is used to manufacture pine oil, dimethyl sulfoxide (DMSO), and other useful chemical products. In addition, the noncellulose portion of the wood is used to provide energy for the pulping process through the combustion of concentrated black liquor.

Over 75 percent of the pulp produced in the United States is manufactured using the kraft process. Because of the predominance of the kraft process, the remainder of this section will address pollution prevention methods for kraft pulp and paper mills. Some of these techniques may be applicable or adaptable to other pulping processes, especially sulfite mills. The major steps in the kraft process are described in the following paragraphs, followed by a discussion of major wastestreams, and proven pollution prevention methods for each of these steps.*

51.2   Major Steps in the Kraft Process

51.2.1   Wood Yard Operations

The kraft process begins with the procurement of wood (and, increasingly, the use of secondary fiber). The wood may be delivered as logs, chips, or sawmill waste. Storing logs in water produces soluble materials, silt, and bark debris to be released to the water. The result may be an increase of 1 to 15 pounds of BOD_5 (five-day biological oxygen demand) per ton of pulpwood in storage ponds, whose water must then be treated.[2] Land storage of logs produces a lower environmental impact.

Solids from debarking logs may comprise the largest quantity of solid waste at a mill, and some debarking methods consume a large quantity of water as well. Bark can be used for garden applications but is commonly burned in hogged-fuel boilers or in combination furnaces for energy recovery. Dry debarking methods are advantageous in eliminating the additional wastewater and in

*Much of the information on pollution prevention options has been adapted from the U.S. Environmental Protection Agency's publication *Model Pollution Prevention Plan for the Kraft Segment of the Pulp and Paper Industry,* EPA 910/9-92-030, U.S. EPA Region 10, Management Division, Program Planning and Evaluation, Seattle, Washington, authored by G. A. Amendola and P. G. Jenkins, September 1992.

retaining a low moisture content in the bark, increasing the energy value of the bark when combusted.

Once the logs are chipped, the wood chips may be stored in large piles at the pulp mill, are then screened to remove pieces that are too large, and conveyed into the digesters. Waste produced from furnished chips is primarily dirt, trash, and fine materials. The furnish is screened and the rejects are usually burned for energy recovery.

51.2.2 Pulping

The digestion process may be either a batch process or continuous, where the wood chips are mixed with white liquor and heated to about 350°F. Figure 51-1 illustrates the kraft process. White liquor contains the active chemicals [sodium hydroxide (NaOH) and sodium sulfide (Na_2S)] that dissolve the lignin from the wood fibers. In a batch process, after digestion, the cooking mixture is conveyed into a blow tank, where vapors are drawn off the material and condensed in a heat exchanger. Pulp-washing water is heated in the heat exchanger by the blow tank vapors and other hot gases. The results of the cooking process are a dark brown pulp and black liquor, which are carefully separated in a controlled process called *brownstock washing*. After washing, the pulp is screened, thickened, and washed.

51.2.3 Deknotting, Washing, and Screening

The cooked pulp is processed in a number of ways, depending on the intended product. Clean, bleachable pulps are typically deknotted before washing. Knots include uncooked chips, overthick chips, and irregularly sized wood pieces that may not have been thoroughly cooked. Deknotting equipment includes the older type of vibrating screen and newer totally enclosed pressure screen knotters. The materials separated from the pulp in the knotters are usually returned to the digesters for cooking or discarded as waste.

Brownstock washing is an important step in the liquor recovery cycle, where residual liquor is removed from the pulp. Standard washing equipment was for years a series of rotary vacuum washers. Today, a variety of washers in addition to rotary drums are being used, including diffusion washers, rotary pressure washers, dilution/extraction equipment, and horizontal belt washers.[1] The efficiency of washing depends upon the number of stages in the washer or washing time, and the quantity of washwater used.

Screening may be performed after washing or between washing stages. The purpose is to rid the pulp of undesirable particles. Screening equipment types include vibrators, gravity centrifugal, and pressure screens. In each, the pulp is passed through a perforated barrier that permits only acceptable fibers to flow through. Rejected material is then collected on the other side of the barrier, and can be repulped or dried and burned in a hogged-fuel boiler.

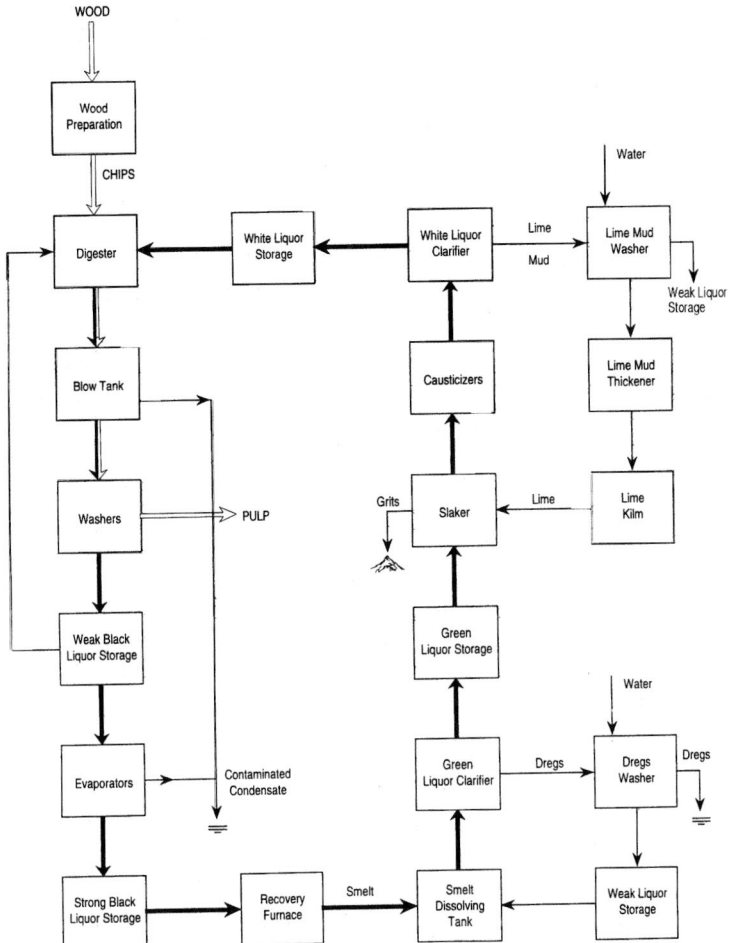

Figure 51-1. Schematic of the kraft process. (*Courtesy of G. A. Smook*, Handbook for Pulp & Paper Technologists, *2d ed., Angus Wilde Publications, 1992.*)

51.2.4 Chemical Recovery

The black liquor is composed of spent cooking liquor and dissolved lignin. Chemical recovery includes the reclamation of chemicals for reuse within the process, recovery of by-products, and utilization of remaining organic materials for heat generation. This recovery process is economically very important to the kraft process, and is an ideal example of pollution prevention, though it was conceived and implemented long before "pollution prevention" became a focus of regulatory interest. Brownstock washing separates the cooked pulp from a weak black-liquor solution containing approximately 15 percent solids. This weak black liquor is then concentrated, through a series of steps involving evaporation and chemical additions, to about 60 to 80 percent solids. The resulting heavy black liquor is combusted in the recovery furnace, forming an inorganic smelt. The smelt, composed of sodium carbonate (Na_2CO_3) and sodium sulfide, is dissolved in water to produce green liquor. The green liquor is then reacted with quicklime (CaO) to convert the sodium carbonate into sodium hydroxide and reconstitute the original white liquor. The lime mud produced in this stage then is burned to recover the quicklime, which completes the recovery cycle.

During the digestion process and various steps in chemical recovery, air and other noncondensible gases are released from the cooking vessels, boilers, and other process units. Turpentine is recovered from digestion vapors, condensed, and decanted. Depending on market prices and demand, turpentine may be sold to chemical manufacturers as an ingredient in such products as solvents, insecticides, camphor, and synthetic resins.[1] Sodium soaps, converted from fatty and resin acids, alcohols, and other compounds, and dissolved in the black liquor, are produced during the kraft digestion process. This material can be sold to chemical producers, or may be processed at the mill to recover raw tall oil and soda.

51.2.5 Pulp Bleaching

The *brightness* (whiteness) of pulp is an important characteristic for both paper and pulp products. Brightness is measured by the reflection of monochromatic light from the pulp, compared with a known standard. Unbleached kraft pulp typically ranges from 15 to 30 in brightness, whereas mechanical, bisulfite, and sulfite pulps range from 50 to 65.[1] In the production of high-quality, stable paper pulps and dissolving pulps, bleaching methods are used to remove the residual lignin from the pulp. Bleaching occurs in a sequence of steps, typically involving chlorination, alkaline extraction, chlorine dioxide, oxygen, hypochlorite, peroxide, and ozone.

Over the past fifteen years, much research has gone into the development of bleaching processes that produce less toxic effluents, particularly chlorinated compounds. Polychlorinated dioxins and furans, highly toxic and persistent compounds, are produced in the chlorine-bleaching process, particularly in the chlorination/extraction sequence. Several of the pollution prevention options presented in Sec. 51.4 achieve significant reductions in the formation of these contaminants.

51.2.6 Pulp Drying and Paper Making

If the pulp is to be used off site, it is usually first dewatered in order to reduce transportation costs. Commonly, pulp deliveries are of pulp in baled sheets, where the pulp is usually 90 percent air-dry. Three methods used for drying pulp are (1) air float drying, (2) flash drying, and (3) techniques developed from the paper machine, utilizing sheet forming, pressing, drying, and usually slitting and cutting. In these processes, the moisture in the pulp is removed either through circulation of preheated air around the pulp web, vaporization in a hot gas stream, or through vacuum, application of pressure, and heating.

In a paper machine, depending on the grade and type of paper being produced, surface sizing, coatings, and other treatments may be added to the product.

White water is the water that drains from wet pulp as it is sequentially dried in the paper machine or pulp dryer. White water contains fiber and may contain a number of other constituents as well, depending on the additives used in the paper-forming process. Discharge of this water directly to the wastewater treatment system would significantly increase the BOD load, and there would be little or no recovery of heat from this wastestream.

51.2.7 Wastewater Treatment

Pulp and paper mills generate a large volume of wastewater containing dissolved lignin, suspended solids, adsorbable organic halides (AOX, which encompasses many chlorinated organic compounds including chloroform), and other organic contaminants. A wealth of information in the existing literature addresses wastewater treatment methods and pollution prevention and process optimization techniques for wastewater treatment; therefore, such alternatives will not be discussed here.

51.3 Major Wastestreams

Various gaseous, particulate, liquid, and solid wastestreams are produced within a kraft mill. Gaseous air emissions include hydrogen sulfide, methyl mercaptan, dimethyl sulfide, dimethyl disulfide (these compounds are usually summed and described as *total reduced sulfur,* or TRS); nitrous oxides, sulfur dioxide, acetone, methanol, acetaldehyde, catechol, ammonia, chlorine, chlorine dioxide, formaldehyde, hydrochloric acid, sulfuric acid, volatile organics, and chloroform. Particulate air emissions include sodium sulfate and sodium carbonate from the recovery boiler, sodium compounds from the smelt tank, calcium and sodium compounds from the lime kiln, and fly ash from wood- or coal-fired boilers. The primary sources of air pollution from a kraft mill are digester gases, washer vents, evaporator gases, condensate water, blow tower vents, tall oil vents, recovery boiler, smelt tank, lime kiln, slaker vent, paper machine, and power boilers.

The major sources of water pollution are log-handling and debarking operations; chip washing; digester and evaporator condensates; white waters from screening, cleaning, and thickening the pulp; filtrates from bleach plant washers; paper machine white water; and liquor spills and fiber losses from all areas. Water contaminants produced by the kraft process include suspended solids, numerous chlorinated compounds including dioxins and furans, AOX, color, reduced sulfur compounds, methanol, and other volatile organic compounds (VOCs).

Solid wastes include boiler fly ash, bottom grate ash, wastewater sludge, lime grits, and slaker grits. Wastewater treatment sludge is often dewatered and burned for energy recovery.

51.4 Pollution Prevention Options

51.4.1 Wood Yard Operations

The following pollution prevention options for wood yard operations are briefly described below:

- Chip quality controls and chip thickness screens
- Dry debarking
- Log flume recycling
- Stormwater control

Chip Quality Controls and Chip Thickness Screens. Chip quality can have a significant impact on wood waste generation and digestion efficiency. Chippers should be maintained and operated to produce chips of uniform dimensions, emphasizing control of chip thickness. Chip thickness screens can be used to ensure that thick chips are not introduced into the digesters. This promotes uniform kraft liquor penetration, higher pulping yield, and reduced rejects and shive content. Uniformly sized chips will be more evenly digested, resulting in lower bleach consumption and subsequent reduction in formation of chlorinated compounds that enter the air and wastewater.

Chip screening for size and thickness can enhance the pulping process by providing more uniform chips that will be completely digested. Chip size screens alone will not do as thorough a job as size screens and thickness screens.

Sawdust and fine rejects generated from screening operations can be used in a hogged-fuel boiler for power generation. Oversize material can be rechipped, screened, and pulped, or also used in the hogged-fuel boiler.

Dry Debarking. Dry debarking methods are preferred over wet methods because wet debarking consumes 2000 to 3000 gallons of water per ton of wood barked, which requires treatment before discharge. Contaminant loadings range from 1 to 10 pounds of BOD_5, and 6 to 55 pounds of total suspended solids per ton of product. In addition, bark from dry debarking operations does not have to be dewatered before use as hogged fuel. Dry debarking can be used at kraft and sulfite mills where roundwood comprises the fiber source.

Log Flume Recycling. The water used to convey logs from log piles to debarkers and chippers can be recycled with minimal blowdown to wastewater treatment facilities. Collected bark and fiber can be burned in hogged-fuel boilers for energy recovery. Treated wastewater may be used as makeup water for the log flumes. This alternative is applicable to any mill that uses log flumes. The benefits of this practice are reduction in water consumption and reduced wastewater discharge loadings of BOD and total suspended solids (TSS).

Stormwater Control. Curbing and diking around chip piles and chip-processing areas and collection of wastewater can prevent chips and dust from entering runoff. Collected stormwater is then processed in the wastewater treatment system. The benefits of stormwater control include the avoidance of contaminated surface water and elimination of the need for a separate stormwater permit.

51.4.2 Pulping and Chemical Recovery

Many pollution prevention options are available for improved performance in the pulping and chemical recovery processes. Brief descriptions of the following methods are provided below:

- Extended delignification
- Closed screen room
- Liquor spill prevention and control
- Improved brownstock washing and new brownstock washing systems
- Use of defoamers and pitch dispersants
- Steam stripping of foul condensates
- TRS controls for high-volume–low-concentration and low-volume–low-concentration vents
- Conversion of recovery boilers to low-odor designs
- Recovery boiler operations for low-odor emissions
- Use of weak wash for scrubbing fluid in air pollution control systems
- Lime mud diversion basin
- Oxygen delignification
- Use of anthraquinone in digesters

Extended Delignification. The purpose of this pulping process is to produce brownstock of lower kappa number, i.e., to remove more of the lignin from the pulp, thereby reducing the demand on the bleaching system and associated generation of contaminants. Extended delignification is accomplished by cooking the fiber in modified time-temperature-alkaline cycles, typically with the same total alkali charge. Continuous digesters can be modified to allow for the addition of cooking liquor at several points, a process called *modified continuous cooking* (MCC) or *extended modified continuous cooking* (EMCC). In batch digesters, extended delignification can be conducted by completing several liquor exchanges during the pulping cycle. This process is called *rapid-displacement heating* (RDH).

The benefits of extended delignification are reduced bleach plant chemical consumption, reduction in chlorinated compounds in effluent, improved energy efficiency through enhanced recovery of liquor solids, reduction of rejects, and increased pulp yield.

Closed Screen Room.   New pulp-screening and deknotting systems can be installed or new units installed to increase the capture and recycling of black liquor into the recovery system. Closed screening systems improve the efficiency of the recovery process and reduce the demand for makeup chemicals in white liquor. Elimination of spills and overflows reduces the BOD loading and concentration of other organic contaminants in raw wastewater. Closed screening systems can be used in sulfite as well as kraft mills.

Liquor Spill Prevention and Control.   A number of maintenance practices and small equipment modifications can be implemented to minimize black-liquor and red-liquor spills and leaks. Pumps, tanks, pipelines, and other equipment can be inspected on a regular, frequent basis and quickly repaired should any leaks or other problems be noted. Engineered systems to collect and recover leaks, spills, and diversions of weak and strong liquor from digester and chemical recovery areas can be installed. These systems include tank level alarms, curbs and dikes, and storage tanks. Sewer conductivity monitoring with frequent equipment inspections can help reduce liquor losses. A spill and emergency response plan that outlines specific procedures for recovering spill material can be developed and workers trained in these procedures.

The benefits of a comprehensive liquor spill prevention and control program accrue from improved efficiency of the chemical recovery process due to increased capture of liquor solids. This reduces the demand for makeup chemicals and decreases the contaminant loading in wastewater. Fugitive emissions containing odorous reduced sulfur compounds and VOCs will also be minimized.

Improved Brownstock Washing and New Brownstock Washing Systems.
Improvements in brownstock washing can lead to a significant improvement in chemical recovery efficiency. This can be accomplished through the addition of one or more washing stages to the existing washing lines or through the installation of newer, more sophisticated washing equipment to replace an existing system. Newer systems may include pulp presses, belt washers, compaction baffle washers, or diffusion washers.

Washing losses in the range of 11 to 22 pounds of sodium sulfate per ton and less are attainable with new washing systems. Increased capture of black liquor reduces the demand for makeup chemicals, and decreases the contaminant load in the effluent. Well-controlled washer vents can reduce emissions of odorous reduced sulfur compounds and VOCs.

Use of Defoamers and Pitch Dispersants.   Defoamers and pitch dispersants that do not contribute to the formation of dioxins and furans during the bleaching process can be substituted for agents that contain dioxin and furan precursors (i.e., chlorine compounds). These substitutes can be water-based or oil-based additives. The advantages of this material substitution are reduced formation of chlorinated compounds in the bleaching process, or reduced toxic contaminant loading in the wastewater sludges and effluent in unbleached-kraft mills.

Steam Stripping of Foul Condensates.   Digester, evaporator, and turpentine condensates can be steam-stripped to remove reduced sulfur compounds (TRS) and BOD (methanol and acetone). A 15 to 20 percent proportion of steam to feed is necessary to achieve efficient BOD removal. Stripper overheads can be burned for energy recovery in the power boilers, and stripper bottoms can be reused for brownstock washing.

Steam stripping can significantly reduce emissions of VOCs to the air, and can also reduce TRS emissions to a lesser extent. In bleached-kraft mills, steam stripping may reduce raw wastewater BOD loadings by up to one-third, and more in unbleached-kraft mills. This method also reduces raw water consumption. The formation of chlorinated compounds contributing to wastewater contamination is also decreased.

TRS Controls for High-Volume–Low-Concentration and Low-Volume–Low-Concentration Vents.   Ten to 20 percent of total mill TRS emissions are contributed by vent streams from brownstock washers, foam tanks, black-liquor filters, oxidation tanks, and storage tanks that are not typically collected in the noncondensible gas (NCG) system. The emissions from these sources can be collected and combusted for energy recovery, reducing the atmospheric emissions of TRS and VOCs.

Conversion of Recovery Boilers to Low-Odor Design.   Old-design recovery boilers may contribute 60 to 80 percent of total mill TRS emissions. These boilers can be converted to low-odor design by elimination of the direct contact evaporator, modification of secondary- and tertiary-combustion air systems, and installation of a new economizer section. Moderate expansion of boiler capacity can be accomplished at the same time. Although expensive, this alternative has significant benefits: improved boiler efficiency and reduction of recovery boiler emissions by more than 90 percent.

Recovery Boiler Operations for Low-Odor Emissions. Reduced TRS emissions can be achieved through the operation of recovery boilers within specific parameters. These include operating within critical loading levels for liquor solids; maintaining secondary- and tertiary-combustion air at greater than 35 percent of total air supply; maintaining excess oxygen levels between 2.0 and 2.5 percent; maintaining liquor sulfidity as low as possible while maintaining pulp quality; maximizing liquor solids within the capability of the evaporator system; minimizing inert contents of the liquor; and maximizing dispersion of the liquor spray by using and maintaining appropriate liquor spray nozzles. Little or no capital investment may be required to implement these practices, which will improve boiler efficiency and reduce atmospheric emissions.

Use of Weak Wash for Scrubbing Fluid in Air Pollution Control Systems. Weak washwater from the causticizing system can be used as scrubber solution for smelt-dissolving-tank emissions, bleach plant and chlorine dioxide plant vents, lime slakers, and lime kilns. The scrubber water is then returned to the causticizing circuit. This operation will reduce fresh water consumption, decrease air emissions and wastewater discharge, and improve chemical recovery efficiency.

Lime Mud Diversion Basin. Lime mud that is usually discharged to the wastewater treatment system from process upsets and lime mud clarifier maintenance activities can be captured in a diversion basin. A relatively inexpensive option, installation of a lime mud diversion basin, would improve resource recovery, reduce wastewater loading and treatment costs, and reduce sludge accumulation.

Oxygen Delignification. Oxygen delignification is a process in which the pulp is treated with oxygen in an alkaline environment prior to bleaching and after a conventional or extended pulping process. In medium-consistency systems (10 to 20 percent), an additional 40 to 45 percent delignification is achievable. Delignification up to 50 percent is attainable in high-consistency operations (25 to 28 percent). This procedure requires a moderate amount of available recovery boiler capacity. If ozone bleaching is used, oxygen delignification is a prerequisite.

The advantages of oxygen delignification are primarily in bleach plant chemical savings and reduced wastewater contamination. Significant reductions in effluent contaminant loadings are achievable, including BOD, chlorinated compounds, AOX, and color. Chlorine consumption may be reduced by 50 percent.

Use of Anthraquinone in Digesters. Anthraquinone can be injected into the digesters to accelerate the pulping process and increase pulp yield. This practice can be used selectively to offset increased recovery boiler loading caused by oxygen delignification. This technique is particularly applicable in bleached-kraft mills where increased pulp yield is desired but which are limited by recovery boiler capacity. Though operating costs are significant with anthraquinone use, the advantages of this practice include improved yield, slight reduction in bleach plant chemical consumption and associated formation of chlorination contaminants, and reduction in effluent color and BOD.

51.4.3 Pulp Bleaching

The bleaching process has been the focus of much research in recent years to develop methods to reduce the production of dioxins, furans, and other chlorinated compounds that are environmentally undesirable. Eight pollution prevention alternatives for the bleaching process are briefly described below, including

- Chemical controls
- Chemical mixing
- Split addition of chlorine
- Chlorine dioxide substitution
- Enhanced extraction
- Replacement of hypochlorites
- Ozone bleaching
- Countercurrent jump-stage washing

Chemical Controls. Special instrumentation can be used to monitor and control the addition of chlorine and/or chlorine dioxide in the first bleaching stage in order to maintain the chlorine multiple or kappa factor within specific ranges. This type of control can minimize the formation of dioxins, furans, and other chlorinated compounds, thereby reducing air and wastewater contaminants.

Chemical Mixing. High-shear mixing equipment can be installed in the bleach plant to ensure efficient and even chemical application. Such mixers minimize the tendency for localized overchlorination, which leads to the formation of undesirable chlorinated compounds. This option is relatively inexpensive, and can assist in maximizing bleaching efficiency and minimizing the generation of chlorine compounds that may be released to the atmosphere and wastewater.

Split Addition of Chlorine. Multiple additions of chlorine can be employed to ensure efficient chemical application and to minimize localized overchlorination. This method requires the installation of multiple addition points in the bleach plant; it is roughly 10 times as expensive as chemical mixing, and from one-fourth to one-half as expensive as high-rate chlorine dioxide substitution. Though not as effective as the latter technique, split addition of chlorine aids in the minimization of dioxins, furans, and other undesirable chlorinated compounds.

Chlorine Dioxide Substitution.   The use of chlorine dioxide as a substitute for some or all of the elemental chlorine in the first bleaching stage is a highly effective method for reducing chlorinated-compound formation. Chlorine dioxide is produced at the mill and applied in solution upstream of the chlorine charge. The substitution rate may range from 5 to 100 percent of the elemental chlorine. A kappa factor of less than 0.15 is maintained through careful control of the application of chlorine dioxide and chlorine. High substitution rates (>70 percent) are most effective for decreasing production of chlorinated compounds. Chlorine use can be drastically reduced, and the formation of associated chlorinated by-products is significantly minimized.

Enhanced Extraction.   Oxygen alone or in combination with hydrogen peroxide or hypochlorite in bleach plant caustic extraction stages is used in many bleached-kraft mills to improve extraction efficiency and delignification. As the pulp enters the extraction stage, a small amount of molecular oxygen is mixed with it. Residual lignin in the pulp reacts with the oxygen, subsequently reducing the demand for chlorine dioxide in the following bleaching stage. Small amounts of hydrogen peroxide or hypochlorite can be added with the oxygen, producing better delignification with finer control of the process. The advantages of this method are its relatively low cost, minimization of chlorinated compounds, and reduced BOD, COD, and color in the effluent.

Replacement of Hypochlorites.   Chlorine dioxide can be used in place of sodium or calcium hypochlorites in the bleaching process. This method is most effective when high-rate chlorine dioxide substitution is used in the first bleaching stage and when hypochlorite is replaced with high-purity chlorine dioxide (i.e., less than 0.2 percent chlorine). The benefits include a significant reduction in the formation of chloroform.

Ozone Bleaching.   Research into methods for elimination or reduction of chlorine-bearing chemicals for bleaching has resulted in the development of several alternatives. One technique recently applied (1992) at commercial scale is ozone bleaching. In this process, ozone is used as the primary bleaching agent instead of chlorine, following oxygen delignification. In addition to eliminating the formation of undesirable chlorinated compounds, ozone bleaching also facilitates the recovery of ozone and extraction stage filtrates in the chemical recovery system. This method significantly reduces wastewater discharge loadings of AOX, BOD, and color from the bleach plant.

Bleach Plant Washing.   Countercurrent jump-stage washing entails the reuse of acid-stage filtrates (from hypochlorite or chlorine dioxide stages) as dilution and washwater on the first bleaching stage (using chlorine, chlorine/chlorine dioxide, or chlorine dioxide/chlorine). Filtrate from the second extraction stage can be used in the first extraction stage on log sequence bleach lines. The advantages of this method include (1) reduction in bleach plant wastewater flow by 2000 to 6000 gallons per ton of product in conventional bleach plants, (2) steam and energy savings, (3) reduced water consumption, and (4) slight reduction in BOD_5 loadings from the bleach plant.

51.4.4 Pulp Drying and Paper Making

Several pollution prevention options for the pulp-drying and paper-making processes are described below, including

- Fiber and white-water recovery with "savealls"
- Reuse of vacuum pump seal water
- Recovery of steam condensates
- Chemical substitutions

Fiber and White Water Recovery with Savealls. Fiber can be recovered from white water or brown water from pulp dryers and paper machines using settling-tank, drum, flotation, or polydisc savealls. Fiber from the savealls is returned to the stock system and subsequently reused. Depending upon the solids loading of the cleaned white water or brown water, it can be reused for stock dilution; wire, headbox, grooved-roll, trim knockdown, wire knockoff, and breast roll showers; consistency regulation; and beaters. The primary benefit of using savealls is an increase in yield. Fiber recovery may exceed 90 percent, depending on the type of paper and type of saveall. In addition, savealls reduce water and energy consumption, wastewater effluent loadings, wastewater treatment costs, and sludge disposal costs. Fresh water reductions from 200 to 10,000 gallons per ton of product have been reported, depending on the type of paper machine and the level of white water reuse.

Reuse of Vacuum Pump Seal Water. Another method that reduces water and energy demand is the use of cascade systems or recirculating water systems for the operation of water ring vacuum pumps. New installations include this feature, and retrofits can also be accomplished. Resulting water savings are estimated at 2000 gallons per ton of product. There is also a reduction in wastewater flow and therefore in treatment costs.

Recovery of Steam Condensates. The steam condensates from pulp dryers and paper machines can be collected and recovered. This method reduces boiler feedwater treatment costs and improves energy and water conservation.

Chemical Substitutions. A number of nonfiber additives may be utilized in paper making, including sizing agents, resins, fillers, dyes or pigments, flocculants, brighteners, and other chemicals for specific paper characteristics. Some additives have been found to produce undesirable or toxic emissions, for example, formaldehyde from urea formaldehyde resins. Adequate substitutes for toxic additives can be used. The benefits are reduced worker exposure to toxic substances, potential product quality improvements, reduced hazardous waste disposal, and reduced levels of toxics in emissions and effluent.

51.5   Pollution Prevention Successes

The entire history of the pulp and paper industry is a success story in pollution prevention. Development of the kraft process enabled the recovery of cooking chemicals and by-products, and the generation of process heat from "waste" cooking liquor, with significant economic advantages. Now, both regulatory pressure and continuing economic progress require further advancement in the capture and recovery of waste materials in the pulp and paper-making processes. Continued research on innovative methods such as biofiltration and black-liquor gasification may result in favorable plant-scale applications and further pollution prevention progress.

Water consumption in newer mills has been cut to levels that were unimaginable twenty years ago. The closed-cycle concept, where maximum utilization of countercurrent washing and water recycling is employed, has been implemented in one Canadian mill.[1] This process provides significant advantages in lower effluent treatment costs, significantly reduced water consumption, and sizable savings in chemical and energy costs.

Success has been attained in other areas of pump mill operations as well. For example, the James River paper mill in Camas, Washington, has achieved significant reductions in plantwide waste disposal through an aggressive waste reduction and recycling effort begun in 1987. A team of Yard Department employees identified solid wastes that could be compacted and recycled, including corrugating from the mill's tissue- and towel-converting operation, spent paper cores, plastic wrappers, parent roll body wrap, and other used materials. Over a three-year period, the mill reduced the number of cans of landfilled waste from 80 to 36 cans per month. Much of the recovered material was sold, including 855 tons of recyclable materials in 1990, which brought the company $65,700. The company's program is basically a grass roots effort, initiated by Yard Department employees who are given the authority to implement waste reduction ideas. James River recycles fiber, office wastepaper, pallets, vehicle batteries, solvents, used oil, scrap ferrous and nonferrous metals, cardboard, paper cores, waxed paper, plastic, and wood scraps.[3] The company works with its suppliers to reduce packaging wastes and maximize utilization of products delivered in containers.

The Simpson Tacoma Kraft Company in Tacoma, Washington, replaced the chlorine bleach plant with a high-substitution chlorine dioxide process in 1989.

Dioxin and furan levels dropped to below detection limits at 75 percent chlorine dioxide substitution. At 85 percent substitution, AOX final effluent levels dropped from 11.0 to 3.3 pounds per ton of pulp bleached, and average chloroform generation dropped to less than 0.1 pounds per oven-dry ton of pulp, a 93 percent reduction. Chlorine emissions were reduced 97 percent.

Continued research and development of new technologies for the manufacture of pulp and paper will, no doubt, further enhance the recovery of materials from raw wood and continue to reduce the environmental impacts from these processes.

References

1. Gary A. Smook, *Handbook for Pulp & Paper Technologists*, 2d ed., Angus Wilde Publications, Vancouver, B.C., 1992.

2. Howard Edde, *Environmental Control for Pulp and Paper Mills*, Noyes Publications, Park Ridge, N.J., 1984.

3. Willson Edgerley, "Recycling/Waste Minimization—One Mill's Success Story," in *Technical Association of the Pulp and Paper Industry Proceedings—1991 Environmental Conference*, TAPPI Press, Atlanta, Ga., 1991.

Further Reading

International Environmental Committee for the Forest Industries, *1992 Compilation of Research Projects on Environmental Problems in the Pulp and Paper Industry in Canada, Finland, Norway, Sweden, and the United States of America*, National Council of the Paper Industry for Air and Stream Improvement, Inc., New York, May 1992.

National Council of the Paper Industry for Air and Stream Improvement, Inc., *1991 Review of the Literature on Pulp and Paper Industry Effluent Management*, Technical Bulletin no. 632, NCASI, New York, May 1992.

———, *Delignification and Bleaching of Chemical Pulps with Ozone: A Literature and Patent Review*, Technical Bulletin no. 619, NCASI, New York, December 1991.

Springer, Allan M., *Industrial Environmental Control—Pulp and Paper Industry*, Wiley, New York, 1986.

U.S. Environmental Protection Agency, *Pollution Prevention for the Kraft Pulp and Paper Industry Bibliography*, EPA 910/9-92-031, EPA Region 10, Management Division, Program Planning and Evaluation, Seattle, Wash., September 1992.

———, *Model Pollution Prevention Plan for the Kraft Segment of the Pulp and Paper Industry*, EPA 910/9-92-030, EPA Region 10, Management Division, Program Planning and Evaluation, Seattle, Wash., September 1992.

———, *Pollution Prevention Opportunity Assessment and Implementation Plan for Simpson Tacoma Kraft Company, Tacoma, Washington*, EPA 910/9-92-027, EPA Region 10, Management Division, Program Planning and Evaluation, Seattle, Wash., August 1992.

52

Pollution Prevention in the Pharmaceutical Industry

E. S. Venkataramani

Merck & Co., Inc.
Rahway, New Jersey

52.1 Introduction

Over the last several years, it has become increasingly apparent to leaders in industry, government, and academia that an environmental protection strategy based on waste minimization and other forms of pollution prevention would be more effective than the historical focus on pollution control, remediation of past problems, and protecting specific environmental media through end-of-pipe treatments and pollution controls.[1–9] While the industry has made great strides in environmental protection over the past twenty years, the costs of pollution control continue to skyrocket. The U.S. Environmental Protection Agency (EPA) estimates that the costs for pollution control in this country will be $155 billion a year by the year 2000.[6] The cost associated with the control and remediation of pollution should be evidence enough to change the way we approach environmental protection. In general, waste minimization results in lower material(s) usage and inventory costs, leading to more efficient and economic processes, and reduces the amount of process wastes that require treatment and/or disposal, which, in turn, lowers the environmental burden and potential liability costs associated with a process. Thus, pollution prevention makes sense economically, environmentally, strategically, and politically.

52.2   Overview of Pharmaceutical Industry and Waste Generation

The primary goal of the pharmaceutical industry is to produce substances that have therapeutic value for humans and animals. There are four primary categories of pharmaceutical products manufactured by the industry:[10] (1) medicinal chemicals and botanical products [Standard Industrial Classification (SIC) code 2833], (2) pharmaceutical preparations (SIC code 2834), (3) in vitro and in vivo diagnostic substances (SIC code 2835), and (4) biological products (SIC code 2836). The activities of the pharmaceutical industry can be classified into three main categories, namely, (1) research and development, (2) primary manufacturing to produce the bulk drugs and vaccines, and (3) secondary manufacturing to produce the dosage-form pharmaceuticals. The pharmaceutical industry, for the most part, relies on complex batch processes and technologies for the manufacture of drug substances and pharmaceutical products. A whole array of other manufacturers that produce solvents and basic raw materials through to excipients, packaging materials, vials, needles, and so forth, support the pharmaceutical industry. In this chapter, only process- and production-related waste generation, disposal, and minimization are discussed in detail.

52.2.1   Research and Development

New drug research and development are complex, costly, and time-consuming endeavors. It has been estimated that, on average, it costs $359 million and takes ten years (industrywide averages) to bring a new drug from discovery through Food and Drug Administration (FDA) approval to market.[11,12] This includes, on average, 18 to 24 months of preclinical animal testing, four to six years of phased clinical testing to establish drug efficacy and safety, and two years for the FDA to review and approve the new drug application. Generally, it is sometime well into the clinical testing program that a new drug process enters the process development phase, in which the process is scaled up from laboratory quantities to a commercial scale. Clinical quantities of the drug substances and dosage forms, ranging from a few kilograms to several hundred kilograms, are prepared in pilot plants. During this phase, significant process development and optimization are usually required to develop a safe, reliable, cost-effective, and environmentally sound manufacturing process.

Research and development in the pharmaceutical industry is a diverse activity that relies on the cooperative efforts of specially trained personnel to screen, isolate, and develop new drug substances, pharmaceutical applications, and products. Distinct areas of research include chemical research, biological research, and pharmaceutical research. Due to the very great diversity of research and development efforts in the industry, the types of raw materials utilized and wastes generated vary greatly. Examples of common wastes resulting from pharmacological research and development include halogenated solvents, nonhalogenated solvents, organic chemicals, natural products, biomass, radionuclides, oxidizers, acids, bases, and a myriad of reagents. In general, a major portion of the raw materials that are used to conduct R&D would potentially end up as waste.[13] These quantities, while significant, are small in comparison to those from manufacturing operations.

52.2.2 Bulk Drug Manufacturing

The main purpose of bulk drug manufacturing is to produce active ingredients. The quantities produced depend on the particular drug substance. The most common methods of bulk drug manufacturing include chemical synthesis, natural product extraction, and fermentation. Each of these methods is briefly described in the following sections with emphasis on the types of raw materials and wastes generated. The main sources of waste from bulk drug manufacturing are liquids, gases, and solids, in that order. Chemical syntheses, extractions, and solvent interchanges all use large volumes of solvents. For example, a study conducted in 1988 indicated that the pharmaceutical industry in the United States used 14,400 metric tons of methylene chloride.[14] Solvent wastes constitute at least 75 percent of the waste load from the pharmaceutical industry.[15] The considerable amount of organic solvents used in synthetic work and the vast amounts of water in fermentation operations end up as waste. One of the primary wastes in terms of volume is "trade effluent," or dirty water resulting from a variety of sources that include boiler and cooling-tower blowdowns. Usually, the volume of nonprocess wastewater in a pharmaceutical manufacturing facility would be 5 to 10 times the volume of process wastewater. Due to reliance on complex batch processes and technologies, equipment cleaning accounts for a large volume of solvent and nonsolvent waste generated by the pharmaceutical industry. There are currently a number of aqueous-based cleaning solutions, nonhazardous and phosphate-free detergents, and sodium bicarbonate products that may be substituted for traditional chlorinated and nonchlorinated solvents. The selection of alternate cleaning materials is highly dependent on the process and type of equipment used. Mechanical assistance (e.g., high-intensity jets, spray nozzles) to enhance contact and cleaning efficiency and minimize solvent use and waste generation is an area of definite interest to the pharmaceutical industry.

Chemical Synthesis. Chemical synthesis is used to produce the majority of drugs on the market today. Chemical synthesis processes typically consist of one or more batch reactions followed by separation and purification steps. The reaction vessels and ancillary equipment can be dedicated to the production of one product line, typically products that are produced in large quantities. Many products, however, are manufactured in "campaigns," which may last a few weeks to several months. At the end of each campaign, all process equipment is thoroughly cleaned to prepare for further processing.

There are a wide variety of chemical reactions, product recovery schemes, and raw materials used to produce drug substances, each with strict product specifications. Chemical synthesis operations use a wide array of organic and inorganic reactants, solvents, and catalysts. Chemical synthesis and downstream processing are solvent-intensive operations. Solvents are used as reaction media as well as for product recovery and purification.

Due to the varied nature of operations employed and raw materials used in chemical synthesis, wastestreams generated are numerous, as well as complex. Nearly every step of organic synthesis generates a mother liquor, which contains unconverted reactants, by-products, and residual product in a solvent or aqueous base. Organic synthesis processes may also generate acids, bases, cyanide, metals, etc. Volatile solvent use results in solvent-rich organic waste as well as air emissions. Equipment cleaning, spills, filtrates, concentrates, wet scrubbers, and miscible solvents each result in the generation of solvent-laden and/or aqueous wastestreams. Some of these waste streams may not be amendable to treatment by conventional wastewater treatment plants.

Natural Product Extraction. Natural product extraction utilizes natural materials such as biomass (e.g., roots, leaves) and animal glands to produce pharmaceutical products. Examples of drugs produced by natural products extraction include taxol, morphine, digitoxin, ajamalicine, pilocarpine, and insulin. Close to 25 percent of the pharmaceutical industry's sales result from plant-derived drugs. Natural product extraction processes are typically characterized by the extraction of a very small amount of product relative to the amount of raw materials used (e.g., product mass may be as low as one-ten-thousandth of the original mass). Product recovery operations utilize a wide variety of solvents. Chlorinated solvents, ketones, and alcohols are frequently used for solvent extraction. Ammonia, acids, and bases are commonly used to control pH. These operations result in the generation of spent raw materials (e.g., spent biomass), organic waste rich in solvents, solvent vapors, wastewater laden with solvents, and wastewater containing natural products and salts.

Fermentation. Fermentation processes are used to produce products such as primary and secondary metabolites (e.g., antibiotics and other bulk drugs), recombinant proteins (e.g., erythropoetin, viral antigens), and vaccines (e.g., HepA, chicken pox). The fermentation process generally consists of batch fermentation, product recovery, and purification. Once the fermentation cycle is complete, the fermentation broth is usually subjected to whole-broth extraction to remove the product, or filtration to remove the solids and products. Solvent extraction, precipitation, ion exchange, and adsorption chromatography are frequently used to further purify the product. Solvent extraction processes utilize organic solvents such as alcohols, toluene, acetone, acetates, or methylene chloride to transfer the fermentation product into the solvent phase. The product is recovered by additional extraction processes, precipitation, or crystallization.

Fermentation operations and downstream processing typically result in the generation of large volumes of aqueous and solvent-laden wastes. Aqueous wastes are produced from the spent fermentation broth consisting primarily of unconsumed raw materials and cell debris. The filtration process produces large quantities of wastes in a solid form (e.g., filter cake). Wastewater is generated as a result of product recovery operations (i.e., spent filtrate), equipment cleaning, and fermenter vent gas scrubbing. The volatile solvents used for product recovery generate solvent waste and air emissions.

52.2.3 Formulation and Dosage-Form Manufacturing

Pharmaceutical formulation is the preparation of dosage forms such as tablets, capsules, liquids, parenterals, creams, and ointments. Tablets and capsules are the most widely used and preferred oral medication. Tablets are made by blending the active ingredients with fillers or excipients and compressing into a tablet form. The form of the tablet (i.e., plain compressed, coated, or molded) depends on the desired release characteristics. The pharmacological release characteristics of the tablet are often controlled by spraying the tablet with a coating material.

Capsules can be produced in hard or soft form. Hard capsules are made by forming two pieces, which are then filled and joined. Each piece is formed by dipping pins into a solution of gelatin; the pieces are then dried and trimmed to form the final product. Soft capsules are formed by placing two gelatin layers between a die plate and injecting the formulation into the capsule.

A liquid dosage is prepared by mixing the solute with a solvent, typically in a glass-lined vessel. The solution is then filtered and transferred to a storage tank for inspection before packaging. Parenteral dosages are produced as solutions, dry solids dissolved before injection, or as suspensions. Ointments are typically produced by melting a base, such as petrolatum, blending the base with the drug, and passing the cooled mixture through a roller mill.

Formulation and dosage-form manufacturing are probably the least wasteful operations in the pharmaceutical industry.[13] Apart from active ingredients, these activities use excipients and packaging materials, substantially converting them to final products. However, as in any process, waste, such as off-specification products or contaminated packaging and solvents and other chemicals, is generated. The tablet-coating operation may be solvent-intensive, depending on the finished tablet presentaiton, and could result in solvent waste, including air emissions. However, aqueous-based coating processes are gaining popularity. In the manufacture of vaccines and sterile products, considerable amounts of water are used for washing and "water for injection" (WFI), which results in the generation of a large volume of wastewater. The most prevalent waste stream generated during formulation operations is contaminated wastewater from cleaning and equipment sterilization. Packaging activities may also generate large quantities of waste, mainly nonhazardous in nature.

52.3 Waste Disposal Practices in the Pharmaceutical Industry

52.3.1 Primary (Bulk Drug) Manufacturing

The front-end as well as downstream processing generates relatively large volumes of waste. Solvent-laden waste constitutes a major fraction of this waste. This is despite the fact that solvents are commonly recycled in dedicated processes, with wash liquors being sent back to earlier parts of the process to become the mother liquor. This form of recycling is difficult when the solvent becomes contaminated with other solvents during a solvent exchange, or with water or other materials such as color compounds, by-products, and suspensions. At this point two options exist: in-house recovery or disposal. The disposal of waste solvent is determined by, among other factors, reusability, price, the hazard it presents, and physical/chemical/biological properties such as volatility and biodegradability. If the material contains highly active substances hazardous to humans and the environment, or if some uncertainty exists as to what it may contain, then incineration is usually the preferred disposal route. Again, recycling is a possibility. Fuel blending is an accepted and increasingly practiced option. Recovery by distillation is commonly practiced. Recycle back to pharmaceutical manufacture must meet quality standards. In such a situation, nonprocess third-party reuse is common. Distillation can clean up the vast majority of solvents to the standard required by alternate recycle options or for beneficial reuse as solvents where high purity is not required; for example, the manufacturers of paint strippers, windshield washer fluids, and the like. Aqueous wastes are often high strength in terms of biological and chemical oxygen demand (BOD >10,000 mg/L and COD >30,000 mg/L), and are almost always subjected to biological treatment. Typical secondary wastes from primary manufacturing are filter aids and carbon. Here the problem is somewhat more intractable, as the reuse of these materials is seldom practiced. Carbon can be regenerated, but the practicality of this is dependent on content, e.g., metals. Also, these materials are used in very small quantities that make the recycling option less practical. In many instances, there are no economic incentives to recycle. This being the case, disposal by incineration or landfilling is a common practice. However, carbon and filter aids containing precious metal catalysts are often recovered and reused.

52.3.2 Secondary (Formulation and Dosage-Form) Manufacturing

The primary wastes to be considered from secondary manufacturing are solvents from tablet-coating operations and water from the production of high-quality water for WFI. Large amounts of water are rejected as waste in the process of WFI production. Reuse and recycling of water is practiced to some extent.

Although the majority of the nonaqueous solvents used during pharmaceutical granulation and coating operations are likely to be released as air emissions, excess formulation solutions are subject to the land disposal restrictions of the Resource Conservation and Recovery Act (RCRA). Significant changes are possible for these operations. In the case of solvent-free formulations, atmospheric releases are minimal. However, excess granulating or coating solutions add a large amount of BOD and COD if discharged as wastewater. Wastage of the excess coating solution can be minimized by making only the required amount. The most effective waste minimization technique for existing solvent-based formulations is to reformulate without using solvents. If this is not possible, reformulating with a solvent that has a lower environmental impact or reducing the amount of solvent used should be considered. Any reformulations aimed at reducing waste, however, need to be carefully evaluated for approval from the FDA for the manufacturing process. When there are specific BOD and/or COD discharge limits for wastewater, reductions can be achieved through work practices (e.g., dry cleanup methods using vacuum) that precede washdown of process equipment.

Materials substitution methods have been successfully applied to tablet-coating and equipment-cleaning operations. One pharmaceutical manufacturing plant reported that the use of a water-based solvent and the development of new spray equipment for tablet coating reduced the need for air pollution control equipment, resulting in cost savings of $180,000.[10] Another facility reduced methylene chloride usage in their tablet-coating operation from 60 to 8 tons per year by switching from a conventional film-coating process to an aqueous film-coating process.[10]

Recycling of packaging, another major waste of secondary manufacturing, is difficult to tackle, but is being considered by the European Economic Community (EEC) regulators. The EEC position on packaging materials is that much more will have to be recycled.[13] There has been considerable movement, in recent years, toward the use of original package dispensing for a variety of sound reasons, not the least of which is improved self-administration by the customer/patient. This form of packaging is wasteful, as it generates large quantities of plastics and cardboard for disposal. Also, original packages occupy 7 times the volume compared to bulk packages, increasing waste in the areas of transport, warehousing, and handling.[13] This is not the case for vaccines, where single-use vials and sterile products are probably the best for the patient and practically the only option for freeze-dried drugs. The use of parenteral products is generally less popular, so the pressure is on the pharmaceutical industry to produce oral versions of the products. From a pollution prevention standpoint, an oral version of the finished product (e.g., tablets) is often favorable as the production of the parenteral is less efficient in terms of water and energy utilization.

52.4 Solvent Usage in the Pharmaceutical Industry

52.4.1 Toxic Release Inventory Data

A representative data set pertaining to waste generation practices in the pharmaceutical industry was obtained by conducting a search of the EPA's Toxic Release Inventory (TRI) database. The search was conducted for all reporting facilities located in the state of New Jersey and classified under SIC code 283 for the reporting year 1991. It is worth keeping in mind here that the state of New Jersey is home to many major pharmaceutical companies. This search yielded a total of 30 reporting facilities. The report provided data on the quantity of production and R&D waste generated by each facility for reporting year 1991, as well as data for 1990 and projected data for 1992. It should be stressed that this is only a sample data set representing the pharmaceutical industry as a whole. However, it does provide useful information on the types and magnitudes of waste generated by the industry. The TRI database does not provide any data on solvent usage. It is to be remembered that TRI tracks only the amount of individual chemical constituents subject to reporting, not the total quantity of the waste generated. Also, facilities reported this information without any guidance from the EPA regarding "reportable recycling," e.g., what is considered recycled.

TRI Data Evaluation. A review of the data compiled revealed that a total of 91 million pounds of production-related waste was generated in 1991, as compared to a total of 93 million pounds generated in 1990, and a projected total of 76 million pounds in 1992 (refer to Table 52-1). The total quantity of waste generated was composed of 60 different chemicals. However, the types of chemical waste generated by each facility generally varied widely, with a few exceptions. Of the total quantity of waste generated in 1991, 89.3 percent came from the following eight chemicals: acetone (26.6 percent), methylene chloride (12.1

Table 52-1. Waste Quantities and Methods of Disposal by the Pharmaceutical Industry in New Jersey in 1990, 1991, and 1992 (In Millions of Pounds)

Waste disposal method	1990 Amount	1990 Percent of total	1991 Amount	1991 Percent of total	1992 Amount	1992 Percent of total
Released	3.96	4.27	4.02	4.42	3.52	4.66
Recycled	45.0	48.52	39.46	43.42	34.75	45.99
Energy recovery	10.7	11.53	10.40	11.45	9.27	12.27
Treated	33.09	35.69	37.0	40.71	28.01	37.08
Total	92.75	100	90.9	100	75.55	100

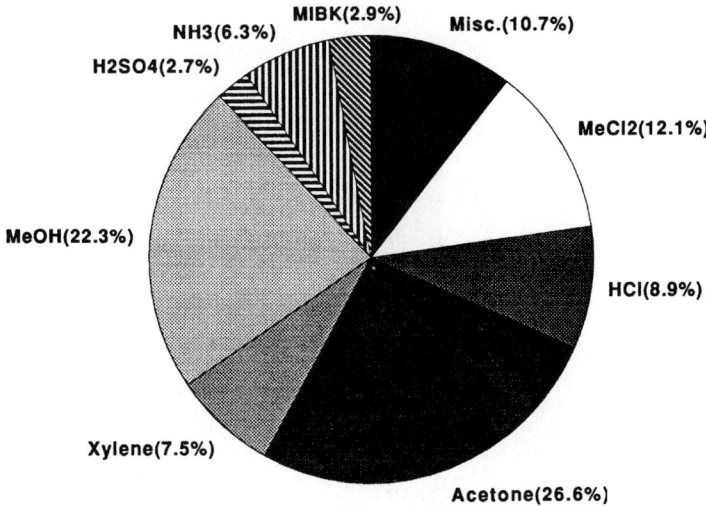

TOTAL = 91 MILLION LBS.

Figure 52-1. Chemicals (percent) contributing to 1991 waste generation by the pharmaceutical industry in New Jersey.

percent), hydrochloric acid (8.9 percent), sulfuric acid (2.7 percent), xylene (7.5 percent), methanol (22.3 percent), ammonia (6.3 percent), and methyl isobutyl ketone, or MIBK (2.9 percent). The contribution from the five major solvents accounted for 71.4 percent of the reportable waste by the pharmaceutical industry in New Jersey (1991). The remaining 52 chemicals each contributed less than 1 percent of the total quantity of waste generated in 1991. A summary of the quantities and percentages of these eight chemicals contributing to the amount of waste generated in 1991 is presented in Fig. 52-1.

The TRI data provide a breakdown of wastes generated according to the ultimate method of disposal. The database indicates whether the waste was released (i.e., released on site or disposed off site), recycled on site, recycled off site, burned for energy recovery on site, burned for energy recovery off site, treated on site, or treated off site. A comparison of the data reported for 1990, 1991, and 1992 indicates that the method of waste disposal is consistent from year to year (refer to Table 52-1). An evaluation of the methods of waste disposal revealed that recycling and treatment were the most common methods of waste disposal (e.g., in 1991, 43 percent of the waste was recycled and 41 percent was treated). The amount of waste released on site or disposed off site generally accounted for only 4 to 5 percent of the total waste generated (refer to Table 52-1 and Fig. 52-2). However, the amount of waste being used for energy recovery ranged from 11 to 12 percent. Recycling and energy recoveries being

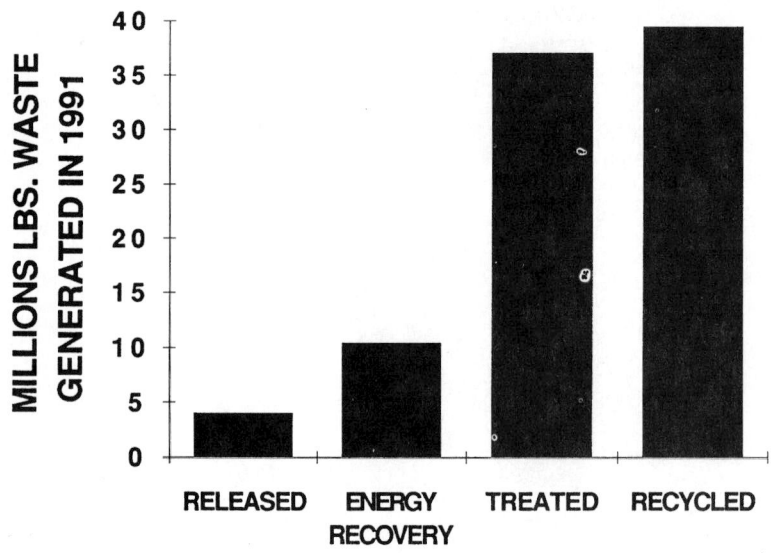

Figure 52-2. Ultimate disposal methods for wastes generated in 1991 by the pharmaceutical industry in New Jersey.

practiced account for approximately 50 to 60 percent of the total reportable waste generated by the pharmaceutical industry in New Jersey.

An analysis of the waste disposal methods for the eight chemicals contributing to the bulk of the waste generated in 1991 (i.e., acetone, methylene chloride, hydrochloric acid, sulfuric acid, xylene, methanol, ammonia, and MIBK) indicates (refer to Table 52-2) that recycling methods have been very successful for acetone (79.4 percent recycled), methylene chloride (81.8 percent recycled), xylene (92.7 percent recycled), and MIBK (76.7 percent recycled). For methanol, 21.9 percent was recycled and 21.6 percent used for energy recovery. Beneficial waste management activities, such as recycling and energy recovery, appear to have been most successful for MIBK, which had only 4.5 percent treated and 0.1 percent released or disposed. The method of disposal for hydrochloric acid, sulfuric acid, and ammonia was almost entirely through treatment (i.e, >97 percent treated).

52.5 Pollution Prevention in the Pharmaceutical Industry

Despite the potential benefits of pollution prevention to industry as a whole, the pharmaceutical industry is faced with a unique regulatory challenge for implementing an effective pollution prevention program. Pharmaceutical oper-

Table 52-2. Waste Disposal Methods for the Eight Chemicals Contributing to the Bulk Quantity of Waste Generated by the Pharmaceutical Industry in New Jersey in 1991

Chemical name	Percent released	Percent recycled	Percent energy recovery	Percent treated
Acetone	2.0	79.4	8.9	9.6
Methylene chloride	4.5	81.8	3.2	10.5
Hydrochloric acid	0.5	0	0	99.5
Sulfuric acid	0.1	0	0	99.9
Xylene	3.4	92.7	3.7	0.2
Methanol	9.5	21.9	21.6	47.0
Ammonia	2.1	0	0	97.9
MIBK	0.1	76.7	18.8	4.5

ations are characterized by a low ratio of finished product to raw materials used, and operations typically rely on batch processing. Batch processes are typified by short production runs, varied product mixes, and frequent equipment cleaning, resulting in a greater challenge to minimize wastes. Most of the waste generated during batch operations results from equipment cleaning, unwanted by-products, and spent solvents and raw materials resulting from frequent start-up and shutdown. New pharmaceutical products are subject to approval by the FDA. The FDA approval process is lengthy. Once approval is obtained, strict adherence to the approved process and product specifications is required, making pollution prevention very difficult for processes that have already received FDA approval. On the same count, these strict regulations and restrictions naturally favor the conception and development of "environmentally benign manufacturing technology." Lately, the FDA is also encouraging and crediting the pharmaceutical industry for pollution prevention.

Pollution prevention is not a new concept for the pharmaceutical industry,[10,16–36]. However, due to the confidential and highly competitive nature of the business, there is a dearth of published information in this area. Process improvements, yield enhancement, and raw material charge reductions are just a few of the many objectives that have been part of the pharmaceutical industry's manufacturing strategy for decades.[20–22] Though waste reduction was not often the driving force for initiating these projects, it is a natural consequence of such efforts. The primary difference in what's been practiced in the past several years is the point in the product and process development cycle at which waste reduction efforts are conceived and implemented.[23]

It is within the process development period that the best opportunities to incorporate waste minimization initiatives are found. Before a new drug process reaches the process development phase, it is generally too ill-defined to expend extensive energies on waste minimization. On the other hand, by the time process development nears completion, clinical efficacy has likely been established and process modifications (including those intended to effect waste minimization) become difficult (but not impossible) to incorporate, due to regulatory constraints. Therefore, it is important that waste minimization initiatives be evaluated early in the process development phase.

A detailed discussion of waste minimization and pollution prevention achieved in six process/facility engineering modification projects, in the different sectors of the pharmaceutical industry, can be found in a publication by the author and colleagues.[24] The projects and their associated challenges were described from technical, regulatory, and economic standpoints. The modification projects included major process chemistry changes, recovery and reuse of solvents and raw materials, substitution with environmentally benign chemicals, hazardous waste elimination, equipment modification, and "at-source" and "end-of-pipe" treatments. Solvent recovery in one antibiotic process alone accounted for an annual savings of greater than $10 million. Methylene chloride emission (160,000 lb in 1989) was eliminated. In another process, isopropyl acetate was substituted for toluene and a recovery scheme was implemented that resulted in 90 percent recovery of the solvent. Yet another example involved replacing a two-solvent extraction system with a single, environmentally more benign solvent. This switch resulted in an environmentally benign process. However, it was not without yield loss. Thus, pollution prevention need not always be attractive from a productivity and process economics point of view. The publication also describes solvent selection, recovery and reclamation of mercury, elimination of granulation and coating solvents, and selection of emission control technologies. Because waste minimization efforts are increasingly focused on product candidates early in the development cycle, the role of R&D in conceiving environmentally sound processes was also discussed in detail.

A discussion of the development and implementation of rational waste minimization strategies during pharmaceutical process development can also be found in other publications.[25-27] The general approach consists of several steps (refer to Fig. 52-3). First, complete process material balances are calculated in order to provide detailed descriptions of all prospective process wastestreams; computerized process simulation is an important tool in this effort. This can be difficult, considering that, especially early in the process development period, reaction stoichiometries and partitioning of various chemical species during mass-transfer operations may be poorly understood. Often, one must rely on "guesstimation" until various process and waste streams have been analyzed and characterized. Second, after obtaining estimated waste stream compositions, an environment assessment is conducted to screen these prospective waste streams for possible waste minimization opportunities. Typical waste minimization initiatives considered include materials substitution, materials use reduction, materials recovery and reuse, and a variety of production technology and operations improvements. Usually, the most important waste minimization technique considered is solvent selection and recovery and reuse, since for the pharmaceutical industry at large, organic solvents account for 75

1. Calculate initial process material balance to estimate prospective wastestreams. *Note:* Computerized batch process simulation is an important tool in this effort.

2. Perform environmental assessment of process; evaluate waste minimization options.

3. Where feasible, incorporate waste minimization options into process design.

4. Update process material balance, incorporating waste minimization initiatives.

5. Iterative strategy: Repeat steps 1 to 4 as necessary.

6. For waste minimization residuals and nonminimized wastes, evaluate treatment and disposal options.

Important Considerations

Requires collaboration among chemists, engineers, and environmental staff.

May involve significant process changes; usually requires lab and pilot validation.

Must maintain product quality; safety and economic concerns must be addressed.

Early intervention is critical, before process becomes "locked in."

Figure 52-3. Waste minimization and process development strategy.

percent or more of reported emissions under the Superfund Amendments and Reauthorization Act (SARA).[15] A primary focus area is to substitute nonhalogenated and environmentally benign solvents for halogenated solvents. The third step in the waste minimization strategy involves testing and evaluation of specific waste minimization initiatives which, if feasible, are incorporated into the process design. This usually involves extensive laboratory and pilot-plant evaluation over several months, and must be demonstrated not to compromise product quality. Finally, the process material balance is revisited and updated to incorporate all process modifications; the update thus reflects and documents waste minimization results. Highlights from two recent case studies illustrating the application of this strategy during process development included the following: for case 1, technology was developed to recover and reuse 85 percent of all process solvents; for case 2, methylene chloride was engineered out of the process and 55 percent of all other solvents recovered and reused within the process.[25]

An important tool in calculating these process material balances is availability of computerized process simulation tools. Expert-system-based environmental assessment systems to assess potential environmental impact and implications of manufacturing processes during the early stages of development are also useful.[28] An evaluation of the disposition of liquid-, gas-, or solid-phase waste such as solvent recovery opportunities and preferred ways of handling toxic, hazardous, biocidal, and malodorous chemicals can be made.[29] Also, simulation of the fate of waste stream components in a typical activated-sludge treatment plant using a treatabilty index to predict toxicity and oxygen demand of the effluent, stripping of volatile organic chemicals (VOCs) , sludge production, and sludge toxicity has also been demonstrated.[30] This systematic way of evaluating process concepts for their potential environmental impact and exploitation of pollution prevention opportunities has been carried out for a number of development projects.[30]

In formulation manufacturing, many pharmaceutical companies have or are currently reformulating several granulating and coating solutions to eliminate chlorinated solvents, acetone, and methanol, and have already reduced chloroform and other chlorinated solvents emissions considerably. Future total elimination of chloroform and other chlorinated solvents will depend on the completion of these formulation research projects and the appropriate approvals from the FDA. Needless to mention, implementation of this will incur considerable capital cost.

The pharmaceutical industry has not totally eliminated wastes generated in processing. Residual wastes still require disposal or treatment and continue to have environmental impacts—air emissions, wastewater, or still bottoms. But the commitment and concerted effort keep the industry moving in the right direction.

Merck is one of the many pharmaceutical companies that has subscribed to the EPA's 33/50 Program. In addition, in 1990, Merck announced a very ambitious corporate goal emphasizing minimization of environmental emissions (refer to Fig. 52-4): reduce all environmental releases of carcinogens and suspected carcinogens by 90 percent (by December 1991), eliminate carcinogens in air emissions or apply best available control technology (by December 1993), reduce SARA-listed toxic chemicals by 90 percent worldwide, using 1987 as a baseline (by December 1995). These ambitious goals have led Merck to closely review all aspects of its operations for pollution prevention opportunities. As such, the process development effort has become a focal point for evaluating waste minimization and pollution prevention initiatives for incorporation into new pharmaceutical processes.

Merck has met the first two corporate goals, and is on target toward achieving the third goal. Ultimately, the company will reduce by 90 percent releases and transfers for treatment and disposal of all toxic chemicals by the end of 1995. Merck has also exceeded EPA's guidelines for its voluntary 33/50 Program. All this could not be achieved without continued emphasis on waste minimization in production operations and pollution prevention in process development. At least 15 percent of the company's capital program for its chemical manufacturing operations for the next five years (1993–1998) is ear-

By December 1991, reduce by 90 percent air emissions of carcinogens and suspect carcinogens.

By December 1993, eliminate emissions of carcinogens and suspect carcinogens or apply best available control technology.

By December 1995, reduce by 90 percent all environmental releases of toxic chemicals.

- Toxic chemicals are defined as those on the SARA 313 list.
- Baselines for reductions are 1987 release levels reported to the EPA under SARA.

Figure 52-4. Waste minimization initiatives at Merck & Co., Inc. (worldwide).

marked for environmental projects. In addition to the process cost savings, Merck is reducing its potential environmental liability by not sending waste off site for disposal.

52.6 Conclusion

Pharmaceutical manufacturing is subject to very strict regulations and restrictions. The FDA approval process is lengthy. Once approval is obtained, strict adherence to the approved process and product specification is required, making pollution prevention very difficult for processes that have already received FDA approval. On the same count, these strict regulations and restrictions naturally favor the conception and development of "environmentally benign manufacturing technology." A clear understanding of the process, reaction pathways, process equipment, operational requirements, and waste stream characteristics are critical for the evaluation, selection, and implementation of pollution prevention. Although pollution prevention opportunities are always preferred over treatment and disposal techniques, consideration of a full range of options—including at-source treatments and disposal—is a practical necessity to ensure protection of the environment using best available technology. General housekeeping can also play a major role in waste minimization.

Waste minimization and pollution prevention are not new concepts for the pharmaceutical industry. But the confidential and highly competitive nature of the business stands in the way of disseminating information regarding specific activities in this area. The pharmaceutical industry could probably do much better in this respect.

Successful implementation of waste minimization in the pharmaceutical industry requires that a process modification not have a negative impact on product quality. Recovered and recycled materials must meet quality specifications that are similar to those for virgin raw materials. Any changes made early in a process have to be verified through to the end product by extensive laboratory tests of all subsequent processing steps to ensure that there has been no compromise in the quality of the final product. In practice, once a process has undergone significant development at the pilot-plant scale, it is generally difficult and costly to make major changes and modifications. Also, regulatory constraints that are unique to the pharmaceutical industry restrict modification of a process once clinical efficacy of the drug is established.

A rational waste minimization strategy for pharmaceutical process development focuses on constructing overall process material balances to estimate the quantity and composition of prospective process wastestreams, and using that information to spur development and implementation of waste minimization initiatives. The major waste minimization accomplishment has been the selection as well as development and incorporation of solvent recovery schemes that resulted in the on-site recovery and reuse of process solvents from the otherwise wastestreams. It is important to stress that a successful waste minimization strategy requires close collaboration among chemists, chemical and biochemical engineers, and environmental professionals, and is most effective when performed early in the development cycle, since process modifications are difficult to implement later on due to regulatory and economic constraints.

It should be highlighted that recent advancements in drug development, a result of refinements in molecular biology, chemistry, and other sciences, have led to the development of very specific and more potent drug substances. This is translating into newer products whose production volumes are a fraction of traditional drug volumes, resulting in minimization of overall waste generation.

The TRI data were used to evaluate the total quantity of waste generated, the types of chemical waste generated, and the ultimate method of waste disposal for the pharmaceutical industry in the state of New Jersey. Although this data set was limited by the search range (i.e., New Jersey facilities only), the chemicals required to be reported by the TRI, and the accuracy of data provided by individual facilities, it provides a valuable sample evaluation of waste generation trends in the pharmaceutical industry. The data indicate that there is high diversity among the types of chemical waste generated. However, acetone, methylene chloride, hydrochloric acid, sulfuric acid, xylene, methanol, ammonia, and MIBK were found to be used by the largest number of facilities and to account for the largest percentage of the total quantity of waste generated. A comparison of data from 1990 to 1992 indicates that the quantity of waste generated by the pharmaceutical industry is declining. This is clear evidence that pollution prevention is happening in the pharmaceutical industry. The TRI data do not provide any information on solvent usage; therefore, no information regarding source reduction methods being utilized by the pharmaceutical industry could be garnered from these data. However, the TRI data do provide information on the method of waste disposal. The TRI data indicate that recycling methods have been used widely for some chemicals, energy recovery has been used to a limited extent, treatment is still a prevalent method of waste management, and on-site release or off-site disposal is used to a very limited extent. Recycling and energy recoveries are practiced on approximately 50 to 60 percent of the total reported quantities.

To a large extent, pharmaceutical manufacturing is solvent-intensive, with solvent use accounting for more than 75 percent of the waste burden. Thus, careful choice of environmentally benign solvents and recovery, recycling, and reuse of these solvents offer the best pollution prevention strategy. The information from the TRI database for 30 pharmaceutical companies in New Jersey substantiates this. The contribution from the five major solvents alone accounted for 71.4 percent of the reportable waste generation in 1991.

Since the late 1980s, the requirements of the Control of Substances Hazardous to Health (COSHH) Regulations[37] have percolated through to educational establishments, where it can now be found as an integral effort in laboratory experiments and science and engineering curricula. Several industry-university consortia actively pursue research in the pollution prevention area. For example, the Emission Reduction Research Center (ERRC), with its headquarters at New Jersey Institute of Technology, is a consortium composed of pharmaceutical companies and universities in the northeastern United States.[38] Some of the projects researched include solvent selection for process and non-process use, aqueous-based and mechanically assisted cleaning techniques, batch process simulation and design, environmental simulation, and waste minimization. Students are now taught to assess as well as conceive and design environmentally benign processes. Teaching the importance of the environmental aspects of process synthesis, design and development from school level onwards will ensure that they are instilled in future generations of scientists and engineers. Also, it is still not be too late to get traditional and veteran industrialists and scientists to take waste minimization and pollution prevention seriously. In fact, the awareness of scientists and engineers of environmental issues has increased considerably in recent years. In the years to come, it is hoped that waste management, including pollution prevention, will be considered automatically and routinely by all those involved, through R&D to full-scale manufacturing.

52.7 Acknowledgment

The author thanks Ms. Jan Baldauf for compiling the TRI data and those in Process R&D, the Merck Manufacturing Division, and Central Environmental Resources at Merck & Co., Inc., who have contributed directly or indirectly to the waste minimization projects described in this paper.

References

1. U.S. Congress, Office of Technology Assessment, *Serious Reduction of Hazardous Waste*, OTA-ITE-318, U.S. Government Printing Office, Washington, D.C., 1986.

2. H. Freeman, *Hazardous Waste Minimization*, McGraw-Hill, New York, 1990.

3. H. Freeman et al., "Industrial Pollution Prevention: A Critical Review," *Journal of the Air and Waste Management Association,* **42:**(5)618–656, May 1992.

4. U.S. Environmental Protection Agency, *Industrial Pollution Prevention Opportunities for the 1990s,* U.S. EPA, EPA/600/8-91/052, Office of Research and Development, August 1991.

5. R. T. McHugh (Ed. Harry Freeman), *Economics of Waste Minimization in Harzardous Waste Minimization,* Chap. 6, McGraw-Hill, 1990.

6. U.S. Environmental Protection Agency, "The Next Environmental Policy: Preventing Pollution," *Communications and Public Affairs,* August 1991.

7. T. Foecke, "A New Mandate for Pollution Prevention," *Pollution Prevention Review,* Winter 1990–1991, Executive Enterprises, Inc., 1991.

8. *Proposal for a Council Regulation* (EEC) allowing voluntary participation by companies in the industrial sector in a Community ECO-audit scheme.

9. BS 7750, Specification for Environmental Management Systems, 1992.

10. U.S. Environmental Protection Agency, *Guides to Pollution Prevention: The Pharmaceutical Industry,* U.S. EPA, EPA/627/7-91/017, Office of Research and Development, October 1991.

11. M. Mathieu, *New Drug Development: A Regulatory Overview,* PAREXEL International Corp., Cambridge, Mass., 1990.

12. N. A. Nichols, "Scientific Management at Merck: An Interview with CFO Judy Lewent," *Harvard Business Review,* January–February 1994, pp. 88–99.

13. L. Andreassen and N. A. Fletcher, "Pharmaceutical Waste Handling and Disposal," *Pharmaceutical Engineering,* September–October 1993, pp. 61–66.

14. Katy Wolf et al., "Chlorinated Solvents: Will the Alternatives Be Safe?" *Journal of the Air and Waste Management Association,* vol. 41, no. 8, August 1991, pp. 1055–1129.

15. U.S. Environmental Protection Agency, *SARA 313 Toxics Release Report,* U.S. EPA, Washington, D.C., 1987.

16. "O'Sullivan, Bayer Targets Process Modification as Approach to Pollution Prevention," *Chemical and Engineering News,* Oct. 21, 1991, pp. 21–25.

17. H. Redwood, *The Pharmaceutical Industry—Trends, Problems, and Achievements,* Oldwicks Press, London, England, 1987.

18. Sarokin, *Cutting Chemical Wastes: What 29 Organic Chemical Plants Are Doing to Reduce Hazardous Wastes,* INFORM, New York, 1985.

19. D. P. Bowers, *Testimony before the Senate Environmental Quality Committee,* New Jersey Department of Environmental Protection, Trenton, New Jersey, May 17, 1990.

20. M. L. King, A. L. Forman, C. Orella, and S. H. Pines, "Extractive Hydrolysis for Pharmaceuticals," *Chemical Engineering Progress,* **81:**(5)36, 1985.

21. E. L. Paul and C. B. Rosas, *Chemical Engineering Progress,* **86:**(12)17, 1990.

22. E. S. Venkataramani, W. Olsen, and S. Bacher, "Waste Minimization in an Ethical Pharmaceutical Company," *Environmental Progress,* **9:**(3)A10, 1990.

23. L. Naldi, "Waste Reduction Via Process Development," paper presented at the 28th Annual Symposium of the North Jersey AIChE, East Brunswick, N.J., May 9, 1988.

24. E. S. Venkataramani, F. Vaidya, W. Olsen, and S. Wittmer, "Create Drugs, Not Waste—Case Histories of One Company's Successes," *Chemtech,* **22:**674–679, 1992.

25. E. Dienemann, and S. Bacher, "Waste Minimization Strategies during Pharmaceutical Process Development—Pollution Prevention Case Studies," in *Environmental Strategies Handbook*, R. V. Kolluru (ed.), McGraw-Hill, New York, 1994, pp. 189–197.

26. E. Dienemann, D. Wolf, K. Larson, and E. S. Venkataramani, "Waste Minimization Strategies during Pharmaceutical Process Development," AIChE National Meeting, Pittsburgh, Aug. 18–21, 1991.

27. E. S. Venkataramani and S. Bacher, "Waste Minimization in the Pharmaceutical Industry," in *Proceedings of the AIChE Pollution Prevention Topical Meeting*, Washington, D.C., 1989, p. 108.

28. E. S. Venkataramani, G. Bamopolous, A. L. Forman, and S. Bacher, "Design of an Expert System for Early Environmental Assessment of Manufacturing Processes," in *Proceedings of the 43rd Purdue Industrial Waste Conference*, J. M. Bell (ed.), Lewis Publishers, Chelsea, Mich., 1989, p. 425.

29. E. S. Venkataramani, "Case Study: Merck Makes It EASY," *Waste Minimization and Recycling Report*, **46:**6, 1990.

30. E. S. Venkataramani, M. J. House, and S. Bacher, "Implementation of an Expert System Based Environmental Assessment System (EASY)," presented at the AIChE Summer National Meeting, Philadelphia, August 20–23, 1989.

31. E. S. Venkataramani, "Waste Reduction in Bulk Drug and Pharmaceutical Formulation Manufacturing Operations," presented at the AIChE National Meeting, Los Angeles, Nov. 19–22, 1991.

32. E. S. Venkataramani and F. Vaidya, "Development of Environmentally Sound Pharmaceutical Manufacturing Processes through Appropriate Solvent Selection," presented at the Special Conference on Technical Solutions for Air Toxics, Chemical & Pharmaceutical Industry Case Histories, Philadelphia, May 21, 1992.

33. E. S. Venkataramani, "Pollution Prevention through Early Intervention During Process Development," presented at Pollution Prevention in the Process Industries Workshop, Newark, N.J., Sept. 10–11, 1992.

34. E. Dienemann, H. Mahadevan, and E. S. Venkataramani, "Incorporating Pollution Prevention into Pharmaceutical Process Development," presented at the AIChE Summer National Meeting, Seattle, Wash., Aug. 15–19, 1993.

35. E. S. Venkataramani and B. Tarantino, "Novel and Practical Approaches to Waste Minimization and Pollution Control in the Pharmaceutical Industry," AIChE Annual Meeting, St. Louis, Nov. 7–12, 1993.

36. E. S. Venkataramani and A. L. Forman, "Environmentally Proactive Pharmaceutical Process Development," presented at the AIChE Summer National Meeting, Denver, Colo., Aug. 21–24, 1994.

37. *Control of Substances Hazardous to Health (COSHH) Regulations, 1988*, S.I. 1657, 1988.

38. Emission Reduction Research Center, *Pollution Prevention Opportunities in the Pharmaceutical Industry*, New Jersey Institute of Technology, Newark, N.J., May 1991.

53

Sources of Pollution Prevention Information

Madeline M. Grulich

Pacific Northwest Pollution Prevention
Research Center
Seattle, Washington

Pollution prevention success is dependent on finding information that responds to an organization's needs, but accessing information on pollution prevention can be challenging. Because pollution prevention programs must be tailored to respond to an organization's technical needs, physical site restrictions, and culture, generic, readily accessible solutions are unusual. Fortunately, over the past few years a wide and growing network of support has developed for those investigating pollution prevention options. These sources range from federal, state, and local government agencies to academic research centers and private consultants. Information available varies from one source to another. Programs offer everything from where and how to start a program to applications of complicated chemistries that enhance process efficiency. A general overview of services commonly available from sources of information on pollution prevention is provided in Table 53-1.

53.1 Introduction

This chapter serves as an introduction to organizations and publications that may be useful in establishing a network of pollution prevention resources that

Table 53-1. Overview of Services Commonly Offered by Sectors Working on Pollution Prevention

	Whole-saler/ retailer	Publish news-letters	Publish case studies/ fact sheets	Offer publicy accessible clearinghouse or hotline	Offer opportunity assessments and/or planning assistance	Host work-shops	Maintain vendor database	Provide external funding	Conduct research/ demon-stration projects
Federal agencies	W/R	x	x	x		x		x	x
State programs	R	x	x	x	x	x	x	x	
University programs and research institutions	W					x			x
Trade associations	R	x				x		x	
Librarians	R			x					
Consulting companies	R				x	x			x
Nonprofit research centers	W	x				x		x	

are responsive to specific needs. To identify an appropriate source of information it is important to consider the organization's perceived role and audience. One way to divide sources of information is from the perspective of wholesalers of information and retailers of information.

Wholesalers. Wholesalers of information conduct research, compile information, and then prepare documents and other materials for use by others who might be closer to a target audience.

Retailers. Retailers of information serve as conduits linking those who generate the information with those who may put the information into practice. Retailers add value to information by identifying target audiences, developing a communication strategy, locating and compiling relevant materials, screening sources, and delivering the materials to the target audience in a form that is acceptable to that particular audience.

An example of a wholesaler/retailer relationship is the relationship between the United States Environmental Protection Agency (EPA) and state pollution prevention technical assistance programs. While the EPA distributes its own materials to target audiences, the states rely on EPA documents to supplement the information they provide to the companies in their states. The one-on-one technical assistance provided by the state programs has been an effective vehicle for the dissemination of EPA materials.

When considering a source of information, consider whether the source is designed to meet your needs. A retailer of information may be most appropriate for customized responses to specific questions, advice on how to start a program, inspiration on technologies or strategies to deploy, or access to information sources. A wholesaler may be most appropriate for advice on research trends, for details of technologies and their deployment in an industry sector, or for long-term analysis of an emerging technology.

Sources of information are provided in this chapter. The tables are provided as an overview of the types of sources generally available. Most of the sources listed in this chapter should be able to provide general information on pollution prevention or, at a minimum, provide a referral to an organization or materials that respond to a general request for information. The more specific the question, the more important it will be to expand a search to include terms such as *clean manufacturing, waste reduction, waste minimization, source reduction,* and to use the appropriate industry jargon.

53.2 Forms of Information

Information on pollution prevention is available in many forms, from hard copy to electronic to one-on-one consultations. These forms are discussed as follows. An overview of services commonly offered by sectors working on pollution prevention is provided in Table 53-1.

53.2.1   Newsletters

These newsletters may provide information on upcoming workshops, conferences, and other events, explanations of new policies or regulations and their impact on companies, case studies, and other relevant information.

53.2.2   Case Studies

Case studies are a popular method of describing successful pollution prevention activities. Many state technical assistance programs document the successes of companies in their state through case studies that are available on file. Typically, case studies describe the technical achievements of the company, as well as a brief economic analysis of the projects highlighted. Increasingly, case studies also either focus on or provide insights into the management techniques deployed to incorporate pollution prevention into the organization's culture. Case studies vary in length and detail. Because many of the case studies focus on successes, they frequently overlook the investments in time and resources that the organization may have spent in unsuccessful pollution prevention ventures.

53.2.3   Fact Sheets

Fact sheets cover topics from general program descriptions to abbreviated forms of case studies. Look to fact sheets for summary information that may be supported by more complete reports or more details available anecdotally from program staff.

53.2.4   Journals and Publications

Several journals and publications on pollution prevention are available and listed under "Further Reading" at the end of this chapter. The free trade press is also an excellent source of industry-specific, timely, responsive information.

53.2.5   Electronic Databases and Hotlines

Numerous electronic databases exist to provide pollution prevention researchers with on-line electronic information or hotline access to knowledgeable individuals trained to deliver advice over the telephone. Internet is emerging as one of the most valuable conduits for information. The EPA operates the Pollution Prevention Information Clearinghouse, a comprehensive on-line database. An example of a highly specific database is the Program for Assisting the Replacement of Industrial Solvents (PARIS), which uses a database of about 800 chemicals to identify solvent mixtures which are less toxic but similar in

Center for Hazardous Materials (800) 334-2467	Toxic Release Inventory User Support (202) 260-1531
Emergency Planning and Community Right-to-Know (800) 535-0202	U.S. Environmental Protection Agency Pollution Prevention Information Clearinghouse 401 M Street SW (PM 211-A) Washington, D.C. 20460 (202) 260-1023
Green Lights Customer Service Center (202) 775-6650	
Hazardous Waste Ombudsman (800) 262-7937	U.S. Environmental Protection Agency Center for Environmental Research Information 26 W. Martin Luther King Drive Cincinnati, OH 45268 (513) 569-7562
RCRA/Superfund (800) 424-9346	
Solid Waste Assistance Program (800) 677-9424	
TSCA Assistance Information Service (202) 554-1404	

Figure 53-1. EPA hotlines and clearinghouses.

function to the original. The Southeast Waste Reduction Resource Center has compiled a summary of Environmental Information Sources. Information on selected environmental information sources available electronically and on hotlines is provided in Fig. 53-1.

53.2.6 Opportunity Assessments

State technical assistance programs commonly offer the services of trained engineers ready to provide on-site assessments of pollution prevention opportunities to any business or organization that requests the service. These assessments are offered free of charge. Companies concerned about having representatives from a state agency with ties to regulatory enforcement programs concerns are frequently allayed by formal or informal agreements between the technical assistance programs and regulatory branches that protect companies participating in an assessment from enforcement action resulting from information provided in the course of the assessment. The level of detail provided in an assessment varies from program to program. Some programs are prepared to provide specific recommendations for housekeeping changes that address operations and maintenance-related releases, equipment, or process modifications, or substitutions for hazardous materials. Other programs concentrate on inspiring a company to explore pollution preventing alternatives through the use of case studies, economic analyses, and management studies. Once a company decides to take action, technical assistance program staff may help the company help itself or use the services of an outside consultant more effectively.

53.2.7 Planning Assistance

A growing number of states are passing legislation that requires companies to develop pollution prevention/toxics use reduction plans. These states offer assistance in developing plans. State planning assistance programs include the Washington Department of Ecology, the Massachusetts Office of Technical Assistance for Toxics Use Reduction, and the Arizona Waste Minimization Program.

53.2.8 Workshops

Workshops are offered by a wide variety of organizations. These workshops offer companies an opportunity to share information on pollution prevention, to get advice from local consulting companies, and to gain access to sources of information and funding.

53.2.9 Vendor Databases

A few organizations have compiled vendor databases that provide information on equipment and consultants active in the field. For a variety of reasons, most of which involve litigation, state programs are reluctant to provide advice recommending one vendor or consultant over another. However, program staff may provide guidance that will narrow a search for a vendor or consultant. Examples of programs with vendor databases include the Washington State Department of Ecology and the New England Waste Management Officials Association.

53.2.10 Grants and Contracts

Government agencies offer grant programs and/or contract pollution prevention activities. For example, the California Department of Toxic Substances Control, the North Carolina Office of Waste Reduction, the Connecticut Technical Assistance Program, and the Minnesota Office of Waste Management have matching grant programs that fund demonstration programs and research. The Center for Waste Reduction Technologies and the National Center for Manufacturing Sciences also fund projects, especially projects involving their membership. Government, trade associations, private foundations, and others are potential funders of training programs for industries, environmental regulatory staff, and other concerned parties.

53.3 Federal Programs

For several years, federal programs focusing on pollution prevention and waste minimization have been operated by the U.S. EPA and the U.S. Department of Energy.

53.3.1 U.S. Environmental Protection Agency (EPA)

The EPA has been supporting research and demonstration projects and working in conjunction with the states on pollution prevention since the mid-1980s. In the area of information, the EPA acts primarily in the role of wholesaler of information to the states; however, the agency also acts as a retailer through its conferences, seminars, and the distribution of its materials. The EPA has produced many fine publications that run the gamut from general to specific, depending on the source within the EPA.

Headquarters. At EPA Headquarters in Washington, D.C., pollution prevention effects are centralized at the Office of Pollution Prevention and Toxic Substances (OPPTS), where staff administer a variety of grants programs and work on intra- and interagency integration efforts, as well as specific areas such as the financial aspects of pollution prevention. Although pollution prevention is being integrated throughout the agency, OPPTS is an important first contact for referrals within the agency. Many of the EPA's programs are listed in tables included in this section. The EPA publishes *Pollution Prevention News*, a monthly newsletter on pollution prevention. The EPA's Pollution Prevention Information Clearinghouse, which is based at EPA Headquarters, is a large electronic database that provides users with on-line access to case studies, a calendar of events, specialized conferences, and other services.

The EPA publication clearinghouse is located in Cincinnati. This clearinghouse houses informative technical materials, from two-page fact sheets to full technical research reports to detailed analyses in fields of study such as life-cycle assessment. The EPA annually publishes the comprehensive *Reference Guide to Pollution Prevention Resources*, a directory of sources of information ranging from state and federal contacts to training videos and publications. This reference guide should be included in any pollution prevention library.

Regional Offices. EPA pollution prevention activities vary significantly from EPA region to EPA region. Most EPA regional offices have a designated pollution prevention coordinator, have a library of EPA pollution prevention publications (with limited copies available on site), and sponsor training, seminars, conferences, and other outreach activities. Regional offices that sponsor regional pollution prevention roundtables are particularly good references for information on activities occurring at the state level within the region.

For the past few years, the EPA has been concentrating on the integration of pollution prevention into all media programs. Consequently, in coming years, technical assistance may be diffused within the regional offices and may be more closely tied to the regulatory programs.

Research Laboratories. The EPA's pollution prevention research is conducted primarily at the Risk Reduction Engineering Laboratory (RREL) in Cincinnati, Ohio, and at the Air and Energy Research Laboratory (AERRL) in

Research Triangle Park, North Carolina. A compilation of RREL summaries is available by writing to "Current Projects," RREL, MS-466, Cincinnati, OH 45268.
   EPA contacts are found in Fig. 53-2.

53.3.2 U.S. Department of Energy

The U.S. Department of Energy (DOE) has several programs that address waste minimization and pollution prevention. Programs offered by the DOE are heavily oriented toward research. The Office of Waste Reduction Technologies funds large research and demonstration projects. The Innovative Technologies Program has invested in inventions that minimize wastes. The Industrial Waste Reduction Program has issued several publications on research needs, a direc-

---

Headquarters

U.S. Environmental Protection Agency
Pollution Prevention Division
401 M Street SW
Washington, D.C. 20460

Region I

U.S. Environmental Protection Agency
Pollution Prevention Program
John F. Kennedy Federal Building
Boston, MA 02203

Region II

U.S. Environmental Protection Agency
Pollution Prevention Coordinator
26 Federal Plaza, Rm. 900
New York, NY 10278

Region III

U.S. Environmental Protection Agency
Environmental Services Division
841 Chestnut Building (3ES43)
Philadelphia, PA 19107

Region IV

U.S. Environmental Protection Agency
Office of Policy and Management
345 Courtland Street, NE
Atlanta, GA 30365

Region V

U.S. Environmental Protection Agency
Policy and Management Division
77 W. Jackson Blvd.
Chicago, IL 60604-3590

Region VI

U.S. Environmental Protection Agency
Office of Planning and Evaluation
1445 Ross Avenue (6M-P)
Dallas, TX 75270

Region VII

U.S. Environmental Protection Agency
Waste Management Division
726 Minnesota Avenue
Kansas City, KS 66101

Region VIII

U.S. Environmental Protection Agency
Policy Office
999 18th Street, Suite 500
Denver, CO 80202-2405

Region IX

U.S. Environmental Protection Agency
Pollution Prevention Program
75 Hawthorne Street (H-1-B)
San Francisco, CA 94105

Figure 53-2. U.S. Environmental Protection Agency.

tory on federal agencies engaged in waste minimization research, and information on research the program sponsors. The Innovative Technologies program is noted for its sponsorship of emerging technologies. Other programs have supported the development of new technologies, processes, and materials. Recently, the agency has been investigating better electronic information distribution systems. DOE programs frequently fund research performed at the national laboratories. For contacts with information on DOE programs, refer to Fig. 53-3.

National Laboratories. The National Laboratories are federally funded research centers located throughout the United States. Several of these laboratories are engaged in waste minimization work, particularly work funded by the U.S. Department of Energy and the U.S. Department of Defense. Until recently they have been primarily wholesalers of information that is distributed through federal agencies. This relationship may change as the National Laboratories change to meet new demands. For more information on national laboratories, refer to Fig. 53-4.

U.S. Department of Energy
Department of Management
Trevion II
Washington, D.C. 20585-0002
(301) 903-7449

U.S. Department of Energy
Inventions and Innovations Division
1000 Independence Avenue SW
Washington, D.C. 20585
(202) 586-1476

U.S. Department of Energy
Office of Industrial Technologies
1000 Independence Avenue, SW
Washington, D.C. 20585
(202) 586-9496

Figure 53-3. U.S. Department of Energy.

Los Alamos National Laboratory
P.O. Box 1663
Los Alamos, NM 87545
(505) 667-5061

Pacific Northwest Laboratory
P.O. Box 999
Richland, WA 99352
(509) 372-4528

Sandia National Laboratory
P.O. Box 5800
Albuquerque, NM 87185
(505) 846-0011

Figure 53-4. National laboratories.

53.4   State Programs

Most states in the United States have technical assistance programs dedicated to pollution prevention. For the most part, these technical assistance programs are located in state environmental agencies or state universities.

Agency-based technical assistance programs are primarily retailers of information, staffed by technical assistance providers trained to respond to a broad range of needs. Resources available from most state technical assistance programs include a library of waste minimization/pollution prevention materials, fact sheets, case studies, access to engineers trained in pollution prevention; on-site audit capabilities, and general assistance. Most technical assistance programs retain a distance from state environmental regulators and have established agreements that distance them from compliance/enforcement responsibilities. State pollution prevention programs work closely with regulatory agencies on the integration of pollution prevention into policies, laws, and regulations. Because of these responsibilities, they may be strong allies for companies seeking permits that incorporate pollution prevention.

The state technical assistance programs are well networked within their state and region, as well as nationally. The National Roundtable of State Pollution Prevention Programs, which hosts conferences and other activities, facilitates information-sharing among the states. Because of this networking, states are able to provide sophisticated responses to questions through their informal relationships throughout the United States and Canada. State program contacts are listed in Fig. 53-5.

Several states operate their technical assistance programs through or in conjunction with a state university. This separation provides client companies with a comfortable distance from the regulatory responsibilities of an agency-based program. The university-affiliated technical assistance programs frequently have a stronger research orientation than the agency-based programs. Among university-based technical assistance programs are the Iowa Waste Reduction Center operated out of the University of Northern Iowa; Kentucky Partners— State Waste Reduction Center at the University of Louisville; and the Massachusetts Toxics Use Reduction Institute at the University of Lowell.

In addition to the pollution prevention technical assistance programs, the Clean Air Act Amendments of 1990 have created a new state-level source of pollution prevention information. The Act requires each state to establish a Small Business Technical and Environmental Compliance Assistance Program. These programs usually operate in cooperation with the state pollution prevention technical assistance programs and may be a source of information on pollution prevention opportunities, especially targeted toward air releases.

53.5   University Programs and Research Institutions

Universities and community colleges throughout the nation have been developing programs in pollution prevention, particularly by establishing centers

Alabama
Department of Environmental Mgmt.
1751 Congressman William L.
 Dickinson Drive
Montgomery, AL 36130
(205) 260-2779

Alaska
Alaska Health Project
1818 West Northern Lights Boulevard
Suite 103
Anchorage, AK 99517
(907) 276-2864

Department of Environmental
 Conservation
Pollution Prevention Office
3601 C Street Suite 1334
Anchorage, AK 99503-1795
(907) 563-6529

Arizona
Department of Environmental Quality
Pollution Prevention Unit
3033 North Central Avenue, Room 558
Phoenix, AZ 85012
(602) 207-4210

Arkansas
Department of Pollution Prevention
 and Ecology
Hazardous Waste Division
P.O. Box 8913
Little Rock, AR 72219-8913
(501) 570-2861

California
California Integrated Waste
 Management Board
8800 Cal Center Drive
Sacramento, CA 95826
(916) 255-2289

California Local Government
 Commission
909 12th Street
Suite 205
Sacramento, CA 95814
(916) 448-1198

Department of Toxic Substances
 Control
Pollution Prevention, Public &
 Regulatory Assistance Division
P.O. Box 806
Sacramento, CA 95812-0806
(916) 322-3670

U.S. Environmental Protection
 Agency
Office of Pacific Island/Native
 American Programs E-4
75 Hawthorne Street
San Francisco, CA 94105
(415) 744-1599

Colorado
Department of Health
Pollution Prevention Waste Reduction
 Program
4300 Cherry Creek Drive S
Denver, CO 80220
(303) 692-3003

Connecticut
Connecticut Hazardous Waste
 Management Service
Connecticut Technical Assistance
 Program (ConnTAP)
900 Asylum Avenue
Suite 360
Hartford, CT 06105-1904
(203) 241-0777

Department of Environmental
 Protection
Bureau of Waste Management
165 Capitol Avenue
Hartford, CT 06106
(203) 566-5217

Delaware
Department of Natural Resources &
 Environmental Control
Pollution Prevention Program
Kings Highway
P.O. Box 1401
Dover, DE 19903
(302) 739-5071

Figure 53-5. Technical assistance programs and state resources.

University of Delaware
Department of Civil Engineering
Newark, DE 19716
(302) 451-8522

Florida
Department of Environmental
  Regulation
Waste Reduction Assistance
  Program
2600 Blair Stone Road
Tallahassee, FL 32399-2400
(904) 488-0300

Georgia
Department of Natural Resources
Environmental Protection
  Division
4244 International Parkway
Suite 104
Atlanta, GA 30334
(404) 362-2537

Hawaii
Department of Health
Office of Solid Waste
5 Waterfront Plaza
Suite 250
500 Ala Moana Blvd.
Honolulu, HI 96813
(808) 586-4373

Idaho
Department of Health and Welfare
1410 North Hilton Street
Boise, ID 83720-9000
(208) 334-5860

Illinois
Department of Energy and Natural
  Resources
Hazardous Waste Research &
  Information Center
One East Hazelwood Drive
Champaign, IL 61820-7465
(217) 333-8569

Environmental Protection Agency
Office of Pollution Prevention
2200 Churchill Road
P.O. Box 19276
Springfield, IL 62794-9276
(217) 785-0533

Indiana
Department of Environmental Mgmt.
Office of Pollution Prevention and
  Technical Assistance
P.O. Box 6015
105 South Meridian Street
Indianapolis, IN 46225
(317) 232-8172

Purdue University
Environmental Management and
  Education Program
2129 Civil Engineering Bldg.
West Lafayette, IN 47907-1284
(317) 494-5038

Iowa
Department of Natural Resources
Waste Management Authority Div.
Wallace State Office Building
Des Moines, IA 50319
(515) 281-8941

University of Northern Iowa
Iowa Waste Reduction Center
75 Biology Research Complex
Cedar Falls, IA 50614-0185
(319) 273-2079

Kansas
Department of Health and
  Environment
State Technical Action Plan
Forbes Field, Building 740
Topeka, KS 66620
(913) 296-1603

Kansas State University
Engineering Extension Programs
133 Ward Hall
Manhattan, KS 66506-2508
(913) 532-6026

Figure 53-5. (*Continued*)

Kentucky

University of Louisville
Kentucky Partners—State Waste
 Reduction Center
Ernst Hall, Room 312
Louisville, KY 40292
(502) 588-7260

Louisiana

Department of Environmental Quality
P.O. Box 82263
Baton Rouge, LA 70884-2263
(504) 765-0720

Maine

Department of Environmental Protection
State House Station #17
Augusta, ME 04333
(207) 287-2811

Maine Waste Management Agency
State House Station 154
Augusta, ME 04333
(207) 287-5300

Maryland

Department of the Environment
Hazardous Waste Program
2500 Broening Highway, Building 40
Baltimore, MD 21224
(410) 631-3344

Maryland Environmental Services
2020 Industrial Drive
Annapolis, MD 21401
(301) 974-7281

University of Maryland
Technical Extension Service
Engineering Research Center
College Park, MD 20742
(301) 454-1941

Massachusetts

Department of Environment
Office of Technical Assistance
100 Cambridge Street
Boston, MA 02202
(617) 727-3260

Department of Environmental Protection
Toxics Use Reduction Act
 Implementation
1 Winter Street
Boston, MA 02108
(617) 292-5870

University of Massachusetts—Lowell
Toxics Use Reduction Institute
1 University Ave.
Lowell, MA 01854
(508) 934-3262

Michigan

Department of Commerce and Natural
 Resources
Office of Waste Reduction Services
Environmental Services Division
P.O. Box 30004
116 West Allegan Street
Lansing, MI 48909-1178
(517) 335-2142

Michigan Technological University
Waste Reduction and Management
 Program
1400 Townsend Drive
Houghton, MI 49931
(906) 487-2098

Minnesota

Minnesota Office of Waste Management
1350 Energy Lane
Suite 201
St. Paul, MN 55108-5272
(612) 649-5744

Minnesota Pollution Control Agency
Environmental Assessment Office
520 Lafayette Road
St. Paul, MN 55155
(612) 296-8643

University of Minnesota
Technical Assistance Program
School of Public Health
1313 5th Street, S.E., Suite 207
Minneapolis, MN 55414
(612) 627-4646

Figure 53-5. (*Continued*)

Mississippi

Department of Environmental Quality
Waste Reduction/Waste Minimization
  Program
P.O. Box 10385
Jackson, MS 39289-0385
(601) 961-5171

Technical Assistance Program and
  Solid Waste Reduction Assistance
P.O. Drawer CN
Mississippi State, MS 39762
(601) 325-8454

Missouri

Department of Natural Resources
Division of Environmental Quality
205 Jefferson Street
P.O. Box 176
Jefferson City, MO 65102
(314) 751-3176

Environmental Improvement and
  Energy Resources Authority
225 Madison Street
P.O. Box 744
Jefferson City, MO 65102
(314) 751-4919

Montana

Department of Health and
  Environmental Sciences
Water Quality Bureau
Room A-206, Cogswell Building
Helena, MT 59620
(406) 444-2406

Montana State U. Extension Service
807 Leon Johnson Hall
Bozeman, MT 59717-0312
(406) 994-5683

Nebraska

Department of Environmental Control
Hazardous Waste Section
301 Centennial Mall South
P.O. Box 98922
Lincoln, NE 68509
(402) 471-4217

Nevada

Bureau of Waste Management
Division of Environmental
  Protection
123 West Nye Lane
Carson City, NV 89710
(702) 687-5872

Nevada Energy Conservation
  Program
Office of Community Services—
  Capitol Complex
201 South Fall Street
Carson City, NV 89710
(702) 885-4420

University of Nevada—Reno
Business Environmental Program
Nevada Small Business Development
  Center
Reno, NV 89557-0100
(702) 784-1717

New Hampshire

Department of Environmental
  Services
Waste Management Division
New Hampshire P2 Program
6 Hazen Drive
Concord, NH 03301-6509
(603) 271-2912

New Hampshire Business and
  Industry Association
New Hampshire Waste Cap
122 North Main Street
Concord, NH 00301
(603) 224-5388

New Jersey

Department of Environmental
  Protection
Office of Pollution Prevention
  CN-402
401 East State Street
Trenton, NJ 08625
(609) 777-0518

Figure 53-5. (*Continued*)

New Jersey Institute of Technology
Technical Assistance Program
Hazardous Substance Management
  Research Center
323 Martin Luther King Blvd.
Newark, NJ 07102
(609) 292-8341

New Mexico

New Mexico Environmental
  Department
Municipal Water Pollution Prevention
  Program
1190 St. Francis Drive
P.O. Box 26110
Sante Fe, NM 87502
(505) 827-2804

New York

Clarkson University
Hazardous Waste/Toxic Substance
  Research/Management Center
Rowley Laboratories
Potsdam, NY 13699
(315) 268-6542

Department of Environmental
  Conservation
Bureau of Pollution Prevention
50 Wolf Road
Albany, NY 12233-7253
(518) 457-7276

New York Pollution Prevention Program
Erie County Office of Pollution
  Prevention
Erie Co. Office Bldg.
95 Franklin St.
Buffalo, NY 14202
(716) 858-6231

New York State Environmental
  Facilities Corp.
Technical Advisory Services Division
50 Wolf Road
Albany, NY 12205
(518) 457-4138

North Carolina

Department of Environmental, Health,
  and Natural Resources
Pollution Prevention Program
P.O. Box 27687
Raleigh, NC 27611-7687
(919) 571-4100

North Dakota

Department of Health and
  Consolidated Laboratories
Environmental Health Section
1200 Missouri Avenue
Rm. 201
P.O. Box 5520
Bismarck, ND 58502
(701) 221-5150

Ohio

Department of Development
Technology Transfer Organization
77 South High Street, 26th Floor
Columbus, OH 43255-0330
(614) 644-4286

Department of Natural Resources
Division of Litter Prevention and
  Recycling
Fountain Square Court—Building F2
Columbus, OH 43224-1387
(614) 265-6333

Environmental Protection Agency
Pollution Prevention Section
Division of Hazardous Waste
  Management
P.O. Box 1049
Columbus, OH 43266-0149
(614) 644-3969

Ohio's Thomas Edison Program
77 South High Street, 26th Floor
Columbus, OH 43215
(614) 466-3887

Figure 53-5. (*Continued*)

Oklahoma

Department of Health
Environmental Quality Council
Environmental Health Adm.—0200
1000 N.E. 10th Street
Oklahoma City, OK 73117-1299
(405) 271-7353

Department of Health
Pollution Prevention Technical
  Assistance Program
Hazardous Waste Management
1000 N.E. 10th Street
Oklahoma City, OK 73117-1299
(405) 271-7047

Oregon

Department of Environmental Quality
Hazardous and Solid Waste Division
811 S.W. 6th Avenue
Portland, OR 97204
(503) 229-5458

Pennsylvania

Department of Environmental Resources
Office of Air and Waste Management
P.O. Box 2063
Harrisburg, PA 17105-8472
(717) 787-7382

Penn State University
Pennsylvania Tech. Assistance Program
110 Barbara Bldg. II
810 North University Drive
University Park, PA 16802
(814) 865-0427

University of Pittsburgh
Center for Hazardous Materials Research
Applied Research Center
320 William Pitt Way
Pittsburgh, PA 15238
(412) 826-5320

Rhode Island

Department of Environmental Mgmt.
Hazardous Waste Reduction Program
83 Park Street
Providence, RI 02903-1037
(401) 277-3434

South Carolina

Clemson University
Continuing Engineering Education
Hazardous Waste Management
  Research Fund
P.O. Drawer 1607
Clemson, SC 29633
(803) 656-3308

Department of Health and
  Environmental Control
Center for Waste Minimization
2600 Bull Street
Columbia, SC 29201
(802) 734-4715

South Dakota

Department of Environment and
  Natural Resources
Waste Management Program
Joe Foss Building
523 East Capitol Avenue
Pierre, SD 57501-3181
(605) 773-4216

Tennessee

Department of Health and
  Environment
Bureau of Environment
14th Floor, L & C Building
401 Church Street
Nashville, TN 37243-0455
(605) 741-3657

Tennessee Valley Authority
Waste Reduction & Management
  Section
311 Board Street
Chattanooga, TN 37406
(615) 751-4574

University of Tennessee
Waste Reduction Assistance
  Program
Center for Industrial Services
226 Capitol Blvd. Building
Nashville, TN 37219-1804
(615) 242-2456

Figure 53-5. (*Continued*)

Texas

Texas Tech University
Center for Hazardous and Toxic Waste
  Studies
P.O. Box 4679
Lubbock, TX 79409-3121
(806) 742-1413

Texas Water Commission
Office of Pollution Prevention and
  Conservation
P.O. Box 13087
Capitol Station
Austin, TX 78711-3087
(512) 463-7869

Utah

Department of Environmental Quality
Office of Executive Director
168 N 1950 West Street
P.O. Box 144810
Salt Lake City, UT 84114-4810
(801) 536-4477

Vermont

Department of Environmental
  Conservation
Pollution Prevention Division
103 South Main Street
Waterbury, VT 05676
(802) 563-8702

Department of Environmental
  Conservation
Recycling and Resource Conservation
  Section
103 South Main Street
Waterbury, VT 05676
(802) 244-8702

Virginia

Department of Waste Management
Waste Minimization Program
Monroe Bldg., 11th Floor
101 North 14th Street
Richmond, VA 23219
(804) 371-8716

Virginia Polytechnic Institute and State
  University
University Center for Environment
  and Hazardous Materials Studies
Blacksburg, VA 24061-0113
(703) 231-7508

Washington

Washington Department of Ecology
Toxics Reduction Section
Mail Stop PV-11
Olympia, WA 98504-8711
(206) 407-6723

West Virginia

Division of Natural Resources
Waste Management Section
Pollution Prevention and Open Dump
  Program
1356 Hansford Street
Charleston, WV 25301
(304) 558-4000

Wisconsin

Department of Development
Hazardous Pollution Prevention Audit
  Grant Program
P.O. Box 7979
132 West Washington Ave.
Madison, WI 53707
(608) 266-3075

Department of Natural Resources
Bureau of Solid and Hazardous Waste
  Management
P.O. Box 7921
Madison, WI 53707-7921
(608) 267-3763

Wyoming

Department of Environmental Quality
Solid Waste Management Program
122 W. 25th Street—Herschler Building
Cheyenne, WY 82002
(307) 777-7752

Figure 53-5. (*Continued*)

dedicated to specialized research. These wholesalers of information are diverse in their orientation. Some, such as Tufts University and the University of California—Los Angeles, have contributed significantly to curricula development. Others, such as the Pollution Prevention Research Center at North Carolina State University, specialize in pollution prevention research through their Chemical Engineering Department. The Tellus Institute is a leader in research on the economics of pollution prevention.

Finding pollution prevention contacts at universities is usually difficult. For technical information chemical engineering programs are a good place to start. Many universities also have a technology transfer office that has an overview of research and activities at the university. Industrial relations program offices at universities help link industry partners with researchers. Environmental sciences or law programs may house experts in policy issues, while some business programs are researching procurement and environmental accounting issues.

A central source of information on university programs is the EPA-funded National Pollution Prevention Center for Higher Education (NPPC) at the University of Michigan. The NPPC was established to develop and disseminate pollution prevention educational materials for colleges and universities in a variety of disciplines. The NPPC publishes the *Directory of Pollution Prevention in Higher Education: Faculty and Programs,* a comprehensive, detailed, and indexed listing of pollution prevention activities occurring at academic institutions nationwide. University directories are available in most libraries. Another resource found in many libraries is research capabilities guides, which are organized by subject and discipline. Selected university contacts are provided in Fig. 53-6.

53.6 Corporate Programs and Trade/Manufacturing Associations

Several companies are willing to share information on their corporate programs. This information may be available from the company or through state technical assistance programs or trade associations. Usually, companies are willing to share their corporate environmental policies. An increasing number of companies are willing to share information on nonproprietary technologies, but these are usually made available through workshops or case studies prepared and disseminated by another organization such as a state program or trade association. If a company is willing to share information on its program, the management aspects are often as valuable as the technical details. Information on procurement practices, employee incentive programs, and integration of pollution prevention into the corporate culture are all management issues that may be transferable.

Information from trade/manufacturing associations frequently has the advantage of being targeted, application-oriented, and peer reviewed by indi-

Alabama

University of Alabama
Hazardous Materials Management and
 Resource Recovery Program
275 Mineral Industries Bldg.
BOX 870203
Tuscaloosa, AL 35487-0203
(205) 348-8403

California

Electric Power Research Institute
P.O. Box 10412
Palo Alto, CA 94303-0813
(415) 855-2000

University of California Extension
Environmental Hazards Management
 Program
Riverside, CA 92521-0112
(714) 787-5804

University of California—Los Angeles
Center for Waste Reduction
 Technologies
405 Hilgard Avenue
Los Angeles, CA 90024
(310) 206-0300

Colorado

Colorado State University
Waste Minimization Assessment
 Center
Mechanical Engineering Dept.
Fort Collins, CO 80523
(303) 491-5317, 294-5317

Connecticut

University of Connecticut
Pollution Prevention Research and
 Development Center
Environmental Research Institute
Box U-120, Route 44
Longley Bldg. 146
Storrs, CT 06269-3210
(203) 486-4015

Waterbury State Technical College
Industrial Environmental Management
750 Chase Parkway
Waterbury, CT 06708-3089
(203) 596-8703

District of Columbia

American Institute of Architects
1735 New York Avenue NW
Washington, D.C. 20006-5292
(202) 626-7463

Howard University
Great Lakes and Mid-Atlantic
 Hazardous Substance Research Center
Dept. of Civil Engineering
Washington, D.C. 20059
(202) 806-6570

Florida

Florida Institute of Technology
Research Center for Waste Utilization
150 W. University Blvd.
Melbourne, FL 32901-6988
(305) 768-8000

University of Florida
Center for Training, Research, and
 Education for Environmental
 Occupations
3900 S.W. 63rd Blvd.
Gainesville, FL 32608-3848
(904) 392-9570

University of Florida
Florida Center for Solid & Hazardous
 Waste Management
3900 S.W. 63rd Blvd.
Gainesville, FL 32608-3848
(904) 392-9570

Illinois

University of Illinois
Hazardous Waste Research and
 Information Center
1 East Hazelwood Drive
Champaign, IL 61820
(217) 333-8940

Figure 53-6. Universities and research organizations.

Indiana

Purdue University
Pollution Prevention Program
2129 Civil Engineering Building
West Lafayette, IN 47907-1284
(317) 494-5038

Iowa

University of Northern Iowa
Iowa Waste Reduction Center
75 BRC
Cedar Falls, IA 50614-0185
(319) 273-2079

Kansas

Kansas State University
Hazardous Substance Research Center
 (HSRC)
Durland Hall, Room 105
Manhattan, KS 66506-5102
(913) 532-5584

University of Kansas
Center for Environmental Education
 and Training
6330 College Boulevard
Overland Park, KS 66211
(913) 491-0810

Kentucky

University of Louisville
Kentucky PARTNERS
State Waste Reduction Center
Ernst Hall, Room 312
Louisville, KY 40292
(502) 588-7260

University of Louisville
Waste Minimization Assessment Center
Department of Chemical Engineering
Louisville, KY 40292
(502) 588-6357

Louisiana

Louisiana State University
Hazardous Waste Research Center
3418 CEBA Building
Baton Rouge, LA 70803
(504) 388-6770

Southern University at Baton Rouge
Center for Energy and Environmental
 Studies
Cottage #8
P.O. Box 9764
Baton Rouge, LA 70813
(504) 771-4723

Maine

University of Maine
Chemicals in the Environment
 Information Center
5737 Jenness Hall
Orono, ME 04469-5737
(207) 581-2301

Massachusetts

Massachusetts Institute of Technology
Center for Technology, Policy and
 Industrial Development
E40-241
Cambridge, MA 02139
(617) 253-7753

Tellus Institute
Risk Analysis Group
11 Arlington Street
Boston, MA 02116-3411
(617) 266-5400

Tufts University
Center for Environmental Management
474 Boston Avenue, Curtis Hall
Medford, MA 02155
(617) 627-3452

University of Massachusetts—Lowell
Toxics Use Reduction Institute
Lowell, MA 01854-2881
(508) 934-3275

Michigan

Grand Valley State University
Waste Reduction and Management
 Program
School of Engineering
301 W. Fulton, Rm. 617
Grand Rapids, MI 49504
(616) 771-6750

Figure 53-6. (*Continued*)

Michigan Technological University
Center for Clean Industrial and
  Treatment Technologies
Environmental Engineering Center
1400 Townsend Drive
Houghton, MI 49931
(903) 487-3143

University of Detroit Mercy
Center of Excellence in Polymer
  Research & Environmental Study
4001 West McNichols Road
Detroit, MI 48219-3599
(313) 993-1270

University of Michigan
EPA Pollution Prevention Center for
  Curriculum Development and
  Dissemination
School of Natural Resources
430 E. University
Ann Arbor, MI 48109-1115
(313) 764-1412

University of Michigan
Great Lakes and Mid-Atlantic
  Hazardous Substance Research Center
Suite 181, Engineering 1-A
Ann Arbor, MI 48109-2125
(313) 763-2274

University of Michigan
National Pollution Prevention Center
430 E. University
Ann Arbor, MI 48109-1115
(313) 764-1412

Minnesota

University of Minnesota
Minnesota Technical Assistance Program
1315 5th Street SE
Suite 207
Minneapolis, MN 55414-4504
(612) 627-4646

Mississippi

Mississippi State University
Technical Assistance Program
P.O. Drawer CN
Mississippi State U., MS 39762
(601) 325-8454

Nevada

University of Nevada—Reno
Business Environmental Program
Nevada Small Business Dev. Center
Reno, NV 89557-0100
(702) 784-1717

New Jersey

New Jersey Institute of Technology
Hazardous Substance Management
  Research Center
323 Martin Luther King Boulevard
University Heights
Newark, NJ 07102
(201) 596-5864

New Mexico

New Mexico State University
Waste Management Education &
  Research Consortium
Box 30001, Dept. 3805
Las Cruces, NM 88003-0001
(505) 646-2038 (6419)

New York

Cornell University
Waste Management Institute
413 Hollister Hall
Ithaca, NY 14853
(607) 255-8674

North Carolina

North Carolina State University
EPA Research Center for Waste
  Minimization and Management
Box 7905
Raleigh, NC 27695-2325
(919) 515-2325

North Carolina State University
Pollution Prevention Research Center
Raleigh, NC 27695-7905
(919) 737-2325

Figure 53-6. (*Continued*).

University of North Carolina
EPA Research Center for Waste
  Minimization and Management
Dept. of Environmental Science and
  Engineering
Chapel Hill, NC 27514
(919) 966-1024

North Dakota

University of North Dakota
Energy and Environmental Resource
  Center
15 N. 23rd Street, Box 8213
University Station
Grand Forks, ND 58202-8213
(701) 777-5131

Ohio

University of Cincinnati
American Institute for Pollution
  Prevention
Office of University Dean for
  Research
Cincinnati, OH 45221-0071
(513) 556-3693

University of Findlay
RCRA Generator Training Program
P.O. Box 538
St. Clairsville, OH 43950
(614) 695-5036

Oklahoma

Oklahoma State University
Center for Resource Conservation and
  Environmental Research
2630 Northwest Expressway
Suite B
Oklahoma City, OK 73112
(405) 943-8989

Pennsylvania

Pennsylvania State University
Environmental Resource and Research
  Institute
Land and Water Research Building
University Park, PA 16802
(814) 863-0291

University of Pittsburgh
Ctr. for Hazardous Materials Research
320 William Pitt Way
Pittsburgh, PA 15238
(412) 826-5320

Rhode Island

University of Rhode Island
Chemical Engineering Dept.
Crawford Hall
Kingston, RI 02881
(401) 792-2443

South Carolina

University of South Carolina
Hazardous Waste Management
  Research Fund
Gambrell Hall, 4th Floor
Columbia, SC 29208
(803) 777-8157

Tennessee

University of Tennessee
Waste Minimization Assessment Center
310 Perkins Hall
Dept. of Engineering, Science, and
  Mechanics
Knoxville, TN 37996-2030
(615) 974-7682

Texas

Lamar University
Gulf Coast Hazardous Substance Research
P.O.Box 10613
Beaumont, TX 77710
(409) 880-8707

Texas A&M University
Center for Waste Management
Box 3376
College Station, TX 77843
(409) 845-4930

Texas Tech University
Center for Environmental
  Technologies
P.O. Box 43121
Lubbock, TX 79409-3121
(806) 742-1413

Figure 53-6. (*Continued*).

University of Texas at Arlington
Environmental Institute for
 Technology Transfer
Box 19050
Arlington, TX 76019
(817) 273-2300

Utah

Weber State University
Center for Environmental Service
Ogden, UT 84408-2502
(801) 626-7559

Wisconsin

University of Wisconsin—Extension
Solid and Hazardous Waste Education
 Center
529 Lowell Hall
610 Langdon Street
Madison, WI 53703
(608) 265-2360

University of Wisconsin at Madison
Engineering Professional Development
 Program
432 North Lake Street
Madison, WI 53706
(608) 263-7429

Figure 53-6. (*Continued*).

viduals working daily with the technology or strategy. Trade associations are generally retailers of general information that benefits their membership. A particularly good reference published by a trade/manufacturing association is the *Pollution Prevention Resource Manual,* published by the Chemical Manufacturer's Association.

Finally, it is necessary to underscore the importance of being professionally networked through local, state, regional, and national professional societies and organizations. The information relationships developed in the course of this networking often lead to opportunities to get candid, detailed responses to requests for advice.

53.7 Consulting Companies

Throughout the United States, consulting companies are beginning to offer a variety of pollution prevention services—especially opportunity assessments and responses to state planning requirements. Consultants are both wholesalers or retailers, depending on the needs of their clients. State and federal technical assistance programs may be able to provide guidance on consultants working in pollution prevention.

Selection of a consultant claiming to do pollution prevention work may require caution. Questions to ask the consultant might include:

- What pollution prevention work has the consultant done in the past? Examples? The use of the federal government's Standard Forms 254 and 255 (SF 254 and SF 255, available at any government bookstore) allows a direct and comprehensive comparison of qualifications of consultants competing for a project.

- What is the range of activities the consultant considers to fall within "pollution prevention"? Is it consistent with the needs of the contracting organization?

- Has there been a specific emphasis in previous work? Consultants that have a strong background with a limited number of technologies may restrict their view of options to what they have worked with in the past.

- Will the consultant be able to provide a broad range of assistance or will the focus be on limited aspects of the project? For example, has the consultant done all or most of his or her work in process design? Hazardous materials substitution? Water and energy conservation? Total quality management? Procurement and hazardous materials management? Waste reduction planning? Or has the work been in control strategies?

- With respect to waste reduction planning, does the consultant use boilerplate language in final plans? How did the consultant integrate the plan with existing activities? Who is responsible for plan implementation? Boilerplate may not be responsive to the company's needs and if the plan was not integrated into existing activities at the facility, it is likely to serve the short-term need to comply with planning requirements, but is unlikely to change the company's operations in the long term. Because pollution prevention plans should establish a sustainable program, the plan champion and an implementation team should be clearly identified and engaged in the development of the plan.

- Who are the consultant's references? It is worthwhile to speak with others who have employed the consultant to determine whether he or she will be able to provide services that are responsive to the specific site and cultural needs of the facility. References should be asked if the consultant's work was worthwhile and whether the company would engage the consultant again.

- Who are the principals assigned to the project? What will their level of activity be?

53.8 Nonprofit Research Organizations

There are a few nonprofit organizations dedicated specifically to pollution prevention and related areas of study. These organizations are separately incorporated, governed by boards of directors. Their funding is derived from multiple sources that may include membership fees, grants, and contracts from public and private sources, publication sales, and other sources. They are not affiliated with universities. Five examples of nonprofit research organizations are listed in Fig. 53-7. Each is unique and should be contacted for program information.

The National Center for Manufacturing Sciences is an industry membership organization that funds research, demonstration projects, and other activities

Center for Waste Reduction Technologies 345 East 47th Street New York, NY 10017 (212) 705-7407	Pollution Prevention Partnership 1099 18th Street Suite 2100 Denver, CO 80202 (303) 294-1200
National Center for Manufacturing Sciences 3025 Boardwalk Ann Arbor, MI 48108 (313) 995-0300	WRITAR 1313 Fifth Street SE Minneapolis, MN 55414 (612) 379-5995
Pacific Northwest Pollution Prevention Research Center 1326 Fifth Avenue Suite 650 Seattle, WA 98101 (206) 223-1151	

Figure 53-7. Nonprofit research organizations.

that further United States manufacturing. Most of the work of NCMS is not identified as pollution prevention; however, their research on manufacturing efficiency and improved processes has contributed heavily to pollution prevention. NCMS publishes an excellent newsletter, *Focus.*

The Center for Waste Reduction Technologies is a part of the American Institute for Chemical Engineering. CWRT has produced several excellent publications that are prepared as a collaborative effort by their members. Two CWRT research topics are "The Reduction and Control of VOC Emissions" and "Environmental Considerations in Process Design and Simulation." They also fund research of interest to their membership.

The Pacific Northwest Pollution Prevention Research Center and the Colorado Pollution Prevention Partnership are public-private partnerships that mobilize resources toward research and other projects that spur the implementation of pollution prevention. The PPRC has funded research projects; acts as a source of information on pollution prevention research; publishes a newsletter; and works closely with industry, public interest groups, academia, and the states in EPA Region 10 and the Province of Canada. The Colorado-based Pollution Prevention Partnership has done some excellent work with the reduction of trichloroethane. Members of the Partnership include industry, public interest groups, EPA, and a public utility.

WRITAR organizes and conducts waste reduction training courses for targeted industries, universities, and government agencies. "Train-the-trainer" programs are also offered. These valuable programs provide pollution prevention advocates with the tools and skills they will need as champions for pollution prevention within their organizations.

Further Reading

The following directories and references are useful in identifying sources of information. Any pollution prevention library should consider obtaining these publications as standard reference materials.

Bakshani, Nandkumar, and David Allen, *Directory of Pollution Prevention in Higher Education: Faculty and Programs,* 1992. Available through the National Pollution Prevention Center for Higher Education, University of Michigan, Dana Bldg., #2540, 430 East University Avenue, Ann Arbor, MI 48109.

Guide to Pollution Prevention Funding Organizations, Pacific Northwest Pollution Prevention Research Center, Seattle, Wash., 1992.

Reference Guide to Pollution Prevention Resources, U.S. EPA, Prevention, Pesticides, and Toxic Substances, (TS 792), EPA/742/B-93-001, published annually.

Toxics Use Reduction Research Directory, Massachusetts Toxics Use Reduction Institute, University of Massachusetts, Lowell, Mass., 1992.

U.S. Department of Energy, *Federal Agencies Active in Waste Minimization and Pollution Prevention,* DOE/CE/40956T-H1, 1992. Available through the National Technical Information Service, 703-487-4968.

Wigglesworth, David, *Pollution Prevention: A Practical Guide for State and Local Government,* Arkansas, 1993. To order call 1-800-272-7737. $58.

Selected Newsletters and Journals

at the source, quarterly newsletter of the Great Lakes Pollution Prevention Centre. To order call 519-337-3423.

In Review, the annual review issue of the *Environment and Resource Management Division Newsletter* of the Special Librarians Association. To order call 318-266-8692. $20.

Focus, monthly publication by the Strategic Development Staff of the National Center for Manufacturing Sciences. For more information about receiving this publication call 313-995-0300.

Pollution Prevention News, published by U.S. EPA, 401 M Street SW (MC 7409), Washington, D.C. 20460

The Pollution Prevention Letter, biweekly publication of Clarity Publishing. To order call 301-495-7747.

Pollution Prevention Northwest, bimonthly publication of the Pacific Northwest Pollution Prevention Research Center, 1326 5th Avenue, Suite 650, Seattle, WA 98101. To order call 206-223-1151.

Pollution Prevention Review, quarterly publication of Executive Enterprises Publications Co., Inc., 22 West 21st St., New York, NY 10010-6990. To order call 1-800-332-8804.

Total Quality Environmental Management, quarterly publication of the Executive Enterprises Publications Co., Inc., 22 West 21st St., New York, NY 10010-6990. To order call 1-800-332-8804.

Waste Minimization and Recycling Report, monthly publication of Government Institutes, Inc., 4 Research Place, #200, Rockville, MD 20850, 301-921-2355.

Pollution Prevention Advisor, quarterly publication of the U.S. Department of Energy, Office of Defense Programs. For subscription information call 615-481-0036.

Index

About the Editor

Harry M. Freeman is the chief of the Pollution Prevention Research Branch at the U.S. EPA's Risk Reduction Engineering Laboratory in Cincinnati, Ohio. He is recognized in the United States and other countries for his speaking and writing related to pollution prevention. He is a member of the American Institute of Chemical Engineers, a diplomate of the American Academy of Environmental Engineers, and a member of the International Juggling Association (an affiliation that has served him well in putting together books such as this one). He is also the editor of the McGraw-Hill books, *Standard Handbook of Hazardous Waste Treatment* and *Disposal and Hazardous Waste Minimization*.